Stem Cells

Stem Cells

Edited by

C. S. Potten
Paterson Institute for Cancer Research,
Christie Hospital NHS Trust,
Manchester, UK

ACADEMIC PRESS
Harcourt Brace & Company, Publishers
London San Diego New York Boston
Sydney Tokyo Toronto

ACADEMIC PRESS LIMITED
24–28 Oval Road
London NW1 7DX

US edition published by
ACADEMIC PRESS INC
San Diego, CA 92101

This book is printed on acid-free paper

A catalogue record for this book is available from the British Library

ISBN 0-12-563455-2

Typeset by P & R Typesetters Ltd, Salisbury, Wiltshire
Printed in Great Britain by The University Press, Cambridge

Contents

Contributors

E. Arciniegas
CRC Department of Medical Oncology, Christie Hospital NHS Trust, Wilmslow Road, Manchester M20 9BX, UK
P. W. Barlow
Department of Agricultural Sciences, University of Bristol, Institute of Arable Crops Research, Long Ashton Research Station, Bristol BS18 9AF, UK
R. Barraclough
Cancer and Polio Research Fund Laboratories, Department of Biochemistry, University of Liverpool, PO Box 147, Liverpool L69 3BX, UK
A-M. Buckle
Leukaemia Research Fund, Cellular Development Unit, Department of Biochemistry and Applied Molecular Biology, UMIST, Sackville Street, Manchester M60 1QD, UK
P. Cameron-Curry
Howard Hughes Medical Institute, 701 West 168th Street, 10th floor, New York, NY 10032, USA
F. M. F. van Dissel-Emiliani
Department of Functional Morphology, Veterinary School, University of Utrecht, Postbus 80.157, 3508 TD Utrecht, The Netherlands
C. Dulac
Institut d'Embryologie Cellulaire et Moleculaire, UMRC9924, 49bis, Avenue de la Belle Gabrielle, 94736 Nogent-sur-Marne Cedex, France
Present address: Harvard University, Cambridge, MA, USA
D. G. Fernig
Cancer and Polio Research Fund Laboratories, Department of Biochemistry, University of Liverpool, PO Box 147, Liverpool L69 3BX, UK
D. Francis
School of Pure and Applied Biology, University of Wales College Cardiff, PO Box 915, Cardiff CF1 3TL, UK
J. W. Grisham
Department of Pathology, CB #7525, University of North Carolina at Chapel Hill, 303 Brinkhaus-Bullitt Building, Chapel Hill, NC 27599-7525, USA
C. M. Heyworth
CRC Department of Experimental Haematology, Paterson Institute for Cancer Research, Christie Hospital NHS Trust, Wilmslow Road, Manchester M20 9BX, UK
G. E. Jones
The Randall Institute, King's College London, 26–29 Drury Lane, London WC2B 5RL, UK
D. B. Kohn
Division of Research Immunology and Bone Marrow Transplantation, Children's Hospital Los Angeles, Departments of Pediatrics and Microbiology, University of Southern California School of Medicine, 4650 Sunset Boulevard, Los Angeles, CA 90027, USA

J. Kummermehr
Institute of Radiobiology, GSF – Forschungszentrum für Umwelt und Gesundheit, Institut für Strahlenbiologie, Ingolstädter Landstr. 1, D-8042 Neuherberg, Germany

R. M. Lavker
Department of Dermatology, University of Pennsylvania School of Medicine, Philadelphia, PA, USA

M. Loeffler
Institüt fur Medizinische Informatik, Statistik und Epidemologie, Universität Leipzig, Liebigstrasse 27, 04103 Leipzig, Germany

B. I. Lord
CRC Department of Experimental Haematology, Paterson Institute for Cancer Research, Christie Hospital NHS Trust, Wilmslow Road, Manchester M20 9BX, UK

S. J. Miller
Department of Dermatology, Johns Hopkins Medical School, 601 North Carolina Street, Baltimore, MD 21287-0900, USA

J. A. Nolta
Division of Research Immunology and Bone Marrow Transplantation, Children's Hospital Los Angeles, Departments of Pediatrics and Microbiology, University of Southern California School of Medicine, 4650 Sunset Boulevard, Los Angeles, CA 90027, USA

C. S. Potten
CRC Department of Epithelial Biology, Paterson Institute for Cancer Research, Christie Hospital NHS Trust, Wilmslow Road, Manchester M20 9BX, UK

D. G. de Rooij
Department of Cell Biology, Medical School, University of Utrecht, Postbus 80.157, 3508 TD Utrecht, The Netherlands

P. S. Rudland
Cancer and Polio Research Fund Laboratories, Department of Biochemistry, University of Liverpool, PO Box 147, Liverpool L69 3BX, UK

A. M. Schor
Department of Dental Surgery and Periodontology, The Dental School, Park Place, University of Dundee, Dundee DD1 4HR, UK

S. L. Schor
Department of Dental Surgery and Periodontology, The Dental School, Park Place, University of Dundee, Dundee DD1 4HR, UK

J. A. Smith
Cancer and Polio Research Fund Laboratories, Department of Biochemistry, University of Liverpool, PO Box 147, Liverpool L69 3BX, UK

T-T. Sun
Epithelial Biology Unit, Ronald O. Perelman Department of Dermatology and Department of Pharmacology, Kaplan Comprehensive Cancer Center, New York University Medical School, New York, NY, USA

N. G. Testa
CRC Department of Experimental Haematology, Paterson Institute for Cancer Research, Christie Hospital NHS Trust, Wilmslow Road, Manchester M20 9BX, UK

S. S. Thorgeirsson
Laboratory of Experimental Carcinogenesis, National Cancer Institute/National Institutes of Health, Building 37, Room 3C 28, Bethesda, MD 20892-0001, USA

K-R. Trott
Department of Radiation Biology, St Bartholomew's Hospital Medical College, Charterhouse Square, London EC1M 6BQ, UK
D. J. Watt
Department of Anatomy, Charing Cross and Westminster Medical School, Fulham Palace Road, London W6 8RF, UK
A. D. Whetton
Leukaemia Research Fund, Cellular Development Unit, Department of Biochemistry and Applied Molecular Biology, UMIST, Sackville Street, Manchester M60 1QD, UK
N. A. Wright
Department of Histopathology, Royal Postgraduate Medical School, Hammersmith Hospital, Du Cane Road, London W12 0NN, UK

Preface

In 1983 I edited a volume, published by Churchill Livingstone, consisting of ten chapters on stem cells in a variety of invertebrate and vertebrate systems, entitled *Stem Cells – their identification and characterisation.* The book was successful and quickly went out of print. I decided in 1993 that perhaps it was time to produce a new, updated book on the subject perhaps covering fields that were not dealt with in 1983. There have been a number of significant developments since that time. Little was known then, for example, about the complex interacting network of cytokines and growth factors that regulate the proliferation and differentiation of stem cells. Much of the complexity of these signals is now well understood in systems such as the mammalian bone marrow. Furthermore the use of a variety of newly developed molecular biological techniques and probes has made possible dramatic advances in our ability to study the lineage development that is derived as a consequence of divisions of the stem cells. Also in the last few years there have been significant increases of interest in identifying, characterizing and isolating stem cell populations from a variety of tissues. Major granting bodies have targeted stem cells and stem cell concepts amongst the topics identified as high priority for funding. A major impetus here has been the desire to identify these all-important cells in tissues so that strategies for targeting these permanent lineage ancestor cells for gene therapy may be developed. Also there has been an increased interest, arising from the fact that many of the regulatory growth factors and cytokines have been identified, in manipulating the tissue stem cells in various clinical situations including, for example, the normal tissue stem cells during cancer therapy.

When I decided to put together this book I contacted all the original contributors to ask whether they felt there were developments that warranted a new review in this 1996 book. Most responded, either indicating that they thought there were new topics to review or suggesting alternative authors and topics, some did not – which I interpreted to indicate a lack of new developments in that particular field. I have also endeavoured to cover some topics in this new book not covered in the previous one, examples of which are obvious from the contents page. I was particularly interested in trying to solicit a chapter describing what is known about precursor cells or stem cells in insect systems. The extensive work on the genetics and morphogenic signals determining tissue development of *Drosophila* suggested that the concept of stem cells in this system would be interesting. Somewhat surprisingly to me, those working in the field tend not to think of these developmental processes in terms of cell lineages and lineage ancestor or stem cells but rather in terms of morphogens and topography (cell tissue interfaces or boundaries) determining development and cell fate. Unfortunately, in spite of contacting successively about ten authors in the field, I could not persuade anyone to put together a chapter. I am not sure how one interprets this reticence. There were a few other fields where I encountered similar difficulties.

I have managed to recruit altogether 15 chapters for this new book. Inevitably, it has taken a considerable time to achieve this and to acquire the promised chapters from the various authors. This was exacerbated by the consecutive chain of prospective authors that I sometimes had to contact as indicated above. To those authors who did respond quickly and were diligent in submitting their chapters by the initial deadlines, I offer my sincere apologies for the length of time that it has taken to collect the remaining chapters and produce the book. I think the only mitigating factor is that inevitably the book is more valuable and comprehensive as a consequence of waiting for important contributions. Most of the initial contributors have had the opportunity to update their chapters in the intervening period.

It is still not possible to identify stem cells in most tissues by either their morphological characteristics or by the use of a specific marker. Numerous attempts have been made to find stem cell specific markers, but the problem here may be that what characterizes these cells is more likely to be the absence of specific features than the presence of something that can be identified by a marker or probe. In the absence of such abilities, finding these elusive cells is somewhat analogous to the needle in the haystack problem. However, there are approaches that can be adopted to enrich these cells. In several systems one of the major problems in studying stem cells is the fact that the techniques available involve perturbation of the system, which inevitably causes changes in the behaviour and characteristics (and possibly the number) of the stem cells which are to be studied, resulting in a situation analogous to the Heisenberg uncertainty principle in the field of subatomic particle and quantum physics.

There is still considerable variability or context dependence in the definition of stem cells and the operation of the definition to identify and study stem cells amongst different investigators. Although stem cells inherently and by definition are proliferative cells, this is a weak and ineffective criterion to use on its own. The ability to divide a large number of times and maintain a tissue throughout the life of an animal is a stronger criterion for defining these cells but difficult to study directly and as a consequence usually inferred indirectly. There are two further criteria which are virtually inherent in the point just raised. Firstly, that if a stem cell is to divide a large number of times and maintain a tissue, it must consequently also maintain its own numbers. This property of self-maintenance is again difficult to study experimentally, usually requiring serial transplantation or subcloning procedures. It is a cardinal property of stem cells that they maintain their numbers and that the self-maintenance probability is subject to control and hence variation. The second feature arising from the ability of stem cells to maintain the cellularity of the tissue, results from the fact that tissues commonly contain a variety of cell types, which implies that the stem cells are capable of generating cells that differentiate down a variety of lineages. This is clearly demonstrable in some of the more extensively studied tissues, such as bone marrow and the gastrointestinal tract. It is a topic that is further discussed in several chapters dealing with other tissues in this book.

Several questions concerning stem cells remain unresolved, for example to what extent their attributes are inherently determined and to what extent they are governed by the environment in which the cell finds itself. It is quite clear that environmental (niche) determinants play an important role in the functional capabilities of stem cells. However, it equally seems likely that there are intrinsic differences between stem cells and cells later in the proliferative lineage; thus it is probable that both intrinsic and extrinsic factors are involved. In spite of many years of discussion and investigation it is still not clear how differentiation of stem cells is achieved and what controls this process, i.e. symmetric versus asymmetric divisions, stochastic versus deterministic processes, environmental versus inherent signals. Another question that is commonly raised is to what extent the differentiation and proliferative capacities (life span) of a stem cell are limited. Bone marrow stem cells can clearly be demonstrated to have a division potential far in excess of that required for day-to-day replacement of haematopoietic cells. Intestinal stem cells seem likely to have a division potential of about 1000 in the mouse – a large number by any criteria. It is surprising that with their large numbers, extensive division potential and short cell cycle, stem cells in the small intestine rarely develop tumours. This must indicate efficient damage detection and protection mechanisms. Bone marrow stem cells seem to have an extraordinarily broad repertoire of differentiation. Whether further unknown elements of their repertoire could be unmasked provided the correct signals were given remains unanswered, but is perhaps unlikely bearing in mind the efforts that have gone into studying this question. This topic also has important implications in terms of understanding the cellular elements involved in ageing. The role that stem cells play in the development of cancers is also an important area, subject to much debate and comment. In tissues such as the gastrointestinal tract with its highly mobile and dynamic cell lineages with their short life expectancy in the tissue, the stem cells would seem to be the only candidates present in the tissue long enough to play a role at least in the initiation stage in the development of cancer. Once the stem cell is initiated it would produce a lineage of similarly altered cells, and it is conceivable that subsequent stages in carcinogenic transformation may involve cells other than stem cells. However, even here in the gastrointestinal tract, the probability would be, based on the rapid turnover and the time it would take for a transformed cell to grow into a micro-tumour that tumours are unlikely to arise in any other cells than the stem cells or very early lineage cells. These and many other topics are raised and discussed in the various chapters to be found in this volume.

I should like to thank all the contributors for their chapters and for the patience shown by those who submitted their chapters some considerable time ago. It would not be possible to put together such a book without dedicated skilled secretarial help, and for that I am very grateful. I should like to thank the publishers for help in producing this volume and finally I should like to thank my wife and family for their support and understanding over the years.

1 Stem cells and cellular pedigrees – a conceptual introduction

Markus Loeffler and Christopher S. Potten*

*Institute of Medical Informatics, Statistics and Epidemiology, Leipzig, Germany; *CRC Department of Epithelial Cell Biology, Paterson Institute for Cancer Research, Manchester, UK*

SUMMARY

In this chapter, we consider some of the problems involved in current discussions on stem cells in adult mammalian tissues. The present concepts involve a number of pitfalls, logical, semantic and classification problems. This indicates the necessity for new and well defined concepts that are amenable to experimental analysis.

One of the major difficulties in considering stem cells is that they are defined in terms of their functional capabilities which can only be assessed by testing the abilities of the cells, which itself may alter their characteristics during the assay procedure; a situation similar to the uncertainty principle in physics. Hence, a proper description requires the measurement i.e. manipulation process itself to be taken into account.

If such context-dependent interactions exist between the manipulation and measurement process and the challenged stem cells, the question of the number of stem cells in a tissue has to be posed in a new way. Rather than obtaining a single number, one might end up with different numbers under different circumstances, all being complementary. This might suggest that stemness is not a property but a spectrum of capabilities from which to choose. This concept might facilitate a reconciliation between the different and sometimes opposing experimental results. Given certain experimental evidence, we have attempted to provide a novel concept to describe structured cell populations in tissues involving stem cells, transit cells and mature cells. It is based on the primary assumption that the proliferation and differentiation/maturation processes are in principle independent in the sense that each may proceed without necessarily affecting the other.

Stem cells may divide without maturation, while cells approaching functional competence may mature but do not divide. In contrast, transit cells divide and mature showing intermediate properties between stem cells and mature functional cells. The need to describe this transition process and the variable coupling between proliferation and maturation leads us to formulate a *screw model of cell and tissue organization.*

This concept is illustrated for the intestinal epithelium. Reference is made also to other tissues including the basal epidermal cell layer and the haematopoietic system.

STEM CELLS
ISBN 0–12–563455–2

INTRODUCTION

At present there is no experimental way to decide if a given cell in a functional mammalian tissue is a stem cell or not. There are also no morphological criteria to identify such cells. This is partly due to lack of appropriate experimental techniques, and partly due to some conceptual problems. At the present stage it is helpful to discuss stemness as a latent variable which cannot directly be observed and which can only be deduced retrospectively on the basis of some indirect evidence based on measurable observable parameters in specific experimental settings. If one accepts stemness as a hidden property, it becomes obvious that one has to talk about stem cells and stem cell properties within the framework of concepts and models. This clearly implies that there cannot be a canonic unique definition but that a wide variety of models and concepts can be imagined, and have in fact to be proposed to describe various features of stem cell systems. Furthermore, it is evident that such models will differ in their attitude, their methodology and the set of phenomena on which they focus. At present there is no generally accepted standard stem cell model.

This introductory chapter is designed to serve as a framework for the stem cell models discussed in subsequent chapters. Starting with some general definitions, we try to set up criteria that should be fulfilled for stem cells. We point out the conceptual distinction between stem cells and transit cells before we discuss the problem of obtaining measurements on stem cells. We further discuss the basic elements of models used to describe cellular hierarchies and stem cells. In undertaking this exercise we will highlight our present reflections on this topic but also try to relate these to concepts suggested by other authors. In this respect we will extend ideas discussed in a previous paper (Potten and Loeffler, 1990).

DEFINITIONS

General definitions and concepts

In order to understand the full meaning and implications of the definition of stem cells, we need to consider some subsidiary definitions. The most important ones are associated with differentiation and maturation.

Differentiation

Differentiation can be defined as a qualitative change in the cellular phenotype that is the consequence of the onset of synthesis of new gene products, i.e. the non-cyclic (new) changes in gene expression that lead ultimately to functional competence (see Lajtha, 1979c). It may be recognized by a change in the morphology of the cell or by the appearance of changes in enzyme activity or protein composition. Since it is a qualitative change, a cell can be said to be differentiated only relative to another cell, and during its life a cell may be capable of undergoing several differentiation events. Differentiation is commonly identified by the detection of a novel protein. The ability to define a cell as differentiated thus clearly depends on the sensitivity of the detection procedures. A few molecules of a novel protein may be detectable, as may the changes in the messenger RNA responsible for these molecules, but ultimately the differentiation event involves a change in the repression/activation of the genome, i.e. in transcription, and this may approximate to a quantal phenomenon. According to this definition cells developing from a primitive stage to functional competence may undergo many, even a series of, differentiation events each linked to a novel change in the gene activation pattern. In many circumstances, it may be practically helpful to consider only some primary key (marker) genes as relevant indicators of differentiation, particularly if secondary genes are activated subsequently.

Maturation

Maturation in contrast can be regarded as a quantitative change in the cellular phenotype or the cellular constituent proteins leading to functional competence (see Lajtha, 1979c). Thus the degree of maturation, in principle, could be measured on a quantitative scale, e.g. of the amount of a specific protein per cell. A differentiated cell matures with the passage of time to form a functionally competent cell for that particular tissue. Its passage through time and space could in principle be mapped, as new differentiation events occur changing the path of the cell. This relationship is illustrated in Figure 1.

The terms differentiation and maturation are often used in a loose fashion, interchangeably, with a consequent potential for confusion. It is also common to see the term *terminal differentiation* used without an adequate definition of its meaning, which is presumably an implicit indication of either the activation of the last differentiation event in the cell's life history (e.g. involucrin synthesis for epidermal keratinocytes) or more likely an indication of terminal maturation, i.e. the accumulation of differentiated

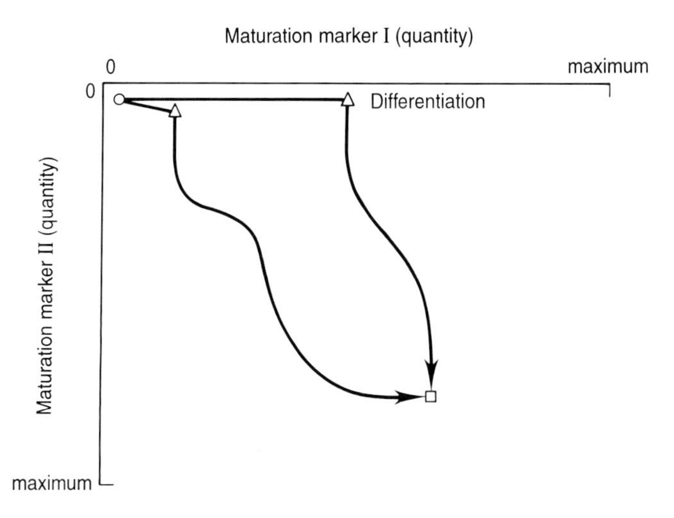

Figure 1 The course of an individual cell can be described in a differentiation–maturation diagram. Acquisition of a qualitatively new marker is defined as differentiation (△), while the trajectory for a given marker (from ○ to △) or, for a set of markers (from △ to □) is defined as maturation. Different maturation/differentiation paths may lead to the same state (□).

product(s) consistent with the final functional role of the cell (e.g. terminal keratinization, or cornification for epidermal keratinocytes).

Proliferation

Proliferation is a process involving a sequential pattern of (cyclic, repeating) changes in gene expression leading ultimately to the physical division of the cells. This is in contrast with cell growth, which involves an increase in cell size or mass. In order to identify a proliferating cell these changes have to be detected, and sensitivity problems similar to those associated with differentiation are encountered. The changes may be represented by discrete step-wise changes in the cellular concentration of, or by sharp peak alterations in, proliferation gene products. Many of these changes can be, and indeed have been, mapped on a time scale represented by the interval in time between two subsequent cell divisions, i.e. mapped in relation to the *cell cycle*. A large number of the gene products of these proliferation-associated genes (which include many cellular oncogenes S-phase and mitotic enzymes, cyclins etc.) have been mapped as transition points in the cell cycle. Traditionally the four major transition points, the onset and termination of DNA synthesis and mitosis have been used to identify proliferative cells, but many other transition points may be equally valid.

There are certain difficulties in distinguishing cells on the basis of our definitions of differentiation and proliferation. The first thing to note about these two processes is that they are not necessarily mutually exclusive. Indeed many cells in the adult body may exhibit differentiation markers, and hence be differentiated relative to cells earlier in tissue development, and yet they also proliferate. Certainly many cells in bone marrow exhibit both properties. Haematopoietic stem cells in the bone marrow are differentiated relative to embryonic stem cells. The stem cells in surface epithelia may be differentiated relative to the bone marrow stem cells and *vice versa*. The characterizations of the state of proliferation and differentiation are dependent upon the ability to identify changing patterns in gene expression and gene products. If these changes are of a cyclical nature they may be associated with proliferation. However, the cells under consideration may divide only once or we may have no knowledge of their previous history, in which case we are unable to tell if a particular gene product has been produced cyclically. Hence, it is more useful to define proliferation on the basis of the appearance of gene products associated with DNA replication, or the cell division process, which are in fact produced in a cyclic fashion. This implies a knowledge of many or all the metabolic processes associated with, and leading to cell division.

The distinction between differentiation, maturation and proliferation appears important as the development from stem cells to functionally competent cells can be viewed as a transition from one extreme (prolif: yes; diff/mat: no) to the opposite extreme (prolif: no; diff/mat: yes). The transition takes place through states of coexistence with some flexibility to accelerate or slow down one or both processes. It is this flexibility that permits cells to be stimulated to differentiate and stop proliferation and vice versa. Below we will introduce the assumption that proliferation, differentiation and maturation are not strictly coupled and in many circumstances should be considered independent of each other.

A special consideration here is to what extent a differentiated cell can dedifferentiate, whether this involves a switching off of the already activated differentiation genes (this process in itself could be regarded as a differentiation step in certain circumstances) or an inhibition of further maturation. Under certain natural circumstances it appears that some limited categories of cells in the bone marrow and in the intestine can dedifferentiate and assume stem cell potential. To what extent this process can be experimentally manipulated by, for example, providing the correct set of signals/growth factors, remains to be seen.

Definition of stem cells

Criteria for actual and potential stem cells

Stem cells are defined by virtue of their functional attributes. This immediately imposes difficulties since in order to identify whether or not a cell is a stem cell its function has to be tested. This inevitably demands that the cell must be manipulated experimentally, which may actually alter its properties. We will return to this circular problem later. The second problem faced in defining the stem cell population is that the definition can only be relative compared with other cell types. We would define the stem cells of a particular tissue as (a) undifferentiated cells (i.e. lacking certain tissue specific differentiation markers), (b) capable of proliferation, (c) able to self-maintain the population, (d) able to produce a large number of differentiated, functional progeny, (e) able to regenerate the tissue after injury, and (f) flexible use of these options (see Lajtha, 1967, 1979a, 1979b, 1979c; Steel, 1977; Potten and Lajtha, 1982; Wright and Alison, 1984; Potten and Morris, 1987; Hall and Watt, 1989; Potten and Loeffler, 1990). Table 1 gives a summary of these criteria. Ideally, in order to categorize a population of cells as containing stem cells, all of these criteria should be satisfied; in practice, there are experimental limitations. This is further complicated by the fact that not all of these functions have the same weighting. For example, it would not be sufficient to characterize a stem cell by virtue of its ability to proliferate alone. Cells or populations of cells actually fulfilling all these criteria at a given instance will be called *actual stem cells,* while those not actually expressing these capabilities at a particular moment in time, though they possess these capabilities, will be termed *potential stem cells.* It may be possible for a stem cell to cease proliferation, i.e. become *quiescent,* in which case it is not an actual stem cell, but since it can re-enter the cycle and it has the potential to be a stem cell. These cells should

Table 1 Stem cell criteria.

Criteria	Stem cells	Transit cells	Maturing cells
(a) Differentiation marker	no	onset	yes
(b) Capable of proliferation	yes	yes	no
(c) Capable of self-maintenance	yes ($p_{sm} \geq 0.5$ possible)	no (but $0.5 > p_{sm} \geq 0$ possible)	no ($p_{sm} = 0$)
(d) Capable of many progeny cells	yes	limited	no
(e) Capable of regenerating tissues after injury	yes (long term)	temporarily	no
(f) Flexibility in options	(b)–(e)	(b), (d)	no

p_{sm}, self-maintenance probability.

perhaps be termed quiescent actual stem cells to distinguish them from other potential stem cells. Likewise a transit cell (see below) may not normally self-maintain, but may do so under special circumstances, thereby representing a potential stem cell. Without going too much into detail here, it is sufficient to point out that we may have two classes of stem cell; those that actually satisfy the requirements of the definition, i.e. actual stem cells, and those that may have the potential to do so under special conditions, i.e. potential stem cells. We choose the word actual in preference to the term functional, which has been used previously (Steel, 1977; Cairnie *et al.*, 1965; Wright and Alison, 1984).

Some terms outlined in our stem cell definition have stronger weight than others and hence some of them could be accepted on their own as a means of identifying stem cells: (1) self-maintenance and the ability to vary self-maintenance; (2) the ability to produce a large family of differentiated functional cells; (3) the ability to regenerate the tissue or elements of it by producing a large family of differentiated functional progeny following injury. However, it is evident that taken alone any of these criteria will select quite different cell populations. Subsequently we discuss the precise meaning of some of the definitions just used.

Self-maintenance and self-renewal

Self-maintenance, self-renewal, self-reproduction, self-replication and self-regeneration are terms that have been used in connection with stem cells, often interchangeably and without definition, to the detriment of clarity. However, these terms have subtle differences in meaning and should be used with care, as discussed in our earlier review (Potten and Loeffler, 1990).

Maintenance means 'keeping at an existing state or level', and when considered in terms of numbers is a meaningful expression to apply to a stem cell population (see Lajtha, 1979a; Potten and Lajtha, 1982). The ability to maintain its own numbers without input from other cell stages, i.e. self-maintenance, is exclusively a property of stem cells. The term *renewal* can be defined as 'to make like new' which implies an element of rejuvenation. We would like to restrict the term renewal to a specific process to be discussed below in which transit cells regain stemness properties.

The term *reproduction* means to 'give rise to offspring' and is thus a property of all proliferative cells. The term self-reproduction, however, implies that the offspring are identical in every sense, including genetically, with the parent, and is therefore a term best restricted to budding or cell cloning processes. It is clear that self-reproduction has a stronger implication than self-maintenance.

Replication implies duplication or repetition and has connotations somewhat similar to reproduction. Self-replication implies production of identical twins, while self-maintenance implies maintenance of a functional ability (e.g. number) irrespective of the identity.

Regeneration implies 'to make again' something that was already pre-existing. It could apply to a tissue or a population of cells and would be more appropriately used in connection with other processes, to be discussed later. So stem cells may be defined as cells capable of self-maintenance and regeneration under certain conditions.

There are basically two ways to describe stem cell self-maintenance conceptually (Figure 2). The first concept relies on a deterministic description of the entire cell

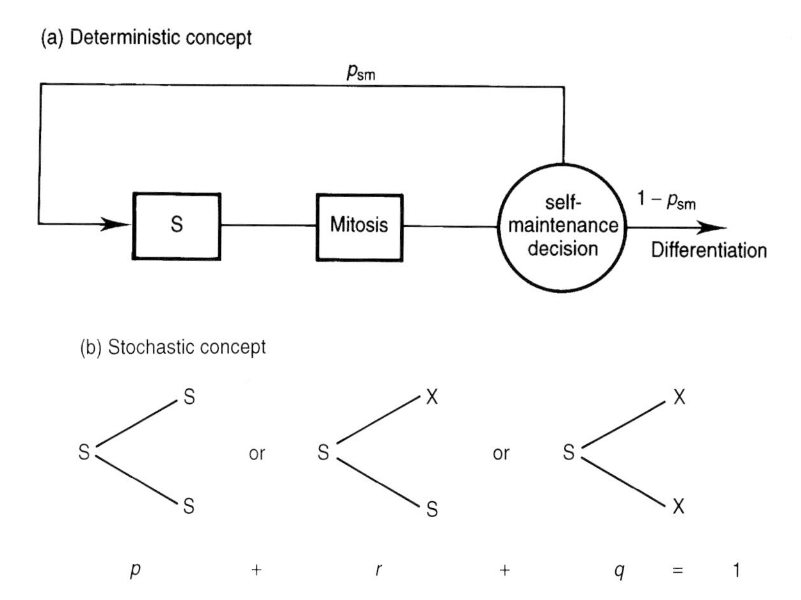

Figure 2 Stem cell concepts. (a) Deterministic concept of population growth; (b) Stochastic concept of single cell growth. (p_{sm}, fraction of postmitotic stem cells remaining in the stem cell pool; steady state condition $p_{sm}=0.5$; p, probability of symmetric cell division generating two stem cells; r, probability of symmetric cell division generating one stem cell; q, probability of stem cell loss by symmetric cell division or extinction; strict steady state $r=1$; note that $p=q\neq 0$ implies long term random extinction for finite populations.)

population (Figure 2a). In this case we consider a compartment, i.e. a pool of many stem cells which are actively cycling and undergo mitosis at a certain rate. The population concept implies that at any given instance in time a fraction p_{sm} of postmitotic stem cells remains in the stem cell pool. The cell population is maintained at a steady state if $p_{sm}=0.5$, indicating that on average half of the daughter cells maintain stem cell properties while the other half leaves the compartment. There is an increase in the stem cell population if p_{sm} becomes larger than 0.5, and there is a decrease if it becomes smaller than 0.5. It should be noted that this quantity has often been termed by us and others 'self-maintenance probability', which is not literally correct, as no random process is involved. The concept describes an average population and does not take into account variations and fluctuations in the population size due to random events. Hence the description is phenomenological as it does not relate the processes of growth of the population to the microscopic events occurring at the single cell level.

In contrast the second frequently used concept relates the population dynamics to the microscopic behaviour of single cells as indicated in Figure 2b. Each stem cell undergoing a cell division can either generate two, one or no daughter stem cells. In order to take our general ignorance about the specific decision in a specific cell into account we describe the processes with probabilities. We distinguish three such probabilities which add up to 1 (i.e. $p+r+q=1$). A strict steady state is maintained if only asymmetric divisions are occurring ($r=1$). On the population scale a stationary state is also possible if r is smaller than 1 provided that $p=q$. However, it should be noted that such a situation leads to long term random extinction of a finite cell population if p and q are constant with time. This is a particular feature of a stochastic process not present in deterministic concepts. Hence, stochastic models will require special mechanisms to assure long term stability.

Consideration of the particular type of division shown in Figure 2b illustrates a particular conceptual problem. If one looks at individual cells, a cell that produces two daughters that are not stem cells cannot itself be considered a stem cell since it does not satisfy the stem cell criteria of self-maintenance. However, the definition is still applicable to a pool of many such cells contributing as a whole to the self-maintenance process. Thus the two concepts of deterministic population growth and stochastic single cell behaviour do not coincide under all circumstances. This problem of compartment size relating to the stem cell definition reappears on numerous occasions in the consideration of stem cells. Only in the case of a large stem cell population can one equate $p_{sm} = p + 1/2(r+1)$. In the case of small stem cell populations the experimental procedures may lead to a serious estimation bias of p_{sm}. This is illustrated in Figure 3. Here, we have two cell lineages, both characterized by predominantly asymmetric microscopic divisions. In the case of lineage (a), whatever compartment size one considers even down to a single cell division, one could conclude that stem cells are involved and a population determination of p_{sm} yields 0.5. However, in lineage (b) the overall p_{sm} equals 0.5 (steady state), but there is a series of inappropriately small subcompartments for which p_{sm}, if valued, would be very different (box E, 0.0; box C, 0.33; box D, 1.0). Hence the determination of self-maintenance properties of a population on a small sample may be very misleading. Furthermore it is necessary to have constant growth conditions over an appropriate period of time. Thus compartment size is important in considering

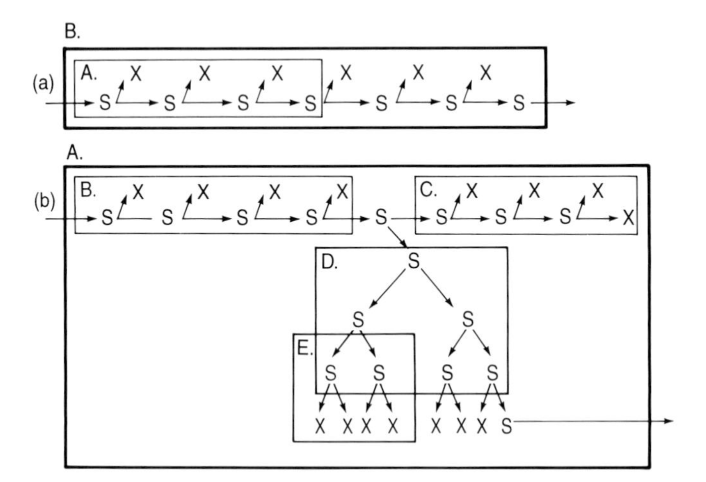

Figure 3 Compartment size considerations. (a) Permanent asymmetric stem cell lineage. Whatever compartment size is considered, self-maintenance is satisfied. (b) A second stem cell lineage. If all the divisions are considered self-maintenance is satisfied (box A). Similarly for a selective smaller box (B) it is satisfied although a similarly sized different compartment (box C) does not satisfy these self-maintenance criteria. Other compartments may show no self-maintenance (box E) or maximum values (box D). The values for p_{sm} in these boxes are: $p_{sm}=0.5$ for A, $p_{sm}=0.5$ for B, $p_{sm}=0.33$ for C, $p_{sm}=1.0$ for D, $p_{sm}=0.0$ for E.

self-maintenance but Figure 3 could also be thought of in terms of time frames. Short time frames are as inappropriate as small compartment sizes.

Relativity

One criterion of stem cells is that they are undifferentiated compared with the functional end cells of the particular tissue to which they give rise. This definition is essentially a relative one as it relates the stem cells to the functional end cells, or relates them to cells at earlier stages of the development, or stem cells in other times. Apparently, this definition is compatible with the existence of various stem cells of different tissues as well as of a hierarchy of stem cells for one particular tissue. It is possible that there are specific differentiation markers which would enable a distinction of stem cells in relation to one another and in relation to the functional cells they are eventually producing.

The relativity of stemness is an essential feature to keep in mind and one has to be specific with respect to the particular experimental circumstances. For example, in the haematopoietic system one is presently inclined to distinguish several stem cell classes out of a continuum of stem cells ranging from long term repopulating cells to colony forming units in spleen (CFU-S) (see Chapter 13). A similar continuum has recently been suggested for intestinal stem cells (Potten and Hendry, 1995).

Pluripotency

In the stem cell definition given above, pluripotency was not requested as a prerequisite of stemness. However, it is clear that most tissues contain a range of different specialized functional cells. These may all originate from a common compartment of stem cells in the tissue, the range of variable different differentiation options being facilitated by the length of the transit compartment. Those tissues with the greatest differentiation potential, for example bone marrow, tend to have the longest transit compartment, although the differentiation of individual lineages may well occur earlier in the transit compartment. The limit to the differentiation potential for individual stem cells is unclear and may well differ from tissue to tissue. The ability to produce progeny that differentiate down various lineages (pluripotency) is not necessarily a property of stem cells *per se*, although it appears that many stem cells possess this capability (see Chapters 10 and 13).

Definition of maturing cells

Maturing cells can be defined as cells with (a) full expression of a differentiation marker, (b) no capability of proliferation, (c) no capability of self-maintenance, (d) no capability to produce any progeny cells after injury and (e) hence no ability to regenerate tissue after injury.

Maturing cells therefore represent cell stages which are close to completing their development and becoming functional end cells. In this context, for example, reticulocytes would be maturing cells, as would segmented neutrophils in the bone marrow, villus cells in the intestine and superficial stratified cells in epidermis.

Definition of transit cells

Transit cells can be defined as a cell stage which is intermediate between stem cells and maturing cells. We define transit cells by the following criteria (see Table 1): (a) they are characterized by the onset of differentiation marker expression during their development which are, however, not mandatory; (b) they are capable of proliferation; (c) they cannot self-maintain. This implies that transit cells may be able to operate as amplifying cell stages generating many maturing cells from the few cells entering the transit cell stage. If one would determine the self-maintenance probability p_{sm} for these amplifying cells, one would obtain a value clearly below 0.5. Therefore transit cells would be capable of producing many progeny cells (criterion d) which are temporarily capable of regenerating a tissue after injury (criterion e). However, no long term regeneration and no functional re-establishment of the tissue would be possible.

The essential feature of a transit cell population is that it irreversibly develops towards maturing cells thereby undergoing several rounds of cell division. This process essentially operates as a cellular amplification machine generating numerous end cells from the few cells entering the system. There are again two ways of describing the transit cell stages (Figure 4). One way would be a population description in which a sequence of

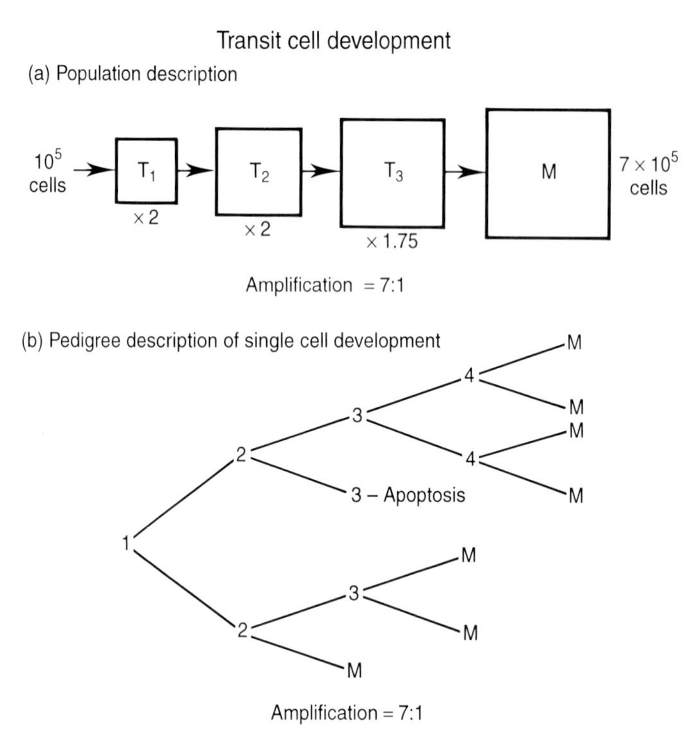

Figure 4 Schemes of transit cell developments. The two diagrams indicate cellular developments which generate numerous mature cells (M) out of less mature and still dividing cells. (a) In a population description the average amplification of many cells can be used as a parameter disregarding single cell development. (b) In a pedigree description the fate of an individual cell is sketched out. Such a pedigree can be quite asymmetric as indicated and may in fact vary between different T_1 cells.

developmental steps (T1, T2, T3, T ...) follow one another. Without looking into the details of the cellular development one could count the cells in each of these stages.

An alternative description relates to the microscopic single cell development and leads to a pedigree description. Figure 4b shows an example of one such pedigree of cells developing out of a primitive transit cell. In this particular pedigree maturing cells originate as early as two cell divisions and as late as four cell divisions after entry into the transit cell stage. This example shows a marked asymmetry in the cellular development. Furthermore, particular processes like apoptosis can be easily introduced into this description. However, at the present stage of cell biological experiments our knowledge is very limited on individual cell development and pedigree descriptions at this level of sophistication are rarely available. Most of our present knowledge is therefore based on the population approach which is an averaging process of the latent pedigree development.

It should be emphasized that a pedigree description or a population description of an age-structured transit cell population as shown in Figure 4 implies a particular distribution of the number of generations cells need to process through. Very little is known at the present stage about this process, its variability and its controls. The question could be rephrased as: 'what determines the number of transit generations?' Is it a clock within the cells counting their divisions or are external signals restricting the division potential? Different systems seem to have selected either the former or the latter, or perhaps some combination of both. It would appear, for example, that in bone marrow the number of transit divisions is under regulation and can be modified, while in some more primitive systems a counting clock may operate. If these processes are under regulatory processes then if the restricting factors are removed transit cells may divide many times (e.g. in culture condition) without exceeding any certain maximum.

A good example of a population description of the cellular developments is given by models of the haematopoietic system discussed by Wichmann and Loeffler (1985) and Loeffler *et al.* (1989). In these examples of the haematopoietic system an entire cell stage (e.g. colony forming unit erythroid (CFU-E) see Chapter 13 for details of cell types) was pooled together into one population compartment. In this compartment a cellular amplification was assumed. Hence, a CFU-E is generating further CFU-E cells which might be misinterpreted as evidence of self-maintenance. In fact there is the mathematical possibility to translate amplification within one compartment into a self-maintenance probability p_{sm}. However, this self-maintenance probability will in no situation exceed the value of 0.5.

PROBLEMS IN MEASURING STEM CELLS

How to check the criteria in practice?

If we consider the stem cell definition presented above, the question arises as to whether it can be used in a practical sense. A stem cell is a proliferative cell but this is the weakest part of the definition. Proliferation can be strictly identified in a population only by determining the future behaviour of the cell in question, i.e. whether it will divide into two cells in the future. In practice it is usually sufficient to identify that the cell, or population of cells, expresses one or more of the many markers of transition through the cell cycle, the commonest of which is whether it enters DNA synthesis, but the simplest is whether it enters or is in the mitotic phase.

On the other hand it has been suggested that stem cells having the potential for division may in fact be in a resting phase for most of their life and can therefore be identified by appropriate techniques such as long term S-phase label retention. However, as is evident from this feature such cells cannot be participating actively in tissue maintenance.

The second aspect of the definition is whether or not the cell is undifferentiated, which is a qualitative and relative term. It would usually be assessed by observing the morphological status of the cell and whether or not it expresses one or more markers for differentiation.

Another aspect of the stem cell definition relates to the ability of these cells to produce a large progeny of differentiated cells. This again is a question related to the future potential of the cells in question and can only be tested by placing the cell or cells in a situation where they can express this potential, e.g. placing the cells in culture or arranging for a situation *in vivo* where the regulators limiting stem cell growth are removed, as would happen in a situation where some time stem cells were killed, i.e. during a regeneration. We will return to the question of regeneration.

Self-maintenance is the cardinal property of stem cells, but here again it can only be assessed in terms of the future of the cell. Can the cell produce other cells like itself and maintain the population over a period of time?

Self-maintenance is perhaps the most difficult property to determine experimentally. There are at present no easy and reliable assay techniques to measure this feature. Perhaps still the best approaches to the problem are modifications of the original recloning experiments originally used in the haemopoietic system. In a series of classical experiments Till and McCulloch (1961) examined whether a colony produced in the spleen (CFU-S) after transplantation of bone marrow into lethally irradiated mice contained cells which would generate secondary colonies if excised and re-transplanted. Thus the basic idea to investigate self-maintenance is to check by recloning experiments whether one can obtain cells which fulfil the same criteria of stem cells successively. This approach has in fact been used to determine p_{sm} for CFU-S in the early 1960s (see below).

Referring to the boxing phenomena discussed above (Figure 3) it is evident that any determinations of self-maintenance have to be conducted on populations of cells. As a consequence of using populations of cells one loses the information about an individual cell. This necessarily implies that one can only make statistical statements. On the other hand, the problem of investigating a cell population is that one cannot control whether the cell population under investigation receives any input from more primitive cell stages which may contaminate the population. It is therefore always possible, and has in fact been suggested for the CFU-S assay, that some very primitive stem cells have been present in the spleen colonies which gave rise to the CFU-S cells which themselves could be transit cell stages. Thus, the examination of cell populations to determine self-maintenance is often open to criticism, and there are very few biological systems like the intestinal crypt where one can conclude from the anatomical construction of the system that stem cells cannot immigrate and must be constitutive to the cell population.

The final aspect of the stem cell definition, which is associated with the property of a large proliferative potential, is whether regeneration can be achieved. Besides being a property associated with the future this is specifically a property associated with disturbance of the system. In practice, clonal growth assays are a frequently used way of assessing stem cell function. For adult tissue stem cells, a variety of clonal regeneration assays have been developed *in vivo* and a number *in vitro* (summarized in Potten and Hendry, 1985a, 1985b). The most effective of these is the spleen colony assay for haemopoietic stem cells (Till and McCulloch, 1961) and the micro and macro colony assays for intestinal epithelium (Withers and Elkind, 1969, 1970; Potten and Hendry, 1985b), and epidermis (Withers, 1967). In these cases, the colonies assessed are large and contain very many cells and, if the necessary conditions are satisfied (high doses of radiation), represent *clones* derived from a single surviving stem cell. The fact that these clones contain many cells demonstrates the large division potential of the originator cell (*clonogenic cell*). The fact that the clones can often contain several differentiated cell lineages indicates that the original stem cell was pluripotent. The self-maintenance element can be assessed by either a second clonal regeneration assay starting with the first clone, or simply from the longevity of the clone in terms of maintenance of its cellularity and differentiation and the fact that it eventually repopulates an entire area of the tissue.

Previous functional assays of stem cells were often clonogenic assays. A *clonogenic cell* is thus a cell that is capable of producing from one cell a large number of progeny, i.e. a clone (see Potten and Hendry 1985a). Clonogenic competence is usually measured by looking at surviving clones and as a consequence clonogenic cells might better be described as those cells in a tissue which if all killed will cause the tissue to degenerate. This is really only a technical aspect of stem cell measurement in some specialized circumstances. If the clone can be demonstrated (usually by secondary clonogenic assays) to contain further clonogenic cells, then self-maintenance has been satisfied. Clonogenic cells thus may satisfy some of the criteria for stemness, e.g. clone formation and self-maintenance. However, it is often not clear whether clonogenic cells are able to regenerate the entire tissue in the long run because they are assayed fairly soon after the damage. This raises the question of whether clonogenic cells are a mixture of stem and early transit cells. It is most likely that clonogenic assays measure all, or a part, of the potential stem cells, which may be a considerable overestimation of the number of actual stem cells.

Ontology, uncertainty and probability

A question frequently posed by biologists is: 'is this particular cell a stem cell?' We refer to this as the ontology question. It implies the idea that one can decide about the capabilities of a given cell without relating it to other cells and without testing the capabilities functionally. We believe that this is a very dogmatic and unrealistic point of view.

As we have seen above, the main attributes of stem cells relate to their potential in the future. These can only be studied effectively by placing the cell, or cells, in a situation where they have the opportunity to express their potential. Here, we find ourselves in a circular situation. In order to answer the question whether a cell is a stem cell we have to alter its circumstances, and in doing so inevitably lose the original cell, and in addition we may only see a limited spectrum of responses. This situation has a marked analogy with *Heisenberg's uncertainty principle* in quantum physics. In simple terms, this states that the very act of measuring the properties of a certain body inevitably alters the characteristics of that body, hence giving rise to a degree of uncertainty in the evaluation of its properties. The analogy holds true for the functional stem cell assay procedures, all of which study the response after a perturbation to the system thereby challenging the different capabilities of the cells in different though complementary ways. Therefore it might be an impossible task to determine the status of a single stem cell without changing it. We hereby postulate a fundamental uncertainty problem for stem cells. This implies that one will not be able to make a definitive statement about whether or not a given cell is a stem cell. It implies that all statements that we can make will be necessarily probabilistic statements about the future behaviour of the cell under consideration. Essentially this has two particular aspects. The first is that we can only make statements about cell populations in the statistical sense of expected values under a given statistical model. This implies that measurements will necessarily be conducted on populations of cells. However, the second essential aspect is that we cannot disregard the experimental procedure by which the stem cell under consideration was challenged. Each particular measurement or perturbation process may induce a different response in one or several of the characteristics of the stem cell. This is constitutive to the stem cell property, as it is thought to be reactive and responsive to various types of perturbations. Hence, all statements about stem cells and their reactions have to be given in the context of the perturbation of the measurement process under which they were obtained.

This type of uncertainty concept makes it obvious that the ontology question can be rather misleading. We therefore advocate focusing research not so much on the question of the causes of effects but much more on the effects of causes. This relates to comparing the behaviour of stem cell populations exposed to different types of manipulation.

Another aspect to consider here is to what extent a stem cell is intrinsically different from a transit cell and to what extent its differences are merely imposed upon it by its environment. It is extremely difficult to make dogmatic statements here but the indications are that at least to some extent some intrinsic properties exist.

MODELS OF CELLULAR HIERARCHIES

In the previous section we have already introduced some theoretical concepts which can be used to describe cellular development. Here we analyse and extend this topic by giving a brief review of model concepts actually used in various descriptions of biological systems.

The essential aspect of the following three types of model is that they all are designed to describe some features of the cellular development from the stem cells to the maturing cells via transit cells.

Compartment models

A frequently used class of models is based on the compartment concept. It is applicable if one considers a large number of cells in which only the average behaviour is the focus of interest and where variability and fluctuations do not play a role. Usually such models imply the assumption that all cells in a compartment behave alike and that the behaviour of the compartment can be described by some deterministic laws.

A simple population compartment in connection with stem cells would be one which was purely expansionary in growth (see Figure 2). Such compartments may exist when a large number of stem cells is placed in culture, during early embryogenesis, and possibly under some conditions of wound or tissue repair and tumour growth. It is possible to control the expansionary growth by removal of some stem cells which could for example be achieved by applying a simple spatial cut-off. As the cells reach a particular point in the tissue, for example the top of the intestinal crypt, they are instructed by some signal(s) to become mature functional cells. Such a cut-off could operate via some chemical signal from outside the crypt or by a chemical gradient of intracellular factors.

To our knowledge there are few (if any) mammalian tissues which have only stem cells and mature cells. It is unclear why this is so. Perhaps the switch from proliferation to differentiation and maturation is not a simple change in genetic programs but requires time and even cell divisions (Holtzer, 1985). In a situation where there are only stem cells (A) and mature cells (M), there is a large population of stem cells at risk from genetic error (see Cairns, 1975). One solution to these problems might have been to use the time that it takes for maturation for some continued cell proliferation. Such a dividing and maturing cell population allows much of the workload in terms of cell production to be removed from the stem cell compartment, which as a consequence becomes much smaller and hence offers a smaller target for genetic and carcinogenic damage. This maturation time also allows for the generation of diversity of function, i.e. additional differentiation events, whether this requires rounds of cell division (quantal cell cycles, Holtzer, 1978, 1979) or some other mechanism remains uncertain (Lajtha, 1979b).

This reasoning introduces a new class of proliferative cells, the *dividing transit cell compartment* (*T*) (see Lajtha, 1979a, 1979b; Gilbert and Lajtha, 1965). The concept implies that, wherever a high cell production rate is required, a T population might be expected and the higher the cell production rate the more cell divisions could be expected in the T population (see Cairns, 1975; Lajtha, 1979b; Potten and Lajtha, 1982). The converse might also be expected. If turnover is extremely slow the tissue could in principle operate effectively with just stem cells and maturing differentiated cells. Convincing examples of this type of organization are lacking. It is clear from the scheme represented in Figure 5 that the cell production rate, i.e. the number of M cells produced per unit time, is determined by the number of stem cells considered, their cell cycle time and the number of cell divisions (amplification) in the T compartment. The advantage of the transit population is that it enables some genetic protection to be afforded to the stem cells. It effectively amplifies each stem cell division thus minimizing the number of stem cell divisions required, and hence conserves the stem cell genetic load. It also allows for diversity of specialization at a low cost in terms of proliferation and genetic risk for the stem cells. The disadvantage of such a hierarchy is that the length of time spent in the T population can result in an instability in cell output following damage, i.e. overshoots and fluctuations. This can be overcome by introducing feedback loops and dampening phenomena, such as a high variability of cell cycle or transit times (Wichmann *et al.*, 1988). The total cell output can be controlled by either the number of cell generations in the T compartment, which might be controlled, for example, by a feedback loop from the M compartment, or by the output from the stem cells, i.e. their cycle time. If there

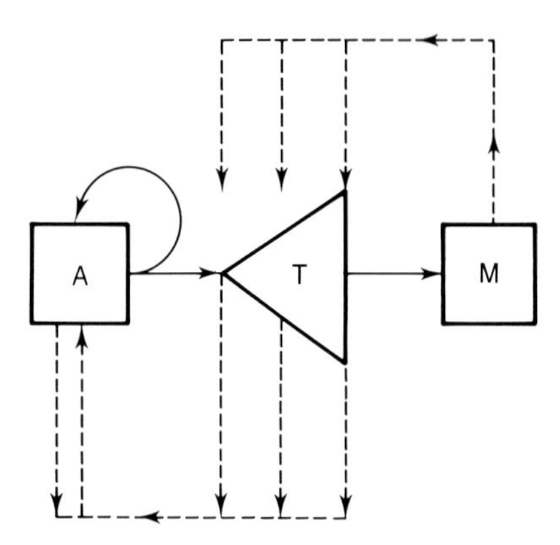

Figure 5 General scheme of an ATM tissue organization with actual stem cells (A), transit cells (T) and mature cells (M). Full arrows indicate cellular development. Dashed arrows indicate possible regulatory feedback. This structure has been proposed for the haematopoietic system (Wichmann and Loeffler, 1985).

are many T generations, there is a logistic problem in terms of the spatial distance in a tissue over which a feedback loop would have to operate from the M population to the stem population. This may be overcome by growth factors that operate over long distances, i.e. hormones, or a breakup of the system into several feedback loops. Such feedback loops are illustrated in Figure 5.

Compartment models are particularly useful to describe cell fluxes and cell production in a dynamic setting to highlight the behaviour with time. A particular advantage is that they are very useful in investigating the assumptions on regulatory control. Such models have in fact been used successfully by many authors who describe the haematopoietic system (e.g. Wichmann and Loeffler, 1985; Mackey, 1978), the intestinal crypt system (Britton *et al.*, 1982; Paulus *et al.*, 1992), the epidermal system and various tumour systems (see also other chapters in this volume). The remarkable technical advantages are that these models can be put into operation using differential equations making them easily manageable by computer simulations.

There are, however, some disadvantages with such models. First of all they are not helpful in taking spatial arrangement of cells and cellular heterogeneity into account. Furthermore, descriptions of systems containing only a few cells will be inadequate, as the stochastic fluctuation in such populations cannot be adequately considered. In addition, compartment models of cellular hierarchies are frequently limited as they include the assumption of a fairly strict link between cellular development towards maturation and the remaining potential for proliferation in the transit cell stage. In other terms such models imply an assumption about a fairly rigid age structure in which the differentiation age and proliferation age are fairly strictly coupled. This implies that cells can only become mature if they have undergone a certain number of transit cell populations. This may not in fact be correct for all tissues under consideration.

Single cell models and pedigrees

In Figure 4 we have already illustrated the principle or idea of a cellular pedigree used to describe the development of a single cell and its progeny. Such a pedigree description would, for example, ideally describe the cellular development seen in an *in vitro* cell culture. Clearly the structure of the pedigree may vary depending on the system, on the growth conditions and on the manipulations previously performed. Furthermore, the temporal development of the cells through the pedigree may be subject to random or systematic variations. It is evident that a cellular pedigree is a much more flexible way of describing cellular development than the compartment approach discussed above. There are various ways of operating such a model technically. One particularly interesting way is to relate the individual cell development to the spatial arrangement of cells. One can, for example, assume that cells are arranged on a two- or three-dimensional lattice and that each cell undergoes a development according to the cellular pedigree. If cells divide on such a lattice, migration processes have to be introduced to rearrange the spatial situation and to remove excess cells from the environment. Such model descriptions usually cannot be described by deterministic rules, and a certain degree of randomness, i.e. stochasticity, has to be introduced. Such models have in fact been successfully developed and applied to the intestinal crypt (Loeffler *et al.*, 1986, 1987, 1988; Potten and Loeffler, 1987; Paulus *et al.*, 1993) and the epidermal basal layer (Loeffler *et al.*, 1987). Such models provide a particular insight into the mechanisms of cellular migration and of development of cells belonging to one pedigree. Similar modelling should also be encouraged for other tissues.

The advantage of a cellular pedigree description is that one can describe systems with only a few cells for which detailed information is available. The conceptual advantage lies in the fact that the pedigrees can vary considerably, become symmetric or become highly asymmetric. Hence the pedigrees provide a microscopic basis for the other models. Furthermore, the pedigree concept permits one to decouple cell proliferation and cell differentiation. It is possible in this concept to develop maturing cells after few transit cell divisions as well as after many. Hence such models are useful in situations where a relative decoupling of the two processes is biologically likely.

Pedigree models technically can be simulated by using stochastic cellular automata. In our view they have not yet been sufficiently exploited.

The screw model

A third class of model is illustrated in Figure 6. Because of its shape we call it the screw model. Its main characteristic is that cell proliferation and maturation/differentiation are assumed to be two independent biological processes. While the cell cycle is represented by a cyclic coordinate on which cells may operate, the maturation/differentiation axis (downwards) indicates the irreversible development towards a mature cell stage. In this cylindrical coordinate system cells travel on a helical downwards spirally. While stem cells are represented by the circle at the top (self-maintenance cycling without differentiation) the transit cell population is characterized by proliferation and a continuous maturation and differentiation downwards. This concept was to our

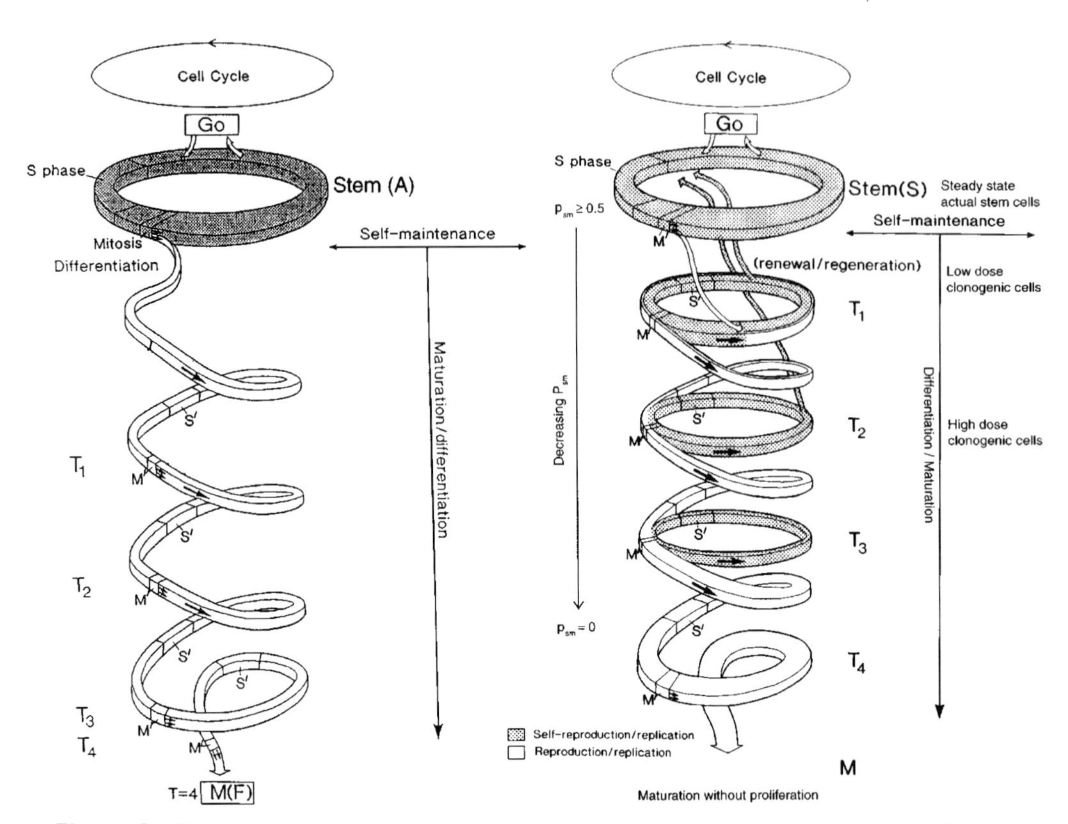

Figure 6 The screw model. Diagrammatic representation of the stem cell (A) and dividing transit (T) and mature (M) cell population. Proliferation is represented by the horizontal cylindrical axis and differentiation/maturation by the vertical axis. The stem cells are characterized by their ability to self-maintain (circle) at the same level of maturation and differentiation with p_{sm} exceeding 0.5 if necessary. The transit cells (T) retain some ability for self-maintenance (horizontal cell cycles as opposed to spiralling cycles). The probability of self-maintenance is always less than that for the stem cells ($p_{sm} < 0.5$) and declines with each transit generation. Thus the probability of a cell progressing down the spiral increases from >0.5 at T_1 generation to 1.0 at the T_4 generation. The ribbons indicate the potential fate of a cohort of cells starting development at the level of A-cells. In a real system the height and width of the ribbons will be much broader and frequently overlapping due to variation in cell cycle times, maturation velocities and self-maintenance processes. Since the T_1 population at least possesses some self-maintenance ability, it can be regarded, when under self-maintenance cell cycle conditions, to be indistinguishable from a stem cell and if a vacant space becomes available in the stem cell environment (niche – see Schofield, 1978) then such a T_1 cell could reoccupy a vacant niche and become an actual stem cell. This represents renewal. Hence renewal is a process that is unique to some transit cells. Having renewed, such a cell could then regenerate the stem compartment, the differentiation spiral(s) and the tissue. Regeneration is thus a property unique to stem cells. The T_1 cells here could be equivalent to the committed stem cells described in some cases. They are also potential stem cells as they have the possibility of renewal. At each level of this diagram, the self-maintenance probability (p_{sm}) changes as does the range of options open to a cell. It is important to realize that since this diagram represents the possible path of a cell in time where there is a bifurcation in the path (cell division in the T_1–T_3 population) the two paths represent the extremes of the options open to individual cells. (a) In the steady state situation self-maintenance at the level of T_1, T_2 and T_3 may not occur and hence the situation would be described by a corkscrew. (b) In regenerative situations self-maintenance and even self-renewal may come into play yielding the picture displayed.

knowledge first proposed by Mackey and Dörmer (1982) and elaborated in a recent review (Potten and Loeffler, 1990).

The situation represented in Figure 6a is one of two alternative possibilities for the T population. It shows a discrete (quantal) change from the stem to the transit population and although the T population in Figure 6a retains the property of division in common with the stem compartment, it cannot self-maintain (stay at the same level of maturation/differentiation). The T population is entirely dependent on an input from the stem compartment. If the stem compartment is removed or destroyed the T population will disappear and has no possibility of maintaining the tissue or itself. In this particular model, only the stem cells have the ability to regenerate the T population and the tissue.

The second and more realistic possibility for the T population is illustrated in Figure 6b, which suggests that the T compartment retains some additional attributes of stemness, i.e. they possess a progressively declining spectrum of stemness and an increasing spectrum of differentiation and maturation. This model suggests that the earlier transit compartments retain a certain ability for self-maintenance, i.e. some do not progress down the screw (i.e. mature) but remain for at least one cycle at the same level. This property may be retained to a lesser extent in the second and third generation (at declining levels). In this case, the feature that distinguishes a transit cell from a stem cell is not so much the question of whether or not it can self-maintain, but its maximum capability of self-maintenance. A transit cell population will by definition always have a p_{sm} value less than 0.5 under steady state conditions. Its ability to vary its p_{sm} may be considerable but restricted, and certainly declines with increasing maturation. The advantage of this model is that some T cells are very similar to the stem cells or indeed indistinguishable from stem cells in some situations but, although these cells may have an ability to behave like stem cells (i.e. $p_{sm} \geqslant 0.5$) in some special circumstances, under normal steady state conditions, they do not. The necessity for considering such a model comes from regeneration experiments *in vivo*, particularly those in the small intestine (see Potten, 1991; Potten *et al.*, 1987; Hendry *et al.*, 1992, and recent developments in techniques for studying bone marrow stem cells (see Chapter 13). It should be noted that the screw model is a much more comprehensive model of the stem–transit–mature (A-T-M) scheme than the compartment concepts in Figure 4 because it inherently allows for a description of a wide variety of different individual cell developments (trajectories). In this respect, it is an extension of the concept of a hierarchical tissue organization (Michalowski, 1981; Gilbert and Lajtha, 1965; Potten, 1974; Loeffler *et al.*, 1987; Clausen and Potten, 1990; Potten, 1991; Potten and Hendry, 1983; Potten and Morris, 1987; Potten *et al.*, 1979, 1982a, 1982b, 1987).

If we consider the scheme in Figure 6b the stem cells within this figure are clearly capable of clone formation and regeneration. A major question is whether any of the T population under conditions of severe cellular depletion could regenerate the epithelium. If this could occur it would involve a sort of rejuvenation process of the T cell, or a renewal of a T cell, i.e. its return to the status of a fully effective stem cell. Thus both stem cells and T_1 cells may be capable of *regeneration* of the tissue but it is only T_1 cells that undergo a process of *renewal*.

Clearly if T_1 cells really represent a spectrum of declining stem cell properties then these questions could in principle be asked concerning the T_2 population and so on.

If we regard the T_1 population as capable of re-entering the stem cell cycle then they constitute a class of *potential stem cells* as distinct from those that are performing stem cell functions which we would term *actual stem cells*. (A second class of potential stem cells would be those that had actually stopped progression through the cell cycle and therefore were in some *quiescent or* G_0 phase that could at any moment be recalled into proliferation.) We would define the *renewal* of a transit cell population as the fraction of T cells that rejuvenates such that it can perpetually self-maintain and retain the capability of tissue regeneration.

However, the difference between stem cells and T_1 transit cells may be very small and it might be difficult to prove the renewal process on a molecular basis because it would involve the demonstration that an activated T_1 differentiation marker disappears under certain circumstances. On the other hand, there is good experimental evidence in haemopoietic and epithelial systems that fully competent tissue regeneration is still possible even after frequent severe damage, e.g. chronic or fractionated irradiation or drug application.

In the small intestinal crypt such an hierarchical stem cell model has been proposed (Potten and Hendry, 1995). This attempts to take into account the following observations: (a) there are a small number of cells (about six per crypt) that occur at the stem cell location that exhibit an exquisite radiosensitivity and die via apoptosis when small amounts of dosage are induced; (b) after moderate doses of radiation, which would kill all of the above mentioned cells, there are about six other cells that can regenerate the crypt, i.e. are clonogenic; (c) after even higher doses where the above clonogenic cells are sterilized, there are a further 24 resistant clonogenic cells (Hendry *et al.*, 1992; Potten and Hendry, 1995). Altogether they constitute a total of about 36 potential 'stem' cells. The crypt then contains about 124 other proliferating transit cells with no stem cell attributes. The crypt stem cell hierarchy might then be structured as follows:

(i) Up to six actual steady state stem cells – responsible for day to day cell replacement but very intolerant of any DNA damage.
(ii) Up to six potential stem cells which are called into regeneration activity (clonogenic cells) after low levels of injury but when all six actual stem cells are killed.
(iii) Up to 24 other potential stem cells which are the most radio-resistant (good repair system) clonogenic cells called into action after high levels of damage.

STEM CELL MODELS

In the previous section we have discussed descriptions of the cellular hierarchies and in particular the development of cells through the transit cell stages towards the maturing cell stage. In this section we will focus on the various aspects of modelling the stem cell population. This section is not intended to be a review of stem cell models published so far but rather presents a selection of topics that have to be considered.

Clonal succession concept – pseudo stem cell models

There have recently been arguments that several cellular systems may be organized without self-maintaining stem cells. This was advocated in particular for the haematopoietic system (Kay, 1965; Abkowitz *et al.*, 1993; Guttorp *et al.*, 1990) and to some extent for the intestinal crypt (Winton *et al.*, 1988; Winton and Ponder, 1990) and possibly for the keratinocyte stem cells found in the hair follice (see Chapter 11). The somewhat simplified basic idea is that a cellular system is maintained by a reservoir of dormant stem cells which are available throughout life time and which can be challenged one after another to enter a transit-cell-like amplification and differentiation/maturation process. Once being triggered such a 'stem cell' would give rise to a large clone of maturing cells. Any of those clones would have a limited life span, as no further input is maintaining the cell flux. After a certain time, which may be long, such a clone would disappear and another 'stem cell' would have to be triggered and recruited to form another clone. This concept is generally considered as the clonal succession theory. The major arguments for these theories stem from observations of somatic mutation experiments in which the occurrence of clones of cells with a specific marker are observed in time. If few such stem cells exhibit the marker it is possible under specific circumstances to observe wide fluctuations in macroscopic clone compositions.

It is evident from these arguments that the clonal succession theory is a concept which is in marked contrast to the stem cell concept of self-maintaining stem cells discussed above. A 'stem cell' in the clonal succession concept is not self-maintaining and can only be viewed as a cell of origin. Once triggered, these cells propagate down the screw of the pedigree development. Hence the clonal succession theory implies that there is a naturally determined age limit for tissues (and organisms in general) which is determined by the exhaustion of the reservoir. It must be noted that in our mind there is so far no clear proof for this concept. As has been analysed for the haematopoietic system and for the intestinal crypt system the majority of findings used to support the clonal succession theory can equally well be integrated into a self-maintaining stem cell concept (e.g. Loeffler *et al.*, 1993). Thus at least for the intestinal system the question on the nature of stem cells is undecided. The basic question of whether or not a small population of self-maintaining stem cells really exists and can regenerate functionally competent stem cells remains to be fully elucidated.

At this point we would like to emphasize that the existence of an age structured stem cell population with a sequence of cells is not by itself an argument for either theory.

The crucial question is whether or not there are cells with self-maintenance capacity in the functional sense. The term 'functional' in this context is important, as it refers to the typical time scale and functional requirements in an organism under stress. For example, it would be sufficient to test whether a stem cell is self-maintaining in the sense that it can reconstitute a tissue for many months, e.g. haematopoiesis, or whether it can sustain this tissue for many years in sequential transplant experiments, which would be a very artificial situation.

Because of the lack of the self-maintaining property in the clonal succession theory we call these concepts pseudo stem concepts. Subsequently we will focus our interests on models which imply functional self-maintenance.

Deterministic models with self-maintenance and self-renewal

As already mentioned above, deterministic models are closely connected to the compartment concept. They represent average cell productions and cell numbers in single- or multi-compartment systems with or without feedback regulation. Perhaps the most simple stem cell model of this type was already described in Figure 2 where we assumed a homogeneous stem cell pool. A certain fraction 'a' is actively propagating in the cell cycle which has an average cycle time of T. Hence per time unit aS/T cells enter mitosis. Of these cells the fraction p_{sm} self-maintains. This in summary leads to a frequently used differential equation to describe the stem cell dynamics:

$$dS/dt = 2p_{sm}aS/T - aS/T = (2p_{sm} - 1)aS/T.$$

This equation has the important property that p_{sm}, a and T are considered to be independent of each other and therefore can be subject to different regulatory mechanisms. In an extremely simplified case (e.g. tumour growth) all three of these parameters can be constant with time, for example in a small homogeneous tumour. In this circumstance the solution of the differential equation will be an exponential growth characteristic. If we assume that either the fraction of cells in the cycle diminishes with time or the cell cycle time becomes longer with time or the self-maintenance probability reduces with time (e.g. increasing cell loss) then this exponential growth characteristic will slow down, and potentially flatten or even decline. Thus this type of equation can be used for a wide variety of growth circumstances, and several special cases of this type of model are used throughout this book.

We have ourselves used this type of model frequently to describe regulatory processes in stem cells in regenerative situations. It was for example utilized in great detail for the haemopoietic CFU-S population (Wichmann and Loeffler, 1985) and for the intestinal crypt (Paulus *et al.*, 1992).

However, it must be re-emphasized that this model can only be applied to large populations of stem cells where single cell fluctuations do not play a role. Furthermore, more sophisticated and detailed models of the stem cell proliferative process can be envisaged and have in fact been suggested.

In an attempt to understand the recovery behaviour of the small intestinal crypt a deterministic model was used in which self-maintenance properties were attributed to actual stem cells as well as to early transit cell stages under a circumstance of severe demand (Paulus *et al.*, 1992). This was necessary in order to reconcile findings on a small number of actual stem cells in the normal steady state crypts with findings of a much larger number of clonogenic cells becoming active after severe radiation damage. Thus we came to the conclusion that one way of explaining this was to assume a gradual loss of self-maintenance capacity as cells develop through the transit cell stage, e.g. Figure 6b. In the normal situation this capacity may not be utilized and hence these cells appear functionally as transit cells while under other circumstances the same cells can operate in a self-maintaining fashion. Whether this can be regarded as a real self-renewal process with transit cells rejuvenating to become actual stem cells or whether this is simple self-maintenance could not be deduced from these data.

Stochastic models

Stochastic models are legitimate if one considers fluctuations in cell populations due to small cell numbers. The classic model was proposed in the 1960s by Vogel *et al.*, (1968, 1969, reviewed by Wichmann and Loeffler (1983)). It has been used to describe the CFU-S assay of haemopoietic stem cells. In its basic variant it was described already in Figure 2. We refer to it as the *p, r, q* model, which indicates that three division processes, generating two, one or no daughter stem cells, are possible. In this model one assumes that at each division all three possibilities exist for a given stem cell, that this decision can be described as a random decision and that previous decisions do not play a role (Markovian property). In the model considered by Vogel *et al.* the probabilities *p, r, q* were always assumed to be constant with time. These models were able to explain the large variation of CFU-S numbers found in reseeding and recloning experiments of isolated CFU-S colonies. In fact Vogel *et al.* analysed such data with a restricted version of this model, assuming that only symmetric cell divisions would take place ($r=0$). He found the probability of producing two stem cells was in the range of 0.6–0.7 under most circumstances. As a consequence one obtains a clonal expansion of the stem cell population.

Recently we have applied the *p, r, q* model to the small intestinal crypt (Loeffler *et al.*, 1991, 1993) based on a microscopic growth process of few actual stem cells in the intestinal crypt and were able to explain a wide variety of macroscopic observations of the crypt's behaviour. In particular it was possible to explain a stem cell somatic mutation experiment in which the conversion of a crypt to a new monoclonality of the mutated marker was observed (Winton *et al.*, 1988, 1989; Winton and Ponder, 1990). After about 100 days a crypt with 300 cells converted from one marker status to another starting with an assumed mutation in one single actual stem cell. We deduced from model simulations that if the probability of asymmetric cell division *r* is in the order of 0.95–0.98 with about six actual stem cells being present the data could also be explained (Loeffler *et al.*, 1993). Thus under most circumstances the stem cell division would be asymmetric but in the remaining circumstances symmetric division would cause stochastic fluctuations. In the published version of these models we assumed that the values for *p, q* and *r* were constant with time. However, a number of recent experimental observations hint to the fact that this may not be the case. We have at present good arguments to believe that the probabilities *p, q* and *r* may in fact be variable depending on the actual status a crypt is in. Thus in a situation of severe stem cell depletion there may be a shift towards increasing the probability *p* of generating new stem cells at the expense of the other two processes.

Miscellaneous problems of stem cell biology

There are a number of problems which have so far not been elucidated sufficiently in stem cell models.

The first area of problems relates to the mechanisms of symmetric and asymmetric division. It is somewhat doubtful whether an asymmetric division really exists. We rather believe that quite a number of different processes can contribute to this asymmetry. It is possible that in particular micro-environmental and spatial effects contribute to this process, rendering two daughter cells unequal. Such reasoning would be implied in the niche concept proposed for the haematopoietic system (Schofield, 1978). There may, however, also be cooperative phenomena involved in which the stem cells or other cells send out signals on a local basis, generating something like a morphogenetic field in which cells depending on their position are subjected to such a decision. If such local processes play a role, the growth situation in an intact *in vivo* situation may be quite different from the *in vitro* environment.

A second area of problems relates to the regulatory processes acting at the stem cell level. Although many people have suggested an autoregulatory process controlling the stem cell population and activity, very little direct biological proof has been gathered. Perhaps one of the most sophisticated stem cell regulatory processes was postulated by our group (Wichmann and Loeffler, 1985) in which the self-maintenance property and the proliferative property of haemopoietic stem cells was assumed to be regulated independently by inter-related feedback loops. One loop relates to the autoregulation of stem cells while the second relates to the demand regulation for differentiated cells. Although this type of regulatory process has shown great potential in explaining a wide variety of experimental situations, still no proof or disproof of this theory has been collected. It should be emphasized that these regulatory concepts are based on the idea of a homogeneous reaction in the tissue which implies that all cells know about each other. In biological terms this mostly implies systemic control processes usually relating to systemically acting growth factor exposure. However, it may well be possible that all these phenomena can also be explained by a different type of concept which originates from the field of synergetics in which self-organization of multiparticle systems is the focus. It will be an interesting area for future research to investigate whether such an idea of cellular next neighbourhood interactions and resulting self-organization leads to new insights into the dynamics of the cellular regenerative tissue. The potential advantage of such an approach is that it takes the spatial arrangement of the cells in their micro-environment constitutively into account.

A third field of problems is the question of how one can describe a hierarchy of stem cells which may develop out of each other and which may gradually lose stemness properties in order to become transit cells. At present one of the strongest limitations of the model discussed so far is that the transition from the stem cell pool to the transit cell pool is a fairly marked transition which may be incompatible with a gradual change as suggested by experiments. Perhaps one has to consider concepts in which the self-maintenance possibility is gradually lost while cells travel down a differentiation/maturation process. It will, however, be a crucial question to investigate whether this is an irreversible process or whether cells can in fact renew to become fully functional stem cells under particular circumstances. Perhaps the best biological system to look at these questions is the intestinal crypt, as we know with great confidence that the crypts are closed systems and all cell regeneration must originate from within the crypt. These crypts live for many hundreds of days with all cells cycling at a high rate. Thus, we assume that stem cells operating in the intestinal crypt not only have self-maintaining properties but may also have self-renewal properties. However, clear proofs for this hypothesis are not present. Similarly in the haematopoietic system it remains to be seen whether long term repopulating cells once having progressed to somewhat more 'mature' cells can become long term repopulating cells again. The basic question is whether the different cell populations develop in a chain fashion or whether they are just different representations of the same type of cells with markers which develop independently of each other.

One of the final considerations relating to stem cells is how they are controlled. How do they know how many stem cells there are and how many there should be? The questions are interesting to consider in relation to the crypt. There are indications that the six actual steady state stem cells somehow monitor their members and know if an extra stem cell is produced, perhaps as a consequence of occasional symmetric divisions. Unless one stem cell is removed either by differentiation or death the crypt architecture would be lost since an entire extra transit lineage of 32–64 cells would be produced. It is believed that the occasional spontaneous apoptosis seen at the stem cell position removes such excess stem cells. Conversely if using low dose radiation a simple stem cell is killed (induced into apoptosis) there is good evidence that proliferation in the stem cell zone is up-regulated to compensate. The mechanism by which this monitoring of numbers is achieved is conceptually complex and completely unknown. The six stem cells are believed to be located in a ring of 16 cells which on average is at the fourth cell position from the crypt base. However, this is an average. In some cases the stem cell might be at the second cell position while in others it might be at the seventh. Even if they were all at the fourth position they would not be touching since other cells would be interspersed. Nevertheless, one way or another they seem to know that six is the correct number and can detect if the number changes to five or seven and respond accordingly.

There are two further hypotheses which have to be mentioned in the context of stem cell biology and models. One is a concept of chromosomal DNA segregation postulated by Cairns (1975). This hypothesis concept postulated that stem cells might be characterized by a selective chromosomal segregation process which would have a specific function in genetic hygiene by effectively maintaining the template DNA strands in the stem cell and transferring all newly synthesized DNA strands with potential replication errors to the transit cells where they would not be as harmful. At present, however, there is no proof or disproof for this speculation.

The other hypothesis is one of ageing of cells where it is assumed that stem cells have only a limited number of possible cell divisions. This hypothesis is closely related to the concept of clonal succession, although it is not identical. A recent variant of this concept is known as the telomere shortening process (e.g. Vaziri *et al.*, 1994). It was observed that the terminal end of DNA usually contains many repeats of a six base pair motive and that this shortens with the age of an organism. It was therefore suggested that all cells have only a limited division potential and that their life span may expire. However, presently no proof is available that this limitation has an actual impact on stem cell biology.

CONCLUDING REMARKS

In this chapter we have summarized some of the present concepts and models of stem cell behaviour and cellular hierarchies. Sime of the topics that we find relevant at present have been highlighted. Furthermore, it was designed to be an introduction to the various stem cell concepts discussed throughout this book and to illustrate how these concepts relate and differ. It was therefore not intended to be an encyclopaedic summary of the various models published so far and we apologize for all models and groups not mentioned in this chapter.

On the other hand we wanted to point out that a standard or unified stem cell hierarchical model is lacking. We also point out why this is the case and why it will remain a matter of active research in the foreseeable future. We believe that the stem cell problem is closely connected to the question of describing the tissue organization from an integrated point of view. It does not appear to be very meaningful to model a stem cell population in an isolated way, but rather its description should be integrated into a comprehensive theory of dynamic regenerative tissues.

ACKNOWLEDGEMENTS

We thank Dr U. Paulus for valuable comments on this manuscript and the Deutsche Forschungsgemeinschaft (Lo 342/411) who enabled work included in this chapter.

REFERENCES

Abkowitz, J.L., Linenberger, M.L., Penik, M., Newton, M.A. and Guttorp, P. (1993). *Blood* **82**:2096–2103.
Britton, N.F., Wright, N.A. and Murray, J.D. (1982). *J. Theor. Biol.* **98**:531.
Cairnie, A.B., Lamerton, L.F. and Steel, G.G. (1965). *Exp. Cell Res.* **39**:528–538.
Cairns, J. (1975). *Nature* **255**:197–200.
Clausen, O.P.F. and Potten, C.S. (1990). *Cutaneous Pathol.* **9**:129–143.
Gilbert, C.W. and Lajtha, L.G. (1965). In *Cellular Radiation Biology* (ed. M. D. Anderson), pp. 474–495. Williams and Wilkins, Houston.
Guttorp, P., Newton, M.A. and Abkowitz, J.L. (1990). *J. Math. Appl. Med. Biol.* **7**:125–143.
Hall, P.A. and Watt, F.M. (1989). *Development* **106**:619–633.
Hendry, J.H., Roberts, S.A. and Potten, C.S. (1992). *Radiat. Res.* **132**:115–119.
Holtzer, H. (1978). In *Stem Cells and Tissue Homeostasis* (eds B. Lord, C. S. Potten and R. Cole), pp. 1–28. Cambridge University Press, Cambridge.
Holtzer, H. (1979). *Differentiation* **14**:33–34.
Holtzer, H. (1985). *Cell Lineages, Stem Cells and Tissues Homeostasis*, pp. 1–28. Cambridge University Press, Cambridge.
Kay, H.G.M. (1965). *Lancet* **2**:418.
Lajtha, L.G. (1967). In *Canadian Cancer Conference*, pp. 31–39. Toronto: Pergamon Press.
Lajtha, L.G. (1979a). *Nouv. Rev. Fr. Hematol.* **21**:59–65.
Lajtha, L.G. (1979b). *Differentiation* **14**:23–34.
Lajtha, L.G. (1979c). *Blood Cells* **5**:447–455.
Loeffler, M. and Grossmann, B. (1991). *J. Theor. Biol.* **150**:175–191.
Loeffler, M., Stein, R., Wichmann, H.E., Potten, C.S., Kaur, P. and Chwalinski, S. (1986). *Cell Tissue Kinet.* **19**:627–645.
Loeffler, M., Potten, C.S. and Wichmann, H.E. (1987). *Virchows Arch. B* **53**:286–300.
Loeffler, M., Potten, C.S., Paulus, U., Glatzer, J. and Chwalinski, S. (1988). *Cell Tissue Kinet.* **21**:247–258.
Loeffler, M., Pantel, K., Wulff, H. and Wichmann, H.E. (1989). *Cell Tissue Kinet.* **22**:51–61.
Loeffler, M., Birke, A., Winton, D. and Potten, C.S. (1993). *J. Theor. Biol.* **160**:471–491.
Mackey, M.C. (1978). *Blood* **51**:941.
Mackey, M.C. and Dörmer, P. (1982). *Cell Tissue Kinet.* 15:381–392.
Michalowski, A. (1981). *Radiat. Environ. Biophys.* 19:157–172.
Paulus, U., Potten, C.S. and Loeffler, M. (1992). *Cell Proliferation* 25:559–578.
Paulus, U., Loeffler, M., Zeidler, J., Owen, G. and Potten, C.S. (1993). *J. Cell Sci.* 106:473–484.
Potten, C.S. (1974). *Cell Tissue Kinet.* 1:77–88.
Potten, C.S. (1991). In *Chemically Induced Cell Proliferation* (eds B.E. Butterworth, T.J. Slagg, W. Farland and M. McClain), pp. 155–171. John Wiley, New York.
Potten, C.S. and Hendry, J.H. (1983). In *Stem Cells: Their Identification and Characterization* (ed. C.S. Potten). Churchill-Livingstone, Edinburgh, pp. 200–232.

Potten, C.S. (1991). In *Chemically Induced Cell Proliferation* (eds B. E. Butterworth, T. J. Slagg,W. Farland and M. McClain), pp. 155–171. John Wiley, New York.
Potten, C.S. and Hendry, J.H. (1985a). *Manual of Mammalian Cell Techniques. Cell Clones.* Churchill-Livingstone, Edinburgh.
Potten, C.S. and Hendry, J.H. (1985b). In *Cell Clones* (eds C.S. Potten and J.H. Hendry), pp. 50–60. Churchill-Livingstone, Edinburgh.
Potten, C.S. and Hendry, J.H. (1995). *In Radiation & Crypt* (eds C.S. Potten and J.H. Hendry) pp. 45–69. Elsevier, Amsterdam.
Potten, C.S. and Lajtha, L.G. (1982). *Ann. NY Acad. Sci.* 397:49–61.
Potten, C.S. and Loeffler, M. (1987). *J. Theor. Biol.* 127:381–391.
Potten, C.S. and Loeffler, M. (1990). *Development 110*:1001–1018.
Potten, C.S. and Morris, R. (1987). *J. Cell Sci. Suppl.* 10:45–62.
Potten, C.S., Schofield, R. and Lajtha, L.G. (1979). *Biochim. Biophys. Acta* 560:281–299.
Potten, C.S., Chwalinski, S., Swindell, R. and Palmer, M. (1982a). *Cell Tissue Kinet.* 15:351–370.
Potten, C.S., Wichman, H.E., Loeffler, M., Dobek, K. and Major, D. (1982b) *Cell Tissue Kinet.* 15:305–329.
Potten, C.S., Hendry, J.H. and Moore, J.V. (1987). *Virchows Arch. B* 53:227–234.
Potten, C.S., Owen, G. and Roberts, S. (1990). *Int. J. Radiat. Biol.* 57: 185–199.
Schofield, R. (1978). *Blood Cells* 4:7–25.
Steel, G.G. (1977). *Growth Kinetics of Tumours*, p. 351. Clarendon Press, Oxford.
Till, J.E. and McCulloch, E.A. (1961). *Radiat. Res.* 14:213–222.
Vaziri, H., Dragowska, W., Allsopp, R.C., Thomas, T.E., Harley, B. and Landsdorp, P.M. (1994). *PNAS* 91:9857–9860.
Vogel, H., Niewisch, H. and Matioli, G. (1968). *J. Cell. Physiol.* 73:221–228.
Vogel, H., Niewisch, H. and Matioli, G. (1969). *J. Theor. Biol.* 22:239–270.
Wichmann, H.E. and Loeffler, M. (1983). *Blood Cells* 9:475–483.
Wichmann, H.E. and Loeffler, M. (1985). *Mathematical modelling of cell proliferation: stem cell regulation in hemopoiesis.* Vols I & II. CRC Press, Boca Raton.
Wichmann, H.E., Loeffler, M. and Schmitz, S. (1988). *Blood Cells* 14:411–425.
Winton, D.J. and Ponder B.A.J. (1990). *Proc. R. Soc. B.* 241:13–18.
Winton, D.J., Blount, M.A. and Ponder, B.A.J. (1988). *Nature* 333:463–466.
Winton, D.J., Peacock, J.H. and Ponder, B.A.J. (1989). *Mutagenesis* 4:404–406.
Withers, H.R. (1967). *Br. J. Radiol.* 40:187–194.
Withers, H.R. and Elkind, M.M. (1969). *Radiat. Res.* 38:598–613.
Withers, H.R. and Elkind, M.M. (1970). *Irit. /. Radiat. Biol.* 17:261–267.
Wright, N.A. and Alison, M. (1984). *The Biology of Epithelial Cell Population,* Vol. 1, p. 536. Clarendon Press, Oxford.

2 Stem cells and founder zones in plants, particularly their roots

Peter W. Barlow

Department of Agricultural Sciences, University of Bristol, Bristol, UK

INTRODUCTION

Compared to the cell theory introduced by Schleiden and Schwann in about 1838, the concept of the stem cell is relatively recent. The concept is based on both theoretical predictions and experimental observations, particularly in the context of mammalian systems (Lajtha, 1963, 1979, 1983; Potten and Lajtha, 1982; Wolpert, 1988; Hall and Watt, 1989; Potten and Loeffler, 1990). In the last few years, many of these systems have been the subject of comprehensive multi-author reviews (Lord and Dexter, 1988; Smith, 1992; Potten, 1993). By contrast, comparatively little has been written on the subject of stem cells in relation to plant structure and development (Clowes, 1967, 1975; Barlow, 1978; Ivanov, 1986), possibly because their importance is not so acutely appreciated. The molecular characterization of the stem cells located in root meristems (meristems being discrete zones in the plant body where cell proliferation is concentrated), and the evidence of their differentiation from surrounding meristematic zones, are obviously topics which merit considerable interest (Sabelli *et al.*, 1993; Barlow, 1994a). One of the aims of this chapter concerns the formal characterization of stem cells in the context of plant root (and shoot) development, because, as Lajtha (1983) has rightly remarked, the term 'stem cell' is often used quite imprecisely with the consequence that the stem-cell concept is rendered trivial and its biological significance obscured. A second aim is to discuss some of the features used empirically to define or to describe stem cells (and these features are remarkably similar in both plant and animal systems – a point to which I shall draw attention from time to time). These stem-cell properties seem to be a consequence of a supracellular control. Therefore, the molecular and cytological properties of stem cells in roots, some of which have been reviewed elsewhere (Barlow, 1994a, 1994b), would be also subordinate to such a type of control. If this view is correct, it is then necessary to enquire into the nature of this more global regulation of stem-cell behaviour which should, in turn, have implications for intra- and inter-cellular signalling mechanisms that relate to stem-cell proliferation. Moreover, if living systems have a logical structure (cf. Jackson *et al.*, 1986; Kauffman, 1987), these mechanisms might

STEM CELLS
ISBN 0-12-563455-2

even be expected to predict, in some general way, the probable locations of stem cells within an organism.

STEM CELLS, BRANCHING STRUCTURES AND FOUNDER ZONES

Omnis cellula e cellula (Virchow)

One element in the *definition* of a stem cell, in terms of the criteria proposed by Potten and Loeffler (1990), is that it is a cell capable of proliferation and self-maintenance, the latter condition indicating an unlimited potential for mitotic division. However, the very *concept* of the stem cell implies that this condition is a relative one: at some stage an immediate descendant of a stem cell must differ from the stem cell itself, for otherwise there could be no further differentiation or development. This principle of relativity, then, should also enter the definition of a stem cell and ascribe the property of self-maintenance to at least one of the two daughters of a stem cell. Moreover, the tissues or organs to which stem cells contribute are usually ones in which there is a relatively high degree of cell loss from the functional cell population, i.e. there is terminal differentiation with or without actual loss of cells. Thus, these tissues are 'self-renewing' and contrast with 'static' tissue populations which, once formed, exhibit little ability to grow further or to respond to the loss of cells. This distinction introduces another property of stem cells – and another element in the definition – namely, their ability to regenerate or compensate for damage to, i.e. unusual and excessive cell loss to, the tissues with which they are associated.

Cellular branching structures

Proliferative cells divide into two daughter cells, hence the growth of a cell population can be considered as a bifurcating branching process. A representation of this process can be by means of a 'branching structure' that consists of nodes and axes. An axis, or axes, on arising from a node, can be assigned a state; and different axes may be assigned different states according to their ancestry. In the present context (Figure 1), an axis represents a cell, its state reflects some cellular property, and a node corresponds to some event in the life of an axis such as cell division, i.e. mitosis and cytokinesis, or a stage of differentiation. State transitions coincide with such an event.

Consider a proliferative branching process in which there are three cellular states that indicate the potentialities of a given cell for further division. Two of the states, A and B, define whether the division potential is (A) unlimited or (B) limited. A third state, C, defines the lack of any active progress towards mitosis and division. Additional states (D, E, . . .) defining differentiating or mature cellular characteristics may be developed subsequently, but obviously do not depend on cell division for the transitions from these states to occur. In the first proliferating system, shown in Figure 1i, none of the cells, all of which are in state A, can be accorded the status of stem cell because, although their capacity for self-maintenance is high, there is no differentiation of states within the expanding cell population as required by the relativity principle implicit to the stem-cell

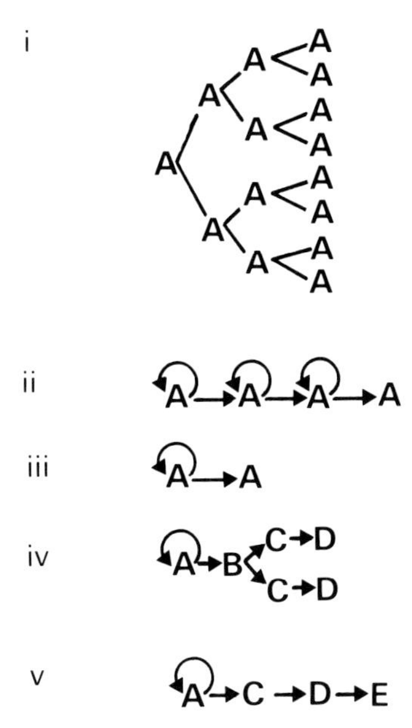

Figure 1 Branching structures consisting of axes and nodes (denoted at the position of upper-case letters) which depict cell genealogies and sequences of cellular state transitions. The latter occur when new axes are developed from nodes. In the five schemes (i–v) there may be up to five different states, A–E. (Although these letters are placed at nodes, they actually indicate the state of the supporting axes that emerge from them.) A new axis (or axes) (—) develops at each developmental step. At the cellular level, the emergence of two axes from each node represents the division of a cell into two daughters; this may (in iv and v) or may not (in i, ii or iii) be accompanied by a change in state. Whether or not a change in state occurs depends upon a probability, P, or upon a counter, λ, of time or some event such as cell divisions. A single axis from a node represents a step in the processes of differentiation, or maturation, unaccompanied by growth or division. The genealogy in (i) represents the multiplication of cells holding state A. No particular physical structure or organization for the group of cells is indicated and they could be free-living. However, in (ii), which is another representation of the multiplication system in (i), the linearity of the scheme, denoted by the straight arrow (→), implies an organized polar structure with a cellular 'source', or initial cell, for the lineage fixed at its left-hand end. A more minimal representation of this system is shown in (iii). The same convention could apply to (iv) and (v), and also in Figure 3. In this way, the structured cellular organization of cell-files in plant meristems can be represented. The branching structures can equally well represent the growth and development of protoplasmic domains (with states **A**–**E**) within an organ. At this organ level, the appearance of new axes represents the growth of a domain, during a developmental step, which may or may not be accompanied by further differentiation of states.

concept. Only in the second and third systems (Figures 1iv and 1v) is state A associated with stem-cell status, for not only is the A state renewed in one of the daughter cells produced by each cell division but the other daughter cell has also acquired a different state (B or C). Thus, because of the differentiation of states during the branching process and the self-renewal of the A state, this latter state can be accorded stem-cell status. Moreover, in terms of the actual anatomical structure of such a lineage of cells within a tissue, an A state cell would occupy a special position at the head of a cellular lineage in which different cellular states were possible. The branching structures in Figures 1i and 1ii, although similar to the extent that the progeny of division have the same state (A), they might be regarded as different if Figure 1i represented an expanding population of physically disconnected cells (as may exist in an *in vitro* culture of animal cells), and if Figure 1ii represented a population with a definite spatial structure, or orientation, in which the sister cells remain connected. In plants, this is effected by the enclosure of cell lineages within a common cellulosic extracellular matrix, or set of cell walls. The placement of the nodes (letters) in Figure 1 could therefore represent the relative positions of related cells within a tissue. It follows that the straight arrows in Figure 1ii, for example, represent both the physical connectedness of sister cells and also the orientation of their growth and mitotic division. The cells here would be organized in a unidirectional chain. A minimal representation of this branching structure is shown in Figure 1iii.

Thus, it becomes apparent that the branching structures in Figure 1 represent two things simultaneously; they can represent (as in Figures 1iii and 1v) cell genealogies, and they also define a set of cellular state transitions within a cell population. Both these aspects are sufficient to represent a growing and differentiating population. A branching structure is, in effect, an algorithm specifying a cellular branching process and its associated sequence of developmentally related state transitions. Each structure could also more explicitly represent the dynamics of a population of cells if both the division probability (P) of the cells associated with each state (i.e. P_A, P_B, P_C, . . .), or, alternatively, the number of divisions allotted to each state (i.e. λ_A, λ_B, λ_C, . . .), as well as the state transition probabilities (p, q) (see Figure 2), were specified. Likewise, in an actual branching system, or genealogy (of cells, for example), the number of axes (cells) associated with each divisionally competent state, i.e. A and B, at any given time would be related to the number of cell generations (x_A, x_B) which had occurred in each of the two respective types of cells; because division is a binary process, this number would be $2^{(xA+xB)}$. In this connection, the structure in Figure 1iv would represent a rapidly expanding population if cells in B state were both numerous (i.e. $x_B \gg 1$, and where $P_B < 1$ but > 0) and proliferated more rapidly than the cells in A state (i.e. $x_B > x_A$, and where $P_A = 1$). This is more explicitly represented in Figure 3i. However, because of the limited division potential of B state cells, the overall rate of increase in cell numbers is dependent on the relative rates of division of A and B state cells. Alternatively, the progeny of a given A state cell may increase in number before any of the daughter cells differentiate to B state (Figures 3ii and 3iii). In these branching structures, an A state cell (A_3) is shown as dividing to give rise to either one (Figure 3ii) or two (Figure 3iii) daughters in state B. The latter situation contradicts the definition of A state as being of unlimited potential since the division potential of cell A_3 is clearly limited if *both* its daughters acquire state B. This contradiction is acceptable only if the branching structures shown in Figures 3ii and 3iii are actually now representations of a level of organization in which the behaviour of the individual cells is, in addition to the dictates of their own state, also subject to the environment, or domain, in which the population of cells resides (as mentioned in the previous paragraph in connection with Figure 1ii). Therefore, whereas at the cellular level of organization (l) it may be sufficient to define a stem cell simply in terms of its division potential within a group of dividing and differentiating cells, at the organ level ($l+1$) it is pertinent to consider the properties of stem cells in relation to the anatomical structure of the organ as a whole. The cell state transition A→B could

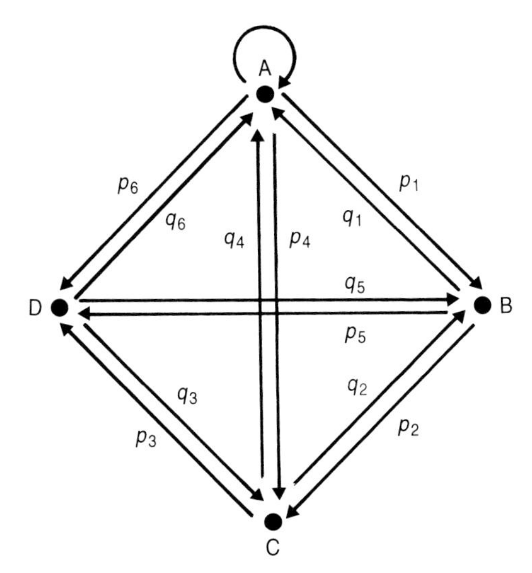

Figure 2 State transition graph for the cells (or the domains which they comprise) of a developing organ such as a root. The sequence of upper case letters (A–D) represents an ontogenetic sequence in which cells develop from a state (●) that is physiologically young to a state that is mature. For the present purposes, state A = stem cell; B = meristematic cell; C = growing but non-meristematic cell; D = mature cell (neither growing nor meristematic). The state transitions (→) are governed by transition probabilities (p, q) that may be under physiological and genetical control. The probabilities are implicitly dependent on some measure of time, which may be measured biologically (in terms of events such as the numbers of cell division accomplished, distance moved, etc.), or non-bioiogically (in terms of elapsed clock time). The 'p' probabilities regulate the transitions from younger to older states, whereas the 'q' probabilities regulate transitions from older to younger states. The normal pathway of root development is regulated by probabilities (p_0–p_3) of high value, with the occasional group (domain) of cells becoming rejuvenated (as in lateral root primordium initiation) by means of a local elevation, within mature D-state tissue, of q_1, q_4 or q_6.

be the result of a special type of mitosis, designated as 'quantal' by Holtzer *et al.* (1975). Such mitoses can be considered to account for divergences in the subsequent pattern of cell differentiation (e.g. $B_2 \rightarrow C_{a1} + C_{b1}$ in Figure 3i). However, it is difficult to envisage how a quantal mitosis, which is an event at level l, could occur at a specific stage in development without additional information being supplied from level $l+1$.

The branching structure in Figure 3i also draws attention to a central problem in the relationship between stem cells and their derivatives. This concerns the balance between numbers of A state and B state cells or, in other words, the rates at which the cells in these states produce new cells. The most straightforward way of achieving this balance is for the number of B state cells to regulate the rate of proliferation of A state stem cells (described in general terms by Prothero (1980), and in more elaborate form for haematopoietic tissue by Wichmann *et al.* (1988)). This hypothesis is supported by results derived from animal systems that range from the body column of the coelenterate *Hydra oligactis* (Holstein and David, 1990) to the cornea and the small intestine of the laboratory mouse (*Mus musculus*) (Cotsarelis *et al.*, 1989; Potten, 1991). In maize (*Zea mays*) roots, there is also supportive evidence (Clowes, 1964; Zezina and Grodzinsky, 1981; Barlow

i $A_1 \rightarrow B_1 \rightarrow B_2 < \begin{matrix} C_{a1} \\ C_{b1} \end{matrix}$

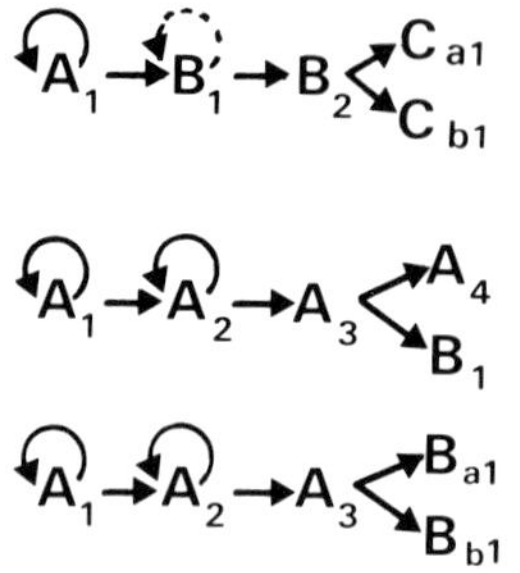

ii $A_1 \rightarrow A_2 \rightarrow A_3 < \begin{matrix} A_4 \\ B_1 \end{matrix}$

iii $A_1 \rightarrow A_2 \rightarrow A_3 < \begin{matrix} B_{a1} \\ B_{b1} \end{matrix}$

Figure 3 Branching structures showing the amplification of cellular populations holding states A, B and C, as shown in Figure 1. The cells also maintain a linear polarized structure, whose growth is from left to right; this also determines the direction of cell movement through the cell-file. The structure in (i) has a self-maintaining A state stem cell from which B state cells are descended. These latter cells can then multiply through additional cell divisions before terminating division in state C. Here, the two C state cells that result from the final division of a B state mother cell are shown existing in two alternative forms, a and b. Cell B_1 is indicated as being partially self-renewing (broken curved arrow-shaft), but its role as a functional stem cell may only be temporary until it is physically displaced from its location in the cell file by a daughter of an A state cell. In (ii) and (iii), state A is amplified. One or both daughters of an A state mother cell acquire state B. In terms of cells and their location within domains, this could be the result of some of the A state cells emigrating beyond the limit of the corresponding **A** state domain, whereupon state B is acquired by the emigrant daughters of the A state mother cells. The transition from A state to B state is interpreted as being due to a quantal mitotic division (Holtzer *et al.*, 1975). In (iii), the quantal division is such that, although two B state cells (B_{a1} and B_{b1}) are produced, they are determined to give two types of differentiated cells, C_{a1} and C_{b1}, as indicated by the different subscript letters. The sequence of subscript numerals indicates the spatial ordering of the cells of different states within a polarized meristematic cell file. Movement of cells within the file is caused by their growth and division.

and Adam, 1989); interaction between A and B state cells is here the basis for the recovery of root growth after sustaining environmental shocks which result in the premature maturation (loss) of proliferating B state cells. The interaction between these two cell populations, demonstrated by various experimental situations that involve disturbance to the usual pattern of cell behaviour, implies also that normal cellular growth and development are governed by homeostasis-generating interactions.

Organs and their domains

Although plant organs are comprised of cells, at a higher, organ, level of organization ($l+1$) the cellular characteristic is of secondary importance. An organ can be regarded as being comprised of a continuum of protoplasmic domains, each domain differing in its state of physiological specialization. Organ development is therefore dependent on the growth and change of state (differentiation) of the domains. Already-existing domains may retain their current state, or they may change to other, already existing states, or acquire completely new states. Thus, the set of state transitions which earlier were applied within a lineage of dividing and differentiating cells, such as represented in the branching structures shown in Figures 1 and 3, may also apply to a set of growing supracellular domains. It may also be more comprehensively illustrated in the state transition graph shown in Figure 2. (Such a graph can apply not only to organs (level $l+1$) but also to cells (level l).) Accordingly, a self-maintaining domain with physiological state **A** and a differentiated domain in state **B** derived from the former domain, would be analogous, at the cellular level of organization, to a stem cell in A state and a B state descendant (cf. Figure 1iv). 'Founder zone' might therefore be a suitable term for the **A** state domain; 'stem-cell domain' would be an inappropriate term since it mixes the terminology of the two levels and, besides, it is the domain, not the cells within it, that is relevant in the context of the organ. Moreover, if the scheme shown in Figure 2 is applicable at different levels of organization, then the defining characteristics of the states at one level, l, may have some correspondence with those of the states at the next level, $l+1$. At the cellular level, for example, there may be relatively low rates of state transition between stem and proliferative cells, dependent upon the rate of division of individual cells. At the organ level there could also be relatively low rates of state transitions between founder and proliferative domains dependent upon the rate at which the sizes of the domains increase.

Because organs are comprised of cells, it is necessary for practical purposes to find correspondences between the two levels. This is largely to enable interpretation of the behaviour of cells in terms of domain properties rather than vice versa; moreover, experimental analyses are usually carried out at the cellular level. For example, the peculiar behaviour of certain A state cells (A_3) in giving rise to B state progeny, as shown in Figures 3ii and 3iii, may be because of their location near a boundary of an **A** state domain. The emigration of cells across this boundary, and their consequent change of state, may be due to their growth. Furthermore, the structure of the domains and of the cell-files which they enclose, regulates the number of cells in any given state. For example, the population of A state cells represented in Figure 3iii would expand indefinitely unless emigration of cells across the domain boundary was also accompanied by a transition from A to B state ($A_3 \rightarrow B$). A quantal mitosis would accompany this transition and as a result lead to other changes associated with the subsequent differentiation of the cells. In discussing the possibility of quantal mitoses for cells emerging from the founder zone of pea (*Pisum sativum*) roots, Gahan and Rana (1985) concluded that cell differentiation – and this could include a change in cell division potential – was more likely to arise from a change in the relative location of a cell due to growth than to be a direct result of a preceding mitosis.

ANATOMY OF THE FOUNDER ZONE

Cell lineages and the quiescent centre

One great advantage of plants for investigating the interrelationships between cell and organ growth is that the former process is often markedly directional and, moreover, is, with few exceptions, symplastic: the cells do not glide over one another, as they may do in animal embryos, for example. Organs such as the root can be considered as a coherent bundle of cellulosic cylinders showing polarity: that is, there is a growth zone

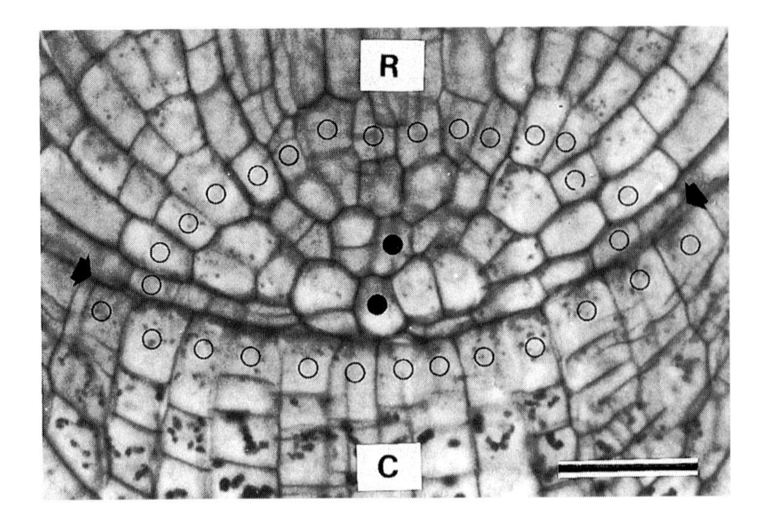

Figure 4 Median longitudinal section through the apex of a tomato (*Lycopersicon esculentum*) root. Note the prominent wall (arrowed) separating the root cap (C) from the rest of the root (R). Cell files in R are branched due to longitudinal cell divisions and converge to a few initial cells (indicated by ●) within what is a proliferatively quiescent centre, whereas in the central columella of the cap, in which longitudinal divisions are absent, there are as many initials (○) as there are cell-file origins. In terms of their position and relative rates of cell production, the ● cells are structural initials, whereas the ○ cells are functional initials. In R, these latter cells are located on the basiscopic face of the quiescent centre, whereas in C they are on its acroscopic face. Scale bar = 25 μm.

at, or near, one end of the bundle. Within the distal part of this zone, mitosis and cytokinesis not only increase the number of cells which comprise each cylinder (or cell-file) but they also permit the files to branch. Because the pattern of growth and division is often stable through time, these files preserve a record of past patterns of cell growth and division and, importantly, the files can be traced to an origin. A stem cell can thus be identified, usually located at the head of a file. Within the root tip, the collective location of many such stem cells traces to a region just behind a distal paraboloidal tissue known as 'root cap' (Figure 4).

Rates and directions of regional growth within a root can be estimated using the transverse cell walls in the cell-files as reference points (Gandar and Hall, 1988), though markers applied to the external boundary wall of the organ can also be used. These methods can be coupled with others that record internal metabolic properties (Erickson and Goddard, 1951; Silk *et al.*, 1986). Together they define not only the location of supracellular physiological domains within the organ but also their kinetics and state transition rates. This kinematic approach reveals that the self-maintaining founder zone (**A** state domain) within the tip of an angiosperm root grows slowly (estimated as an elemental rate of volume increase) compared to the neighbouring zones (**B** state domains) (Barlow, 1994a). Elements (cells) of the **B** state domain later make a transition to **C** and then **D** states, the latter being comprised of non-growing but functionally mature cells (Figure 2). Importantly, when coupled with anatomical studies, kinematics also reveal that many of the stem cells of roots of angiosperms and gymnosperms lie within this founder zone, or 'quiescent centre' (QC), which is so called (Clowes, 1956) because of its relatively low metabolic activity (including low rates of nucleic acid synthesis). Recently, refined methods for studying the replication of sub-cellular organelles reveal that, although the QC may have low levels of nuclear DNA synthesis, this need not apply to the mitochondrial and plastid nucleoids housed within this zone (Fujie *et al.*, 1993a; Suzuki *et al.*, 1995). These important findings indicate that cells in and around the QC are active sites of organelle biogeneses (Kuroiwa *et al.*, 1992; Fujie *et al.*, 1993b). The QC may also be a site with relatively stable, or long-lived, pools of certain metabolic precursors (e.g. thymidine, tubulin), as is indicative of certain animal stem-cell domains (Hume and Potten, 1982), although this remains to be tested experimentally.

As indicated elsewhere (Barlow, 1994b), the QC has been examined in detail in only a relatively few species, so its general status as housing a stem-cell population requires further validation. The quiescent property is a relative one – labelling of nuclei with ^{3}H-thymidine or the incidence of mitoses is infrequent relative to these features in neighbouring cells. Such 'infrequent' proliferative activities, however, might also include cases where there is complete inactivity in the QC. Inert QCs might then be accorded the same status (qualitatively speaking) as QCs where there is a low degree of proliferative activity. The distinction between inactivity and low activity of QC cells with respect to cell division could be important because in the former case, given the continuous growth of the root, the QC could not be legitimately regarded as a stem-cell population. In fact, the QC of *Arabidopsis thaliana* seems to be proliferatively inert: it consists of four cells which, once they have been established during embryogeny (for details of this and other embryonic root systems see Barlow, 1996) have not been seen to progress through either the DNA synthetic (S) phase or mitosis (Dolan *et al.*, 1993). Nor, apparently, can they be stimulated to proliferate by destruction of neighbouring cap cells (B. Scheres and C. van den Berg, personal communication), as would occur in the QC of maize primary roots, for example (Barlow and Rathfelder, 1985). Whether or not proliferatively inert cells occur within the larger QCs of other species remains an open question, but at present there is little definite evidence for this. In view of the situation in *Arabidopsis*, however, it is necessary to keep open this possibility. The lack of self-renewal of inert QC cells may be crucial in regulating whether root growth is determinate or indeterminate (see page 41). With these considerations in mind, it is possible to restate, as shown in Table 1, the schemes for QC evolution presented earlier (Barlow, 1994a, 1994b) and to use it as a basis for research into the various cellular conditions in and around the QC of plant roots.

Open and closed types of root meristems

Roots of higher plants show two types of disposition of the cell-files in and around the QC. The terms 'open' and 'closed' are used to describe the corresponding apical constructions. In roots with a closed construction (typical of grasses, for example), there is a thickened cell wall that visibly sets the root cap apart from the more proximal portion of the root (which includes the QC), whereas in roots of open construction such a wall is not evident and the cell-files seem to have continuity between cap and root. Consequently, in the closed type of root, the stem cells of the root cap lineages appear to lie just outside the distal, acroscopic surface of the QC (Figure 4). However, the root/cap boundary, although it often appears unbroken in histological preparations, can

Table 1 The presence of a quiescent centre within a root apex is the outcome of a differentiation process in which there is redundancy of information.

Degree of redundancy	R_p	R_c	Q	Q′	Type of meristem
0	–	–	–	–	Those with an apical cell (e.g. *Azolla*)
1	–	–	+	–	Determinate. *Arabidopsis* (?)
1	–	+	–	+	Indeterminate. At present, it is uncertain which type of Q cells are present in higher plant quiescent centres.
2	+	+	–	+	
2	–	+	+	+	
3	+	+	+	+	

If quiescence is due to specific gene action, perhaps regulated by 'positional information', then the QC may be viewed as an autonomous, or constitutive, characteristic (Q) of the root. But if quiescence is maintained by continually-operating physical (R_p) or chemical (R_c) constraints (these constituting another class of positional information) that interfere with processes which occur in otherwise mitotically-active cells, then quiescence is an induced, or facultative, characteristic (Q′). Cells with either Q or Q′ (or even both) characteristics could theoretically co-exist within a QC, although this would be a doubly redundant condition.

be broached from time to time, whereupon descendants of cells in the QC pass into the cap (Clowes and Wadekar, 1989). There they re-establish new stem cells and, hence, new cell-files within the cap. The rate k_a (cells.day^{-1}) at which cells make the transition from the QC to the cap determines whether the apex is regarded as open or closed (Clowes, 1981); the two types of construction are associated with high and low values of k_a, respectively (Figure 5i). Another value, k_b, is assumed to apply on the proximal, basiscopic surface of the QC. Unless there are distinct differences between the values of k_a in different species, openness or closedness will be a relative, rather than an absolute, property of their apices. That k_a and k_b always have a positive value also suggests that the true stem cells for all root tissue, including the cap, are located in the QC or founder zone.

The parameters k_a and k_b, although they reflect the behaviour of cells in the founder zone, have counterparts, K_a and K_b, at the organ level. Respectively, they define the movement of the founder zone's distal and proximal borders relative to the centre of the zone. Thus, rates K_a and K_b (μm.day^{-1}, where values >0 or <0 indicate movement of the borders away from or towards the centre of the founder zone, respectively) determine the size of the founder zone at any given time, and are the outcome of a correlative control – synonymous with 'positional information' – which defines the location of the founder zone within the root tip. Moreover, positional information, by regulating the locations and the states of the domains, also regulates the behaviour of the cells within them. For example, if cells formerly within the founder zone (**A** state domain) become relocated outside it (i.e. in a **B** state domain) by shrinkage of the zone, they will assume a more rapid rate of cell growth and mitotic division (Figure 5ii). The

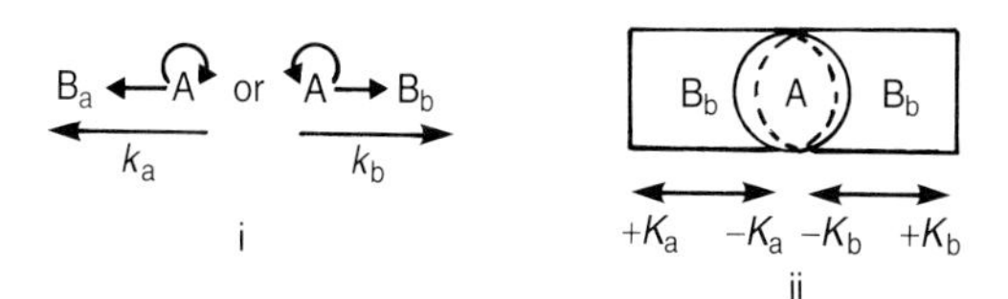

Figure 5 (i) Cellular branching systems in a tissue can have directionality. Here, there are two directions in which an A state cell can branch, apically or basally (denoted by subscripts a and b, respectively). The rate at which it does so in either direction is regulated by the parameters k_a and k_b, respectively. In the context of the root, this is analogous to the release of a cell from the quiescent centre, with the concomitant production of new cells, from either its acroscopic or basiscopic face. In roots of open construction, the differential between the two rates is less than in roots of closed construction. In the latter case, the relatively low value of k_a favours the formation of a prominent boundary between the root and the surmounting cap tissue. (ii) In terms of domains, the scheme in Figure 5i can be reinterpreted as a shrinkage of the **A** state domain (to give a new domain defined by broken lines), at rates $-K_a$ and $-K_b$ on its apical and basal borders, respectively, and the corresponding capture of former **A** state space by the **B** state domain. The apico-basal polarity of this structure is denoted by subscripts a and b. In some cases, apices appear to have a stochastic structure in which the limits of the founder zone (**A** state domain) appear to wander (as, for example, in root meristems that alternate between an open and closed structure). This is brought about, in part, by temporal variations in the values of k_a and k_b, but may also be due to oscillation between positive and negative values of K_a and K_b, particularly K_a.

rates k and K may also be self-regulatory. Should a low value of k_a favour the development of a thickened boundary wall between root and cap, this may then restrict further growth of QC cells in the distal direction.

Periodic expansion and contraction of the QC has been described during root development (Miksche and Greenwood, 1966; Alfieri and Evert, 1968). In cultured roots of a tomato (*Lycopersicon esculentum*) variety bearing the *gib-1* allele (which greatly decreases gibberellin biosynthesis), the QC undergoes similar cycles of growth and shrinkage. Apparently, when some critical volume is attained by the QC, its contraction ensues owing to the basipetal movement of its acroscopic surface (i.e. $K_a < 0$, $K_b = 0$). Consequently, a new meristem for the root cap is formed from formerly quiescent cells (Barlow, 1992). Possibly, some of the subsequent expansion of the QC (i.e. $K_a = 0$, $K_b > 0$) is due to the inheritance of the quiescent property by daughter cells. This, when coupled with a compensatory positional control determining where the quiescent property should be located, confers the episodic behaviour on the QC. This behaviour of the QC might be interpreted as evidence of a temporary mismatch of controls that emanate from two successive levels of organization (l and $l+1$): between inheritance of slow proliferative behaviour at the cellular level (l), and positional controls that regulate this behaviour at the organ level ($l+1$). The controls inherent to each level may operate on different aspects of the proliferative process. Instability of the QC in roots of the *gib-1* mutant, and the contrasting stable behaviour of wild-type QCs, have been simulated and discussed by Nakielski and Barlow (1995) in terms of a growth field and an associated growth tensor that governs cell division rates and morphogenesis at the root apex (see also Barlow, 1996).

Structured and stochastic apices

Shifting domain boundaries are characteristic of a 'stochastic' type of root apex in which no cell occupies a permanent position within the apex relative to those boundaries (Figure 4). It contrasts with a 'structured' type of apex (found in roots of leptosporangiate ferns and horsetails, for example) where cell position does remain constant (Figure 6). The two types of apices also contrast in their size. In a structured root apex, which can be minimally represented by the branching structure in Figure 3i, for example, a single cell (A_1) is both stem cell and sole occupant of the **A** state founder zone. In a stochastic apex, the founder zone encompasses many more stem cells whose positions relative to each other and to some other reference point may change with time. Such an apex may be represented by the branching structures in Figures 3ii and 3iii. In both cases, the domain boundaries may be as indicated in Figure 5ii: in structured apices $k_a = k_b = 0$, but in stochastic apices $k_a \neq k_b \neq 0$.

Stochastic behaviour is also evident in some, but not all, shoot apices, and in the cambium. In the latter tissue, it might account for the inability of anatomists to identify a stem-cell layer with any certainty (see Gahan, 1989). In shoot tissue it might even underlie the apparently spontaneous changes in the phyllotactic pattern that are especially evident in transitions in the Fibonacci series which describe leaf positions on shoot apices (Zagórska-Marek, 1994). In both the shoot and the cambial systems, no cell conserves a constant pattern of behaviour because the spatial controls that govern such behaviour

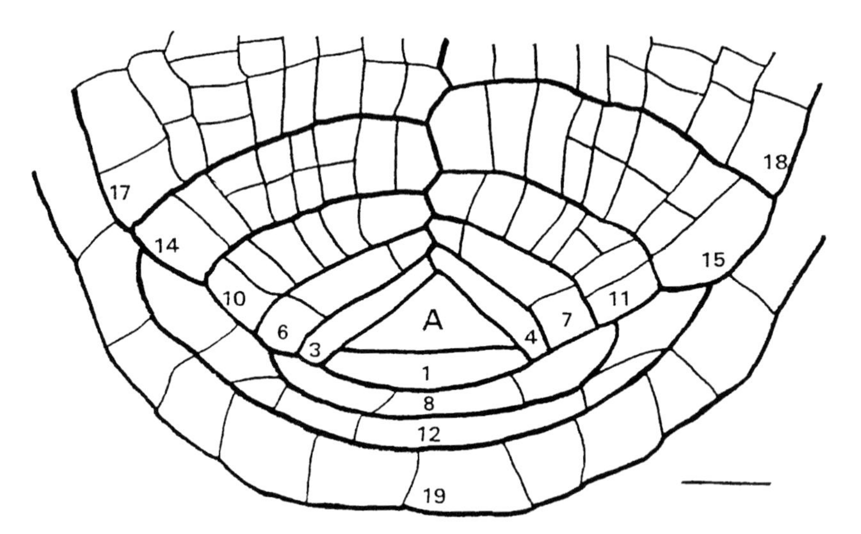

Figure 6 Drawing of a median longitudinal section through a young root tip of the leptosporangiate fern, *Ceratopteris thalictroides*, with a structured apex. Note the single triangular apical cell (A) (tetrahedral, when seen in three dimension) which serves as a stem cell. The heavy lines represent the walls bounding groups (merophytes) of related cells generated by successive divisions of the apical cell. The numbers refer to one possible order in which these merophytes were produced by the apical cell, the lowest number referring to the most recently formed merophyte. Scale bar = 25 μm. (From an original micrograph kindly provided by Dr Alexander Lux.)

are inherently unstable. However, there are also shoot apices which are structured and where, like the roots mentioned in this connection, there is a single apical cell (A_1) at their summits.

The founder zones in structured and stochastic root apices differ in behaviour. For instance, it is by no means general that all roots have a QC (Barlow, 1994a, 1994b). A QC is present only in the stochastic roots of angiosperms and gymnosperms; it can be absent from the structured roots of some ferns (Table 1). In the leptosporangiate water-fern, *Azolla pinnata*, for example, the founder zone has a high growth rate in contrast to the low rate associated with the QC of an angiosperm root. The single apical cell which occupies the founder zone of the *Azolla* root divides at least three times faster than any of its immediate cellular descendants (Gunning *et al.*, 1978). In the earliest stages of root development, the apical cell, which has a tetrahedral form, produces a new daughter cell at each of its facets in sequence. The three basiscopic facets contribute cells to the root; the fourth, acroscopic facet provides cells to the cap (Figure 6). The rate of cell production, k_a, for the acroscopic facet equals the rate k_b for any one of the three basiscopic facets; thus, the rates at all facets are equal. Later in development, the acroscopic facet ceases cell production, i.e. $k_a \rightarrow 0$, and the cap is henceforth maintained by its own autonomous set of meristematic cells, whereas the basiscopic facets remain productive. In this respect, the apex behaves similarly to apices of angiosperm roots with closed construction in which the meristem of the cap is also independent of the QC (Figure 4). As in these latter roots, the fern root apical cell can, under certain circumstances, resume cell production from its acroscopic face ($k_a > 0$). In roots of all types of structure or construction, at least some of the cells enclosed in the founder zone continue to contribute to the proximal portion of the root by maintaining a positive value of k_b (Figure 5i).

Determinate apical growth

Roots

Some roots have a limited life span. In the case of *Azolla pinnata*, for example, the roots grow to a length of only 5 cm. This is associated with the completion of about 55 division cycles in the apical cell whereupon this activity ceases. (Whereas $x_A \approx 55$ in the apical cell, in the descendent merophytes $x_B \leqslant 7$; i.e. the division counter, λ, never exceeds 7 in these latter cells.) The loss of 'stemness' (and here it is assumed that initially the apical cell was in A state and was not a B state cell descended from an A state cell present at an earlier stage in ontogeny) may be a result of the physiological isolation of this cell as the root lengthens, since a conspicuous decrease in the number of cytoplasmic connections (plasmodesmata) between the apical cell and surrounding cells accompanies root growth (Gunning, 1978; Overall and Gunning, 1982). Some angiosperm roots also have limited, or determinate, growth (Varney and McCully, 1991, and references therein), sometimes accompanied by unusual orientations; e.g. the determinate upward-growing roots of *Myrica gale* (Torrey and Callaham, 1978). The reason for this limited growth is not known. It may be because of the inactivity of the QC ($k_{a,b} = 0$), perhaps as a result of its insensitivity to the behaviour of meristematic cells in B state which consequently decrease in number since they continue to make the transition to a differentiated C state. Or it could be, as results of Raju *et al.* (1964)

suggest, that a quiescent zone is absent, all cells of the root thus having a determinate proliferative life span.

There can also be a genetic component to determinate growth. For instance, in *Arabidopsis thaliana*, the mutation *short-root* is associated with the cessation of division in the root meristem some time during the post-germination phase of growth (Benfey *et al.*, 1993). Within the apex, differentiated, non-growing cells eventually take the place of the former meristematic cells. The mutants *elf* (Scheres *et al.*, 1994) and *rml1* and *rml2* (*root meristemless*) (Cheng *et al.*, 1995) produce even shorter roots as a result of the non-occurrence of mitoses in the immediate post-germination phase. As indicated earlier (page 37), the QC of the wild-type *Arabidopsis* primary root could be regarded as a D-state cell, as a result of $p_0 \rightarrow 0$ immediately after its cells are formed (see Figure 2). In the mutant roots mentioned, the transition probability p_1, is also reduced to zero (though at different times after embryogeny in the different mutants) and, since p_2 and p_3 need not be affected, the remainder of the developing meristematic system runs through to the D state (Figure 2). Actually, the wild-type *Arabidopsis* root also eventually ceases growing; so, in accordance with the suggestions above, this might be the result of the inability of its QC to divide and to produce descendants which replenish the B state meristematic cell population, p_1 reducing to zero at a still later stage in development. This could be more closely related to the physiology of the whole plant rather than being due to the imposition of a genetic block affecting the meristematic cells.

Shoots

Shoots, also, can become determinate in the course of their growth. The formation of a spine, for example, is the result of a switch from a relatively indeterminate to a determinate mode of growth and division by the cells comprising an axillary or terminal bud. An inflorescence is another case where a shoot apex becomes determinate as a result of a major switch in developmental fate. In the case of the shrub, *Carissa grandiflora*, spine and inflorescence development represent alternative determinate options for the shoot apex (Cohen and Arzee, 1980). In contrast to the example of the spine, more cell proliferation is usually allotted to the florally determined apex due to either higher division probabilities, P_B, of B state cells or an increased value of λ_B, the number of divisions allocated to these cells. However, not all inflorescences are determinate, beautiful examples being found, for example, in the Australian genera *Callistemon* and *Malaleuca* where a new vegetative shoot grows out from the tip of an otherwise floral apex. Pea (*Pisum sativum*) also has an indeterminate inflorescence in which a terminal vegetative apex is retained. Here, the indeterminism of this apex is under genetic control; the mutation *determinate* (*det*) results in the apex becoming non-proliferative whereupon it proceeds to a differentiated, mature condition (Singer *et al.*, 1990). The *det*$^+$ gene may be associated with the maintenance of the A state of the stem cells (or with the **A** state of the founder zone). A fault (mutation) in this maintenance system results in loss of the A (or **A**) state. Possibly, at some stage in the ontogeny of the shoot apex of the mutant *Pisum*, all daughter cells make a transition to state B and thence, inevitably, to state C, until a final mature state is reached. This latter is state D in the graph shown in Figure 2. In the switch to determinacy, the probability, p_0, associated with the A state collapses to zero and, consequently, there is a progression to states B, C and D at a rate governed by the values of p_1, p_2 and p_3, or by the associated values of λ. Extreme determinism of a shoot apex is seen in the *shoot meristemless* mutant of *Arabidopsis* (Barton and Poethig, 1993) which may be the shoot counterpart of the root mutants *rml* and *elf* mentioned earlier. Again, it may be that transition probabilities p_0 and p_1, are reduced and that p_6 (see Figure 2) is elevated.

A basis for determinate growth

In connection with the state transitions of cells (A→B, B→C, etc.) which regularly occur in root growth, it is worth considering the possibility that they could be accompanied by progressive quantitative and qualitative alterations in the genomic DNA and that these transitions could accordingly lead to new patterns of gene activity. Bassi *et al.* (1984), for example, discovered that the ontogenetic maturation of cells (B→C, C→D, as in Figure 2) of bean (*Vicia faba*) roots was accompanied by a depletion of a repetitive fraction of DNA (buoyant density, ρ, of 1.677 g cm^{-3} in a Cs_2SO_4 gradient). Presumably, the loss of the DNA sequences making up this fraction is not total for they can be re-amplified in circumstances where a new meristem regenerates from mature cells (i.e. D→B, or D→A, as in Figure 2). In fact, when dedifferentiation of mature root cells was provoked by the removal of younger, distal zones (see Cionini *et al.*, 1985), there was synthesis of a satellite DNA (but with $\rho = 1.708$ g cm^{-3}) not normally found in mature tissue (Natali *et al.*, 1986). The same group of authors (Frediani *et al.*, 1992) have also found region-specific patterns of DNA methylation in *Vicia* roots which, in the mature zone, could be reversed when the cells were caused to dedifferentiate. Progressive epigenetic modification of DNA, including its loss and/or methylation, could provide cells of a growing system, such as a root apex, with a counting mechanism (the division counter, λ) that could govern the timing and rate of state transitions. However, this has yet to be investigated in detail, and so remains a speculation. Nevertheless, something analogous has already been proposed for human cell cultures (Harley, 1991) where the progressive loss of telomeric DNA in each cell cycle is relevant to cell ageing. Epigenetic changes have also been discussed in relation to cell differentiation and quantal mitoses in developing embryonic roots by Barlow (1996).

STEM CELLS VERSUS INITIAL CELLS

The term 'initial' is sometimes used to designate cells at the origin of a cell lineage (see Figure 1). However, with the discovery of the QC in roots came the realization that some of the cells which comprise this zone would also be initials (Clowes, 1961). But because they were now also found to divide infrequently, their importance in terms of actively providing cells to the root seemed compromised. In the light of this contradiction, 'initials' might actually be a more appropriate term for the cells lying on the surface of the QC since here they would be a more direct source of the cells required continually to supply the cell-files. To avoid these terminological difficulties, 'structural initial' has been suggested (Barlow, 1994a, 1994b) as a term to denote a self-maintaining, slowly dividing cell at the summit of a lineage (e.g. cell A_1 in Figure 3i), whereas 'functional initial' designates a more rapidly dividing cell from which most of the cells in a lineage would in practice be descended (e.g. cell B_1 in Figure 3i). In certain circumstances, either type of initial may be regarded as the stem cell of the lineage. The functional initial (B_1) in Figure 3i is, for example, denoted as partially self-maintaining. Logically, this initial cell is self-maintaining only until its division potential is exhausted; it must then be replaced by the descendant of an A state cell if the lineage is to continue growing. When cell A_1 divides, one daughter becomes a new B_1 cell, and the former B_1 cell becomes cell B_2. In the structured root apices of leptosporangiate ferns (Figure 6), where there is no zone exactly comparable to the QC, the actively dividing, single apical cell would be both a functional initial and a structural initial. It would also be a stem cell. Moreover, it corresponds to a founder zone of state **A**. It is theoretically possible, however, that, because of its finite division potential (recall the 55 divisions allotted to the *Azolla* root apical cell), the apical cell of such root apices is really a cell in state B (B_1 in Figure 3i), the A state cell from which it is descended actually being the apical cell of the shoot. (The way in which such a root apical cell descends from a shoot apical cell in the fern *Ceratopteris thalictroides* has been described by Howe (1931).) A putative B state of the apical cell might account not only for its faster cell cycle (all A state cells could then be unequivocally slowly proliferating) but also for the determinate growth of the roots in which they are housed. Further research at the molecular level may be able to distinguish between cells in A and B states.

Because the QC tends to conserve a constant size, even though its cells do grow and divide at appreciable rates, the functional initials will, as mentioned, be replaced from time to time by cells issuing from the QC. In closed meristems, the rate of this occurrence is greater at the basiscopic surface than at the acroscopic surface ($k_b > k_a$), but the differential may be less in open meristems (cf. Figure 5i). Thus, the division probability (P_B) of B state, functional initial cells varies according to their location with respect to the acroscopic or basiscopic surfaces of the QC, P_B being high at the former surface (where k_a is low) but lower at the basiscopic surface (where k_b is high).

CONTROLS OF STEM-CELL AND FOUNDER-ZONE PROPERTIES

The foremost problem of stem-cell regulation is to understand the means of maintaining the **A** state domain in which stem cells reside. This is true of both structured or stochastic apices with their contrasting patterns of proliferative behaviour. The central issue is one of feedback control and also of domain polarity (expressed through the fact that $k_a \neq k_b$ in closed meristems). Ultimately, it reduces to a question of what provides the positional controls for ordered development and differentiation, a topic poorly understood in plants, especially in relation to the behaviour and maintenance of meristematic zones (Barlow, 1984; Racusen and Schiavone, 1990). If it is true that the differentiated state requires continuous regulation for its maintenance (Blau and Baltimore, 1991; Blau, 1992), then each domain in the root actively maintains either its own state or that of a neighbouring domain, or both. In the example of the structured fern root apex, the postulated regulatory factors maintain rapid growth of the founder zone and hence rapid cell cycling, whereas in the stochastic apex of the higher plant the controlling factors are apparently negative and induce slow growth and slow cell-cycling (Barlow, 1994a). The assumption made here is that cell growth and cell cycle are coupled by means of intracellular titration of cycle-related regulator molecule(s) (see, for example, Novak and Tyson, 1993).

Hormonal controls

Because of the possibility that mobile plant hormones might provide one type of supracellular control over the behaviour of the quiescent centre, experiments have been performed to test hypotheses of their involvement as regulatory factors of **A** state. Abscisic acid (ABA) was found to maintain the quiescent condition of the QC in maize roots in circumstances where cells of this zone would normally have been proliferatively active (Müller *et al.*, 1993), apparently by bringing about a decrease in the rate of cell growth (Barlow and Pilet, 1984). However, this hormone has rather a general inhibitory effect on cellular metabolism in plants. Therefore, if ABA is a regulator of the QC in intact roots either it must be particularly concentrated in this zone, or QC cells must be especially sensitive to its presence, in which case its concentration could be relatively unimportant. Another class of hormones, the gibberellins, may also be implicated in regulating the QC, at least in *in vitro*-cultured tomato roots, since a decrease of their levels by means of the *gib-1* mutation increases the growth and proliferative rate of the QC cells (Barlow, 1992). Remarkably, the effect of this mutation is specific to the QC since rates of cell proliferation elsewhere in the tomato root meristem are undisturbed and are little different from the rates in wild-type roots. Cytokinins have also been postulated to control the QC (Feldman, 1975, 1979a). However, immunoassays performed on tissue sections (Zavala and Brandon, 1983; Sossountzov *et al.*, 1988) have failed to reveal any superabundance of cytokinins in the QC zone. The relatively low levels that are evident in the QC of maize roots are consistent with this zone's low metabolic rate. Thus, at present, the chemical regulation of quiescence by means of some of the major classes of hormones does not look too promising as a hypothesis, except perhaps for the involvement of gibberellins.

More recently, ascorbic acid has emerged as a regulator of plant cell proliferation (Citterio *et al.*, 1994) and has been found to induce divisions in the QC of onion roots (Innocenti *et al.*, 1990). However, not all cells of the QC could be brought to division, suggesting that there may be two classes of cells within this zone, tentatively assigned to a constitutive (Q) and facultative (Q′) quiescent condition (also see Table 1). Given the widespread occurrence of ascorbic acid in plants, it would be interesting to examine its distribution, the regulation of its amounts and activity in root apices and its effects on cellular pH (a possible regulator of cell division cycle in plants), particularly in relation to maintenance of the QC.

Positional signalling

The belief in positional signalling within roots has underlain studies on the regenerative potential of their stem cells. Maize roots have again provided a model system. The distal portion of its apex can be thought of as being made up of three domains in states **A**, **B** and **C**. For the sake of the present argument, these states correspond respectively, at the level of cellular domains, to QC cells, dividing cells of the root cap, and maturing cap cells. One effective way of inducing regeneration, and hence of demonstrating the regenerative capacity of the QC, is to remove part, or all, of the cap from the root tip (Barlow, 1974). This can be achieved without evident damage to the QC. Upon decapping, the QC (founder zone) is triggered to metabolic activity and its cells lose their proliferative quiescence. New cells are rapidly produced at the exposed root apex and, after a few days, a new cap, similar in structure to the lost tissue, is developed (Barlow, 1974; Barlow and Sargent, 1977). At the same time, a new QC is reformed. The initial loss of quiescence by the QC cell population should not necessarily be regarded as a loss of 'stemness' (A state) since this is actually a feature common to stem-cell populations in situations where derivative cells in other states (B or C) have been removed from the system. Possibly, a number of A state cells are triggered to make the transition to B state (regulated by p_1 in Figure 2) more rapidly than usual as a result of a diminished feedback from the depleted B state population. However, it is also likely that a few A state cells remain unaffected and continue to behave as stem cells.

The regeneration process suggests that there is feedback of information from the **B** domain to the **A** domain; the **C** domain may also be involved since partial removal of cap (non-proliferative tissue only) can also activate weak proliferation in the QC (Clowes, 1972). Cap regeneration might therefore be formalized as follows:

$$[\mathbf{A}+\mathbf{B}+\mathbf{C}]-[\mathbf{B}+\mathbf{C}]\rightarrow[\mathbf{A}]\rightarrow[\mathbf{A}+\mathbf{B}]\rightarrow[\mathbf{A}+\mathbf{B}+\mathbf{C}]$$

where the '+' indicates not only 'with the additional presence of' but also the orientation of the domain. In the latter context, the '+' connects a domain that is proximally located in the root tip to an adjacent, more distal domain. The '−' denotes 'removal of'. The parentheses [ ] denote a set of domains. An arrow (→) denotes a time-step in the regeneration process. From an electron microscopic study of QC reactivation and cap regeneration in maize roots (Barlow and Sargent, 1977), there was some evidence that features characteristic of **B** and **C** state domains could co-exist in the region of the tip that was formerly an **A** state domain. Thus, characters normally associated with maturing, **C** state tissue, such as hyperactive Golgi bodies and starch-filled plastids, could be found in dividing cells of the former QC zone. In this circumstance, therefore, a more realistic formalization of regeneration could be:

$$[\mathbf{A}+\mathbf{B}+\mathbf{C}]-[\mathbf{B}+\mathbf{C}]\rightarrow[\mathbf{A}]\rightarrow[\mathbf{A},\mathbf{B}]\rightarrow[\mathbf{A},\mathbf{B},\mathbf{C}]\rightarrow[\mathbf{A}+\mathbf{B}+\mathbf{C}]$$

where the comma indicates 'coincident with'.

It is likely that regeneration, at the tip of the root, of the domains with correct relative positions (i.e. [**A**+**B**+**C**]), is the result of a re-formation of correct positional information within the apex. This is usually accompanied by cell growth and division and, intriguingly, regeneration seems also to require an input from the Earthly 1 g gravitational field since, according to observations from experiments on board the space shuttle *Columbia*, cap regeneration does not occur normally in microgravity (10^{-4} g) (Moore *et al.*, 1987). Decapping the root seemingly disturbs an informational coordinate system, constructed of positional values, by removing reference points for its specification. One reference point may be the boundary of the root tip with the external environment. Indeed, as cap regeneration proceeds, a small quiescent zone is at first seen close to the tip, and it then appears to move proximally, away from the tip (Bednara, 1974; Feldman, 1975, 1976), as though sensing its distance from the decapped surface. If there is genuine movement of the quiescent property, rather than there being an apparent relocation of this zone due to the formation, by cell division, of new, non-quiescent cells at the tip, then this means that a wave of quiescence passes through cells on or near the proximo–distal axis of the root until the locality of the quiescent state is stabilized as a result of the restoration of the correct positional values in the regenerating tip, coupled with an ability to interpret these values. Positional values, then, determine the **A** state domain and consequently the behaviour of the cells included within it. Moreover, relocation of the quiescent condition seemed also to occur in decapped root tips which had been exposed to inhibitors of DNA synthesis and then allowed to recover in their absence (Barlow, 1981). Here, a quiescent zone appeared at the usual distance from the tip immediately the recovery phase began, even though little new cell division had previously occurred during the inhibition phase. It was as though the appropriate positional values for its specification had been regenerated during the latter period. A partial representation of the regeneration of the proposed informational coordinate system has been simulated by Nakielski (1992) for decapitated root apices of radish (*Raphanus sativus*) by means of a basipetal drift of coordinates specifying a growth tensor. A consequence of the migration of the quiescent zone would be that formerly B state cells revert to state A in response to the positionally controlled re-specification of an **A** state domain. The B state cells are subject to the state transition probability q_1 in Figure 2 which, in undisturbed apices, is assumed to have a value of zero.

It should be remarked that because the QC has the capacity to regenerate neighbouring tissue following injury (Clowes, 1964; Barlow, 1974), its cells, or the **A** state domain which encloses them, thereby conform to another important and general characteristic of stem cells (or founder zones). However, this regenerative potential of stem cells is an empirical *observation* and it might be argued that, of itself, it is not a necessary formal element for their definition (cf. Potten and Loeffler, 1990). Moreover, other regions of the root (and other parts of the plant body), including mature root tissue, are able to regenerate parts lost by amputation (Francis, 1978; Feldman, 1979b; Rost and Jones, 1988). (This latter situation again admits the possibility that, as mentioned above, cells in B or C state can, under certain circumstances, make the transition to A and B state, respectively. They do this by elevation of the probabilities q_1, q_4 and q_6 (see Figure 2).) A similar situation holds in many animal systems, though there are cases (e.g. regeneration in the planarian *Dugesia lugubris*) where stem cells (from the male gonad, in this case) actually migrate to a wounded site and there participate in the regeneration process (Gremigni and Miceli, 1980). Therefore, although regeneration, like low cell production rate, is a trait often associated with stem cells (e.g. Cotsarelis *et al.*, 1989), it may not be a feature unique to them. However, the regenerative trait confers on the cells an important biological function in relation to the vitality of the organism (at level $l+2$) in addition to their having simply a geometric or genealogical role in the structuring of an organ (at level $l+1$). Responsiveness to cell loss could also be a basis for distinguishing between self-renewing and static cell populations with indeterminate and determinate growth, respectively. All the systems mentioned give evidence for feedback of information between differentiated cells and stem cells which, in turn, regulates the p and q probabilities shown in Figure 2.

The nature of positional signals

Hormonal controls of stem cell behaviour were briefly discussed above. They comprise one type of positional control. An entirely different, non-chemical perspective on positional information can come from consideration of the biophysical aspects of root development. Interestingly, the connection is made here through the putative chemical controls of quiescence discussed earlier, and is again illustrated by the *gib-1* mutant roots of tomato and also by the analogous *d-5* mutant in maize. (Both mutant genes, *gib-1* and *d-5*, interfere with neighbouring steps in the gibberellin biosynthetic pathway.) In these two plants, low endogenous levels of gibberellins disturb the intracellular organization of the cortical microtubules (MTs) (Baluška *et al.*, 1993; P. W. Barlow and J. S. Parker, unpublished); it will also be recalled that, in the QC of the *gib-1* mutant tomato roots, there is an enhanced rate of growth and proliferation (Barlow, 1992). Moreover, the wholesale disruption of MTs by means of cold treatment and by agents such as colchicine and oryzalin also stimulates cell proliferation in the QC (Baluška and Barlow, 1993). On the other hand, taxol, which stabilizes MTs, can negate these stimulatory effects (F. Baluška and P. W. Barlow, unpublished). Differential behaviour of MTs in the various domains of the root may thus provide an alternative basis from which to form hypotheses about the relationship between the founder zone and positional information. This differential behaviour may also provide explanations for the respective fast and slow growth rates of founder zones in structured and stochastic apices since the growth process is intimately connected with (though not solely dependent upon) the orientation and dynamics of MTs. In addition, MTs and other components of the cytoskeleton might provide the link for some of the 'mechanical' explanations ventured for quiescence (Barlow, 1974), such as that it is due to a particular pattern of forces which focus onto the QC domain (Clowes, 1975).

The integration of biomechanics and cell biology into a system of structural epigenetics is of topical interest (Lintilhac, 1984), particularly since it has now been firmly established that applied mechanical forces can re-orient a pre-existing direction of growth (Lintilhac, 1984) by action upon the cytoskeleton (Hush and Overall, 1991; Cleary and Hardham, 1993; Wang *et al.*, 1993). Moreover, the electric fields that surround and run within roots also has some importance for cytoskeletal orientation (Hush and Overall, 1991). Presumably, the cytoskeleton has sufficient sensitivity to respond to the physical inter- and intracellular stresses associated with the normal pattern of tissue tensions which exist within plant organs (Kutschera, 1992). Thus, reference points for the informational and positional coordinates that specify the location of a stem-cell population might be tissue or organ boundaries whose shape and pattern of extensibility govern intra-organ patterns of stress.

Once the position of the founder zone has been specified according to such principles – and in the root this specification would occur in the hypophysis zone of the early embryo – certain patterns of gene activity may follow, including the apparent inactivity of the *cdc2* gene (Martinez *et al.*, 1992) essential for the G_1-S transition in proliferating plant cells. Some of these patterns of activity could, in part, be related to MTs since these elements of cytoplasmic structure are able to control the degree of chromatin dispersion in the nuclei of QC cells (Baluška and Barlow, 1993) which in turn relates to transcriptional activity (Barlow, 1985). Differential patterns of gene action may also be related to properties of the extracellular matrix (cell wall). In this regard, it is of interest that a matrix glycoprotein, or 'differentiation inhibiting factor', has been suggested as being involved in stem cell differentiation in mammalian embryos (Rathjen *et al.*, 1990). An extracellular glycoprotein has been identified at the site of lateral root primordia in tobacco (*Nicotiana tabacum*) by Keller and Lamb (1989) and seems to be carried into the apical zone (the QC?) of the emerging root.

The quiescence of the **A** state domain in the maize root is not only characterized by a certain pattern of gene activity (Sabelli *et al.*, 1993) but may also be determined by that pattern of activity. In this situation, however, there is the practical difficulty of establishing the sequence and interdependence of the various ontogenetic processes. Although there is evidence for the active maintenance of the quiescent condition of potentially proliferative individual cells of animals (Epifanova *et al.*, 1982; Epifanova and Brooks, 1994) and plants (Lelievre *et al.*, 1987; Kodama *et al.*, 1991), as would be predicted by Blau's (1992) hypothesis of continual regulation of the differentiated state, this condition, as it applies to cells within organs, may nevertheless be the product of a higher level of regulation which, for want of more precise knowledge, we call positional information. It is the sequence of ontogenetic processes mentioned above that needs to be analysed more carefully, making full use, as Ingber (1993) suggests, of concepts from a range of biological, and even engineering, disciplines.

STEM CELLS AND FOUNDER ZONES IN RELATION TO PLANT DEVELOPMENT

Stem cells, besides being responsible for the generation and maintenance of cell lineages in tissues and organs, are involved in the branching of organ axes and hence for the development of the root and shoot systems of the whole plant. In the root system, new founder zones, with their stem cells, rise in cellular descendants of pericycle tissue of an axis of branch order N and generate the meristem of a new axis of order $N+1$. Similarly, in the shoot system, new founder zones destined to form new axes develop upon the flanks of the apex of a parental shoot axis. Here, each new founder zone is eventually located at the summit of an axillary bud, which itself is part of a reiterated set of structures, the metamers (Barlow, 1994c). The inception of each metamer is suggested as being due to a quantal event at the boundary of the **A** state domain in the apex: [**A**]→[**A**+**B**]. Then, within the **B** domain, a new **A** domain is differentiated: [**B**]→[**B**+**A**]. Thus, one founder zone begets the precursor of another and, hence, at the organ level ($l+1$), makes possible the large-scale branching of root and shoot systems. This situation is mirrored at the cellular level (l) where stem cells are also responsible for the cell lineages which support cell differentiation, including the differentiation of new groups of stem cells: [A]→[A+B] and [B]→[B+A]. At this level, there are evident similarities with the models of Bailey (1986) and Brown *et al.* (1988) who have discussed, respectively, the ontogenetic programming of the human immunohaemopoietic system as the consequence of a sequential development, or differentiation, of successive stem-cell populations. At the organ level ($l+1$), various types of plant axis branching that contribute to distinctive shoot morphologies have been discussed by Bugnon (1971) and, for mammals, the related problem of stem-cell zone creation (and loss) as been discussed by Loeffler and Grossmann (1991).

Diversity of organ axes

Just as either a cell lineage or the developing domains within an organ can be represented as a branching structure, so the whole plant, consisting of root and shoot systems, can

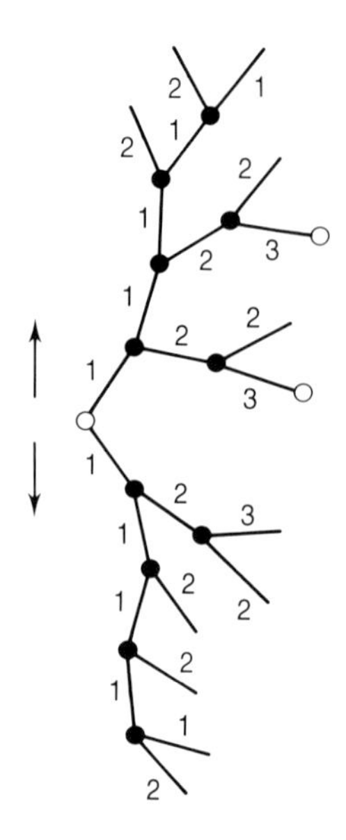

Figure 7 Plants can be represented as multi-order branching systems (or branching structures) consisting of nodes (○, ●) and axes (—). Node ○ represents the zygote from which both primary shoot and root organ axes emerge. These are, respectively, in the upper, ↑, and lower, ↓, parts of the diagram, though other details of axis orientation have no special significance. The outgrowth of one of the two axes from a node is usually delayed relative to the other (i.e. dichotomy of an axis apex is uncommon). Thus, axes of order N (where $N=1, 2, 3, \ldots$ corresponding to organ axes holding states **A**, **B**, **C**, . . .), when they branch, propagate themselves and also generate new primary axes of order $N+1$. Occasionally, flowering occurs at the apex of a shoot axis (on axis $N=3$, which would hold state **C**, in this scheme) enabling a new node (○), or zygote, to develop. In viviparous species, the branching system continues beyond this node (i.e. two generations, N and $N+1$, of branching systems co-exist), but in seminiferous species, as shown here, the branching system of generation N is dislocated, at the zygotic node, from that of generation $N+1$.

be considered in a like manner (Figure 7). As before, the branching structure consists of nodes and axes. Relating this branching structure to a developing plant shows that the zygote (the proximate stem cell or founder domain) represents a node from which arise two first-order axes (N) corresponding to the shoot and root poles of the early embryo. These two axes then branch further, each giving multi-order branching structures ($N+1$, $N+2, \ldots$) that represent the entire systems of root and shoot and the state transitions which accompany the morphological differentiation of their component organ axes. The state transition diagram for cell-types shown in Figure 2 can be similarly reinterpreted in terms of organ- or axes-types within the context of the whole organism. Some axes produced later in the ramification develop a zygote from which a new set of organ axes is reiterated. In viviparous plants, the growth of the branching system of axes is continuous through successive generations of plants (**N**, **N + 1**, . . .), whereas in seminiferous plants the axes of one generation (**N**) become physically disconnected from those of the next (**N + 1**) through the shedding of seed. The disjuncture of the various generations of axes ($\Sigma[N, N+1, \ldots]$) at the node which supports the zygote (Figure 7) defines a branching system (or a set of branches) known as a 'genet' and is symptomatic of sexual reproduction. Where the disjuncture occurs at nodes other than the zygotic node, the separated branching system defines a 'ramet' and is typical of vegetative reproduction.

Each branching event in a root and shoot axis which is relevant to the development of a new generation of organ axes occurs in a *primary meristem*. Importantly, primary meristems usually possess a founder zone containing stem cells. It is this character that confers on a primary axis its state of potentially unlimited growth. However, there is another type of branching event, also involving the primary meristem, which generates the determinate and relatively ephemeral axis structure of the leaf. In this case, the primary meristem branches to produce a localized *secondary meristem* of limited growth potential. This branching event may depend upon a modification of the pattern of gene activity within the apex. Recent research into the action of the *KNOTTED-1* gene in both maize shoot apices (Jackson *et al.*, 1994) and in transgenic tobacco plants (Sinha *et al.*, 1993) suggests that this gene is associated with a state of indeterminate growth (A state), and that when its activity is repressed in certain zones of the apex the state of these latter zones is switched to determinate B state. From such a zone a leaf meristem develops. This new secondary meristem subsequently becomes located either in the distal portion of the secondary axis which it produces, and has there a fragmented condition (as in the leaves of dicots), or it is proximal and unfragmented (as in the leaves of monocots). The limited growth potential of leaf meristems, to which there are exceptions, e.g. the spectacular leaves of the ancient gnetalian species *Welwitschia mirabilis* (though in this case the early abortion of the shoot apical meristem (Rodin, 1953) could have something to do with leaf longevity) may be because they do not possess a set of self-renewing stem cells. The *KNOTTED-1* gene, therefore, may govern the A→B and the B→A state transitions in the shoot by maintaining, at the least, the transition probability of q_1 at zero (Figure 2). Perhaps in the *Welwitschia* leaf meristem the probability q_1 is >0 and consequently a stem cell population is permitted to form.

Intercalary meristems (Fisher and French, 1976), associated with the base of stem internodes in certain tropical monocotyledonous plants, are another example of a localized secondary meristem. Here, however, no organ axis branching is involved and the meristem continues to contribute to the growth of the primary axis. Again, the limited growth potential of this meristem can be ascribed to the lack of a stem-cell population and may also be associated with a certain pattern of gene action. Fruits, because they develop from carpels, are homologous to leaves. However, they initially possess a diffuse secondary meristem which is also of limited growth potential. The anatomy of fruits is specialized to assist the dispersal of the next generation (**N**+1) of primary meristems (and their associated axes) contained within the seeds.

Diversity of meristems

In the context of meristems, primary and secondary are adjectives that describe their state of differentiation, particularly in relation to the tissues and organs that develop from them (Figure 8). One of the products of a primary meristem, when it branches, is a new primary meristem and an axis of similar state and behaviour (i.e. it is a self-maintaining axis with state **A**). Consequently, the plant's branching systems

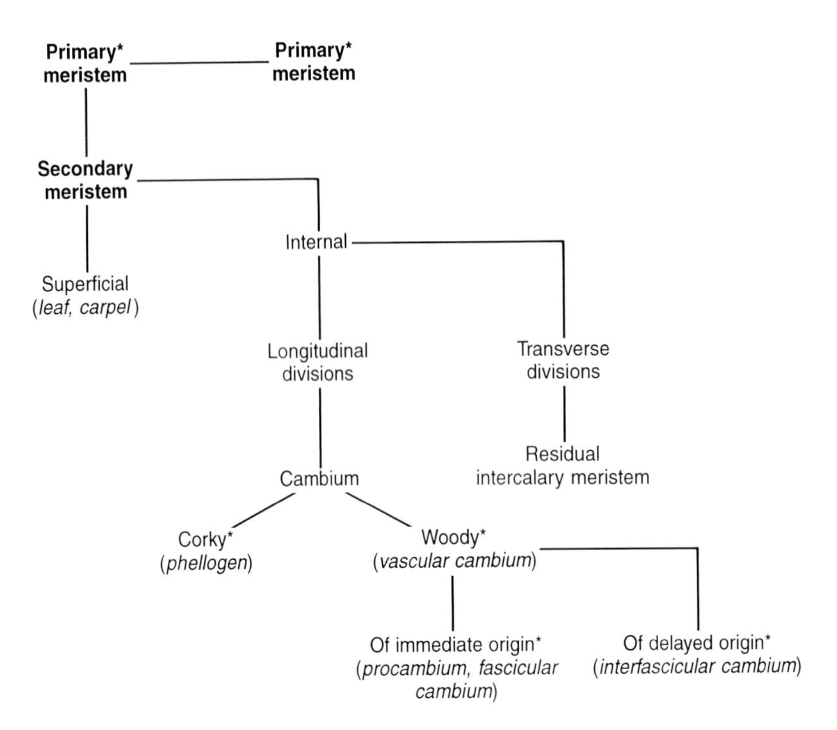

Figure 8 Differentiation of primary and secondary meristems in the shoot of higher plants. Meristems which support cell lineages that trace to cells fulfilling the criteria of a stem cell are marked (*). Each meristem may have a structure that is stochastic to a varying degree, where no stem cell is necessarily permanent. Roots conform to a similar scheme, though they may lack the residual intercalary meristem and superficial meristem types. Also, in the root, cambium does not derive directly from the primary meristem but seems to originate some distance behind it, and then to propagate acropetally towards the tip.

(Figure 7) ramify by the production of morphologically equivalent organ axes; primary root and shoot axes of order N produce, respectively, lateral root and shoot axes of order $N+1$ (Figure 7), each of which is developed by a primary meristem. However, as already mentioned in relation to leaves, many primary meristems also have the ability to produce secondary meristems associated with the development of axes having a different, or secondary, state (state **B**). In contrast to primary meristems, secondary meristems usually do not generate additional organ axes, though there are exceptions (e.g. some leaves generate adventitious shoots). The secondary meristems from which leaves arise initially have discrete *superficial* locations on the primary meristem, whereas secondary meristems which develop *internally* are more extensive and long-lived. The latter are known as cambia (*sing.* cambium) of which there are two types, vascular cambium and cork cambium (phellogen) (Figure 8). The vascular cambia of shoots are generated continuously from the primary meristems (excepting periods of dormancy when all meristems are inactive). In contrast, leaf meristems are initiated discontinuously on the shoot apex, but with a certain temporal rhythm (Barlow, 1994d). As suggested elsewhere (Barlow, 1994d), this rhythm may be related to a quantal event in the shoot apex, which could be interpreted as a set of quantal mitoses (Holtzer *et al.*, 1975) whereby a set of daughter cells emigrates from the founder zone to occupy a new niche within the more proximal portion of the meristem. As a result, the cells become committed to make one of the structural units (a metamer) of which a shoot axis is comprised. This quantal event leads, at a higher level of organization, to the branching of the axis with a corresponding state transition, [**A**]→[**A**+**B**].

There is also variation in the timing of cambium formation, but once established it is continually active, suggesting that it is maintained by its own set of stem cells. In shoots, the fascicular portion of the vascular cambium differentiates almost immediately from the primary meristem, whereas the interfascicular cambium differentiates later from non-dividing cells. Roots of herbaceous plants often possess interfascicular cambium only. Thus, the root apex, in terms of meristem productions, is a bifurcating branching system, if only the primary meristems are considered; if the secondary meristem of the interfascicular cambium is included, it is a trifurcating system. Shoot apices are often more complex quadri- or quinquefurcating branching systems since leaf and phellogen meristems might also be included. The branching system in Figure 7 is therefore a simplification and represents only the branching of primary meristems (represented as occurring at the nodes) and the subsequent development of organ axes.

Temporal-spatial displacements apply to the development of new primary meristems during the branching of axes. In roots of most angiosperms and gymnosperms, new primary meristems of an $(N+1)$th-order axis develop *behind* the meristem of the parental Nth-order axis, whereas in some ferns the $(N+1)$th-order axis meristem develops *within* the meristem of the parental axis. Occasionally, branching of the primary meristem is not delayed and the apex undergoes dichotomy (see Piché *et al.* (1982) and Tomlinson and Posluszny (1977) for descriptions of dichotomy in roots and shoots, respectively). The relative timing (or spacing) of branching events in the primary meristem of an axis, and especially the timing of the subsequent outgrowth of the new primary axes – and here the shoots of sea-grasses in the genus *Halophila* provide spectacular examples of having up to eight orders of axis branching compressed into a diminutive (<1 mm) bud (Posluszny and Tomlinson, 1991) – are features acted on by selection in the course of the evolution of plant forms. The relevant determinants constitute a class of heterochronic genes (Conway and Poethig, 1993) which may influence the timing of the onset of stem-cell activity in the newly developed axis primordia.

CONCLUSION

Stem-cell populations in plants show many properties common to those of animal stem-cell systems (these latter are mostly mammalian, but some are from invertebrates). These properties are:

(i) Proliferation, which is usually at a lower rate than in other dividing somatic cells;
(ii) Self-maintenance;
(iii) Occupation of a particular anatomical site from which recognizable cell lineages originate;
(iv) Asymmetric division, where one daughter cell may, given certain conditions, embark on a differentiation pathway;
(v) Capacity to respond, by an activation of more rapid proliferation, to the loss or destruction of neighbouring differentiating cells.

The last-mentioned property is not strictly necessary for the definition of a stem cell because other cells in roots and shoots, even those which are differentiated non-proliferating cells, can respond in a similar way. This proliferative response is, however, evidence of the sensitivity of stem cells to their immediate environment and, importantly, it enables proliferating populations to adopt a particular cell production rate commensurate with, and even governed by, the needs of the organism. Moreover, it emphasizes that stem cells have more than just a geometrical relationship with other cells in the same lineage, but are integrated with the growth of entire tissues and organs.

There are two ways in which stem cells can be regarded: one pertains to the cellular level of organization, the second pertains to the organ level. It follows that properties 1 and 2 correspond to the cellular level, and properties 3, 4 and 5 to the organ level. With respect to the cellular viewpoint, the stem cell has a unique context within a lineage of cells which, with time, acquire different states. At the organ level, stem cells also comprise a founder domain. This domain has properties similar to the five enumerated above for cells, and are vital to the growth zone of an organ.

Just as the regenerative response of stem cells is a property of a more integrated organ level of organization, so the property of asymmetric division is a feature more satisfactorily interpreted at the organ level than at the cellular level. (Exceptions, perhaps, are the rather special cases where asymmetry of genetic structure following DNA replication is associated with cell division, as in yeast, for example (Klar, 1992).) Divisional asymmetry represents the movement of a cell from one physiological milieu (or domain) to another; in the process, a division counter is set running in one daughter cell to restrict its division potential. Hence, the properties of stem cells, or founder domains, and their respective differentiating derivatives are the responses to different bits of positional information, the totality of which contributes to a field where growth and morphogenesis occur. Positional information thus leads to the definition of the 'second anatomy' of an organism, which Slack (1982) suggests is made up of a system of codings and decisions, superimposed upon the familiar visible cellular anatomy.

Plants are modular constructions. Their organs and their founder zones beget further organs, also with founder zones. And within organs themselves, various types of meristems are developed, some with stem cells (as defined above) and others without. The latter, as a consequence, contribute to determinate, as opposed to indeterminate, growth. With such variety of meristem types in plants, it is obviously important to understand not only the details of their structure and of how one type of meristem can give rise to another, but also the means by which any given meristem is maintained, since this may influence the life span of the associated tissue or organ. It is in all these areas that the concept of the stem cell and founder zone can be applied since it is their behaviour, and that of their descendants, which enable the root and shoot systems to develop. Moreover, for the plant sciences, an appreciation of these concepts is of practical importance because meristems, and the stem cells which support them, are the bases of productivity in terms of consumable biomass.

ACKNOWLEDGEMENTS

I am grateful to Drs Jacqueline Lück and Hermann B. Lück, and also to Dr Hanna E. Zieschang, for their helpful discussions of certain aspects of this chapter.

REFERENCES

Alfieri, I.R. and Evert, R.F. (1968). *New Phytol.* **67**:641–647.
Bailey, D.W. (1986). *Differentiation* **33**:89–100.
Baluška, F. and Barlow, P.W. (1993). *Europ. J. Cell Biol.* **61**:160–167.
Baluška, F., Parker, J.S. and Barlow, P.W. (1993). *Planta* **191**:149–157.
Barlow, P.W. (1974). *New Phytol.* **73**:937–954.
Barlow, P.W. (1978). In *Stem Cells and Tissue Homeostasis* (eds B. I. Lord, C. S. Potten and R. J. Cole), pp. 87–113. Cambridge University Press, Cambridge.
Barlow, P.W. (1981). In *Structure and Function of Plant Roots* (eds R. Brouwer, O. Gašparíková, J. Kolek and B. C. Loughman), pp. 85–87. M. Nijhoff/Dr W. Junk Publishers, The Hague.
Barlow, P.W. (1984). In *Positional Controls in Plant Development* (eds P. W. Barlow and D. J. Carr), pp. 281–318. Cambridge University Press, Cambridge.
Barlow, P.W. (1985). *J. Exp. Bot.* **36**:1492–1503.
Barlow, P.W. (1992). *Ann. Bot.* **69**:533–543.
Barlow, P.W. (1994a). In *Plant Molecular Biology. Molecular Genetic Analysis of Plant Development and Metabolism* (eds G. Coruzzi and P. Puigdomènech), NATO-ASI Series, Vol. H81, pp. 17–30. Springer-Verlag, Berlin.
Barlow, P.W. (1994b). *Plant and Soil* **167**:1–16.
Barlow, P.W. (1994c). In *Growth Patterns in Vascular Plants* (ed. M. Iqbal), pp. 19–58. Dioscorides Press, Portland, OR.
Barlow, P.W. (1994d). *Biol. Revs* **69**:475–525.
Barlow, P.W. (1996). In *Plant Roots. The Hidden Half* (eds Y. Waisel, A. Eshel and U. Kafkafi), pp. 77–109. Marcel Dekker, New York.
Barlow, P.W. and Adam, J.S. (1989). *J. Exp. Bot.* **40**:81–88.
Barlow, P.W. and Pilet, P.-E. (1984). *Physiol. Plant.* **62**:125–132.
Barlow, P.W. and Rathfelder, E.R. (1985). *Envir. Exp. Bot.* **25**:303–314.
Barlow, P.W. and Sargent, J.S. (1977). *Ann. Bot.* **42**:791–799.
Barton, K.M. and Poethig, R.S. (1993). *Development* **119**:823–831.
Bassi, P., Cionini, P.G., Cremonini, R. and Seghizzi, P. (1984). *Protoplasma* **123**:70–77.
Bednara, J. (1974). In *Structure and Function of Primary Root Tissue* (ed. J. Kolek), pp. 23–32. Veda, Bratislava.
Benfey, P.N., Linstead, P.J., Roberts, K., Schiefelbein, J.W., Hauser, M.-T. and Aeschbacher, R.A. (1993). *Development* **119**:57–70.
Blau, H.M. (1992) *Ann. Rev. Biochem.* **61**:1213–1230.
Blau, H.M. and Baltimore, D. (1991). *J. Cell Biol.* **112**:781–783.
Brown, G., Bunce, C.M., Lord, J.M. and McConnell, F.M. (1988). *Differentiation* **39**:83–89.
Bugnon, F. (1971) *Mem. Soc. Bot. Fr.* **1971**:79–85.
Cheng, J.-C., Seeley, K.A. and Sung, Z.S. (1995). *Plant Physiol.* **107**:365–376.
Cionini, P.G., Zolfino, C. and Cavallini, A. (1985) *Protoplasma* **124**:213–218.
Citterio, S., Sgorbati, S., Scippa, S. and Sparvoli, E. (1994). *Physiol. Plant.* **92**:601–607.
Cleary, A.L. and Hardham, A.R. (1993). *Plant Cell Physiol.* **34**:1003–1008.
Clowes, F.A.L. (1956). *J. Exp. Bot.* **7**:307–312.
Clowes, F.A.L. (1961). *Apical Meristems.* Blackwell Science, Oxford.
Clowes, F.A.L. (1964). *Brookhaven Symp. Biol.* **16**:46–58.
Clowes, F.A.L. (1967). *Phytomorphology* **17**:132–140.
Clowes, F.A.L. (1972). *Nature New Biol.* **235**:143–144.
Clowes, F.A.L. (1975). In *The Development and Function of Roots* (eds J. G. Torrey and D. T. Clarkson), pp. 3–19. Academic Press, London.
Clowes, F.A.L. (1981). *Ann. Bot.* **48**:761–767.
Clowes, F.A.L. and Wadekar, R. (1989). *New Phytol.* **111**:19–24.
Cohen, L. and Arzee, T. (1980). *Bot. Gaz.* **141**:258–263.
Conway, L.J. and Poethig, R.S. (1993). *Seminars Devel. Biol.* **4**:65–72.
Cotsarelis, G., Cheng, S.-Z., Dong, G., Sun, T.-T. and Lavker, R.M. (1989). *Cell* **57**:201–209.
Dolan, L., Janmaat, K., Willemsen, V., Linstead, P., Poethig, S., Roberts, K., Scheres, B. (1993). *Development* **119**:71–84.

Epifanova, O.I. and Brooks, R.F. (1994). *Cell Prolif.* **27**:373–394.
Epifanova, OI., Setkov, N.A., Polunovsky, V.A. and Terskikh, V.V. (1982). In *Cell Function and Differentiation* (ed. G. Akoyunoglo), Part A, pp. 231–242. Alan R. Liss, New York.
Erickson, R.O. and Goddard, R.O. (1951). *Growth* **15 (Suppl)**, 89–116.
Feldman, L.J. (1975). In *The Development and Function of Roots* (eds J. G. Torrey and D. T. Clarkson), pp. 55–72. Academic Press, London.
Feldman, L.J. (1976). *Planta* **128**:207–212.
Feldman, L.J. (1979a). *Planta* **145**:315–321.
Feldman, L.J. (1979b). *Ann. Bot.* **43**:1–9.
Fisher, J.B. and French, J.C. (1976). *Amer. J. Bot.* **63**:510–525.
Francis, D. (1978). *New Phytol.* **81**:357–365.
Frediani, M., Cremonini, R., Sassoli, O. and Cionini, P.G. (1992). *Chromatin* **1**:79–88.
Fujie, M., Kuroiwa, H., Kawano, S. and Kuroiwa, T. (1993a). *Planta* **189**:443–452.
Fujie, M., Kuroiwa, H., Suzuki, T., Kawano, S. and Kuroiwa, T. (1993b). *J. Exp. Bot.* **44**:689–693.
Gahan, P.B. (1989). *Bot. J. Linn. Soc.* **100**:319–321.
Gahan, P.B. and Rana, M.A. (1985). *Ann. Bot.* **56**:437–442.
Gandar, P.W. and Hall, A.J. (1988). *Planta* **175**:121–129.
Gremigni, V. and Miceli, C. (1980). *W. Roux' Arch.* **188**:107–113.
Gunning, B.E.S. (1978) *Planta* **143**:181–190.
Gunning, B.E.S., Hughes, J.E. and Hardham, A.R. (1978). *Planta* **143**:121–144.
Hall, P.A. and Watt, F.M. (1989). *Development* **106**:619–633.
Harley, C.B. (1991). *Mutat. Res.* **256**:271–282.
Holstein, T.W. and David, C.N. (1990). *Dev. Biol.* **142**:392–400.
Holtzer, H., Rubinstein, N., Fellini, S., Yeoh, G., Chi, J.D., Birnbaum, J. and Okayama, M. (1975). *Q. Rev. Biophys.* **8**:523–557.
Howe M.D. (1931). *Bot. Gaz.* **92**:326–329.
Hume, W.J. and Potten, C.S. (1982). *Cell Tissue Kinet.* **15**:49–58.
Hush, J. M. and Overall, R.L. (1991). *Cell Biol. Int. Rep.* **15**:551–560.
Ingber, D.E. (1993). *Cell* **75**:1249–1252.
Innocenti, A., Bitonti, M.B., Arrigoni, O. and Liso, R. (1990). *New Phytol.* **114**:507–509.
Ivanov, V.B. (1986). *Tsitologiya* **28**:295–302 (in Russian).
Jackson, D., Veit, B. and Hake, S. (1994). *Development* **120**:405–413.
Jackson, E.R., Johnson, D and Nash, W.G. (1986). *J. Theor. Biol.* **119**:379–396.
Kauffman, S.A. (1987). *BioEssays* **6**:82–87.
Keller, B. and Lamb, C.J. (1989). *Genes Devel.* **3**:1639–1646.
Klar, A.J.S. (1992). *Trends Genet.* **8**:208–213.
Kodama, H., Ito, M., Hattori, T., Nakamura, K. and Komamine, A. (1991). *Plant Physiol.* **95**:406–411.
Kuroiwa, T., Fujie, M. and Kuroiwa, H. (1992). *J. Cell Sci.* **101**:483–493.
Kutschera, U. (1992). *Physiol. Plant.* **77**:157–163.
Lajtha, L.G. (1963). *J. Cell. Comp. Physiol.* **62 (Suppl. 1)**:143–145.
Lajtha, L.G. (1979). *Differentiation* **14**:23–34.
Lajtha, L.G. (1983). In *Stem Cells. Their Identification and Characterization* (ed. C. S. Potten), pp. 1–11. Churchill Livingstone, Edinburgh.
Lelievre, J.-M., Balague, C., Pech, J.-C. and Meyer, Y. (1987). *Plant Physiol.* **85**:400–406.
Lintilhac, P.M. (1984). In *Positional Controls in Plant Development* (eds P. W. Barlow and D. J. Carr), pp. 83-105. Cambridge University Press, Cambridge.
Loeffler, M. and Grossmann, B. (1991). *J. Theor. Biol.* **150**:175–191.
Lord, B.I. and Dexter, T.M. (eds) (1988). *J. Cell Sci.* **Suppl. 10**:288 pp.
Martinez, M.C., Jørgensen, J.-E., Lawton, M.A., Lamb, C.J. and Doerner, P.W. (1992). *Proc. Natl Acad. Sci. USA* 89:7360–7364.
Miksche, J.P. and Greenwood, M. (1966). *New Phytol.* 65:1–4.
Moore, R., McClelen, C.E., Fondren, W.M. and Wang, C.-L. (1987). *Am. J. Bot.* 74:218–223.
Müller, M.L., Pilet, P.-E. and Barlow, P.W. (1993). *Physiol. Piant.* 87:305–312.
Nakielski, J. (1992). In *Mechanics of Swelling* (ed. T. K. Karalis), NATO ASI Series, vol. H64, pp. 179–191. Springer-Verlag, Berlin.

Nakielski, J. and Barlow, P.W. (1995). *Planta* **196**:30–39.
Natali, L., Cavallini, A., Cremonini, R., Bassi, P. and Cionini, P.G. (1986). *Cell Differ.* **18**:157–161.
Novak, B. and Tyson, J.J. (1993). *J. Theoret. Biol.* **165**:101–134.
Overall, R.L. and Gunning, B.E.S. (1982). *Protoplasma* **111**:151–160.
Piché, Y., Fortin, J.A., Peterson, R.L. and Posluszny, U. (1982). *Can. J. Bot.* **60**:1523–1528.
Posluszny, U. and Tomlinson, P.B. (1991). *Can. J. Bot.* **69**:1600–1615.
Potten, C.S. (1991). In *Chemically Induced Cell Proliferation: Implications for Risk Assessment* (eds B. E. Butterworth *et al.*), pp. 155–171. Wiley-Liss, New York.
Potten, C.S. (ed.) (1993). *Epithelial Stem Cells* (Parts 1 and 2). Seminars Develop. Biol. 4 (parts 4 and 5): 207–259, 261–312.
Potten, C S. and Lajtha, L.G. (1982). *Ann. NY Acad. Sci.* **397**:49–61.
Potten, C.S. and Loeffler, M. (1990). *Development* **110**:1001–1020.
Prothero, J. (1980). *J. Theoret. Biol.* **84**:725–736.
Racusen, R.H. and Schiavone, F.M. (1990). *Cell Differ. Devel.* **30**:159–169.
Raju, M.V.S., Steeves, T.A. and Naylor, J.M. (1964). *Can. J. Bot.* **42**:1615–1628.
Rathjen, P.D., Nichols, J., Toth, S., Edwards, D.R., Heath, J.K. and Smith, A.G. (1990). *Genes Devel.* **4**:2308–2318.
Rodin, R.J. (1953). *Am. J. Bot.* **40**:371–378.
Rost, T.L. and Jones, T.J. (1988). *Ann. Bot.* **61**:513–523.
Sabelli, P.A., Burgess, S.R., Carbajosa, J.V. *et al.* (1993). In *Molecular and Cell Biology of the Plant Cell Cycle* (eds J. C. Ormrod and D. Francis), pp. 97–109. Kluwer Academic Publishers, Dordrecht, The Netherlands.
Scheres, B., Willensen, V., Janmaat, K., Wolkenfelt, H., Dolan, L. and Weisbeek, P. (1994). In *Plant Molecular Biology. Molecular Genetic Analysis of Plant Development and Metabolism* (eds G. Coruzzi and P. Puigdomènech), NATO ASI Series Vol. H81, pp. 41–50. Springer-Verlag, Berlin.
Silk, W.K., Hsiao, T.C., Diedenhofen, U. and Matson, C. (1986). *Plant Physiol.* **82**:853–858.
Singer, S.R., Hsiung, L.P. and Huber, S.C. (1990). *Am. J. Bot.* **77**:1330–1335.
Sinha, N.R., Williams, R.E. and Hake, S. (1993). *Genes Devel.* **7**:787–795.
Slack, J.M.W. (1982). In *Developmental Order: Its Origin and Regulation* (eds S. Subtelny and P. B. Green), pp. 423–436. Alan R. Liss, New York.
Smith, A. (ed.) (1992). *Stem Cells.* Seminars Cell Biol. 3 (Part 6): 383–456.
Sossountzov, L., Maldiney, R., Sotta, B. *et al.* (1988). *Planta* **175**:291–304.
Suzuki, T., Sasaki, N., Sakai, A., Kawano, S. and Kuroiwa, T. (1995). *J. Exp. Bot.* **46**:19–25.
Tomlinson, P.B. and Posluszny, U. (1977). *Am. J. Bot.* **64**:1057–1065.
Torrey, J.G. and Callaham, D. (1978). *Can. J. Bot.* **56**:1357–1364.
Varney, G.T. and McCully, M.E. (1991). *New Phytol.* **118**:535–546.
Wang, N., Butler, J.P. and Ingber, D.E. (1993). *Science* **260**:1124–1127.
Wichmann, H.E., Loeffler, M. and Schmitz, S. (1988). *Blood Cells* **14**:411–429.
Wolpert, L. (1988). *J. Cell Sci.* **Suppl. 10**:1–9.
Zagórska-Marek, B. (1994). *Acta Soc. Bot. Polon.* **63**:117–137.
Zavala, M.E. and Brandon, D.L. (1983). *J. Cell Biol.* **97**:1235–1239.
Zezina, N.V. and Grodzinsky, D.M. (1981). *Fiziol. Biokhim. Kul't. Rast.* **13**:642–647 (in Russian).

3 The stem cell concept applied to shoot meristems of higher plants

Dennis Francis

School of Pure and Applied Biology, University of Wales College of Cardiff, Cardiff, UK

INTRODUCTION

For botanists, the title of this paper is an unfortunate play on words because in plants the stem bears the shoot apex. Indeed, the meristem is sometimes described as the stem apex. For animal systems, the term has been used to describe cells from which all others in a tissue stem. For shoot meristems, founder cell(s) is perhaps more useful and I shall use the words founder and stem interchangeably throughout this chapter.

Stem cell properties

A recent definition listed the following properties expected of stem cells:

(i) proliferation;
(ii) self-maintenance;
(iii) the production of a large number of differentiated functional progeny;
(iv) the regeneration of the tissue after injury;
(v) flexibility of the use of these options.

(Potten and Loeffler, 1990.)

These criteria prove exacting even for intensively studied systems such as those in the mammalian ileum crypt. As such, the progeny of stem cells are regarded as transit cells which, through proliferation, lead to the production of differentiated cells. This is a useful let-out clause because, as will be described, the cells of the shoot meristem are capable of exhibiting these properties but rarely do so as an identifiable cohort. Moreover, unlike the root meristem, where the quiescent centre exists as a distinct population of potential stem cells (see Chapter 2), it is far less obvious where founder cells reside in the shoot meristem; perhaps all shoot meristem cells are founders. This would conform to the

STEM CELLS
ISBN 0-12-563455-2

definition of the stem cell in its widest context. In a rather narrower sense, a true stem cell is one which can undergo self-renewal. Identifying such cells in the shoot meristem is infinitely more difficult.

A non-cycling cell in G1 is a so-called G0 cell. Potten and Loeffler (1990) interpreted such cells in the ileum crypt as potential stem cells. Again, the quiescent centre of the root meristem possesses G0 cells (see Chapter 2) and although non-cycling cells have been demonstrated in shoot meristems (Gonthier *et al.*, 1987; Taylor and Francis, 1989), they tend to be dispersed throughout.

A cornerstone of the concept as applied to the ileum is the assignment of a defined stem cell number; 16 such cells were deduced for the ileum crypt (Potten and Loeffler, 1990). Moreover, in the ileum cells are arranged so that lineages give rise to a precise pattern of differentiated cells which then perform a precise function. In plant meristems, the fate of cells can be shown to be very plastic and it is not uncommon to find one lineage of cells contributing to two tissues as a result of branching of the files (Smith and Hake, 1992). In animals, clones of cells can often contain several cell lineages which indicate that the original stem cell had the ability to provide cells that differentiate in a variety of ways. This response is much closer to that of shoot meristem cells.

Thus, the main problems to grapple with are: Do shoot meristems have founder cells? Where are they? What is their molecular identity? How do they function? Is the concept in keeping with the functioning of the shoot meristem during vegetative or floral growth or both? Can the concept be applied usefully to shoot meristems?

The aim will be to describe the function of the vegetative shoot meristem and to show how its cellular properties change upon the transition to floral growth. Special attention will be given to cell lineage analysis and the extent to which fate maps can locate founder cells in the shoot meristem. Some commentary will also be provided on the expression of organ identity genes in floral shoot meristems and I shall attempt to link the idea of founder cells in the meristem to the expression of these homeotic genes.

This review will not so much be paraphrased as 'animal developmental biologists do, plant biologists re-do', more 'animal developmental biologists do, should plant biologists follow?'

THE VEGETATIVE SHOOT APEX

Morphology

It is worth bearing in mind that there are about 600,000 species of angiosperm in the world today. Hence, the following account of apical morphology, centring on a mere handful of 'model' species, hardly takes account of the enormous diversity of higher plant form.

A term that is useful in clarifying shoot apical morphology is the 'apical dome' or 'true shoot apical meristem'. This is the region above the youngest leaf primordia and has a basal diameter anywhere between 50 and 3,000 μm (Mauseth, 1991). It is to be distinguished from the major shoot meristematic region which extends proximally from the apical dome over several centimetres of tissue. These meristematic cells are themselves descendants from the shoot meristem. Founder cells must lie within the dome since it is a source of mitotic cells whose descendants are 'differentiated functional progeny'. Three well-researched examples to illustrate the range of shapes are:

(i) *Chrysanthemum segetum* which has a dome 1,400 μm in diameter during vegetative growth and which enlarges to 6,400 μm at the start of floral growth (Nougarède, 1967).

(ii) *Helianthus annuus*, which has a flattened apex spanning about 70 μm (Langenauer *et al.*, 1974, Figure 6); dome is a false description for this species.

(iii) *Arabidopsis thaliana*, the plant geneticist's *Drosophila*, also has a flat meristem. At the dry seed stage it only measures about 20 μm in diameter and comprises 110 cells (Irish and Sussex, 1992) and during vegetative growth the diameter is no more than 50 μm (Furner and Pumfrey, 1992).

Histologically, three separate areas can be recognized in the apical dome (Figure 1): the central or axial zone (CZ), the peripheral zone (PZ) and the pith rib meristem (PRM). The zones are largely distinguished on the basis of cytochemical staining and reflect the presence of larger, more vacuolated cells in the central zone compared with smaller, less vacuolated cells in the peripheral zone, and cells of intermediate or yet larger size in the pith rib meristem (PRM) (Lance, 1957; Nougarède, 1967). Hence, a gradient of cell size exists from the larger cells of the central zone to the smaller cells of the peripheral zone. The gradient of cell size is often matched by a gradient of relative growth rates in the dome. For example, in *Pisum sativum*, the fastest relative growth rates are on the flanks (peripheral zone) of the apical dome (Lyndon, 1970a).

Throughout the life of the plant, the apical dome constantly changes its volume and initiates leaf primordia. Between the initiation of successive leaves, the apex increases in size exponentially by cell division until it reaches a critical size prior to the initiation of the next leaf or leaf pair. This period of exponential growth between the initiation of successive leaves is termed the plastochron.

Much recent work has focused on putative meristem specific gene expression but as reviewed and noted by Medford (1992), neither specific genes nor their gene products

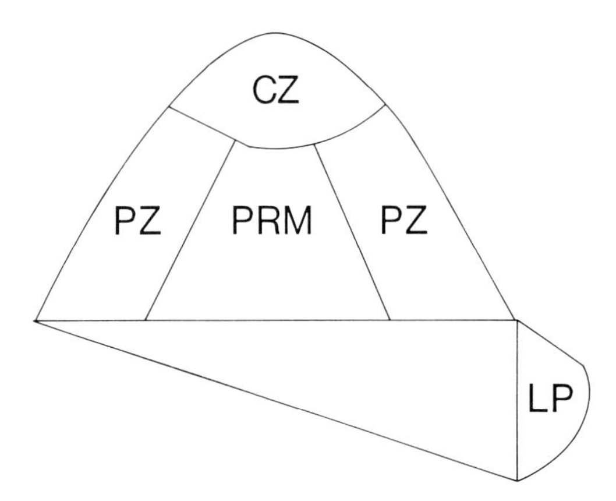

Figure 1 Generalized layout of the central zone (CZ), peripheral zone (PZ), pith-rib meristem (PRM) and the youngest leaf primordium (LP) in an idealized median longitudinal section of a vegetative shoot apex of an angiosperm. In angiosperms the diameter of the apical dome ranges from 50 μm to 3,000 μm (see text for details).

can be assigned exclusively to the shoot apical meristem. It has come as no great surprise that cell division cycle genes and histone gene expression have been reported in the shoot apical meristems just as they have been in other meristematic regions of the plant. Moreover, expression of histone H3 and H2B in the peripheral zone but not in the central zone (Medford, 1992; Kohler *et al.*, 1992) represent molecular confirmation of cytological observations of the relatively low frequency of cell division in the axial zone (see Nougarède, 1967).

Cell cycle heterogeneity in vegetative shoot meristems

Clearly, cell proliferation forms a very important component of the stem cell concept at two levels. First, a cell must be proliferating in order to be a stem cell, or second, a cell must have the potential to divide (the potential stem cell). Hence, a survey of the heterogeneity of cell cycles that have been reported in shoot meristems may be worthwhile in evaluating the potential locations of founder cells within the meristem.

Characteristically, different rates of cell division typify each zone described above. Most often, the slowest cell cycles are in the axial zone, the fastest are in the peripheral zone, whilst the duration of the cell cycle in the PRM tends to be variable (Table 1). The magnitude of the gradient differs from species to species. In the apex of *Helianthus annuus* and *Chrysanthemum segetum*, very few dividing cells occur in the central zone during vegetative growth (Langenauer *et al.*, 1974; Steeves *et al.*, 1969; Nougarède and Rembur, 1978). In fact, most cells are in G1 and only during floral growth is there an activation

Table 1 Cell doubling times in the central zone (CZ), peripheral zone (PZ), pith rib meristem (PRM), and youngest leaf primordium (LP), in the shoot meristem of a range of angiosperm species. The measurements for *Helianthus* and *Pharbitis* were at 28°C and 25°C, respectively and for *Chrysanthemum segetum* the day, night temperatures were 22°C and 16°C, respectively; all others were at 20°C. Data for *Dactylis* were from three natural populations collected from different sites within mainland Europe, and for *Lolium* were from the wild type (Ba3081) and *slow-to-green* mutant.

	Cell doubling time (hours)				
	CZ	PZ	PRM	LP	Reference
Dactylis glomerata					
6265	210	383	412	316	
5971	313	329	396	255	E. A. Kinsman, M. S. Davies,
5401	635	447	343	261	D. Francis and H. J. Ougham, unpublished work
Lolium temulentum					
WT	76	34	73	73	
Slow-to-green mutant	82	41	177	107	L. Moses, D. Francis and H. J. Ougham, unpublished work
Chrysanthemum segetum	139	48	70		Nougarède and Rembur, 1976
Helianthus annuus	83	37	118		Marc and Palmer, 1984
Sinapis alba	287	157			Bodson, 1975
Pisum sativum	69	29			Lyndon, 1970a
Pharbitis nil	100	70			Herbert *et al.*, 1992

of cell division in this zone; this fits the stem cell concept very well and upholds a much older albeit much disputed theory that the central zone is inactive during vegetative growth and active only during floral growth (Buvat, 1953). The difficulty is that not all meristems of higher plants exhibit this behaviour; there are just as many examples that refute the theory as support it. For example, in the shoot meristem of *Pisum sativum* faster rates of cell division have been recorded on the flanks compared with the central zone but the central zone is not devoid of cell division. Moreover, the lower rate of division in the central zone is not consistent with it being a reservoir of cells in G1/G0; cells of the central zone are apportioned equally, and at random, in both G1 and G2 (Lyndon, 1976). The gradient of cell division found in shoot meristems may be more to do with the shape of the apex than indicating founder cells within the apex. A stronger gradient is common with a flatter apex compared with a more conical apex (Francis, 1992). Hence, it seems more likely that founder cells are distributed throughout the shoot apex but this does not lessen the importance of the transition from G0 to G1 for founder cell activation, but merely means that founder cells are more difficult to locate.

In the yeasts, the G0 to G1 transition coincides with START of the cell cycle. Cells have to pass through the start window in order to be competent for cell division. Hunt (1991) likens this transition to a checkpoint which ensures that the cell is of an adequate size to go through an entire division cycle. In other words, cell growth drives the cell cycle. Critical cell size for division is well documented for a range of unicellular organisms including both budding and fission yeast and *Chlamydomonas* (Lorincz and Carter, 1979; Fantes and Nurse, 1981; John *et al.*, 1993). The virtues of studying cell size requirements in such unicellular organisms are obvious. However, exploring the cell size requirements of cells in meristems may provide some insight into the basic property expected of a founder cell, its ability to re-enter the cell cycle. John *et al.* (1993) also recognized this as the acquisition of a prerequisite cell size prior to both cell division and the onset of differentiation. Hence, the acquisition of a minimum cell size, which may well be nutrition dependent, regulates the subsequent participation of that cell in a cell cycle and perhaps contributes in a developmental context.

Leaf initiation

Leaf initiation begins by both an increase in the number of mitotic figures and a change in the plane of cell division in the part of the peripheral zone of the apex destined to form the next primordium. In *Pisum sativum*, these changes were first observed about halfway through a plastochron (Lyndon, 1970a, b). Much has been published about the relative contributions of cell division and cell expansion to leaf initiation and development. One line of evidence argues for cell expansion as the primary control of growth and comes from treatments which eradicate cell division. For example, γ-irradiated wheat seeds were no longer capable of cell division but the gamma plantlets initiated a new leaf (Foard, 1971). One conclusion drawn from these experiments is that cell division has a minor or non-existent role in the initiation of the primordium (Smith and Hake, 1992). If the meristem's cell division machinery is incapacitated, the cells may be capable of little else but expansion, but this should not be a reason for emphasizing the importance of one process rather than the other. In normal circumstances, the meristem increases the cell population by cell division and those progeny expand. Indeed, expansion growth contributes mostly to the rate of elongation in a root (Erickson and Sax, 1956) and cell expansion is the primary driving force for establishing leaf shape (Dale, 1988). In other words, cell division and cell expansion are equal partners which in combination achieve balanced growth. In the case of the γ-irradiated wheat plantlets, the new primordium had an abnormal shape and, in my view, the phyllotaxy, or leaf pattern, of the plants was altered compared with the controls (Figure 1 of Foard, 1971).

An increased rate of cell division in the peripheral zone, coupled with a change in the plane of cell division acts as a prerequisite for the emergence of the leaf primordium as a bump on the side of the apex. However, controversy exists concerning the primary event in leaf initiation. Is it the increased rate of division, the change in the plane of division (Lyndon and Cunninghame, 1986) or a preferential weakening of cell walls in the epidermal region which bulges out (Selker and Green, 1984)? It is probably, all three.

In a classic review of shoot morphogenesis, Nougarède (1967) assessed leaf initiation in *Leucanthemum parthenium, Coleus blumei, Teucrium scordonia, Lupinus albus* and *Tropaeolum majus*. The overwhelming conclusion was that in each case, founder cells for leaf initiation resided in the peripheral zone (anneau initial). However, she did not discount the possibility that through less frequent division, cells of the central zone could contribute to lateral growth, but the carefully assembled data were more consistent with progeny from the central zone contributing to the peripheral and pith rib zones and the latter giving rise to the leaves and sub-axial tissues, respectively (Nougarède, 1967).

The molecular identity of those cells which become destined to form a leaf primordium and the molecular controls which govern the localized change in the polarity of growth which leads to primordium formation are unknown. However, one line of evidence from work on maize suggests that when cells on the side of the apex become founder cells they do not express a particular gene. Specifically, the *KNOTTED-1* (*KN1*), a gene well characterized and best understood in functional maize leaves (Freeling and Hake, 1985), is not expressed in founder cells on the side of the shoot apex nor in young primordia although it is expressed at the protein level in other regions of the shoot apex (Smith *et al.*, 1992). Smith and Hake (1992) suggest that the down-regulation of the *KN1* gene is central to the events that lead to leaf initiation in maize. Whether the complex changes that occur during leaf initiation are due to the suppression of one gene remains to be seen but its absence at the protein level is undeniably correlated with the first events of leaf initiation. A functional homologue of a key cell division cycle gene product, $p34^{cdc2}$, binds to the preprophase band (PPB) in maize (Colasanti *et al.*, 1993). The latter is a vestigial array of microtubules which mark where a new cell wall forms (Pickett-Heaps and Northcote, 1966; Wick, 1991). In other words, the PPB 'predicts' the plane of cell division. Colasanti *et al.* (1993) concluded that the *cdc2* protein kinase could contribute to the establishment of the division site. Hence, it could be involved with the events of leaf initiation. The difficulty with this interpretation is that all dividing cells form PPBs, and why the binding of $p34^{cdc2}$ should be different in those cells which lead to a change in morphogenesis is unanswered.

A newly initiated leaf primordium can be deflected from forming a mature leaf by explanting it to an *in vitro* medium. For example in the fern, *Osmunda cinnamorea*, the youngest of 10 primordia (P_1 and P_2), when explanted on to culture medium, gave rise to plantlets. However, when the oldest (P_{10}) was explanted it developed as a leaf (Steeves, 1961). The response was related to size; those primordia greater than 1 mm in length developed into leaves (Steeves and Sussex, 1957). In other words, initiation and determination of a leaf are separate events although the interval between the two may be quite short (see Steeves and Sussex, 1987). The extent to which a lack of *KN1* expression is linked to both events is unknown.

Cell lineages

Unequivocal evidence that would qualify specific apical dome cells as founders would be knowledge both that a particular cell gave rise to a particular lineage of cells and that the fate of those cells at the end of the lineage were functional differentiating cells. This brings into play a vast amount of literature on chimeric shoot meristems, both natural and artificially created (see, for example, Tilney-Bassett, 1986). Clearly, the object is to track back from a particular sub-set of differentiated cells in an organ to a particular cell or cells in the shoot apex. To what extent do fate maps and cell lineages lead us to identify founder cells in the apical dome of a higher plant? In *Datura stramonium* clonal analysis first demonstrated the existence of three cell layers: L1, L2 and L3, responsible for the subepidermal, cortex and vascular tissues of the plant, respectively (Satina *et al.*, 1940). Moreover, cells in the apical dome have been marked by chemical mutagens or by irradiation and these studies have mapped marked cells from the dome through cell proliferation, to the differentiated tissues of the body of the plant. Critically, however, this type of study has shown that cell fate depends less on lineage and more on positional effects (Stewart, 1978).

An elegant study of irradiated seeds of *Arabidopsis* established fate maps or, what the authors described as 'probability maps', extending from the apical meristem to mature leaves (Irish and Sussex, 1992). They concluded that, in *Arabidopsis*, a single layer of the embryonic apical meristem contains 36–38 cells. Their data supported the idea that one rank of cells within the subepidermal layer of the shoot apical meristem acted as founders for each leaf but there was a somewhat variable number of cells within the rank. Note, they were unable to define a sub-set of unique founder cells which underlines the plasticity of cells within the shoot meristem. In a similar, but longer-term study of *Arabidopsis*, cell fates were mapped through to the stage when the meristem developed into the inflorescence meristem; the pattern did not support the idea of specific founder cells for a germ line (Furner and Pumfrey, 1992). Moreover, similar studies in maize shoot apices indicated that separate, lineage restricted apical initials do not exist (McDaniel and Poethig, 1988; Poethig *et al.*, 1990).

Interestingly, Irish and Sussex (1992) suggested that cells acquire a fate as they enter the flanks of the meristem; this fits well with Nougarède's (1967) conclusion (see above). If one accepts the notion that more cells are in G0 in the central zone, at least for some species, then the transition from G0–G1–S-phase would parallel this suggested acquisition of cell fate.

The stem cell concept for vegetative meristems

How rigorously do the cells of the apical dome conform to the criteria set up by Potten and Loeffler? Steeves and Sussex (1987) drew attention to the remarkable regenerative properties of shoot meristems. For example, dissection of the shoot meristem of *Vicia faba* resulted in two regenerating apices forming side-by-side (Pilkington, 1929), and regeneration of entire shoot apices from cultured halves also occurred in *Trachymena coerulea* (Ball, 1980). Remarkably, multiple dissections (six) of *Lupinus* yielded plantlets although eight cuts proved too much (Ball, 1948, 1952). Even more remarkable was regeneration of a potato plantlet from $\frac{1}{20}$ of the original meristem (Sussex, 1952). Clearly, in a normal shoot meristem, not all cells are founders; some are programmed to exhibit other developmental fates. Nevertheless, the plasticity and versatility of shoot meristem cells is evident; all apical dome cells are either founder cells or potential founder cells.

The dividing cohorts which are invested in the new primordium, or into the incipient internode below, could be regarded as transit cells. They fulfil this criterion quite well, being dividing cells but ultimately giving rise to differentiated cells in the leaf and internode. However, among these descendent cells there exists a tissue from which xylem and phloem ultimately form. This is the procambium and, ultimately, the cambium. The cambium could be a true stem cell population dividing asymmetrically to give rise to a new phloem cell on one side and a new xylem cell on the other. On closer examination, this fascinating tissue proves to be complex often consisting of more than one layer of initials which may be referred to as the cambial zone (Newman, 1956). Unfortunately, the cambium is an intractable tissue to work on because when removed and cultured *in vitro* it may stay alive but it ceases to proliferate (Bailey and Zirkle, 1931). In effect, removal of cambial zones from their normal niche perturbs the organization of cells which are necessary for ordered pattern formation.

Since cambial cells owe their origin to the shoot (and root) meristems, there exists a cell lineage from the meristem to the cambium with the inherent molecular properties of stem cells. Incidentally, a cambial-mediated wound response is central to the regenerative capacity of tissues when explanted to culture media (Warren-Wilson, 1978). So, cambial cells have renewal capacity, they proliferate and they can replace lost tissues.

THE FLORAL SHOOT APEX

Morphology

One major problem in attempting to analyse the stem cell concept in shoot meristems is that typically, the meristem's fate changes from an indeterminate growth habit during vegetative growth to a determinate growth habit during floral growth. The meristem changes shape and function as new arrangements of primordia form on the apex, new patterns are created and, ultimately, gametes are produced. The begging question is which cells do what, and when, as a result of phase change? A simple answer could be that all founder cells are used up in making a flower. My view is that several domains of cells in the floral meristem respond in a coordinated way to produce a flower. If so, in a determinate meristem cells should exist which can proliferate, be self-maintained and produce a large number of differentiated progeny.

We know that the leaves synthesize a floral stimulus and that it is exported to the apex. The initial apical response to the stimulus is an increased rate of metabolism, an increased rate of cell division and faster growth often leading to an increase in the volume of the apical dome (reviewed by Bernier, 1988; Lyndon and Francis, 1992). A shortening of the cell cycle typically precedes the formation of floral parts, and in one well-documented case (the long day plant, *Sinapis alba*), there is an increase in the proportion of cycling cells in the meristem. Hence, more cells are activated into shorter cycles (Gonthier *et al.*, 1987). This coupled with a synchronization of cell division just prior to the morphological appearance of flowers (Bernier *et al.*, 1967; Francis and Lyndon, 1979) led to the notion that such cell cycle changes may be restricted to the activation of particular subsets of cells destined to form floral primordia (Francis, 1987). Indeed earlier, it was suggested that such an activation of the cell cycle occurs in the central zone of the apex (Nougarède, 1967). The latter idea is supported by the greatest activation of cell division occurring in the central zone prior to flowering (e.g. Nougarède *et al.*, 1991; Bernier, 1988), but synchrony and a general activation of cells into the cell cycle seems to be spread throughout the apex (Gonthier *et al.*, 1987; Nougarède *et al.*, 1991; Francis, 1992).

These unseen changes are collectively part of floral evocation, events which commit the apical meristem to a floral mode of growth whereas the so-called floral realization stage begins with the initiation of the flower (Evans, 1969). Apart from well-characterized increases in the rRNA synthesis concomitant with, or immediately preceding the increase in the growth rate of the evoked apex (Arzee *et al.*, 1975; Gressel *et al.*, 1978) nothing is known about the molecular basis of floral evocation which is in rather stark contrast to the wealth of information now available on the molecular biology of floral realization. Floral realization begins with the initiation of the first whorl of the flower, the sepals (the calyx), followed by petals (corolla), then the stamens in which the male gametes form (androecium) and finally, the carpels in which the female gametes form (gynoecium). The whole is the hermaphrodite flower. Once again, in angiosperms there is enormous diversity of floral form. It is not uncommon to observe stamens forming at the same time as, or ahead of, the petals (Lyndon, 1978; Donnison and Francis, 1993) and commonly there are 4 or 5 organs which comprise each whorl, but there are exceptions, e.g. in *Arabidopsis* two carpels form a fused gynoecium. Sometimes sepals and petals are indistinguishable morphologically (perianth parts) and some flowers can be male (carpels are suppressed) and some female (the stamens are suppressed). A monoecious species exhibits both male and female flowers on the same plant, a dioecious species exhibits separate male and female plants.

Simultaneous ripening of unisexual flowers leads to self-pollination while asynchronous ripening facilitates cross-pollination. The grass flower forms differently for although containing the same type of organs, small flowers or florets are grouped into spikelets. Each spikelet is separated from its neighbour by two small modified leaves known as glumes. Each floret is ensheathed by two additional bracts, the lemma and palea. The latter is frequently the smaller organ and enclosed by the former. Inside these modified leaves there are normally three stamens and two fused carpels. Hence, the sepal and petal whorls have been suppressed and small structures at the base of the flower, the lodicules, may represent vestigial remnants of the perianth parts (see Weier *et al.*, 1982 for a particularly well-explained account of floral structure).

Homeotic genes, meristem identity genes and floral founder cells

The above botanical sermon is noteworthy when focusing on founder cells for floral organs. Any generalizations must be tempered by the stunning diversity of floral form of which the above description only gives a superficial insight. However, the above will help when discussing the vast array of homeotic mutants of flowering plants where a normal organ forms at an abnormal site; molecular genetic analysis of these mutants has led to the discovery of homeotic or organ identity genes that direct normal development. This burgeoning research area has been expertly reviewed elsewhere (Coen, 1991; Coen and Meyerowitz, 1991). I shall provide some commentary and try to extrapolate back to the location and function of founder cells during floral organ initiation.

Two types of gene have been identified: genes which regulate the identity of meristems and homeotic genes which determine the identity of organs.

One of the best characterized meristem identity genes in plants is *FLORICAULA* (*FLO*) from *Antirrhinum* (for the details of its isolation and characterization see Coen, 1991). Plants carrying this mutation exhibit inflorescence meristems which cannot change to floral meristems (Carpenter and Coen, 1990). *FLO* is expressed transiently in wild type inflorescences, and then sequentially, in bract, sepal, petal and carpel but not in stamen primordia. Coen (1991) suggested that *FLO* acts as a master switch between an inflorescence and floral meristem and may facilitate the expression of other genes downstream. In particular, *FLO* may be a key player interacting with other whorl or organ identity genes but perhaps becoming repressed in the founder cells for the stamen whorl.

Detailed analyses of various homeotic flowering mutants of *Arabidopsis* and *Antirrhinum* exist whereby an organ type in one whorl is replaced by organs of another whorl. For example, the *apetala2* mutant of *Arabidopsis* makes flowers in which the outermost whorl (4 sepals) is replaced by 4 carpels, and the next whorl (4 petals) is replaced by 4 stamens. The number and pattern of organs is the same in the mutant but organ identity is different. Similar observations in the green pistillate mutant of tomato led to the conclusion that the wild type and mutant showed identical patterns and timing of initiation of organs but that mechanisms governing organ initiation are distinct from those governing organ identity (Rasmussen and Green, 1993). This may be so, but how should this idea be reconciled with conclusions drawn from the floral patterns exhibited by other mutants? For example, the *apetala2* mutant of *Arabidopsis* grown at 29°C formed carpelloid leaves instead of sepals, but skipped the petal whorl and went on to form normal third (stamens) and fourth (carpel) whorls. This observation was consistent with the idea that the *AP2* gene in the wild type is involved in organ (petals) initiation as well as organ identity (Bowman *et al.*, 1989).

Bowman *et al.* (1989) took the view that *AP2* (together with three other homeotic genes) enables cells to sense their position in the flower. How is this achieved? Does this mean that such homeotic genes exert changes upstream of a series of changes which, in effect, isolate a domain of primordium cells? That one set of primordia take their cue from an outer whorl (Heslop-Harrison, 1964; McHughen, 1980) is dismissed as a likely mechanism since some mutants can exhibit altered outer whorls but normal inner ones (see, for example, Irish and Sussex, 1990). The currently accepted view is that molecular domains exist as concentric spheres of activity which overlap and regulate the initiation and development of each whorl.

The homeotic genes which regulate the identity of floral whorls have been apportioned to 3 distinct classes:

Class A affects whorls 1 and 2
Class B affects whorls 2 and 3
Class C affects whorls 3 and 4.

Hence the model describes overlapping molecular domains which regulate floral morphogenesis (Coen and Meyerowitz, 1991). Organ identity genes in both *Arabidopsis thaliana* and *Antirrhinum majus* (which also exhibits a range of homeotic mutants) may encode transcription factors. *AG* (in *Arabidopsis*) and *DEFICIENS* (*DEF* in *Antirrhinum*), encode proteins with a high degree of homology to conserved DNA binding domains of two transcription factors, *serum response factor* (*SRF*) and *mini chromosome maintenance protein* (*MCM1*); collectively referred to as *MADS*-box genes (Schwarz-Sommer *et al.*, 1990).

The recessive homeotic *apetala* (*ap1-1*) mutation of *Arabidopsis* exhibits first-whorl organs as bract-like structures instead of the sepals found in the wild type (Irish and Sussex, 1990). Moreover, in the mutant, flower buds develop in the axil of each bract-like organ. Hence, the mutants exhibit a quartet of flowers (which lack petals), each one arising in the axil of the bract which has developed instead of the sepal. Since petal-like epidermal cells were observed in *ap1-1* plants, Irish and Sussex (1990) concluded that the mutation did not disrupt processes required for petal differentiation but did affect the formation of second whorl primordia. Hence, here is an example of a homeotic gene affecting the initiation of organs, in this case the petal primordia. They concluded that *AP1* and *AP2* gene products interact with *AG* to establish a determinate floral meristem, whereas other homeotic gene products are required for cells to differentiate according to their position. It would follow that expression of *AP1* and *AP2* are part of a programme which gives floral character to the meristem (Irish and Sussex, 1990).

Weigel and Meyerowitz (1993) noted that the expression of homeotic genes is governed by negative interactions. For example, in *Arabidopsis* the class C gene, *AGAMOUS* (*AG*), (the *ag* mutant exhibits more petals instead of stamens and a variable number of inner carpelloid organs) is repressed in the outer whorls by the class A gene *APETALA2* (the *ap2* mutant makes carpelloid leaves in the outer whorl instead of sepals (see above and Bowman *et al.*, 1991; Yanofsky *et al.*, 1990; Drews *et al.*, 1991). Moreover, the expression of the class B genes *APETALA3* and *PISTALLATA* (both *ap3* and *pi* mutants make two whorls of sepals and two whorls of carpels) is repressed by the whorl identity gene *SUPERMAN* (Bowman *et al.*, 1992) (Table 2). In *Arabidopsis*, homeotic genes are activated at least partly by two meristem identity genes *LEAFY* and *APETALA1* (Weigel and Meyerowitz, 1993). This conclusion was drawn by considering the latter two genes as controllers and considering how the expression patterns of homeotic genes were affected by *lfy* and *ap1* mutations. As mentioned above, homeotic genes have been assigned to a particular class affecting overlapping whorl identity. For example, the expression pattern of *AG* gives a domain which encompasses both whorls 3 and 4. In *lfy-6* mutants, the expression pattern of *AG* is abnormal; the detection of

Table 2 A selection of homeotic genes in *Arabidopsis thaliana* and the corresponding arrangement of whorls in the flowers of the mutant phenotypes (see text for details and references).

Class	Homeotic gene	Mutant phenotype
A	*AP2*	carpelloid leaves, stamens, stamens, carpels
B	*AP3*	sepals, sepals, carpels, carpels
C	*AG*	sepals, petals, petals, 'carpels'

AG RNA was confined to a smaller domain than in the wild type. Later, they observed a more normal pattern of *AG* expression. Moreover, in the double mutants, *ap1-lfy-6*, *AG* expression did not accumulate in the centre of the flower. Hence, *AG* (and, *AP3* and *PI*) require the activity of the meristem identity genes, *LFY* and *AP1* (see Weigel and Meyerowitz, 1993, for more details). Hence:

This apparent digression into some aspects of homeotic gene activity can be considered in relation to the location of founder cells for the spectacular changes observed during floral morphogenesis. The expression patterns reported by Weigel and Meyerowitz using *in situ* hybridization revealed whole primordia lighting up in the wild type but in the mutants the expression patterns were more restricted. For example, in the wild type flowers, *AG* was expressed throughout the stamens and in the gynoecium but not in the sepals and petals. In the *lfy6* mutant, the *AG* expression pattern was confined to a smaller region (see above). It seems to me that the expression patterns in the wild type indicate a general molecular competence to express a certain homeotic gene (under the control of a meristem identity gene), but only a certain number of cells are demarcated as a domain of primordial cells. Hence, another component may exist to cause a sub-population of cells to develop as a floral primordium, given that the essential prerequisite is a general level of molecular competence of an entire ring of cells. In other words, the establishment of a molecular domain in which a ring of cells is expressing the right homeotic gene at the right time, is an essential prerequisite for the correct spacing of primordia in a given whorl. However, some other mechanism may demarcate a domain of cells to do this. A gene which regulates meristem and flower development in *Arabidopsis*, *CLAVATA1* (*CLV1*), acts independently of A, B, C-type genes. *CLV1* affects organ number but not organ identity (Clark *et al.*, 1993) and hence may be a gene which is part of a demarcation mechanism. Work is in progress examining this possibility (E. M. Meyerowitz, personal communication). Francis and Herbert (1993) argued that a cell size control may be a component of such a mechanism. Remarkably, transgenic tobacco plants expressing a fission yeast mitotic inducer gene (*cdc25*) under the 35S constitutive promoter exhibited petalless flowers alongside normal ones (Bell *et al.*, 1993). Perturbation of normal cell size in the petalless flowers may have contributed to the aberrancy. We are now testing this hypothesis.

Models which explain homeotic gene function feature cells recognizing their position in the flower (Bowman *et al.*, 1990) or, incorporate cells as part of a coordinated mechanism which imposes a floral character on a meristem (Irish and Sussex, 1990). Notably, these models have led to testable predictions for the phenotypes of mutants and double mutants etc. Valuable as these models are, they still leave the unanswered questions of which cells do what and when. A mutant of tomato (*lateral suppressor*) which suppresses some of the axillary buds that normally grow out in the wild type may offer new clues (these mutants also fail to develop petals (Williams, 1960)). Recently, chimeras were generated by grafting the *lateral suppressor* and wild type plants (Szymkoviak and

Sussex, 1993). The chimera comprised an L1 layer of mutant cells and L2 and L3 layers of wild type cells. Such a chimera formed normal flowers. This implies that one whorl of the flower (in this case, the petals) takes a developmental cue from cells within the meristem. The nature of the signalling molecules involved is unknown but this developmental response is not dissimilar to that of maize plants carrying the *KNOTTED* mutation where the pattern of cell division which gives rise to knot formation takes its cue from underlying mesophyll cells (see Freeling and Hake, 1985). Whether internal cells provide a diffusible signal is unknown; Szymkoviak and Sussex (1993) prefer a model in which the inner cells represent part of a mechanism that gives a developmental cue to cells in an outer whorl. Hence, another component of primordium initiation may well reside in cells within the floral meristem which results in signals that interact with cells, already demarcated into a molecular domain, on the outside of the meristem.

Coen and Meyerowitz (1991) commented on how a greater understanding of the relationship between the timing of cell division and the regulation of homeotic gene expression will bring us nearer to understanding the developmental regulation of floral morphogenesis. Substantial progress has been made on the way that homeotic gene expression is regulated (see, for example, Weigel and Meyerowitz, 1993). Now we need to know more about plant genes which affect the timing of cell division. Whether a sizer control is an important component may well be worth further exploration and would focus more clearly, on the identity of founder cells within floral meristems.

CONCLUSIONS

In deciding where founder cells are located in the shoot meristem the plasticity of development of the vegetative meristem comes to the fore. The remarkable regenerative ability of mere slices of the meristem are consistent with an 'all potential founder cell' concept in the vegetative meristem. Clearly, in normal development not all cells in the meristem are founders. More likely is the notion that G0 cells in the central or axial zone of the shoot meristem are true founders and their activation into the cell cycle, as they transit from the central to the peripheral zone, is a key regulatory event but, paradoxically, in some species (e.g. *Pisum sativum*) such cells cannot be located.

If, and how, cells are programmed to be leaf primordial cells has yet to be explained although such cells exhibit a change in the plane of cell division which then results in a change in the polarity of growth for those cells destined to be the next leaf primordium. The primary events controlling planes of cell division would also help to uncover the identity of leaf founder cells.

The task of locating founder cells in floral shoot meristems may, at first glance, seem easier in a determinate developing structure. Indeed, the major advances in our understanding of the function of meristem identity genes and homeotic genes has led to the concept of molecular concentric domains in the floral apex. Clearly, not all cells within a molecular domain develop as a floral organ. One of the key interactions must be molecular controls of the cell cycle which predispose cells to divide as a cohort or cellular domain coupled to the expression of the organ identity genes. One component of this interaction could be a sizer control which could demarcate a cellular domain within an already defined molecular domain. I suggest that it is this interaction which defines a founder cell population in floral morphogenesis.

ACKNOWLEDGEMENTS

I thank Elliot Meyerowitz (Cal. Tech., Pasadena) and Arlette Nougarède (Univ. P. M. Curie, Paris) for critically reviewing the manuscript but I take responsibility for its eventual form. Unpublished data listed in Table 1 is from AFRC (BAGEC programme)-funded work.

I wrote this paper in 1993 – hence, papers which have appeared since then, notably on homeotic genes, seem to be unnoticed. However, the concepts discussed then still seem to be applicable to papers published in the interim.

REFERENCES

Arzee, T., Zilberstein, A. and Gressel, J. (1975). *Plant Cell Physiol.* **16**:505–511.
Bailey, I.W. and Zirkle, C. (1931). *J. Gen. Physiol.* **14**:363–383.
Ball, E.M. (1948). *Symp. Soc. Exp. Biol.* **2**:246–262.
Ball, E.M. (1952). *Growth* **16**:151–174.
Ball, E.M. (1980). *Ann. Bot.* **45**:103–112.
Bernier, G. (1988). *Ann. Rev. Plant Physiol.* **39**:175–219.
Bernier, G., Kinet, J.M. and Bronchart, R. (1967). *Physiol. Veg.* **5**:311–324.
Bell, M.H., Halford, N.G., Ormrod, J.C. and Francis, D. (1993). *Plant Mol. Biol.* **23**:445–451.
Bodson, M. (1975). *Ann. Bot.* **39**:547–554.
Bowman, J.L., Smyth, D.R. and Meyerowitz, E.M. (1989). *Plant Cell* **1**:37–52.
Bowman, J.L., Smyth, D.R. and Meyerowitz, E.M. (1991). *Development* **112**:1–20.
Bowman, J.L., Sakai, H., Jack, T., Weigel, D., Mayer, U. and Meyerowitz, E.M. (1992). *Development* **114**:599–615.
Buvat, R. (1953). *Ann. Nat. Sci. Bot.* **11**:199–300.
Carpenter, R. and Coen, E.S. (1990). *Cell* **63**:1311–1322.
Clark, S.E., Running, M.P. and Meyerowitz, E.M. (1993). *Development* **119**:397–418.
Coen, E.S. (1991). *Ann. Rev. Plant Physiol. Plant Mol. Biol.* **42**:241–279.
Coen, E.S. and Meyerowitz, E.M. (1991). *Nature* **353**:31–37.
Colasanti, J., Cho, S.-K., Wick, S. and Sundaresan, V. (1993). *Plant Cell* **5**:1101–1111.
Dale, J.E. (1988). *Ann. Rev. Plant Physiol. Plant Mol. Biol.* **39**:267–295.
Donnison, I.S. and Francis, D. (1993). *Physiol. Plant.* **89**:315–322.
Drews, G.N., Bowman, J.L. and Meyerowitz, E.M. (1991). *Cell* **65**:991–1002.
Erickson, R.O. and Sax, K.B. (1956). *Proc. Am. Phil. Soc.* **100**:499–514.
Evans, L.T. (1969). In *The Induction of Flowering; Some Case Histories* (ed. L. T. Evans), pp. 1–30. Macmillan, Melbourne.
Fantes, P.A. and Nurse, P. (1981). In *The Cell Cycle* (ed. P. C. L. John), pp. 11–33. Cambridge University Press, Cambridge.
Foard, D.E. (1971). *Can. J. Bot.* **49**:1601–1603.
Francis, D. (1987). In *Manipulation of Flowering* (ed. J. G. Atherton), pp. 289–300. Butterworths, London.
Francis, D. (1992). *New Phytol.* **122**:1–20.
Francis, D. and Herbert, R.J. (1993). In *Molecular and Cell Biology of the Plant Cell Cycle* (eds J. C. Ormrod and D. Francis), pp. 201–210. Kluwer Academic Publishers, Dordrecht.
Francis, D. and Lyndon, R.F. (1979). *Planta* **145**:151–157.
Freeling, M. and Hake, S. (1985). *Genetics* **111**:617–634.
Furner, I.J. and Pumfrey, J.E. (1992). *Development* **115**:755–764.
Gonthier, R., Jacqmard, A. and Bernier, G. (1987). *Planta* **170**:55–59.
Gressel, J., Zilberstein, A., Strausbach, L. and Arzee, T. (1978). *Photochem. Photobiol.* **27**:237–240.
Herbert, R.J., Francis, D. and Ormrod, J.C. (1992). *Physiol. Plant.* **86**:85–92.
Heslop-Harrison, J. (1964). *Brookhaven Symp. Biol.* **16**:109–125.
Hunt, T. (1991). In *The Cell Cycle Cold Spring Harbor Symposium on Quantitative Biology* (eds J. Watson, D. Beach and B. Stillman). Cold Spring Harbor, Long Island.
Irish, V.F. and Sussex, I.M. (1990). *Plant Cell* **2**:741–753.

Irish, V.F. and Sussex, I.M. (1992). *Development* **115**:745–753.
John, P.C.L., Zhang, K. and Dong, C. (1993). In *Molecular and Cell Biology of the Plant Cell Cycle* (eds J. C. Ormrod and D. Francis), pp. 9–34. Kluwer Academic Publishers, Dordrecht, Boston, London.
Kohler, S., Coraggio, I., Becker, D. and Salamini, F. (1992). *Planta* **186**:227–235.
Lance, A. (1957). *Ann. Sci. Nat. Bot. Biol. Vég.* **Sér 11, 18**:91–421.
Langenauer, H.D., Davis, E.L. and Webster, P.L. (1974). *Can. J. Bot.* **52**:2195–2201.
Lorincz, A. and Carter, B.L.A. (1979). *J. Gen. Microbiol.* **113**:287–295.
Lyndon, R.F. (1970a). *Ann. Bot.* **34**:1–17.
Lyndon, R.F. (1970b). *Ann. Bot.* **34**:19–28.
Lyndon, R.F. (1976). In *Cell Division in Higher Plants* (ed. M. M. Yeoman), pp. 285–314. Academic Press, London.
Lyndon, R.F. (1978). *Ann. Bot.* **42**:1349–1360.
Lyndon, R.F. and Cunninghame, M.E. (1986). In *Plasticity in plants, 40th Symp. Soc. Exp. Biol.* (eds A. J. Trewavas and D. H. Jennings), pp. 233–255. Company of Biologists, Cambridge.
Lyndon, R.F. and Francis, D. (1992). *Plant Mol. Biol.* **19**:51–68.
Marc, J. and Palmer, J.H. (1984). *Am. J. Bot.* **71**:588–595.
Mauseth, J.D. (1991). *Botany, an Introduction to Plant Biology.* W. B. Saunders, Florida.
McDaniel, C.N. and Poethig, R.S. (1988). *Planta* 175:13–22.
McHughen, A. (1980). *Bot. Gaz.* 141:389–395.
Medford, J.I. (1992). *Plant Cell* **4**:1029–1039.
Newman, I.V. (1956). *Phytomorphology* **6**:1–19.
Nougarède, A. (1967). *Int. Rev. Cytol.* **21**:203–251.
Nougarède, A. and Rembur, J. (1978). *Z. Pfl. Physiol.* **90**:379–389.
Nougarède, A., Francis, D. and Rembur, J. (1991). *Protoplasma* **169**:1–10.
Picket-Heaps, J.D. and Northcote, D.H. (1966). *J. Cell Sci.* **1**:109–120.
Pilkington, M. (1929). *New Phytol.* **28**:37–53.
Poethig, R.S., McDaniel, C.N. and Coe, E.H. (1990). In *Genetics of Pattern Formation and Growth Control* (ed. A. Mahowald), pp. 197–208. Wiley-Liss Inc., New York.
Potten, C.S. and Loeffler, M. (1990). *Development* **110**:1001–1020.
Rasmussen, N. and Green, P.B. (1993). *Am. J. Bot.* **80**:805–813.
Rembur, J. and Nougarède, A. (1977). *Z. Pfl. Physiol.* **81**:173–179.
Satina, S., Blakeslee, A.F. and Avery, A.G. (1940). *Am. J. Bot.* 27:985–1005.
Schwarz-Sommer, Z., Huijser, P., Nacken, W., Saeder, H. and Sommer, H. (1990). *Science* 250:931–936.
Selker, J.M.L. and Green, P.B. (1984). *Planta* **160**:289–297.
Smith, L.G. and Hake, S. (1992). *Plant Cell* **4**:1017–1027.
Smith, L.G., Greene, B., Veit, B. and Hake, S. (1992). *Development* **116**:?.
Steeves, T.A. (1961). *Phytomorphology* **11**:346–359.
Steeves, T.A. and Sussex, I.M. (1957). *Am. J. Bot.* **44**:665–673.
Steeves, T.A. and Sussex, I.M. (1987). *Patterns in Plant Development*, 2nd edn. Cambridge University Press, Cambridge.
Steeves, T.A., Hicks, M.A., Naylor, J.M. and Rennie, P. (1969). *Can. J. Bot.* **47**:1367–1375.
Stewart, R.N. (1978). In *The Clonal Analysis of Development* (eds S. Subtelney and I. M. Sussex), pp. 131–160. Academic Press, New York.
Sussex, I.M. (1952). *Nature* **170**:755–757.
Szymkoviak, E.J. and Sussex, I.M. (1993). *The Plant Journal* **4**:1–7.
Taylor, M. and Francis, D. (1989). *Ann. Bot.* **64**:625–633.
Tilney-Bassett, R.A.E. (1986). *Plant Chimeras.* Arnold, London.
Warren-Wilson, J. (1978). *Proc. R. Soc. Lond. Series B* **203**, 153–176.
Weier, T.E., Stocking, C.R., Barbour, M.G. and Rost, T.L. (1982). *An Introduction to Plant Biology*, pp. 269–283, 6th edn. John Wiley & Sons, New York, Chichester, Brisbane, Toronto, Singapore.
Weigel, D. and Meyerowitz, E.M. (1993). *Science* **261**:1723–1726.
Wick, S.M. (1991). *Curr. Op. Cell Biol.* **3**:253–260.
Williams, W. (1960). *Heredity* **14**:285–296.
Yanofsky, M.F., Ma, H., Bowman, J.L., Drews, G.N., Feldman, K.A. and Meyerowitz, E.M. (1990). *Nature* **346**:35–39.

4 Skeletal muscle stem cells: function and potential role in therapy

Diana J. Watt and Gareth E. Jones*

*Department of Anatomy, Charing Cross & Westminster Medical School, London, UK; *The Randall Institute, King's College London, UK*

ORIGIN AND IDENTIFICATION OF SKELETAL MUSCLE STEM CELLS

The concept of a stem cell in relation to skeletal muscle has to be viewed both in terms of development and in terms of the mature muscle fibre. The differentiated skeletal muscle fibre is a multinucleate cell, having formed by the fusion of many mononuclear cells termed 'myoblasts' (Figure 1). It is this fusion event which gives rise to the differentiating myotube which then elaborates contractile protein elements into myofilaments to mature into the fully differentiated skeletal muscle cell or fibre. Although it is nowadays generally accepted that the mechanism of vertebrate skeletal muscle formation is a result of myoblast fusion, earlier theories of myogenesis favoured amitotic division of myoblast and myotube nuclei, unaccompanied by a corresponding cytoplasmic division, as the mechanism responsible for the increase in nuclear number within the developing muscle fibres (Adams *et al.*, 1954; Altschul and Lee, 1960; Boyd, 1960). Evidence of fusion of myoblasts to form multinucleate fibres came from the studies of Capers (1960), Holtzer *et al.* (1957, 1958), Konigsberg *et al.* (1960), Lash *et al.* (1957) and the tritiated thymidine studies of Stockdale and Holtzer (1961) which categorically showed that myotubes were the result of fusion of mononuclear myoblasts.

Vertebrate skeletal muscle takes its origin, during embryonic development, from the paraxial mesoderm situated on either side of the developing neural tube (Theilar, 1989; Ott *et al.*, 1990). In turn, the paraxial mesoderm segments into blocks or somites (Bellairs, 1963) and from part – the dermamyotome – of each of these somites, one pair of which exist for each segment of the body, both the axial and limb musculature of the future body form is derived by migration of cells from the myotome component (Kaehn *et al.*, 1988) of the dermamyotome. It is, however, only relatively recently that experimental evidence using chick/quail grafting has conclusively shown the contribution of the somite to the limb musculature. Prior to such grafting work, earlier studies failed to show the contribution of somitic mesoderm to limb myogenesis (Hamburger, 1938; Saunders, 1948; Seichert, 1971; Gumpel-Pinot, 1974). Quail/chick grafting using biological labelling methods (Le Douarin and Barq, 1969; Christ *et al.*, 1974; Chevallier *et al.*, 1976) has now

STEM CELLS
ISBN 0–12–563455–2

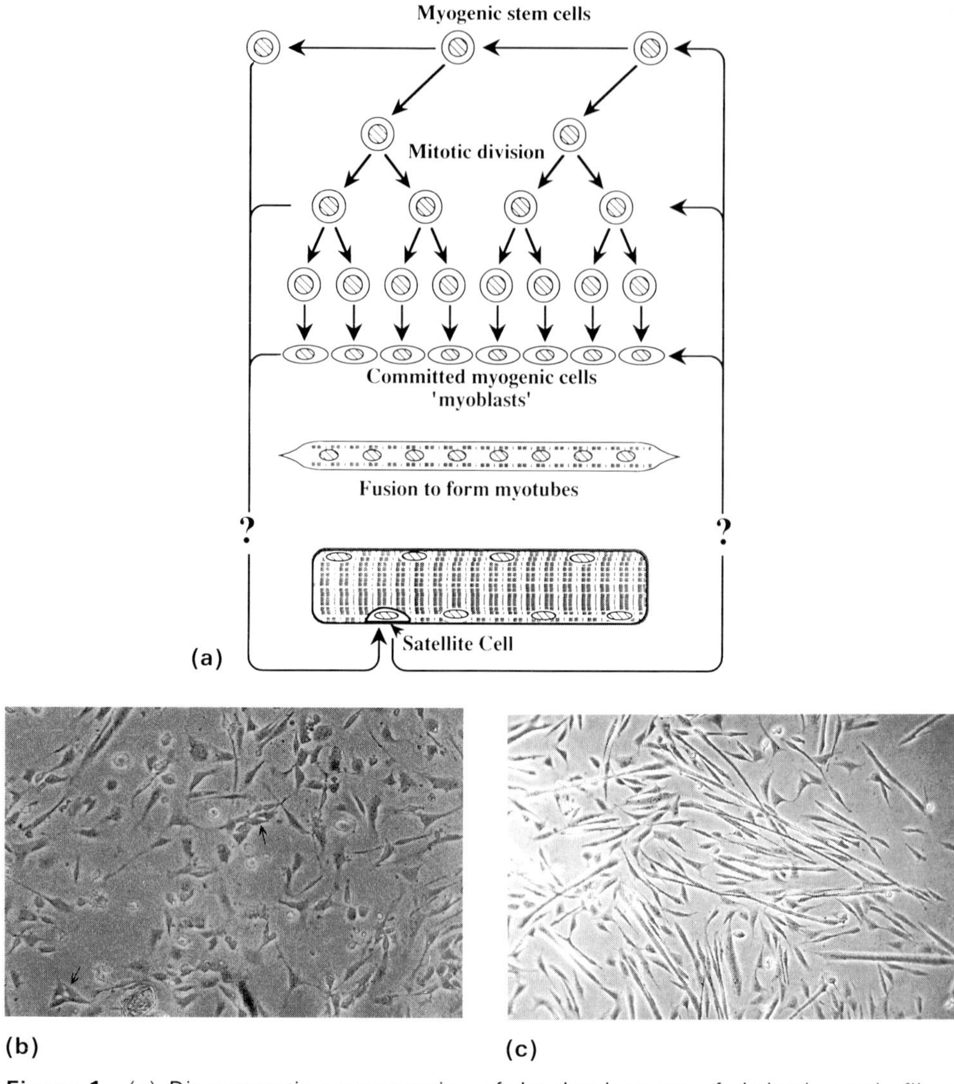

Figure 1 (a) Diagrammatic representation of the development of skeletal muscle fibres. The stem cells increase their number by mitosis, some of which become committed and differentiate into myoblasts. Mononuclear myoblasts fuse to form the multinucleate myotube which by elaboration of contractile elements mature to form the muscle fibre. Some stem cells remain between the muscle plasma membrane and the basal lamina and constitute the 'satellite cells' which will form new muscle fibres during growth and regeneration of the fibre. (After Morgan and Watt, 1993.) (b) A primary culture of mouse myogenic cells. Cells were obtained by enzymatically disaggregating neonatal mouse muscle and enriching for myogenic cells by incubating the mixed cell population obtained from the disaggregate with anti N-CAM (neuronal cell adhesion molecule) antibody. The myogenic cells bearing N-CAM could thus be selected for and a population enriched for myogenic cells grown *in vitro* (Jones *et al.*, 1990). 72 hours after plating out, the panning procedure has reduced the number of fibroblastic cells present in the culture. Myogenic cells are present at this stage as mononuclear myoblasts (arrowed). Magnification ×480. (c) Nine days after plating out many of the cells shown in (b) had fused to form multinucleate myotubes. Magnification ×480.

firmly established the origin of the myogenic component of the muscle belly in the limb to have migrated from a somitic origin (Chevallier *et al.*, 1977; Christ *et al.*, 1977a, 1977b; Jacob *et al.*, 1978; Wachtler *et al.*, 1982). This pattern of formation of the axial and limb musculature also holds true for mammalian systems (Milaire, 1976; Ede and El-Gadi, 1986) although studies have not been so extensively carried out in mammalian systems as compared with avian ones.

During the myogenic programme, the myoblast itself differentiates from a precursor cell. The initial events along the pathway from somitic mesoderm to the emergence of precursors of skeletal muscle have been studied using morphological criteria and *in vitro* clonal analysis. In studies using developing chick wing buds as a model system, colonies of mesodermal cells have been identified which are destined to give rise to multinucleate muscle fibres of the myogenic regions of these limbs (Bonner and Hauschka, 1974; White *et al.*, 1975). Such colonies have been designated into 'early-' or 'late'-muscle-forming colonies depending on the stage they appeared in the developing limb and the percentage of their cells capable of fusing to the multinuclear state. Grafting studies between early and late chick embryos further confirmed that these different myogenic colonies arose from separate precursor cell populations (Seed and Hauschka, 1984), predestined to such fates prior to the migration of the myogenic cells into the wing buds. Similar distinct clonal morphologies were observed to occur during human myogenesis, elucidated using *in vitro* techniques (Hauschka, 1974).

FACTORS INFLUENCING THE DIFFERENTIATION OF MUSCLE STEM CELLS

The work, predominantly of Hauschka and colleagues in the 1970s, identified muscle precursor cells by their ability to form multinucleate fibres *in vitro*. Within the past decade, identification of the muscle precursor cell and the progeny it gives rise to have centred on the discovery of a family of genes and their products (Olson, 1991). The myogenic transcription factors, as they have been so termed, are first expressed within the early precursor cell, each stage of myogenesis being characterized by the expression of one or more of such muscle regulatory genes. Such transcription factors play a central role in the differentiation of muscle precursor cells and their subsequent fate and differentiation beyond this stage. The family consists of 4 members – myogenin, MyoD1, Myf-5 and MRF-4. The role of these DNA binding proteins in myogenesis began to be elucidated following the work of Taylor and Jones (1979) when treatment of the fibroblastic 10T1/2 cell with the anti-cancer agent 5-azacytidine converted such cells along various phenotypic pathways, including myogenesis. The consequence of incorporation of this agent into the genome of 10T1/2 fibroblasts was the production of stable myogenic cells (Konieczny and Emerson, 1984) which could remain as proliferating myoblasts or which could enter terminal differentiation to give rise to multinucleate muscle cells. These two studies opened up the debate for expression of a muscle regulatory gene capable of converting these fibroblasts to a myogenic lineage. Soon after this Davis *et al.* (1987), employing differential cDNA screening techniques, were the first to isolate a gene which was accredited with the role of a muscle determination gene. Forced expression of this gene, termed MyoD1 – the Myogenic

Determination factor, was capable of converting approximately 50% of 10T1/2 cells to myoblasts, indicating a role for the MyoD1 gene product in establishment of the myogenic lineage. The remaining three muscle regulatory factors – myogenin (Wright *et al.*, 1989; Edmonson and Olson, 1989), Myf-5 (Braun *et al.*, 1989) and MRF-4 (Rhodes and Konieczny, 1989) – the latter also given the name herculin or Myf-6 by other laboratories (Miner and Wold, 1990; Braun *et al.*, 1990) were also found capable of directing 10T1/2 cells down the myogenic pathway. The proteins contained an evolutionary conserved core consisting of a basic region, which is involved in binding to DNA, and a helix-loop-helix (HLH) domain, which is important for heterodimer formation with other HLH proteins in the nucleus, such as E12 (Murre *et al.*, 1989). A consensus sequence for the binding of the myogenic factor heterodimers, CANNTG, usually occurs in tandem in muscle specific enhancers, which regulate the high level expression of muscle structural protein genes. This motif is also present in the promoters of many muscle genes.

Each of the genes governing the four myogenic regulatory factors are expressed at very specific stages throughout the myogenic programme (see Figure 2), the most extensive studies of their expression having been carried out in the mouse by Buckingham and colleagues (Buckingham *et al.*, 1992). Before secondary muscle fibres form, muscle masses in the embryo do not show fibre type specialization. Thus muscle genes are expressed uniformly in any given muscle at any given time (Lyons *et al.*, 1990). At later stages there is striking asynchrony in temporal expression of muscle genes (Buckingham 1992). The only muscle regulatory factor to be expressed prior to myotome formation in the mouse is Myf-5, appearing in this animal during genesis of the first somites. At this stage the distribution of Myf-5 is patchy and confined to distinct regions known to be sites of muscle cell determination (Buckingham *et al.*, 1992). This once suggested that Myf-5 may be responsible for specialization of somitic cells to a myogenic lineage (but see results of Myf-5 knock-out mice, page 79). During mouse myogenesis, the other muscle regulatory genes express their products later, in relation to Myf-5, but all have a specific pattern of expression and these differ in the myotome as compared with

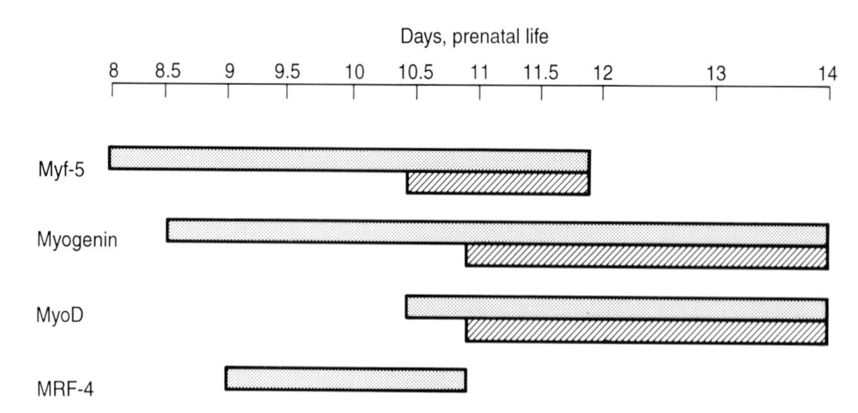

Figure 2 Expression of mRNA for myogenic transcription factors in the mouse somite/myotome and forelimb bud between 8 and 14 days embryonic life. Expression of MRF-4 appears at a later stage in the forelimb, i.e. 16 days (after Buckingham, 1992). Unusually, myogenin is detectable 2 days after the mRNA is produced. (▨ somite/myotome; ▩ forelimb bud).

the limb. In the former, myogenin mRNA follows Myf-5 expression (Sassoon *et al.*, 1989), followed in turn by MRF-4/Myf-6 (Bober *et al.*, 1991) and then MyoD1. In the mouse limb, MRF-4 is the last to be expressed (Bober *et al.*, 1991), Myf-5 being first (Ott *et al.*, 1991), closely followed by MyoD1 and myogenin (Sassoon *et al.*, 1989; Buckingham *et al.*, 1992). While it is not clear what physiological factors cause different genes to be activated at different times, at the cell level it seems likely that the gradual build up of different levels of regulatory factors results in asynchronous activation of genes that differ in their threshold sensitivity. Despite the fact that there are four regulatory factors, one is sufficient to convert 10T1/2 cells to a myogenic lineage and an as yet unanswered question remains as to what the precise role of each is and if each can totally functionally replace the others. Buckingham *et al.* (1992) suggest that although there is no correlation of expression of regulatory factors and muscle specific genes, the latter's expression may require differing activation levels of the four myogenic factors. However, with the advent of muscle regulatory factor 'knock-out mice' further insight into the role of these factors is being realized. Although MyoD1 was the first regulatory factor to be identified and termed 'the' muscle determination factor, the phenotype of the MyoD1 knock-out mouse has shown skeletal muscle development to be little affected in these animals (Rudnicki *et al.*, 1992). This too has been the case for Myf-5; Myf-5 knock-out mice suffer from perturbation of derivatives of sclerotome origin and present a phenotype where the axial skeleton is severely distorted (Braun *et al.*, 1992). The champion's role for myogenic determination is to date pointing more in the direction of myogenin, whose knock-out results in mice where skeletal muscle development is severely affected (Hasty *et al.*, 1993; Nabeshima *et al.*, 1993). These mutants seem to have a deficiency, not of myoblast generation, but rather of differentiation. Thus it is possible that transcription factors such as myogenin have the capacity to prevent cells entering a new cycle of mitosis, in a similar manner to that of the tumour suppressor proteins. Evidence for such a role is corroborated by the work of Gu *et al.* (1993) who showed that myogenin binds to the unphosphorylated gene product of the retinoblastoma tumour suppressor gene. Moreover, the differentiation of myoblasts can be regulated by growth factor-dependent phosphorylation of myogenin (Li *et al.*, 1992), an observation that strongly suggests that myogenin is involved in cell cycle control. Recent work on myogenic differentiation suggests other positive regulators of myogenesis. Muscle LIM protein (MLP) is enriched in striated muscle and coincides with myogenic differentiation. In the absence of MLP, cells of the C2 mouse myogenic cell line, grown in conditions to induce the differentiation of these cells to the multinucleate state, were found to express myogenin but failed to exit from the cell cycle and enter terminal differentiation (Arber *et al.*, 1994). This, and other data, suggests that MLP is an essential promoter of myogenesis distinct from the established factors described above. For the future, therefore, myogenin and the other HLH muscle transcription factors are sure to be at the centre of studies of how muscle cells make the decision between proliferation, quiescence and differentiation (Buckingham, 1994).

THE MUSCLE SATELLITE CELL

Thus myogenic transcription factors allow the muscle precursor cell to enter into and continue along the myogenic pathway and even take it to its terminally differentiated state, that of the multinucleate muscle fibre. Once fusion between myogenic cells has occurred and the muscle fibre is established, the muscle cell has been accredited, even up to recent times, with very limited powers of regeneration (Taussig, 1984), despite the treatises of scientists throughout the last two centuries indicating the possibilities of albeit, limited, yet functional, regeneration of skeletal muscle fibres (Gunsberg, 1848; Bottcher, 1858; Kuttner and Landois, 1913). It is certainly true, as seen from the work of Stockdale and Holtzer (1961), that once incorporated into the muscle fibre, the myoblast nucleus rapidly loses its ability to divide and give rise to further precursor cells and further, that following synthesis of contractile proteins, DNA synthesis does not occur in the maturing muscle fibre. From such evidence the powers of regeneration of the muscle cell would indeed appear to be limited. However, the work of Enesco and Puddy (1964) and MacConnachie *et al.* (1964) shows that there is a phenomenal increase in DNA content in muscle fibres during postnatal development in the rat which seems contradictory to the evidence of Stockdale and Holtzer (1961) of no myonuclear mitosis and DNA synthesis in the chick. The apparent enigma here is due to the role of another cell in the formation of skeletal muscle fibres. Muscle formation not only gives rise to cells destined to form the differentiated fibre, but also to a second population of cells. These 'muscle satellite cells' occupy a unique position and role in myogenesis and indeed it is these cells which have generated so much interest for its continued ability to re-enter the cell cycle and give rise to new muscle fibres.

Save for uninjured adult newt muscle (Hay, 1979; Popiela, 1976), the satellite cell is present in all vertebrate skeletal muscle so far studied, occupying a position interposed between the muscle plasma membrane and the basal lamina. It is this compartmentalization of the cell – between plasma membrane and basal lamina – that establishes the cell as a satellite cell (Figure 3). The satellite cell is thus distinct from the myonuclei of the fibre and is reputed to have no direct communication between it and the fibre (Cull-Candy *et al.*, 1980; Schmalbruch, 1978) although desmosome-like specializations between the satellite cell and myofibre were observed in the craniovelar muscle of the Atlantic hagfish (Sandset and Korneliussen, 1978). Satellite cells were first observed independently by Geberg (1884) and Waldeyer (1865) as early as the nineteenth century but without the advent of electron microscopy further study of this cell was not possible at this time. Katz (1961) was the first to describe the appearance of these cells in relation to intrafusal muscle fibres of the frog, followed a few months later by more extensive observations in the extrafusal muscle fibres of varied animal species by Mauro (1961). The satellite cell is not readily viewed by light microscopy, for at such resolution distinction between it and the muscle fibre nuclei, occupying their normal subsarcolemmal position, is difficult, although Ontell (1974) and Brotchie *et al.* (1995) have described a method for their identification at the light microscope level. Confirmation of cells as satellite cells has been achieved using electron microscopy (Snow, 1977) although various other techniques are now employed for their identification and verification (see Mazanet and Franzini-Armstrong, 1980). The ultrastructure of the satellite cell has been adequately covered in two reviews, by Mazanet and Franzini-Armstrong (1980) and Campion (1984) as has the variation in shape and size of satellite cells derived from different species and will not be further discussed in this chapter.

The question which arises is how the satellite cell came to take up its unusual position during myogenesis. To answer this it might be pertinent to look at the origin of the cell and to identify from where it arose. Using chick/quail interspecific grafting methods,

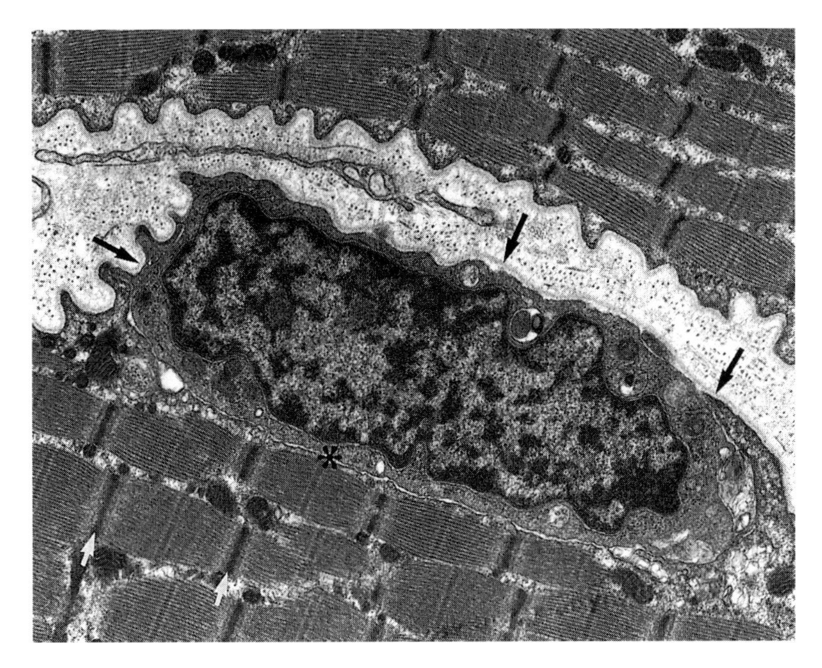

Figure 3 Electron micrograph of normal human muscle. A satellite cell with a prominent nucleus is seen lying under the basal lamina (arrows) of the muscle fibre. Note the closely opposed plasma membranes of the satellite cell and the muscle fibre (*). Myofibrils of the muscle fibre show characteristic banding, with each sarcomere delineated by dense 'Z' lines (white arrow). Magnification ×15,000. (Courtesy of Dr Jill Moss, Department of Histopathology, Charing Cross & Westminster Medical School.)

Armand *et al.* (1983) showed satellite cells to be derived from the same somitic cell lineage as the muscle precursor cells which are destined to fuse and form the myonuclei of the multinucleate muscle fibre. The view that satellite cells are merely muscle precursor cells that have failed to fuse with the muscle fibre and are a remnant of embryogenesis (Tennyson *et al.*, 1973) is not a sufficient explanation for their presence (Grounds and Yablonka-Reuveni, 1993) in the light of several studies in both chick and mammals where the nature and behaviour of satellite cells have been compared with that of embryonic myoblasts. Studies with chick cells, principally from the laboratory of Yablonka-Reuveni, indicate that satellite cells differ in a number of parameters from embryonic myoblasts; their time to fusion to the multinucleate state in culture is greater in comparison with embryonic myoblasts (Yablonka-Reuveni *et al.*, 1987); in the mononuclear state they express desmin more frequently than embryonic myoblasts (Yablonka-Reuveni and Nameroff, 1990); and they have more receptors to platelet-derived growth factor (Yablonka-Reuveni *et al.*, 1990), a cytokine which promotes myoblast proliferation and inhibits their differentiation to the multinucleate state (Jin *et al.*, 1991). Culturing of mouse satellite cells, even when treated with a tumour promoter which reversibly inhibits differentiation of chick myoblasts to myotubes (Cohen *et al.*, 1977), resulted in satellite cells entering terminal differentiation, in contrast to embryonic myoblasts which did not (Cossu *et al.*, 1983). Further, embryonic myoblasts do not express acetylcholine receptor channels whereas satellite cells do (Eusebi and Molarino,

1984; Cossu *et al.*, 1987). During development of muscle in various species, muscle fibres form which express different forms of the myosin heavy chain isoform and hence are classified as fast, slow, or fast/slow fibres (for review see Stockdale, 1992). This is also true at the mononuclear stage where several groups have shown the expression of different myosin heavy chains (MHC) in both avian and mammalian embryonic myoblasts (Crow and Stockdale, 1984, 1986a, 1986b; Lyons *et al.*, 1983; Miller and Stockdale 1986a, 1986b; Vivarelli *et al.*, 1988). Recently DiMario *et al.* (1993) showed that avian embryonic myoblasts of one particular MHC isoform fused to form skeletal fibres of this same type when introduced into developing avian limb buds. In relation to satellite cells, or 'adult myoblasts' as they are sometimes called (Stockdale, 1992), Hughes and Blau (1992) found that clonally related satellite cells were capable of fusing with, and taking on the myosin heavy chain expression pattern of, each of the various muscle fibre types seen in postnatal animals. These results therefore show that satellite cells are not restricted in their fusion to particular types of muscle fibres and further that the myosin heavy chain isoform expressed by the satellite cell nucleus could be determined by the muscle fibre into which it fused. The studies of DiMario *et al.* (1993) and Hughes and Blau (1992) are not however at variance with each other, the former identifying the formation of different fibre types during development and the other the incorporation of 'adult myoblasts' into already established multinucleate muscle fibres. *In vitro* studies using avian and mammalian satellite cells show that the fibre type formed *in vitro* is dependent of the characteristics of the muscle from which the satellite cells are derived (Matsuda *et al.*, 1983; Feldman and Stockdale, 1990; Hartley *et al.*, 1991; Dusterhoft *et al.*, 1992).

All these lines of evidence suggest that the satellite cell is most definitely a different cell from that of the embryonic myoblast whose principle task during development is that of fusion with other such cells, or with existing muscle fibres, resulting in their entering terminal differentiation and forming the end point of the myogenic pathway – the multinucleate muscle fibre. However, the quail/chick grafting experiments of Gulati (1986) did ascertain that satellite cells are derived from the same somitically derived myogenic lineage as are embryonic muscle cells. Although the embryonic myoblast and the satellite cell are similar in nature (Jones, 1982; Campion, 1984; Wright, 1985; Quinn *et al.*, 1988) it is however unclear whether all of the different types of embyronic myoblasts which arise during development (Bonner and Adams, 1982; Seed and Hauschka, 1984; Miller and Stockdale, 1986b; Mouly *et al.*, 1987) are capable of giving rise to satellite cells.

ROLE OF THE MUSCLE SATELLITE CELL

But what of the satellite cell? What is the role of this cell, imprisoned in its own compartment between the muscle fibre plasma membrane and the basal lamina which ensleeves each muscle fibre? Indeed should we continue to call the cell by this name, thereby stressing its unique location, or should we take note of the similarities between these cells and those of embryonic myoblasts and assign the terminology 'adult myoblasts' to these cells, as suggested by Stockdale (1990) and Stockdale and Feldman (1990)?

Whichever terminology is adopted, during normal development, the role of the satellite cell is to continue to allow the muscle fibre to accrue more myonuclei and hence continue to grow at a controlled rate during postnatal life, when, in many species, fibres are still continuing to grow, albeit at a reduced pace, in relation to embryonic development (Cardasis, 1979). This postnatal accumulation of myonuclei from a satellite cell source was first shown by Moss and Leblonde (1971). Approximately 50% of tritiated-thymidine labelled rat satellite cells fused with muscle fibres 48 hours after their injection into growing muscle of juvenile rats. The interpretation of this result was that the introduced cells were capable of undergoing one division prior to their incorporation into host myofibres. Certainly there is evidence of mitosis of satellite cells *in vivo* (Shafiq *et al.*, 1968; Allbrook *et al.*, 1971; Hellmuth and Allbrook, 1973) and as postnatal life ensues the number of satellite cells has been observed to decrease in some species, although Campion (1984) argues that this is only truly the case in mouse where the studies of Schultz (1974), Cardasis and Cooper (1975) and Young *et al.* (1978) provide evidence for this phenomenon, whereas in his own studies on pig muscle (Campion *et al.*, 1979, 1981) there was an increase in satellite cell number with longevity. The decrease in satellite cell number in rodent muscle (Snow, 1977; Schmalbruch and Hellhammer, 1976) results in the satellite cells contributing only 2–5% of all muscle nuclei within mature muscle. In addition to a decrease in number, the satellite cell has also been shown to become less metabolically active with maturity (Schultz, 1976). As the role of the satellite cell in normal postnatal growth is to increase the number of myonuclei, it seems reasonable to assume that their numbers would decline as the animal matures. Although there are numerous studies which stress structural (Cossu *et al.*, 1983, 1987, 1988; Eusebi and Molarino, 1984; Senni *et al.*, 1987) and behavioural differences between muscle precursor cells and the satellite cell (Bischoff, 1979, 1990a; Yablonka-Reuveni *et al.*, 1987; Allen and Boxhorn, 1989; Le Moigne *et al.*, 1990), many of these reports also show that satellite cells can, and do, give rise to muscle precursor cells.

It is therefore to the satellite cell that we have to look as a muscle precursor cell for the maintenance and repair of skeletal muscle. Such a cell is obviously present in early embryogenesis but remains distinct from the other precursor cells which are destined to continue down the myogenic lineage to the differentiated cell. Following damage to the 'adult' (postnatal) muscle fibre, the satellite cells are activated, as evidenced by their increase in number and also by the increased number of cytoplasmic organelles they contain, a second feature that attests to their activated state (Teravainen, 1970; Schultz, 1978). Thus the second major function of the satellite cell in skeletal muscle is in repair of the damaged fibre, whether it be in regeneration following traumatic injury or damage due to pathological processes. Regenerating muscle fibres often show a stump or 'bud' of sarcoplasm at the ends of the damaged myofibres populated by a variable number of nuclei. It was once thought that muscle was repaired by an outgrowth of the sarcoplasm with migration of myonuclei from the parent myofibre into these budded areas (Ali, 1979). This is no longer accepted as the way in which skeletal muscle regenerates, for it certainly could not account for the regeneration of muscle fibres exhibited following the total mincing of muscle fibres as performed by Studitsky (1964). Instead it is now accepted that mononuclear satellite cells fuse with the existing myofibre or fuse together to form myotubes which then fuse with the ends of myofibres. The 'sarcoplasmic stumps' previously viewed are thus the result of fusion of such myogenic cells with damaged myofibres (Snow, 1978). This 'Discontinuous Theory' of myofibre repair has recently been verified by Robertson *et al.* (1993) in murine muscle, where fusion between satellite cell progeny and myotubes was readily seen up to 5 days after injury.

A wealth of literature attests to activation of satellite cells following damage to muscle fibres. Regeneration of the fibres after such insults which include the use of toxins (Kouyoumdjian *et al.*, 1986; Maltin *et al.*, 1983), anaesthetics (Foster and Carlson, 1980) and mechanical damage (Carlson, 1986; McGeachie and Grounds, 1987; Grounds and McGeachie, 1989) is brought about by the initial death of the injured fibre. In other studies, satellite cell activation has been shown to occur without the differentiation of the fibre. Stimuli which produce this sort of activation have varied from degeneration of the muscle fibre (Ontell, 1974; Murray and Robbins, 1982), muscle fibre overwork (Schiaffino *et al.*, 1976; Taylor and Wilkinson, 1986; Darr and Schultz, 1987; Appell *et al.*, 1988) and compression of the fibres (Teravainen, 1970). Many *in vitro* studies have been carried out where mitogens and growth factors known to stimulate myogenic cells have been added to cultures of myogenic cell lines and satellite cells derived from primary explants (see Florini, 1987). The limitations of studies using immortalized cell lines such as the rat L6 myogenic cell line or the mouse C2 myogenic cell line is that they do not exhibit cell senescence which is a feature of primary derived cultures. Further, most work has been executed on differentiating muscle rather than that undergoing regeneration, and there may well be inherent differences between the two. The work of Bischoff in 1990 showed very elegantly the effect of growth factors on both intact and injured muscle fibres in an *in vitro* study where the normal spatial relationship between muscle fibre and satellite cell had been retained (Bischoff, 1990b).

EFFECT OF GROWTH FACTORS ON MUSCLE STEM CELLS

Many reports indicate that growth factors such as basic fibroblast growth factor (bFGF) (Gospodarowicz *et al.*, 1976; Linkhart *et al.*, 1981; Allen *et al.*, 1984; Lathrop *et al.*, 1985) and insulin-like growth factors (IGFs) (Florini *et al.*, 1977; Ewton and Florini, 1980, 1990) stimulate the proliferation of myogenic cells, including satellite cells (Dodson *et al.*, 1985) but suppress their differentiation to multinucleate muscle fibres, whereas transforming growth factors, specifically TGF-β, depress satellite cell proliferation (Allen and Boxhorn, 1987) and inhibit differentiation of both myoblasts (Evinger-Hodges *et al.*, 1982; Massague *et al.*, 1986; Olson *et al.*, 1986) and satellite cells (Allen and Boxhorn, 1987). The role of such growth factors in the regeneration of muscle has also been postulated (see Allen and Boxhorn, 1989). Basic FGF and IGFs are the growth factors best characterized in muscle regeneration. IGFs can stimulate differentiation in chick embryo muscle cells (Schmid *et al.*, 1983) and in rat satellite cells (Allen and Boxhorn, 1989). Recent work by Florini *et al.* (1991) has linked IGF-stimulated muscle differentiation to a sixty-fold increase in myogenin mRNA and they conclude that increased myogenin gene expression is the primary mechanism by which IGF-I stimulates muscle differentiation. In contrast, bFGF inhibits the expression of MyoD1 and myogenin (Vaidya *et al.*, 1989) which is consistent with the assumption that bFGF is acting more as an inhibitor of differentiation rather than directly as a stimulator of mitogenesis. Although bFGF is presumed to mediate its cellular responses by binding to cell surface receptors (Olwin and Hauschka, 1988), it lacks the classic signal peptide sequence believed to be necessary for protein secretion via the exocytotic pathway. Thus the mechanism by which bFGF is released into the extracellular environment by producing cells has been unclear. One possibility for secretion is that bFGF is released from cytoplasmic storage sites into the extracellular environment via diffusion through mechanically induced plasma membrane disruptions. In support of this hypothesis it has been shown that bFGF can be detected in the cytoplasm of myofibres *in vivo*, and that wounded or disrupted fibres contain significantly reduced levels of bFGF which is released to the endomysium (Clarke *et al.*, 1993). Loss of cytoplasmic bFGF also occurs from physiologically wounded myofibres of normal and dystrophic muscle, and following its release from these fibres it can be stored in bound form to the extracellular matrix (DiMario *et al.*, 1989).

In addition to bFGF and IGF, there are numerous reports from culture studies of mitogenic effects on satellite cells induced by a series of less well characterized cytokines. The best known of these is platelet-derived growth factor (PDGF) where L6 rat myoblasts were shown to be directly stimulated to proliferate in the presence of the PDGF-BB isoform. In addition, PDGF-BB is a potent inhibitor of L6 myoblast differentiation, being able to counteract the differentiation-promoting effects of insulin (Jin *et al.*, 1991). The same authors have also shown that developing skeletal muscle and myoblasts *in vitro* express the gene encoding the PDGF receptor. As L6 myoblasts differentiate to myotubes, the expression of mRNA coding for PDGF receptor decreases, a pattern which is replicated *in vivo* (Jin *et al.*, 1990). These and other studies appear to identify PDGF-BB and its receptor as a potent mitogen which has yet to be fully explored in muscle regeneration. Recent data on the regeneration of the X-linked muscular dystrophic (mdx) mouse muscle, the genetic homologue of human DMD, tends to confirm PDGF-mediated mechanisms for regeneration, but more work is necessary to confirm this proposal (Tidball *et al.*, 1992). Even more tenuous is the present data on a group of proteins, leukaemia inhibitory factor (LIF), transforming growth factor-α (TGF-α) (Austin *et al.*, 1992) and colony-stimulating growth factor 1 (CSF-1) on myogenic cells. In the latter case, we (Jones *et al.*, 1991), and others (Leibovitch *et al.*, 1989) have provided some evidence for the role of CSF-1 and its receptor in regulating murine satellite cell proliferation. More recently it has been shown that an autocrine mechanism of CSF-1 stimulated mitogenesis exists in rat muscle cultures (Borycki *et al.*, 1993), a possibility we have suggested also for murine C2 myoblasts (in preparation).

Whatever their role(s) in controlling cell proliferation or differentiation, it is worth noting in the context of this review that apart from the IGFs, data exists to suggest that the factors described above are also chemoattractants for myoblasts and satellite cells, serving to attract muscle precursor cells to sites of high concentrations of the agents in question (Austin *et al.*, 1992; Robertson *et al.*, 1993). Since PDGF-BB, bFGF, LIF and CSF-1 are all secreted by exudate macrophages present in great numbers at the sites of muscle fibre injury and differentiation, it seems valid to speculate that macrophages may play a pivotal role in stimulating mitogenesis of quiescent satellite cells. Recent data from *in vitro* experiments have provided supporting evidence for a role for macrophages in activation of satellite cells (Cantini and Carraro, 1995).

ROLE OF THE MUSCLE STEM CELL IN THERAPY

The fact that satellite cells can be activated in damaged muscle makes them a strong candidate for use in a therapy to alleviate myopathic conditions of skeletal muscle, this therapy having been termed myoblast transfer therapy. This form of therapy makes use of the fact that muscle grows and regenerates by the fusing of mononuclear precursor/satellite cells with each other and existing muscle fibres. Thus gene products deficient in a myopathic muscle may be incorporated by the fusion of normal precursor cells into the multinucleate fibre (see review by Partridge, 1991). This approach has been suggested for the alleviation of Duchenne muscular dystrophy (DMD). This is an X-linked progressive muscle wasting disease which leads to immobility and death of the affected boy due to the severe muscle wasting that ensues throughout childhood and the teenage years, affecting both locomotor and eventually respiratory muscles. Momentum for such a therapy for DMD gained increased pace in the late 1980s when it was discovered that DMD patients have sequence defects in a very large gene located on the short arm of the X-chromosome and that they are also deficient in the protein product of this gene, given the name dystrophin (Hoffman *et al.*, 1987; Hoffman and Kunkel, 1989). Three animal models for this disease exist: the X-linked muscular dystrophic (*mdx*) mouse (Bulfield *et al.*, 1984); the X-linked muscular dystrophic (*xmd*) dog (Cooper *et al.*, 1988; Kornegay *et al.*, 1988); and the muscular dystrophic cat (Carpenter *et al.*, 1989). Little work has been carried out on the cat model, due to insufficient animals which have been identified with the abnormality. Most studies to test the efficacy of myoblast transfer have been carried out using the *mdx* mouse which, arising as a spontaneous mutant from the C57Bl/10ScSn strain of mouse (Bulfield *et al.*, 1984), has been shown to be genetically homologous to DMD (Sicinski *et al.*, 1989), and like its human counterpart lacks the protein dystrophin normally located in a sub-sarcolemmal position in the muscle fibre (Figure 4). As with the muscle pathology of the DMD patient, the muscle fibres of the *mdx* mouse undergo widespread degeneration (Carnwath and Schotton, 1987; Coulton *et al.*, 1988a), but unlike DMD, they are capable of extensive fibre regeneration to such an extent that in older mice the muscles grow to a greater size and strength than is apparent in the normal C57Bl/10ScSn strain (Anderson *et al.*, 1987; Coulton *et al.*, 1988b). In contrast to the *mdx* mouse, the histopathology of the *xmd* dog shows greater similarity to human DMD (Cooper, 1990). As such the dog is a valuable model to investigate the use of muscle precursor cells to alleviate the genetic defect, which, as with the *mdx* mouse, resides in the homologous gene to that in DMD patients and results in absence of the dystrophin protein. Prior to the discovery of *mdx* and *xmd*, studies indicated that implantation of mononuclear muscle precursor cells into regenerating muscle does result in their incorporation into host muscle fibres (Watt *et al.*, 1984a, 1984b; Law *et al.*, 1988a, 1988b). Implantation of normal muscle stem cells into *mdx* mice showed their incorporation into these fibres and the expression of the missing dystrophin gene product within them (Partridge *et al.*, 1989). These results indicated that precursor cell implantation to introduce a missing gene product into dystrophic muscle was feasible. Introduced stem cells could therefore either fuse with the host muscle fibres into which they were introduced and/or with host satellite cells activated within DMD muscles whose fibres undergo a limited but unsustained regenerative event after the fibre degeneration characteristic of this disease. Either way, the goal of this therapy is to produce new

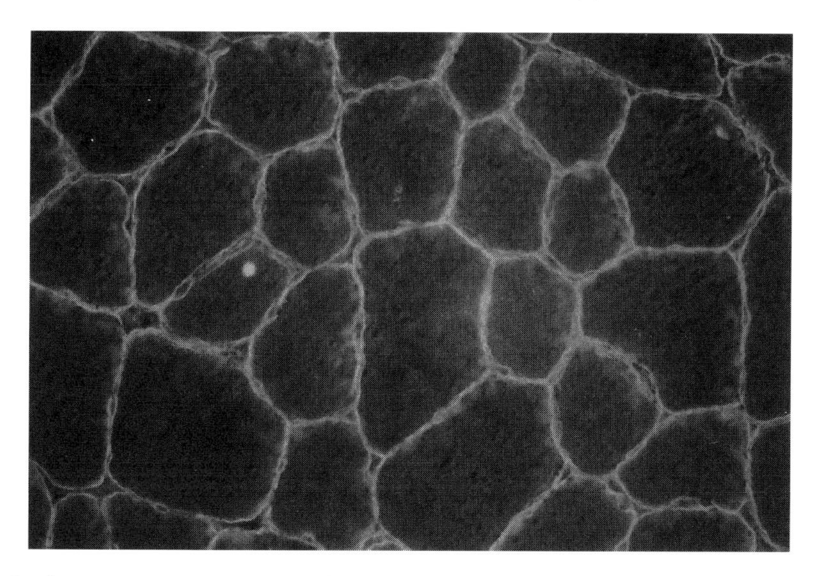

Figure 4 Cryostat section cut from C57Bl/10ScSn tibialis anterior muscle and immunocytochemically stained for dystrophin. Dystrophin is visualized in its normal sub-sarcolemmal position in all muscle fibres. In the *mdx* mouse, which arose as a spontaneous mutation from the C57Bl/10ScSn strain, dystrophin is absent from all fibres, save for a very low number of revertant fibres. Magnification ×1600.

muscle fibres in which gene products, either defective or deficient in the diseased state, are subsequently expressed. To achieve this goal many factors have to be addressed before successful therapy can even be contemplated.

Immune tolerance of implanted stem cells

A readily available source of muscle stem cells for implantation is required for myoblast transfer therapy. This could be achieved by harvesting muscle stem cells from muscle biopsies of unaffected individuals (Webster *et al.*, 1988). The major problem in generating such supplies of stem cells is that of mismatch at the major histocompatibility (MHC) locus between donor cell and the recipient muscle. Although neither normal murine or human muscle stem cells express either MHC class I or class II antigens (Ponder *et al.*, 1983; Appleyard *et al.*, 1985; Karpati *et al.*, 1988), they are expressed on muscle fibres in several human muscle diseases of an inflammatory or degenerative nature (Appleyard *et al.*, 1985; Karpati *et al.*, 1988; Emslie-Smith *et al.*, 1989). MHC class I antigens have been shown to be expressed, under normal conditions, on muscle precursor cells and myotubes *in vitro* (McDouall *et al.*, 1989; Hohlfeld and Engle, 1990). Other cell types which are resistant to killing by primed cytotoxic T-cells exhibit low expression of MHC class I antigen (Zuckerman and Head, 1987; Main *et al.*, 1988) and there is evidence that accessory proteins which are necessary for adhesion of cytotoxic T-cells to its target cells (Krensy *et al.*, 1984; Spits *et al.*, 1986) are lacking in cells in which expression of MHC class I is low (Lampson and Tucker, 1988). Such evidence suggests that 'privileged' cells do exist which are not subject to the immune rejection phenomenon encountered by the majority of cells and tissue types. On the other hand, however, is the reported killing of cultured human myotubes by cytotoxic T-cells, this reaction being mediated via recognition of class I MHC antigens (Hohlfeld and Engel, 1991). In some of our own implantation experiments in mice, we have acceptance of histocompatible and even non-histocompatible muscle stem cells in animals which were not tolerized to the implanted cells (Watt, 1990; Watt *et al.*, 1991). In other studies, however, implantation of MHC-compatible stem cells into mouse muscle resulted in either their rejection or at the very least a florid infiltration of lymphocytes into the grafted areas which suggested the presence of minor histocompatibility antigens (Morgan *et al.*, 1989). However, one group have carried out extensive studies following myoblast implantation in mice and have concluded that effective immunosuppression is essential for successful myoblast transfer in humans (Huard *et al.*, 1994). These authors have observed immune reactions in DMD patients implanted with donor myoblasts, even where the host and donor have been compatible for HLA class I and II. In the *mdx* mouse model this same group have reported a very high success rate in terms of dystrophin-positive fibres when the host animals were suppressed with the immunosuppressant FK506 (Kinoshita *et al.*, 1994).

A second approach to the implantation of muscle stem cells into myopathic muscle is to harvest the patient's own muscle to obtain the satellite cells from this tissue, introduce the missing or defective gene into these cells and then re-implant them back into the recipient's own muscle (Partridge, 1991). The advantage of harvesting stem cells from damaged muscle which is undergoing regeneration is the increased mitotic activity of satellite cells in fibres following injury leading potentially to greater numbers of donor cells into which the normal genes could be implanted. However, caution is necessary here for a study conducted by Schultz and Jaryszak (1985) has indicated that stem cells harvested from regenerated muscle are less competent in their proliferative capacity in comparison to stem cells which have not previously undergone regeneration. Corroborating this is the work of Webster and Blau (1990) who found the proliferative capacity of muscle stem cells harvested from DMD patients to be reduced. The ideal for such a therapy is that implanted cells should be proliferation-competent in order to increase the number of cells carrying the introduced gene product into the recipient muscle. In this respect therefore, is the muscle stem cell of the DMD boy a good candidate for therapy, bearing in mind the findings of Shultz and Jaryszak (1985) and Webster and Blau (1990)? Its candidature for this role is based on the fact that it will fuse readily with myoblasts or immature myotubes and that it expresses muscle-specific genes whose products are often those that are deficient in many of the primary muscle disorders. Is there any other possible cell which could be used as a muscle stem cell which is not functionally compromised by the disease, but yet is capable of incorporating into host muscle fibres and expressing muscle-specific gene products which are deficient within the recipient muscle? In an *in vitro* study, Chaudhari *et al.* (1989) reported that dermal fibroblasts, marked with the lacZ reporter gene, fused with muscle myoblasts derived from the muscular dysgenic (*mdg*) mouse, for they observed β-galactosidase (encoded lacZ) positive myotubes within their cultures. Muscle fibres of the *mdg* mouse are deficient in a functional copy of the skeletal muscle dihydropyridine receptor (Tanabe *et al.*, 1989) and hence cannot contract. The contraction of some *mdg* myotubes following co-culture of their myoblasts with normal dermal fibroblasts, coupled with the fact that normal dihydropyridine-sensitive Ca^{2+} currents were observed in these rescued myotubes

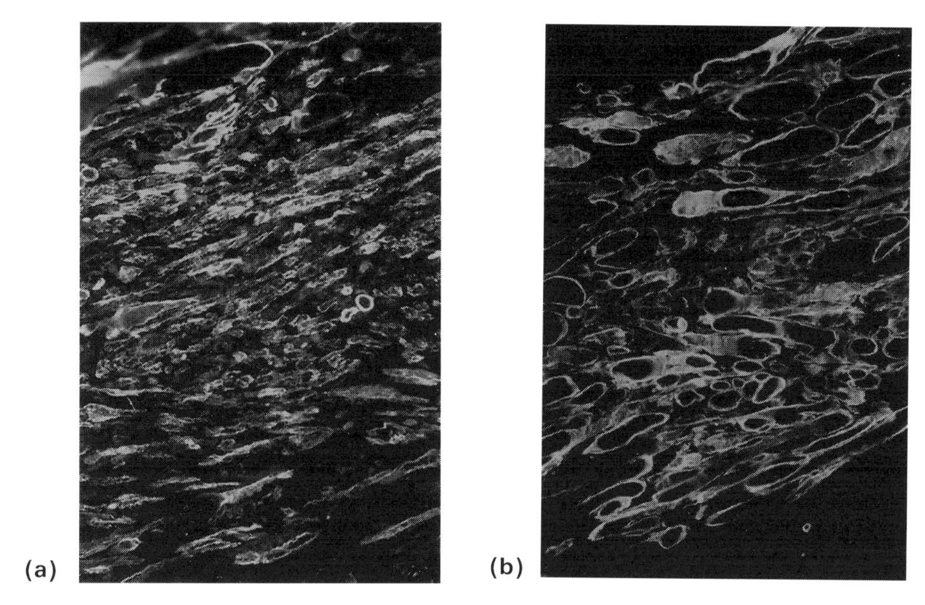

Figure 5 (a) Implantation of cloned dermal fibroblasts into irradiated *mdx* muscle where host muscle precursor cells had been depleted. High numbers of dystrophin-positive fibres were observed three weeks after cell implantation. (b) Non-irradiated *mdx* muscle implanted with cloned dermal fibroblasts also yielded high numbers of dystrophin-positive fibres. (Micrographs taken from Gibson *et al.*, 1995.) Magnification ×400.

(Courbin *et al.*, 1989), led to the hypothesis that the muscle cytoplasm was inducing the fibroblast nuclei to express muscle-specific genes, a phenomenon first described by Blau *et al.* (1985), who induced artificial fusion between myoblasts and fibroblasts. In a recent *in vivo* study we injected cloned normal mouse dermal fibroblasts into the muscles of the *mdx* mouse (Gibson *et al.*, 1995). The result was the formation of immature myotubes which were found to express the protein dystrophin, normally absent in *mdx* muscle fibres (Figures 5a and 5b). Further, analysis of isoenzyme types specific for host and donor tissue showed that where host muscle precursor cells had been depleted by subjecting the host muscle to X-irradiation (Wakeford *et al.*, 1990), the newly-formed dystrophin-positive fibres were of donor origin. Where dermal fibroblasts were implanted into non-irradiated *mdx* muscle, this muscle containing a normal complement of host myoblasts, dystrophin-positive mosaic muscle fibres formed by the fusion of donor dermal fibroblasts with host muscle precursor cells. These results suggest that the implanted dermal fibroblasts have converted to a myogenic lineage within the environment of the host *mdx* muscle, which could have significant consequences for therapy for primary muscle disease, such as DMD. The dermal fibroblast, a cell not functionally compromised by the muscle disease, could be readily harvested from the patient's own skin, hence avoiding immune rejection complications, the missing gene transfected into these cells and their numbers expanded *in vitro*, prior to their implantation into the patient's diseased muscle where they would participate in fibre regeneration.

Ability of implanted cells to remain as a stem cell

For long term benefit in myoblast transfer it is desirable, if not imperative, that some of the implanted stem cells should remain as such, i.e. capable of retaining their capacity to undergo division in order to introduce their gene products into myopathic fibres during subsequent degenerative and regenerative events which may occur within the fibres throughout the course of the myopathic disease. That muscle stem cells do indeed retain the ability to continuously contribute to host myofibres has been shown in an excellent study devised by Yao and Kurachi (1993). These authors were able to recover, as viable myoblasts, the myogenic cells they implanted into host muscle several months previously. Further, on re-implantation of these cells into a second host, they fused with recipient fibres, some of them even retaining the capacity to remain as muscle stem cells. In the *mdx* mouse there is some evidence that implanted cells remain as stem cells after their introduction into the myopathic muscle. Morgan *et al.* (1991) were able to extract mononuclear myogenic cells shown to be of donor origin by virtue of their carrying a donor form of the glucose-6-phosphate isomerase (GPI) isoenzyme, used as a marker to distinguish between host and donor tissue, and also because dystrophin within myotubes subsequently formed when the extracted cells were grown *in vitro*. A similar study carried out by the same group (Morgan *et al.*, 1990) also indirectly suggests the capabilities of some muscle stem cells to retain their proliferative capacity following implantation into *mdx* muscle. Donor *mdx* muscle stem cells, carrying a different isoenzyme form from that normally expressed within *mdx* mice, were injected into *mdx* muscle. Hence both donor and host cells were myopathic but donor cells were engineered to carry a different GPI isoform from the host and therefore the relative contribution of donor and host cells to the muscle could be assayed. The proportion of donor GPI was found to be

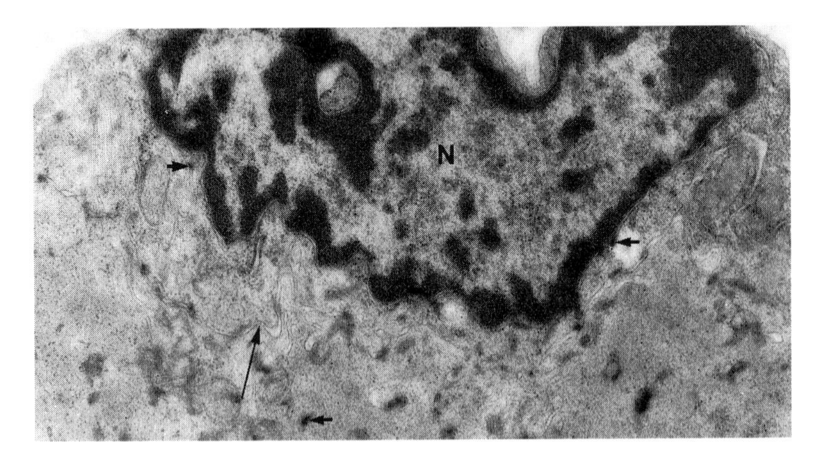

Figure 6 Myoblasts derived from the C2 mouse cell line were infected with the mouse moloney leukaemia virus carrying the lacZ gene. Ultrastructurally the product of this gene, β-galactosidase, is seen as a dense electron deposit. We detected the lacZ gene product (short arrows) both in the cytoplasm of the muscle fibre and that of satellite cells. Note the undulating closely apposed plasma membranes of the satellite cell and muscle fibre (long arrow) N: nucleus of satellite cell. EM section of cryostat material, $\times$29,000 courtesy of Dr J. Moss, Department of Histopathology, CXWMS.

much greater in later stages of the experiment, i.e. 67 days after implantation as opposed to 29 days, and was interpreted as the continued presence at later stages of the experiment of donor mononuclear stem cells still capable of fusing with host muscle fibres. In studies where we have been tracing the extent of migration of muscle precursor cells through the muscle into which they have been implanted (see below), we have noted that implanted myoblasts carrying the lacZ reporter gene have in some cases located to a satellite cell position (Figure 6) when observed at the ultrastructural level (Watt *et al.*, unpublished data). Although we have as yet no evidence that these β-galactosidase expressing cells are re-utilized in further bouts of degeneration and regeneration in the *mdx* muscle into which they were implanted, their very presence below the basement membrane of the muscle fibre indicates that they are in the correct location for reactivation in such an event.

Migration of implanted stem cells throughout the target muscle

Given that sufficient stem cells could be introduced into the muscles of DMD, or other conditions where the defect is due to a deficiency or defect in a myogenic gene product, it is imperative for success in implantation therapy that the introduced gene product is efficiently disseminated throughout the target tissue and expressed within a high percentage of recipient fibres in order to achieve the maximum beneficial effect of implantation. The extent of myoblast migration in experimental systems is, however, controversial. In the rat, precursor cells have been reported to migrate between individual fibres within one particular muscle (Jones, 1979; Lipton and Schultz, 1979; Ghins *et al.*, 1984, 1985; Schultz *et al.*, 1985; Hughes and Blau, 1992; Phillips *et al.*, 1990). There is also evidence of movement of cells from one muscle belly into one adjacently placed, such movement being reported in cases where the muscle into which the cells migrate is regenerating after injury (Watt *et al.*, 1987; Morgan *et al.*, 1987, 1990) or where the epimysia of the adjacent muscles have been disrupted (Schultz *et al.*, 1986; Watt *et al.*, 1993). In relation to DMD, the disease affects multiple muscle bellies. In early clinical trials to test the efficacy of myoblast transfer for DMD, multiple delivery of muscle precursor cells into one muscle has been undertaken (Karpati, 1991). In other cases, the efficiency of myoblast transfer has been low, with the amount of dystrophin transcript expressed in DMD fibres after such injection requiring amplification by polymerase chain reaction techniques to detect its presence (Gussoni *et al.*, 1992). For diseases affecting multiple muscle bellies the migration of muscle stem cells into several such muscles would be advantageous. In the *mdx* mouse we have reported the migration of muscle precursor cells from one muscle belly into those adjacently placed. In these experiments the introduced myogenic precursor cell carried the lacZ marker coding for β-galactosidase (β-gal). If the implanted cell is incorporated into the host muscle fibre, the fibre stains blue, due to the expressed β-gal enzyme. We detected blue muscle fibres within fibres of the adjacent muscle (Figures 7a and 7b), particularly when the neighbouring muscle was undergoing extensive regeneration, suggesting some chemotactic influence on the muscle cells (Watt *et al.*, 1993). More recently (Watt *et al.*, 1994), we have detected the preferential movement of the injected lacZ carrying muscle stem cell towards areas of muscle injury and regeneration. These latter experiments demonstrate the movement of muscle precursor cells into other muscles, but also supply further evidence, as

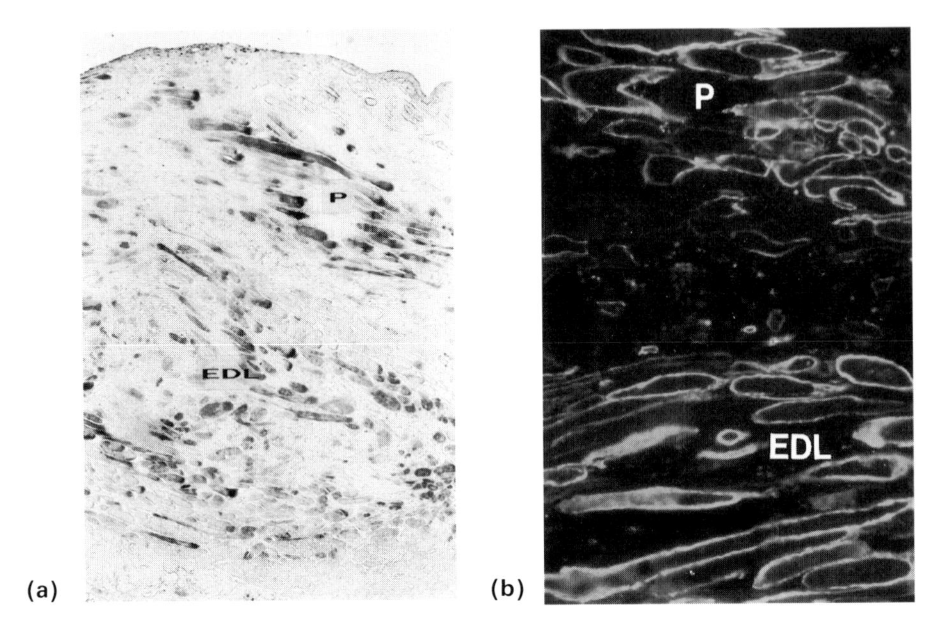

Figure 7 Following implantation of β-gal marked myogenic cells into the extensor digitorum longus (EDL) muscle of the *mdx* mouse, cells migrated from the muscle into which they were implanted and were detected in the neighbouring muscle, particularly if the muscle fibres were undergoing regeneration following crush injury and hence capable of incorporating muscle stem cells. The incorporation of the implanted cells into fibres of the adjacent muscle and expression of gene products present within these cells is shown by the presence of (a) β-gal-positive fibres – Magnification ×160; and (b) dystrophin-positive fibres within the neighbouring peroneus (P) muscle. Magnification ×800.

discussed above, of the ability of introduced cells to retain their proliferative status. Such cells were incorporated into fibres of an *mdx* extensor digitorum longus (EDL) muscle following their implantation into this muscle, for EDL fibres were β-gal positive. The EDL muscle was isografted into a second *mdx* host, ensuring a unique supply of lacZ-muscle stem cells. Yet when neighbouring muscles were subsequently injured, the EDL muscle was capable of surrendering some of the implanted cells to neighbouring tibialis anterior and peroneal muscles of the final host, for β-gal positive fibres were present within them only when their fibres were undergoing regeneration following injury. Cells were not attracted to adjacent muscles if fibre degeneration followed by regeneration was not occurring. This experiment satisfies two criteria for use of muscle stem cells in transfer therapy; migration of cells and their continued presence as stem cells capable of being incorporated into host muscles at varied times after the initial implantation. It also shows the necessity for implanting cells into myopathic muscle at a time when the recipient muscle is 'responsive' to them, i.e. when the fibres are regenerating and capable of incorporating stem cells within them. Thus in terms of stem cell therapy to alleviate diseases such as DMD, it is imperative to introduce the stem cells to the myopathic muscle when it is in a state where the introduced cells can be readily incorporated.

USE OF MUSCLE STEM CELLS IN THERAPY FOR NON-MUSCLE DISEASES

A novel approach to drug delivery in the treatment of other types of diseases involves using cells to introduce into the body genes that express therapeutic proteins continuously. Muscle stem cells appear to be well suited to this purpose because they can become an integral part of the muscle into which they are injected. Once the muscle stem cells have fused with muscle fibres, their myonuclei become post-mitotic and remain an integral part of the myofibre. Any gene products they express have therefore access to the vascular circulation of the recipient. Using this approach, a recombinant gene encoding human growth hormone was stably introduced into mouse myoblasts of the C2 cell line via a retrovirus vector and the genetically engineered myoblasts injected into mouse muscle. Human growth hormone was subsequently detected in the serum of the mice for a period of three months (Dhawan *et al.*, 1991; Barr and Leiden, 1991). This encouraging data has been confirmed using Factor IX protein where stable production and secretion was observed for six months (Yao and Kurachi, 1992).

In conclusion therefore, the muscle stem cell, whether it be of the embryonic myoblast lineage, or of the adult satellite cell status, may well turn out to be a cell with far greater importance to tissues other than its tissue of origin and may well hold the key to future therapies for diseases other than those of a myogenic nature.

ACKNOWLEDGEMENTS

The authors wish to thank the Muscular Dystrophy Group of Great Britain and Ireland and the Leverhulme Trust, charities who have supported their work.

REFERENCES

Adams, R.D., Denny-Brown, D. and Pearson, C.M. (1954). In *Diseases of Muscle, A Study in Pathology* (ed. P. B. Hoeber), pp. 1–735. Harper & Row, New York.
Ali, M.A. (1979). *J. Anat.* **128**:553–562.
Allbrook, D.B., Han, M.F. and Hellmuth, A.E. (1971). *Pathology* **3**:233–243.
Allen, R.E. and Boxhorn, L.A. (1987). *J. Cell Physiol.* **133**:567–572.
Allen, R.E. and Boxhorn, L.A. (1989). *J. Cell Physiol.* **138**:311–315.
Allen, R.E., Dodson, M.V. and Luiten, L.S. (1984). *Exp. Cell Res.* **142**:154–160.
Altschul, R. and Lee, J.C. (1960). *Anat. Rec.* **136**:153.
Anderson, J.W., Ovalle, W.K. and Bressler, B.H. (1987). *Anat. Rec.* **219**:243–257.
Appell, H.J., Forsberg, J.S. and Hollmann, W. (1988). *Intl J. Sports Med.* **9**:297–299.
Appleyard, S.T., Dunn, M.J., Dubowitz, V. and Rose, M.L. (1985). *Lancet* **i**:361–363.
Arber, S., Halder G. and Caroni, P. (1994). *Cell* **79**:221–231.
Armand, O., Boutineau, A-M., Mauger, A., Pautou, M-P. and Kieny, M. (1983). *Arch. d'Anat. Microsc.* **72**:163–181.
Austin, L., Bower, J., Kurek, J. and Vakakis, N. (1992). *J. Neurol. Sci.* **112**:185–191.
Barr, E. and Leiden, J.M. (1991). *Science* **254**:1507–1509.
Bellairs, R. (1963). *J. Embryol. Exp. Morphol.* **11**:697–714.
Bischoff, R. (1979). In *Muscle Regeneration* (ed. A. Mauro), pp. 13–29. Raven Press, New York.

Bischoff, R. (1990a). *Development* **109**:943–952.
Bischoff, R. (1990b). *J. Cell Biol.* **111**:201–207.
Blau, H.M., Pavlath, G.K., Hardeman, E.C. *et al.* (1985). *Science* **230**:758–766.
Bober, E., Lyons, G., Braun, T., Cossu, G., Buckingham, M. and Arnold, H. (1991). *J. Cell Biol.* **113**:1255–1265.
Bonner, P.H. and Adams, T.R. (1982). *Dev. Biol.* **90**:175–184.
Bonner, P.H. and Hauschka, S.D. (1974). *Dev. Biol.* **37**:317–328.
Borycki, A.G., Lonormand, J-L., Guillier, M. and Liebovitch, S.A. (1993). *Biochem. Biophys. Acta* **1174**:143–152.
Bottcher, A. (1858). *Virchows Arch.* **13**:227–392.
Boyd, J.D. (1960). In *The Structure and Function of Muscle* (ed. G. H. Bourne), pp. 63–109. Academic Press, New York.
Braun, T., Buschhausen-Denker, G., Bober, E., Tannich, E. and Arnold, H.H. (1989). *EMBO J.* **8**:701–709.
Braun, T., Bober E., Winter, B., Rosenthal, N. and Arnold, H.H. (1990). *EMBO J.* **9**:821–831.
Braun, T., Rudnicki, M.A., Arnold, H-H. and Jaenisch, R. (1992). *Cell* **71**:369–382.
Brotchie, D., Davies, I., Ireland, G. and Mahon, M. (1995). *J. Anat.* **186**:97–102.
Buckingham, M. (1992). *Trends in Genetics* **8**:144–148.
Buckingham, M. (1994). *Curr. Opin. Genet. Dev.* **4**:745–751.
Buckingham, M., Biben, C., Catala, F., Lyons, G. and Ott, M-O. (1992). In *Neuromuscular Development and Disease* (eds A. M. Kelly and H. M. Blau), pp. 59–72. Raven Press, New York.
Bulfield, G., Siller, W.G., Wight, P.A.L. and Moore, K.J. (1984). *Proc. Natl Acad. Sci. USA* 81:1189–1192.
Campion, D.R. (1984). *Int. Rev. Cytol.* 87:225–251.
Campion, D.R., Richardson, R.L., Kraeling, R.R. and Reagan, J.O. (1979). *J. Anim. Sci.* **48**:1109–1115.
Campion, D.R., Richardson, R.L., Reagan, J.O. and Kraeling, R.R. (1981). *J. Anim. Sci.* **52**:1014–1018.
Cantini, M. and Carraro, U. (1995). *Exp. Neurol.* **54**:121–128.
Capers, C.R. (1960). *J. Biophys. Biochem. Cytol.* **7**:559.
Cardasis, C.A. (1979). In *Muscle Regeneration* (ed. A. Mauro), pp. 155–166. Raven Press, New York.
Cardasis, C.A. and Cooper, G.W. (1975). *J. Exp. Zool.* **191**:347–358.
Carlson, B.M. (1986). *J. Morphol.* **125**:447–472.
Carnwath, J.W. and Schotton, D.M. (1987). *J. Neurol. Sci.* **80**:39–54.
Carpenter, J.L., Hoffman, E.P., Romanul, F.C.A. *et al.* (1989). *Am. J. Pathol.* **135**:909–919.
Chaudhari, N., Delay, R. and Beam, K.G. (1989). *Nature* **341:**445–447.
Chevallier, A., Kieny, M. and Mauger, A. (1976). *C. R. Acad. Sci. Hebd. Seannc, Paris D* **282**:309–311.
Chevallier, A., Kieny, M. and Mauger, A. (1977). *J. Embryol. Exp. Morphol.* **41**:245–258.
Christ, B., Jacob, H.J. and Jacob, M. (1974). *Experientia* **30**:1449–1451.
Christ, B., Jacob, H.J. and Jacob, M. (1977a). *Anat. Embryol.* **150**:171–186.
Christ, B., Jacob, H.J. and Jacob, M. (1977b). *Verh. Anat. Ges.* **71**:1231–1237.
Clarke, M.S.F., Khakee, R. and McNeil, P.L. (1993). *J. Cell Sci.* **106**:121–133.
Cohen, R., Pacifici, M., Rubinstein, N., Biehl, J. and Holtzer, H. (1977). *Nature* **266**:538–540.
Cooper, B.J. (1990). In *Myoblast Transfer Therapy* (eds R. C. Griggs and G. Karpati), pp. 279–284. Plenum Press, New York.
Cooper, B.J., Winand, N.J., Stedman, H., Valentine, B.A. *et al.* (1988). *Nature* 334:154–157.
Cossu, G., Molinaro, M. and Pacifici, M. (1983). *Dev. Biol.* **98**:520–524.
Cossu, G., Eusebi, F., Grassi, F. and Wanke, E. (1987). *Dev. Biol.* **123**:43–50.
Cossu, G., Ranaldi, G., Molarino, M. and Vivarelli, E. (1988). Development **102**:65–69.
Coulton, G.R., Morgan, J.E., Partridge, T.A. and Sloper, J.C. (1988a). *Neuropath. Appl. Neurobiol.* **14**:53–70.
Coulton, G.R., Curtin, N.A., Morgan, J.E. and Partridge, T.A. (1988b). *Neuropath. Appl. Neurobiol.* **14**:299–314.
Courbin, P., Koenig, J., Ressouches, A., Beam, K.G. and Powell, J.A. (1989). *Neuron* **2**:1341–1350.
Crow, M.T. and Stockdale, F.E. (1984). *Exp. Biol. Med.* **9**:165–174.
Crow, M.T. and Stockdale, F.E. (1986a). *Dev. Biol.* **113**:238–254.
Crow, M.T. and Stockdale, F.E. (1986b). *Dev. Biol.* **118**:333–342.

Cull-Candy, S.G., Miledi, R., Nakajima, Y. and Uchitel, O.D. (1980). *Proc. R. Soc. Lond. (Biol.)* **209**:563–568.
Darr, K.C. and Schultz, E. (1987). *J. Appl. Physiol.* **63**:1816–1821.
Davis, R.L., Weintraub, H. and Lassar, A.B. (1987). *Cell* **51**:987–1000.
Dhawan, J., Pan, L.C., Pavlath, G.K. *et al.* (1991). *Science* **254**:1509–1514.
DiMario, J.X., Buffinger, N., Yamada, S. and Strohman, R.C. (1989). *Science* **244**:688–690.
DiMario, J.X., Fernyak, S.E. and Stockdale, F.E. (1993). *Nature* **362**:165–167.
Dodson, M.V., Allen, R.E. and Hossner, K.L. (1985). *Endocrinology* **117**:2357–2363.
Dusterhoft, S., Dieterle, R. and Pette, D. (1992). *Eur. J. Cell. Biol.* **57 (S36)**:16.
Ede, D.A. and El-Gadi, A.O.A. (1986). In *Somites in Developing Embryos* (eds R. Bellairs, D. A. Ede and J. W. Lash), pp. 209–224. Plenum Press, New York.
Edmonson, D.G. and Olson, E.N. (1989). *Genes Dev.* **3**:628–640.
Emslei-Smith, A.M., Arahata, K. and Engel, A.G. (1989). *Hum. Pathol.* **20**:224–231.
Enesco, M. and Puddy, D. (1964). *Am. J. Anat.* **114**:235–244.
Eusebi, F. and Molarino, M. (1984). *Muscle & Nerve* **7**:488–492.
Evinger-Hodges, M.J., Ewton, D.Z., Seifert, S.C. and Florini, J.R. (1982). *J. Cell Biol.* **93**:395–401.
Ewton, D.Z. and Florini, J.R. (1980). *Endocrinology* **106**:577–583.
Ewton, D.Z. and Florini, J.R. (1990). *Proc. Soc. Exp. Biol. Med.* **194**:76–80.
Feldman, J.L. and Stockdale, F.E. (1990). *Dev. Biol.* **143**:320–334.
Florini, J.R. (1987). *Muscle & Nerve* **10**:577–598.
Florini, J.R., Nicholson, M.L. and Dulak, N.C. (1977). *Endocrinology* **101**:32–41.
Florini, J.R., Ewton, D.Z. and Magri, K.A. (1991). *Ann. Rev. Physiol.* **53**:201–216.
Foster, A.H. and Carlson, B.M. (1980). *Anesth. Analg.* **59**:727–736.
Geberg, A. (1884). *Int. Monatschr. Anat. Histol.* **1**:7.
Ghins, E., Colson-van-Schloor, M. and Marechal, G. (1984). *J. Mus. Res. Cell Motil.* **5**:711–722.
Ghins, E., Colson-van-Schloor, M., Maldauge, P. and Marechal, G. (1985). *Arch. Int. Phys. Biochim.* **93**:143–153.
Gibson, A.J., Karasinski, J., Relvas, J.B., Moss, J., Sherratt, T.G., Strong, P.N. and Watt, D.J. (1995). *J. Cell Sci.* **108**:207–214.
Gospodarowicz, D., Weseman, J., Moran, J.S. and Lindstrom, J. (1976). *J. Cell Biol.* **70**:395–405.
Grounds, M.D. and McGeachie, J.K. (1989). *Exp. Cell Res.* **180**:429–439.
Grounds, M.D. and Yablonka-Reuveni (1993). In *Molecular and Cell Biology of Muscular Dystrophy* (ed. T. A. Partridge), pp. 210–256. Chapman & Hall, London.
Gu, W., Schneider, G.W., Condorelli, G., Kauschal, S. *et al.* (1993). *Cell* **72**:309–324.
Gulati, A.K. (1986). *J. Embryol. Exp. Morphol.* **92**:1–10.
Gumpel-Pinot, M. (1974). *C. R. Acad. Sci. Hebd. Seannc, Paris D* **279**:1305–1308.
Gunsberg, F. (1848). In *Studien zur Spediallenpathologie*, pp. 1–412. Brockhaus., Leipzig.
Gussoni, E., Pavlath, G.K., Lanctot, A.M., Sharma, K.R. *et al.* (1992). *Nature* **356**:435–438.
Hamburger, V. (1938). *J. Exp. Zool.* **77**:379–397.
Hartley, R.S., Bandman, E. and Yablonka-Reuveni, Z. (1991). *Dev. Biol.* **148**:249–260.
Hasty, P., Bradley, A., Morris J.H. *et al.* (1993). *Nature* **364**:501–506.
Hauschka, S.D. (1974). *Dev. Biol.* **37**:345–368.
Hay, E.D. (1979). In *Muscle Regeneration* (ed. A. Mauro), pp. 73–81. Raven Press, New York.
Hellmuth, A.E. and Allbrook, D.B. (1973). *Proc. Int. Cong. Muscle Dis.* pp. 343–345.
Hoffman, E.P. and Kunkel, L.M. (1989). *Neuron* **2**:1019–1029.
Hoffman, E.P., Brown, R.H. and Kunkel, L.M. (1987). *Cell* **51**:919–928.
Hohlfeld, R. and Engel, A.G. (1990). *Am. J. Pathol.* **136**:503–508.
Hohlfeld, R. and Engel, A.G. (1991). *J. Clin. Invest.* **86**:370–374.
Holtzer, H., Marshall, J. and Finck, H. (1957). *J. Biophys. Biochem. Cytol.* **3**:705–723.
Holtzer, H., Abbot, J. and Lash, J. (1958). *Anat. Rec.* **131**:567.
Huard, J., Verrault, S., Roy, R., Tremblay, M. and Tremblay, J.P. (1994). *J. Clin. Invest.* **93**:586–599.
Hughes, S.M. and Blau, H.M. (1992). *Cell* **68**:659–671.
Jacob, M., Christ, B. and Jacob, H.J. (1978). *Anat. Embryol.* **153**:179–193.
Jin, P., Rahm, M., Claesson-Welsh, L., Heldin, C.H. and Sejersen, T. (1990). *J. Cell Biol.* **110**:1665–1672.
Jin, P., Sejersen, T. and Ringertz, N.R. (1991). *J. Biol. Chem.* **266**:1245–1249.

Jones, P.H. (1979). *Exp. Neurol.* **66**:602–610.
Jones, P.H. (1982). *Exp. Cell Res.* **139**:401–404.
Jones, G.E., Murphy, S.J. and Watt, D.J. (1990). *J. Cell. Sci.* **97**:659–667.
Jones, G.E., Murphy, S.J., Wise, C. and Watt, D.J. (1991). *J. Cell. Biochem.* **Suppl. 15C**:39.
Kaehn, K., Jacob, H.J., Christ, B., Hinricksen, K. and Poelmann, R.E. (1988). *Anat. Embryol.* **177**:191–201.
Karpati, G. (1991). In *Muscular Dystrophy Research; From Molecular Diagnosis Towards Therapy* (eds C. Angelini, G. A. Danielli and D. Fontanari), pp. 101–108. Excerpta Medica, Amsterdam, New York, Oxford.
Karpati, G., Pouliot, Y. and Carpenter, S. (1988). *Ann. Neurol.* **23**:64–72.
Katz, F.R.S. (1961). *Philos. Trans. R. Soc. Lond. (Biol.)* **243**:221–240.
Kinoshita, I., Vilquin, J-T., Guerette, B., Asselin, I., Roy, R. and Tremblay, J.P. (1994). *Muscle & Nerve* **17**:1407–1415.
Konieczny, S.F. and Emerson, C.P. (1984). *Cell* **38**:791–800.
Konigsberg, I.R., McElvain, N., Tootle, M. and Herrmann, H.J. (1960). *Biophys. Biochem. Cytol.* **8**:333.
Kornegay, J.N., Tuler, S.M., Miller, D.M. and Levesque, D.C. (1988). *Muscle & Nerve* **11**:1056–1064.
Kouyoumdjian, J., Harris, J.B. and Johnson, M.A. (1986). *Toxicon* **24**:575–583.
Krensy, A.M., Robbins, E., Springer, T.A. and Burakoff, S.J. (1984). *J. Immunol.* **132**:2180–2182.
Kuttner, H. and Landois, F. (1913). *Deutsche Chirurgie* **25a**:1–303.
Lampson, L. and Tucker, M. (1988). *FASEB J.* **2**:A879.
Lash, J., Holtzer, J. and Swift, H. (1957). *Anat. Rec.* 128:679–693.
Lathrop, B.K., Olson, E.N. and Glaser, L. (1985). *J. Cell Biol.* 100: 1540–1547.
Law, P.K., Goodwin, T.G. and Li, H-J. (1998a). *Trans. Proc.* XX (3) Suppl. 3:1114–1119.
Law, P.K., Goodwin, T.G. and Wang, M.G. (1988b). *Muscle & Nerve* 11:525–533.
Le Dourain, N. and Barq, G. (1969). *C. R. Acad. Sci. Hebd. Seannc, Paris D* 269:1543–1546.
Leibovitch, S.A., Leibovitch, M.P., Borycki, A.G. and Havel, J. (1989). *Oncogene Res.* **4**:157–162.
Le Moigne, A., Martelly, I., Barlovatz-Meimon, G. *et al.* (1990). *Int. J. Dev. Biol.* **34**:171–180.
Li, L., Zhou, J., James, G., Heller-Harrison, R. *et al.* (1992). *Cell* **71**:1181–1194.
Linkhart, T.A., Clegg, C.H. and Hauschka, S.D. (1981). *Dev. Biol.* **86**:19–30.
Lipton, B.H. and Schultz, E. (1979). *Science* **205**:1292–1294.
Lyons, G.E., Haselgrove, J., Kelly, A.M. and Rubinstein, N.A. (1983). *Differentiation* **25**:1–8.
Lyons, G.E., Ontell, M., Cox, R., Sassoon, D. and Buckingham, M. (1990). *J. Cell Biol.* **111**:1465–1476.
MacConnachie, H.F., Enesco, M. and Leblonde, C.P. (1964). *Am. J. Anat.* **114**:245–251.
McDouall, R.M., Dunn, M.J. and Dubowitz, V. (1989). *J. Neurol. Sci.* **89**:213–226.
McGeachie, J.K. and Grounds, M.D. (1987). *Cell Tissue Res.* **248**:125–130.
Main, E.K., Monos, D.S. and Lampson, L.A. (1988). *J. Immunol.* **141**:2493–2950.
Maltin, C.A., Harris, J.B. and Culen, M.J. (1983). *Cell Tissue Res.* **232**:565–577.
Massague, J., Cheifetz, S., Endo, T. and Nadal-Ginard, B. (1986). *Proc. Natl Acad. Sci. USA* **83**:8206–8210.
Matsuda, R., Spector, D.H. and Strohman, R.C. (1983). *Dev. Biol.* **100**:478–488.
Mauro, A. (1961). *J. Biophys. Biochem. Cytol.* **9**:493–495.
Mazanet, R. and Franzini-Armstrong, C. (1980). In *Myology* (eds A. G. Engel and B. Q. Banker), pp. 285–307. McGraw-Hill, New York.
Milaire, J. (1976). *Arch. Biol. (Bruxelles).* **86**:177–221.
Miller, J.B. and Stockdale, F.E. (1986a). *Proc. Natl Acad. Sci. USA* **83**:3860–3864.
Miller, J.B. and Stockdale, F.E. (1986b). *J. Cell Biol.* **103**:2197–2208.
Miner, J.H. and Wold, B. (1990). *Proc. Natl Acad. Sci. USA* **87**:1089–1093.
Morgan, J.E. and Watt, D.J. (1993). In *Molecular & Cell Biology of Muscular Dystrophy* (ed. T. A. Partridge), pp. 303–331. Chapman & Hall, London.
Morgan, J.E., Coulton, G.R. and Partridge, T.A. (1987). *J. Mus. Res. Cell Motil.* **8**:386–396.
Morgan, J.E., Coulton, G.R. and Partridge, T.A. (1989). *Muscle & Nerve* **12**:401–409.
Morgan, J.E., Hoffman, E.P. and Partridge, T.A. (1990). *J. Cell Biol.* **111**:2437–2449.
Morgan, J.E., Pagel, C.N. and Partridge, T.A. (1991). In: *Muscular Dystrophy Research; From Molecular Diagnosis Towards Therapy* (eds C. Angelini, G. A. Danielli and D. Fontanari), pp. 235–236. Excerpta Medica, Amsterdam, New York, Oxford.

Moss, F.P. and Leblond, C.P. (1971). *Anat. Rec.* **170**:421–435.
Mouly, V., Toutant, M. and Fiszman, M.Y. (1987). *Cell Diff.* **20**:17–25.
Murray, M.A. and Robbins, N. (1982). *Neuroscience* **7**:1823–1833.
Murre, C., McGaw, P., Vaessin, H. *et al.* (1989). *Cell* **58**:537–544.
Nabeshima, Y., Hanaoka, K., Hayasaka, M. *et al.* (1993). *Nature* **364**:532–535.
Olson, E.N. (1991). *Genes Dev.* **4**:1454–1461.
Olson, E.N., Sternberg, E., Hu, J.S., Spizz, G. and Wilcox, C. (1986). *J. Cell Biol.* **103**:1799–1805.
Olwin, B.B. and Hauschka, S.D. (1988). *J. Cell Biol.* **107**:761–769.
Ontell, M. (1974). *Anat. Rec.* **178**:211–228.
Ott, M-O., Robert, B. and Buckingham, M. (1990). *Medicine/Science* **6**:653–663.
Ott, M-O., Bober, E., Lyons, G., Arnold, H. and Buckingham, M. (1991). *Development* **111**:1097–1107.
Partridge, T.A. (1991). *Muscle & Nerve* **14**:197–221.
Partridge, T.A., Morgan, J.E., Coulton, G.R., Hoffman, E.P. and Kunkel, L.M. (1989). *Nature* **337**:176–179.
Phillips, G.D., Hoffman, J.R. and Knoghton, D.R. (1990). *Cell Tissue Res.* **262**:81–88.
Ponder, B.A.J., Wilkinson, M.M., Wood, M. and Westwood, J.H. (1983). *J. Histochem. Cytochem.* **31**:911–919.
Popiela, H. (1976). *J. Exp. Zool.* **198**:57–64.
Quinn, L.S., Norwood, T.H. and Nameroff, M. (1988). *J. Cell. Physiol.* **134**:324–336.
Rhodes, S.J. and Konieczny, S.F. (1989). *Genes Dev.* **3**:2050–2061.
Robertson, T.A., Papadimitriou, J.M. and Grounds, M.D. (1993). *Neuropath. Appl. Neurobiol.* **19**:350–358.
Rudnicki, M.A., Braun, T., Hinuma, S. and Jaenisch, R. (1992). *Cell* **71**:383–390.
Sandset, P.M. and Korneliussen, H. (1978). *Cell Tissue Res.* **195**:17–27.
Sassoon, D., Lyons, G., Wright, W. *et al.* (1989). *Nature* **341**:303–307.
Saunders, J.W. (1948). *Anat. Rec.* **100**:756.
Schiaffino, S., Pierobon Bormioli, S. and Aloisi, M. (1976). *Virchows Arch. B Cell Pathol.* **21**:113–118.
Schmalbruch, H. (1978). *Anat. Rec.* **191**:371–376.
Schmalbruch, H. and Hellhammer, U. (1976). *Anat. Recs.* **185**:279–288.
Schmid, C., Steiner, T. and Froeschh, E.R. (1983). *FEBS Lett.* **161**:117–121.
Schultz, E. (1974). *Anat. Recs.* **180**:589–596.
Schultz, E. (1976). *Am. J. Anat.* **147**:49–70.
Schultz, E. (1978). *Anat. Recs.* **190**:299–312.
Schultz, E. and Jaryszak, D.L. (1985). *Mech. Ageing Dev.* **30**:63–70.
Schultz, E., Jaryszak, D.L. and Valliere, C.R. (1985). *Muscle & Nerve* **8**:217–222.
Schultz, E., Jaryszak, D.L., Gibson, M.C. and Albright, D.J. (1986). *J. Mus. Res. Cell Motil.* **7**:361–367.
Seed, J. and Hauschka, S.D. (1984). *Dev. Biol.* **106**:389–393.
Seichert, V. (1971). *Acta Morph. Neerl. Scand.* **9**:129.
Senni, M.I., Castrignano, F., Poiana, G., Cossu, G., Scarsella, G. and Biagioni, S. (1987). *Differentiation* **36**:194–198.
Shafiq, S.A., Gorycki, M.A. and Mauro, A. (1968). *J. Anat.* **103**:135–141.
Sicinski, P., Geng, Y., Rider-Cook, A.S. *et al.* (1989). *Science* **244**:1578–1580.
Snow, M.H. (1977). *Anat. Rec.* **188**:201–218.
Snow, M.H. (1978). *Cell Tissue Res.* **186**:535–540.
Spits, H., Schooten, W. van., Keizer, H. *et al.* (1986). *Science* **232**:403–405.
Stockdale, F.E. (1990). *Proc. Exp. Biol. Med.* 194:71–75.
Stockdale, F.E. (1992). *Dev. Biol.* 154:284–298.
Stockdale, F.E. and Holtzer, H. (1961). *Exp. Cell Res.* **24**:508–520.
Stockdale, F.E. and Feldman, J.L. (1990). In *The Dynamic State of Muscle Fibers* (ed. D. Pette), pp. 641–649. Walter de Gruyter, New York.
Studistsky, A.N. (1964). *Ann. N.Y. Acad. Sci.* **120**:789–801.
Tanabe, T., Beam, K.G., Powell, J.A. and Nume, S. (1989). *Nature* **336**:134–139.
Taussig, M.J. (1984). *Processes in Pathology and Microbiology*. 2nd edn. Blackwell Science, Oxford.
Taylor, S.M. and Jones, P.A. (1979). *Cell* **17**:771–779.
Taylor, N.A. and Wilkinson, J.G. (1986). *Sports Med.* **3**:190–200.
Tennyson, V.M., Brzin, M. and Kremzner, L.T. (1973). *J. Histochem. Cytochem.* **21**:634–652.

Teravainen, H. (1970). *Z. Zellforsch. Mikros. Anat.* **103**:320–327.
Theilar, K. (1989). *The House Mouse; Atlas of Embryonic Development*. Springer-Verlag, New York.
Tidball, J.G., Spencer, M.J. and St. Pierre, B.A. (1992). *Exp. Cell Res.* **203**:141–149.
Vaidya, T.B., Rhodes, S.J., Taparowski, E.J. and Konieczny, S.F. (1989). *Mol. Cell. Biol.* **9**:3576–3579.
Vivarelli, E., Brown, W.E., Whalen, R.G. and Cossu, G. (1988). *J. Cell Biol.* **107**:2191–2197.
Wachtler, F., Christ, B. and Jacob, H.J. (1982). *Anat. Embryol.* **164**:369–378.
Wakeford, S., Watt, D.J. and Partridge, T.A. (1990). *Muscle & Nerve* **14**:42–50.
Waldeyer, W. (1865). *Virch. Arch.* **34**:473–514.
Watt, D.J. (1990). In *Myoblast Transfer Therapy* (eds R. C. Griggs and G. Karpati), pp. 35–39. Plenum Press, New York.
Watt, D.J., Morgan, J.E. and Partridge, T.A. (1984a). *Muscle & Nerve* **7**:741–750.
Watt, D.J., Morgan, J.E. and Partridge, T.A. (1984b). *Clin. Exp. Immunol.* **55**:419–426.
Watt, D.J., Morgan, J.E., Clifford, M.A. and Partridge, T.A. (1987). *Anat. Embryol.* **175**:527–536.
Watt, D.J., Morgan, J.E. and Partridge, T.A. (1991). *Neurological Disorders* **5**:345–355.
Watt, D.J., Karasinski, J. and England, M.A. (1993). *Muscle Res. Cell Motil.* **14**:121–132.
Watt, D.J., Karasinski, J., England, M.A. and Moss, J. (1994). *Nature* **368**:406–407.
Webster, C. and Blau, H.M. (1990). *Somat. Cell Mol. Genet.* **16**:557–565.
Webster, C., Pavlath, G.K., Parks, D.R. *et al.* (1988). *Exp. Cell Res.* **174**:252–265.
White, N.K., Bonner, P.H., Nelson, D.R. and Hauschka, S.D. (1975). *Dev. Biol.* **44**:346–361.
Wright, W.E. (1985). *Exp. Cell Res.* **157**:343–354.
Wright, W.E., Sassoon, D.A. and Lin, V.K. (1989). *Cell* **56**:607–617.
Yablonka-Reuveni, Z. and Nameroff, M. (1990). *Differentiation* **45**:21–28.
Yablonka-Reuveni, Z., Quinn, L.S. and Nameroff, M. (1987). *Dev. Biol.* **119**:252–259.
Yablonka-Reuveni, Z., Bowen-Pope, D.F. and Hartley, R.S. (1990). In *The Dynamic State of Muscle Fibres* (ed. D. Pette), pp. 693–706. Walter de Gruyter, New York.
Yao, S-N. and Kurachi, K. (1992). *Proc. Natl Acad. Sci. USA* **89**:3357–3361.
Yao, S-N. and Kurachi, K. (1993). *J. Cell Sci.* **105**:957–963.
Young, R.B., Miller, T.R. and Merkel, R.A. (1978). *J. Anim. Sci.* **46**:1241–1249.
Zuckerman, F.A. and Head, J.R. (1987). *J. Immunol.* **139**:2856–2864.

5 Cell progenitors in the neural crest

Catherine Dulac and Patrizia Cameron-Curry*†

Institut d'Embryologie Cellulaire et Moleculaire, Nogent-sur-Marne, France; (Present address: Harvard University, Cambridge, MA, USA)
**Howard Hughes Medical Institute, New York, USA*
†Deceased

INTRODUCTION

The neural crest is a transient structure of the vertebrate embryo composed of migratory precursor cells which leave the neural primordium at a definite stage of development, i.e. when the neural plate folds and closes on its mediodorsal aspect. Neural crest cells spread over the embryo and settle in various tissues where they differentiate into a large variety of cell types. Among these are neurons and glial cells of the peripheral nervous system (PNS), melanocytes, and certain endocrine and paraendocrine cells. In addition, the neural crest originating from the cephalic region of the neural axis yields mesectodermal derivatives that form the connective and skeletal structures of the face.

Owing to these characteristics, the development of neural crest cells has provided a unique system in vertebrate embryos, both for investigating the process by which a highly diversified progeny is generated from a multipotent population of progenitor cells, and for delineating the relative contributions of intrinsic and extrinsic factors to cell fate determination.

The choice of the avian system, whose embryo is accessible to experimentation in the egg during the entire period of development, has allowed the migration and the fate of crest cells to be extensively monitored and manipulated *in vivo*. The nuclear marker provided by quail cells in quail/chick chimeras (Le Douarin, 1969, 1973, and 1982 for a review) and more recently iontophoretic injections of fluorescent dyes into single neural crest cells *in situ* (Bronner-Fraser and Fraser, 1988, 1989) have been particularly instrumental in this respect. Quail/chick constructs in which neural primordia have been transplanted to heterotopic sites have revealed that the embryonic micro-environment plays an essential role in the differentiation of the cells that issue from the neural crest and form the PNS.

Although *in vivo* studies clearly demonstrate the pluripotence of the neural crest and the importance of cell interactions in cell fat determination, the nature of the progenitor cells that compose the migrating neural crest, and the factors which influence their fate, remain elusive. Is there a totipotent precursor for all crest cell derivatives whose fate is

STEM CELLS
ISBN 0-12-563455-2

influenced by the micro-environment encountered during or after migration? An alternative hypothesis would be that, from the beginning of neural crest ontogeny, at the premigratory stage, precursors are already committed to distinct lineages.

These questions have been directly addressed by *in vitro* studies. Recent attempts have been made to analyse the phenotypes generated by individual cells taken at different time points during their migration. This led to the discovery of a small subpopulation of highly multipotent progenitors, suggesting the existence of a neural crest stem cell. Some developmental decisions made by crest precursors have been recently documented in the case of the glial lineage and will be described here.

THE DEVELOPMENTAL POTENTIAL OF NEURAL CREST PRECURSORS: *IN VIVO* STUDIES

Fate map of premigratory neural crest

Work performed in the 1970s and 1980s on avian embryonic chimeras led to a precise account of the spatial and temporal migration patterns of neural crest in two avian species, the quail (*Coturnix coturnix japonica*) and the chick (*Gallus gallus*). The basic experimental paradigm was to exchange defined segments of the chick neural primordium with their exact counterpart from quail (or vice versa), at stages preceding the migration of the neural crest. The migration thus takes place in the host tissues, enabling the establishment of a neural crest fate map and the determination of the level of origin of the different crest derivatives. A structural difference in the interphase nucleus of quail and chick cells (Le Douarin, 1969, 1971, 1973) makes it easy to distinguish cells from host and donor. Moreover, the stability of the quail marker allows cells to be followed until they have reached a fully differentiated state.

Quail/chick chimeras were systematically constructed at all levels of the neural axis (reviewed by Le Douarin, 1982) in order to establish the fate map of the premigratory neural crest (Figure 1). More recent experiments have refined the analysis by grafting the neural tube and associated neural crest corresponding to a single somite (Teillet *et al.*, 1987). Two main results emerge from these experiments:

(1) Crest cells display a wide-ranging capacity for differentiation (Table 1). Crest cell precursors give rise to PNS neurons and glial cells, some endocrine and paraendocrine cells (adrenomedullary cells, calcitonin-producing cells and carotid body type I and II cells) and melanocytes (except pigment cells of the retina). In addition, the crest originating from the cephalic level of the neural axis plays an important role in head morphogenesis. Its contribution to that process has recently been shown to be more important than previously supposed (Couly *et al.*, 1993). The meninges of the cerebral hemispheres, the facial skeleton and dermis, the mesenchymal cells of the thymus and of the thyroid, parathyroid, pituitary, lacrymal and salivary glands, the musculo-connective wall of the arteries arising from the aortic arches, as well as the skull roof (comprising the frontal, parietal and squamosal bones), are all of crest origin.

(2) The fate map of the various neural crest derivatives shows that the crest is regionalized in different areas of the neural axis (Figure 1). While melanocytes arise from all levels of the crest, mesenchymal cells are produced in birds only by the cephalic and

Table 1 Neural crest derivatives.

Mesectoderm (from cephalic crest only)
facial and precordal skull skeleton
contribution to head and neck connective tissues
prosencephalic meninges
odontoblasts (mammals)
Melanocytes
Endocrine cells
carotid body type I and II cells
thyroid C cells
adrenal medulla chomaffin cells
Peripheral nervous system
Glial cells
satellite cells
Schwann cells
enteric glia
Neurons
spinal sensory
cranial sensory (some are of placodal origin)
sympathetic autonomic
parasympathetic
enteric

Data from: Le Douarin, 1982; Teillet *et al.*, 1987; Couly *et al.*, 1993.

vagal crest extending from the level of the epiphysis (Couly and Le Douarin, 1987) down to the fifth somite pair (Le Lièvre and Le Douarin, 1975). The adrenergic sympathetic chain is generated from cells leaving the neural tube at the trunk level, whereas the majority of the parasympathetic enteric system was found to derive from axial levels corresponding to somites 1–7. These so-called 'vagal' neural crest cells migrate long distances while vigorously proliferating and colonize the whole gut from the pharyngeal entry point down to the cloaca. A minor stream of cells migrates to the post-umbilical gut from the lumbosacral level, caudal to the 28th somite level. Similar results were recently obtained in chick and mouse by vital dye and retroviral labelling (Serbedzija *et al.*, 1991; Epstein *et al.*, 1994).

Therefore the destiny of premigratory crest cells at each level of the neural axis seems to be restricted to a particular set of phenotypes. This could be interpreted in two opposite ways. Either the developmental capacity of the crest cells arising from different segments of the neural primordium is already fixed to a subset of the crest potentials, or each segment of the neural crest is equally pluripotent and the appropriate differentiation relies on specific environmental cues encountered by the precursor during migration or in the site to which it finally lies.

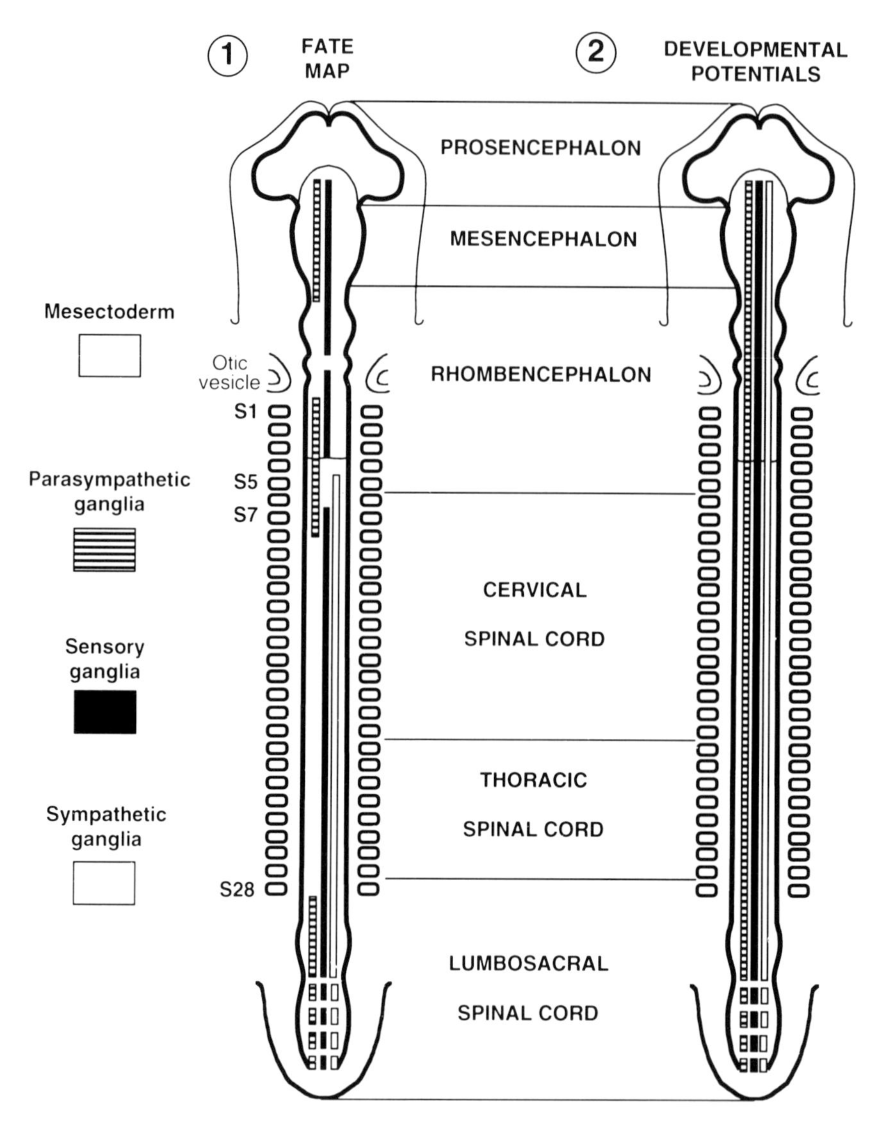

Figure 1 Fate and developmental potential maps of the avian neural crest. (1) Fate map. Schematic representation of the distribution along the neural axis of the presumptive territories giving rise during normal development to different neural crest derivatives: mesectoderm and PNS ganglia. (2) Developmental potential. When transplanted to a different axial level, neural crest cells from any level have the capacity to give rise to the PNS derivatives appropriated for their new location. However, only cephalic crest, rostral to the level of the fifth somite, can give rise to mesectoderm. S = somite.

Development potential of the neural crest along the neural axis

The level of commitment of premigratory crest cells has been documented by experiments in which fragments of the quail neural primordium were implanted heterotopically along the neural axis of chick embryos. Morphogenesis of peripheral ganglia and the adrenomedullary gland as well as the neurotransmitters synthesized by the various types of neurons that differentiated following the operation were subsequently examined (Le Douarin and Teillet, 1974; Le Douarin *et al.*, 1975; Fontaine-Pérus *et al.*, 1882).

It appeared that when displaced along the neural axis prior to the onset of migration crest cells adopted the fate of their new position in the body. The migratory pathways, the patterning of peripheral ganglia as well as the neurotransmitter phenotype of individual cells display characteristics corresponding to their new position rather than to their fate in normal development (Figure 1). For example, trunk neural crest cells, which normally give rise to catecholamine-synthesizing cells, are able to colonize the gut and differentiate into enteric acetylcholine-producing neurons when transplanted to the vagal level.

Therefore it seems that the premigratory neural crest displays a certain degree of pluripotency along the neural axis. This implies that crest cells must respond to environmental cues in order to select or to permit the development of the cell type appropriate to a particular location. An exception to this is the confinement of mesectodermal potential to the cephalic region, suggesting an early segregation of this particular lineage. The distribution of the various potentials is quantitatively non-uniform. For instance, trunk crest is a less abundant potential source of enteric ganglia than cranial crest, while conversely, trunk crest has a greater aptitude than cranial crest to differentiate into adrenergic cells and melanocytes (Le Douarin and Teillet, 1973; Smith *et al.*, 1977; Newgreen *et al.*, 1980).

Role of environment cues in neural crest differentiation

The role of the environment in the process of crest cell differentiation was further emphasized by experiments in which forming peripheral ganglia were heterotopically and heterochronically back-transplanted into the neural crest migratory pathway. Quail dorsal root ganglia (DRG) and sensory ganglia of cranial nerves (e.g. nodose or petrosal) were grafted into the neural crest migration pathway of a younger (E2) chick embryo at the adrenomedullary level of the neural axis. Following this procedure, the grafted ganglia lose their cohesiveness and their component cells become dispersed in the host tissue. Strikingly, adrenergic cells of quail origin were subsequently found in the chicken sympathetic ganglia and adrenal medulla demonstrating that the capacity to yield cells of the sympatho-adrenal lineage exists in sensory ganglia although it is repressed during the normal course of development (Le Lièvre *et al.*, 1980; Schweizer *et al.*, 1983). Further experiments showed that adrenergic cells differentiating in backtransplants of either DRG or ciliary ganglia originate from undifferentiated precursors present in a quiescent state in the non-neuronal population of the ganglia, rather than from a phenotypic switch of postmitotic neurons (Schweizer *et al.*, 1983; Dupin, 1984; Le Douarin, 1986).

Thus, transplantation experiments have clearly established that neural crest cell fate is determined, at least in part, by the local environment. However, they do not address the issue of how the lineages which give rise to the various crest derivatives become segregated and whether the demonstrated pluripotency of the neural crest population corresponds to the existence of individual multipotent precursors.

Development fate of individual neural crest cells

By injecting lysinated rhodamine-dextran, a vital fluorescent dye, into single cells of the dorsal neural tube of E2 chick embryos, Bronner-Fraser and Fraser (1988, 1989) were able to follow the fate of single neural crest cells *in ovo*. Two days after the injection, clones of cells that had inherited the dye were seen in multiple structures and expressed several phenotypes: melanocytes, neurons in DRG and sympathetic chain ganglia, satellite and Schwann cells, and adrenomedullary cells. These observations provided direct evidence that at least some neural crest cells are multipotent at the time of emigration from the neural tube. Some clones were also found to be quite restricted, limited to DRG for example, suggesting that some crest cell precursors might have a smaller repertoire of potentials. This approach enabled the range of cell fates adopted by the descendants of an individual crest cell to be visualized for the first time. However, these results do not necessarily reflect the actual range of potentials for this cell since the local environment met by the migrating cell after the injection may restrict an initial multi- or even totipotency.

THE DEVELOPMENTAL POTENTIAL OF INDIVIDUAL CREST CELL PROGENITORS: *IN VITRO* STUDIES

Cultures of single neural crest cells

A decisive breakthrough in the analysis of the commitment and the developmental capacities of individual neural crest cells came from the use of a culture system allowing single cells to grow and produce colonies in which virtually all crest-derived cell types are represented (Baroffio *et al.*, 1988; Dupin *et al.*, 1990). This allowed the differentiation capacity of individual neural crest cells to be tested at different time points during their migration.

One would expect multipotent cells to reveal the full extent of their repertoire if provided with a permissive environment. The culture conditions in which single cells are grown should provide a large array of growth and differentiation factors so that there are no restrictions on the expression of their development potential. In pioneer experiments, clonal cultures of trunk neural crest cells were made using the limit dilution method (Cohen and Konigsberg, 1975; Sieber-Blum and Cohen, 1980). Three phenotypes were subsequently considered: melanocytes, adrenergic cells and cells expressing neither trait. This study has been extended to an additional phenotype, sensory neurons (Sieber-Blum, 1989, 1991). Homogeneous as well as heterogeneous clones were found, demonstrating the coexistence in the crest cell population of committed and pluripotent precursors.

Another culture system has allowed the analysis of a large range of phenotypes generated *in vitro* by individual crest cells taken at different time points during their migration (Figure 2) (Baroffio *et al.*, 1988, 1991; Dupin *et al.*, 1990; Sextier-Sainte-Claire Deville *et al.*, 1992, 1994). The use of a feeder layer of growth-inhibited mouse 3T3 fibroblasts, a culture system devised to clone human keratinocytes (Barrandon and Green, 1985), has proved remarkably efficient in supporting the survival and the growth of

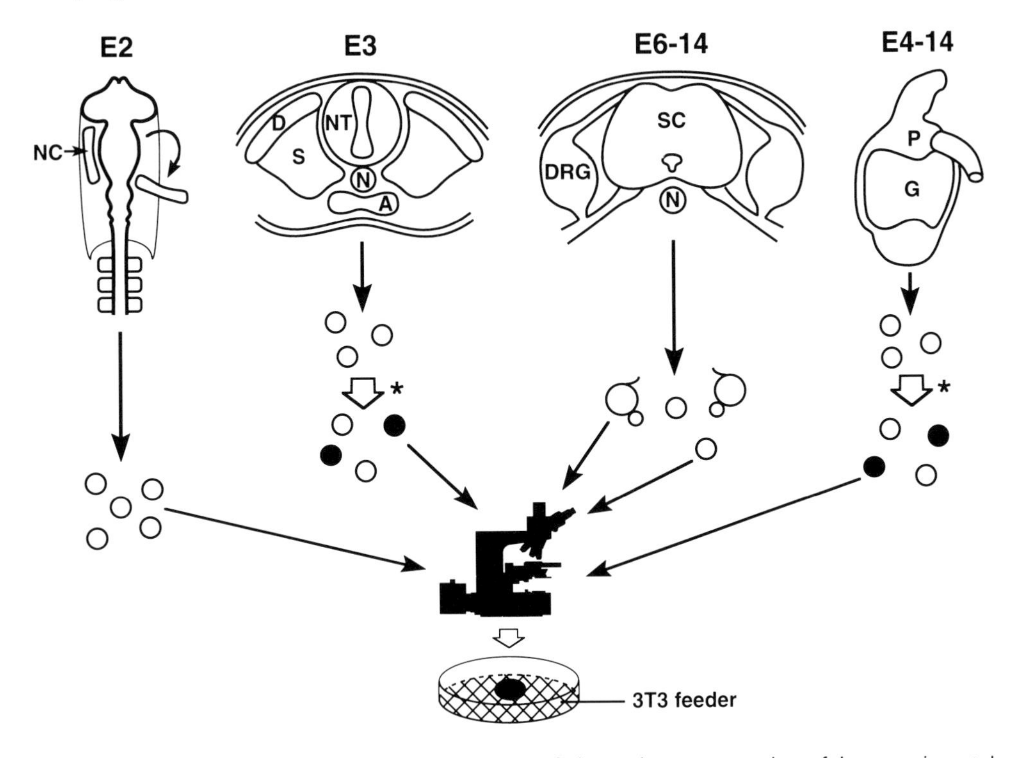

Figure 2 Procedure used for cloning experiments. Schematic representation of the experimental procedure used for cloning experiments. Neural crest and DRG were dissected microsurgically, sclerotomes enzymatically with pancreatin, and gizzard plexuses with collagenase treatment. Cell suspensions were obtained by trypsin digestion, followed by immunostaining with HNK1 and a fluorescent secondary antibody where indicated (*). Single cells were then manually selected under the microscope and seeded onto a feeder layer of growth-inhibited 3T3 murine fibroblasts. NC = neural crest; NT = neural tube; DRG = dorsal root ganglia; SC = spinal cord; D = dermomyotome; S = sclerotome; G = gizzard; P = enteric plexus; N = notocord; A = aorta.

avian neural crest cells. All the major crest-derived cell types differentiate under the conditions of this culture system. Cloning was accomplished by isolating single cells from a suspension of neural crest cells under microscopic control (Figure 2). Individual cells were then seeded with a micropipette into culture wells on top of mitomycin-treated 3T3 cells in a complex culture medum.

Development potentials of isolated cephalic neural crest cells

The clone-forming ability and developmental potential of single neural crest cells were then analysed. Cephalic crest cells were first studied (Baroffio *et al.*, 1988, 1991; Dupin *et al.*, 1990) since the pluripotency of this crest cell population seems to be the highest. After 7–10 days of culture, the single cell-derived colonies were easily recognized on top of the large 3T3 cells either by simple phase contrast or after DNA staining with the Hoechst reagent. The small nuclei of the quail cells are clearly distinguishable from

3T3 nuclei. The cultured individual cells generated colonies that varied considerably both in size and in phenotypic expression, thus reflecting the heterogeneity of early migrating neural crest. The cell numbers in the clones ranged from one or two to several thousands.

The majority of the clones contained cells expressing two or more phenotypes: neurons, glial cells, cartilage or melanocytes. A few clones were composed of cells exhibiting only one phenotype. Among these were colonies of glial cells in which all the cells expressed the marker Schwann cell myelin protein (SMP; see the paragraph on glial cell lineage below), and some clones consisting of one or two neurons, or only of cartilage.

These experiments revealed the coexistence in the cephalic neural crest of largely pluripotent cells and of precursor cells with more restricted developmental potential.

A special feature of the cephalic neural crest is its ability to generate mesectodermal tissue, from which the craniofacial mesenchyme (including cartilage, bone, muscles and connective tissue) develops. As described above, it has been proposed, on the basis of transplantation experiments, that mesectodermal precursors diverge from the other precursors at the premigratory stage. In fact, Baroffio *et al.* (1991) showed that 2.6% of migrating cephalic neural crest consists of precursors able to produce colonies in which mesectodermal cells, revealed by the presence of cartilage islets, are found together with other crest cell derivatives, in particular neurons and glia. Therefore it is clear that, at the cephalic level, the migrating crest still contains cells with dual mesectodermal and neural potential.

Do neural crest stem cells exist?

In this study, a single clone was also found that contained the entire spectrum of phenotypes identifiable by the available markers, *i.e.* neurons, adrenergic cells, Schwann cells, melanocytes and cartilage. The identification in the progeny of a single cell of all phenotypes that may be expected from the cephalic neural crest suggests the existence of a multipotent neural crest stem cell, comparable to the stem cell described in the haemopoietic system.

The isolation in the mammalian system of pluripotent neural crest cells that yield neurons and glia and are endowed with self-renewal capacity (Stemple and Anderson, 1992) further demonstrates the existence of a true crest stem cell.

These cloning experiments revealed the striking heterogeneity of migrating neural crest cells in terms of their capacity for differentiation and proliferation. On the basis of these results, it may be proposed that neural crest cells become progressively committed to different lineages during their migration process (Figure 3).

Using a descriptive mathematical method to analyse the likelihood of occurrence of combination of the different cell types among the clones, Baroffio and Blot (1992) showed that, in contrast to the haemopoietic system, the transitions between pluripotent, oligopotent and monopotent progenitors do not proceed in a sequentially ordered manner. Rather, it appears from this study that a stochastic mechanism restricts the developmental potential of pluripotent progenitors to give rise to the major neural crest-derived lineages.

Similar mechanisms are likely to be conserved in other vertebrate species: single cell tracing experiments done in Xenopus and zebrafish embryos with fluorescent dye have

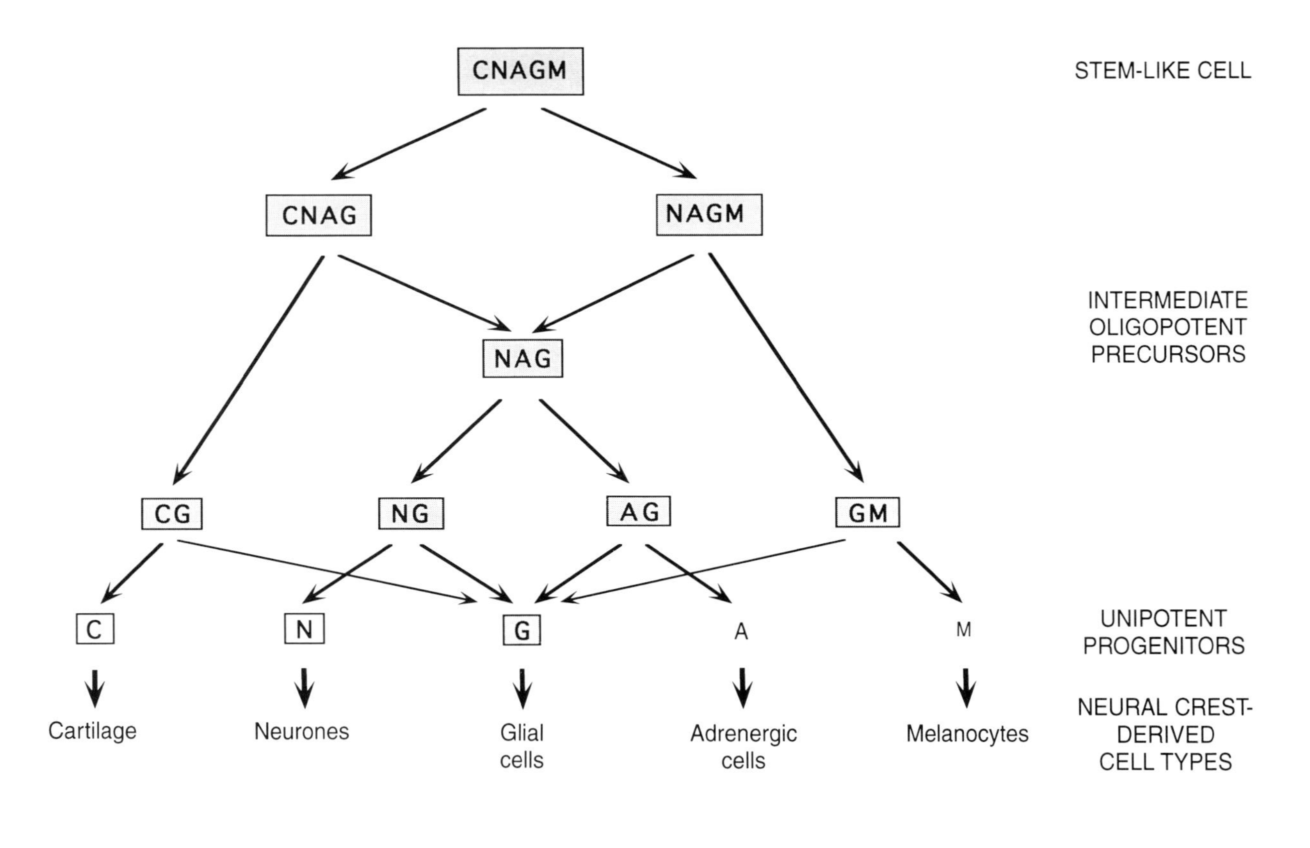

Figure 3 Progenitors in the cephalic neural crest. The diagram summarizes data from analysis of more than 500 clones in which different phenotypes were found. The putative progenitors are classified according to the number and types of cells in their progeny. Progenitors with gliogenic potential are shaded. The arrows indicate hypothetical filiations between precursors. In the proposed model, totipotent neural crest stem-like cells generated committed unipotent progenitors through intermediate oligopotent cells. A = adrenergic cell; C = cartilage; G = glial cell; M = melanocyte; N = neuron.

demonstrated the temporal coexistence of both pluripotent and committed cells, even though specification appears to be more precocious in zebrafish (Collazo *et al.*, 1993; Schilling and Kimmel, 1994; Raible and Eisen, 1994).

Restriction of developmental potential in neural crest derivatives

The progressive commitment of neural crest-derived cells during gangliogenesis was further analysed in clonal cultures of cells originating from the trunk level and found in developing somites at E3 and in DRGs at E6–E14 (Figure 2) (Sextier-Sainte-Claire Deville *et al.*, 1992). The migration pathway through the somite is used by neural crest cells homing to the DRG, autonomic sympathetic ganglia and adrenal medulla (Teillet and Le Douarin 1983).

Crest cells migrating in the sclerotomal part of the somites were analysed by selecting the immunofluorescence-labelled HNK1-positive cells under microscopic control and seeding them individually onto the 3T3 feeder layer. HNK1 is not a lineage marker, (Baroffio and Blot, 1992), and is not restricted to the nervous system. However, it recognizes the majority of migrating neural crest cells, and later in development, glial cells and some neurons of the PNS (Abo and Balch, 1981; Tucker *et al.*, 1984). The epitope recognized by the HNK1 mAb is found on many surface glycolipids and glycoproteins, among which is the avian SMP.

The developmental potency of the cloned cells was found to be higher for E3-derived cells than for cells from older embryos. DRG-derived clones were mostly small (< 100 cells) while cones from E3 migratory HNK1-positive cells were often large (> 1,000 cells), although never as large as the biggest clones from E2 cells (> 10,000 cells). Thus it appears that the proliferation capacity of crest-derived cells decreases with increasing age of the embryo. The phenotypic diversity of the cells in the clones appears also to be smaller: neuronal and non-neuronal cells were found in E3-derived clones, while only non-neuronal phenotypes were expressed in those derived from E6–14 DRGs. It is striking that no neurons were ever generated when single DRG cells from E6 and older embryos were cultured. This result was unexpected, as 2% of the non-neuronal population of the DRG was previously demonstrated in mass cultures to comprise resting precursors for adrenergic cells (Xue *et al.*, 1987). The absence of this particular development potential in clonal cultures suggests that cell–cell interactions between appropriate cells and occurring at a certain cell density are necessary to support some developmental choices.

Similar progressive restriction of developmental potential of trunk crest-derived cells, was shown by Duff *et al.* (1991) in an *in vitro* clonal analysis of progenitor cell in DRG and sympathetic ganglia. In particular, the proportion of pluripotent cells endowed with melanogenic potential decreases rapidly with increasing age of embryos.

The developmental potential of the crest cells colonizing the gut was also examined in clonal cultures (Sextier-Sainte-Claire Deville *et al.*, 1994; Serbedzija *et al.*, 1991). Crest derived cells located in the gizzard were isolated by labelling with the HNK1 antibody as before (Figure 2). Large clones (> 10^3 cells) were numerous when generated by cells derived from E4 and E5 gizzard, while E7 and E8 gizzard crest cells produced very small colonies. Likewise, the phenotypic diversity of the cells composing the clones decreased during the time period under scrutiny. Moreover, two phenotypes that are never encountered *in situ* in the enteric plexus were found in these cultures: cells belonging to the adrenergic type identified by tyrosine hydroxylase (TH) immunoreactivity and glial cells expressing SMP. However, the capacity to yield cells possessing neuronal features (TH or neurofilament) in clonal cultures disappears when cells were derived from later stages than E6. Therefore, proliferation as well as differentiation potential seems to progressively decrease in the developing enteric system.

The developmental capacities of neural crest cell populations that migrated to the trunk ectoderm (Richardson *et al.*, 1993), or to the posterior visceral arches (Ito and Sieber-Blum, 1993) were also investigated. In both cases the number of colonies containing multiple phenotypes declined with the age of the embryos, evidencing a progressive restriction of developmental potential.

Cultures of single neural crest cells demonstrated that, at or soon after initiation of migration from the neural tube, the neural crest is an heterogeneous population composed of pluripotent cells and of cells with restricted developmental capacity. The relative proportion of precursors with limited development potential dramatically increases with later embryonic stages.

THE PROGRESSIVE SEGREGATION OF THE PERIPHERAL GLIAL CELL LINEAGE

Glial cells entertain an intimate relationship with neurons in peripheral ganglia, fibres, and at the synapse level, providing an essential support for neuronal function and survival. It is therefore particularly interesting to try to understand how the glial cell lineage individualizes from other crest cell derivatives and how different factors provided by various cell types dramatically affect glial cell fate.

The different types of PNS glial cells and their origin in the embryo

In the PNS, glial cells surround the soma and neurites of all neurons, isolating them from each other and from the environment. These glial cells are classified into four types according to distinctive morphological and molecular characteristics. Myelinating Schwann cells are associated with a segment of large-diameter axon, while non-myelinating Schwann cells surround a few to many small axons. Each axon/Schwann cell(s) unit is further isolated by a basal lamina largely produced by the Schwann cell at its outer surface. All Schwann cells are able to make a myelin sheath when in contact with the right axonal signal, both *in vitro* and *in vivo* (Mezei, 1993).

The cell body of each neuron is similarly surrounded by a glial sheath formed by one to several glial satellite cells. In some special cases, a myelin sheath is elaborated by the satellite cells around the neuronal cell body in the ganglia (Hess, 1965; references in Peters *et al.*, 1976). A basal lamina is formed at the outer satellite cell membrane that is continuous with the one formed by Schwann cells along the axon (Pannese, 1974). Satellite cells can also form myelin *in vitro*.

A fourth type of PNS glial cell is represented by enteric glia. Enteric glia cells are stellate and although they contact different neurons and/or neurites, the ensheathment of each neuronal unit is incomplete even in the adult animal. No basal lamina is formed inside the ganglia and nerve tracts of the Auerbach's (myenteric) and Meissner's (submucosal) plexuses, but is rather assembled at the neural–mesenchymal interface. The glial cell organization contributes to the definition of the ENS as a 'CNS-like' compartment of the PNS (Gabella, 1971; Gershon and Rothman, 1991).

As demonstrated with quail/chick embryonic chimeras all peripheral glial cells derive from the neural crest. Moreover neurons of peripheral ganglia and the associated satellite and Schwann cells were found to derive from the same axial level of the neural tube (Le Douarin and Teillet, 1973; Teillet *et al.*, 1987). The only exceptions to this rule are the distal ganglia of the cranial V, VIII, IX and X nerves (trigeminal, otic, petrose and nodose ganglia), as their neurons are of placodal origin (Narayanan and Narayanan, 1980; Ayer Le Lièvre and Le Douarin, 1982; D'Amico-Martel and Noden, 1983).

New glial markers

To follow gliogenesis in the avian system, we first defined new molecular markers, specific for glial cells and appearing sequentially during embryogenesis. We used the monoclonal antibody (mAb) strategy and injected high molecular weight glycoproteins extracted from adult quail peripheral nerves, selecting for hybridomas producing mAbs against antigens expressed early in embryogenesis. Two of these were studied in more detail, enabling us to characterize the 4B3 epitope, a pan-glial marker in the avian system, and the SMP glycoprotein, a Schwann cell specific surface molecule.

The 4B3 epitope is a carbohydrate moiety associated with several unidentified surface molecules of the PNS and of the CNS and with the SMP protein (Cameron-Curry *et al.*, 1991). We demonstrated by double staining *in vitro* that oligodendrocytes and astrocytes from the brain and spinal cord express the 4B3 epitope. Its expression is somewhat down-regulated in culture. In the PNS, the surfaces of all Schwann cells and satellite cells are stained, while neurons are never positive. 4B3 is also present in the ENS.

A very interesting aspect of the distribution of the 4B3 epitope during embryogenesis is its first appearance around E3.5, when neural cells aggregate to form the ganglia but glial or neuronal cell types are still not recognizable. Thus the 4B3-positive phenotype can be considered as a very early marker of gliogenesis. At this point it is not possible to assess whether the cells expressing 4B3 at that stage represent fully-committed glial precursors or still retain multiple developmental capacities.

The pattern of expression of the second marker used, the SMP protein, is more restricted *in vivo*: oligodendrocytes in the CNS and myelinating and non-myelinating Schwann cells are the only immunoreactive cells (Dulac *et al.*, 1988; Cameron-Curry *et al.*, 1989). Satellite cells and enteric glia are SMP-negative, and do not express the SMP mRNA. The protein is first expressed in the sciatic nerve of the quail around E6, when Schwann cells are aligned along the axon fascicles and axon segregation has not yet begun. Myelination does not start until around E11 in quail. Thus, in the PNS, the first appearance of the SMP-positive phenotype corresponds to a very early step during the process of Schwann cell differentiation. In the CNS, SMP expression also precedes myelination by one to a few days, depending on the region considered.

A striking feature of SMP expression is its presence in culture at the surface of Schwann cells in non-myelinating conditions and in the absence of neurons. Moreover, a sub-population of both truncal and cranial neural crest cells in culture spontaneously follows the glial cell differentiation pathway and acquires the SMP marker (Dupin *et al.*, 1990). Therefore it seems that the emergence of the SMP-positive phenotype is early enough in the glial cell differentiation pathway to be independent from axonal contact, a major signal for terminal Schwann cell differentiation and myelination (Mezei, 1993).

SMP is a surface glycoprotein, belonging to the immunoglobulin superfamily (Dulac *et al.*, 1992). Its extracellular domain consists of five Ig-like domains, one of the C2 type and four of the V type. As for the intracellular domain, the eight terminal amino acids form a conserved motif of unknown functional significance, found in a few other molecules of the same superfamily. The highest homology, 60% at the amino acid level, is found between SMP and the myelin associated glycoprotein (MAG), but in contrast to SMP, MAG has been shown to be expressed at later stages of embryogenesis and only by myelinating PNS and CNS cells. The regulation of the expression of the two proteins *in vivo* and *in vitro* as well as their functional properties are also different.

The anti-SMP mAb was of crucial use in two sets of experiments aimed at studying: (1) the segregation of the glial lineage in the avian system; and (2) the influence exercised by the environment on the glial phenotype.

Glial cell precursors in the neural crest

How and when is the glial lineage generated from the neural crest population? How many different types of precursor with gliogenic potential coexist and what is their degree of determination?

An important improvement in the study of glial lineage was made possible thanks to the clonal culture system developed in our laboratory (Baroffio *et al.*, 1988; Dupin *et al.*, 1990). The use of the anti-SMP mAb permitted the first accurate study of the emergence of the peripheral glial lineage.

Different kinds of precursors with gliogenic potential were scored in a study of 185 clones obtained from cephalic crest (Dupin *et al.*, 1990). These can be grouped into five categories:

(1) We demonstrated the existence of a population of fully committed glial precursors which give rise to homogeneous clones of SMP-positive cells (Table 2). These glial clones

Table 2 Clones containing SMP-positive cells.

	Clones with SMP+ cells, %	Clones with SMP+ cells only,* %	Clones with SMP+ cells and neurons,* %
Neural crest (1)	87	13 (15)	37 (42.5)
Sclerotome (2)	69	23 (33)	5.5 (8)
DRG E6-14 (2)	37	17 (46)	0 (0)
DRG E8 (N+nN) (3)	100	92.5 (92.5)	/

*Figures in brackets are expressed as a percentage of the total number of clones containing SMP+ cells.

Data from: (1) Dupin *et al.*, 1990; (2) Sextier-Saint-Claire Deville *et al.*, 1992; (3) Cameron-Curry *et al.*, 1993.

which represent 13% of the total number of clones analysed, clearly demonstrate that the presence of neurons is not necessary for the emergence of the PNS glial phenotype.

(2) In 37% of the clones, glial cells and neurons coexisted together with unidentified unpigmented cells. These demonstrate the existence of neurogenic precursors. Statistical analysis of the distribution of the different phenotypes in clonal cultures suggests that neuronal, adrenergic and Schwann cell phenotypes are not randomly associated. In contrast, these three neural cell types differentiate in the clones independently of cartilage and melanocytes, by a process of stochastic restrictions of their developmental potentialities (Baroffio and Blot, 1992).

In vivo lineage tracing in the chick was undertaken by Bronner-Fraser and Fraser (1988, 1989) using lysinated-rhodamine–dextran labelling, and by Frank and Sanes (1991) using retroviral insertion. These studies, which enabled us to follow the progeny of individual neural crest cells also demonstrated the existence of both common (40%) as well as separate precursors for neurons and glial cells.

(3) SMP-positive cells were also found in a third type of clone, together with non-neuronal, unpigmented HNK1-positive and HNK1-negative cells whose identity remains unclear but could represent still undifferentiated precursors. These clones represent 34.5% of the total.

(4) A minority of the clones (1%) contained glial cells and melanocytes, with HNK1-positive and negative cells. This glio/melanogenic precursor was also found in the *in vivo* labelling study, as well as in cultures of E5 quail embryonic nerves, which were found to generate melanocytes in particular conditions (Stoker *et al.*, 1991; Sherman *et al.*, 1993).

(5) Even more interesting is another rare (1%) precursor that can generate glial cells and cartilage, thus having dual mesectodermal and neurogenic potential. This demonstrates that the segregation of the two lineages is still incomplete at the migration stage of the neural crest, as was confirmed by a later study (Baroffio *et al.*, 1991) involving 305 clones. In this series, SMP-positive cells were scored in seven out of the 10 cartilage-containing clones. The two largest of these clones also contained neurons, confirming the full neurogenic potentiality of the founder cells.

To summarize, we showed that differently committed Schwann cell precursors coexist in the neural crest during the migration stage (Figure 3). The presence in our culture system of the 3T3 feeder layer plays a critical role for the development of the glial precursors. The gliogenic potential seems to be very high in the migrating neural crest, as SMP-positive cells were found in 87% of the clones. At the same time, a fully-committed glial precursor can already be demonstrated which is at the origin of 13% of the total number of analysed clones, that is 15% of the SMP-containing clones (Table 2). A highly multipotent progenitor, the putative neural crest stem cell, represents an extremely rare glial cell progenitor.

Although statistical analysis suggests a process of stochastic restriction of crest cell potentials, some studies have found that environmental signals influence to a large extent the phenotypic choice of neural crest-derived cells. In a study of the rat glial lineage (Stemple and Anderson, 1992) most of the neural crest cells cloned on a substrate of fibronectin and poly-D-lysine gave rise to mixed clones composed of neurons and glial cells, while on fibronectin only glial cells were generated. The neurogenic potential of these precursors was none the less retained, as after subcloning in neurogenic conditions, on fibronectin and poly-D-lysine, mixed clones were obtained. This is a clear example of the control of the fate of neural crest cells by extrinsic factors, in this case the substrate molecules.

Gliogenic potential in neural crest-derived cell populations

The gliogenic capacities of neural crest cells were investigated at different time points of gangliogenesis at the trunk level (Figure 2). Neural crest cells migrating at E3 in the sclerotomal part of the somites were examined. Of the obtained clones, 69% contained SMP-positive cells. A high (23%) proportion of clones was homogeneously SMP-positive, representing 33% of the SMP-containing clones. In contrast, neurogenic precursors giving rise exclusively to neurons and glia represented only 5.5% showing that the segregation of the two lineages is more advanced at E3 than at the beginning of crest migration (E2).

SMP-positive cells were found in 37% of the DRG-derived clones, and 17% (representing 46% of the SMP-containing clones) were homogeneous clones composed exclusively of glial cells. Thus determination of the glial lineage further continues.

The developmental potentials of neural crest cells that had migrated to the gut to form the enteric plexuses were also examined (Figure 2) (Sextier-Sainte-Claire Deville *et al.*, 1994). Clones containing both neurons and glial cells were found in E4 (14.5%) to E6 (10%) derived cultures, but the neuronogenic potential was lost by cells from older embryos.

Therefore we were able to demonstrate that the determination of the glial lineage, and in particular the segregation from neuronal fate, is a progressive event that takes place from early E2 (beginning of crest cell migration), to E7, when all the ganglionic neurons are postmitotic. In fact, fully determined restricted glioblasts represent 15% of the precursors cloned from E2 migrating neural crest that can generate SMP-positive cells, 33% of the E3 sclerotome-derived cells, and later on 46% of the non-neuronal, non-satellite cells of the DRG (Table 2). When satellite cells were cloned, all clones contained almost exclusively SMP-positive cells, as discussed below (Cameron-Curry *et al.*, 1993).

Moreover, the restriction of the developmental potential of each type of precursor, for example the glial/melanocyte bipotential progenitor, does not occur simultaneously. As a result, at all times the neural crest derived populations are a mixture of precursors at different stages of their determination.

The genetic mechanisms responsible for the restriction of potential of the crest cells are largely unknown: however, transcription factors which were shown to control similar events in invertebrate PNS development might be interesting candidates (Uemura *et al.*, 1989; Jan and Jan, 1990).

Molecular mechanism of glial cell determination

Recently, molecular and genetic studies have attempted to elucidate the basis of the events described above. The most spectacular result achieved in this field is the recent report by the group of Anderson on the action of the glial growth factor (GGF) on rat crest cells (Shah *et al.*, 1994). This factor has recently been cloned (Marchionni *et al.*, 1993) and belongs to a group of proteins obtained by alternative splicing of the same transcript and which includes secreted and membrane-bound forms previously known as neuregulins or heregulins, neu differentiation factor, or ARIA. These molecules belong to the EGF/TGFa superfamily, and their receptors are the tyrosine kinases $p185^{erbB2}$/HER2/c-Neu and $p180^{erbB4}$/HER4. Both c-Neu and the ligand GGF have been previously shown to be important for Schwann cells (Lemke and Brockes, 1984). C-Neu is expressed by the majority of migrating neural crest cells, and *in vivo* neurons express GGF2 during early gangliogenesis (Shah *et al.*, 1994).

When recombinant human GGF2 was added to the medium in the clonal culture system, neurogenesis was abolished in most of the clones, while gliogenesis was not affected. Moreover, the expression of MASH1 (Mammalian Achaete-Scute Homologue 1), the earliest known marker of neuronogenesis, was abolished. MASH1 is a basic helix-loop-helix (HLH) transcription factor that is thought to have an essential role during PNS neurogenesis. It is expressed by subsets of neuroepithelial cells and neural crest cells several days before detectable neuronal differentiation, and targeted mutagenesis in the mouse affects subclasses of neural-crest derived neurons (Lo *et al.*, 1991; Anderson, 1993; Guillemot *et al.*, 1993). Since GGF2 inhibits the expression of this neuronogenic lineage determination gene, it was deduced that GGF2 exercises an instructive action on cell lineage determination, suppressing the differentiation of neurons from a bipotential neural progenitor, and thus promoting gliogenesis. This is the first report to suggest strongly that an extrinsic signal can control the choice of fate of a neural crest cell. As GGF2 is produced by neurons *in vivo*, a model of lateral inhibition is proposed in which GGF, produced by differentiated neurons, inhibits neuronal differentiation of nearby precursors (which express the GGF receptor) and instead pushes them along the glial differentiation pathway.

Regulation of the expression of the SMP phenotype *in vivo*

As mentioned above, SMP is an early marker of the Schwann cell lineage *in vivo*, so we used the anti-SMP mAb in a second set of experiments to address the question of the role of the environment on the differentiation of the four glial cell types. We noted a discrepancy between the high number of precursors that have gliogenic potential (as revealed by our *in vitro* cloning experiments in cranial and trunk neural crest populations), and the distribution of SMP protein and mRNA *in vivo*: these are restricted to the two types of Schwann cells and are absent from satellite cells and enteric glia.

However, when E8 to E11 quail DRG were dissociated and cultured *in vitro*, all the glial cells switch from an SMP-negative to an SMP-positive phenotype (Cameron-Curry *et al.*, 1993). The phenotypic conversion occurs very rapidly: the first satellite cells become positive after only 4 hours of culture, when they are still in contact with the neuron with which they were associated *in vivo*. A negative signal seems to act *in vivo*, such that the competent satellite cells are not permitted to express SMP.

After partial enzymatic dissociation of E8 DRGs, doublets of cells, each composed of a neuron and a satellite cell, were cloned on 3T3 cells. The glial-derived clones obtained were almost exclusively composed of SMP-positive glial cells, demonstrating that all the satellite cells of a ganglion have the capacity to give rise to cells of the Schwann cell phenotype. Thus, an active repression of SMP expression is exerted *in vivo* in the particular environment of the ganglia. The mechanism responsible for this striking negative regulation, that does not concern the Schwann cells present in the same ganglia, is unknown. Phenotypic heterogeneity and high plasticity of avian PNS glial cells was also revealed by others with another set of mAbs (Rüdel and Rohrer, 1992).

The SMP protein and mRNA are also absent *in vivo* from the ENS, which represents a major peripheral compartment of neural crest origin. When enteric plexuses were dissected out from E8 to E15 quail bowel and cultured, the glial cells acquired the Schwann cell, SMP-positive phenotype in a few days (Dulac and Le Douarin, 1991). The cultured cells also produced laminin, an extracellular matrix component that is expressed by Schwann cells but not by enteric ganglia *in vivo*. The gut environment is responsible for the SMP-negative phenotype of the enteric glia *in vivo*. This was demonstrated by culturing fragments of chick embryonic gut organotypically, *in vitro* or on the chick chorioallantoic membrane, in association with quail neural crest from different embryonic levels. The crest cells invaded the gut tissues, but never became SMP-positive as they did in association with other embryonic control tissues such as skin and muscle. This is due to an active repression of the expression of this particular phenotype. In fact, when fragments of SMP-positive embryonic quail nerves were used, the glial cells lost the SMP marker.

In conclusion (Figure 4), peripheral glial cells show a high degree of phenotypic

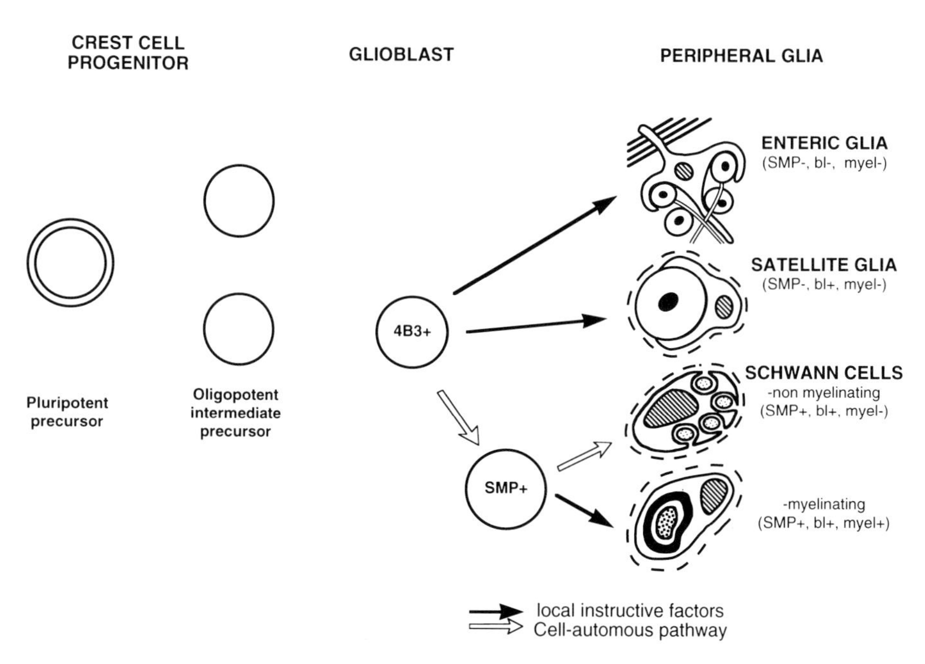

Figure 4 Model of gliogenesis from the avian neural crest cell. Developmental potential of precursor cells in the neural crest are progressively restricted. During crest cell migration and gangliogenesis glioblasts are specified and the glial lineage segregates from the neuronal and the other crest derived lineages. A fraction of the crest cells starts to express the 4B3 epitope, a glial phenotype that later in development characterize all peripheral glial cells. The final phenotype that peripheral glial cells show in the different peripheral compartments are specified by yet unknown local cues. Peripheral glial cells show a high degree of plasticity: positive and negative signals are necessary to maintain the different phenotypes. SMP = Schwann cell myelin protein; myel = myelin; bl = basal lamina.

plasticity, and in the avian system their constitutive differentiation pathway seems to be the non-myelinating, laminin-positive, SMP-positive Schwann cell phenotype. Positive and negative environmental interactions are necessary to stabilize the other phenotypes.

CONCLUSION

In vivo and *in vitro* studies described in this chapter have revealed that, from the beginning of its migration, the neural crest is a globally pluripotent, heterogeneous population of cells that contains progenitors endowed with different individual proliferation and developmental potentials. Highly pluripotent stem-like cells, that in the rat have been demonstrated to have self-renewal capacity, coexist with more restricted precursors and with cells already committed to specific lineages.

The proportion of oligopotent and committed cells increases as development progresses, by a mechanism that seems to be at least in part stochastic. However, several experimental data point to the importance, at all stages of development of the neural crest, and of local environmental signals in guiding the process of crest cell differentiation.

For example, the growth factor GGF2 has been suggested to exercise an instructive action on the glial cell lineage determination (Shah *et al.*, 1994). Moreover, we showed that the stability of the non-Schwann cell glial phenotype, as defined by the absence of SMP expression, depends upon continued environmental control (Figure 4).

REFERENCES

Abo, T. and Balch, C.M. (1981). *J. Immunol.* **127**:1024–1029.
Anderson, D.J. (1993). *Annu. Rev. Neurosci.* **16**:129–158.
Ayer-Le Lièvre, C.S. and Le Douarin, N.M. (1982). *Dev. Biol.* **94**:291–310.
Baroffio, A. and Blot, M. (1992). *J. Cell Sci.* **103**:581–587.
Baroffio, A., Dupin, E. and Le Douarin, N.M. (1988). *Proc. Natl Acad. Sci. USA* **85**:5325–5329.
Baroffio, A., Dupin, E. and Le Douarin, N.M. (1991). *Development* **112**:301–305.
Barrandon, Y. and Green, H. (1985). *Proc. Natl Acad. Sci. USA* **82**:5390–5394.
Bronner-Fraser, M. and Fraser, S.E. (1988). *Nature* **335**:161–164.
Bronner-Fraser, M. and Fraser, S.E. (1989). *Neuron* **3**:755–766.
Cameron-Curry, P., Dulac, C. and Le Douarin, N. (1989). *Development* **107**:825–833.
Cameron-Curry, P., Dulac, C. and Le Douarin, N.M. (1991). *Eur. J. Neurosci.* **3**:126–139.
Cameron-Curry, P., Dulac, C. and Le Douarin, N.M. (1993). *Eur. J. Neurosci.* **5**:594–604.
Cohen, A.M. and Konigsberg, I.R. (1975). *Dev. Biol.* **46**:262–280.
Collazo, A., Bronner-Fraser, M. and Fraser, S.E. (1993). *Development* **118**:363–376.
Couly, G.F. and Le Douarin, N.M. (1987). *Dev. Biol.* **120**:198–214.
Couly, G.F., Coltey, P.M. and Le Douarin, N.M. (1993). *Development* **117**:409–429.
D'Amico-Martel, A. and Noden, D.M. (1983). *Am. J. Anat.* **166**:445–468.
Duff, R.S., Langtimm, C.J., Richardson, M.K. and Sieber-Blum, M. (1991). *Dev. Biol.* **147**:451–459.
Dulac, C. and Le Douarin, N.M. (1991). *Proc. Natl Acad. Sci. USA* **88**:6358–6362.
Dulac, C., Cameron-Curry, P., Ziller, C. and Le Douarin, N.M. (1988). *Neuron* **1**:211–220.
Dulac, C., Tropak, M.B., Cameron-Curry, P., Rossier, J., Marshak, D.R., Roder, J. and Le Douarin, N.M. (1992). *Neuron* **8**:323–334.
Dupin, E. (1984). *Dev. Biol.* **105**:288–299.
Dupin, E., Baroffio, A., Dulac, C., Cameron-Curry, P. and Le Douarin, N.M. (1990). *Proc. Natl Acad. Sci. USA* **87**:1119–1123.

Epstein, M.L., Mikawa, T., Brown, A.M.C. and McFarlin, D.R. (1994). *Dev. Dyn.* **201**:236–244.
Fontaine-Pérus, J., Chanconie, M. and Le Douarin, N.M. (1982). *Cell Diff.* **11**:183–193.
Frank, E. and Sanes, J.R. (1991). *Development* **111**:895–908.
Gabella, G. (1971). *Z. Naturforsch. B.* **26**:244–245.
Gershon, M.D. and Rothman, T.P. (1991). *Glia* **4**:195–204.
Guillemot, F., Lo, L.C., Johnson, J.E., Auerbach, A., Anderson, D.J. and Joyner, A.L. (1993). *Cell* **75**:463–476.
Hess, A. (1965). *J. Cell Biol.* **25**:1–19.
Ito, K. and Sieber-Blum, M. (1993). *Dev. Biol.* **156**:191–200.
Jan, Y.N. and Jan, L.Y. (1990). *Trends Neurosci.* **13**:493–498.
Le Douarin, N. (1969). *Bull. Biol. Fr. Belg.* **103**:435–452.
Le Douarin, N. (1971). *C. R. Acad. Sci.* **272**:1402–1404.
Le Douarin, N.M. (1973). *Exp. Cell Res.* **77**:459–468.
Le Douarin, N.M. (1982). *The Neural Crest*, p. 259. Cambridge University Press, Cambridge, UK.
Le Douarin, N.M. (1986). *Science* **231**:1515–1522.
Le Douarin, N.M. and Teillet, M.-A. (1973). *J. Embryol. Exp. Morphol.* **30**:31–48.
Le Douarin, N.M. and Teillet, M.-A. (1974). *Dev. Biol.* **41**:162–184.
Le Douarin, N.M., Renaud, D., Teillet, M.-A. and Le Douarin, G.H. (1975). *Proc. Natl Acad. Sci. USA* **72**:728–732.
Le Lièvre, C.S. and Le Douarin, N.M. (1975). *J. Embryol. Exp. Morphol.* **34**:125–154.
Le Lièvre, C.S., Schweizer, G.G., Ziller, C.M. and Le Douarin, N.M. (1980). *Dev. Biol.* **77**:362–378.
Lemke, G. and Brockes, J.P. (1984). *J. Neurosci.* **4**:75–83.
Lo, L.C., Johnson, J.E., Wuenschell, C.W., Saito, T. and Anderson, D.J. (1991). *Genes Dev.* **5**:1524–1537.
Marchionni, M.A., Goodearl, A.D.J., Chen, M.S. *et al.* (1993). *Nature* **362**:312–318.
Mezei, C. (1993). In *Peripheral Neuropathy* (eds P. J. Dyck *et al.*), pp. 267–281. W. B. Saunders Company, Philadelphia.
Narayanan, C.H. and Narayanan, Y. (1980). *Anat. Rec.* **196**:71–82.
Newgreen, D.F., Jahnke, I., Allan, I.J. and Gibbins, I.L. (1980). *Cell Tiss. Res.* **208**:1–19.
Pannese, E. (1974). *Adv. Anat. Embryol. Cell. Biol.* **47**:7–97.
Peters, A., Palay, S.L. and Webster, H.D.F. (1976). *The Fine Structure of the Nervous System: the Neurons and Supporting Cells* (ed. A. Peters), p. 406. W. B. Saunders Company, Philadelphia.
Raible, D.W. and Eisen, J.S. (1994). Development **120**:495–503.
Richardson, M.K. and Sieber-Blum, M. (1993). *Dev. Biol.* **157**:348–358.
Rüdel, C. and Rohrer, H. (1992). *Development* **115**:519–526.
Schilling, T.F. and Kimmel, C.B. (1994). *Development* **120**:483–494.
Schweizer, G., Ayer-Le Lievre, C. and Le Douarin, N.M. (1983). *Cell Diff.* **13**:191–200.
Serbedzija, G.N., Burgan, S., Fraser, S.E. and Bronner-Fraser, M. (1991). *Development* **111**:857–866.
Sextier-Sainte-Claire Deville, F., Ziller, C. and Le Douarin, N.M. (1992). *Dev. Brain Res.* **66**:1–10.
Sextier-Sainte-Claire Deville, F., Ziller, C. and Le Douarin, N.M. (1994). *Dev. Biol.* **163**:141–151.
Shah, N.M., Marchionni, M.A., Isaacs, I., Stroobant, P. and Anderson, D.J. (1994). *Cell* **77**:349–360.
Sherman, L., Stocker, K.M., Morrison, R. and Ciment, G. (1993). *Development* **118**:1313–1326.
Sieber-Blum, M. (1989). *Science* **243**:1608–1611.
Sieber-Blum, M. (1991). *Neuron* **6**:949–955.
Sieber-Blum, M. and Cohen, A.M. (1980). *Dev. Biol.* **80**:96–106.
Smith, J., Cochard, P. and Le Douarin, N.M. (1977). *Cell Diff.* **6**:199–216.
Stemple, D.L. and Anderson, D.J. (1992). *Cell* **71**:973–985.
Stocker, K.M., Sherman, L., Rees, S. and Ciment, G. (1991). *Development* **111**:635–645.
Teillet, M.-A. and Le Douarin, N.M. (1983). *Dev. Biol.* **98**:192–211.
Teillet, M.-A., Kalcheim, C. and Le Douarin, N.M. (1987). *Dev. Biol.* **120**:329–347.
Tucker, G.C., Aoyama, H., Lipinski, M., Tursz, T. and Thiery, J.P. (1984). *Cell Diff.* **14**:223–230.
Uemura, T., Shepherds, S., Ackerman, L., Jan, L. and Jan, Y.N. (1989). *Cell* **58**:349–360.
Xue, Z.G., Smith, J. and Le Douarin, N.M. (1987). *Brain Res.* **431**:99–109.

6 Phenotypic diversity and lineage relationships in vascular endothelial cells

Ana M. Schor, Seth L. Schor and Enrique Arciniegas*

*Department of Dental Surgery and Periodontology, University of Dundee, Dundee, UK; *CRC Department of Medical Oncology, Christie Hospital NHS Trust, Manchester, UK*

INTRODUCTION

The circulatory system consists of the heart and an interconnected network of blood vessels which differ in size, structure and function; organ- and age-specific differences contribute to this vascular diversity (Wagner, 1980; Palade, 1988; Machovich, 1988; Simionescu and Simionescu, 1988). Venules and capillaries, the simplest type of vessels, range in diameter from 5 to 50 μm and are composed of the following three structural elements: (a) *the endothelial cells* lining the lumen, (b) *the pericytes* forming a periendothelial cellular network and (c) *the basement membrane*, a complex extracellular matrix encircling and supporting the endothelial cells and pericytes. The vascular cells are partly defined by their position with respect to the basement membrane, i.e. endothelial cells are attached to it on their abluminal side and pericytes entirely surrounded by it, except at points of contact between these cells and the endothelium. The composition of the basement membrane varies as a function of vessel type, age and pathological state. The larger and structurally more complex vessels contain additional concentric layers of cells and extracellular matrix. In these vessels, the role of the pericyte appears to be replaced by smooth muscle cells; however, pericytes or pericyte-like cells are also present in large vessels in a characteristic sub-endothelial location.

Blood vessels are not static structures. The vasculature undergoes continuous remodelling as part of its normal ageing and physiological function. The extensive plasticity of the vasculature is further indicated by the rapid regeneration of damaged endothelium and by the formation of new vessels (angiogenesis) which characterize various pathological processes, such as wound healing, tumour progression and inflammatory conditions. These processes involve the coordinated activity of endothelial and peri-endothelial cells. The microvasculature infiltrates most tissues and consequently achieves close anatomical proximity with a variety of other (tissue-specific) cell types. Interactions amongst the different microvascular cell populations and the surrounding

STEM CELLS
ISBN 0-12-563455-2

tissue cells play a major role in determining vascular structure and function. These cellular interactions are mediated by matrix macromolecules, cytokines and direct cell–cell contacts.

An in-depth understanding of vascular growth, regeneration and remodelling is dependent upon detailed information regarding phenotypic diversity and lineage relationships amongst vascular cells. Such information regarding the internal dynamics of vascular cell populations will also provide the framework in which to understand their interactions with surrounding tissue cells. The principal objective of this chapter is to address the specific question of phenotypic diversity and lineage relationships in vascular endothelial cells and the resultant implications for the existence of a putative endothelial stem cell population in the adult. This question has rarely been explicitly reviewed in the literature. The structural and functional heterogeneity of the vasculature, taken in conjunction with the lack of unambiguous histological, functional and molecular markers of (endothelial) stem cells, has contributed to our current poor state of knowledge. Another confounding factor is the inconsistency with which key terms such as *differentiation* have been used to describe the changes in gene expression which accompany vessel remodelling. We consequently begin by defining a number of such terms and then review the available experimental evidence which (often only indirectly) relates to the existence of distinct endothelial cell populations, including stem cells, and their complex dependency upon peri-endothelial cells. We conclude by summarizing our views relating to lineage relationships amongst vascular cells.

DEFINITIONS

Prior to discussing the evidence relating to the existence of endothelial cell stem cells and their lineage relationships, it is necessary to define a number of terms which are commonly used to describe alterations in cell phenotype and pattern of gene expression. Ambiguities in meaning have often resulted in the same term being used to describe quite distinct phenomenon. In order to avoid such difficulties and facilitate a critical review of the literature, we propose to use the following definitions.

The term **phenotype** refers to the characteristic morphology and functional activity of a cell population; the phenotype therefore reflects the specific pattern of gene expression under a particular set of environmental circumstances. **Differentiation** is the process whereby a cell changes phenotype in an irreversible and directional way, i.e. proceeding along a certain **cell lineage**. The differentiated cell may proliferate and/or undergo further differentiation until a phenotype incapable of further differentiation is eventually reached. This final process has been described by a number of terms, including **terminal differentiation**, and the resultant cell population referred to as **mature**, **terminally differentiated** or **end** cells. End cells may or may not be capable of proliferation. At the molecular level, we suggest that differentiation should be defined as the process by which certain genes become irreversibly incapable of expression; this is commonly accompanied by alterations in the pattern of gene expression which facilitate the identification of the differentiated phenotype. In this regard, it should be noted that all cells express only a proportion of the total number of genes they contain; the zygote and (in certain organisms) a small number of its immediate progeny, retain the potential

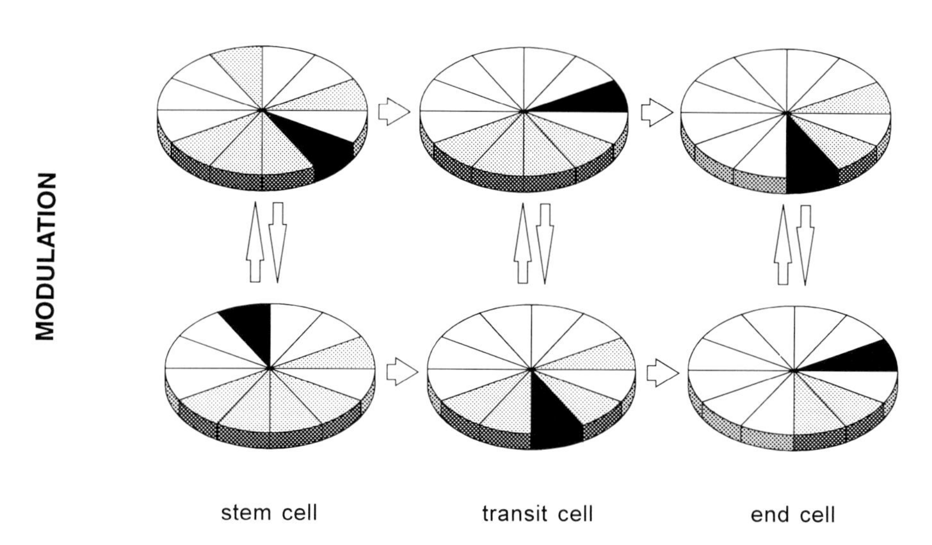

Figure 1 Scheme depicting distinction between differentiation and modulation. Cells move from left to right along a particular lineage by a process of differentiation. This is defined as an irreversible event in which there is a progressive reduction in the sets of genes which can be expressed. In contrast, modulation (depicted in the vertical direction) is a reversible process characterized by an alteration in the particular set of potentially expressible genes which are expressed. White, non-expressible genes; grey, potentially expressible genes; black, set of genes actually expressed.

of expressing the largest subset of these genes. Differentiation is characterized by the progressive reduction in the potential of gene expression. We suggest that reversible alterations in the pattern of potentially expressible genes (these defined by the state of differentiation) be referred to as **modulation** (Young, 1964). Accordingly, both modulation and differentiation lead to an alteration in the pattern of gene expression, the former being *reversible*, whilst the latter is *irreversible* and ultimately defines the phenotypic potential of the cell population (Figure 1).

Stem cells are defined by their capacity of unlimited self-renewal during the life span of the organism (Lajtha, 1979, 1983). Under steady state condition, stem cells usually represent a small population of slow-cycling cells; they are, however, capable of rapid turnover as required for their own self-renewal and maintenance of end cell numbers. Stem cells are already differentiated (the zygote being the only undifferentiated cell according to the above definition), but usually maintain the capacity to undergo further differentiation along one or several distinct lineages and, in this case, are referred to as **ancestral stem cells**. Differentiated progeny of the ancestral stem cell capable of undergoing further differentiation are referred to as **transit** or **transitional** cells. Transit cells are more abundant than stem cells (and consequently easier to identify) and have also been referred to by various other names, including **primitive**, **precursor** or **maturing** cells. As the life span of transit cells is limited, maintenance of their numbers is ultimately dependent upon replenishment from the stem cell population. Both stem and transit cells may be unipotential, bipotential or pluripotential and their differentiation may occur with or without cell proliferation.

Occasionally cells believed to belong to a certain lineage have been observed to undergo an abrupt alteration in phenotype to one characteristic of a different cell lineage. This process has been referred to as **transdifferentiation** (Beresford, 1990; Eguchi and Kodama, 1993). The distinction between transdifferentiation and differentiation is difficult to define at the molecular level; accordingly, transdifferentiation should be regarded as an operational definition and implies a digression from the 'expected' lineage pathway. As such, it may represent a deficiency in our knowledge of cell lineages, rather than a distinct process.

The above definitions can be divided into two broad categories: those which describe *processes* (such as differentiation and modulation) and those which describe the resultant cell populations (such as transit). Experimentally, different cell populations are identified on the basis of their characteristic phenotypes and subsequent developmental fates. Once distinct cell populations have been distinguished, additional information is required to identify the precise process or processes leading to their formation.

In the following sections we shall review the available data pertaining to the existence of distinct endothelial cell populations and the subsequent lineage relationships of their progeny. This information has been gleaned from studies concerned with (a) the development of the vasculature during embryogenesis, (b) endothelial phenotype under 'resting' or steady state conditions and (c) alterations in the resting phenotype which characterize endothelial regeneration and angiogenesis. Since blood vessels must be viewed as functional units consisting of several distinct cell types, this review will necessarily include discussion of peri-endothelial cell populations and their potential lineage relationships with endothelial cells.

DEVELOPMENT OF THE VASCULATURE

The development of the vasculature has been extensively studied. Endothelial and haemopoietic cells are derived from a common pluripotential precursor, the haemangioblast. Commitment of haemangioblast progeny to the endothelial cell lineage occurs early in embryogenesis, prior to gastrulation (von Kirschofer *et al.*, 1994). Initially, clusters of haemangioblasts form blood islands in the extraembryonic membranes (the yolk sac); subsequent differentiation of these cells gives rise to precursors of the haematopoietic lineage in the centre of the blood islands and endothelial precursors (angioblasts) in the periphery. Receptor tyrosine kinases, such as *tie-2* and the VEGF receptor *flk-1*, are expressed in the blood islands and in mature blood vessels; these molecules are therefore early markers of the endothelial cell lineage (Schnurch and Risau, 1993; Yamaguchi *et al.*, 1993). *Extra*embryonic blood vessels form in the yolk sac by the fusion of the blood islands; *intra*embryonic blood vessels derive *de novo*, both by the progressive association of single endothelial cells and their morphogenesis (vasculogenesis and intussusception), as well as from the sprouting or branching of already formed vessels (angiogenesis) (Pardanaud *et al.*, 1987; Noden, 1989; Kadokawa *et al.*, 1990; Burri, 1992).

Experiments involving the transplantation of quail embryo grafts or cell lines into chick embryos (or vice versa) have provided much information regarding the mechanism of vascular development and its control. The resulting chimeric embryos may be examined at different stages of development and cells derived from the quail or chick transplants unambiguously identified using specific antibodies. This powerful experimental approach has demonstrated that angioblasts are produced in most types of mesoderm, including those which are vascularized by angiogenesis. In contrast, both neural crest and neural tube tissue are devoid of angioblasts. During the course of normal development, the mesoderm-derived angioblasts migrate to all tissues, displaying a high degree of invasive behaviour. The emergence of structural and functional heterogeneity in the endothelium is not determined by the endothelial cells of the graft, but rather by the micro-environment provided by the mesenchyme of the host tissue. For example, quail mesenteric vessels growing into grafts of chick brain formed capillaries with structural and functional characteristics specific to the brain microvasculature, whereas brain capillaries growing into grafts of mesodermal tissue resembled mesenteric vessels (Stewart and Wiley, 1981). The identities of the tissue-specific signalling molecules are not known; however, in analogy with other developing systems, both soluble growth factors (e.g. bFGF) and matrix macromolecules (e.g. fibronectin and laminin) have been demonstrated to play an important role in the development and maturation of the blood vessels (Risau and Lemmon, 1988; Noden, 1989; Flamme and Risau, 1992). Endothelial cells within embryonic vessels have been reported to display heterogeneity with respect to expression of von Willebrand factor (Coffin *et al.*, 1991). Vessel regression and remodelling, as well as growth, occur during development. Transdifferentiation of endothelial cells into cells expressing smooth muscle cell or fibroblast phenotypic characteristics has been suggested to occur during programmed vessel regression (Latker *et al.*, 1986) and aortic remodelling (Arciniegas *et al.*, 1989). Embryonic stem cells have the capacity to differentiate in culture into embryo-like structures which contain angioblasts. Under the appropriate culture conditions, these angioblasts can differentiate into vascular structures (Doetschman *et al.*, 1993). In the context of the specific objectives of this review, it is important to note that the phenotypic plasticity of the angioblast continues to be manifested by the mature endothelium in the adult.

In contrast to the relative abundance of data relating to the endothelium, few studies have focused on the developmental origin of pericytes. Le Lièvre and Le Douarin (1975) noted that pericytes were of neural crest origin in all tissues examined (e.g. dermis, thymus, thyroid).

HETEROGENEITY IN ENDOTHELIAL CELL PHENOTYPE IN THE RESTING VASCULATURE OF THE ADULT

The resting phenotype is defined by a set of functional and morphological characteristics displayed by endothelial cells under steady state conditions. Endothelial cells and pericytes are commonly identified by their anatomical position. Endothelial cells form a flat, monolayer lining the vessel lumen. These cells display an apical–basal polarity, with the apical surface facing the lumen and the basal surface attached to the basement membrane. Pericytes are located within the basement membrane (Haudenschild, 1980; Simionescu and Simionescu, 1988). These anatomical definitions may, however, be somewhat ambiguous, as individual cells have been observed to span both locations

(Schoefl, 1963). Large vessels consist of several concentric layers of anatomically distinct tissue. For example, the intima layer of the aorta consists of the endothelium and a small number of intimal subendothelial cells, these variously referred to as subendothelial smooth muscle cells (Taura *et al.*, 1979; Collatz Christensen *et al.*, 1979) or pericytes (Haudenschild, 1980; Bostrom *et al.*, 1993). The intima is surrounded by a layer rich in smooth muscle cells, the media. Media and intima are separated by the internal elastic lamina. In large vessels, the endothelium may be in close contact with both pericytes and smooth muscle cells, as the latter may send extensions through the internal elastic lamina (Haudenschild, 1980). The ratio of endothelial cells to pericytes depends on vessel type and generally increases with age (Engerman *et al.*, 1967; Bär and Wolff, 1972; Sims, 1991). Morphometric studies of skeletal muscle vasculature have indicated that pericytes are randomly positioned in capillaries and concentrated at endothelial cell junctions in venules (Sims *et al.*, 1994). Certain vessels contain cells which have been described as intermediate between pericytes and smooth muscle cells (Sims, 1991; Shepro and Morel, 1993; Nehls *et al.*, 1992).

In addition to anatomical location, endothelial cells are also defined by certain common functional activities, such as providing a non-thrombogenic vessel lining, interacting with leukocytes, and maintaining vessel tone and permeability (Haudenschild, 1980; Machovich, 1988; Pearson, 1991; Reinhart, 1994). Endothelial cells which fulfil these common functions are none the less heterogeneous with respect to a number of morphological, biosynthetic and immunological criteria (Gerritsen, 1987; Belloni and Nicolson, 1988; Fajardo, 1989; McCarthy *et al.*, 1991; Augustin *et al.*, 1994).

Endothelial heterogeneity is dependent upon age, organ and vessel size. Using morphological criteria, three main types of endothelium have been described: continuous (e.g. large vessel, neural and lung microvessels), fenestrated (e.g. kidney and endocrine gland microvessels) and discontinuous (e.g. liver, bone marrow and spleen sinusoids) (Simionescu and Simionescu, 1988). Evidence of endothelial heterogeneity has also been provided by diverse patterns of antibody and lectin binding. For example, Page *et al.* (1992) reported that heart capillary endothelial cells expressed both class I and II major histocompatibility complex molecules (MHC). In contrast, class II MHC antigens were absent from various large vessels examined, but present in the coronary artery. Factor VIII-related antigen (von Willebrand factor, vWF) is commonly used as a marker of large vessel and microvessel endothelial cells; several reports, however, have indicated that vWF is not uniformly present in all capillary endothelial cells (Yamamoto *et al.*, 1988; Page *et al.*, 1992) and conflicting observations have been published regarding its presence in other tissues, such as lymph nodes and liver sinusoids (McCarthy *et al.*, 1991; Page *et al.*, 1992; Fajardo, 1989). Capillary endothelial cells have also been reported to share antigens with a peripheral blood monocyte/macrophage subset; these antigens are either weakly expressed or absent in larger vessels (Yamamoto *et al.*, 1988; Page *et al.*, 1992). In contrast, antibodies that recognize other endothelial markers (EN4 and CD31) have been found to stain all endothelial cells within all vessels examined, regardless of size or organ (McCarthy *et al.*, 1991; Page *et al.*, 1992).

As is the case during development, at least some of the observed heterogeneity appears to result from signals coming from the micro-environment (Schor and Schor, 1983). Thus, Hirano and Zimmerman (1972) reported that the blood vessels present in renal carcinoma metastases to the brain displayed the typical fenestrated morphology of renal vessels, and not the continuous morphology of the invading brain vessels.

Several processes may contribute to the observed diversity in endothelial cell phenotype. For example, heterogeneity could result from the co-existence of cells at various stages of differentiation along a common endothelial cell lineage and/or the modulation of phenotype in cells at a given stage of differentiation. Alternatively, it is possible that functional endothelial cells arise from several distinct lineages and that these exhibit vessel- and organ-specific differences in distribution. We view this latter possibility as a less likely alternative, since endothelial phenotypes are labile and interchangeable (e.g. capillaries become arterioles, large vessels give rise to capillaries).

The pattern of intermediate filament molecule expression is commonly used to assign cells to a particular lineage (Osborn and Weber, 1986; Sappino *et al.*, 1990). However, cells classified as endothelial have been reported to express a range of such molecules. Fujimoto and Singer (1986) observed that capillary and vein endothelial cells in adult chick tissues could express only vimentin, only desmin or both desmin and vimentin. In contrast, large vessel endothelial cells expressed only vimentin. These and related observations on cultured endothelial cells (see below) suggest that: (a) endothelial cells may arise from distinct precursor cell lineages; (b) different intermediate filament molecules may be expressed by cells in the same lineage; or (c) cells classified as 'endothelial' on the basis of their anatomical location may in fact be peri-endothelial cells transiently lining the lumen.

In addition to the extensive documentation of vessel- and organ-specific heterogeneity in endothelial phenotype, a more limited number of observations have indicated the existence of intra-vessel heterogeneity. For example, endothelial cells in large vessels expressed a patchy distribution of class I MHC (Page *et al.*, 1992). Similarly, only a subset of capillary endothelial cells in an atherosclerotic plaque were observed to express the B-chain of PDGF (Wilcox *et al.*, 1988). The A-chain of PDGF was present in some mesenchymal-appearing intimal cells (pericytes?) and, occasionally, on luminal (endothelial?) cells (Wilcox *et al.*, 1988). In bovine corpus luteum, IGF-1 was demonstrated in only a limited number of endothelial cells (Amselgruber *et al.*, 1994). Such heterogeneity in endothelial cell phenotype within a given vessel is not likely to be due to the influence of the micro-environment, although it is possible that local factors (e.g. contact with pericytes) may still play a decisive role.

Consideration of phenotype heterogeneity of the endothelium has generated a diversity of opinion regarding the classification of the cells themselves (e.g. stem cells, differentiated cell) and the nature of the processes responsible for their formation. In this regard, care must be taken to distinguish between processes operative at the cellular and vessel levels. For example, newly formed capillaries may become arterioles or venules (Clark and Clark, 1939; Bär and Wolff, 1972; Simionescu and Simionescu, 1988). In skeletal muscle, pre-existing capillaries can similarly undergo arterialization by the addition of a smooth muscle layer. The transition from capillary to arteriole or venule is normally referred to as 'differentiation', although several processes may be operative at the cellular level: e.g. the migration of smooth muscle cells from terminal arterioles or their differentiation from a precursor cell (Nehls *et al.*, 1992; Price *et al.*, 1994) and the endothelial cell population may not undergo any type of differentiation.

Nehls and Drenckhahn (1991) reported that the pericytes in midcapillaries expressed desmin, but not α-smooth muscle (α-sm) actin; in contrast, pericytes in venules and arterioles were positive for α-sm actin. Other studies have similarly demonstrated heterogeneity amongst pericytes with respect to staining with antibodies to α-sm actin, desmin and a high molecular weight melanoma associated antigen (Nehls *et al.*, 1992; Schlingermann *et al.*, 1991). As is the case with endothelial cells, these cells are classified as pericytes primarily on the basis of their anatomical location and consequently may comprise cells of different lineages (Schor *et al.*, 1990). It is possible, for example, that endothelial cells may be translocated to a subendothelial position. Such translocation could explain the reported presence of Weibel–Palade bodies (an endothelial marker) in pericytes (Zelickson, 1966).

Stem and transit cells are commonly characterized by their ability to proliferate, whereas end cells (as defined by their lack of potential to differentiate further) may or may not proliferate. The resting endothelium is appropriately named regarding its labelling index (LI), which has been estimated to vary between 0% and 2.4%, usually falling below 1% (Tannock and Hayashi, 1972; Fishman *et al.*, 1975; Hobson and Denekamp, 1984). Endothelial labelling indices may display considerable inter-organ variation; the labelling index of pericytes has been reported to be lower than that of the endothelium in the same vessel (Engerman *et al.*, 1967). The labelled cells are generally homogeneously distributed throughout the resting vessel, although focal concentrations of labelled endothelial cells are occasionally seen (Fishman *et al.*, 1975; Schwartz *et al.*, 1990); these are believed to result from local damage to the endothelium. Taken together, these observations indicate that cells capable of proliferation (be they stem, transit or end cells) are located randomly throughout the flat endothelial layer.

In conclusion, it is important to note that the term 'resting' endothelial cell phenotype is commonly used in contradistinction to the phenotypes displayed by 'activated' or 'angiogenic' endothelial cells *in vivo* and *in vitro* (see following sections). In this regard, a true 'resting' phenotype may not exist, as the endothelial cells are continuously responding to external stimuli as part of their functionality.

HETEROGENEITY IN THE RESTING ENDOTHELIAL CELL PHENOTYPE *IN VITRO*

Endothelial cells from various sources have been isolated and maintained in culture. The plating efficiency of freshly isolated endothelial cells in primary culture is high (70–80%) (Antonov *et al.*, 1986). Endothelial cells cultured on two-dimensional substrata generally form a quiescent monolayer of closely apposed cells displaying a clearly defined apical–basal polarity; the cells may form gap junctions, tight junctions and fenestrations similar to those observed *in vivo*. These cells are metabolically active, provide a non-thrombogenic surface, express specific endothelial markers and continue to synthesize effectors of coagulation, fibrinolysis and vessel tone (Huttner and Gabbiani, 1982; Schor *et al.*, 1984; Antonov *et al.*, 1986; Schor and Schor, 1986; Pearson, 1991; Gerritsen, 1987; Belloni and Nicolson, 1988; Milici *et al.*, 1985).

Cultured endothelial cells continue to display phenotypic diversity comparable to their *in vivo* counterparts and exhibit reversible alterations in phenotype in response to culture conditions (Figure 2) (Schor *et al.*, 1983; Schor and Schor, 1986; Canfield *et al.*, 1986, 1990; Canfield and Schor, 1995). In spite of the continued expression of certain common characteristics, these cells also exhibit significant heterogeneity in phenotype with respect to such fundamental aspects of cell behaviour as morphology, production of and response to cytokines and matrix macromolecules, and ability to undergo tubular morphogenesis (Gerritsen, 1987; Fajardo, 1989; Canfield *et al.*, 1992; Canfield and Schor, 1994; Augustin *et al.*, 1994). Such variations in phenotype have been observed amongst endothelial cell lines derived from different tissues and vessel types, as well as from donors of different ages. A similar diversity in phenotype has also been observed amongst endothelial lines derived from the same vessel. For example, primary endothelial cell cultures contained sub-populations of cells which were multinucleate (Antonov *et al.*, 1986) or expressed certain cytokeratins (Spanel-Borowski *et al.*, 1994), as well as vWF. This type of 'clonal' or intra-vessel heterogeneity may also contribute to the different characteristics displayed by endothelial cells derived from the same vessel of different individuals. For example, cloned and uncloned lines of bovine endothelial cells derived from different aortas have been reported to synthesize distinct mixtures of collagenous molecules; such differences remained constant upon repeated subculture (Schor *et al.*, 1984; Canfield *et al.*, 1992).

Heterogeneity in endothelial cell phenotype *in vitro* has been extensively reviewed (Gerritsen, 1987; Fajardo, 1989; Pearson, 1991; McCarthy *et al.*, 1991; Canfield *et al.*, 1992; Canfield and Schor, 1994; Augustin *et al.*, 1994). Several examples will serve to illustrate the breadth of such phenotypic diversity and the factors which influence their expression. Cells are normally cultured on a plastic tissue culture substratum, which may be coated with a variety of matrix macromolecules. Endothelial cells further modify the substratum by synthesizing and depositing a complex extracellular matrix. The nature of the substratum used for tissue culture has a decisive effect on the phenotype expressed by the cells, including the composition of the extracellular matrix that they synthesize and their response to cytokines (Schor *et al.*, 1984; Canfield *et al.*, 1986, 1990; Sutton *et al.*, 1991; Canfield and Schor, 1995). Several studies have demonstrated that the expression of organ-specific characteristics by endothelial cells is determined by stromal components, such as peri-endothelial cells and extracellular matrix. Thus, aortic endothelial cells were induced to express the lung capillary-specific adhesion molecule Lu-ECAM-1 when cultured on lung-derived extracellular matrices (Augustin *et al.*, 1994). Similarly, adrenal cortex endothelial cells lost their fenestrations when cultured on plastic but reformed them when cultured on a matrix deposited by renal epithelial cells (Milici *et al.*, 1985). The co-culture of liver endothelial cells with hepatocytes (Modis and Martinez-Hernandez, 1991) or brain endothelial cells with glial cells (De Bault and Cancilla, 1980) also induced the expression of the corresponding organ-specific endothelial phenotype. Porcine aortic endothelial cells retained their ability to produce angiotensin converting enzyme when cultured in homologous serum, but not in bovine serum (Chesterman *et al.*, 1983).

We have reported that large vessel (aortic) and microvessel (retinal, brain-derived) endothelial cells differ in their proliferative response to the low molecular weight angiogenic factor ESAF (endothelial cell-stimulating angiogenesis factor): when cultured on malleable three-dimensional gels of type I collagen, the proliferation of retinal endothelial cells was stimulated by ESAF, whereas aortic endothelial cells were unaffected. In contrast, neither type of endothelial cells responded to ESAF when cultured on gelatin-coated plastic tissue culture dishes. Pericytes, however, responded to ESAF irrespective of the substratum (Schor *et al.*, 1980; Keegan *et al.*, 1982; and Schor *et al.*, unpublished results). These results indicate a clear difference between large vessel and microvessel endothelial cells with respect to their response to a particular angiogenic factor and further indicate that the nature of the substratum exerts a profound regulatory effect upon this responsiveness. It should be noted, however, that aortic endothelial cells may display responsiveness to ESAF if cultured on the appropriate (as yet unknown) substratum or in the presence of other factors.

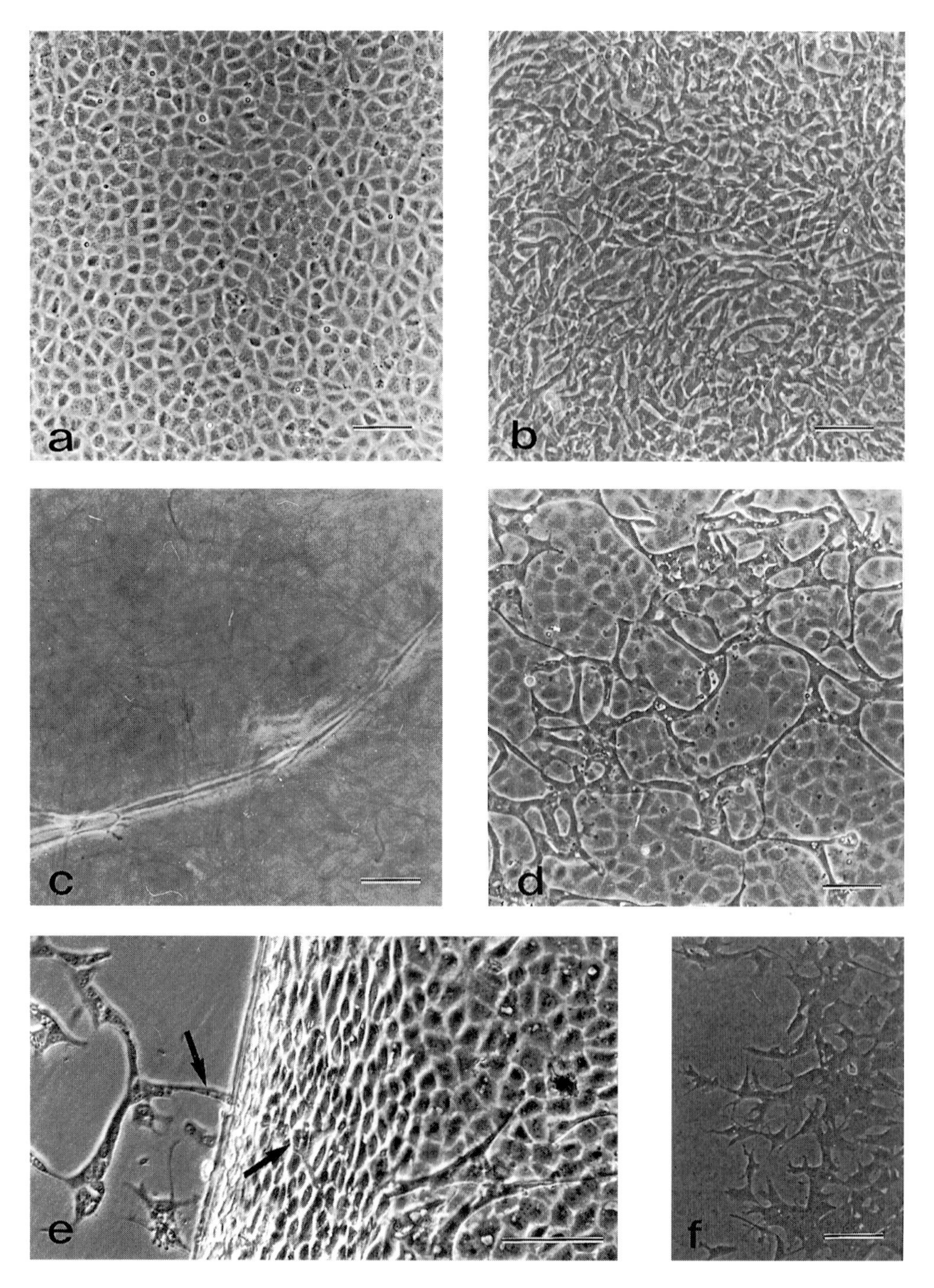

Figure 2 Modulation of endothelial cell morphology by culture conditions. Cloned bovine aortic endothelial cells were maintained in medium containing 20% calf serum and ascorbic acid, unless otherwise stated (Schor *et al.*, 1983). (a) Cobblestone monolayer at confluence on gelatin-coated dishes. (b) Disorganized, overlapping, multilayer formed when cells were

In addition to the substratum, various other environmental factors have been shown to play an important role in modulating endothelial cell phenotype *in vitro*; these include soluble factors (e.g. cytokines, hormones, vitamins, enzymes, coagulation factors), physical factors (e.g. pressure, hydrodynamic flow, malleability of the substratum), the composition of the atmosphere (e.g. oxygen tension), extracellular matrix components, and cell–cell social interactions. These various components of the microenvironment interact with one another and affect the cell in a mutually interdependent fashion; the cells respond to the micro-environment and modify it, thus creating a complex network of iterative regulatory interactions. Virtually all aspects of endothelial phenotype may be modulated by the micro-environment, including the expression of endothelial cell markers, such as vWF (Paleolog *et al.*, 1990) and smooth muscle/pericyte markers, such as α-sm actin (Madri *et al.*, 1989). The complex nature of these cell–environment interactions has been the subject of several recent reviews (Nathan and Sporn, 1991; Aggarwal and Gutterman, 1992; Adams and Watt, 1993; Davies and Tripathi, 1993; Reinhart, 1994; Schor, 1994). It is important to note that many of these micro-environmental factors (e.g. cytokines, matrix macromolecules) are commonly used in cell culture regimes, whereas others are usually omitted (e.g. hydrodynamic flow); as a consequence, it is unlikely that all of the characteristics of the resting endothelial phenotype *in vivo* will be displayed by 'resting' endothelial cells *in vitro*. As a result, certain of the differences between endothelial cells *in vivo* and *in vitro* attributed to cell differentiation (Shima *et al.*, 1995) may be due to modulation of phenotype by the micro-environment; further studies are required to clarify this distinction.

Consideration of the above examples clearly indicates that endothelial heterogeneity results from a combination of intrinsic differences between cells of diverse origin (since these differences are manifested when the cells are maintained under identical conditions and persist following repeated subculture) (Belloni and Nicolson, 1988) and the influence of the culture micro-environment.

Various terminologies have been used to define endothelial cells displaying distinct phenotypes in culture (e.g. resting, activated, differentiated, dedifferentiated) and the processes responsible for their production (e.g. differentiation, dedifferentiation, modulation). Since the majority of alterations in phenotype induced by micro-environmental factors *in vitro* are reversible, they are best referred to as modulation, according to the definitions discussed above. Indeed modulation of endothelial phenotype by the micro-environment may be responsible for many of the differences between endothelial cultures and the endothelium *in vivo*, as well as organ-specific and vessel-specific examples of endothelial heterogeneity *in vitro*. On the other hand, intra-vessel 'clonal' heterogeneity of cells cultured under identical conditions is difficult to understand in terms of micro-environmental modulation. Even in this case, however, Longenecker *et al.* (1983) demonstrated that the micro-environment may still play an important role in defining cell phenotype; they found that cloned populations of aortic endothelial cells displayed

plated at low density and grown in the absence of ascorbic acid. (c) Sprouting cells embedded within a three-dimensional collagen gel. (d) The formation of a sprouting cell network below monolayer in a postconfluent culture. (e) Removal of the cobblestone monolayer reveals the sprouting cells underneath (arrows), which will revert to a cobblestone morphology upon subsequent culture. (f) Stellate morphology obtained by overlaying cells attached to gelatinized dishes with a collagen gel. Bar = 150 μm.

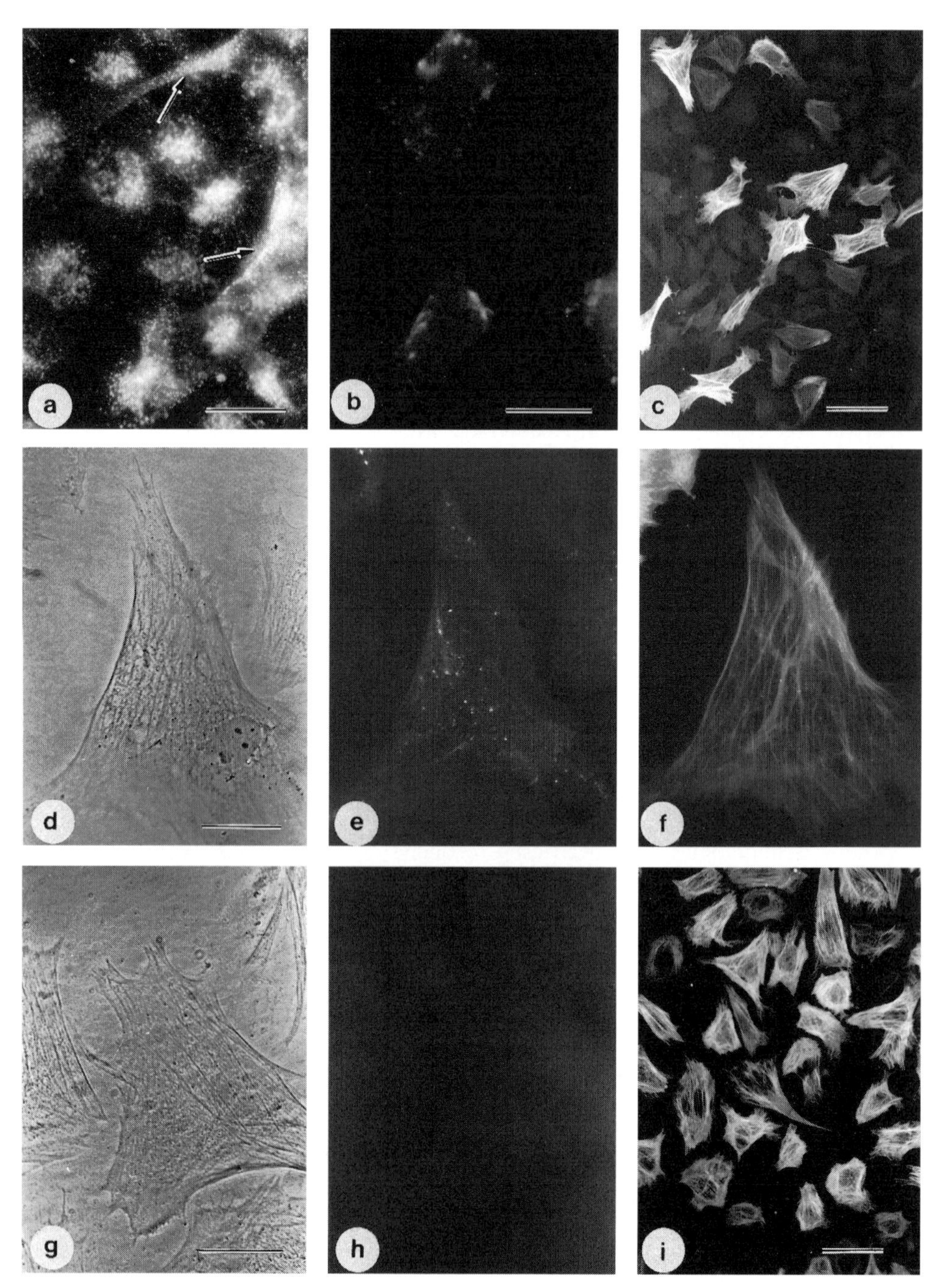

Figure 3 Endothelial cell differentiation induced by TGFβ-1: expression of α-smooth muscle actin and loss of von Willebrand factor. Cloned bovine aortic endothelial cells were plated at low density and maintained in the absence (controls) or the presence of 1 ng ml^{-1} TGFβ-1 for up to 10 days (Arciniegas *et al.*, 1992). (a) Control postconfluent culture containing cobblestone and sprouting cells (arrows) stained for von Willebrand factor (vWF). All cells were positive for vWF and negative for α-smooth muscle actin (α-sm actin) in control

variability in their basal proliferative rate and response to FGF when cultured on plastic tissue culture dishes, but this clonal variability was not manifest when the cells were maintained on an extracellular matrix deposited by vascular smooth muscle cells.

With rare exceptions (Canfield *et al.*, 1986), modulation of endothelial cell phenotype usually involves quantitative, rather than qualitative, alterations in gene expression. Few studies have unambiguously demonstrated the irreversible nature of alterations in endothelial cell phenotype *in vitro*. In this case, the irreversible nature of these alterations justifies the use of the term 'differentiation'. For example, Arciniegas *et al.* (1992) demonstrated that bovine aortic endothelial cells *in vitro* can be induced to differentiate into non-proliferating cells resembling smooth muscle cells or pericytes in terms of their expression of α-sm actin and muscle myosin, and lack of vWF (Figures 3 and 4). The expression of α-sm-actin is characteristic of smooth muscle cells and myofibroblasts. Some fibroblasts also express this actin isoform, whereas endothelial cells (with the exception of rat epididymal fat pad-derived cells: Madri *et al.*, 1989) do not. Bovine aortic endothelial cells in culture express vWF, lack α-sm actin and display a typical cobblestone morphology when confluent (Figure 3a). When exposed to TGFβ-1, the cells acquire a spread morphology, they gradually lose vWF (Figure 3b) and begin to synthesize α-sm actin (Figure 3c). After 5 days' exposure, transitional cells which synthesize both can be seen (Figures 3d, 3e and 3f); after 20 days, over 95% of the cells lack vWF and synthesize α-sm actin (Figures 3g, 3h and 3i) as well as smooth muscle myosin (Figure 4). These changes have been observed with both cloned and uncloned cells and are irreversible. Although expressing these smooth muscle- and pericyte-like characteristics, the TGFβ-treated endothelial cells differ from smooth muscle cells and pericytes in other aspects; they have therefore been described as differentiated endothelial cells. Interestingly, these cells maintain the ability to form tubular structures (Figure 5). Cells of similar characteristics have been observed lining the lumen of large blood vessels *in vivo* (Arciniegas *et al.*, 1992).

Lipton *et al.* (1991, 1992) found that human dermal microvascular endothelial cells irreversibly acquired certain fibroblastic characteristics upon exposure to histamine and withdrawal of cAMP; these cells lost their Weibel–Palade bodies, ceased expressing vWF, and underwent a patterned reorganization of the vimentin cytoskeleton. These changes were described as a transdifferentiation from an endothelial to a mesenchymal phenotype and occurred without cell proliferation; transitional cells were also observed. Cells similar to the latter are observed during inflammation response *in vivo* (Lipton *et al.*, 1991, 1992).

The phenotype expressed by pericytes and smooth muscle cells is also modulated by culture conditions, including the extracellular matrix (Wren *et al.*, 1986; Canfield *et al.*, 1990; Newcomb and Herman, 1993). Although clearly distinct from endothelial and smooth muscle cells (Schor *et al.*, 1992), pericytes share many phenotypic characteristics

---

cultures, irrespective of their density or time in culture. (b)–(f) Cells incubated with TGFβ-1 for 5 days. (b) vWF is expressed weakly in approximately 40% of the cells; (c) α-sm actin is expressed by 20–60% of the cells; (d)–(f) cells showed a ragged, stellate morphology. Double labelling for vWF (e) and α-sm actin (f) of the same cell shown in phase contrast in d. (g)–(i) Cells incubated with TGFβ-1 for 10 days. (g)–(h) Phase contrast (g) and immunofluorescence (h) micrographs of the same field showing stellate morphology (g) and lack of staining with vWF antibody (h); (i) 80–90% of the cells now express α-sm actin. Bar = 150 μm in (c) and (i); 50 μm in the rest.

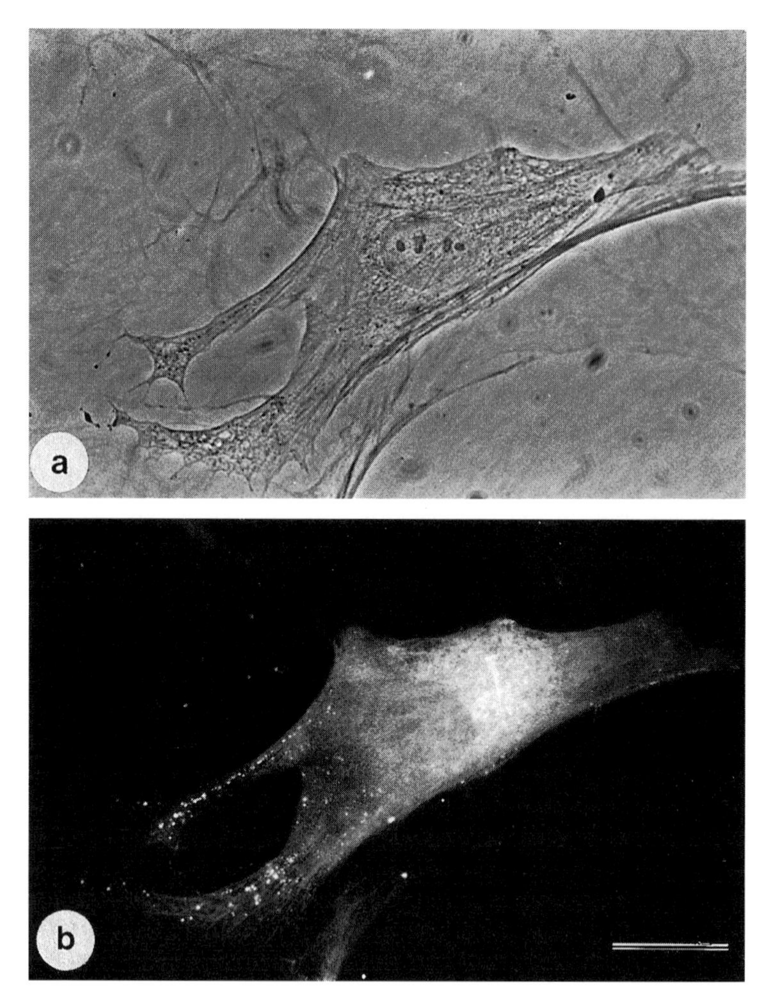

Figure 4 Endothelial cell differentiation induced by TGF-β: expression of muscle myosin. Phase (a) and immunofluorescence (b) micrographs showing the presence of smooth muscle myosin by endothelial cells that had been incubated with TGFβ-1 for 25 days. Control cultures (not shown) did not express this protein. Bar = 50 μm.

with both. In common with endothelial cells, pericytes are able to synthesize various basement membrane components (Canfield *et al.*, 1990; Schor *et al.*, 1991, 1992) and may adopt an endothelial-like morphology under appropriate culture conditions (Figure 6). These alterations in phenotype are reversible and should therefore be considered to result from a modulation in gene expression. In contrast, differentiation (or transdifferentiation) of pericytes into osteogenic cells has been demonstrated both *in vivo* (Diaz-Flores *et al.*, 1992) and *in vitro* (Figure 7) (Schor *et al.*, 1990, 1991, 1992, 1995). Pericytes may also differentiate (or transdifferentiate) into other cell types, including smooth muscle cells (reviewed in Schor *et al.*, 1990; Sims, 1991).

THE RESPONSE OF ENDOTHELIAL CELLS TO INJURY

Vascular damage elicits a number of cellular responses which may provide information regarding lineage relationships amongst vascular cells. For example, damage due to hypertension results in vessel remodelling. Herman and Jacobson (1988) reported an increase in the number of pericytes present in brain capillaries of hypertensive rats. Similarly, hyperplastic capillaries containing multilayered endothelial cells have been observed in venous stasis leg ulcers and in chronically rejected kidney transplants. These endothelial cells have been postulated to become perivascular fibroblasts (Beranek, 1989).

Thyroid hyperplasia is accompanied by vasodilation, and hence a decrease in the number of endothelial cells per unit area of luminal surface. During the first 2 days following thyroid stimulation, vasodilation occurs without cell division; following this initial period, vasodilation correlates with the LI of the endothelium (Smeds and Wollman, 1983). In all vessels examined, a peak in the LI (5–20%) was observed between 2 and 5 days after stimulation, depending on type of vessel and area within the vessel. The LI decreased to near baseline levels afterwards, remaining higher in capillaries than in veins and arteries. Labelling of pericytes and smooth muscle cells was also observed at 36 hours and later throughout the experimental period. By comparison to the number of labelled endothelial cells, the number of labelled peri-endothelial cells was lower and remained constant.

Denudation of the endothelial layer of large vessels by various types of mechanical means activates mechanisms which result in re-endothelialization of the damaged area. The specific nature of the induced response depends on the extent of damage. A small denuded area may be regenerated by endothelial cell migration alone. More extensive denudation elicits a migratory and proliferative response from the endothelial cells, as well as from the peri-endothelial cells of the intima and media. Various studies have suggested that the endothelium may be regenerated both by endothelial cells at the periphery of the wound, as well as by precursor cells in the media variously described as 'undifferentiated', 'smooth muscle-like' and 'fibroblastoid' (Schwartz *et al.*, 1990; Ferns *et al.*, 1992). Fishman *et al.* (1975) found that between 2 and 4 days after injury, sheets of new endothelium spread out from the edges of the denuded area until confluence was achieved. The pattern of tritiated thymidine labelling mirrored the movement of the endothelial sheets, suggesting that the new endothelium was derived from endothelial cells at the periphery of the denuded area. However, these authors also observed a striking thickening of the intimal layer between the endothelium and the internal elastic lamina due to the appearance of smooth muscle-like cells in the deeper zones and a layer of 'undifferentiated' cells closer to the lumen. Undifferentiated cells labelled with tritiated thymidine were also observed in the media. In a similar study, Taura *et al.* (1979) observed that the frequency of mitotic luminal cells was less than 1% on day 3 after injury and reached a maximum of 3% on day 7; in contrast, the frequency of mitotic medial cells reached a peak of 3.6% on day 3. On the basis of these observations, the authors suggested that the damaged intimal cells were initially replaced by the proliferation of medial cells which migrated into the intima, where they continued to proliferate and eventually formed a pseudoendothelium. The formation of a pseudoendothelium has been described by several authors. Although providing a non-thrombogenic surface, the pseudoendothelium does not appear to become a fully functional endothelium (Haudenschild, 1980; Ferns *et al.*, 1992). It has been suggested

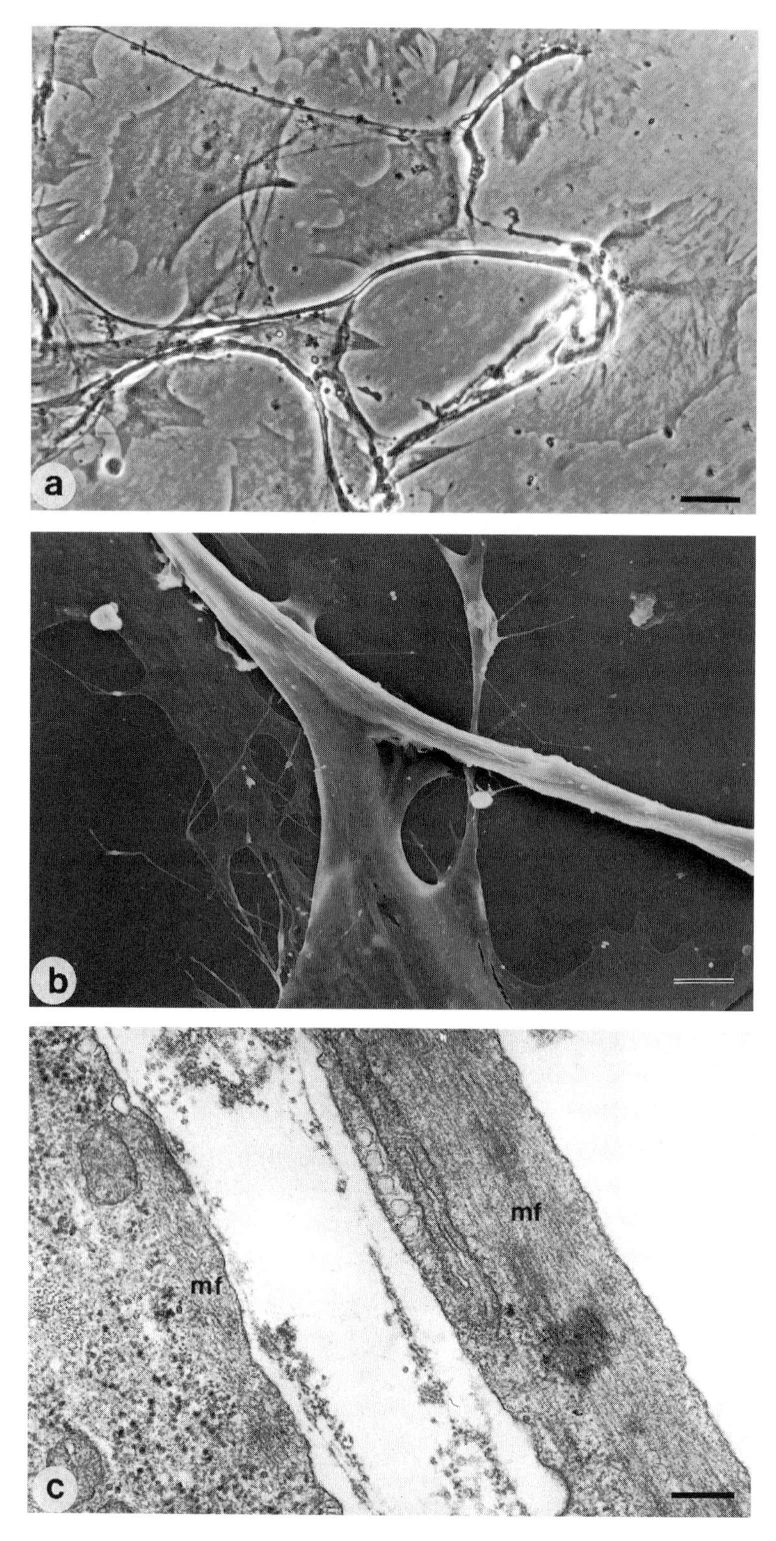
a
b
mf
mf
c

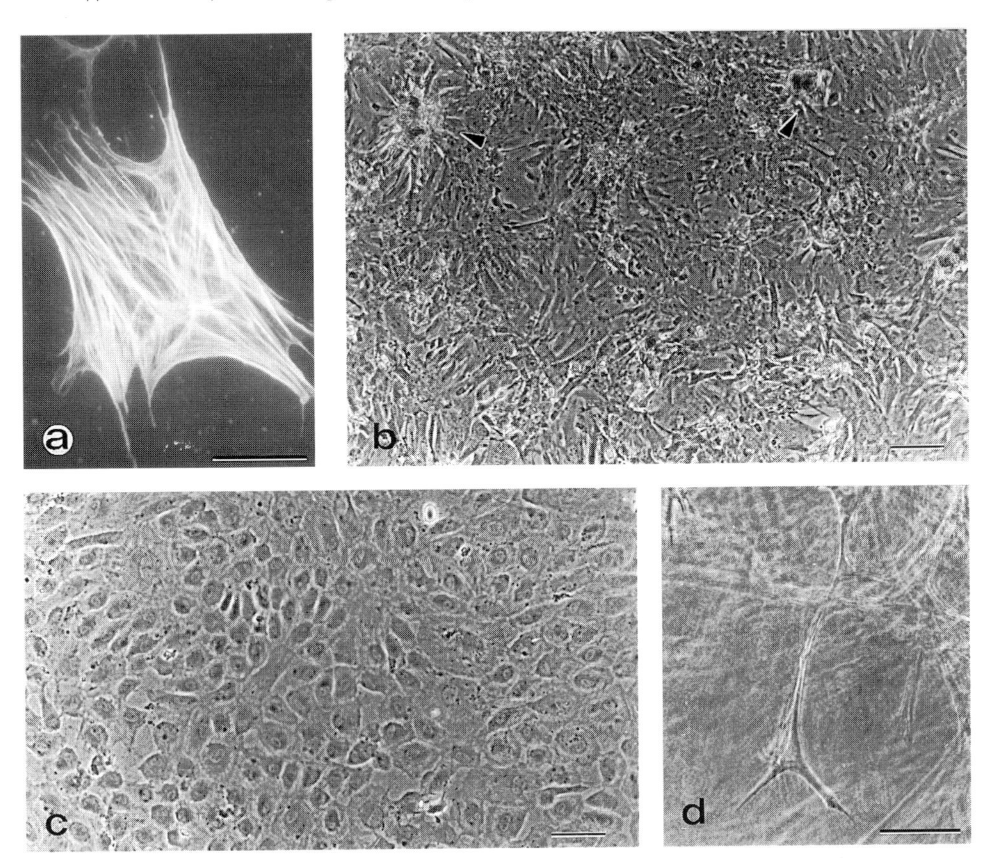

Figure 6 Modulation of pericyte morphology by culture conditions. Cloned pericytes were cultured in medium containing 20% calf serum and ascorbic acid (Schor *et al.*, 1990). (a) Stellate morphology displayed by sparse cells on plastic, glass or gelatinized substrata, showing staining with α-smooth muscle actin antibody. (b) Postconfluent culture on plastic substratum, showing multilayering and the formation of nodules (arrowheads). (c) Cobblestone monolayer formed when the cells were grown on an extracellular matrix deposited by retinal endothelial cells. (d) Sprouting morphology displayed when the cells are embedded within a three-dimensional matrix of fibrin or collagen. Bar = 50 μm in (a); 150 μm in (b), (c) and (d).

Figure 5 Formation of tubular structures by endothelial cells following their (TGFβ-1 induced) differentiation. Aortic endothelial cells were incubated with TGFβ-1 for 20 days. At this point the cells had lost their typical endothelial characteristics and expressed α-smooth muscle actin and smooth muscle myosin. This differentiated phenotype was maintained following removal of TGFβ-1. The cells formed tubular structures 5–10 days later, both in the presence and in the absence of TGFβ-1. (a) Phase micrograph showing stellate morphology of the cells attached to the substratum and tubular structures connecting and in between the cells. Bar = 50 μm. (b) Scanning electron micrograph showing the connection between a tubule and a stellate cell. Bar = 10 μm. (c) Transmission electron micrograph of a longitudinal section of a tubule showing a luminal space and the presence of microfilaments (mf) and abundant cytoplasmic organelles. Bar = 0.17 μm.

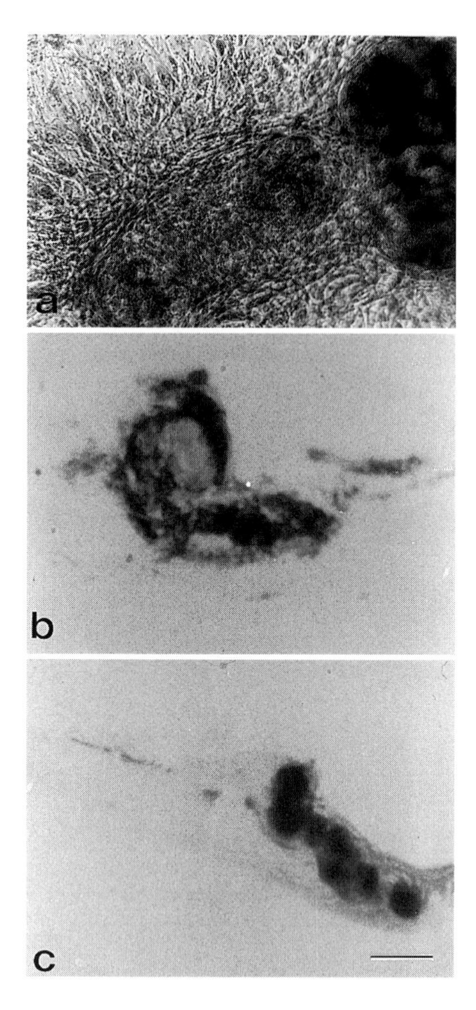

Figure 7 Differentiation of pericytes in culture along the osteogenic pathway. Cloned pericytes were maintained in culture for 30 days in the presence of medium containing 20% calf serum and ascorbic acid. At this point, the culture contained multicellular nodules rich in extracellular matrix in which hydroxyapatite crystals had deposited (Schor *et al.*, 1990). (a) On the tissue culture dish, the nodules and multilayered areas stained with alkaline phosphatase, whereas the surrounding monolayer was negative. (b), (c) Sections cut through a pericyte nodule showed the presence of phosphate and calcium, detected by staining with von Kossa (b) and Alizarin Red (c). Bar = 150 μm.

that the authentic endothelial lining is subsequently re-formed by the displacement of the pseudoendothelium by the ingrowth of endothelial cells from the periphery of the lesion (Collatz Christensen *et al.*, 1979).

Re-endothelialization may occur by a third mechanism. Following denudation, various blood-borne cells adhere to the exposed (thrombogenic) basement membrane. The hypothesis that endothelium may originate from a putative stem cell present in the circulation has been repeatedly proposed, rejected and re-proposed (Taura *et al.*, 1979; Scott *et al.*, 1994). This hypothesis has remained difficult to substantiate since it is possible that damage to the endothelium may result in detachment of viable endothelial cells capable of re-attaching and colonizing a denuded basement membrane or an inert material present in the vessel.

Relatively few studies have documented the events following denudation of microvessels. Ishibashi *et al.* (1987) noted two types of response following laser photocoagulation of the retinal and choroid microvasculature; the re-endothelialization of damaged microvessels by the migration of endothelial cells along the exposed basement membrane, and angiogenesis from both undamaged and re-endothelialized vessels. Mitoses were not observed during re-endothelialization, whilst cell proliferation and budding were commonly noted in both endothelial cells and pericytes involved in angiogenesis. Similarly, denuded vessels in grafts become re-endothelialized by the ingrowth of endothelial cells from the host tissue (Schor and Schor, 1983).

Complementary data may be obtained from the study of endothelial cell response to injury *in vitro*. As discussed in the preceding section, the phenotype expressed by endothelial cells in culture is modulated by the microenvironment. Since conventional culture conditions provide some of the factors associated with wounding *in vivo* (e.g. certain cytokines), cultured endothelial cells may express a mixture of 'resting' and 'injured' phenotypic characteristics. Endothelial cells *in vitro* establish a confluent, quiescent monolayer, the density (cells cm^{-2}) of which depends on culture conditions (Schor *et al.*, 1983; Schor and Schor, 1986). When part of the confluent monolayer is removed, the denuded area is repaired by cells migrating and proliferating from the periphery of the wound until confluence is restored; if culture conditions do not support cell proliferation, confluence may still be achieved, albeit at a lower cell density, by cell migration alone (Ryan *et al.*, 1982; Schor and Schor, 1986).

Endothelial cells plated at subconfluent density proliferate and migrate until a confluent monolayer is formed. 80–100% of the cells may be cycling during the proliferative phase. After confluence there is a small but continuous loss of cells and regeneration of the monolayer; under these conditions, the labelling index is low, with values in the region of 5% commonly observed. Such observations suggest that the majority of cultured endothelial cells retain the ability to proliferate (Ryan *et al.*, 1982; Schor and Schor, 1986).

The clonogenic potential of endothelial cells and pericytes *in vitro* is dependent upon culture conditions, such as the presence of cytokines and oxygen tension (Table 1) (Gospodarowicz *et al.*, 1976; Rhee *et al.*, 1986; Schor and Schor, 1986); values between 3% and 30% have been reported. These data further indicate that the clonogenicity of the various vascular cell populations is differentially affected by culture conditions.

VASCULAR CELL BEHAVIOUR DURING ANGIOGENESIS

Angiogenesis is the process whereby pre-existing blood vessels give rise to new ones. The cellular and molecular events which contribute to this process have been extensively reviewed (Schor and Schor, 1983; Mahadevan and Hart, 1990; Folkman and Shing, 1992). In the context of the present discussion, it is important to emphasize that angiogenic potential is not an exclusive property of the microvasculature; although new blood vessels most commonly arise from capillaries and venules, it is clear that arteriole and large

Table 1 Clonogenic potential of endothelial cells and pericytes. Effects of culture conditions.

Cells	O_2, %	No CM	RP-CM
REC	19	5.5±1	16.0±3
REC	3	5.0±1	14.0±3
RP	19	8.0±2	9.0±1
RP	3	16.5±3	15.5±2

Cloned bovine retinal endothelial cells (REC) and pericytes (RP) were plated at 25 cells cm^{-2} and incubated in growth medium, either in the absence (no CM) or presence of 50% medium conditioned by pericytes (RP-CM) and in an atmosphere containing either 19% or 3% oxygen (Schor and Schor, 1986). The number of colonies was counted 10 days later. Results are expressed as the mean±SD of triplicate dishes. The cloning efficiency of pericytes was significantly increased by low oxygen concentration but not by soluble factors present in the conditioned medium. The reverse was observed with the endothelial cells.

vessel vascular cells are equally capable of producing new capillaries (Schor and Schor, 1983; Gown *et al.*, 1986; Sueishi *et al.*, 1990; Eisenstein, 1991).

New microvessels are composed of endothelial cells and pericytes. Angiogenesis occurs through a series of steps which have been defined better for the endothelial cells than for the pericytes. 'Activation' of both cell types has been reported to be the first step; this involves blebbing of the endothelial cell surface and loosening of cell junctions (Ausprunk and Folkman, 1977), as well as cell enlargement (Paku and Paweletz, 1991). Alterations to the basement membrane (lysis, fragmentation, reduplication) have been observed soon afterwards. Angiogenesis may begin with a bulging of the parental vessel; in this situation there is a direct continuity between basement membranes and lumens of parental and new vessel from the outset (Clark and Clark, 1939; Paku and Paweletz, 1991). Alternatively, angiogenesis may be initiated by the migration of endothelial cells and pericytes through their own basement membrane into the surrounding connective tissue (Ausprunk and Folkman, 1977; Ishibashi *et al.*, 1987). Endothelial cells and pericytes within the perivascular connective tissue matrix adopt an 'angiogenic' phenotype which is distinguishable from their previous 'resting' states in terms of morphology and matrix biosynthesis. In this new environment, these angiogenic cells self-associate to form solid cords or 'sprouts'. The sprouts anastomose, give rise to new sprouts and eventually form new patent vessels or immature capillaries. The subsequent process of capillary maturation involves the formation of an inner (endothelial) layer, an outer (pericytic) layer and intervening basement membrane, thus restoring the cell–cell and cell–matrix relationships characteristic of the parent vessel. During vessel maturation, perivascular fibroblastoid cells have been reported to migrate towards the new vessel and become incorporated in it as pericytes (reviewed in Schor and Schor, 1983; Nehls *et al.*, 1992).

New vessels may originate from several sites along the parent vessel (Sholley *et al.*, 1984). It therefore appears that endothelial cells capable of adopting the angiogenic phenotype may originate from any location within a given blood vessel. The same seems to apply to angiogenic pericytes.

The development of new blood vessels displays a clear directionality, with sprouts growing towards the source of the angiogenic stimulus. Both cell migration and proliferation normally take place during sprouting. The LI in forming vessels may be as high as 15–38% (Tannock and Hayashi, 1972; Hobson and Denekamp, 1984). Although a clear distinction between sprouting endothelial cells and pericytes is not generally possible, it has been suggested that pericytes are present at the leading end of the growing sprout; these leading cells do not proliferate, whilst the more proximal lying cells do so (reviewed in Schor and Schor, 1983; Nehls *et al.*, 1992).

Several features of vascular cell behaviour during angiogenesis merits special comment. Firstly, new blood vessel formation has been observed to take place in the absence of cell proliferation, albeit at a slower rate compared to situations in which proliferation occurs (Sholley *et al.*, 1984). Capillary maturation does not occur simultaneously along the newly formed sprout and different extracellular matrix components are synthesized at different stages during this process (Paku and Paweletz, 1991). These observations indicate a progressive change in the phenotype of the sprouting cells, this possibly being modulated by alterations in the macromolecular composition of the microenvironment.

Yamaura and Matsuzawa (1984) have reported that growing capillaries display two thresholds of radiosensitivity, thereby suggesting the existence of two angiogenic populations with different radiosensitivity. They suggest that one of these may be a stem cell population.

Complementary data have been obtained from numerous *in vitro* studies. Endothelial cells derived from both large vessels and microvessels are capable of adopting a sprouting phenotype and forming tubules. This expression of the angiogenic phenotype may occur within three-dimensional matrices (Figure 2) (Schor *et al.*, 1983; Schor and Schor, 1986; Montesano, 1992; Nicosia *et al.*, 1992) or on the surface of two-dimensional substrata (Figure 5) (Ingber and Folkman, 1989; Grant *et al.*, 1991; Vernon *et al.*, 1995).

As is the case *in vivo*, modulation between the resting and angiogenic phenotypes occurs. For example, large vessel and microvessel endothelial cells cultured on a two-dimensional substratum form a contact-inhibited monolayer which exhibits many characteristics of the resting endothelium *in vivo* (Figure 2a). These 'resting' cells rapidly adopt a spindle-shaped morphology resembling sprouting cells when they are trypsinized and replated within a three-dimensional macromolecular matrix (Figure 2c). Sprouting endothelial cells also arise in postconfluent cultures of resting cells maintained in the presence of serum (Figure 2d); in this case, the sprouting cells assume a sub-monolayer position and are embedded within the three-dimensional cell-deposited matrix. The overlying cobblestone monolayer can be mechanically removed, thus exposing the network of sprouting cells (Figure 2e), which are thereby induced to proliferate and reform a cobblestone, resting monolayer (Schor *et al.*, 1983).

Resting endothelial cells (growing on a two-dimensional surface) and sprouting cells (growing within a three-dimensional environment) differ with respect to a number of criteria, including the synthesis of certain matrix macromolecules, migratory activity, response to cytokines, production of cytokines and proliferative activity. Interchange between the angiogenic and resting phenotypes is completely reversible and repeated alternations in phenotype may be reproducibly achieved by the appropriate changes in culture conditions (Schor and Schor, 1986; Canfield *et al.*, 1986, 1990, 1992; Sutton *et al.*, 1991).

Pericytes also display a sprouting cell phenotype when embedded within a three-dimensional matrix (Figure 6d). Sprouting pericytes and sprouting endothelial cells are indistinguishable in terms of morphology and synthesis of certain matrix macromolecules (Schor *et al.*, 1992). In spite of their similarities, these cells display distinctive proliferative

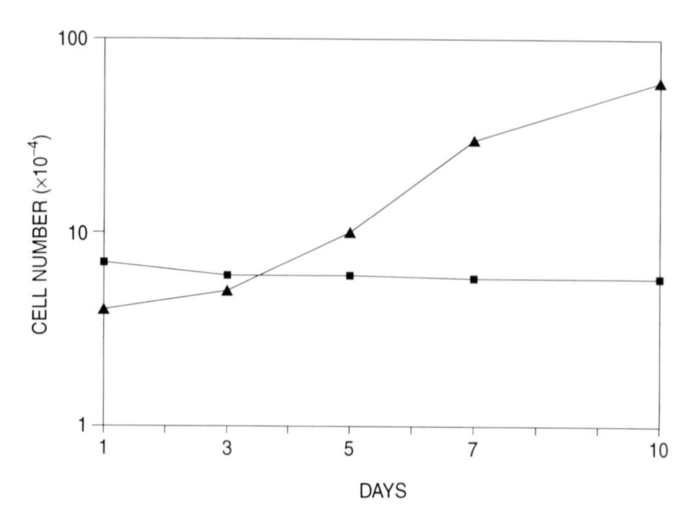

Figure 8 Growth characteristics of sprouting retinal endothelial cells (■) and pericytes (▲). Endothelial cells and pericytes were induced to express a sprouting phenotype by plating them within three-dimensional collagen gels. Cell numbers were determined at various points after plating (Schor and Schor, 1986). Initial plating densities were 2×10^5 endothelial cells and 5×10^4 pericytes per culture.

behaviour (Schor and Schor, 1986): when cultured within a three-dimensional macromolecular matrix, sprouting pericytes proliferate whilst endothelial cells do not (Figure 8).

Resting endothelial cells growing on a two-dimensional substratum may also be induced to form 'tubules' by other means; these include the deposition of matrix fibrils on top of the resting monolayer and various procedures which reduce the relative strength of cell–substratum adhesion compared to cell–cell adhesion. This latter situation may be achieved by using a poorly adhesive substratum, when cell–substratum interactions are weakened by exogenous factors and by increasing the contractability of the cells (Ingber and Folkman, 1989; Grant *et al.*, 1991; Vernon *et al.*, 1995). Pericytes are also able to form tubular structures on two-dimensional substrata (Schor *et al.*, 1990, 1991). The formation of endothelial tubules is stimulated by the presence of pericytes and other peri-endothelial cell types (Sato *et al.*, 1987; Miyazato *et al.*, 1991).

The process of tubule formation *in vitro* has often been referred to as differentiation. We suggest that use of this term is not appropriate, as tubule formation in this situation is reversible and there is consequently no basis for believing that an irreversible alteration in the potential for gene expression has occurred; 'modulation' and 'morphogenesis' would be more accurate descriptions of the processes involved, the former describing changes in cell phenotype and the latter the generation of a new multicellular structure. The ability to form tubular structures *in vitro* is not an exclusive property of cells capable of proliferation and further differentiation; thus, non-dividing endothelial cells which have differentiated into a pericyte/smooth muscle cell phenotype (end cells?) are still able to form tubular structures *in vitro* (Figure 5).

The phorbol ester PMA has been shown to induce angiogenesis *in vivo* and the

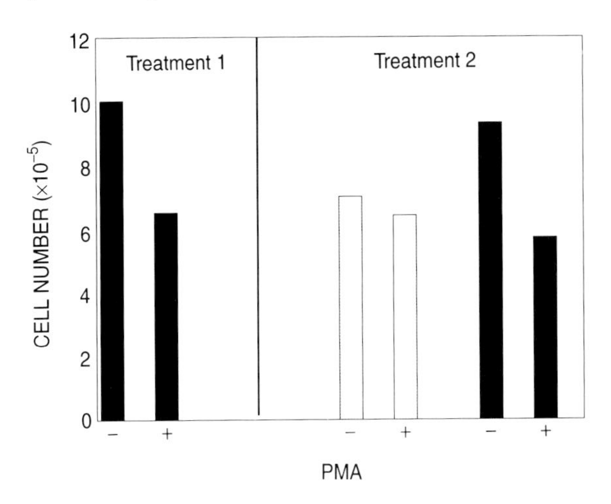

Figure 9 Differences in the toxicity of PMA for two subpopulations of endothelial cells. (Treatment 1) Cloned bovine aortic endothelial cells were plated at various densities and grown in the presence of medium containing 20% calf serum for 3–5 days. They were then incubated with (+) or without (−) PMA (4β-phorbol-12-myristate-13-acetate) (100 ng ml^{-1}), in serum-free medium, for 48 hours. At the end of the incubation, the number of cells present was determined. As shown in the figure, approximately 40% of the cells were killed by PMA. The same results were obtained irrespective of whether the cells were confluent, sparse, quiescent or proliferating. (Treatment 2) Cultures that had been treated once with PMA ('+' in 'Treatment 1' box) were extensively washed and maintained in growth medium (without PMA), supplemented with either 1% or 20% calf serum for 10 days. After this time, half of the cultures were treated again with PMA (+) for 48 hours and the number of cells present determined at the end of this period. Cells maintained in 1% serum (white bars) had undergone minimal or no proliferation between the first treatment and the second; approximately 10% of these cells were killed by PMA. Cultures maintained in 20% serum (black bars) had proliferated during the 10 day interval; approximately 40% of these cells were killed by the second treatment with PMA. These results indicate the presence of distinct PMA-resistant and PMA-sensitive populations, with the former giving rise to the latter upon proliferation.

expression of a sprouting endothelial cell phenotype *in vitro* (reviewed in Montesano, 1992). The effects of PMA on endothelial cells are complex: changes in gene expression may be detected 4 hours after its addition (Hla and Maciag, 1990) and the induction of a sprouting phenotype is apparent 24 hours later. We have observed that approximately half of the cells die following a 24–48 hour incubation with PMA (Figure 9). As a result, PMA appears to separate two distinct subpopulations: one sensitive and the other resistant. The resistant population, if allowed to proliferate (e.g. in the presence of high serum concentrations), gives rise to a new PMA-sensitive population; if the PMA-resistant population is maintained under non-proliferating conditions (e.g. low serum), formation of the PMA-sensitive population does not occur. These observations indicate a proliferation-dependent lineage relationship between two distinct subpopulations of endothelial cells. Further studies are required to characterize these cells, as well as the processes involved in their generation (differentiation or modulation).

SUMMARY AND CONCLUSIONS

The objective of this chapter has been to review the available data regarding lineage relationships amongst vascular endothelial cells, with particular reference to the possible existence of an endothelial stem cell population. Identification of stem cells in other tissues has been facilitated by (a) the short lifespan of the end cell populations, (b) the location of stem cells in recognizable anatomical sites and the directionality of cell flow from them (e.g. intestinal crypt, skin), and (c) the existence of markers for the various transit and end cells (e.g. haematopoietic). Unfortunately, no comparable handles exist for the endothelial cell lineage.

Inferences regarding endothelial cell lineage relationships in the adult have been derived from studies of phenotype diversity in cells under steady state conditions, as well as from endothelial cell behaviour during regeneration and angiogenesis. In considering such data, however, it is important to bear in mind that the vessel should be viewed as a functional unit composed of several cell types and a distinctive extracellular matrix. The significant size- and organ-specific diversity in vessel structure and function further complicate the issue and raise the possibility that conclusions regarding endothelial cell lineage relationships in one vessel type may not apply to another.

Cell turnover in the endothelium is generally low under steady state conditions, thereby suggesting that cell flow from a putative stem compartment to its derivative populations must occur rather slowly. Various factors stimulate endothelial cell proliferation, including a reduction in endothelial cell density (due to vessel dilation or denudation), cytokines and alterations in the extracellular matrix. Irrespective of the stimulus and the final outcome (regeneration, angiogenesis), the vessel responds as a unit, cell migration and proliferation taking place in endothelial and periendothelial locations, and changes occurring in the composition and amount of extracellular matrix. Cell migration alone (i.e. without cell proliferation) is often sufficient to achieve endothelial regeneration and angiogenesis. Consequently, there are no clonogenic assays *in vivo* which provide information regarding vascular cell lineages (i.e. a capillary sprout is not clonally derived, but a structure formed by the cooperation of at least two cell types).

According to the definitions we have adopted, differentiation along a particular cell lineage is characterized by the progressive and irreversible reduction in the potential of gene expression. In order to study such lineage relationships it is necessary to identify the different cell populations unambiguously and demonstrate the irreversible nature of phenotype alterations. A survey of the literature reveals that, in the majority of cases, phenotypic heterogeneity in endothelial cells has not been demonstrated to be the consequence of differentiation. In contrast, several examples of such heterogeneity have been shown to result from modulation of phenotype. Intra-vessel heterogeneity in endothelial phenotype has been documented and, although difficult to prove unambiguously, is less likely to result from modulation by the micro-environment. Such heterogeneity may result from the presence of populations of endothelial cells at different stages of differentiation. Further information regarding the directionality of phenotypic changes from one population to another within the same vessel is required before inferences can be made regarding lineage relationships. Endothelial differentiation *in vitro* is indicated by the loss of typical endothelial cell characteristics (morphology, vWF,

Weibel–Palade bodies) and the adoption of phenotypic characteristics of other cell types (e.g. expression of α-sm actin, smooth muscle myosin, fibroblastoid morphology).

Endothelial cells defined on the basis of their location lining the vessel lumen, staining with presumed pan-endothelial markers (such as vWF or CD31) and functionality (e.g. provision of a non-thrombogenic apical surface), have been observed to proliferate under a variety of situations both *in vivo* (regeneration of the endothelium, angiogenesis) and *in vitro* (regeneration of confluent monolayer). These observations indicate that these cells have the capacity of self-renewal. There is no clear evidence that endothelial cells may be replenished from a non-endothelial source.

Endothelial cells and pericytes share certain phenotypic characteristics *in vivo* and *in vitro*. Their identities are often defined by their position within the vessel. Location within the vessel may not, however, be sufficient to distinguish between pericytes and endothelial cells as they may exchange position under a number of circumstances (e.g. the formation of pseudoendothelium, during angiogenesis, within hyperplastic vessels). Possible exchange also occurs between pericytes and perivascular fibroblasts (e.g. during the maturation of new blood vessels). Cells present in the stroma, described as primitive mesenchymal cells are believed to differentiate into fibroblasts, osteoblasts, smooth muscle cells and others. It has been suggested that such primitive cells may be the pericytes or the endothelial cells. Indeed, the plasticity of the stromal cell lineages has long been recognized (Owen and Friedenstein, 1988; Beranek, 1989).

Taken together, these various observations have led us to conclude that:

(i) Endothelial cells appear to be able to differentiate into cells displaying phenotypic characteristics of pericytes and smooth muscle cells. The differentiation of pericytes and smooth muscle cells into endothelial cells has not been observed.
(ii) The majority of phenotypic variations amongst endothelial cells is due to modulation of gene expression, although a certain degree of differentiation may also occur.
(iii) Functional endothelial cells are capable of self-renewal and may therefore be considered stem cells. Future studies are required to determine whether different cell populations (stem, transit, end) are present within the functional endothelium.

ACKNOWLEDGEMENTS

The authors gratefully acknowledge the financial support of the Cancer Research Campaign, Medical Research Council and Consejo Nacional de Investigaciones Cientificas y Tecnologicas de Venezuela.

REFERENCES

Adams, J.C. and Watt, F.M. (1993). *Development* **117**:1183–1198.
Aggarwal, B.B. and Gutterman, J.U. (1992). *Human Cytokines. Handbook for Basic and Clinical Research.* Blackwell Scientific Publications, Oxford.
Amselgruber, W., Sinowatz, F., Schams, D. and Skottner, A. (1994). *J. Reprod. Fertil.* **101**:445–451.
Antonov, A.S., Nikolaeva, M.A., Klueva, T.S. *et al.* (1986). *Atherosclerosis* **59**:1–19.
Arciniegas, E., Servin, M., Arguello, C. and Mota, M. (1989). *Atherosclerosis* **76**:219–235.
Arciniegas, E., Sutton, A.B., Allan, T.D. and Schor, A.M. (1992). *J. Cell Sci.* **103**:521–529.

Augustin, H.G., Kozian, D.H. and Johnson, R.C. (1994). *BioEssays* **16**:901–906.
Ausprunk, D.H. and Folkman, J. (1977). *Microvascul. Res.* **14**:53–65.
Bär, T. and Wolff, J.R. (1972). *Z. Zellforsch.* **133**:231–248.
Belloni, P.N. and Nicolson, G.L. (1988). *J. Cell Physiol.* **136**:398–410.
Beranek, J.T. (1989). *Med. Hypotheses* **28**:271–273.
Beresford, W.A. (1990). *Cell Differ. Dev.* **29**:81–93.
Bostrom, K., Watson, K.E., Horn, S., Wortham, C., Herman, I.M. and Demer, L.L. (1993). *J. Clin. Invest.* **91**:1800–1809.
Burri, P.H. (1992). In *Angiogenesis: Key Principles–Science–Technology–Medicine* (eds R. Steiner, P. B. Weisz and R. Langer), pp. 32–39. Birkhäuser Verlag, Basel.
Canfield, A.E. and Schor, A.M. (1994). *J. Cell Physiol.* **159**:19–28.
Canfield, A.E. and Schor, A.M. (1995). *J. Cell Sci.* **108**:797–809.
Canfield, A.E., Schor, A.M., Schor, S.L. and Grant, M.E. (1986). *Biochem. J.* **235**:375–383.
Canfield, A.E., Boot-Handford, R.P. and Schor, A.M. (1990). *Biochem. J.* **268**:225–230.
Canfield, A.E., Allen, T.D., Grant, M.E., Schor, S.L. and Schor, A.M. (1990). *J. Cell Sci.* **96:**159–169.
Canfield, A.E., Wren, F.E., Schor, S.L., Grant, M.E. and Schor, A.M. (1992). *J. Cell Sci.* **102**:807–814.
Chesterman, C.N., Ager, A. and Gordon, J.L. (1983). *J. Cell Physiol.* **116**:45–50.
Clark, E.R. and Clark, E.L. (1939). *Am. J. Anat.* **64**:251–301.
Coffin, J.D., Harrison, J., Schwartz, S. and Heimark, R. (1991). *Dev. Biol.* **148**:51–62.
Collatz Christensen, B., Chemnitz, J., Tkocz, I. and Kim. C.M. (1979). *Acta Pathol. Microbiol. Scand. Sect. A.* **87**:265–273.
Davies, P.F. and Tripathi, S.C. (1993). *Circ. Res.* **72**:239–245.
De Bault, L.E. and Cancilla, P.A. (1980). *Science* **207**:653–654.
Diaz-Flores, L., Gutierrez, R., Lopez-Alonso, A., Gonzalez, R. and Varela, H. (1992). *Clin. Orthop.* **275**:280–286.
Doetschman, T., Shull, M., Kier, A. and Coffin, J.D. (1993). *Hypertension* **22**:618–629.
Eguchi, G. and Kodama, R. (1993). *Curr. Opin. Cell Biol.* **5**:1023–1028.
Eisenstein, R. (1991). *Pharmacol. Ther.* **49**:1–19.
Engerman, R.L., Pfaffenbach, D. and Davis, M.D. (1967). *Lab. Invest.* **17**:738–743.
Fajardo, L.F. (1989). *Am. J. Clin. Pathol.* **92**:241–250.
Ferns, G.A.A., Stewart-Lee, A.L. and Änggård, E.E. (1992). *Atherosclerosis* **92**:89–104.
Fishman, J.A., Ryan, G.B. and Karnovsky, M.J. (1975). *Lab. Invest.* **32**:339–351.
Flamme, I. and Risau, W. (1992). *Development* **116**:435–439.
Folkman, J. and Shing, Y. (1992). *J. Biol. Chem.* **267**:10931–10934.
Fujimoto, T. and Singer, S.J. (1986). *J. Cell Biol.* **103**:2775–2786.
Gerritsen, M.E. (1987). *Biochem. Pharmacol.* **36**:2701–2711.
Gospodarowicz, D., Moran, J., Braun, D. and Birdwell, C. (1976). *Proc. Natl Acad. Sci. USA* **73**:4120–4124.
Gown, A.M., Tsukada, T. and Ross, R. (1986). *Am. J. Pathol.* **125**:191–207.
Grant, D.S., Lelkes, P.I., Fukuda, K. and Kleinman, H.K. (1991). *In Vitro Cell. Dev. Biol.* 27A:327–336.
Haudenschild, C.C. (1980). *Adv. Microcirc.* 9:226–251.
Herman, I. and Jacobson, S. (1988). *Tissue Cell* **20**:1–12.
Hirano, A. and Zimmerman, H.M. (1972). *Lab. Invest.* **26**:465–468.
Hla, T. and Maciag, T. (1990). *J. Biol. Chem.* **265**:9308–9313.
Hobson, B. and Denekamp, J. (1984). *Br. J. Cancer* **49**:405–413.
Huttner, I. and Gabbiani, G. (1982). *Lab. Invest.* **47**:409–411.
Ingber, D.E. and Folkman, J. (1989). *J. Cell Biol.* **109**:317–330.
Ishibashi, T., Miller, H., Orr, G., Sorgente, N. and Ryan, S.J. (1987). *Invest. Ophthalmol. Vis. Sci.* **28**:1116–1130.
Kadokawa, Y., Sueromi, H. and Nakatsuji, N. (1990). *Cell Differ. Develop.* **29**:187–194.
Keegan, A., Hill, C., Kumar, S., Phillips, P., Schor, A. and Weiss, J. (1982). *J. Cell Sci.* **55**:261–276.
Lajtha, L.G. (1979). *Differentiation* **14**:23–34.
Lajtha, L.G. (1983). In *Stem Cells. Their Identification and Characterisation* (ed. C. S. Potten), pp. 1–11. Churchill Livingstone, New York.
Latker, H.C., Feinberg, R.N. and Beebe, D.C. (1986). *Anat. Rec.* **214**:410–417.

Le Lièvre C.S. and Le Douarin N.M. (1975). *J. Embryol. Exp. Morphol.* **34**:125–154.
Lipton, B.H., Bensch, K.G. and Karasek, M.A. (1991). *Differentiation* **46**:117–133.
Lipton, B.H., Bensch, K.G. and Karasek, M.A. (1992). *Exp. Cell Res.* **199**:279–291.
Logenecker, J.P., Kilty, L.A., Ridge, J.A., Miller, D.C. and Johnson, L.K. (1983). *J. Cell Physiol.* **114**:7–15.
McCarthy, S.A., Kuzu, I., Gatter, K.C. and Bicknell, R. (1991). *TiPS* **12**:462–467.
Machovich, R. (1988). In *Blood Vessel Wall and Thrombosis*, Vol. 1 (ed. R. Machovich), pp. 115–140. CRC Press Inc., Boca Raton, Florida.
Madri, J.A., Kocher, O., Merwin, J.R., Bell, L. and Yannariello-Brown, J. (1989). *J. Cardiovasc. Pharmacol.* **14 (suppl 6)**:S70–75.
Mahadevan, V. and Hart, I.R. (1990). *Acta Oncol.* **29**:97–103.
Milici, A.J., Furie, M.B. and Carley, W.W. (1985). *Proc. Natl Acad. Sci. USA* **82**:6181–6185.
Miyazato, M., Fukuda, M. and Iwamasa, T. (1991). *Acta Pathol. Jpn* **41**:133–142.
Modis, L. and Martinez-Hernandez, A. (1991). *Lab. Invest.* **65**:661–670.
Montesano, R. (1992). *Eur. J. Clin. Invest.* **22**:504–515.
Nathan, C. and Sporn, M. (1991). *J. Cell Biol.* **113**:981–986.
Nehls, V. and Drenckhahn, D. (1991). *J. Cell Biol.* **113**:147–154.
Nehls, V., Denzer, K. and Drenckhahn, D. (1992). *Cell Tissue Res.* **270**:469–474.
Newcomb, P.M. and Herman, I.M. (1993). *J. Cell Physiol.* **155**:385–393.
Nicosia, R.F., Bonanno, E. and Villaschi, S. (1992). *Atherosclerosis* **95**:191–199.
Noden, D.M. (1989). *Am. Rev. Respir. Dis.* **140**:1097–1103.
Osborn, M. and Weber, K. (1986). *TIBS* **11**:469–472.
Owen, M.E. and Friedenstein, A.J. (1988). In *Cell and Molecular Biology of Vertebrate Hard Tissues* (eds D. Evered and S. Harnett), pp. 42–60. Wiley, Chichester.
Page, C., Rose, M., Yacoub, M. and Pigott, R. (1992). *Am. J. Pathol.* **141**:673–683.
Paku, S. and Paweletz, N. (1991). *Lab. Invest.* **65**:334–346.
Palade, G.E. (1988). In *Endothelial Cell Biology in Health and Disease* (eds N. Simionescu and M. Simionescu), pp. 3–22. Plenum Press, New York.
Paleolog, E.M., Crossman, D.C., McVey, J.H. and Pearson, J.D. (1990). *Blood* **75**:688–695.
Pardanaud, L., Altmann, C., Kitos, P., Dieterlen-Lièvre, F. and Buck, C.A. (1987). *Development* **100**:339–349.
Pearson, J.D. (1991). *Radiology* **179**:9–14.
Price, R.J., Owens, G.K. and Skalak, T.C. (1994). *Circ. Res.* **75**:520–527.
Reinhart, W.H. (1994). *Experientia* **50**:87–93.
Rhee, J.G., Lee, I. and Song, C.W. (1986). *Radiat. Res.* **106**:182–198.
Risau, W. and Lemmon, V. (1988). *Dev. Biol.* **125**:441–450.
Ryan, U.S., Absher, M., Olazabal, B.M., Brown, L.M. and Ryan, J.W. (1982). *Tissue Cell* **14**:637–649.
Sappino, A.P., Schurch, W. and Gabbiani, G. (1990). *Lab. Invest.* **63**:144–161.
Sato, N., Sawasaki, Y., Senoo, A., Fuse, Y., Hirano, Y. and Goto, T. (1987). *Microvasc. Res.* **33**:194–210.
Schlingermann, R.O., Rietveld, F.J.R., Kwaspen, F., van de Kerkhof, P.C.M., de Waal, R.M.W. and Ruiter, D.J. (1991). *Am. J. Pathol.* **138**:1335–1347.
Schnurch, H. and Risau, W. (1993). *Development* **119**:957–968.
Schoefl, G.I. (1963). *Virchows Arch. Pathol. Anat.* **337**:97–141.
Schor, A.M. and Schor, S.L. (1983). *J. Exp. Pathol.* **141**:385–413.
Schor, A.M. and Schor, S.L. (1986). *Microvasc. Res.* **32**:21–38.
Schor, A.M., Schor, S.L., Weiss, J.B., Brown, R.A., Kumar, S. and Phillips, P. (1980). *Br. J. Cancer* **41**:790–799.
Schor, A.M., Schor, S.L. and Allen, T.D. (1983). *J. Cell Sci.* **62**:267–285.
Schor, A.M., Schor, S.L. and Allen, T.D. (1984). *Tissue Cell* **16**:677–691.
Schor, A.M., Allen, T.D., Canfield, A.E., Sloan, P. and Schor, S.L. (1990). *J. Cell Sci.* **97**:449–461.
Schor, A.M., Canfield, A.E., Sloan, P. and Schor, S.L. (1991). *In Vitro Cell. Dev. Biol.* **27A**:651–659.
Schor, A.M., Canfield, A.E., Sutton, A.B., Allen, T.D., Sloan, P. and Schor, S.L. (1992). In *Angiogenesis: Key Principles–Science–Technology–Medicine* (eds R. Steiner, P. B. Weisz and R. Langer), pp. 167–178. Birkhäuser Verlag, Basel.

Schor, A., Canfield, A.E., Sutton, A.B., Arciniegas, E. and Allen, T.D. (1995). *Clin. Orthop.* **313**:81–91.
Schor, S.L. (1994). *Prog. Growth Factor Res.* **5**:223–248.
Schwartz, S.M., Heimark, R.L. and Majesky, M.W. (1990). *Physiol. Rev.* **70**:1177–1209.
Scott, S.M., Barth, M.G., Gaddy, L.R. and Ahl E.T. Jr. (1994). *J. Vasc. Surg.* **19**:558–593.
Shepro, D. and Morel, N.M. (1993). *FASEB J.* **7**:1031–1038.
Shima, D.T., Saunders, K.B., Gougos, A. and D'Amore, P. (1995). *Differentiation* **58**:217–226.
Sholley, M.M., Ferguson, G.P., Seibel, H.R., Montour, J.L. and Wilson, J.D. (1984). *Lab. Invest.* **51**:624–634.
Simionescu, N. and Simionescu, M. (1988). In *Cell and Tissue Biology. A Textbook of Cytology*, 6th edn (ed. L. Weiss), pp. 355–400. Urban and Schwarzenberg, Inc., Baltimore.
Sims, D.E. (1991). *Can. J. Cardiol.* **7**:431–443.
Sims, D.E., Miller, F.N., Horne, M.M. and Edwards, M.J. (1994). *J. Submicro. Cytol. Pathol.* **26**:507–513.
Smeds, S. and Wollman, S.H. (1983). *Lab. Invest.* **48**:285–291.
Stewart, P.A. and Wiley, M.J. (1981). *Dev. Biol.* **84**:183–192.
Spanel-Borowski, K., Ricken, A.M. and Patton, W.F. (1994). *Differentiation* **57**:225–234.
Sueishi, K., Yasunaga, C., Castellanos, E., Kumamoto, M. and Tanaka, K. (1990). *Ann. NY Acad. Sci.* **598**:223–231.
Sutton, A.M., Canfield, A.E., Schor, S.L., Grant, M.E. and Schor, A.M. (1991). *J. Cell Sci.* **99**:777–787.
Tannock, I.F. and Hayashi, S. (1972). *Cancer Res.* **32**:77–82.
Taura, S., Taura, M., Hideshige, I. and Kummerow, F.A. (1979). *Tohoku J. Exp. Med.* **129**:25–39.
Vernon, R.B., Lara, S.L., Drake, C.J. *et al.* (1995). *In Vitro Cell. Dev. Biol.* **31**:120–131.
von Kirschofer, K., Grim, M., Christ, B. and Wachtler, F. (1994). *Dev. Biol.* 163:270–278.
Wagner, R.C. (1980). *Adv. Microcirc.* 9:45–7.
Wilcox, J.N., Smith, K.M., Williams, L.T., Schwartz, S.M. and Gordon, D. (1988). *J. Clin. Invest.* 82:1134–1143.
Wren, F.E., Schor, A.M., Grant, M.E. and Schor, S.L. (1986). *J. Cell. Physiol.* 127:297–302.
Yamaguchi, T.P., Dumont, D.J., Conlon, R.A., Breitman, M.L. and Rossant, J. (1993). *Development* 118:489–498.
Yamamoto, M., Shimokata, K. and Nagura, H. (1988). *Virchows Arch. A. Anat. Histopathol.* 412:479–486.
Yamaura, H. and Matsuzawa, T. (1984). *J. Radiat. Res.* 25:296–304.
Young, R.W. (1964). In *Bone Biodynamics* (ed. H. M. Frost), pp. 117–139. J. & A. Churchill Ltd, London.
Zelickson, A.S. (1966). *J. Invest. Dermatol.* 46:167–171.

7 Mammary stem cells in normal development and cancer

Philip S. Rudland, Roger Barraclough, David G. Fernig and John A. Smith

Cancer and Polio Research Fund Laboratories, Department of Biochemistry, University of Liverpool, Liverpool, UK

PREFACE

The mammary gland of rodents and humans consists of three major cell types, epithelial, myoepithelial and in pregnancy and lactation alveolar cells, and these can be identified by immunohistochemical and ultrastructural techniques. In growing pubertal ducts, particularly terminal end buds, cells intermediate between the epithelium and myoepithelium can be identified, and at a later developmental stage, intermediates are also seen between epithelial and secretory alveolar cells. These results suggest that two separate pathways of differentiation exist. Transplantation of dissected mammary glands in syngeneic rats suggests that the intermediates along the epithelial–myoepithelial pathway are repopulating stem cells for regeneration of the entire mammary gland. Epithelial cell lines have been isolated from normal mammary glands and benign lesions of rats and humans that can interconvert, in a stepwise fashion, into myoepithelial-like and alveolar-like cells. The cellular intermediates between epithelial and myoepithelial cells can be isolated as clonal cell lines and these and the transitional intermediates between epithelial and alveolar-like cells bear strong similarities to those seen *in vivo*. Thus cellular interconversions *in vitro* probably reflect the two differentiation pathways *in vivo*, and the clonable intermediate cells are candidates for mammary stem cells.

Of the main mammatrophic hormones, prolactin synergizing with oestradiol stimulates the epithelial cell lines to differentiate *in vitro* into casein-secretory, alveolar-like cells, but these hormones alone fail to stimulate appreciable proliferation of this mammary stem cell system. However, both locally produced growth factors, e.g. transforming growth factor α (TGFα) from the myoepithelial cells and prostaglandin E_2 (PGE_2) from the mammary adipocytes, and potentially systemic growth factors, e.g. pituitary mammary growth factor, are mitogenic for the epithelial and intermediate cell lines. Moreover, the growth-promoting effects of oestrogen are believed to be mediated through growth factors, possibly TGFα and/or insulin-like growth factor-1 (IGF-1).

STEM CELLS
ISBN 0–12–563455–2

Differentiation of epithelial cell lines along the myoepithelial-like pathway stimulates production of basic fibroblast growth factor (bFGF) and its receptors, but this factor is probably sequestered by extracellular glycosaminoglycans, except when the intermediate stem cells are proliferating in the growing gland. External agents which trigger differentiation along the myoepithelial pathway have not yet been identified. One protein, called p9Ka, which has been identified as a calcium-binding regulatory protein, arises at an early stage in this pathway and binds intracellularly to the actomyosin cytoskeleton. This binding may trigger the major observable change in differentiation along this pathway, the rearrangement of the cytoskeleton to that of a smooth muscle-like cell.

As with the normal gland, benign lesions of rats and humans contain myoepithelial-like and, under suitable hormonal conditions, alveolar-like cells, and epithelial/intermediate stem cell lines have been isolated that can undergo these differentiation pathways *in vitro*. However, malignant carcinomas of rats and humans contain only epithelial cells and the resultant cell lines fail to differentiate completely along either pathway. Limited expression of early markers of the myoepithelial pathway are sometimes seen in rat and human invasive carcinomas. This result is consistent with the malignant cell bearing some similarity to the normal intermediate stem cells, but with a much more limited differentiating ability. Loss of the myoepithelial cell, in particular, in some human invasive carcinomas may account, in part, for compensatory changes in the malignant cells. Thus overexpression of c-erbB-2 receptor may compensate for reduction of TGFα, ectopic production of bFGF may compensate for its loss, and ectopic production of p9Ka may help in metastasis. Thus compensation for or retention of molecules potentially involved in growth and/or differentiation of the mammary stem cell system by some human invasive carcinomas may be a mechanism by which a malignancy progresses.

DEVELOPMENT OF THE NORMAL MAMMARY GLAND *IN VIVO*

Anatomical structures in developing glands

The development of the mammary gland in rodents and humans occurs in three distinct phases that are largely controlled by the levels of circulating hormones. The first phase occurs during fetal development. The mammary glands arise as buds from the epidermal layer of the skin along two 'milk lines', which run from the shoulder to the inguinal region on either side of the body (Dawson, 1934; Raynaud, 1961). The mammary buds elongate and form simple, branched ducts during interuterine life, initially under the influence of maternal hormones (Myers, 1917; Dawson, 1934).

The second phase occurs prior to puberty and results in the progressive growth of the mammary parenchyma within the surrounding fat pad (Dawson, 1934; Hollman, 1974). This phase is characterized by a rapid extension of the ductal tree and generation of its branching pattern by lengthening of existing ducts, by dichotomous branching of the growing ductal tips and by monopodial branching from the sides of existing ducts (Dawson, 1934; Vorherr, 1974). During their period of rapid growth, the ducts sometimes terminate in globular structures called terminal end buds (TEBs) (Figure 1) (Dawson, 1934; Russo *et al.*, 1982), which are observed more frequently in rodent than in human mammary glands (Russo and Russo, 1987; Rudland, 1991a). The terminal ducts/TEBs

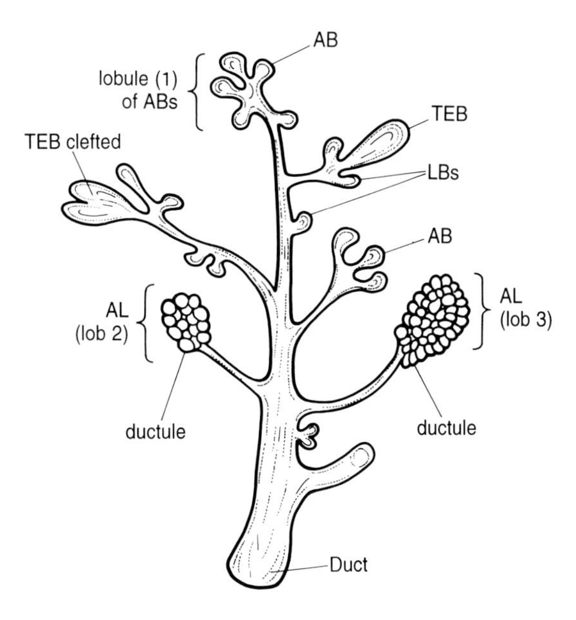

Figure 1 Diagrammatic representation of the structures present in the developing mammary gland. TEB, terminal end bud; TEB clefted, terminal end bud subdivided by a cleavage furrow at its tip, further cleavage results in alveolar buds; LB, lateral bud; AB, alveolar bud; AL, alveolar lobule (Ormerod and Rudland, 1984; Rudland, 1993). In humans there is a wider range of lobules than in rodents and these have been classified as type 1 (lob 1), synonymous with lobule of ABs, and types 2, 3, 4 (4 not shown, present only in lactating alveoli) which correspond to ALs containing increasing numbers of smaller ductules (Russo and Russo, 1987).

contain the most actively proliferating cells in the mammary parenchyma (Russo *et al.*, 1982; Russo and Russo, 1987), and are responsible for the elongation of the ducts and probably for their dichotomous branching by bifurcation (Ormerod and Rudland, 1984; Rudland, 1991a). The number of terminal ducts/TEBs reaches a maximum in rats of about 20 days old and in humans of about 13 years old (Dawson, 1934; Russo *et al.*, 1982). Thereafter the number rapidly decreases, although the rate of decrease is much more variable in humans. Monopodial branching of ducts is produced by the growth of small lateral buds (LBs) from the sides of existing ducts (Figure 1) (Myers 1919; Dawson, 1934). After puberty TEBs/terminal ducts/LBs differentiate to form lobules of alveolar buds (ABs) which consist of 3–5 lobes (Figure 1), although this transition is not as sharp in humans as in rodents (Russo and Russo, 1978, 1987). Thereafter, sprouting of new ABs in rodents and humans can gradually occur during puberty, and the ABs can subdivide further at each oestrous cycle, giving rise to alveolar lobules (ALs) consisting of progressively larger numbers of smaller lobes (Figure 1) (Dawson, 1934; Russo and Russo, 1978). In rodents the terminal ductal structures that include ABs and ALs correspond to the terminal ductal lobular units (TDLUs) normally described in humans (Russo *et al.*, 1982). In humans, as opposed to rodents, there is much variation in glandular structure both within the same mammary gland and between mammary glands of females of the same age; larger and more complex ALs are observed in humans.

The third phase of mammary development occurs during pregnancy and lactation. During the initial period of pregnancy the ABs and ALs subdivide even further, giving rise to large clusters of ALs, some of which can contain up to 200 individual lobes in humans. They become distended and form secretory alveoli during the period of lactation (Russo and Russo, 1978; Salazar and Tobon, 1974; Russo *et al.*, 1988) and produce milk predominantly under the influence of pituitary prolactin and other peptide hormones as well as circulating steroids. Milk itself consists predominantly of milk fat globules pinched off from the surface of the alveolar cells, caseins and α-lactalbumin (Vorherr, 1974).

Identification of different cell types in mature glands

The ducts and lobes or ductules are separated from the stroma by a basement membrane. These structures consist of two cell types, an inner lining of epithelial cells and an outer, sometimes discontinuous layer of myoepithelial cells (Figure 2) (Ozzello, 1971; Radnor, 1972; Salazar and Tobon, 1974). A third cell type, the alveolar cell, lines distended ductules or alveoli during pregnancy and lactation, and is responsible for production and secretion of milk (Ozzello, 1971; Radnor, 1972). Within the ducts, ductules and alveoli these cell types can be distinguished in histological sections both by position and by their ultrastructure (Vorherr, 1974). Epithelial cells possess apical microvilli and specialized intercellular junctions including desmosomes. The myoepithelial cells usually possess an irregular nucleus with peripheral heterochromatin, myofilaments, pinocytotic vesicles which line the inner plasma membrane and basement membrane often connected by hemidesmosomes to the outside of the plasma membrane. The alveolar cell possesses a developed secretory apparatus (Ozzello, 1971).

More recently, histochemical and immunocytochemical reagents have been used to distinguish epithelial cells, myoepithelial cells and alveolar cells (Table 1). In rats and humans, peanut lectin (Newman *et al.*, 1979), antiserum to epithelial membrane antigen (EMA) (Sloane and Ormerod, 1981; Warburton *et al.*, 1982), monoclonal antibodies (MAbs) to human (or rat, Dulbecco *et al.*, 1983) milk fat globule membranes (MFGM) (Foster *et al.*, 1982a; Taylor-Papadimitriou *et al.*, 1983) and MAbs to cytokeratin 18 (Lane, 1982) stain some epithelial cells throughout the parenchymal structures; the myoepithelial cells are largely unstained (Table 1). Antisera and monoclonal antibodies to MFGM stain the luminal membrane of all epithelial cells, irrespective of their position in the gland, whereas peanut lectin and MAbs to keratin 18 stain relatively few epithelial cells in main ducts but more epithelial cells in terminal lobular structures (Rudland and Hughes, 1989). Removing terminal sialic acid residues from histological sections with neuraminidase enables many more luminal epithelial cells to react with peanut lectin (Rudland and Hughes, 1989). None of the above reagents stains the stromal cells (Table 1).

In contrast in both rats and humans, antisera to the cytoskeletal components of smooth muscle actin (Bussolati *et al.*, 1980), myosin (Gusterson *et al.*, 1982; Warburton *et al.*, 1982), MAbs to cytokeratins 5 and 14 (Dairkee *et al.*, 1985; Nagle *et al.*, 1986), to the intermediate filamental protein vimentin (Warburton *et al.*, 1989; Rudland and Hughes, 1989) and to cell surface determinants such as the common acute lymphoblastic leukaemia antigen (CALLA) (Gusterson *et al.*, 1986) stain the myoepithelial but not the epithelial cells (Table 1). Some reagents, e.g. those MAbs to a 135 K_d cell-surface glycoprotein are specific for human cells (Gusterson *et al.*, 1985), whilst others, e.g. MAbs to complex

(a) Epithelial markers

luminal epithelial cells
cortical cells
lumen
peripheral cap cells
peripheral cells
myoepithelial cells
epithelial cells
subtending duct

(b) Myoepithelial markers

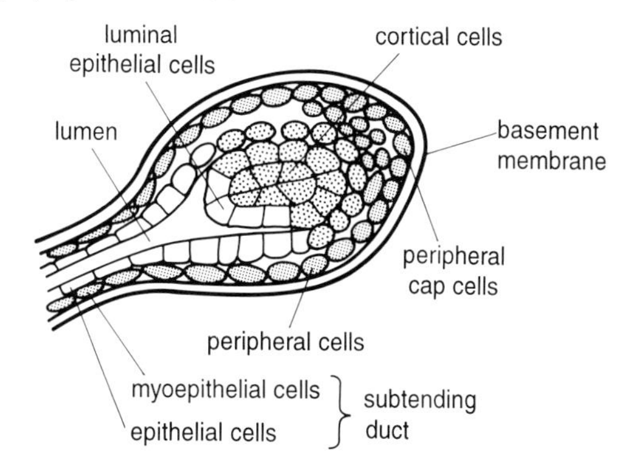

Figure 2 Schematic representation of the staining patterns for epithelial and myoepithelial markers in terminal end buds. (a) Epithelial markers: histochemical staining with MAbs to MFGM, anti-EMA and peanut lectin after treatment of the sections with neuraminidase. The loosely packed peripheral cells are stained weakly. This staining becomes stronger in the central or cortical epithelial cells where it is predominantly cytoplasmic, and becomes intense along the apical surfaces of the epithelial cells which line the lumina in both the duct and TEB. Gradations in morphology and intensity of staining are observed. The proximal peripheral cells and the myoepithelial cells of the subtending duct are unstained. (b) Myoepithelial markers: histochemical staining for smooth muscle actin/myosin and basement membrane proteins. The peripheral cells are stained moderately for actin/myosin. The elongated morphology and intensity of staining increases for the peripheral cells closer to the subtending duct, and they merge eventually with those of the myoepithelial cells of the duct. The cortical epithelial cells show usually only a weaker staining for actin/myosin and this staining is usually not apparent in the luminal epithelial cells of the TEB and duct. Staining for the basement membrane proteins laminin and Type IV collagen encircles completely the TEB, producing a thin band at its distal tip and gradually increasing in thickness in the neck regions of the TEB, where it merges with that of the subtending duct (Ormerod and Rudland, 1984; Rudland, 1993).

keratins (LP34, PKK2) (Warburton *et al.*, 1982; Taylor-Papadimitriou *et al.*, 1983) and the lectins *Griffonia simplicifolia-1* (GS-1) and pokeweed mitogen (PWM) (Hughes and Rudland, 1990a), show selective reaction with myoepithelial cells only in the rat. In addition, antisera to laminin, Type IV collagen and Thy-1, the last only in the rat, stain

Table 1 Summary of immunocytochemical staining patterns of rat and human mammary cells *in vivo*.

Reagent[a]	Ductal epithelial cells	Myoepithelial cells	Alveolar cells
Anti-MFGM[b], anti-EMA[c], MAbs to EMA[c]	+	−	+
MAb to keratin 18	+	−	±[d]
PNL alone	−	−	+
PNL plus neuraminidase	+	−	+
Anti-caseins[b,c]	−	−	+
Anti-smooth muscle actin/myosin	−	+	−
MAb to keratin 14	−	+	−
MAbs to CALLA	−	+	−
MAb to 135 K_d glycoprotein	−	+[c]	−
Anti-vimentins[e]	−	+	−
GS-1, PWM lectins	−	+[b]	−
Antibodies to complex keratins[f]	−[b] (±)[c]	+	−
Anti-Type IV collagen/laminin/Thy-1	−	+[g]	−

[a] Abbreviations as used in text; MAb, monoclonal antibody; ± weakly stained; + strongly stained; [b] Rat; [c] Human; [d] Variably stained; [e] Also stains stromal fibroblasts; [f] Includes polyclonal to human callus keratin and MAbs PKK2 and LP34; [g] Basement membrane adjacent to myoepithelial cells (Warburton *et al.*, 1982; Rudland, 1987a, 1987b, 1993; Rudland and Hughes, 1989).

the basement membrane which is usually associated with myoepithelial cells (Table 1) (Warburton *et al.*, 1982; Gusterson *et al.*, 1982; Barsky *et al.*, 1983; Monaghan *et al.*, 1983). However, MAbs to vimentin also stain stromal fibroblasts, and these antibodies, together with those of the smooth muscle-specific actin/myosin, of the 135 K_d surface glycoprotein and of laminin, Type IV collagen and Thy-1 in the rat also stain stromal blood vessels (Rudland, 1987a; Rudland and Hughes, 1989), thus necessitating a combination of suitable antibodies to identify unambiguously myoepithelial cells (Table 1).

The secretory alveolar cell is characterized by staining with peanut lectin alone (Newman *et al.*, 1979) and with antisera to rat (Herbert *et al.*, 1978) or human (Earl and McIlhinney, 1985; Rudland and Hughes, 1989) caseins.

Identification of transitional cells

Although the majority of rodent mammary glands consist of three discrete cell types, the TEBs, LBs, ABs and terminal ducts of the developing gland are composed of a heterogeneous collection of cells (Williams and Daniel, 1983; Ormerod and Rudland, 1984). These include not only epithelial and myoepithelial cells but also irregular loosely-adherent cells and the occasional undifferentiated basal clear cells. The irregular, loosely packed cells, or cap cells, are situated mainly around the periphery of the TEBs and LBs and appear undifferentiated. They show gradations in ultrastructure to the epithelial cells within the cortex and also to the myoepithelial cells of the subtending duct (Williams and Daniel, 1983; Ormerod and Rudland, 1984). The gradations to cortical epithelial cells are accompanied by an increase in staining with anti-MGFM (Ormerod and Rudland, 1984), with an MAb to rat epithelial cells (Dulbecco *et al.*, 1983) and with an MAb to keratin 18 (Rudland, 1987a) (Figure 2a). The gradations to myoepithelial cells are accompanied by an increase in staining with antisera to smooth muscle actin, myosin, complex keratins, vimentin, laminin, Type IV collagen (Dulbecco *et al.*, 1983; Ormerod and Rudland, 1984; Rudland, 1987a) and the lectins GS-1 and PWM (Hughes and Rudland, 1990b) (Figure 2B).

The above results suggest that the undifferentiated cap cells do not represent a discrete cell type, but show transitional forms to epithelial cells on the one hand and to myoepithelial cells on the other hand (Figure 2), and that the tendency towards the myoepithelial phenotype predominates in the more differentiated budded structures, the ABs (Ormerod and Rudland, 1984). Results from pulse-chase experiments with DNA precursors (Dulbecco *et al.*, 1982) and additional monoclonal antibodies to cytokeratins specific for luminal epithelial and myoepithelial cells (Allen *et al.*, 1984) are consistent with this interpretation. Cells intermediate in binding characteristics for the lectins GS-1 and PWM can also be identified in mammary ducts of postnatal rats up to 5 days old (Hughes and Rudland, 1990b), consistent with a similar interpretation for the appearance of the differentiated outer layer of ductal myoepithelial cells in rats of 6–7 days of age (Radnor, 1972; Warburton *et al.*, 1982). In growing TEBs and in growing ducts up to 5 days old the intermediate/basal cells are actively dividing whilst in non-growing ducts the epithelial and particularly the myoepithelial cells show little cell division (Dulbecco *et al.*, 1982; Joshi *et al.*, 1986a, 1986b).

Basal clear cells which possess few cytoplasmic organelles but form junctions with neighbouring epithelial cells have also bee proposed as precursors of epithelial and myoepithelial cells (Radnor, 1971; Salazar and Tobon, 1974; Smith and Medina, 1988). However, unlike the cap cells, basal clear cells identified by the above criteria are infrequently seen to divide (Joshi *et al.*, 1986b) and fail to bind either lectin (Hughes and Rudland, 1990b). These results suggest the basal clear cells represent a non-growing cell population and that when GS-1/PWM receptors are used as markers, the basal clear cells are not involved in the myoepithelial lineage, at least in its later stages.

In the human mammary gland the globular structures at the ductal termini of peripubescent, 13-year-old girls have a similar histological appearance to TEBs in rodents (Figure 3A). Peripheral cap cells show morphological and immunocytochemical staining gradations to cortical epithelial cells in the centre of TEBs (antisera to EMA, PNL, MAbs to MFGM, Figure 3B) and to myoepithelial cells of the subtending duct (MAbs to vimentin and to smooth muscle actin/myosin, Figures 3C and 3D). Furthermore, some of the more immature AB structures of young women also contain a heterogeneous collection of cells which include epithelial cells, myoepithelial cells and peripheral cells of a more intermediate ultrastructure (Stirling and Chandler, 1976; Smith *et al.*, 1984a) and staining pattern with antisera to MFGM, actin and myosin (Rudland and Hughes, 1989; Rudland, 1991a). Although these peripheral cells in ABs show gradations in ultrastructure and staining patterns to the myoepithelial cells of the subtending terminal duct, in these respects they are still closer to the myoepithelial cells, as in the rat (Stirling and Chandler, 1976; Smith *et al.*, 1984a; Rudland and Hughes, 1989). Antibodies to the basement membrane proteins, Type IV collagen and laminin form a continuous stained band round such terminal ductal structures, but its thickness is reduced round the distal tips of TEBs and some peripheral cells exhibit cytoplasmic staining (Rudland, 1991a)

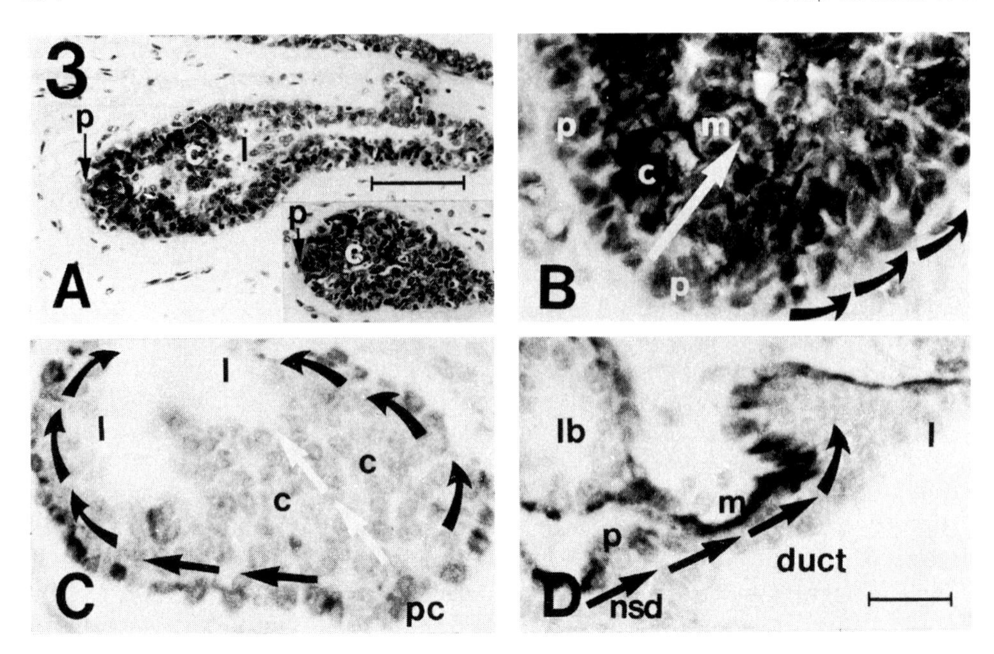

Figure 3 Histochemical and immunocytochemical staining of terminal end bud structures in prepubescent developing human mammary glands. (A) Longitudinal section of terminal end bud (TEB) stained by haematoxylin and eosin with central lumen (l). Inset, TEB without central lumen. Peripheral cells are seen at the distal tip (p) and within the centre or cortex (c) of the cellular mass. (B) Section of TEB incubated with MABs to MFGM showing the distal tip at higher magnification with weak staining of the loosely packed, peripheral cells (p), which is gradually reduced the nearer the cells become to the subtending duct (curved arrows). Staining of the cells increases towards the centre (white arrow), first cytoplasmically (c) and then membranously (m). (C) Section of TEB incubated with MAb to smooth muscle actin showing the distal tip at higher magnification with the loosely packed peripheral cells (pc) stained moderately; this staining increases gradually the further such cells are positioned from the distal tip (black arrows). The staining decreases for the close-packed cells (c) in the cortex of the TEB (white arrows), and is lost completely from the luminal epithelial cells (l). (D) Section of TEB incubated with MAb to smooth-muscle actin showing its proximal region at higher magnification with the now more-closely-packed peripheral cells (p) increasing their staining to form a continuum with the myoepithelial cells (m) of the neck region of the subtending duct (nsd). The luminal epithelial cells are unstained and a lateral bud (lb) is also shown. Original magnification ×230 and ×730. Bars = 75 μm and 20 μm for A and B, C, D, respectively (Rudland, 1991a).

suggesting the site of synthesis. Finally there is increased staining for the epithelial-specific keratin 18 and myoepithelial-specific keratin 14 for similar cells in the terminal budded structures (Rudland and Hughes, 1989; Rudland, 1991a). These results, therefore, suggest that staining for both keratins is not reciprocally related in such structures, as it is in main ducts, and may relate more to such cells possessing some properties of both epithelial and myoepithelial cells (Rudland, 1993). Basal clear cells have also been identified in human mammary glands, particularly in lactating alveoli (Salazar and Tobon, 1974).

A heterogeneous collection of epithelial cells has also been identified in both rats and humans using ultrastructural and immunohistochemical techniques in all structures of the mammary gland, but in varying proportions (Hagueneau and Arnoult, 1959; Vorherr, 1974; Russo *et al.*, 1976; Du Bois and Elias, 1984; Dulbecco *et al.*, 1983; Edwards and Brooks, 1984). In the rat, histochemical binding of peanut lectin to the developing gland shows that most epithelial cells in ducts fail to stain, even when the cells are treated with neuraminidase to remove terminal sialic acid residues. Many desialylated cells in TEBs are stained only weakly and cytoplasmically by peanut lectin; in ABs the majority of cells are stained along their luminal membranes; whilst in casein-secreting and non-secreting alveolar cells, this lectin stains the cells' luminal membranes without the necessity to remove terminal sialic acid residues (Rudland, 1992). Previous studies have shown that peanut lectin reacts primarily with carbohydrate residues of EMA (Ormerod *et al.*, 1984, 1985; Warburton *et al.*, 1982, 1985). Thus in normal mammary development, the carbohydrate receptor for peanut lectin is probably absent from most ductal epithelial cells, and first occurs capped by terminal sialic acid in the cytoplasm of the majority of epithelial cells within TEBs, then in its sialylated form on the luminal surface of the majority of epithelial cells in ABs, and finally as terminal sugar residues on the cell surface of most of the alveolar cells prior to their secreting casein in pregnant and lactating rats (Figure 4A). Minor populations of epithelial cells, however, that are unstained by peanut lectin, also exist in TEBs and ABs, and minor populations of desialylated stainable cells also exist in ducts. Thus the major peanut lectin-interacting types of mammary epithelial cell can be assigned to different mammary structures (Figure 4a) (Rudland, 1992).

The above assignments are not immutable, however, and with the appropriate stimulus of hormonal agents, ductal epithelial cells can be converted into the major type of epithelial cell in ABs and eventually in alveoli, and those in TEBs can be converted into the major cell type in alveoli, with respect to their interaction with peanut lectin. These results suggest that the casein-secretory alveolar cells *in vivo* are generated by successive interconversions between the major epithelial cell types present in the different mammary structures in the order: ducts, TEBs, ABs, alveoli and secretory alveoli (Rudland, 1992). Results with peanut lectin and with certain MAbs to human MFGM in human mammary glands at different developmental stages support the concept of successive unmasking and desialylating steps along a potential alveolar cell pathway (Newman *et al.*, 1979; Edwards and Brooks, 1984; Rudland and Hughes, 1989).

Demonstration of stem cell populations by transplantation of rodent glands

Whether any one of the three discrete types or the not-so-discrete types of mammary cell can regenerate directly the other cell types *in vivo* is largely unknown. Transplantation studies of mosaic tissue from two inbred strains of mice show that ductal and alveolar structures can breed true and that neoplastic mammary tissue in similar transplant lines can be categorized into one of the three discrete cell types (Slemmer, 1974). However, in the former experiment the ductal and alveolar growths are always in the same proportion in a given pair of mouse strains, suggesting an alternative explanation based on hormonal differences between strains, and in the latter case the argument is complicated by other mammary cell types being found in some of the resultant tumours (Bennett, 1979). Moreover, evidence is now mounting against completely immutable

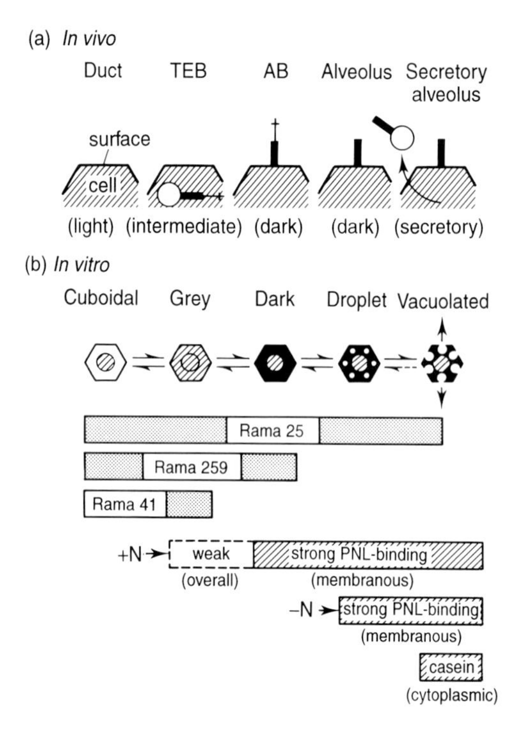

Figure 4 Schematic representation of the changes observed on differentiation to alveolar-like cells. (a) Histochemical binding of PNL to the predominant cell type in different structures of the rat mammary gland. The carbohydrate receptor for PNL (■) is capped by sialic acid (†), possibly on milk fat globule-like structures within the cytoplasm of the major epithelial cell type within TEBs. It is observed in its sialylated form on the cell surface of the major epithelial cell type in ABs; and is then found in its desialylated form on the cell surface of the major epithelial cell type in alveoli (Rudland, 1992). It also appears in its desialylated form on extruded milk fat globule structures in the lumina of casein-secretory alveoli. The major ultrastructural types of epithelial cell observed within these structures are shown in parentheses (Radnor, 1971; Russo *et al.*, 1976). (b) The linear pathway of morphologically intermediate cell types between the cuboidal epithelial and casein-secretory, vacuolated cells of Rama 25 is shown, together with the extent of binding of peanut lectin (PNL) with (+N) or without (−N) treatment of the cultures by neuraminidase (hatched lines). The solid lines represent the extent of the pathway traversed by Rama 25, Rama 259, and Rama 41 cultures. The weak overall binding of PNL to desialylated grey cells is probably due to a cryptic receptor, possibly located within the cell.

mammary cell types *in vivo*. As outlined earlier, mammary ductal epithelial cells may eventually give rise to myoepithelial cells, since myoepithelial cells are absent from the ducts of embryonic and neonatal rats and only appear 6–7 days after birth (Radnor, 1972; Warburton *et al.*, 1982). Moreover, different parts of the mammary gland, e.g. ducts and alveoli, have been dissected out and transplanted to other suitable sites, usually white fat pads, in syngeneic rodents. In all cases fully-developed mammary glands are generated which will secrete milk products in isologous, pregnant hosts (Hoshino, 1963, 1964).

Table 2 Summary of outgrowths with different transplants in interscapular fat pads of rats.

Transplant[a]	Treatment of rat	Outgrowth morphology[b]
DUCT	None	Ductal, LB, TEB, AB
TEB	None	Ductal, LB, TEB, AB
AB	None	Ductal, LB, TEB, AB
DUCT	Perphenazine	Distended ductal, AB/lobule
TEB	Perphenazine	Distended ductal, AB/lobule
AB	Perphenazine	Distended ductal, AB/lobule
DUCT	Mated and lactating	Distended ductal, alveoli
TEB	Mated and lactating	Distended ductal, alveoli
AB	Mated and lactating	Distended ductal, alveoli

[a] 'DUCT' indicates 2–3 segments of duct without any lateral buds, approximately 200 μm in length. 'TEB' and 'AB' indicate 2–3 terminal end-buds and alveolar buds, respectively, both approximately 200 μm in length which were removed from the subtending duct at the neck of the bud. [b] 'Ductal' indicates a branching system of ducts; LB, lateral buds; TEB, terminal end buds; AB, alveolar buds (Rudland, 1991b).

Recently, defined segments of the rodent mammary gland have been excised and transplanted with full regenerating ability, suggesting that the full regenerative capacity of glandular tissue is distributed throughout the mammary parenchyma (Smith and Medina, 1988). Since LBs resemble TEBs in appearance and cell types (Ormerod and Rudland, 1984), ductal fragments have to be carefully selected to exclude those with LBs. When this is done, implants of bud-free ducts, as well as TEBs and ABs, can generate the entire mammary ductal tree, including the ducts and budded structures (Ormerod and Rudland, 1986), and, in hormone-stimulated recipient rats, secretory alveoli as well (Table 2) (Figure 5) (Rudland, 1991b). The different structures generated from outgrowths of the transplants contain the same arrangement of cell types as seen in normal mammary glands, including the cellular intermediates in terminal structures (Ormerod and Rudland, 1986; Rudland, 1991b). However, although secretory alveolar cells are present in implants from lactating hosts, the number of alveoli in each lobulated cluster is increased in the order: bud-free ducts < TEBs < ABs. This order for the different implants is also reflected in the increase in the percentage of peanut lectin-staining and casein-staining cells. (Rudland, 1991b). These results suggest that although the complete cell-differentiating ability for generating lactating mammary glands is present in bud-free ducts, this ability is enhanced with respect to the alveolar cell in TEBs and ABs. However, prolonged exposure to the hormones of pregnancy and lactation enable the outgrowths from bud-free ducts to catch up with those of TEBs and ABs. These results suggest that this enhancement is only of a temporal nature (Rudland, 1991b) and is consistent with the temporal sequence described in the previous section for the generation of secretory alveoli.

Results of the majority of transplant studies in rodents therefore imply that similar stem cell populations capable of generating the fully differentiated lactating mammary gland exist in all the structures discussed. Morphological studies of these developing implants using immunocytochemical and ultrastructural techniques have suggested the

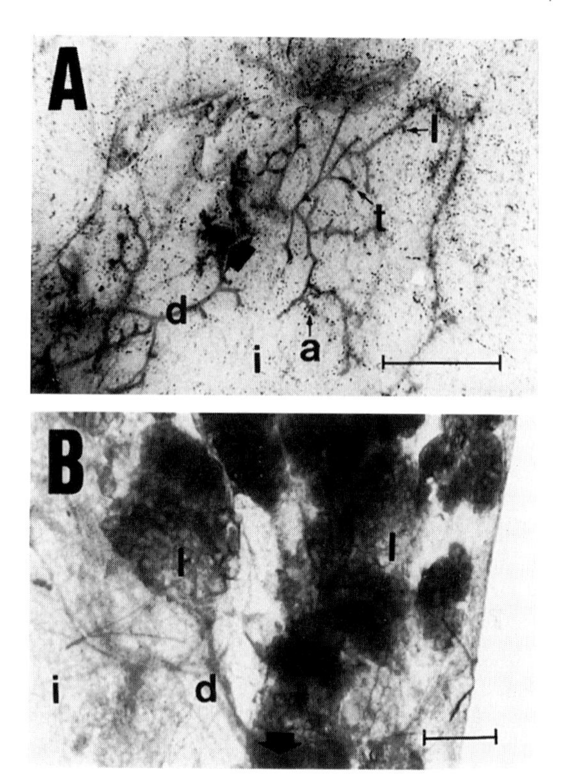

Figure 5 Whole mounts of transplants of rat mammary glands. (A) 16 weeks after implantation of DUCTs within the interscapular fat pad (i), two main branching systems originated from the transplant site (arrow). This ductal system consists of ducts (d) as well as lateral buds (l), terminal end buds (t), and alveolar buds (a). (B) 12 weeks after implantation of TEBs within the interscapular fat pad (i), the rat was mated and examined 3 days postpartum. The transplantation site is shown by the arrow. The ductal system now consists of ducts (d) connecting small and large lobulated structures (l). Original magnifications: A ×19; B ×12. Bars = 1000 μm (Rudland, 1991b).

presence of such stem cell population(s) in the basal cell layers of ducts and end-buds. These cell population(s) may be the basal clear cells (Radnor, 1971; Smith and Medina, 1988) or alternatively a cap cell type intermediate in morphology between epithelial and myoepithelial cells (Williams and Daniel, 1983; Ormerod and Rudland, 1986; Rudland, 1991b). The basal clear cell, however, occurs in the greatest proportions in lactating alveoli and least in TEBs (Joshi *et al.*, 1986b), the reverse order of their regenerative capacities when these two structures are transplanted into the fat-pads of non-mated rodents (Ormerod and Rudland, 1986; Smith and Medina, 1988). The more likely conclusion is that many of the so-called intermediate cell types are also capable of generating the other cell types under suitable conditions. This model is also more consistent with the observation that the proliferation of epithelial cells at each stage of pregnancy and lactation appears to be sufficient to account for their increase in number at that particular period of mammary growth (Joshi *et al.*, 1986a, 1986b).

DIFFERENTIATION OF THE MAMMARY GLAND *IN VITRO*

Isolation of immortalized cell lines of a potential stem cell nature

To investigate the relationship between the three individual cell types within the mammary gland, normal and benign tissue from rats and humans has been maintained in short term culture (Rudland *et al.*, 1977a; Easty *et al.*, 1980; Stampfer *et al.*, 1980) and immortalized, either spontaneously for rat and benign hyperplastic human tissue (Bennett *et al.*, 1978; Rudland, 1993) or with simian virus 40 (SV40) for normal human mammoplasty specimens (Rudland *et al.*, 1989a) to obtain single-cell-cloned, epithelial-like cell lines (Figure 6A, B). Examples of such epithelial-like cell lines are rat mammary (Rama) 704 from normal rat tissue (Ormerod and Rudland, 1985), Rama 25 and Rama 37 from DMBA-induced benign adenomas (Dunnington *et al.*, 1983), SVE3 and Huma 7 from SV40-immortalized normal human glands and Huma 121 and 123, from the benign hyperplastic human cell line HMT-3522 (Rudland, 1993) (Table 3). All these single-cell-cloned epithelial cell lines repeatedly give rise to more-elongated cells at a frequency of 0.1% (human) to 3% (rat). These more-elongated cells can be isolated as clonal cell lines (Figure 6C), representatives of which are Rama 711 from Rama 704, Rama 29 from Rama 25, Huma 25 from SVE3, Huma 62 from Huma 7 etc. (Table 3). These cloned cell lines are completely stable. In addition, two further morphological cell types are observed within the cultures of the epithelial-like cell lines, large flat cells (Figure 6E) (Edwards *et al.*, 1984; Warburton *et al.*, 1985) and dark droplet/vacuolated cells (Figure 6F) (Bennett *et al.*, 1978; Rudland *et al.*, 1989b). During the early stages of culture the epithelial-like cells require the more-elongated cells for growth, but this requirement is partially diminished after immortalization and extended culture (Bennett *et al.*, 1978; Rudland *et al.*, 1989a, 1989b). All these cell lines may also be grown on floating gels of Type I collagen to mimic, to a certain degree, a stromal matrix (Emerman *et al.*, 1977; Ormerod and Rudland, 1982; Rudland *et al.*, 1991). In addition the cell lines derived from benign tumours or immortalized with SV40 can also be grown as tumours in syngeneic rats (Dunnington *et al.*, 1983) or as small tumour nodules in nude mice (Bennett *et al.*, 1978; Rudland *et al.*, 1982a, 1989a). The tumour nodules regress after about 10 days. All three types of growth environment have been used for subsequent analysis.

Immunofluorescent and immunocytochemical staining with reagents against MFGM/EMA stain all the epithelial-like cells in a peripheral manner and the larger, flat cells very intensely confirming both cells' epithelial characteristics (Table 4) (Rudland, 1987a, 1987b, 1993). The epithelial-like cells also stain with epithelial-specific MAbs to keratin 18, but the large flat cells fail to do so appreciably (Table 4). The flat cells, however, stain intensely with MAbs to more complex keratins, suggesting a squamous form of the epithelial cells (Rudland, 1987a, 1993). Immunostaining with antibodies to smooth muscle actin, myosin, vimentin and CALLA in rats and humans, with MAbs to the 135 K_d glycoprotein in humans or with GS-1/PWM lectins in rats stains predominantly the more-elongated cells, albeit in a rather heterogeneous fashion. These results, at first sight, suggest that they may have a mesenchymal origin (Table 4) (Rudland, 1987a, 1987b, 1993). However, the fact that the more-elongated cells produce the basement membrane proteins laminin and Type IV collagen and in the rat Thy-1 (Rudland, 1987a), and shortly after conversion from epithelial cells they also produce keratin intermediate filamental

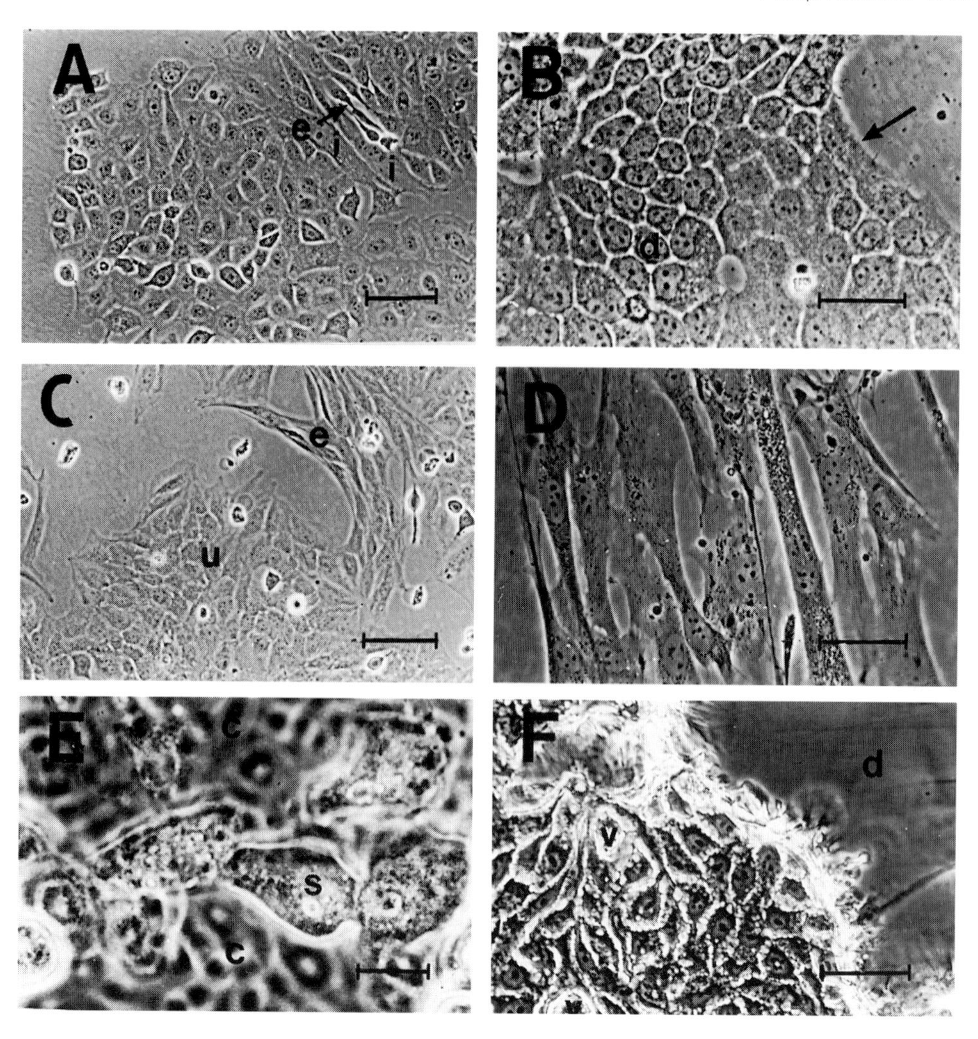

Figure 6 Morphology of normal human mammary cells in culture. SV40-immortalized normal human epithelial cells have been photographed with phase-contrast optics; the morphology of the rat cells are identical. (A) Colony of cuboidal epithelial cells, SVE3 showing the very occasional intermediate cell (i) and more-elongated cell (e). (B) Cuboidal epithelial cells of Huma 7 showing more compact grey cells (←) and dark cells (d) in places. (C) More-elongated, myoepithelial-like cells, Huma 25 which when sparse assume an elongated morphology (e) but when confluent appear pseudocuboidal (u) due to the over- and underlapping of cellular processes. (D) Mammary fast-sticking stromal fraction of fibroblastic cells. (E) SVE3 cuboidal epithelial cells (c) undergoing processes of desquamation by shedding thin, enucleated cellular residues or squames (s) into the medium; this happens much less frequently in the corresponding rat cells. (F) Small dark, droplet cells with the occasional vacuolated cell (v) and associated hemispherical blister or dome (d) in cultures of SVE3. Original magnifications $\times 110$, $\times 280$, $\times 110$, $\times 280$, $\times 550$, and $\times 280$. Bars = 100, 50, 100, 50, 20 and 50 μm for A, B, C, D, E and F, respectively (Rudland *et al.*, 1989a).

Table 3 Origins and differentiating ability of rat and human mammary cell lines.

Mammary tissue	Cell line	Identity	Differentiate[a]
Normal rat	Rama 704	epithelial	Yes
	Rama 711/712	myoepithelial-like	No
	Rama 401	myoepithelial-like	No
Benign DMBA rat tumour	Rama 25	epithelial	Yes
	Rama 259	epithelial, truncated alveolar pathway	Yes
	Rama 25-I	epithelial/myoepithelial intermediate cells	Yes
	Rama 29	myoepithelial-like	No
Benign DMBA syngeneic rat tumour	Rama 37	epithelial	Yes
	Rama 41	epithelial, truncated alveolar pathway	Yes, partially
Weakly-metastasizing rat tumour, TR2CL	Rama 600	epithelial, no alveolar and greatly truncated myoepithelial pathways	Very incompletely
Moderately-metastasizing rat tumour TMT-081	Rama 800	anaplastic epithelial	No
Strongly-metastasizing rat tumour SMT-2A	Rama 900	anaplastic epithelial	No
Normal human immortalized by SV40	SVE3	epithelial	Yes
	Huma 7	epithelial	Yes
	Huma 25	myoepithelial-like	No
	Huma 62	myoepithelial-like	No
Benign breast disease parent line HMT-3522	Huma 121	epithelial	Yes
	Huma 123	epithelial	Yes
	Huma 101	myoepithelial-like	No
	Huma 109	myoepithelial-like	No
	Huma 131	myoepithelial-like	No
Human ductal carcinoma	Ca2-83, KM1	malignant epithelial	No
Metastatic pleural effusion	MCF-7[b], ZR-75[b], T-47D[b], SK-Br-3[b], MDA-MB-231[b]	malignant epithelial	No

[a] Differentiate to both myoepithelial-like and alveolar-like cells, all epithelial cell lines differentiate to squamous-like cells, particularly those of human origin (Rudland 1987a, 1987b, 1993). [b] Established cell lines obtained from other laboratories.

proteins, which may or may not be retained on long-term passaging, suggests that such cells are not simply fibroblastoid cells (Boyer *et al.*, 1989) but are more closely related to myoepithelial cells (Table 4) (Rudland, 1987a, 1987b, 1993). Ultrastructural analysis confirms that some of the more-elongated cells possess a phenotype consistent with that of immature myoepithelial cells, although dense networks of myofilaments are encountered only rarely in such growing cultures (Warburton *et al.*, 1981; Ormerod and Rudland, 1982, 1985; Rudland *et al.*, 1989a, 1991), as observed in an immortal myoepithelial cell line from a human salivary adenoma (Shirasuna *et al.*, 1986).

Table 4 Summary of histochemical staining of different cell types in culture.

Antibody or reagent[a]	Cell type[b]			
	Cuboidal	Large, flat	More-elongated	Vacuolated
Epithelial-related				
Anti-EMA, PNL+N, MAbs to MFGM	+	++	−	+
MAbs to keratin 18	+	±	−	±
Epithelial and myoepithelial-related				
Antibody to complex keratins, PKK2 and LP34	±	++	+	±
Myoepithelial-related				
Anti-smooth muscle actin, myosin[c], MAb to keratin 14[d], Anti-vimentins[c,e]	−	−	+	−
MAbs to CALLA, 135 K_d glycoprotein[c,f]	±	±	+	−
GS-1, PWM lectins[g]	−	−	+	−
Anti-Type IV collagen/laminin/Thy-1[c,h]	−	−	+	−
Alveolar-related				
PNL alone	−	+	−	++
Anti-caseins	−	−	−	+
Conclusions	Epithelial	Squamous-like	Myoepithelial-like	Alveolar-like

[a] Abbreviations and reagents as in text and Table 1, +N with neuraminidase treatment; [b] Identical results obtained for normal and benign rat and human primary cultures and cell lines. Key: ± weak and diffuse staining; + most cells stain; ++ extremely strong staining; [c] Also stains blood vessels in stroma; [d] Staining with MAb to cytokeratin 14 has only been performed on human cell lines as yet; [e] Also stains stromal fibroblasts; [f] MAb to 135 K_d glycoprotein not stain rat system; [g] GS-1, PWM lectins not show differential staining on human systems; [h] Antibodies to Thy-1 not used on human systems.

The epithelial-like, squamous-like and myoepithelial-like cell types have also been observed in primary cultures of rat (Warburton *et al.*, 1985) and human (Rudland *et al.*, 1989b) mammary glands. The most strongly anti-EMA/MFGM-staining and anti-CALLA-staining cell types in primary culture have been separated by fluorescence-activated cell sorting (O'Hare *et al.*, 1991). The former cell type probably corresponds more to the squamous epithelial-like cells and the latter more to myoepithelial-like cells. The peripherally, more-moderately anti-EMA-stained epithelial cells which proliferate in early cultures (Warburton *et al.*, 1985; Rudland *et al.*, 1989b) and which correspond to the immortalized epithelial cell lines (Bennett *et al.*, 1978; Rudland *et al.*, 1989a) do not seem to have been separated by this technique.

When confluent cultures of the epithelial cell lines become densely packed they form small, dark polygonal cells with small vacuoles or droplets at their peripheries (hence the term droplet cells) and hemispherical blisters or domes in the cell layer (Figure 6F).

Conversion of an initially homogeneous culture of epithelial cells to droplet/vacuolated cells can be accelerated with agents that stimulate Friend erythroleukaemia cells, notably DMSO (Bennett *et al.*, 1978), PGE_1 (Rudland *et al.*, 1982b) or retinoic acid (Rudland *et al.*, 1983a) and mammotrophic hormones. These rat cultures synthesize increased amounts (20–40 fold) of immunoreactive casein which has been authenticated as the p42 kDa component present in rat milk by peptide mapping techniques (Warburton *et al.*, 1983). They also demonstrate an increase in staining with peanut lectin (Newman *et al.*, 1979). Based on the above criteria, the droplet/vacuolated cells have been adjudged to be related to alveolar cells, although, since the amount of casein synthesized is only 1–2% of that found in lactating rat mammary gland explants (Warburton *et al.*, 1983), they are classified as alveolar-like cells (Table 4). The formal identification of human casein in human cell lines by similar biochemical criteria has not yet been attempted.

When grown on floating collagen gels (Ormerod and Rudland, 1982, 1988; Rudland *et al.*, 1991) or as tumour nodules in syngeneic rats (Dunnington *et al.*, 1983) or nude mice (Rudland *et al.*, 1982a, 1989a), the epithelial cell lines form rod- or cord-like structures, a few of which, particularly those grown on gels, are hollow and superficially resemble ducts. A few other structures have more bulbous, solid ends like TEBs, others resemble ABs, whilst a further few consist of grape-like structures, some of which superficially resemble ALs. A few of the duct-like structures and many of the sheets of cells on gels consist of two or more cellular layers; the outer/basal layer possessing fewer epithelial and more myoepithelial properties and the inner/upper layer consisting of epithelial cells correctly polarized with respect to the lumen/medium (Ormerod and Rudland, 1982, 1988; Rudland *et al.*, 1991). Lactating, nude mice bearing rat- or human-tumour nodules produced by the epithelial cell lines can form hollow, sac or grape-like structures that secrete rat-specific or human-specific forms of casein, respectively (Rudland *et al.*, 1983b, 1989a). These casein-secretory cells may then be related to alveolar cells and also to the droplet/vacuolated cells observed in primary cultures (Rudland *et al.*, 1977a, 1989b) and in epithelial cell lines (Bennett *et al.*, 1978; Dunnington *et al.*, 1983; Ormerod and Rudland, 1985; Rudland *et al.*, 1989a).

The epithelial cell lines can therefore give rise to myoepithelial-like and to alveolar-like cells (Figure 7), although the full expression of the differentiated phenotype is difficult to achieve in either case. Since the same epithelial cell lines can also give rise, at least superficially, to many of the mammary parenchyma, they may indeed be part of a stem cell system (Ormerod and Rudland, 1982, 1988; Rudland *et al.*, 1991). However, these epithelial cell lines, particularly those of human origin, produce cells and structures not normally encountered in the mammary gland, the most notable being the flat, squamous-like cells (Figure 7) and corresponding sheets and whorls of keratinizing epithelium on collagen gels (Rudland *et al.*, 1991) and in regressing tumour nodules in nude mice (Rudland *et al.*, 1982a, 1989a). However, it is unlikely that immortalization has resulted in this extra phenotype, since much larger numbers of squamous-like cells are produced in primary cultures (Edwards *et al.*, 1986; Rudland *et al.*, 1989b), and their production is probably due to the non-physiological conditions of tissue culture with their emphasis on cell growth. The senescence of primary human cultures is predominantly due to the rapid production of such near-terminal squamous-like cells from the growing epithelial cells (Rudland *et al.*, 1989b), and transformation with SV40 yields epithelial cells with a considerably reduced ability to produce the squamous-like cells (Rudland and Barraclough, 1990), thereby probably effecting immortalization.

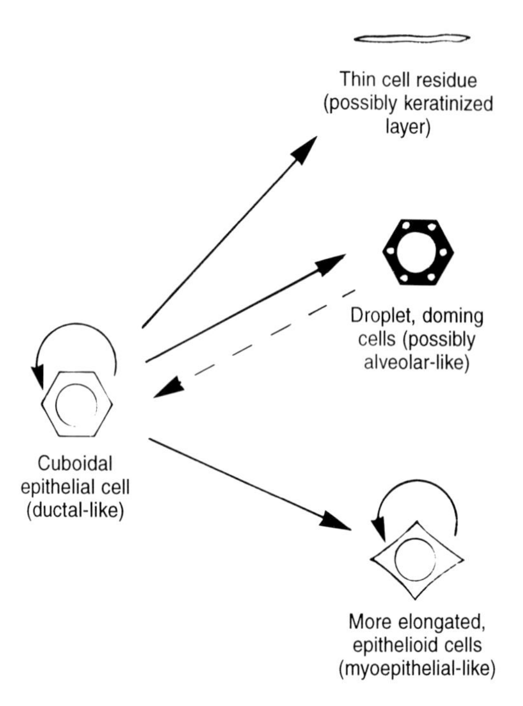

Figure 7 Summary of intercellular conversions of normal and benign epithelial cell lines. The cuboidal epithelial cell lines of rat and human origin can either replicate (curved arrow) or give rise to different cell types (straight arrow). Conversion to thin cellular residues of squamous-like cells or to more-elongated, myoepithelial-like cells is irreversible but that to the droplet/vacuolated/doming, alveolar-like cells can be reversed (discontinuous lines). The epithet 'possibly' has been used, since the production rat caseins has been formally established by biochemical means for the rat lines but only as yet by immunohistochemistry for human lines (Rudland, 1987a; Rudland *et al.*, 1989a).

Identification of discrete differentiation stages of a potential stem cell system *in vitro* and their consequences *in vivo*

Differentiation of the benign rat mammary epithelial cell line Rama 25 along pathways to both myoepithelial-like and to alveolar-like cells occurs in discrete stages. Clonal cell lines that are intermediate in morphology between Rama 25 epithelial cells and more-elongated myoepithelial-like cells have been isolated and they form a morphological series in the order: Rama 25 cuboidal cells, Rama 25-I2, Rama 25-I1, Rama 25-I4, and more-elongated myoepithelial-like cells (Figure 8a). This same order is maintained for increasing frequency of conversion to more-elongated cells, increased binding of antisera to laminin, vimentin, Thy-1, increasing abundance of 7 polypeptides characteristic of the more-elongated myoepithelial-like cells, increasing myoepithelial ultrastructural features (Rudland *et al.*, 1986) and increased binding to the lectins GS-1 and PWM (Rudland and Hughes, 1991) (Figure 8a). Similarly, the same order is maintained for decreasing conversion to cuboidal epithelial cells, decreased binding of antisera to MFGM, decreased abundance of four polypeptides characteristic of cuboidal epithelial cells and decreasing epithelial ultrastructure (Rudland *et al.*, 1986) (for further details see page 183).

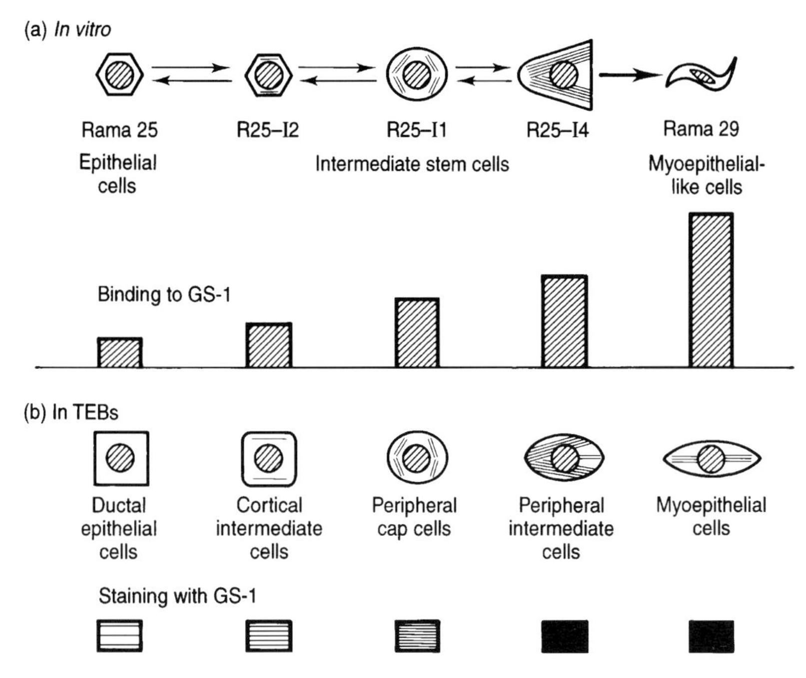

Figure 8 Schematic representation of the changes observed on differentiation to myoepithelial-like cells. (a) The linear pathway of cloned morphologically intermediate cell types between the Rama 25 cuboidal epithelial and more-elongated myoepithelial-like cells (e.g. Rama 29) is shown. This order is based on levels of markers and conversion frequencies of the individual rat cell lines (Rudland *et al.*, 1986) as well as the time of appearance of such markers after triggering differentiation with colchicine (Paterson and Rudland, 1985a). Only the last step is irreversible. The levels of one such marker, the receptor for the lectin GS-1 is shown (Rudland and Hughes, 1991). Since the intermediate cells can give rise to epithelial and to myoepithelial cells they may be considered as mammary stem cells. (b) The morphologically intermediate cell types between the ductal epithelial and the myoepithelial cells in rat terminal end buds are shown. This order is based on marker characteristics (Ormerod and Rudland, 1984). The appearance of one such marker, the receptor for GS-1 is shown (Hughes and Rudland, 1990b).

That the above cell lines of intermediate characteristics represent true intermediate stages along the more-elongated, myoepithelial pathway *in vitro* is suggested by the results of pretreatment of Rama 25 epithelial cells with the microtubule-disrupting agent colchicine. This treatment progressively increases the percentage of more-elongated-cell colonies after removal of the drug (Paterson and Rudland, 1985a). When followed by time-lapse cinematography, this conversion process is seen to occur by sequential morphological changes, similar to those of the cell lines described previously (Figure 8a). The last stage is the only irreversible one. Increases in binding of Thy-1 antiserum, changes in most of the 11 characteristic peptides (Paterson and Rudland, 1985a) and binding to the lectins GS-1 and PWM (Rudland and Hughes, 1991) are also consistent with this model. These cellular intermediates have also been identified by immunocytochemical and morphological criteria in primary cultures of normal rat (Warburton *et al.*, 1985) and human (Rudland *et al.*, 1989b) mammary glands, and similar cell lines to those derived from Rama 25 have been isolated from normal rat (Ormerod and Rudland, 1985) and SV40-immortalized human (Rudland *et al.*, 1989a) mammary epithelial cells. Several of the polypeptides characteristic of Rama 29 cells have also been found in different more-elongated myoepithelial-like cells and cell lines from normal mammary glands (Barraclough *et al.*, 1984a) (page 183). These results suggest that the intermediate cells are not unique to a single cell line from a benign rat mammary tumour.

When grown on floating collagen gels, the intermediate cell line Rama 25-I2 forms more mature duct-like structures than the parental Rama 25 epithelial cells, and these structures possess both epithelial and myoepithelial cells organized correctly within the tubules (Rudland *et al.*, 1986). The other two intermediate cell lines fail to form such structures, but yield instead giant, multinucleated cells which have ultrastructural characteristics of skeletal muscle and which produce striated-specific myoglobin (Rudland *et al.*, 1984a). This result has been confirmed when cells are grown in nude mice as tumours and when similar cells from benign neoplastic Rama 37 (Rudland *et al.*, 1984a) or normal Rama 704 (Ormerod and Rudland, 1985) cell lines are grown on gels *in vitro* or, for Rama 37-derived cells, in syngeneic rats *in vivo*. Ultrastructural and immunocytochemical analysis of cells on gels (Rudland *et al.*, 1986) and quantitative binding of the lectins GS-1 and PWM to cells on plastic (Rudland and Hughes, 1991) suggest that the intermediate Rama 25-I1 cells resemble the peripheral cap cells in TEBs, Rama 25-I2 cells resemble the cortical cells intermediate between the cap cells and the ductal epithelial cells which line lumina, and Rama 25-I4 cells resemble the peripheral cells intermediate between the cap cells and the myoepithelial cells of the subtending duct in TEBs (Figure 8B). The Rama 25-I4 cells resemble the peripheral cap cells in ABs more closely than Rama 25-I1 (Rudland *et al.*, 1986; Rudland and Hughes, 1991). Since the intermediate cell lines can give rise to epithelial and to myoepithelial-like cells as well as to well-differentiated mesenchymal elements such as skeletal muscle (Rudland *et al.*, 1984a), they are better candidates than the closely-related epithelial cell lines such as Rama 25 for both normal stem cells in mammary TEBs *in vivo* and for the neoplastic stem cells in the mixed tumours of glandular origin which produce similar differentiated mesenchymal elements (Hamperl, 1970).

Time-lapse cinematographic analysis of Rama 25 epithelial cells stimulated to differentiate to the droplet/vacuolated alveolar-like cell, indicates that a linear sequence of morphological stages exists and that it is triggered, directly or indirectly, by the pumping action of the Na^+/K^+ ATPase (Paterson *et al.*, 1985a). The linear sequence of morphological changes or stages; cuboidal epithelial (Figure 6A), grey, dark (Figure 6B), droplet and vacuolated cells (Figure 6F) (Paterson *et al.*, 1985b) is associated with the increased production of novel polypeptides at each stage *in vitro* (Paterson and Rudland, 1985b). There is also an increase in the binding of peanut lectin observed for each morphological stage: desialylated cuboidal epithelial cells fail to bind this lectin; desialylated grey cells bind this lectin weakly and cytoplasmically; desialylated dark cells bind this lectin strongly to their cell surfaces; whilst droplet cells and vacuolated cells bind this lectin to their cell surfaces without the necessity to remove terminal sialic acid residues. With the requisite mammatrophic hormones and inducing agent, only the vacuolated cells produce casein (Figure 4B) (Rudland, 1992). Results for the production of the characteristic polypeptides and for the binding of peanut lectin by variant epithelial cell lines, Rama 259 and Rama 41 are consistent with this linear model of cellular interconversions between the different morphological cell types (Figure 4B) (Paterson and Rudland, 1985b; Rudland, 1992). DMSO or retinoic acid in the presence of prolactin, oestrogen, hydrocortisone and insulin accelerates the overall pathway predominantly by increasing the rate of droplet-formation and this can be reversed by removing the inducers (Paterson *et al.*, 1985b). Prolactin is the most important hormone for production of casein in this system.

Similar morphological stages to those in the Rama 25 system have been observed in normal primary cultures and cell lines of rat (e.g. Rama 704) (Rudland *et al.*, 1977a; Ormerod and Rudland, 1985) and human (Rudland *et al.*, 1989a, 1989b) mammary epithelial cells showing that they are not unique to an epithelial cell line from a benign rat mammary tumour. By the criterion of binding peanut lectin discussed on page 155, the cuboidal epithelial cells correspond to the major cell type present in ducts, the grey cells with that present in TEBs, the dark cells with that in ABs, the droplet cells with that in ALs and the vacuolated cells with that in secretory ALs (Figure 4). Thus the different interconverting cell types of Rama 25 which form a linear pathway to casein-secretory cells *in vitro* can be equated with recognizable epithelial cell types *in vivo* (Rudland, 1992). These results support the suggestion on page 155 that casein-secretory cells *in vivo* are generated by successive interconversions between the major epithelial cell types present in the different mammary structures in the order: ducts, TEBs, ABs, ALs, and secretory ALs. The direct conversion of ductal epithelial cells to casein-secretory ALs under the influence of the same mammatrophic hormones has also been accomplished in primary cultures of normal mouse mammary glands (Hamamoto *et al.*, 1988) and clonal mouse epithelial cell lines RAC (Sonnenberg *et al.*, 1986) and COMMA-D (Danielson *et al.*, 1989). Therefore this cellular model for the generation of the secretory alveolar cell would appear to be generally applicable.

NON-HEPARIN-BINDING GROWTH FACTORS IN NORMAL MAMMARY DEVELOPMENT

Hormonal agents and micronutrients

Growth and differentiation of the mammary gland, both in rats and humans, is controlled by systemic agents secreted by the pituitary, ovary and adrenal glands. *In vivo* experiments in which various hormones were used to replace ablated glands have identified oestrogen, progesterone, glucocorticoids, prolactin, placental lactogen and growth hormone as having mammogenic activity (Lyons *et al.*, 1958; Nandi, 1958; Topper and Freeman, 1980). In addition, local trophic agents produced in the vicinity of the mammary glands are also involved in growth control, since rodent mammary tissues and cells will grow when transplanted into the mammary fat pad, but not when transplanted subcutaneously (page 157) (DeOme *et al.*, 1959; Beuving *et al.*, 1967; Gould *et al.*, 1977).

However, *in vitro*, when primary and secondary mammary cell cultures derived from normal tissue or benign tumours, or cell lines developed from them are grown in medium depleted of serum, the established mammatrophic hormones are less effective than might be anticipated in promoting cell growth (Ben-David, 1968; Oka and Topper, 1972;

Hallowes *et al.*, 1977; Kano-Sueoka *et al.*, 1977; Rudland *et al.*, 1977a, 1979; Mittra, 1980; Sirbasku *et al.*, 1982; Smith *et al.*, 1984d; Dembinski *et al.*, 1985; Newman *et al.*, 1987). Instead, three routes have been followed to identify those agents that stimulate DNA synthesis and cell division directly in mammary cells. First, growth-promoting activity has been purified from tissue extracts, notably from the pituitary gland which is believed to be the physiological centre of control of mammary growth. Second, medium that has been used to grow mammary cells in culture, particularly myoepithelial and stromal cells, has been used as a source of autocrine or paracrine growth factors. Third, conventional hormones, and growth-promoting agents for other types of cells, have been tested empirically, alone or in combinations, to discover their effect, if any, on mammary epithelial cells. In practice, the three approaches have given overlapping results, so the various mammatrophic agents will be considered separately before an attempt is made to piece together the overall pattern of growth control of the mammary gland.

One type of growth factor, as identified by the use of cultured cells in defined media, may be characterized as essential nutrients. One such example is phosphoethanolamine (Kano-Sueoka *et al.*, 1979; Barnes *et al.*, 1984), another is the iron-binding protein(s) (Rudland *et al.*, 1977b). Both are isolatable from bovine pituitaries (Riss and Sirbasku, 1987b). In our rat mammary model system, some cell lines (Rama 25, Rama 29), but not others, grow faster in the presence of phosphoethanolamine (Smith *et al.*, 1984d). This idiosyncratic effect may suggest that dependency on phosphoethanolamine for growth is probably an artefact of the isolation of cell lines and may have no significance *in vivo*.

In the case of the iron-binding proteins, transferrin is essential for the growth of many cultured cells in defined media (Barnes *et al.*, 1984; Riss and Sirbasku, 1987a, 1987b), including our mammary epithelial stem cell lines (Smith *et al.*, 1984d). In addition haemoglobin and ferritin promote growth of our Rama 37 epithelial stem cell system (Wilkinson *et al.*, 1996). In the latter case, the optimal concentration of iron ions for cell division is four times that normally present in serum-free medium. Thus the growth-promoting effect of iron-free transferrin may well be to increase the availability of sub-optimal amounts of iron in the medium (Rudland *et al.*, 1977b). This result suggests that the iron requirement also may be an artefact of the culture system. However, a high number of transferrin receptors has been reported in rapidly proliferating cells (Trowbridge and Omary, 1981; Sutherland *et al.*, 1981; Neckers and Trepel, 1986) including those in pregnant and lactating mammary glands of mice (Schulman *et al.*, 1989) and humans (Walker and Day, 1986). Transferrin and its mRNA are also found within the developing rat mammary gland; transferrin itself constitutes 1% of cytosolic protein in early pregnancy, rising to 2% of cytosolic protein in late pregnancy and lactation (Keon and Keenan, 1993). This rise is consistent with increases in its mRNA (Grigor *et al.*, 1990). The finding of mRNA for transferrin suggest that at least some of the transferrin in the mammary gland is produced locally. Surprisingly, there is a sudden increase in the content of transferrin to 15% of cytosolic protein when suckling is prevented and this increase may provide a signal for involution of the gland (Grigor *et al.*, 1990). Both normal transferrin and lactoferrin are found in the milk of sheep; radioactive labelling experiments with explants suggest an extra-mammary origin for some of this transferrin (Sanchez *et al.*, 1992). Moreover, in rabbit mammary epithelial cells endocytosis of added transferrin has been observed, and therefore the observation of transferrin within a cell does not necessarily imply that the cell is its site of synthesis (Seddiki *et al.*, 1992). Thus the possibility exists in the case of transferrin that it may be acting like a growth factor, either in an autocrine/paracrine manner if produced locally, and/or in an endocrine manner if its levels in blood vary appreciably.

Although prolactin has little growth-promoting effect *in vitro*, pituitary extracts are mitogenic for mammary cells (Sirbasku *et al.*, 1982). The pituitary-derived activity that promotes the growth of our rat mammary epithelial stem cell lines has still only been partially purified. None the less, it is separable from bovine prolactin, growth and other pituitary hormones and growth factors including fibroblast growth factors (Smith *et al.*, 1984d). It is also distinct from phosphoethanolamine and transferrin (Smith *et al.*, 1984d) and other iron-containing proteins found in the pituitary gland (Wilkinson *et al.*, 1996), and its stimulatory effect is apparent even in the presence of fetal calf serum. Pituitary mammary growth factor (PMGF) stimulates the growth of our rat epithelial stem cell lines, but not their myoepithelial-like derivatives, nor fibroblasts. Thus the ability to respond to the growth-promoting effects of PMGF is lost when the epithelial stem cell system differentiates to myoepithelial-like cells in culture (Smith *et al.*, 1984d). Although PMGF is not yet chemically defined, it is one of the few growth factors implicated in the process of mammary development *in vivo*. Thus the activity of PMGF from pituitary glands of early lactating or perphenazine-treated rats (Ben-David, 1968), when the mammary glands are growing rapidly, is four to ten times greater than the activity found in untreated virgin females (Smith *et al.*, 1984d). Pure rat or bovine prolactin does not stimulate the growth of our rat epithelial cell lines (Smith *et al.*, 1984d), but does stimulate production of casein-secreting, alveolar-like cells in confluent epithelial cultures and cell lines, as outlined earlier (page 163). Thus, although prolactin and PMGF may be under similar hypothalamic control in the pituitary gland, prolactin acts on the epithelial cells of the mammary gland to induce differentiation, whereas PMGF may serve to promote the proliferation of the epithelial/stem cells in glandular growth.

Epidermal growth factors

The most abundant growth factor of the epidermal growth factor (EGF) family in the mammary gland is TGFα. This growth factor has been studied extensively, primarily as a locally-produced trophic factor. Thus, both epithelial (Rama 25) cells and, particularly, myoepithelial (Rama 29) cells in culture have been shown to secrete mature, bioactive TGFα (Smith *et al.*, 1989), and it is found also in the rat mammary gland *in vivo*. In the rat system, TGFα is easily separated from EGF by reversed-phase HPLC as the rat EGF is significantly less hydrophobic than mouse (or human) EGF (Smith *et al.*, 1984b). This change is due to the truncation of the C terminus of the rat molecule and the consequent loss of the tryptophan residues (Simpson *et al.*, 1985). The identity of the growth-promoting activity of the conditioned medium of the rat mammary cells as TGFα has been confirmed by its competition with radiolabelled EGF for binding to receptors, and by the presence of mRNA for TGFα mRNA in the secreting cell lines (Smith *et al.*, 1989) (Table 5).

The activity of rat TGFα (Smith *et al.*, 1989), like that of EGF (Smith *et al.*, 1984d) is not specific for a particular cell type; it stimulates the growth of fibroblastic, myoepithelial-like and epithelial cell lines from normal mammary glands and from benign tumours. The amounts of TGFα secreted from the Rama 29 myoepithelial-like cells in culture are

Table 5 Summary of data on the presence of EGF/TGFα-related material in normal mammary glands.

Location	Material identified	Reference
Human mammary epithelium		
cultured organoids	mRNA for TGFα	Zajchowski *et al.*, 1988; Bates *et al.*, 1990
epithelial cells in situ from developing glands	mRNA for TGFα	Liscia *et al.*, 1990
epithelial cells in situ	immunoreactive EGF	Mori *et al.*, 1989
epithelial and myoepithelial cells in situ	immunoreactive TGFα	McAndrew *et al.*, 1994b
Rat mammary gland		
tissue extract	bioactive TGFα	Smith *et al.*, 1989
epithelial and myoepithelial-like cloned cell lines	mRNA for TGFα; TGFα protein by receptor and bioactivity	Smith *et al.*, 1989
epithelial and myoepithelial cells *in situ*	immunoreactive TGFα	McAndrew *et al.*, 1994b
Mouse mammary gland	mRNA for TGFα	Snedeker *et al.*, 1991
	mRNA for EGF	Fenton and Sheffield, 1991; Snedeker *et al.*, 1991
Human milk	EGF – immunoreactive	Okada *et al.*, 1991; Corps and Brown, 1987
	– receptor competition	Corps and Brown, 1987
	– bioactive	Corps and Brown, 1987; Shing and Klagsbrun, 1984; Dai *et al.*, 1987; Petrides *et al.*, 1985
	– sequence	Petrides *et al.*, 1985
	TGFα – immunoreactive	Okada *et al.*, 1991
	– receptor competition	Zwiebel *et al.*, 1986
	– bioactive	Petrides *et al.*, 1985
Rat milk	EGF – immunoreactive	Raaberg *et al.*, 1990
Swine milk	EGF – bioactive; – immunoreactive	Tan *et al.*, 1990
Goat colostrum	EGF – receptor competition; – bioactive ($>$20 kDa)	Brown and Blakeley, 1983

sufficient to stimulate the growth of the myoepithelial-like cells themselves, or of epithelial or stromal cell lines. On the face of it, this suggests a paracrine or autocrine role for TGFα in the mammary gland *in vivo*. Apparent confirmation of the pattern, in which differentiation from the epithelial to myoepithelial phenotype is accompanied by an increase in TGFα production, came from immunofluorescent and immunocytochemical studies (McAndrew *et al.*, 1994b). Again, immunofluorescence with anti-TGFα serum is greater in Rama 29 myoepithelial than Rama 25 epithelial cells, whilst immunocytochemical staining of the rat mammary gland shows the presence of immunoreactive (ir)-TGFα primarily in the myoepithelial and to a lesser extent in the epithelial cells during development. The intermediate stem cells in growing TEBs are stained at an intermediate level. In lactating glands alveolar secretions are also stained strongly (Figure 9).

The story is more complicated, however. Western blotting of ir-TGFα in extracts of the rat mammary glands and cell lines shows that ir-TGFα consists mainly of a 50 kDa

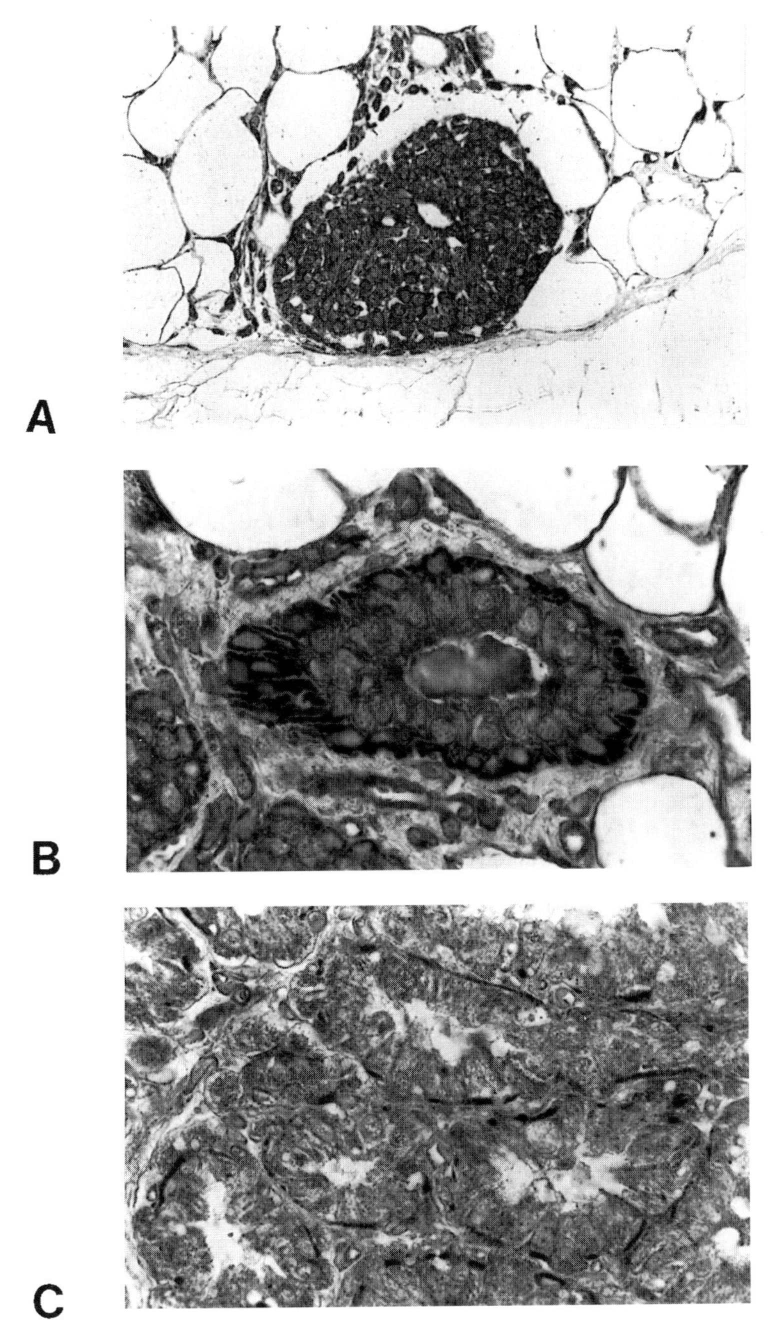

Figure 9 Immunocytochemical staining of the rat gland by anti-TGFα. (A) Terminal end bud in cross-section from 50-day-old rat showing moderate to strong staining of all growing parenchymal cells and associated blood vessels and fibroblasts. (B) Duct from 50-day-old rat showing strong staining of outer (myoepithelial) cells of the duct and little or no staining of the inner epithelial cell layer. The blood vessels and fibroblasts are also stained. (C) Lactating mammary gland showing intense but fragmented staining of the distended myoepithelial cells, but only weak or no staining of alveolar cells. Original magnifications $\times 230$ for A and C; $\times 730$ for B. Bars as for Figure 3.

protein and the amounts of ir-TGFα are considerably in excess of the amounts of bioactive TGFα observed in the same extracts. Moreover, ir-TGFα is also found in rat mammary fibroblasts *in vivo* and ir-TGFα (McAndrew *et al.*, 1994b) and mRNA for TGFα are found in the equivalent Rama 27 cell line *in vitro*. However, this cell line does not secrete bioactive TGFα into the culture medium (Smith *et al.*, 1989). These results and others (Shing and Klagsbrun, 1984; Brown *et al.*, 1986; Bates *et al.*, 1990; Madsen *et al.*, 1992; McAndrew *et al.*, 1994a) (Table 5) warn against too simplistic interpretations in relation to growth control. On balance, though, it is likely that TGFα is a normal product of mammary cells and this suggestion is reinforced by the finding of immunoreactive (Okada *et al.*, 1991) and bioactive TGFα in human (Petrides *et al.*, 1985; Zwiebel *et al.*, 1986) and rat (McAndrew *et al.*, 1994b) milk (Table 5). There is also preliminary but less substantial evidence implicating other EGF-like growth factors, amphiregulin and heregulin-α in autocrine regulation of normal human mammary epithelial cells (Li *et al.*, 1992; Normano *et al.*, 1993). EGF itself is largely absent from the mammary gland except when transported from circulating blood into milk (Brown *et al.*, 1986) (Table 5).

As outlined earlier the development of the mammary gland is controlled, in part, by systemic steroids. There is now some evidence that TGFα and/or EGF can substitute for oestrogens. Thus, when introduced in plastic implants, oestrogen, TGFα and EGF can all stimulate local ductal growth in ovariectomized mice. Local implants of EGF or TGFα in 5-week-old female mice supplemented with oestrogen and progesterone cause local lobuloalveolar development. Moreover, the gland is five times more sensitive to TGFα than to EGF and the TGFα response can be seen even in the absence of the steroid supplements (Vonderhaar, 1987) suggesting that TGFα is the more potent molecule. EGF/TGFα form new TEBs, restore their histomorphology and stimulate the reappearance of the intermediate stem cell layer in ovariectomized mice. EGF/TGFα also reinitiate DNA synthesis and increase the ductal diameter, although no lobuloalveolar or hyperplastic growth is seen in this case. EGF binding is observed to the intermediate stem cells of the TEBs, to the myoepithelial cells of the ducts and to the stromal cells adjacent to the TEBs and subtending ducts (Coleman *et al.*, 1988). These results coupled with *in vitro* evidence with human breast carcinoma cell lines on page 203 suggest that TGFα and to a lesser extent EGF may be mediators of some of the mitogenic effects of oestrogen in the mammary gland.

TGFα normally binds to the EGF receptors (EGFR) in many different cell types. All our rat mammary cell lines possess high-affinity receptors for EGF with dissociation constants ranging from 0.4 nM to 1.3 nM (Smith *et al.*, 1989; Fernig *et al.*, 1990). Furthermore, both epithelial and myoepithelial-like cell lines possess about 22,000 high-affinity receptors per cell with identical K_d values (Fernig *et al.*, 1990). Thus there is no alteration in the EGF receptors when epithelial stem cells differentiate along the myoepithelial-like pathway *in vitro*. These results suggest that locally produced TGFα may stimulate the growth of epithelial, intermediate stem and myoepithelial cell types *in vivo*. These results also suggest that all mammary gland cell types *in vivo* possess cell-surface high-affinity receptors for EGF, in accordance with direct-binding studies in mice (Coleman *et al.*, 1988). In contrast, the SV40-immortalized human mammary epithelial and myoepithelial-like cell lines express much greater numbers (10^5 to 10^6 per cell) of high-affinity receptors for EGF (McAndrew *et al.*, 1994b) than either their rat counterparts above or human mammary cell lines not immortalized by SV40. A fourfold increase in cell-surface EGF receptors upon immortalization by SV40 is also observed in a separate human mammary system (Valverius *et al.*, 1989). These results suggest that one of the mechanisms of immortalization by SV40 may involve increasing the number of cell-surface receptors for EGF such that terminal differentiation (keratinization) to squamous formations in primary culture of human mammary epithelium, the probable cause of their senescence (Rudland *et al.*, 1989a), is largely suppressed (Rudland and Barraclough, 1990) (page 163).

Insulin-like growth factors

The insulin-like growth factors (IGFs) have been studied in connection with the mammary gland in large part because of the widespread use of growth hormone (GH) in the rearing of cattle, and the assumption that the effects of pituitary GH are mediated by IGF-1 (Hauser *et al.*, 1990; Turner and Huynh, 1991; Glimm *et al.*, 1992). However, much of the effect of GH on milk production is probably due to an increase in general metabolic rate (Breier *et al.*, 1991). What is clear, though, is that high levels of IGF-1, IGF-2 and insulin are found in bovine mammary secretions just before and after parturition and these levels fall rapidly after a few days of milking (Skaar *et al.*, 1991; Schams and Einspanier, 1991; Einspanier and Schams, 1991).

Growth effects of IGFs on mammary tissue have been observed. Thus Ruan *et al.* (1992) have shown that administration of IGF-1 induced the development of TEBs and the formation of alveolar structures in the mammary gland of hypophysectomized, castrated and oestradiol-treated immature male rats. These effects are similar to those found with GH. In tissue culture IGF-1 can stimulate the formation of colonies of NMU-induced mammary (benign) tumour cells in soft agar (Manni *et al.*, 1990). Moreover, both IGF-1 and insulin stimulate [^{3}H] DNA synthesis of tissue slices of normal, pregnant and lactating mammary glands (Baumrucker and Stemberger, 1989). The cells radioactively labelled include alveolar and ductal epithelial cells, myoepithelial cells, fibroblasts and white blood cells. That IGF-1 and IGF-2 are mitogenic for mammary epithelial cells has also been confirmed in a number of other systems *in vitro* (Zhao *et al.*, 1992; Winder *et al.*, 1993).

With respect to receptors, the bovine mammary gland expresses high-affinity receptors for IGF-1 and IGF-2 (Hadsell *et al.*, 1990; Winder *et al.*, 1993), although those for IGF-1 unlike the GH receptors are mainly found in the stromal cells (Hauser *et al.*, 1990). The possibility exists, therefore, of an autocrine/paracrine loop. In support of this Lavandero *et al.* (1991) have shown that the mammary glands of pregnant and lactating rats have both high- and low-affinity binding sites for IGF-1, and that explants of lactating mammary glands secret IGF-1 into the culture medium; an inverse relationship is found between the production of IGF-1 and the number of binding sites. On the other hand Manni *et al.* (1992) have found mRNA for IGF-2 but not for IGF-1, in normal and pregnant rat mammary glands. Thus the role of the IGFs in mammary development *in vivo* is best described as open at present.

Other growth factors

Three other families of growth factors need to be considered: platelet-derived growth factors (PDGF), interleukins and prostaglandins (PG). PDGF is secreted into the conditioned medium of human mammary epithelial cells (Bronzert *et al.*, 1990) and mRNAs for both A and B chains have been found in all normal human breasts examined (Coombes *et al.*, 1990). In benign breast lesions the mRNA for PDGF-B chain is located in the epithelial cells (Ro *et al.*, 1989). A growth effect of PDGF-B chain is observed on HCll cells (derived from pregnant mouse mammary glands) (Taverna *et al.*, 1991). This growth stimulation differs from that induced by EGF in that the latter also enables prolactin to stimulate casein synthesis. Interleukins (ILs) have mainly been of interest in relation to attempts at immunotherapy of breast cancer. At this point it is worth noting that IL1 and IL6 have been found in human milk (Saito *et al.*, 1991b; Rudloff *et al.*, 1993) and breast cyst fluid (Reed *et al.*, 1992b), and a peptide immunologically related to IL6 is secreted by human breast fibroblasts in culture (Adams *et al.*, 1991). In the cow, bioactive IL2 has been found in prepartum secretions (Sordillo *et al.*, 1991) and administered IL2 accelerates involution of lactating bovine mammary gland (Nickerson *et al.*, 1992). The significance *in vivo* of PDGF or of the interleukins in the development of the normal mammary gland is still unclear.

Finally there is evidence from *in vivo* studies for locally produced growth factors in the mammary gland. The mammary epithelium does not exist in isolation, but is embedded in a fatty stroma (page 156), and this stroma is required for normal growth and development of the mammary gland, both *in vitro* and *in vivo* (Kratchwil, 1969; Sakakura *et al.*, 1976). Thus the embryonic mesenchyme, being the precursor of the mouse mammary fat pad, promotes organogenesis of the mammary epithelium *in vivo*, and the growth of mammary epithelial cells *in vitro* (Sakakura *et al.*, 1987; Oka *et al.*, 1987; Kanazawa and Hosick, 1992). The molecular mechanisms for these interactions are far from clear. However, mitogenic activity found in conditioned medium of primary cultures of rat stromal cells, or of fibroblastic cell lines such as Rama 27, has been identified as prostaglandin E_2 (PGE_2) (Rudland *et al.*, 1984b). The amount of PGE_2 secreted by rat stromal cells correlates with their ability to differentiate into lipocytes which in turn is related to the levels of GH in the medium. Similar levels of PGE_2 stimulate the growth of epithelial cells from normal rat (Rama 704) and mouse (Bandyopadhyay *et al.*, 1988) mammary glands and from benign mammary tumours (Rama 25) (Rudland *et al.*, 1984b). These mitogenic effects are enhanced by insulin and EGF. PGE_2 is thus associated with cells from normal fatty stroma, and may thus be one of the permissive stromal mitogens *in vivo*. The growth-promoting activity from embryonic mesenchyme has been identified with material of high (100 kDa) molecular weight (Taga *et al.*, 1989), and IGFs, particularly IGF-1, can also be produced by stromal fibroblasts. The latter are considered in more detail on page 173. Any combination of the above or other as yet unidentified growth factor may constitute the local growth environment produced by the normal mammary stroma.

HEPARIN-BINDING GROWTH FACTORS IN NORMAL MAMMARY DEVELOPMENT

Appearance of receptors for fibroblast growth factors on formation of myoepithelial-like cells *in vitro*

The fibroblast growth factors (FGFs) are a family of heparin-binding growth factors and oncogenes with eight members (Figure 10), five of which, acidic FGF (aFGF), basic FGF

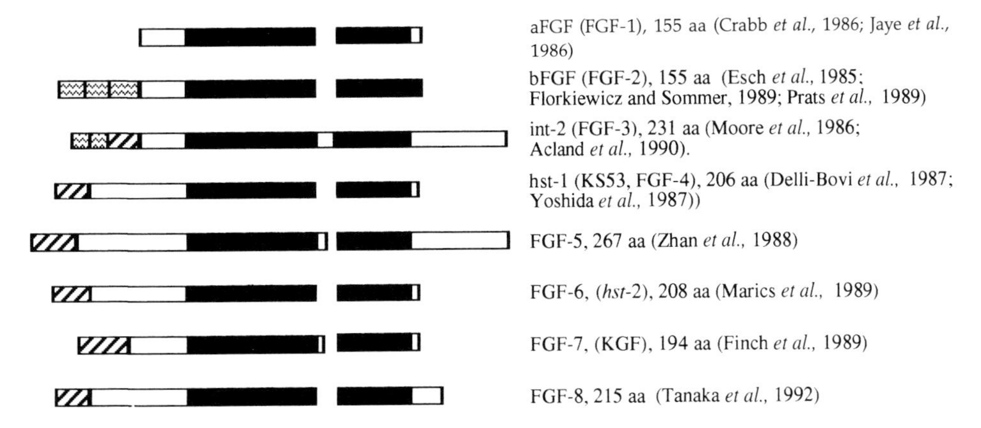

Figure 10 Structural features of the FGF family. ■ Core structure of FGFs that contains the regions of the molecules required for binding to cell-surface receptors. This core structure does not, however, contain a secretory signal sequence and so the mechanism whereby the prototypic members of the family, aFGF and bFGF, are secreted by cells is unclear. ▨ Secretory signal sequence. ▤ N-terminal extensions translated from upstream CUG codons which provide a nuclear localization sequence that overrides other subcellular localization signals and thus directs the extended forms of bFGF or *int*-2 to the nucleus (Acland *et al.*, 1990; Florkiewicz and Sommer, 1989; Prats *et al.*, 1989). The lengths of the primary translation products initiated from the AUG codons are also shown.

(bFGF), *int*-2, *hst*/KS53 and FGF-8, have been associated with both normal and abnormal growth of mammary cells. The archetypal FGFs, bFGF and aFGF were first isolated from pituitary and brain (Crabb *et al.*, 1986; Esch *et al.*, 1985; Gospodarowicz, 1974; Jaye *et al.*, 1986; Smith *et al.*, 1984c). The name FGF is derived from the assay used to identify their growth-promoting activity – the stimulation of DNA synthesis in fibroblasts (Gospodarowicz, 1974; Rudland *et al.*, 1974). Subsequently it became apparent that these growth factors are able to promote the growth of mesodermal and neuroectodermal cells (Gospodarowicz *et al.*, 1978, 1986). Thus in the breast-derived cell systems bFGF has been found to stimulate not only the growth of rat fibroblasts (Rudland *et al.*, 1977a) and cell lines, derived from the stroma, but also the growth of rat myoepithelial-like cell lines derived from the epithelium (Smith *et al.*, 1984d). However, bFGF fails to stimulate the growth of pure rat epithelial cells (Rudland *et al.*, 1977a) and cell lines (Smith *et al.*, 1984d). aFGF also stimulates the growth of the myoepithelial-like and fibroblastic cell lines, but about ten times more material than bFGF is required for maximal stimulation; the pure epithelial cell lines fail to be stimulated (e.g. Ke *et al.*, 1990). These results are consistent for cell lines isolated from normal rat mammary glands and also from benign rat mammary tumours (Smith *et al.*, 1984d).

Receptor-binding studies have allowed the molecular basis of the selective stimulation of the growth of myoepithelial and fibroblastic cells but not pure epithelial cells by bFGF to be understood. The FGFs have two types of receptors: tyrosine kinase receptors (FGFRs) (Figure 11) that are ultimately responsible for the generation of primary intracellular signals and heparan sulphate proteoglycans (HSPGs). FGFs bind to specific sequences of oligosaccharides in the heparan sulphate (HS) chains of the HSPGs (Turnbull *et al.*, 1992) which activates the FGFs and enables their high-affinity binding to the FGFRs (Figure 12).

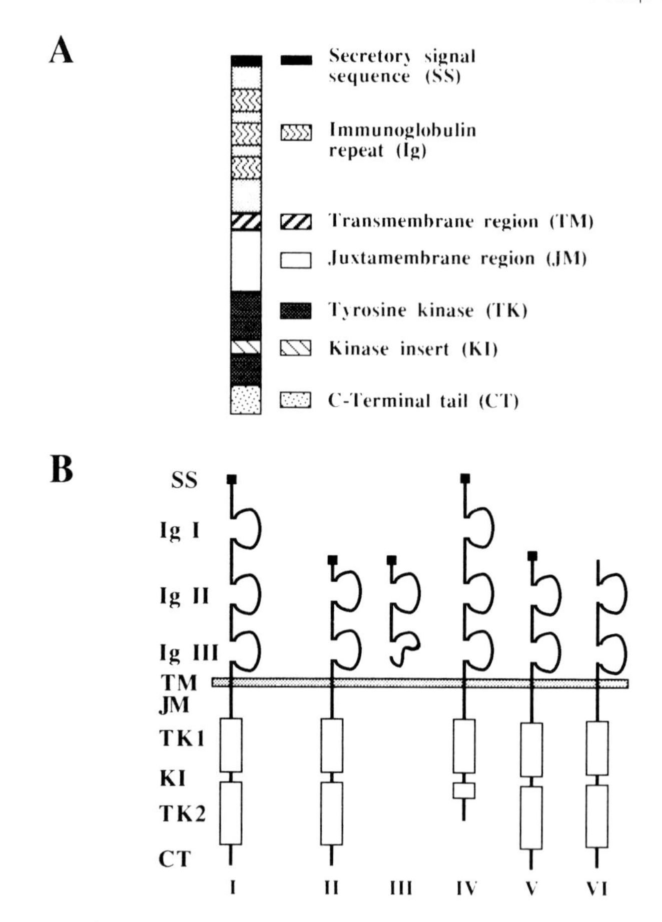

Figure 11 Structure of the tyrosine kinase receptors for FGFs, the FGFRs. (A) Schematic structure of the FGFRs, the fourth family of tyrosine kinase receptors (Ullrich and Schlessinger, 1990). The FGFRs are encoded by four distinct genes, FGFR-1 (*flg*), FGFR-2 (*bek*), FGFR-3 and FGFR-4 (Pasquale and Singer, 1989; Houssaint *et al.*, 1990; Keegan *et al.*, 1991; Partanen *et al.*, 1991). (B) Generation of FGFR-1 receptor subtypes by alternative splicing. The binding of the eight FGFs to the different FGFRs is apparently highly promiscuous, e.g. aFGF, bFGF and hst/KS53 bind to FGFR-2 (Mansukhani *et al.*, 1992), although some degree of specificity is imparted by the alternative RNA splicing (e.g. Championarnaud *et al.*, 1991) which, in the case of the FGFR-1 gene product, can generate six distinct receptors (I–VI) (Hou *et al.*, 1991). Splice forms II and V differ by just two amino acids in their extracellular domains.

The rat mammary fibroblastic and myoepithelial-like cell lines possess both specific high-affinity receptors for bFGF (K_d, 30 pM to 280 pM) and specific low-affinity receptors (K_d, 1–20 nM) (Fernig *et al.*, 1990, 1992, 1993). The high-affinity receptors give rise to complexes of M_r 180 kDa and 160 kDa when (^{125}I)-bFGF is specifically affinity-cross-linked to them on the fibroblastic or the myoepithelial-like cell lines (Fernig *et al.*, 1990). On the basis of their sensitivity to competition by heparin and digestion by heparinase, the low-affinity receptors have been identified as HS (Figure 12), one half to two thirds of which may be associated with the extracellular matrix (Fernig *et al.*, 1992). In addition, whilst the HS receptors for bFGF on the myoepithelial-like Rama 401 cells have a single K_d value of about 20 nM, they possess three distinct dissociation rate constants, suggesting that these receptors have heterogeneous binding characteristics (Fernig *et al.*, 1992). However, it is not known whether these represent three distinct receptors associated with specific sequences of oligosaccharides within HS (e.g. Turnbull *et al.*, 1992) or with the different functions of HS, namely storage and activation of FGFs (Figure 12). In marked contrast to the myoepithelial-like cells, the pure parental rat epithelial cell lines derived from the normal mammary gland or benign tumours do not express cell surface bFGF-receptors (Fernig *et al.*, 1990, 1993). Therefore the failure of bFGF to stimulate the growth of pure epithelial cells (Rudland *et al.*, 1977a; Smith *et al.*, 1984d) is due to an absence of detectable cell-surface receptors rather than the failure at a subsequent step in the stimulation of cell division.

A similar pattern of expression of cell-surface receptors for bFGF has been observed in a human mammary cell system derived from a benign lesion. Thus the epithelial Huma 123 and 121 cells do not possess detectable receptors for bFGF, whilst the myoepithelial-like Huma 109 cells possess both high- and low-affinity receptors for bFGF (Ke *et al.*, 1993). However, our SV40-transformed human mammary cells derived from the normal mammary gland fail to bind bFGF (Ke *et al.*, 1993). Thus neither high- nor low-affinity receptors for bFGF are detectable on any of these cells, regardless of their epithelial or myoepithelial phenotype. Since the immortalization of these cells with SV40 has resulted in a large increase in the number of cell-surface receptors for EGF (McAndrew *et al.*, 1994b) (page 172), it is possible that the immortalization of these cells with SV40 has interfered with the pattern of expression of cell-surface receptors normally associated with cells from the mammary gland.

In the rat cell lines intermediate in characteristics between the epithelial and myoepithelial-like cells (page 164), it appears that both high- and low-affinity receptors for bFGF are first expressed at the cell surface at a very early step in the differentiation pathway of Rama 25 epithelial cells to myoepithelial-like cells, and this step precedes the stage represented by Rama 25-I2 cells. Thereafter, the number of both classes of receptors for bFGF increases, in the same order as the cells' stage in differentiation to the myoepithelial phenotype, up to a maximum value for the myoepithelial-like Rama 29 cells (Fernig *et al.*, 1990). Thus, epithelial cells that progress along the myoepithelial-like differentiation pathway *in vitro* express higher levels of receptors for bFGF and the appearance of both the high- and low-affinity receptors for bFGF may be a characteristic of the differentiation pathway of myoepithelial cells *in vivo*.

In primary cultures of mouse mammary epithelial cells bFGF has been found to stimulate growth and inhibit the production of casein (Levay-Young *et al.*, 1989). An interpretation of this result that is consistent with the results obtained with the Rama and Huma mammary cell systems is that the primary cultures contain a mixture of cells from the mammary epithelium including bFGF-responsive intermediate cells. Similarly, the murine mammary epithelial COMMA-D cell line is responsive to bFGF (Riss and Sirbasku, 1987a, 1989), although the growth of NMuMG mouse mammary epithelial cells is not affected by exogenously-added bFGF (Elenius *et al.*, 1992). bFGF has also been observed to stimulate cell growth and anchorage-independent growth of benzo[a]pyrene- and subsequently SV40-immortalized A1N4 normal human mammary cells (Valverius *et al.*, 1990; Venesio *et al.*, 1992), although the related HMEC 164 cell line does not respond to exogenously-added bFGF (Li and Shipley, 1991). The former

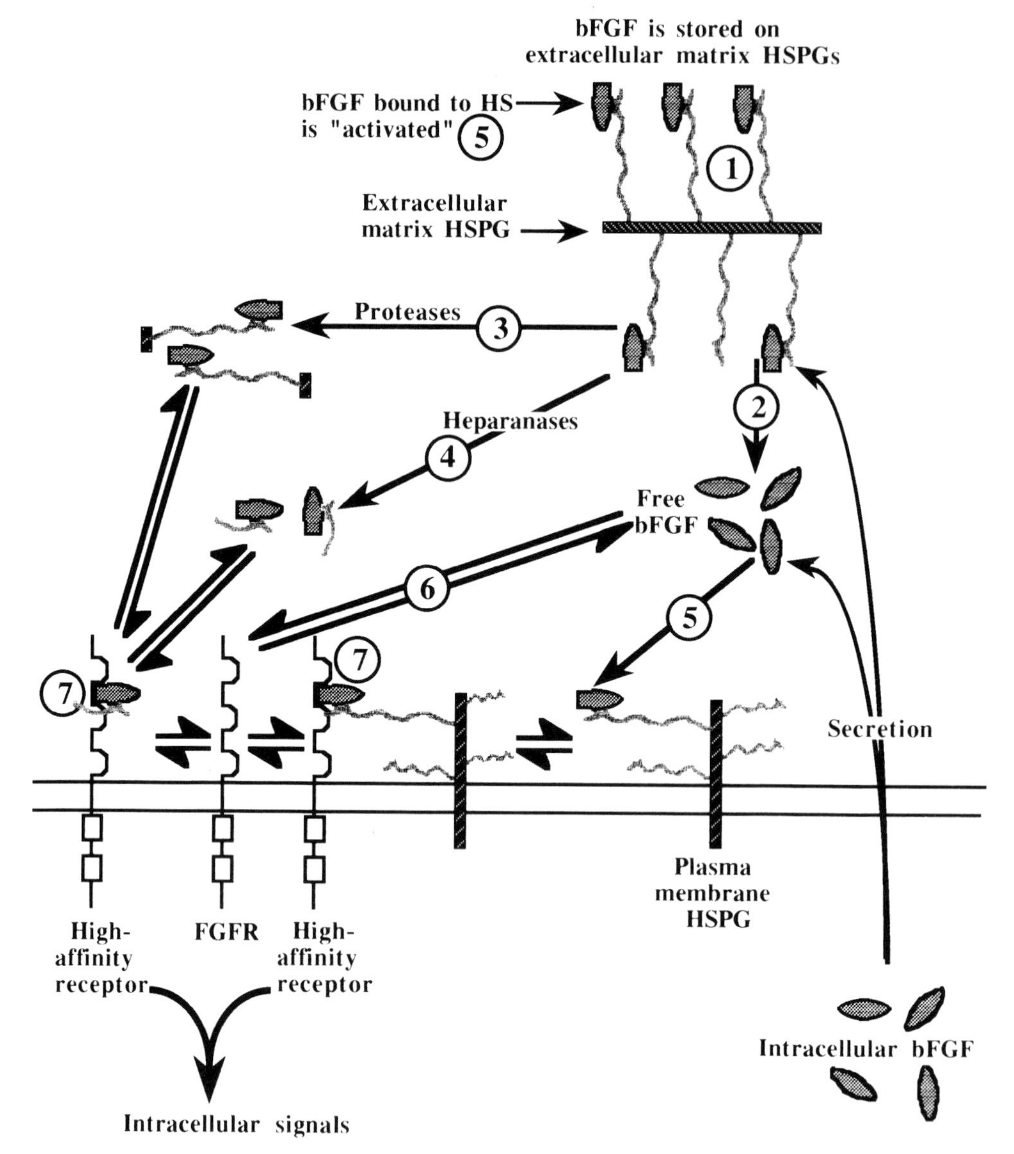

Figure 12 Activation and storage of bFGF. HSPGs consist of heparan sulphate (HS) polysaccharide chains (black) covalently-linked to core proteins (hatched box) (Gallagher, 1989; Gallagher *et al.*, 1992). Six different families of HSPG core protein have been identified. One is perlecan, a proteoglycan (PG) of the extracellular matrix and the five others are associated with plasma membranes, two of which, syndecan and glypican, have been identified in mammary cells (Gallagher, 1989; Gallagher *et al.*, 1992). bFGF secreted by cells is stored on the HSPG receptors which act as a storage depot or sink for extracellular bFGF (1). Under suitable conditions the HSPG storage receptors may modulate the delivery of bFGF to the tyrosine kinase receptors by the continuous release of bFGF (2) (e.g. Flaumenhaft *et al.*, 1989; Presta *et al.*, 1989; Fernig *et al.*, 1992). bFGF may also be released by the action of proteases (3) (Briozzo *et al.*, 1991) or heparanases (4) (Vlodavsky *et al.*, 1991). It is likely that both these mechanisms release bFGF bound to intact HS chains or short oligosaccharides, respectively. The binding of bFGF to HS induces a conformational

possess high-affinity receptors for bFGF. Thus these observations may be accounted for by the responsive mouse and human cell lines being capable of differentiating to cells intermediate between epithelial and myoepithelial cells and the non-responsive mouse and human cells representing a pure epithelial cell that fails in this respect.

At the protein and mRNA level little is known about the FGFRs in mammary cell systems. A recent study suggests that mRNAs encoding FGFR-1 and FGFR-2 are produced in normal human breast tissue (Luqmani *et al.*, 1992), although neither the spliced form(s) (Figure 11) nor the subcellular localization of these mRNAs is known.

Several HSPGs that are potential low-affinity receptors for FGFs have been described in mammary cell systems. Syndecan has been discovered in the NMuMG mouse mammary epithelial cells and in dense cultures it is restricted to the basolateral surface of the cells (Rapraeger *et al.*, 1986), although it is also shed into the culture medium as a consequence of the action of a trypsin-like enzyme (Jalkanen *et al.*, 1987). *In vivo* the core protein of syndecan has been localized immunocytochemically to epithelial cells in ducts and TEBs of the mouse mammary gland (Hayashi *et al.*, 1987). However, the basolateral distribution of syndecan probably precludes its being detected in bFGF-binding experiments on intact cells. Thus, although under suitable conditions (e.g. EDTA detached cells) the syndecan of NMuMG cells can bind bFGF, the NMuMG cells still do not respond to the bFGF, underlining the necessity for cells to have both HSPG and FGFR receptors (Elenius *et al.*, 1992). A recent study has demonstrated the presence of the core proteins and their corresponding mRNAs for a number of HSGPs in mammary cells. Thus both syndecan and the phosphatidyl inositol-anchored glycipan are found in NMuMG and human HBL-100 mammary epithelial cells, although fibroglycan and a novel 125 kDa HSPG core protein are not detected either at the protein or at the mRNA level in these mammary epithelial cells (Lories *et al.*, 1992). In addition a HSPG that is distinct at the level of its core protein from syndecan has been described in the extracellular matrix of NMuMG cells (Jalkanen *et al.*, 1988). Thus four HSPGs have been found in mammary cells but only one, syndecan, has been shown directly to bind bFGF (Bernfield and Sanderson, 1990; Kiefer *et al.*, 1990), and the relationship between these HSPGs and the differentiation of mammary cells has yet to be determined.

Thus the majority of the evidence suggests that the breast cells that are able to respond

---

change in bFGF (5) (Prestelski *et al.*, 1992) that enables it to bind to the high-affinity receptors (Klagsbrun and Baird, 1991; Li and Bernard, 1992; Rapraeger *et al.*, 1991; Yayon *et al.*, 1991). This activation mechanism extends to other members of the FGF family (Mansukhani *et al.*, 1992). FGF-binding data commonly reveal the existence of a high-affinity binding site, K_d of 10–300 pM, and a low-affinity binding site, K_d of 1–30 nM (e.g. Fernig *et al.*, 1992, 1993). It is now apparent that the high-affinity receptor is generated as a consequence of the prior activation of FGFs (5) by either heparin (Ishihara *et al.*, 1993; Nugent and Edelman, 1992; Ornitz and Leder, 1992; Ornitz *et al.*, 1992; Tyrrell *et al.*, 1993) or the more physiologically relevant HS (Walker *et al.*, 1994). Thus in isolation the FGFRs have an affinity for FGFs that is comparable to that of the HSPGs (6) (K_d of 1–30 nM). Since the binding sites for bFGF on HSPGs outnumber the FGFRs by over 50-fold in most normal cell systems, including those of the breast (Fernig *et al.*, 1992, 1993), the low-affinity binding sites can be equated with the HSPGs. However, until the mechanism(s) of activation of FGFs by HS is elucidated at the molecular level, it is not clear whether the high-affinity binding sites are a binary complex of HS and FGFR (7) (Kan *et al.*, 1993) or FGFRs (Klagsbrun and Baird, 1991).

to bFGF *in vitro* are stromal fibroblasts, the intermediate cells of the TEBs and the myoepithelial-like cells. The ability of these cells to mount a response to bFGF is due not only to their possessing FGFRs, but also activating HS receptors for bFGF. In addition these cells also possess storage HSPG receptors for bFGF, and are thus able to sequester relatively large amounts of exogenous bFGF. The storage receptors and the activating receptors for bFGF may be structurally equivalent at the level of their oligosaccharide sequences, although the detection of HS receptors with three distinct rates of dissociation (Fernig *et al.*, 1992) suggests that this interpretation may be too simplistic.

Appearance of fibroblast growth factors on formation of myoepithelial cells

Activity of bFGF is readily detected in extracts of rat myoepithelial-like cells and a single species of bFGF mRNA of 5.7–6 kb is observed in such cells upon Northern hybridization of their mRNA (Barraclough *et al.*, 1990a; Fernig *et al.*, 1993). Thus myoepithelial-like cells derived from the normal rat mammary gland such as Rama 711 (Fernig *et al.*, 1993) and from benign rat mammary tumours such as Rama 29 (Barraclough *et al.*, 1990a) contain both bFGF mRNA and 6–7 ng bFGF per 10^6 cells. However, bFGF mRNA and activity are virtually absent from the epithelial cell lines derived from the normal mammary gland, such as Rama 704, and from benign mammary tumours, such as Rama 25 and Rama 37 (Barraclough *et al.*, 1990a; Fernig *et al.*, 1993). Furthermore, in the cell lines of intermediate characteristics between epithelial and myoepithelial-like cells that are representative of the cells found in TEBs, the levels of bFGF mRNAs and activity increase as the myoepithelial characteristics of the cells increase. Substantial amounts of bFGF mRNA and bFGF activity are observed even in the Rama 25-I2 cells that are closest to the epithelial phenotype, indicating that the production of these molecules is associated with an early step in this differentiation pathway (Barraclough *et al.*, 1990a). Comparison of the level of bFGF activity and bFGF mRNA in the rat mammary epithelial, intermediate and myoepithelial-like cells indicates that their levels of bFGF activity reflects the relative levels of their bFGF mRNA and that there is a greater than 30-fold increase in the amount of bFGF associated with the cells upon differentiation to the myoepithelial phenotype. In the human mammary cell lines a similar pattern of expression of bFGF activity and bFGF mRNA is observed both in the SV40-immortalized Huma cells lines isolated from the normal gland and in the Huma cell lines isolated from a benign lesion. Thus the myoepithelial-like cells possess 80-fold higher levels of bFGF activity and at least 20-fold higher levels of bFGF mRNA than their parental epithelial cells. bFGF mRNA and bFGF (4 ng per 10^6 cells) are also present in the fibroblastic Rama 27 cells indicating that it is also produced by cells derived from the mammary stroma (Barraclough *et al.*, 1990a).

bFGF and its receptors represent a potent autocrine/paracrine growth stimulatory loop in the intermediate cells and the myoepithelial-like cells, since their level of bFGF activity is higher than that required for the maximal stimulation of growth (Barraclough *et al.*, 1990a; Ke *et al.*, 1993). However, these cells do not grow in an uncontrolled manner and their bFGF activity is cell-associated and virtually no activity has been detected in the culture medium (Barraclough *et al.*, 1990a; Fernig *et al.*, 1993; Ke *et al.*, 1993). Moreover, crude cell extractions using a variety of agents such as 2M NaCl, Triton-X-100 and trypsin have suggested that a portion of the cellular bFGF is associated with the extracellular matrix, and with HSPGs in particular, whilst the remainder is intracellular (Barraclough *et al.*, 1990a). Thus the autocrine/paracrine loop that exists in the intermediate cells and the myoepithelial-like cells is probably not usually activated, since the bFGF is probably sequestered in a compartment (extracellular matrix or cytoplasm) separate from the cell-surface receptors.

Analysis of the bFGF activity that is recovered from the human and rat mammary cells by immunoblotting has demonstrated the presence of an immunoreactive band of M_r 18 kDa and three other immunoreactive bands, usually of similar intensity to the 18 kDa band and of M_r values of 22, 23 and 25 kDa (Fernig *et al.*, 1993; Ke *et al.*, 1993). The former corresponds to the translation product of the bFGF mRNA initiated at the AUG codon, and the latter three products probably correspond to the products of translation of the bFGF mRNA which are initiated from each of the CUG codons upstream of the AUG codon (Figure 10) (Florkiewicz and Sommer, 1989; Prats *et al.*, 1989) similar to those observed in another human mammary epithelial cell system (Li and Shipley, 1991). Other bands of immunoreactivity are also observed in some cell lines (Ke *et al.*, 1993; Li and Shipley, 1991). Whilst their origin is uncertain, since covalent multimers of bFGF and aFGF occur *in vitro* upon prolonged storage (Houssaint *et al.*, 1990; Engleka and Maciag, 1992), the immunoreactive bands of molecular weight greater than 30 kDa may be due to artefacts of the extraction procedure.

One major difference between the rat and human cell lines is found at the level of their bFGF mRNA. Thus a single species of bFGF mRNA of 5.7–6 kb is observed in the rat cells (Barraclough *et al.*, 1990a) compared to four discrete mRNA species, corresponding to 6.6, 3.4, 1.9 and 1.3 kb, in the human cells (Ke *et al.*, 1993). These mRNAs arise from transcription of the same strand of DNA (Ke *et al.*, 1993), a result which indicates the absence of any stable reverse transcript, such as that of 1.5 kb identified in Xenopus oocytes (Kimelman and Kirschner, 1989; Volk *et al.*, 1989). bFGF mRNA has also been detected in a number of other human mammary epithelial cells (Luqmani *et al.*, 1992) and in one study three species of bFGF mRNA of 7.5, 4.4 and 2.2 kb were detected (Li and Shipley, 1991). The multiple mRNAs may be due to the utilization of different polyadenylation sites (Kurokawa *et al.*, 1988).

Extracts of virgin rat mammary gland contain 50 ng per gram of tissue of bFGF-like activity (Barraclough *et al.*, 1990a), and immunochemical analysis of these extracts indicates that all four translation products of the bFGF mRNA are present (Rudland *et al.*, 1993a). bFGF mRNA has been detected in normal human mammary gland and in benign lesions including neoplasms (Anandappa *et al.*, 1994). Immunocytochemical localization of bFGF in the rat mammary gland using four independently-isolated antisera demonstrates that within the more-mature, resting ducts and secretory alveoli, bFGF is located predominantly in the basement membrane/myoepithelial cell region (Rudland *et al.*, 1993a), in agreement with results obtained in human glands (Gomm *et al.*, 1991). In the ducts and probably the alveoli, the majority of the staining for anti-bFGF serum occurs in the basement membrane (page 151). The weaker staining for antisera to bFGF on ductal myoepithelial cells suggests that there is less bFGF in the myoepithelial cell than in the basement membrane. During the development of the mammary gland, within the growing mammary ducts of 1-day-old neonatal rats, in growing TEBs, and to a lesser extent in growing ABs, the inner and outer cells are moderately stained to a similar extent by anti-bFGF serum, whilst the staining for bFGF in the basement membranous region is not apparent. At later developmental stages, in non-growing structures, the majority of the staining for bFGF reverts to the basement membrane/myoepithelial cell region (Rudland *et al.*, 1993a). Thus the presence of immunoreactive bFGF in the basement membrane/myoepithelial cell is associated with a relatively quiescent cell population. In contrast the presence of immunoreactive bFGF in the proliferating stem cell population of epithelial/myoepithelial intermediate cells and the absence of immunoreactive bFGF in the basement membrane may be linked to the enhanced growth rate of these cells.

Surprisingly, when antisera to bFGF are incubated with bFGF the immunocytochemical staining produced by the antisera above is not abolished. In addition to the failure of bFGF to inhibit binding of anti-bFGF sera to the basement membranous region of non-growing ducts, mixtures of bFGF and anti-bFGF serum increase the staining of the myoepithelial cell and to a lesser extent that of the epithelial cell, all of which is inhibited by heparin (Rudland *et al.*, 1993a). These results may reflect the cells' possession of unoccupied binding sites for bFGF in the resting ducts which probably correspond to HSPGs of the cellular plasma membrane and of the extracellular matrix/basement membrane (Figure 12). These results strongly suggest that in non-growing areas of the mammary gland a considerable extracellular sink exists for bFGF whereby it may be sequestered without promoting the proliferation of those cells possessing the appropriate receptors. The cellular staining of proliferating intermediate cells and its failure to be enhanced by mixtures of bFGF and anti-bFGF serum may then reflect either that similar cellular binding sites are now occupied by bFGF in growing mammary structures, particularly in the TEBs, or that the HSPGs produced by cells in growing mammary structures contain bFGF-activating but not bFGF-sequestering HS. In support of this hypothesis, the implantation of bFGF-containing slow-release pellets in the mammary glands of mice (Daniel and Silberstein, 1991) has no effect on mammary gland growth, and it is probable that the bFGF released from the implant simply bound to storage HSPG receptors and is thus unable to activate FGFRs on potential targets such as the intermediate cells of the TEBs.

The other members of the FGF family have been investigated in much less detail. aFGF mRNA is readily detected in a rat mammary fibroblastic cell line, but not in rat epithelial- and myoepithelial-like cell lines, suggesting that aFGF may be produced by the mammary stromal cells (Barraclough *et al.*, 1990a). *int*-2, a member of the FGF family that possesses a secretory signal sequence (Figure 10), is not a normal product of the mammary gland. However, it is produced by the epithelial cells of the largely benign tumours found in mice infected by mouse mammary tumor virus (MMTV) (Moore *et al.*, 1986) both in the medium, and also associated with HSPGs on the surface of the cells (Dickson, 1990; Dixon *et al.*, 1989). Mice possessing an *int*-2 transgene, under the transcriptional control of the MMTV long terminal repeat and hormone-responsive element, produce mammary hyperplasia, indicating that the FGF family of growth factors, presumably through their receptors, is able to stimulate the growth of mammary cells *in vivo* (Muller *et al.*, 1990). Even though *int*-2 possesses a secretory signal sequence, it acts only over very short distances in tissue-explants from such mice (Ornitz *et al.*, 1992). Thus, these results support both the existence of an extracellular sink for the FGF family of growth factors and a physiological role for bFGF in mammary gland development.

In conclusion, bFGF is a product of myoepithelial-like, intermediate and fibroblastic cells *in vitro* and *in vivo* and, apart from aFGF which may be a stromal-specific product, the other FGFs are not yet in our systems strong candidates for physiological products of the mammary gland. Since bFGF is abundant in the pituitary, the source of the bFGF that can activate the receptors for bFGF on mammary cells may be either systemic from the pituitary or local, produced constitutively by cells within the mammary gland. However, the presence of bFGF in serum has been a controversial issue (Burgess and Maciag, 1989), although it is certainly associated with some classes of blood cells. Moreover, the distance bFGF can diffuse *in vivo* is likely to be relatively short due to the large number of storage HSPG receptors for bFGF that are found in the mammary gland. Thus it seems likely that bFGF is a local, short range, growth factor produced by the cells of the mammary gland. The physiological role of bFGF may include not only the growth and differentiation of the intermediate cells of TEBs, but also its abstraction by HSPGs may be associated with quiescent myoepithelial cells of the ducts and alveoli.

PROTEINS ASSOCIATED WITH DIFFERENTIATION OF MYOEPITHELIAL-LIKE CELLS

Polypeptide changes associated with the formation of myoepithelial-like cells and their control *in vitro*

The conversion in culture of our rat epithelial cell lines to myoepithelial-like cells is accompanied by changes in a small number of specific, abundant polypeptides separable by two-dimensional polyacrylamide gel electrophoresis (Barraclough *et al.*, 1982). Essentially the same changes are found for the conversion of epithelial cell lines derived from normal rat glands or from benign rat tumours to the myoepithelial-like cells (Barraclough *et al.*, 1984a). Only 6% of the total number of polypeptides resolved show any quantitative or qualitative changes (Barraclough *et al.*, 1982) and of these only nine polypeptides show major changes. Three polypeptides show a marked decrease in the myoepithelial-like cells, whereas six are more abundant. Seven polypeptides are components of the intermediate filaments within the cells (Barraclough *et al.*, 1982). An acidic polypeptide of M_r 9,000, termed p9Ka, is present in abundance in the myoepithelial-like cells, but is barely detectable in the parental epithelial cell lines. Other proteinaceous markers of the myoepithelial-like cells, Thy-1, laminin, Type IV collagen, the receptors for GS-1 and pokeweed mitogen (page 150), bFGF and receptors for bFGF (pages 176 and 180) are of much lower abundance, and are therefore not readily detected using the technique of two-dimensional gel electrophoresis.

The appearance of p9Ka in the development of the myoepithelial-like cells has been compared with the appearance of other markers of differentiation *in vitro* and *in vivo*, namely laminin, Type IV collagen, and Thy-1. In the myoepithelial-like cell line, Rama 29, Thy-1 antigen, p9Ka, Type IV collagen and laminin are respectively 13-fold, 16-fold, 3.5-fold and 3.7-fold more abundant than in the parental, epithelial cell lines (Barraclough *et al.*, 1987a; Rudland *et al.*, 1982c, 1986; Warburton *et al.*, 1986). The levels of the specific mRNAs for these markers have been estimated by hybridizing poly(A)-containing RNA from the different cell lines to radioactive, cloned cDNAs corresponding to the mRNAs for Thy-1, p9Ka (Barraclough *et al.*, 1984b), laminin, and Type IV collagen. Thus, a 13-fold increase in the amount of Thy-1 in the myoepithelial-like cells is accompanied by an equivalent increase in Thy-1 mRNA, as measured by hybridization to Thy-1 cDNA (Barraclough *et al.*, 1987a). In contrast, the 16-fold increase in p9Ka accumulation in the myoepithelial-like cells (Barraclough *et al.*, 1984a) is accompanied by only a tenfold increase in cytoplasmic mRNA for p9Ka (Barraclough *et al.*, 1984b). These results suggest that translational control may play a facilitating role in the increase in p9Ka. On the other hand, the 3.5-fold increases in laminin (Rudland *et al.*, 1986) and Type IV collagen (Warburton *et al.*, 1986) are accompanied by only a very small increase in their respective mRNAs. The increase in Type IV collagen seen in the myoepithelial-like cells relative to the epithelial cells arises predominantly from a fivefold decrease in the rate of intracellular breakdown of Type IV collagen, as estimated from the rate of release of hydroxyproline (Warburton *et al.*, 1986). These results strongly suggest that in the same cells, the steady-state levels of different marker proteins can arise by controls acting at different molecular levels, namely at transcriptional, post-transcriptional processing or post-translational levels.

The idea of controls acting at different molecular levels on the production of marker proteins for the myoepithelial-like cell has been extended by examining the appearance of markers of the myoepithelial-like cells in cloned cell lines intermediate in morphology between the epithelial and myoepithelial-like cells, the potential mammary stem cells. These form a series as described on page 164 in the order Rama 25, Rama 25-I2, Rama 25-I1, Rama 25-I4, Rama 29. As described previously this same order is maintained for increases in myoepithelial characteristics including the marker proteins of the myoepithelial-like cell. However, the individual changes do not occur to the same degree in the same intermediate cell line. Thus, the majority of the increase in p9Ka occurs in the cell line Rama 25-I2, whereas Thy-1 antigen increases predominantly in Rama 25-I1 and I4, and laminin increases almost equally in Rama 25-I2, I1, and I4 cells i.e. on average is I1. Our results show that the production of these proteins in our stem cell lines is not only regulated at different molecular levels, but also that the production is not synchronous, and therefore that the regulation of their production is executed in an asynchronous manner in this development pathway *in vitro*.

Three of the markers of the myoepithelial cells *in vitro* and *in vivo*, Thy-1, laminin and Type IV collagen (page 151) are well characterized proteins. However, the fourth marker protein p9Ka was much less well understood. Progress towards understanding the nature of p9Ka came from a study of its gene.

p9Ka, a marker of the formation of myoepithelial-like cells *in vitro*, is a calcium-binding protein

A cloned cDNA corresponding to p9Ka mRNA has been obtained from a cDNA library constructed with mRNA from the myoepithelial-like cell line, Rama 29 (Barraclough *et al.*, 1984b). It is complementary to 400 nucleotides at the 3′ end of the p9Ka mRNA. In addition, an 18-kb fragment of rat genomic DNA that contains the p9Ka gene and its flanking regions has been isolated from a normal rat genomic library (Barraclough *et al.*, 1987b). The nucleotide sequence contains a potential coding region of 101 amino acids, including the initiating methionine residue, and ends with two termination codons (Barraclough *et al.*, 1987b). A search of computerized data bases of protein sequences shows limited homology (43%) of this potential amino-acid sequence for p9Ka to both the α and β chains of bovine S-100 protein (Isobe and Okuyama, 1978, 1981), to rat S-100 protein (Kuwano *et al.*, 1984) and similar homology to other proteins of the same family, including calcyclin (Calabretta *et al.*, 1986), macrophage migration-inhibitory factor-related proteins 8 and 14 (Odink *et al.*, 1987) and P11, the small subunit of annexin II (Gerke and Weber, 1985).

The protein p9Ka also shows a weak homology (34%) with the well-characterized bovine vitamin D-dependent, intestinal calcium-binding protein (Fullmer and Wasserman, 1981). This protein has two potential calcium-binding sites, one of which conforms to the EF-hand structure of known calcium-binding proteins (Moews and Kretsinger, 1975; Szebenyi *et al.*, 1981). When only those amino acid residues thought to be involved in calcium binding by the two sites of bovine intestinal calcium-binding protein (Szebenyi *et al.*, 1981) are compared with those in similar sequence positions in p9Ka, 11 out of 14 of these residues are identical in the two polypeptides. The protein p9Ka also contains two potential calcium-binding loops between residues 33 and 40 and residues 62 and 73 (Figure 13). The C-terminal loop (residues 62–73) of p9Ka corresponds to an almost perfect EF-hand sequence, with five residues containing carboxylic acid derivatives in their side chains in the exact positions of the loop region that are thought to be important in calcium binding by the vitamin D-dependent, intestinal calcium-binding protein (Szebenyi *et al.*, 1981; Szebenyi and Moffat, 1986).

The p9Ka gene contains two intervening sequences of 1172 nucleotide pairs (np) and 675 np. The mRNA for p9Ka, encoded by this gene is derived from 3 exons. The exon encoding the 5′ end of the mRNA is only about 37 nucleotides in length (Figure 13), there being some apparent heterogeneity in the precise start site of transcription over a 2–3 nucleotide region using primer extension techniques (B. R. Barraclough, unpublished observation). This short exon gives rise to an untranslated region of the mRNA. Exons

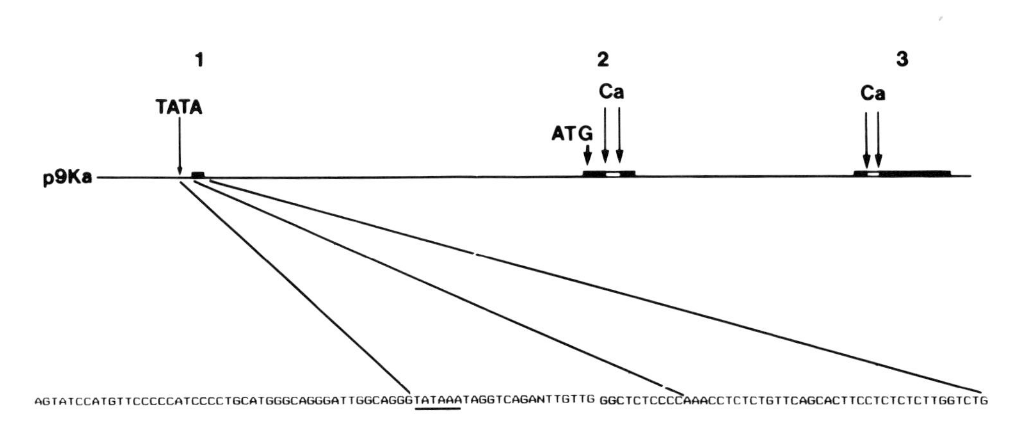

Figure 13 The structure of the gene for p9Ka. The regions corresponding to mRNA (exons) are shown as black boxes. Non-exon DNA, including the intervening sequences, is shown as a line. The potential calcium-binding regions (Ca) are shown as white boxes within the exon regions. The position of the triplet for the initiating methionine residue (ATG), and the location of the TATA box (TATA) are indicated. The nucleotide sequence of the first exon of the gene and adjacent non-transcribed DNA is shown.

2 and 3 (of 156 np and 295 np, respectively) encode the translated region of the p9Ka mRNA, each exon contributing to a functional calcium-binding domain (Figure 13). The calculated size of the p9Ka mRNA (488 nucleotides) is somewhat smaller than the apparent size of the mRNA obtained using Northern blotting procedures (600–700 nucleotides). Part of this discrepancy may be accounted for by the poly(A)-tail (Barraclough *et al.*, 1984b). The p9Ka gene is located on two cloned *Eco*RI fragments derived from normal rat genomic DNA (Barraclough *et al.*, 1987b). These cloned fragments have been linked together to provide a cloned p9Ka gene, which together with a partial cDNA (Barraclough *et al.*, 1984b), has been used to provide information on the physical and biological properties of p9Ka.

Recombinant p9Ka (rp9Ka) has been produced in *Escherischia coli* cells using a high-level expression vector system (Ke *et al.*, 1990), and purified using a rapid, two step, purification procedure (Gibbs *et al.*, 1994) the yield of rp9Ka being 40–50 mg l^{-1} of bacterial culture. Natural p9Ka (np9Ka) has been purified from the myoepithelial-like cell line, Rama 29, using the same purification procedure (Gibbs, 1993). The purified rp9Ka consists of a mixture of two proteinaceous components which can be separated analytically by two-dimensional gel electrophoresis and preparatively by reverse-phase high-performance liquid chromatography. These two forms of rp9Ka differ only in the presence or absence of the N-terminal methionine (Gibbs, 1993).

Both the purified rp9Ka and np9Ka bind calcium ions (Barraclough *et al.*, 1990b; Gibbs, 1993), with K_ds of 34±0.3 to 38±0.6 (over a tenfold range of rp9Ka concentrations) and 76±14 μM respectively (Barraclough *et al.*, 1990b; Gibbs *et al.*, 1994). Notably, p9Ka contains a single class of binding sites, at a concentration of two sites per molecule and these correspond to the two EF-hand binding sites in the polypeptide chain. In p9Ka one potential binding site contains asparagine at position 67 instead of the smaller glycine residue (Figure 13). Although the glycine was thought to play a role in maintaining the EF-hand structure in both S-100 protein and intestinal calcium-binding protein (van Eldik *et al.*, 1982), its substitution in p9Ka does not seem to prevent the binding of calcium (Gibbs *et al.*, 1994).

The affinity of p9Ka for calcium ions *in vitro* is less than that for the regulatory calcium-ion-binding protein, calmodulin (3–20 μM) (Crouch and Klee, 1980), the calcium storage protein, parvalbumin (0.1 μM) (Pechere, 1977), and a calcium-ion transport protein, the intestinal vitamin D-dependent calcium-ion-binding protein (2 μM) (van Eldik *et al.*, 1982). It has a similar affinity for calcium ions to most other closely-related, small calcium-ion-binding proteins of the S-100 protein family (Hilt and Kligman, 1991) except p11, the regulatory subunit of annexin II, which does not bind calcium ions at all (Gerke and Weber, 1985). Furthermore, the binding of calcium by p9Ka is strongly antagonized by physiological concentrations of sodium, potassium or magnesium ions (Gibbs *et al.*, 1994), a property which p9Ka shares with S-100 protein (Baudier *et al.*, 1986).

Cellular targets and tissue distribution of p9Ka

Since the binding of calcium ions by p9Ka is reduced in the presence of physiological concentrations of potassium and magnesium ions, one possibility is that p9Ka binds calcium in a subcellular compartment in which calcium ions are either elevated above normal levels, or where the potassium ion concentration is low. This idea has been

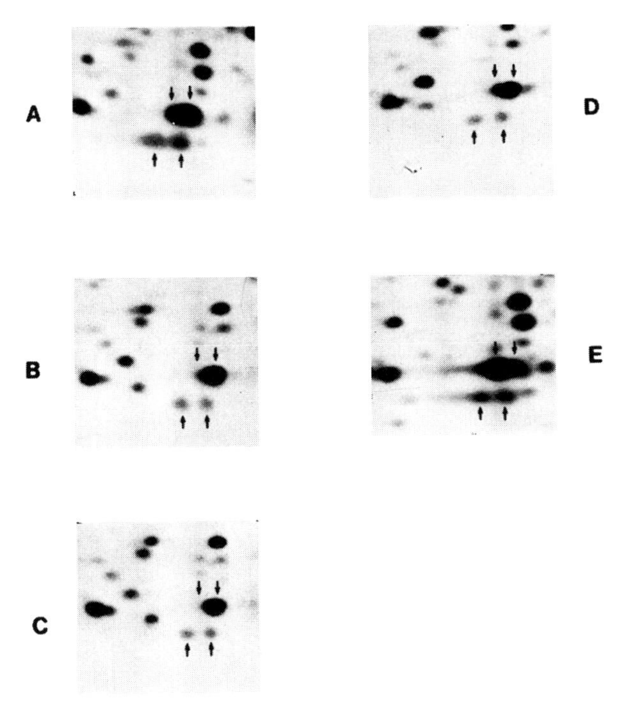

Figure 14 The subcellular distribution of p9Ka. Myoepithelial-like cells, Rama 29, were radioactively-labelled overnight with [^{35}S]methionine and the cells were homogenized in an aqueous buffer. The extract was subjected to differential centrifugation and the pellets from each centrifugation step were dissolved in electrophoresis sample buffer. Samples of the homogenate (A), pellet fractions resulting from centrifugation at 770 *g*(av) (B), 12,000 *g*(av) (C), 100,000 *g*(av) (D) and the supernatant resulting from centrifugation at 100.000 *g*(av) (E) were subjected to isoelectric focusing in the first dimension and SDS gel electrophoresis in the second dimension. The autoradiographs of the relevant portions of the two-dimensional gels are shown. The arrows point down to the acidic (right) and basic (left) isoelectric focusing variants of p9Ka, and point up to the two isoelectric focusing variants of p6Ka (Barraclough and Rudland, 1991).

investigated by experiments to examine the distribution of p9Ka in subcellular fractions of the myoepithelial-like cell lines. When proteins extracted from subcellular fractions are subjected to two-dimensional polyacrylamide gel electrophoresis, p9Ka is found in the nuclear, mitochondrial/lysosomal, and microsomal fractions, as well as in the soluble supernatant fraction (Figure 14). However, in some experiments, the protein spot corresponding to p9Ka, which is usually elongated in the isoelectric focusing direction (Figure 14), splits into two separate, but closely-spaced, isoelectric focusing variants (Figure 14) that are present in about equal amounts in total extracts from the myoepithelial-like cells (Figure 14A). The separation of these variants during two-dimensional gel electrophoresis is somewhat less than was found with the N-terminal methionine variants of rp9Ka (see above). The more-acidic variant predominates in the membrane/cytoskeletal/microsomal fractions (Figures 14B–D), whilst the more-basic variant predominates in the soluble fraction (Figure 14E). These results suggest that a proportion of the p9Ka in these cells may be modified in some as-yet unidentified manner, and that the modified form has a subcellular location different from that of the unmodified form.

In order to clarify the subcellular distribution of p9Ka in the cultured cells using sensitive immunofluorescent techniques, and to explore immunocytochemically the distribution of p9Ka in normal tissues, an antiserum to rp9Ka has been raised in rabbits. In the past, antisera to small calcium-binding proteins have not been specific to one calcium-binding protein. The antiserum to rp9Ka reacts specifically with p9Ka and with no other cellular proteins in Western blotting procedures, and does not react with closely-related, calcium-binding proteins, calcyclin, S-100 protein subunit, MRP8 or MRP14 (Gibbs *et al.*, 1995).

Immunofluorescent staining of the p9Ka-containing myoepithelial-like cell line, Rama 29, shows it to be concentrated on bundles of cytoplasmic filamental structures (Figure 15A), and the staining is completely abolished by prior incubation of the antiserum with rp9Ka. The pattern of immunofluorescent staining obtained with the unblocked p9Ka antiserum is identical to that observed with phalloidin (Davies *et al.*, 1993b), which forms a complex

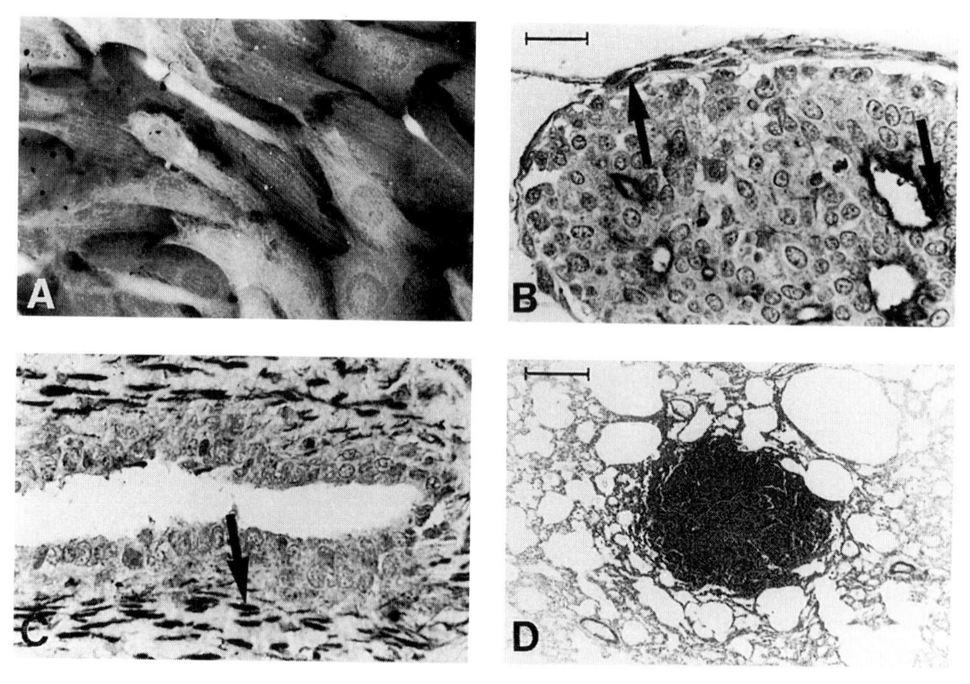

Figure 15 Immunofluorescent and immunocytochemical detection of 9pKa. (Panel A) Rama 29 myoepithelial-like cells were labelled with fluorescein-conjugated anti-p9Ka serum and photographed in fluorescent light. Final magnification ×578. (Panel B) A section through a TEB from the mammary gland of a 21 day old rat was stained with peroxidase-conjugated anti-p9Ka serum. The arrows point to staining of a fibroblastic cell (↑) and staining within the TEB (↓). Final magnification ×578, bar=20 μm. (Panel C) A section through the mammary gland of an adult rat stained with anti-p9Ka serum as for panel B. The arrow points to the unidentified staining for p9Ka; magnification is the same as for panel B. (Panel D) Immunocytochemical staining of a section through a cannon-ball metastasis in the lung arising from Rama 37 cells transfected with the p9Ka gene being injected into the mammary fat pad of the recipient rat. Final magnification ×58, bar=200 μm.

with F-actin (Lengsfield *et al.*, 1974). These results suggest that in those cells in which it is expressed at a high level, p9Ka associates, at least in part, with the actin cytoskeleton. This idea of interaction between p9Ka and the actin cytoskeleton is supported by recent findings in which bovine p9Ka (calvasculin) isolated from aorta has been shown to interact in a calcium-dependent manner with actin using a co-sedimentation assay (Watanabe *et al.*, 1993). It remains to be seen whether the interaction of p9Ka with F-actin alters its affinity for calcium or reduces the effect on calcium-binding of potassium and magnesium ions. Perhaps more importantly it also remains to be seen whether p9Ka serves as a trigger to drive the epithelial cells along the myoepithelial-like cell pathway *in vitro*. If the latter were true, then the levels of intracellular p9Ka could determine the relative proportions of epithelial and intermediate stem cells described on page 164.

Experiments in different laboratories have sought to establish the tissue distribution of the mRNA for p9Ka in rats and mice. These experiments employing cloned cDNA molecules, 18A2 (Jackson-Grusby *et al.*, 1987), *mts1* (Ebralidze *et al.*, 1989) or the cDNA to p9Ka mRNA (Barraclough *et al.*, 1984b), have failed to show a consistent pattern of expression in normal tissues. In the mouse, mRNA corresponding to *mts1* has been detected in spleen, bone marrow, thymus and lymphocytes (Ebralidze *et al.*, 1989). However, others, using a mouse p9Ka cDNA, designated 18A2, detected p9Ka mRNA in uterus, placenta, and kidney with very low levels being detected in the thymus and testes (Jackson-Grusby *et al.*, 1987). In the rat, the mRNA for p9Ka has been detected in the spleen, and low levels have been found in the mammary gland, and uterus with variable levels being reported in liver (Barraclough *et al.*, 1984b). These variable levels of the mRNA for p9Ka reported by different laboratories may be a consequence of its presence in some blood cells and vessels which may contaminate isolated normal tissues (Ebralidze *et al.*, 1989; B. R. Barraclough, unpublished observation).

Immunocytochemically, in normal rat tissues, p9Ka is widely expressed in epithelial, neuronal and muscle cells. Generally, staining is confined to discrete regions within tissues. For example, in the stomach, staining is present in the parietal/oxyntic cells but there is no staining of the chief/peptic cells. In the intestine, staining is mainly in the columnar absorptive cells and neuronally-derived Auerbach's plexuses. In the developing rat mammary gland, p9Ka is present within the dividing cells of the parenchymal TEBs, particularly the peripheral cap or stem cells (Figure 15B), and at an apparently higher level in fibroblasts surrounding these TEBs. All the above staining is intracellular, as expected from the staining obtained with mammary cell lines (Gibbs *et al.*, 1995). However, in the fully-developed rat mammary gland, and in some other tissues, the p9Ka-specific antiserum showed a filamentous staining adjacent to, but separate from the parenchyma, and this staining was not associated with identifiable cells (Figure 15C). The exact extracellular structures which are staining have not yet been identified although both capillaries and nerves are candidates. However, in cultures of certain bovine fibroblasts or smooth muscle cells, p9Ka has been detected in the extracellular medium, where it is reported to be bound to an extracellular matrix protein, a 36 kDa microfibril-associated glycoprotein (Watanabe *et al.*, 1992). These results suggest that p9Ka, in addition to binding to the intracellular cytoskeleton of dividing intermediate stem and myoepithelial-like cells, may be capable, in non-dividing cells, of interaction with extracellular targets, possibly those connected in turn with the actin filamental systems of the cell. The widespread, cell-type-specific distribution, both inside and outside the cell, and its variously-reported interactions with calcium ions and microfilamental systems/associated proteins, points to a highly sophisticated role for p9Ka in the area of cell–cell interaction and cell motility, particularly in the development of the ductal systems of the mammary gland (Gibbs *et al.*, 1995).

MULTIPLE INTERACTIONS DETERMINE THE LEVELS OF PUTATIVE MAMMARY STEM CELLS

In conclusion, the most likely candidates for stem cells in the mammary gland are those cells intermediate between epithelial and myoepithelial cells which occur in growing ductal structures, particularly TEBs (Figure 16) (page 152). They can give rise, by stepwise intercellular conversions, to both ductal epithelial and myoepithelial cells, the latter by an irreversible process (page 164). The ductal epithelial cells, under certain pathological conditions can give rise to squamous cells, but normally differentiate to alveolar cells in a reversible manner (pages 162 and 163). Pituitary prolactin is largely responsible for the production of casein-secretory alveolar cells and its action inhibits formation of squamous or myoepithelial-like cells (pages 162 and 163). In contrast pituitary PMGF (page 169) acts on the ductal and intermediate stem cells causing them to grow and largely prevents differentiation to squamous or myoepithelial cells. Oestrogens are thought to act indirectly through TGFα and/or IGFs (pages 172, 204 and 206) in causing epithelial cell growth along this pathway and preventing differentiation to squamous or myoepithelial cells (Figure 16).

External agents which trigger differentiation to myoepithelial cells have not yet been identified, but the position of the cells within the mammary gland abutting a stromal

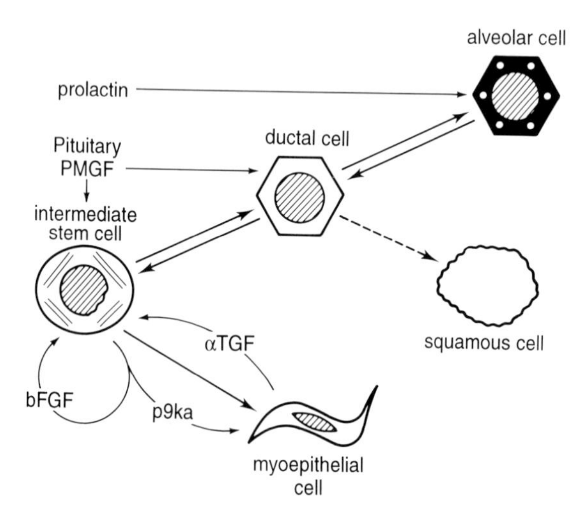

Figure 16 Possible molecular model for mammary development. The intermediate stem cells in TEBs and elsewhere in the mammary gland can differentiate to ductal and then to alveolar-like cells along one reversible pathway (⇆) and to myoepithelial-like cells along another irreversible pathway (→). Ductal epithelial cells can also give rise to squamous cells in certain pathological conditions (--→). The growth-promoting ability of some growth factors and the differentiating ability of prolactin is indicated on cells within this developmental model. The roles of other hormonal agents, e.g. oestrogen and IGFs are not indicated.

matrix seems to play a role (page 166). One of the most likely internal candidates is the calcium-binding regulatory protein p9Ka. This probably causes an alteration, directly or indirectly, in the cytoskeleton of the epithelial/intermediate stem cells (Figure 16) (page 188). Differentiation along the myoepithelial pathway is also accompanied by production of extracellular/basement membranous materials as well as growth factors like bFGF (Figure 16) (page 180). These growth factors may possibly act in a paracrine and/or autocrine fashion, since the receptors for bFGF are also induced on the stem cells and along this pathway. However, a complicating factor in the case of the FGF family of growth factors is their sequestration by HSPG low-affinity receptors on the myoepithelial cell surface/basement membrane/extracellular matrix (page 177). Hence production of bFGF would have to be accompanied by loss and/or destruction of many of these low-affinity receptors to enable cell proliferation to occur. Such circumstances may be encountered in intermediate stem cells in growing or regenerating ducts, particularly in the TEBs (pages 178 and 183), so enabling a negative control to be exerted by largely the extracellular matrix on the production of stem cells.

In addition to the myoepithelial cells and associated basement membrane/extracellular matrix exerting control on the production of stem cells, stromal fibroblasts and fat cells also provide conditions for apparent expansion of the intermediate stem cell population in TEBs and presumably thereby ductal growth (page 167). One of the most powerful mitogens identified so far that is produced by the adipocytes is not a peptide growth factor, but a prostaglandin – PGE_2. Thus differentiation of preadipocytic fibroblasts to adipocytes releases a powerful local stimulus for all mammary cell growth, which is not dependent on the problems of sequestration like aFGF or bFGF. Hence differentiation within the mesenchymal compartment may also influence the production of intermediate stem cells. Thus a variety of factors, both extrinsic, like circulating hormones and growth factors, and intrinsic to the mammary gland in the form of locally-produced diffusable and non-diffusable substances can alter the balance of growth and differentiation within the mammary parenchyma, and thereby affect the levels of the intermediate stem cells. In addition it should be noted that the biological function of the mammary gland is to produce milk and that milk contains a number of growth factors whose function is presumably related to the growth of the offspring and not necessarily to the growth of the mammary gland *per se*. One thing is certain. Many more controls on growth and differentiation have yet to be discovered in the mammary system. On the basis of the systems discussed here, the development of the mammary gland can be seen to resemble, albeit to a more limited extent, development of the haemopoietic system in the bone marrow. Whether this is a useful concept in understanding the process of abnormal development, particularly those of cancer, will be the subject of the remainder of this review.

DEVELOPMENT OF THE MALIGNANT MAMMARY GLAND

Identification of different cell types in benign and malignant mammary lesions

The susceptibility of the rat mammary gland to rapid chemical carcinogenesis is correlated with the presence of TEBs and terminal ducts and declines with the appearance of ABs after 50 days (Russo *et al.*, 1977). Although the degree of glandular differentiation of DMBA- (Huggins *et al.*, 1961) and nitrosomethylurea (NMU)- (Gullino *et al.*, 1975) induced tumours varies, the vast majority in our hands are benign. They are delineated by a fibrous capsule, show little evidence of local invasion and metastases, and in many tumours the cells are cytologically benign by standard human histopathological criteria (Williams *et al.*, 1981). They are best described as adenomas showing varying degrees of atypia (Rudland, 1987a, 1987b). Serial transplantation of these immunogenic (Kim, 1984) tumours in syngeneic rats can yield weakly-metastasizing tumours (e.g. TR2CL type) (Williams *et al.*, 1982), but chemical induction in partially immune-deprived rats that are then immunostimulated non-specifically produces non-immunogenic tumours of much higher metastatic ability (e.g. TMT, MT and SMT types) (Kim, 1979). The former (TR2CL) tumours metastasize to lungs and lymph nodes only, primarily via the bloodstream producing micrometastases (Williams *et al.*, 1985), whilst the latter (TMT, MT, SMT) tumours produce gross metastases ranging from those in the lungs and lymph nodes for TMT-081 which spread via both bloodstream and lymphatics, to extensive metastases in lymph nodes, lungs, bone, spleen and kidney for SMT-2A which spread initially only via the lymphatics (Table 6) (Kim, 1979). The TR2CL tumours are thus classified as weakly-metastatic, the TMT-081 as moderately- and the SMT-2A as strongly-metastatic.

As with the carcinogen-induced tumours of the rat, the terminal structures of the human mammary gland are the most likely initial sites for much of the hyperplastic benign breast disease and many of the benign and malignant mammary neoplasias (Wellings *et al.*, 1975; Ishige *et al.*, 1991). Indeed the incidence of breast cancer increases predominantly in those survivors of the atomic bombs at Hiroshima and Nagasaki whose mammary glands were likely to contain TEBs and terminal ducts at the time of irradiation (McGregor *et al.*, 1977). Human mammary carcinomas are thought to arise mainly from carcinomas-*in-situ* (Page *et al.*, 1987) and they in turn are thought to arise predominantly from epithelial hyperplasias (Page *et al.*, 1987) that are similar to those encountered in hyperplastic benign breast disease, particularly that associated with atypia (Simpson *et al.*, 1982; Wellings and Misdrop, 1985; Norris *et al.*, 1988). The major benign neoplasms of the breast, the fibroadenomas and adenomas, however, appear to arise by proliferation of the entire ductal/ductular unit and rarely develop into carcinomas (Table 6) (Fechner, 1987). Thus both benign and malignant mammary tumours are thought to arise predominantly in terminal ductal structures, and in this respect are more similar to those of the rat than to those of the mouse mammary tumour virus (MMTV)-infected mouse (DeOme *et al.*, 1959). The patterns of metastatic spread of the human carcinomas are also similar, in certain respects to those of the non-immunogenic rat mammary tumours (Kim, 1979) with lymphatic permeation being the favoured route of dissemination, at least in the early stages of the disease (Lee, 1985).

Analysis of the cell types present in both rat and human lesions using the immunocytochemical and ultrastructural criteria (see pages 150–152), which are based predominantly on the presence of a smooth-muscle actomyosin cytoskeleton and particular cell-surface determinants, give similar results to each other. The human benign hyperplastic (fibrous changes, fibrocystic disease, lobular hyperplasia), the benign rat (DMBA-induced or MT-W9 transplanted) and human neoplasias (papilloma, fibroadenoma, phyllodes tumour) contain both epithelial and myoepithelial-like cells, many in duct-like arrangements (Table 6) Azzopardi, 1979; Bussolati *et al.*, 1980;

Table 6 Summary of cell types present in rat and human mammary lesions.

Mammary lesion	Number of females	(Pregnant/ lactating)	Cell types[a]
Rat			
Benign adenomatous			
DMBA-induced	20	(5)	epithelial
NMU-induced	20	(6)	myoepithelial-like
MT-W9	10	(2)	(alveolar-like)[g]
Invasive carcinoma[b]			
TR2CL	20	(5)	epithelial, undifferentiated myoepithelial-like cells
TMT-081	20	(5)	
MT-450	12	(3)	anaplastic
SMT-077	14	(4)	epithelial
SMT-2A	12	(3)	
Human			
Benign diseases			
Fibrous changes	31		
Fibrocystic disease	242[d]	(1)	
Lobular hyperplasia	5		epithelial,
Papilloma	6		myoepithelial-like
Fibroadenoma	76	(1)[f]	(alveolar-like)[g]
Phyllodes tumour	8		
Carcinoma-in-situ			
Ductal	8	(2)	epithelial
Lobular	4		
Invasive carcinoma			
Tubular	4		
Ductal[c]	137[e]		
Medullary	3		epithelial
Colloid	7	(2)	
Lobular	11		

[a] Immunocytochemical identification in parenchymal tissue only, carcinomas-*in-situ* also possess host myoepithelial cells surrounding the lesions; [b] Invasive carcinomas of rat in increasing malignant (as defined by metastatic) potential, TR2CL borderline malignant being only weakly metastatic; [c] Not otherwise specified; [d] This includes cystic changes (22), duct ectasia (9), apocrine metaplasia (16), adenosis (133), blunt duct adenosis (15), sclerosing adenosis (25), epitheliosis (49); the last four contain areas associated with some of the other types of benign breast disease; [e] This includes 6 with ductal carcinoma-*in-situ* with comedo (3) or cribriform (3) pattern and 6 with scirrhous reaction; [f] Lactating adenoma; [g] Secretory alveolar-like cells detected only in appreciable quantities in lesions in pregnant/lactating females, otherwise 5–10% of the lesions in non-pregnant females contained small numbers (<5%) of secretory cells.

Dunnington *et al.*, 1984a; Gusterson *et al.*, 1982, 1985, 1986; Nagle *et al.*, 1986; Dairkee *et al.*, 1985, 1988; Rudland *et al.*, 1993b). However, many of the myoepithelial-like cells, particularly in the rat tumours, possess a more variable and undifferentiated appearance than the myoepithelial cells of mature mammary ducts (Dunnington *et al.*, 1984a), and are therefore more similar to the basal or intermediate cells of TEBs and ABs (Ormerod and Rudland, 1984). In human mammary lesions as well, cells intermediate in ultrastructure

(Ozzello, 1971) and immunocytochemical staining (Rudland *et al.*, 1993b) characteristics between epithelial and myoepithelial cells are observed in areas of hyperplasia and of fibroadenoma, and such cells also bear at least a superficial resemblance to the intermediate cells growing in terminal ductal structures in normal mammary glands (Rudland, 1993).

In contrast to the above, in both rat and human carcinomas, no malignant cells with the same ultrastructural and immunocytochemical characteristics of fully differentiated myoepithelial cells are observed, apart from the host myoepithelial cells surrounding carcinoma-*in-situ* (Table 6) (Dunnington *et al.*, 1984a; Williams *et al.*, 1985; Bussolati *et al.*, 1980; Gusterson *et al.*, 1982; Nagle *et al.*, 1986; Dairkee *et al.*, 1988; Rudland *et al.*, 1993b). Thus only the epithelial cell type is present (Sloane and Ormerod, 1981; Foster *et al.*, 1982b; Taylor-Papadimitriou *et al.*, 1983). The loss of the myoepithelial cell has since been used as a diagnostic marker for separating apparently morphologically similar benign and malignant lesions in humans (Rudland, 1993).

Similarly the benign tumours of the rat (DMBA-induced and MT-W9 transplantable) as well as the rather rare cases of benign hyperplastic and benign neoplastic lesions of humans in hormonally-primed hosts can produce small amounts of the appropriate immunoreactive caseins in alveolar-like cells. In contrast, rat or human carcinomas do not (Table 6) (Herbert *et al.*, 1978; Supowit and Rosen, 1982; Rudland *et al.*, 1983b, 1993b; Rudland, 1987a; Bartkova *et al.*, 1987; Earl *et al.*, 1989). Thus complete and definite markers of fully differentiated myoepithelial cells and, under suitable hormonal conditions, of secretory alveolar cells are found in benign hyperplastic and neoplastic disease but are absent in malignant carcinomas (Rudland *et al.*, 1993b).

Although there is usually a complete loss of myoepithelial and alveolar features in most of the rat and human carcinomas, some residual features of the myoepithelial cell, but probably not the alveolar cell, are retained. For example, in rats, the weakly-metastasizing tumour, TR2CL contains undifferentiated more-elongated cells and patches of fragmented basement membrane (Williams *et al.*, 1985). In humans 11–15% of invasive carcinomas (Albrechstein *et al.*, 1981; Nielson *et al.*, 1983; Aung *et al.*, 1993) produce a fragmentary basement membrane, and there is sometimes an inverse relationship between its appearance and the histological grade or degree of malignancy (Gusterson *et al.*, 1982; Nielson *et al.*, 1983; Aung *et al.*, 1993). Moreover, cytokeratin 14, a specific marker for myoepithelial cells in normal mammary glands (page 150) is found in 9–14% of cases of malignant carcinoma in humans (Dairkee *et al.*, 1988; Wetzel *et al.*, 1989; Aung *et al.*, 1993). An even higher proportion of 50% of human carcinomas stains strongly with an MAb that also stains myoepithelial cells in mature normal ducts (Skilton *et al.*, 1990). These results suggest that, although fully differentiated myoepithelial cells are absent in carcinomas as judged primarily by their smooth muscle-like cytoskeleton, some marker proteins of the myoepithelial cell are retained by 9–50% or more of malignant carcinomas. These myoepithelial-related proteins are expressed to varying degrees by the malignant epithelial cells, and there is sometimes an inverse correlation between their expression and the degree of malignancy (Rudland, 1993).

Isolation and properties of immortalized cell lines from mammary carcinomas

In the rat, only the weakly malignant tumour TR2CL has been cultured by conventional means using collagenase digestion (Williams *et al.*, 1982, 1985). Other strategies employed include culturing ascitic versions of the transplantable tumour, e.g. TMT-081 (Ghosh *et al.*, 1983) and SMT-2A (Rudland *et al.*, 1989c) or selecting the most metastatic variants *in vitro* from cultured metastases (Nehri *et al.*, 1982), usually with other cell types present as feeders. The ascitic and solid forms of the tumour and the cell lines developed from them give the same histological appearance and patterns of spread in the syngeneic rat (Kim, 1979; Ghosh *et al.*, 1983; Dunnington *et al.*, 1984b; Rudland *et al.*, 1989c). Our cell lines, the Rama 600, Rama 800, and Rama 900 series have been isolated from transplantable tumours of increasing malignancy as measured by metastatic potential, i.e. the TR2CL, TMT-081 and SMT-2A tumours, respectively (Table 3). There is a corresponding increase in the malignant cells' ability to grow in suspension as loosely-adherent aggregates, a decrease in growth rate and a greater dependency for support on feeder cells (Williams *et al.*, 1985; Dunnington *et al.*, 1984b; Rudland *et al.*, 1989a). In fact the most malignant, metastatic Rama 900 cell line from the SMT-2A ascites tumour grows only in culture as aggregates with a feeder layer of normal mesothelial cells from the same ascites fluid.

The above behaviour of malignant metastatic rat mammary cells has been used as the basis for the isolation of human malignant cells. Thus collagenase digests of primary tumours of human invasive ductal carcinomas yield on collagen gels a fast-adherent cellular fraction similar to that of benign hyperplastic and neoplastic human mammary lesions (Figure 17A) and a fraction of relatively loosely-adherent cellular aggregates (Figure 17B). The former fraction contains both epithelial and myoepithelial cells. The latter fraction can eventually attach to plastic petri dishes and forms colonies of slow-growing malignant epithelial cells without any characteristic myoepithelial-like cells (Figure 17C) (Rudland *et al.*, 1985; Peterson and van Deurs, 1987; Peterson *et al.*, 1990). Only the loosely-adherent, slow-growing cellular fraction of epithelial cells can be isolated from cultured metastases (Figure 17D) (Rudland *et al.*, 1985), so confirming that these cells do indeed represent the malignant-cell fraction in the primary invasive ductal carcinomas (Hallowes *et al.*, 1983) (Table 7). These normally die out on repeated transfer *in vitro*.

Spontaneously immortalized cell lines from metastatic rat primary carcinomas (Figure 17E) (Rama 600, Rama 800, Rama 900), from human pleural effusions (MCF-7; ZR-75; T-47D; SK-Br-3; and MDA-MB-231: Soule *et al.*, 1973; Engel *et al.*, 1978; Keydar *et al.*, 1979; Trempe and Fogh, 1973; Cailleau *et al.*, 1974; respectively) and from loosely-adherent cellular fractions of human primary carcinomas (Figure 17F) (Ca2-83: Rudland *et al.*, 1985; KM1, K. McCarthy and P. S. Rudland, unpublished) have been subjected to clonal analysis. However, no myoepithelial-like or alveolar-like cells have been detected using the immunocytochemical and ultrastructural criteria outlined on page 152, at frequencies where one cell in a million would have been capable of detection (Table 3) (Rudland, 1987a, 1993). Moreover, different malignant epithelial cell lines selected from other rat carcinogen-induced tumours (Nehri *et al.*, 1982) or developed from primary invasive human carcinomas (Lasfargues and Ozzello, 1975; Nordquist *et al.*, 1975; Hackett *et al.*, 1977; Minafra *et al.*, 1989; Peterson *et al.*, 1990) fail to show any definitive features of myoepithelial cells (Table 3). All malignant cell lines of human and some of rat origin show evidence of squamous-like formations but to varying degrees. Thus whilst benign hyperplastic and benign neoplastic lesions in rats and humans contain epithelial stem cells capable of differentiating to myoepithelial-like and to alveolar-like cells (page 161), malignant epithelial cells fail to differentiate to these two mature cell types in culture, in agreement with the earlier pathological analyses on page 193.

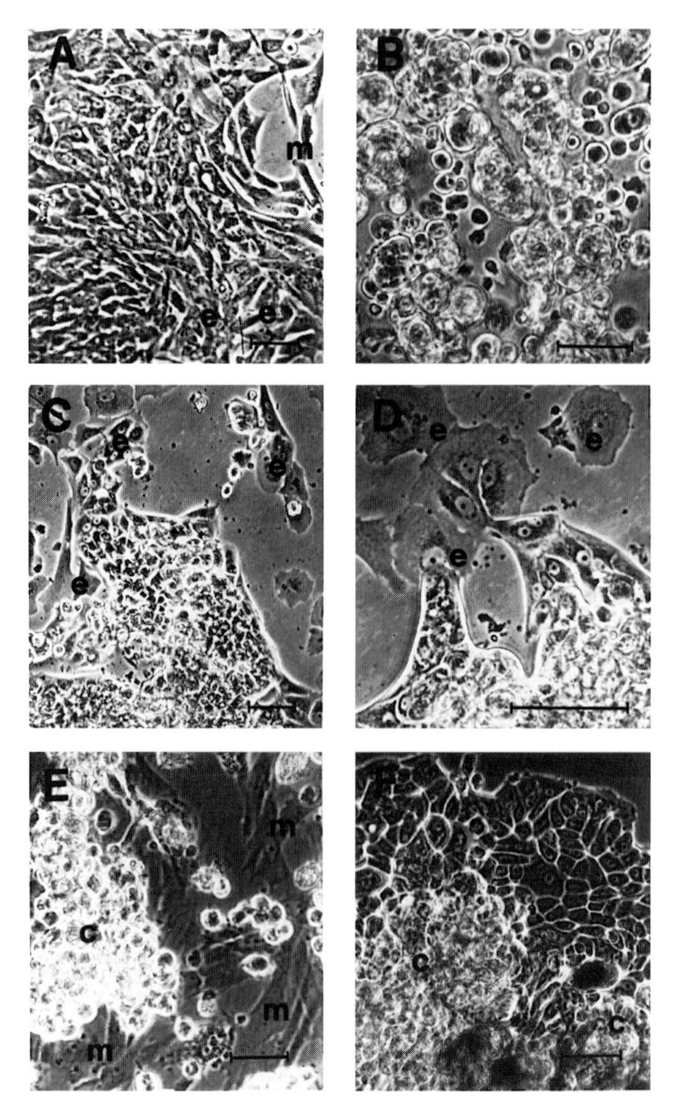

Figure 17 Phase-contrast micrographs of cultured invasive ductal carcinoma. (A) Collagenase-digested human primary carcinoma, fast-adherent fraction to collagen gel showing some epithelial (e) but many more-elongated, myoepithelial-like cells (m). (B) Collagenase-digested human primary carcinoma, loosely-adherent fraction 72 hours after plating on collagen-coated dishes showing no attached cells. (C) Collagenase-digested human primary carcinoma, loosely-adherent fraction now plated on plastic dishes showing large epithelial-like cells (e), but no more-elongated myoepithelial-like cells. (D) Collagenase-digested human secondary carcinoma now plated on plastic dishes showing large pleiomorphic epithelial-like cells (e), but no more-elongated myoepithelial-like cells (Rudland *et al.*, 1985). (E) Permanently growing rat carcinoma cell line Rama 900 growing with mesothelial-like feeder cells (m) (Rudland *et al.*, 1989c). (F) Permanently growing human carcinoma cell line Ca2-83 (Rudland *et al.*, 1985). Both the last two lines show limited attachment to plastic with the formation of loosely-adherent clumps (c), but no more-elongated, myoepithelial-like cells. Original magnifications $\times 160$ for A, C; $\times 250$ for B; $\times 400$ for D, $\times 300$ for E, F. Bars $= 75\,\mu m$ for A–D, $50\,\mu m$ for E and $75\,\mu m$ for F.

Table 7 Adherence of cultured digests of different human mammary tissues to collagen gel.

Mammary tissue[a]	Total no. of patients	No. of patients with different types of epithelium in cultures		Identifiable cell-types
		Fast-adherent[b]	Loosely-adherent[c]	
Normal (reduction mammoplasty)	40	40	0	Epithelial, myoepithelial-like
Benign hyperplastic (benign breast disease)	20	20	0	Epithelial, myoepithelial-like
Benign neoplastic (fibroadenoma)	10	10	0	Epithelial, myoepithelial-like
Primary carcinoma (invasive ductal)	30[d]	25	27	Epithelial, (myoepithelial-like)[e]
Secondary carcinoma in lymph node	10[d]	0	9	Epithelial
Secondary carcinoma in as pleural effusion	2	0	2	Epithelial

[a] Mammary tissue has been digested with collagenase, except pleural effusions, and grown on collagen gel-coated dishes; [b] Fast-adherent colonies stick within 24–48 hours, they always contain epithelial and myoepithelial-like cells as determined by the markers described in the text; [c] Loosely-adherent colonies take greater than 72 hours to stick, if they do at all, and then are easily washed off; they contain only epithelial and not myoepithelial-like cells; [d] A few samples contain no visible epithelium after digestion; [e] Myoepithelial-like cells only present in the fast-adherent and not in the loosely-adherent fraction.

The closest malignant epithelial cell lines to the benign tumour cell lines are the weakly-metastasizing rat cell lines of the Rama 600 series. These epithelial cell lines can generate undifferentiated more-elongated cells in culture at frequencies of the order of 10^{-5} to 10^{-6}. These more-elongated cells retain components of the basement membrane and PWM and GS-1 receptors, but have lost the smooth muscle-like actomyosin cytoskeleton (Williams *et al.*, 1985; Hughes and Rudland, 1990a). They do not metastasize on their own, however, and their presence in the parental strain may serve to enhance the growth rate of the Rama 600 cell type, similar to the effect of the myoepithelial-like cells on the growth of normal and benign epithelial cells in culture. The partial differentiation of Rama 600 epithelial cells towards a myoepithelial phenotype may still be sufficient to generate the fragmented basement membrane which is sometimes seen in the Rama 600 tumours *in vivo*. It would appear then that, at least in the rat, vestiges of the ability of the normal and/or benign stem cells to differentiate to myoepithelial-like cells is probably retained in the less malignant/metastatic cell lines.

Most of the usual properties associated with viral transformation of cultured fibroblasts (Rudland and Jimenez de Asua, 1979a, 1979b) are not applicable to our rat mammary cell lines of increasing metastatic potential or to our human malignant cell lines. The properties of the metastatic cells include reduced not increased growth rates, reduced not increased ability to grow in semi-solid medium, reduced not increased ability for autonomous growth in isolation, reduced (North and Nicolson, 1985) not increased ability to grow in T-cell deficient rodents (Dunnington *et al.*, 1984b; Williams *et al.*, 1985; Rudland *et al.*, 1989c). The properties which do change radically when these cells become highly metastatic are those involved with increasing instability (Nowell, 1976; Nicolson *et al.*, 1983; Poste, 1983) or plasticity (Rudland, 1987b) of genetic expression and hence an increased adaptive ability. Thus, in addition to the loss of some or all of both differentiation pathways of the normal mammary epithelial stem cell, there is a large increase in chromosomal number up to near tetraploid and a large increase in resistance to the cytotoxic effects of drugs such as ouabain and trifluorothymidine (Rudland, 1987a) and, in other rat systems, to chemotherapeutic agents (Welch and Nicolson, 1983). One final example of the ability of the highly metastatic mammary carcinoma cells to adapt to their environment is the growth of the Rama 900 cells in nude mice. Although the non-metastasizing epithelial stem cell lines Rama 25, Rama 37 (Rudland *et al.*, 1982a; Dunnington *et al.*, 1983) and the weakly-metastasizing Rama 600 cells (Williams *et al.*, 1985) grow readily in nude mice, the moderately-metastasizing Rama 800 (Dunnington *et al.*, 1984b) and the strongly metastasizing Rama 900 (Figure 10E) (Rudland *et al.*, 1989c) grow poorly from subcutaneous sites and are rejected like their parental tumours. However, if Rama 900 cells are inoculated interperitoneally, about half the nude mice develop ascites tumours, and these contain mainly mouse mesothelial feeder cells and Rama 900 tumour cells. The mouse-derived tumour cells can now produce tumours and metastases in nude mice, but not in syngeneic rats (Rudland *et al.*, 1989c). Results with the primary cultures and cell lines of human malignant carcinomas are similar to but not so extensive as those in the rat (P. S. Rudland, unpublished). These results, therefore, tend to reduce the importance of simple molecular mechanisms based solely on autostimulation of cell growth for malignant, metastatic properties and tend to emphasize the increasing capability of the malignant, metastatic cell to adapt and utilize normal control mechanisms in new environments (Rudland, 1987b).

Cellular model for the development of breast cancer

Previous sections have shown that certain epithelial cells of normal mammary glands and of benign/hyperplastic and neoplastic lesions of rats and humans have the properties of stem cells. Such epithelial stem cells can give rise to all other cell types, including the alveolar and the myoepithelial cell, whereas the malignant epithelial cells fail to do so. Two simple models may explain this phenomenon: (1) there are two types of epithelial cell in normal mammary glands, the one which can differentiate gives rise to benign lesions and the one which cannot gives rise to malignant lesions; (2) the ability of the epithelial stem cells to differentiate is impaired when they become malignant (Figure 18).

There is some evidence for the first model. Thus immunocytochemical staining patterns for cytokeratins, particularly cytokeratin 19, has suggested that two populations of epithelial cells exist within the normal human mammary gland; epithelial cells in benign tumours usually show a reduced expression whilst those in malignant carcinomas show an increased expression of this cytokeratin (Taylor-Papadimitriou *et al.*, 1983; Taylor-Papadimitriou and Lane, 1987) in support of different cellular origins for benign tumours and carcinomas. However, this first model cannot readily accommodate the observation that 9–50% of carcinomas retain some markers of the myoepithelial cell (page 194).

The second model is supported by evidence obtained from experimental carcinomas

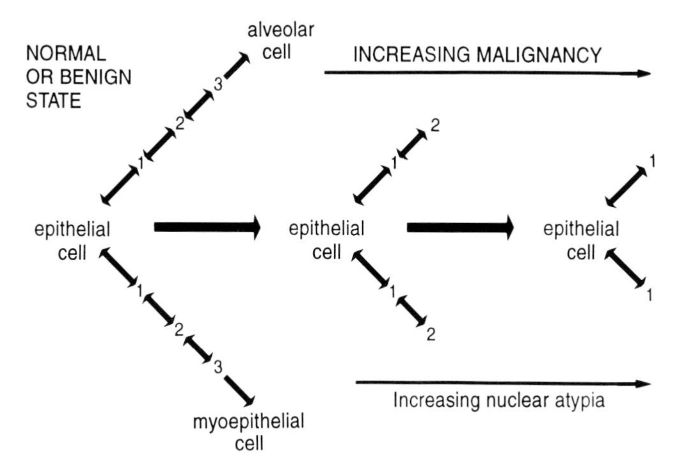

Figure 18 Possible cellular model for the development of carcinoma cells of the mammary gland. Normal epithelial cells in terminal ductal structures can differentiate into myoepithelial cells and into secretory alveolar cells by a series of cellular intermediates here represented by numerals. Details of such steps have only been worked out at present in rats (Figures 4 and 8). It is suggested that many benign hyperplastic and neoplastic lesions arise directly from such epithelial/closely-related intermediate cells which then still retain much of their differentiating abilities to myoepithelial and to alveolar cells. Insults leading to malignant changes produce epithelial/related intermediate cells with a decreasing ability to differentiate into either cell type, and result at best in a retention of only limited expression of markers characteristic of myoepithelial and of alveolar cells in invasive carcinomas. The fact that more vestiges of myoepithelial cell rather than alveolar cell markers are retained in invasive ductal carcinomas, even under hormonal environments conducive to alveolar cell differentiation (Rudland *et al.*, 1993b), suggests that the malignant insult occurs/is expressed in epithelial/intermediate cells along the myoepithelial rather than the alveolar cell pathway.

in rats and from observations made on human carcinomas. Thus, in the rat, the benign tumours, which contain predominantly only the one major type of epithelial cell, the differentiating stem cell (Rudland, 1987a), can develop into malignant metastasizing tumours by serial transplantation, particularly against an immunological barrier (Kim, 1979). This change appears to be accompanied by a gradual loss of differentiating ability to the myoepithelial cell as the malignant potential increases (page 195). Moreover, direct transfection of rat clonal epithelial stem cell lines with fragmented DNA of either rat (Jamieson *et al.*, 1990a) or human malignant, metastatic origin (Davies *et al.*, 1993a) can induce specifically the malignant metastatic state (Table 8) (page 200). Such transformants also fail to differentiate to myoepithelial-like, or to alveolar-like cells. In human mammary cancers detailed studies on markers of the normal myoepithelial cells (page 194), cytokeratin 14, antigenic determinants on the MCF-7 mammary carcinoma cell line and the basement membrane proteins laminin and Type IV collagen, show that 9–50% of human mammary carcinomas retain a marker of the myoepithelial cell and that such markers are now expressed by the malignant epithelial-like cells.

Since the differentiation of epithelial cells to myoepithelial-like cells proceeds in discrete steps (pages 152 and 164) in rat systems *in vivo* and *in vitro* and is likely to do so in humans (pages 153 and 167) it is possible then that variable truncations of this

Table 8 Summary of metastases produced by transfectants of a benign rat mammary cell line.

Transfecting DNA[a]	Latent period of primary tumour (days)	Incidence of metastases
–	44	0/36
Salmon sperm	43	0/20
EJ-*ras*	26[b]	0/15
Large T Antigen	28[b]	0/20
p9Ka	11[b]	16/29[c]
Rama 37-benign	46	0/65
Rama 800-metastatic	50	9/62[c]
Huma DNA-non-metastatic	23[b]	0/16
Huma DNA-metastatic	26[b]	18/51[c]

[a] Abbreviations as in text; [b] Significantly shorter than the parental Rama 37 cells ($P<0.01$, Mann Whitney U-Test); [c] Significantly greater than the parental Rama 37 cells ($P<0.01$, Fisher Exact Test).

differentiation pathway may accompany the malignant process. Thus the malignant, myoepithelial-like cells would retain the vestiges of this pathway, thereby allowing ectopic expression of a limited number of myoepithelial-related markers which, perhaps, increase the ability of the malignant cells to adapt to new environments (Figure 18). Immunocytochemical studies with antibodies to complex and simple keratins in different human mammary diseases are consistent with this model (page 194). Furthermore, these studies show that benign hyperplastic, benign neoplastic and malignant neoplasms largely have a common origin in the growing ductal buds and lobules (Rudland *et al.*, 1993b), a suggestion consistent with conclusions drawn from the more detailed examination of their subgross anatomies (Wellings *et al.*, 1975). The most likely cellular candidates in these growing buds and lobules are those peripheral cells intermediate between epithelial and myoepithelial cells. Whether malignant transformation preserves preferentially the intermediate types closer to the epithelial end of the epithelial–myoepithelial developmental pathway or truncates the epithelial stem cells' ability to differentiate along this developmental pathway is unclear at present. However, the ectopic expression of a limited number of markers related to the myoepithelial cell by a good proportion of human invasive carcinomas, particularly and including carcinoma-*in-situ*, the direct precursor of invasive carcinoma, suggests that the insulted cell originally bears some similarities to the epithelial precursor or stem cells described for normal developmental processes of the mammary gland.

There are several consequences of the above model of sequential progression from benign to malignant epithelial stem cells. The epithelial stem cells from benign tumours with their capacity to differentiate to the other major cell types *in vitro* should be capable of generating all the pathological forms of glandular tumours when re-inserted *in vivo* (Hamperl, 1970). This is indeed the case for Rama 25, SVE3, Huma 7, Huma 121, and Huma 123 in nude mice and Rama 37 in syngeneic rats (Rudland, 1987b, 1993). Complete differentiation of the rat benign neoplastic epithelial cell lines along either pathway often reduces their tumorigenic potential (Rudland, 1987a). Thus most of the myoepithelial-like cell lines from Rama 25 fail to induce tumours, but spontaneous transformants from Rama 29 yield benign spindle-cell tumours reminiscent of myoepitheliomas (Rudland *et al.*, 1982a). This suggests that subsequent and perhaps different transformation events are required to generate the spindle-cell tumour.

A large reduction in rates of DNA synthesis *in vitro* and in rates of tumour formation *in vivo* is also caused by the differentiation of Rama 25 cells to alveolar-like cells triggered by DMSO, PGE_1 (Rudland *et al.*, 1982b), or retinoids (Rudland *et al.*, 1983a) in the presence of prolactin, oestradiol, hydrocortisone and insulin. Moreover, the variant of Rama 25, Rama 259, which fails to complete the later stages of the alveolar-like pathway (page 156) also fails to reduce its rate of DNA synthesis and tumour-forming ability under the same conditions. Thus the effect of mammatrophic hormones, pregnancy (Dao *et al.*, 1960; Moon, 1969) and the retinoids (Sporn and Newton, 1979) in protecting the mammary gland from carcinogenic insult may be due to their ability to induce differentiation. This ability may be exerted either at an early stage (McCormick *et al.*, 1980) converting normal epithelial stem cells in TEBs to less susceptible, slow-growing differentiated cells or at a later stage in the carcinogenic process converting benign tumour stem cells to differentiated cells having lower tumorigenic potential (Rudland, 1987a). On this model such agents are unlikely to be effective against the strongly malignant non-differentiating carcinoma cell. These results in the rat are consistent with the effect of pregnancy (MacMahon *et al.*, 1973) and dietary vitamin A (Peto *et al.*, 1981) in reducing the incidence of human mammary cancer and as such give further validity to this model *in vivo*.

NON-HEPARIN-BINDING GROWTH FACTORS IN BREAST CANCER

Hormonal agents and micronutrients

About one quarter of all human breast carcinomas will regress if any one of the major endocrine organs, the ovaries, the adrenal glands or the pituitary is removed, or if their identified hormones are inhibited in some way, e.g. tamoxifen inhibiting the binding of oestrogen to its cellular receptor or aminoglutethimide inhibiting its synthesis. These so called hormone-sensitive carcinomas invariably recur, but are still capable of inhibition by removing the second and eventually the third type of endocrine gland. Eventually these hormone-sensitive carcinomas recur in a state that is no longer responsive to any further hormonal manipulations, the so called hormone-insensitive state. Thus, as breast cancer progresses, it is believed to pass from a hormonally sensitive to a hormonally-insensitive state (Vorherr, 1980). Since the major hormones of the ovaries, adrenals and pituitary glands are thought to exert their growth-promoting effects indirectly on the normal gland (page 167), the mechanisms by which circulating hormones influence the growth of the breast carcinoma cells and how this influence is lost is largely unknown.

One candidate hormone that has been isolated from, *inter alia*, the pituitary gland is the iron-transporting protein transferrin (page 168). In the MCF-7 breast carcinoma cell line, transferrin is secreted into the culture medium; its secretion is stimulated by oestrogen and reduced by tamoxifen (Vandewalle *et al.*, 1989). The transferrin receptor is also present in the MCF-7 cell lines, being expressed in a cell-cycle specific manner (Kute and Quadri, 1991). Thus the possibility exists of an hormonally-controlled autocrine loop involving transferrin. However, there may be some doubt as to whether the secreted transferrin is available to form an autocrine loop *in vivo*, since it is secreted only when the basolateral surface is accessible (Vandewalle *et al.*, 1991), and it then binds to membrane-associated proteoglycans (Vandewalle *et al.*, 1989, 1991). It is probable, too, that, as found for the rat mammary tumour cell line 64-24, transferrin does not promote growth if the concentration of iron in the medium is optimal (van der Burg *et al.*, 1990). Nevertheless, transferrin receptors are overexpressed in the majority of cells in 71% of invasive breast carcinomas, and in some cells of all breast carcinomas (Walker and Day, 1986). No correlation has yet been found with nodal status or tumour grade, but there is a correlation with a marker of cell proliferation, Ki-67 (Betta *et al.*, 1989; Wrba *et al.*, 1989). The concentration of transferrin receptors in serum, however, is not raised (Raaf *et al.*, 1993). This situation may be somewhat confused by the presence in such tumours of infiltrating T lymphocytes which themselves express transferrin receptors (Whitford *et al.*, 1990).

That the iron metabolism of the cell is important in breast carcinoma cells may be indicated by the following. There is overexpression of, *inter alia*, the ferritin H chain in tumorigenic, but not non-tumorigenic clones of the SW 613-5 breast cancer cell lines (Lavialle *et al.*, 1989). The ratio of placental isoferritin to 'normal' ferritin is greater in cell lines derived from breast carcinomas than from normal breast epithelial cells (Shterman *et al.*, 1991). The expression of the superheavy chain p43 of the placental isoferritin shows an inverse correlation with growth rate in MCF-7 and T-47D cells (Rosen *et al.*, 1992a), and with poor prognosis in patients (Rosen *et al.*, 1992b). In invasive breast carcinomas, the tissue concentrations of ferritin correlate with poor prognosis and histopathological dedifferentiation (Weinstein *et al.*, 1989; Guner *et al.*, 1992). Moreover, the presence of differing antigenic forms of ferritin and ferritin-carrying lymphocytes (Moroz *et al.*, 1989; Rosen *et al.*, 1992c, 1993) may be the cause of disputes as to its utility as a tumour marker in blood (Weinstein *et al.*, 1989; Williams *et al.*, 1990a, 1990b; Monti *et al.*, 1990; Stierer *et al.*, 1991; Guner *et al.*, 1992). Thus the enhanced requirement for iron is an important fact in breast carcinoma cells. The iron transporter transferrin may be supplied locally or systemically to the invasive carcinoma cells. Hence the supply of iron bound to transferrin may be governed by the levels of the transferrin receptor on the surface of the carcinoma cells.

Another pituitary-derived mammary growth factor is PMGF (page 169). Unlike the situation with cell lines isolated from normal tissue or benign lesions, PMGF fails to stimulate the growth of highly malignant, metastatic carcinoma cell lines of rat (Rama 800, Rama 900) or human origin (Ca2-83) (Table 3). This result is consistent with those results *in vivo*, whereby hypophysectomy of rats or humans bearing such advanced tumours largely fails to cause their regression (Kim, 1979; Vorherr, 1980). In contrast, benign tumours of rats and humans largely regress after systemic use of antipituitary agents (Segaloff, 1978; Mansel *et al.*, 1978; Hinton *et al.*, 1986), and epithelial cells from such tumours are still responsive to the growth-promoting effects of PMGF in culture (Smith *et al.*, 1984d; P. S. Rudland, unpublished). Thus, in addition to its failure to differentiate, the highly malignant, metastasizing mammary carcinoma cell is no longer under control of agents released from the pituitary gland. One of the simplest interpretations is that the malignant cells are refractory to the growth-promoting effects of PMGF. The loss of differentiating ability and failure to respond to pituitary-related growth factor(s), however, are changes that may not occur simultaneously in the malignant cell, since weakly metastasizing rat cell lines are responsive to PMGF in culture (Smith, *et al.*, 1984d), and about a quarter of all human breast cancers, when first diagnosed, are responsive to hormone (including pituitary-ablative) therapy (Vorherr, 1980). Thus initiation of the loss of the differentiating ability of the neoplastic epithelial cell may occur at an earlier stage in the malignant process than does the loss of response to pituitary (possibly PMGF) control of cell growth.

Epidermal growth factors

The overexpression of TGFα or of its cognate receptor, EGFR has been implicated in the abnormal growth of the malignant breast. Thus TGFα or EGF-related molecules are overexpressed in a number of cell lines derived from human breast carcinomas as well as in a number of mammary tumours (Salomon *et al.*, 1984; Kidwell, 1986; Perroteau *et al.*, 1985; Travers *et al.*, 1988). The synthesis of EGFR can be stimulated by TGFα, TGFβ progestins, EGF, glucocorticoids and phorbol esters in human breast cancer cell lines (Kudlow *et al.*, 1986; Murphy *et al.*, 1986; Fernandez-Pol *et al.*, 1989; Ewing *et al.*, 1989; Lee *et al.*, 1989). Some express both TGFα and EGFR, and inhibition of EGFR in these cell lines results in reduced growth rates. These results may be examples of autocrine growth regulation in operation (Ennis *et al.*, 1989; Bjorge *et al.*, 1989; Bates *et al.*, 1990). However, high levels of EGF actually inhibit the growth of certain human breast cancer cell lines which possess high levels of EGFR (Murayama, 1990; Prasad and Church, 1991).

In support of an autocrine model involving TGFα/EGFR, *in vivo*, immunoreactive TGFα (ir-TGFα) has been detected in 30–70% of human breast carcinomas, and its presence correlates with the tumour burden (Lundy *et al.*, 1991; Parham and Jankowski, 1992; Umekita *et al.*, 1992; McAndrew *et al.*, 1994b). The major component of ir-TGFα, however, appears to be a 50 kDa protein, and the relationship of this to mature, bioactive TGFα is unknown (McAndrew *et al.*, 1994b). Ir-TGFα has also been found in the urine and effusions of patients with invasive breast carcinomas. Similarly EGFR is also present in some 40–50% of human breast carcinomas and its presence can be correlated with the number of metastases, the histological grade of the tumours, the involvement of lymph nodes and poor prognosis for the patient. This correlation is greater if ir-TGFα is also overexpressed in the same carcinomas (Umekita *et al.*, 1992). EGFR is detected in invasive ductal and medullary carcinomas, but not in lobular or colloid carcinomas (Skoog *et al.*, 1986; Spitzer *et al.*, 1987; Reubi and Torhorst, 1988; Nicholson *et al.*, 1988a, 1988b, 1989; Harris *et al.*, 1989; Llorens *et al.*, 1989; Grimaux *et al.*, 1989; Tauchi *et al.*, 1989; Barker *et al.*, 1989; Moller *et al.*, 1989; Toi *et al.*, 1989; Bolufer *et al.*, 1990) (summarized in Klijn *et al.*, 1992), suggesting at best its involvement in the former but not the latter type of breast carcinomas.

About a quarter of breast cancers are still sensitive to the growth-promoting effects of oestrogen (Vorherr, 1980). In a previous section oestrogen stimulation of cell growth has been suggested to be mediated by increased synthesis of TGFα, which in turn binds to EGFR (page 172). In support of this idea in experimental systems, oestrogen has been reported to induce the expression of TGFα in the MCF-7 breast carcinoma cell line, and ovariectomy to decrease the expression of the TGFα mRNA in rat mammary tumours. These results suggest that, as in the normal epithelium (page 172), production of TGFα is regulated by ovarian steroids in certain neoplastic breast tissues (Dickson *et al.*, 1986). Indeed the 5′ flanking region of the TGFα gene in MCF-7 cells contains a potential oestrogen-responsive element(s) in the region upstream of the sac-II restriction site (Dickson *et al.*, 1986; Liu *et al.*, 1987; Saeki *et al.*, 1991). Moreover, retinoic acid inhibits the proliferation of these MCF-7 cells, antagonizes oestrogenic stimulation of cell growth and inhibits secretion of TGFα. A direct antagonist of oestradiol, oestradiol–chlorambucil also inhibits the oestrogen-induced proliferation of MCF-7 cells and this inhibition is preceded by inhibition of the secretion of TGFα; proliferation is restored by addition of exogenous TGFα to the cells (Fontana *et al.*, 1992; Kosano *et al.*, 1992). Similar evidence has been obtained from the human malignant cell line T-47D and from oestrogen-stimulatable cell lines derived from chemically-induced rat mammary tumours (Eppstein *et al.*, 1989; Murphy and Dotzlaw, 1989; Kosano *et al.*, 1990; Manni *et al.*, 1990; Ahmed *et al.*, 1991). Not everybody agrees, however, that TGFα mediates oestrogen-stimulated growth in malignant breast cell lines. Thus Osborne *et al.* (1988) have found that their oestrogen-responsive MCF-7 cell line does not secrete TGFα; Leung *et al.* (1991) have found that TGFα is not the mediator of oestrogen-stimulated growth of the CAMA-1 breast carcinoma cell line; and Arteaga *et al.* (1988) have found that the use of antibodies to inhibit EGFR in hormone-dependent human breast carcinoma cell lines has no effect on oestrogen-stimulated cell growth. These results suggest that there is an alternative route for oestrogen-stimulated growth in breast cancer cell lines and, by implication, in human breast cancer. Progesterone-stimulated growth is probably not due to production of TGFα.

The majority of human breast carcinomas (75%) are insensitive to hormone, usually oestrogen-based, therapy. Those that are recur eventually in a form that is no longer sensitive to such therapy (Vorherr, 1980). It has been suggested that part of the pathway leading to the oestrogen-insensitive state is due to an increase in the expression of TGFα/EGFR. In support of this, the EGF-induced increase in DNA synthesis in mice was found to mimic the effects of oestrogen on its nuclear receptor (Ignar-Trowbridge *et al.*, 1992). Moreover, a percentage of human carcinomas overexpresses both TGFα and more often EGFR. This overexpression is more common in carcinomas which do not express the oestrogen receptor and hence are less likely to respond to oestrogen treatment (Klijn *et al.*, 1992). In some human breast carcinoma cell lines this change to an oestrogen-insensitive state can be apparently reproduced in culture. Thus withdrawal of oestrogens from the medium results eventually in an increase in the cells' basal growth rate, until this rate reaches the oestrogen-stimulable rate and the cells then become insensitive to oestrogens. In these cells TGFβ is induced by withdrawal of oestrogens, suggesting that it and not TGFα plays a role in the progression of carcinomas to the oestrogen-insensitive state (King *et al.*, 1989; Daly *et al.*, 1990).

Notwithstanding the prominence of the TGFα/EGFR pathway, the idea that it represents an autocrine loop that is out of control is probably simplistic in relation to breast cancer. Thus transfection of the ZR-75 human breast carcinoma cell line with the cDNA for EGFR results in its overexpression but fails to induce an oestrogen-independent state (Valverius *et al.*, 1990). Moreover, the expected outcome of an uncontrolled TGFα/EGFR autocrine loop would be a rapidly growing tumour, such as is found in squamous carcinomas of the head and neck, in which EGFR is invariably overexpressed (Cowley *et al.*, 1984). In contrast, breast carcinomas, at least in the early stages, tend to be slow growing, and frequently still under a degree of hormonal control.

Thus a more gradual evolution to an oestrogen-refractory state is required in which changes in TGFα/EGFR may well contribute, but other changes may be of equal or of more importance. If the malignant epithelial cell bears some resemblance to the intermediate stem cells of the normal breast, then partial differentiation along the myoepithelial pathway would automatically generate an oestrogen-refractory state (page 165), part of which may arise from enhanced activity of the TGFα/EGFR autocrine loop (page 170).

A related system to EGFR is that of the c-erbB-2 protooncogene. It codes for a 185 kDa tyrosine kinase receptor, and is structurally related to EGFR. However, it behaves differently in ZR-75 breast cancer cell lines in that its expression is reduced by oestrogen, and correlates inversely with growth rate (Warri *et al.*, 1991). The situation is found in which a breast cancer cell line (MDA-MB-231) possesses the c-erbB-2 protein and secretes a cognate ligand, namely gp30. When gp30 is administered to such cells, the p185 c-erbB-2 is phosphorylated, but their growth rate is reduced (Lupu *et al.*, 1990). The ligands for p185 c-erbB-2 include a family of proteins (heregulins) related to the EGF family (Holmes *et al.*, 1992) and some ligands also interact directly in some measure with EGFR (Lupu *et al.*, 1990). In a normal mouse mammary epithelial cell line (HC-II), transfection of the gene for the activated rat equivalent, the *neu* receptor, enables prolactin to induce casein secretion (Taverna *et al.*, 1991). Thus the evidence overall from work with cell lines indicates that the c-erbB-2 system represents, perhaps, a differentiation rather than a growth-inducing pathway, at least in fully differentiated mammary epithelial cells.

None the less, in invasive breast carcinomas expression of c-erbB-2 is observed in 20–30% of tumours (McCann *et al.*, 1989; Parkes *et al.*, 1990; Marx *et al.*, 1990; Winstanley *et al.*, 1991; Delvenne *et al.*, 1992; Zoll *et al.*, 1992) which in some cases (about 20% of tumours) is accompanied by gene amplification (Parkes *et al.*, 1990; Borg *et al.*, 1991; Wolman *et al.*, 1992; Zoll *et al.*, 1992). Normal and benign tissues show virtually no expression. Correlations of expression with EGFR, absence of oestrogen receptor, histological grade and nodal status have been reported (Marx *et al.*, 1990; Delvenne *et al.*, 1992), although correlations of detection using antibodies to the protein, cDNA to its mRNA (Zoll *et al.*, 1992) and gene amplification (Wolman *et al.*, 1992) may be poor. There is also evidence of mutated c-erbB-2 in some breast carcinomas (Lofts and Gullick, 1992; Zoll *et al.*, 1992). More importantly there is an inverse correlation between patient survival and expression of c-erbB-2, particularly in patients with no involved lymph nodes (Slamon *et al.*, 1987; Winstanley *et al.*, 1991).

It would appear that almost 50% of early X-ray-screened breast lesions of the ductal/lobular carcinoma-*in-situ* type express c-erbB-2 (Barnes *et al.*, 1988; Gusterson *et al.*, 1988) and that this protein may be subject to loss in progression to invasive breast carcinomas. Its expression occurs in large cell, largely comedo type *in situ* lesions and often in Paget's disease of the nipple (Barnes *et al.*, 1988; Gusterson *et al.*, 1988). As such the cells expressing c-erbB-2 may reflect a more primitive cell than the differentiated mammary epithelial cell. This more primitive cell may be capable of differentiating both to mammary epithelial cells and to squamous, including Paget's, cells of the skin, similar to that described earlier for our mammary stem cell system (page 164). Its biological response to expression of c-erbB-2 may also be different from that of differentiated mammary epithelial cells.

Another related gene, c-erbB-3, is expressed in some mammary carcinoma cell lines

(Krauss *et al.*, 1989). Its expression is stronger than normal in 22% of primary breast carcinomas, as measured by detection of mRNA or immunoreactive protein; gene amplification is not, however, observed (Lemoine *et al.*, 1992). There is a positive correlation with lymph node metastases, but not with poor prognosis. However, unlike c-erbB-2 which is virtually an all-or-none phenomenon in normal malignant tissues, changes in c-erbB-3 are largely quantitative, and hence more difficult to assess.

Insulin-like growth factors

Most of this work in breast cancer cell lines has been performed with MCF-7 and T-47D. As with normal cells these two malignant epithelial cell lines can be stimulated to grow with IGF-1, IGF-2 and insulin (Fernandez and Safe 1992; Osborne *et al.*, 1989; Peyrat *et al.*, 1989; Wakeling *et al.*, 1989; Stewart *et al.*, 1990; Thorsen *et al.*, 1992; McGuire *et al.*, 1992; Adamo *et al.*, 1992; Fontana *et al.*, 1991; Wosikowski *et al.*, 1993). However, relatively high levels may be required and IGF-2 and insulin probably act through the IGF-1 receptor (Osborne *et al.*, 1989; Peyrat *et al.*, 1989; Cullen *et al.*, 1992b). IGF-2 and/or IGF-1 are probably secreted into the culture medium (Osborne *et al.*, 1989; Freed and Herington, 1989; Brunner *et al.*, 1993; van der Burg *et al.*, 1990). Thus the possibility exists in breast carcinoma cells, at least *in vitro*, for an autocrine loop, bypassing systemic control of cell growth through GH and the release of IGFs from the liver. There also appears to be a complex set of interactions between oestradiol and IGF-1/IGF-2 on their mutual synthesis and on cell growth (Cullen *et al.*, 1992a; Brunner *et al.*, 1993; Daly *et al.*, 1991; Wakeling *et al.*, 1989; Wosikowski *et al.*, 1993; Noguchi *et al.*, 1991, 1993). One more plausible scenario is that oestradiol stimulates growth of these human breast carcinoma cell lines, at least in part, by stimulating the production of the IGF-1 receptor, thereby allowing IGFs in the medium and/or those produced by the cells to work more effectively (Stewart *et al.*, 1990). Whether oestradiol mediates cell growth through both IGFs and TGFα working simultaneously in the same breast cancer cell line or whether different culture conditions and/or strain of cell line favour one or other of these mediators is unclear at present.

The evidence for an autocrine loop based on the IGFs conferring a degree of autonomous growth on breast carcinoma cells in the clinical situation is much more doubtful. Thus in patients with breast carcinomas, mRNA for IGF-1 is absent from primary breast tumours, but is found in the normal breast, where it occurs only in the stromal cells (Yee *et al.*, 1989). mRNA for IGF-2 is similarly found in all normal breast tissues, but in only 50% of breast carcinomas (Coombes *et al.*, 1990). Plasma levels of IGF-1, though, are higher in patients with primary breast carcinoma than in unaffected women (Colletti *et al.*, 1989; Bonneterre *et al.*, 1990; Peyrat *et al.*, 1990, 1993) and their treatment with tamoxifen (but not aminoglutethimide) reduces these levels (Lonning *et al.*, 1992; Reed *et al.*, 1992a; Lien *et al.*, 1992; Huynh *et al.*, 1993). The reduction in plasma levels of IGF-1 by tamoxifen is probably not all due to its ability to inhibit the release of GH from the pituitary (Malaab *et al.*, 1992), but may reflect a reduction in the production of IGF-1 at other sites, including that of the cancer (Huynh *et al.*, 1993).

The evidence for the IGF system mediating some of oestrogen's control over malignant growth in the clinical situation is better, however. Thus IGF-1 receptors have been detected in most (87–100%) primary breast carcinomas, and their amounts correlate with the presence of oestrogen and progesterone receptors and survival of the patients (Foekens *et al.*, 1989a, 1989b; Peyrat *et al.*, 1990; Bonneterre *et al.*, 1990; Klijn *et al.*, 1992; Berns *et al.*, 1992). Increases in these receptors are not due to gene amplification, since the IGF-1 gene is increased in only 2% of breast carcinomas (Berns *et al.*, 1992). The proteins which transport the IGFs to their receptors, the so called binding proteins, are also synthesized in greater amounts in the malignant than in the adjacent normal tissue and are also under hormonal control (Pekonen *et al.*, 1992; Lonning *et al.*, 1992; Reed *et al.*, 1992a).

In summary the evidence is generally against the idea of an autocrine system involving IGFs in malignant breast tissue, although many of the ingredients are present. Possible sources of IGF-1 *in vivo* include the breast fibroblasts (page 173). Fibroblasts from benign breast tumours mostly express mRNA for IGF-1, whilst those from malignant tissue express mostly mRNA for IGF-2 (Cullen *et al.*, 1991). The major source of systemic IGFs is generally the liver, however, which makes the investigations *in vitro* more difficult to interpret. The idea that an autocrine loop involving IGF may develop when oestrogen responsiveness is lost is attractive, and in this context it may be significant that different patterns of expression of the IGF binding proteins have been observed between oestrogen receptor positive and negative cell lines (Sheikh *et al.*, 1992a, 1992b). On this model the lower levels of IGF-1 receptors in patients with poor prognosis may be explained by their down-regulation by locally produced IGF-2. Thus, although investigations on IGFs arose from their potential systemic role in mammary growth, those in breast cancer are mainly arising as a result of their local production.

Other growth factors

As before three other growth factors need to be considered, PDGF, interleukins and stromal factors. PDGF has only slight, if any, effect on the growth of human breast carcinoma cell lines (Elliott *et al.*, 1992; Stewart *et al.*, 1992), and that at a high dose of 125 ng/ml^{-1} (Ginsburg and Vonderhaar, 1991). Moreover, *in vivo* mRNA for PDGF has been found in the epithelial cells of all normal, benign and malignant breast tissues examined (Ro *et al.*, 1989; Coombes *et al.*, 1990) tending to negate an autocrine role exclusive to breast carcinomas. On the other hand, plasma concentrations of PDGF are reported to be high in 13 of 41 stage 4 breast cancer patients compared to 2 of 17 stage 2 patients and 0 of 9 controls (Ariad *et al.*, 1991). Moreover, Leal *et al.* (1991) have found that PDGF is the only factor tested that promotes the growth of human medullary carcinoma cells in culture. Both results suggest some correlation with the stage of breast carcinoma.

With respect to the direct effects of interleukins and breast carcinomas, the growth of MCF-7, T-47D, SK-BR-3 and ZR-75 cell lines are inhibited by IL-1 and IL-6 in an additive fashion (Danforth and Sgagias, 1991, 1993; Chen *et al.*, 1991; Sehgal and Tamm, 1991; Tamm *et al.*, 1991; Verhasselt *et al.*, 1992) and IL-4 can inhibit MCF-7 cells (Toi *et al.*, 1992). In the case of IL-1 and MCF-7 cells, TGFβ is induced. This induction is inhibited by oestradiol, yet oestradiol-stimulated growth is inhibited by IL-1, so that the induction of TGFβ does not completely account for the inhibitory effect. Both interleukins affect the motility of the cells (Chen *et al.*, 1991; Sehgal and Tamm, 1991; Tamm *et al.*, 1991; Verhasselt *et al.*, 1992), so that the interleukins, while reducing growth rates may promote the dispersal of the tumour. *In vivo*, a locally produced source of IL-6 is available in the 'fibroblasts' associated with breast carcinomas (Reed *et al.*, 1993). Measurements of aromatase in close and distant parts of the mammary gland suggest that the IL-6 which arises from the mass of the tumour can, in fact, act locally in the unaffected tissue (Reed *et al.*, 1993). Thus the interleukins may affect cell dispersal and provide local signals beyond the immediate carcinomatous area.

In addition to changes in defined growth factors in breast carcinomas, the stroma of the breast of patients with breast carcinoma shows differences from that of the normal breast. For instance, Schor *et al.* (1988) have found a migration stimulating factor (MSF) in fetal fibroblasts, and in fibroblasts from women with breast carcinoma, that is absent from normal adult fibroblasts. This MSF is associated with higher amounts of hyaluronic acid (Schor *et al.*, 1989), binds to heparin, has a molecular weight of 70 kDa and a high proline content (Grey *et al.*, 1989). The MSF is reported to have been found in serum and in skin fibroblasts of cancer patients and their close relatives (Schor *et al.*, 1991; Picardo *et al.*, 1991), and also in wound fluid of otherwise normal patients (Picardo *et al.*, 1992).

The extracellular matrix glycoprotein, tenascin, has also been considered as a stromal marker for breast carcinoma (Mackie *et al.*, 1987). Again this is a fetal product, found in the 10th to 16th day of gestation in the mouse at the commencement of morphogenesis, but is induced in adult nude mice by the injection of cells derived from mammary carcinoma (Inaguma *et al.*, 1988; Chiquet *et al.*, 1989). However, this is not an all-or-none phenomenon, as lesser amounts of tenascin are found in normal human breast, particularly round ducts and ductules, in a manner dependent on the menstrual cycle (Ferguson *et al.*, 1990; Howeedy *et al.*, 1990; Jones *et al.*, 1992; Shoji *et al.*, 1992). Furthermore tenascin synthesis can be induced *in vitro* in MCF-7 cells that do not normally produce it by co-culture with embryonic mesenchyme (Hiraiwa *et al.*, 1993). The existence of an alternatively spliced isoform of tenascin that can induce loss of focal adhesion in cultured cells, facilitates cell migration, and is found preferentially in invasive breast carcinomas may have significance in generating a permissive environment for invasion and metastasis (Borsi *et al.*, 1992). Once again an extracellular matrix product, tenascin, that is probably produced by the myoepithelial cells in the normal mammary gland may be ectopically produced by malignant epithelial cells, this time perhaps in a form more conducive to bestowing invasive properties on the malignant cells.

In addition to potential autocrine loops contributing to uncontrolled growth of the carcinoma cells, paracrine loops produced from the stromal cells of the carcinoma may therefore assist in cell growth and autonomy from systemic control. One such example is the IGFs produced from the stromal cells of breast epithelium (pages 173 and 207). Possibilities of paracrine growth stimulation via stromal IGFs arise from the observation that some breast carcinomas secrete PDGF which enhances IGF-1 production by stromal fibroblasts (Yee *et al.*, 1989). How loss of the myoepithelial cell's interface between the epithelial cell and the surrounding stroma (page 193) alters epithelial–stromal interactions, including any paracrine loops, in malignant breast disease is largely unknown at present, but yet may be of great importance in the progress of breast cancer.

HEPARIN BINDING GROWTH FACTORS IN BREAST CANCER

Ectopic expression of fibroblast growth factors and their receptors

Receptors for bFGF have been found on four, independently isolated, malignant rat epithelial mammary cell lines of differing metastatic potential. These cell lines possess cell-surface, high affinity receptors for bFGF with K_d values of 20 pM to 80 pM (Fernig *et al.*, 1993) and these values are comparable to the K_d values of these receptors found on the myoepithelial-like cells (Fernig *et al.*, 1990, 1992, 1993). However, only the weakly metastatic Rama 600 cells possess low-affinity receptors for bFGF. The moderately metastatic cell lines do not possess detectable levels of low-affinity receptors for bFGF (Fernig *et al.*, 1993).

The very unusual androgen-dependent malignant mouse Shionogi carcinoma (SC-115) and its clonal cells SC-3 and CS-2 also possess receptors for FGFs. The stimulation of growth in SC-3 cells by androgens is a consequence of the activation of an autocrine/paracrine loop: steroids stimulate the production of FGFs, which in turn activate their receptors and cause cell growth (Nakamura *et al.*, 1989). The addition of exogenous androgens or FGFs to SC-3 cells results not only in the stimulation of growth, but also in a morphological change, whereby the cells lose their epithelial characteristics and become elongated and fibroblastoid in appearance (Nakamura *et al.*, 1989). Thus transfection of the *hst* gene into the clonal CS-2 cell line results in the androgen-independent stimulation of both cell growth and morphological change. Moreover, the transfectants are able to form tumours in male but not female mice (Furuya *et al.*, 1991). SC-3 cells possess high- and low-affinity receptors for bFGF and the number of these receptors and their K_ds (Nonomura *et al.*, 1990) is similar to that observed in rat mammary myoepithelial-like cells or in the Rama 600 cells (Fernig *et al.*, 1990, 1992, 1993). The FGF-like growth-stimulatory activities produced in response to androgens are also able to compete with bFGF for the receptors for bFGF on the SC-3 cells (Nonomura *et al.*, 1990). A cDNA encoding the fourth splice variant of the FGFR-2 receptor (Figure 11) has been cloned from SC-3 cells (Kouhara *et al.*, 1991), and both bFGF and androgens upregulate the mRNA for this receptor (Saito *et al.*, 1991a). In contrast the stimulation of SC-3 cells with androgens leads to the suppression of the expression of syndecan (Leppa *et al.*, 1991). However, the transfection of a syndecan gene under the control of the MMTV-LTR into SC-3 cells results in the cells retaining their epithelial morphology after stimulation by steroids, suggesting that the morphological changes elicited by androgens are a consequence of changes in syndecan expression (Leppa *et al.*, 1992). The stimulation of the growth of SC-3 cells by bFGF or by the androgen-dependent FGF autocrine loop is dependent on endogenous HS (Sumitani *et al.*, 1993), and thus it would seem unlikely that syndecan is responsible for the activation of FGFs in this system.

The growth of MCF-7, BT-20 and T-47D cells is stimulated by bFGF whilst that of MDA-MB-321 cells is not (Karey and Sirbasku, 1988; Peyrat *et al.*, 1991, 1992a, 1992b; Stewart *et al.*, 1992). In addition bFGF has been shown to stimulate the production of plasminogen activator in T-47D, but not MCF-7, BT-20, SK-Br-2-II and UCT-Br cells (Mira-y-Lopez *et al.*, 1989). The presence of high-affinity receptors for bFGF has also been demonstrated in the MCF-7, BT-20, T-47D and MDA-MB-231 cells, although low-affinity receptors have been only detected in the latter cell line (Peyrat *et al.*, 1991,

1992a, 1992b). The mRNA for FGFR-1 has been detected in ZR-75, MCF-7, T-47D and MDA-MB-231, and that for FGFR-2 in the ZR-75 and MCF-7 cell lines. Moreover, evidence has been found for expression of different alternative transcripts of FGFR-1 in the ZR-75 cells than those observed in normal human mammary epithelial cells (Luqmani *et al.*, 1992). In agreement with these results the MCF-7 and ZR-75 cells have been found to express mRNAs encoding all four FGFRs (Lehtola *et al.*, 1992). Therefore, the balance of evidence would suggest that malignant human mammary epithelial cell lines possess cell-surface receptors for bFGF and thus bFGF is able to stimulate the growth of these cells, as well as other events, such as the production of proteases.

The genes for FGFR-1 and FGFR-2 are amplified in some invasive breast carcinomas (Adnane *et al.*, 1991). The mRNAs for FGFR-1 and FGFR-2 are also detected in human mammary carcinomas and there is evidence for the differential expression of alternatively spliced transcripts of the FGFR-1 gene between normal breast tissue and human carcinomas (Luqmani *et al.*, 1992). Furthermore, since high- but not low-affinity receptors for bFGF have been detected in 19/39 human breast cancer biopsies (Peyrat *et al.*, 1992a) and the growth of a primary culture of cells from a human mammary carcinoma is stimulated by bFGF (Takahashi *et al.*, 1989), functional FGFRs are likely to be expressed by cells in invasive carcinomas.

FGF-like activity is found in extracts of four, independently isolated, rat mammary carcinoma cell lines of differing metastatic potential. The levels of FGF-like activity in these cells are similar to those found in the myoepithelial-like cells. Immunoreactive bFGF of molecular weight 18, 22, 23, and 25 kDa is observed in extracts of Rama 800 cells, indicating that all four translation products of the bFGF mRNA are produced; the increase in the level of mRNA accounts largely for the increase in bFGF protein, at least in this one cell line analysed (Fernig *et al.*, 1993). However, in human carcinoma cell lines the mRNA for bFGF has been detected in MDA-MB-231 and Hs578T cells but not in BT-474, MCF-7, SK-Br-3, T-47D and ZR-75 cells (Anandappa *et al.*, 1994; Li and Shipley, 1991; Luqmani *et al.*, 1992). Whilst the mRNA for bFGF has been detected in human mammary carcinomas, the level of expression relative to that in the benign/normal tissue is generally lower (Anandappa *et al.*, 1994; Luqmani *et al.*, 1992). This result is consistent with the loss of bFGF-producing myoepithelial cells in invasive carcinomas (page 180). However, approximately 25% of these tumours express bFGF mRNA at levels which are equivalent to, or higher than, those of the benign tissue. Since a similar proportion of the human carcinoma cell lines express relatively high levels of bFGF mRNA (Anandappa *et al.*, 1994; Li and Shipley, 1991; Luqmani *et al.*, 1992), these *in vivo* results are consistent with the notion that in some breast cancers there is a population of malignant cells which acquire the ability to express high levels of bFGF mRNA. bFGF has also been isolated from mammary carcinomas (Rowe *et al.*, 1986) so it is likely that the expression of bFGF mRNA by human carcinomas results in the synthesis of bFGF polypeptides.

aFGF mRNA has also been detected in human mammary carcinomas (Anandappa *et al.*, 1994). However, since epithelial and myoepithelial-like cells derived from benign lesions fail to express detectable levels of aFGF mRNA (page 180), it is possible that much of this mRNA is either ectopically expressed by the malignant cells or is derived from the stromal elements within the tissue samples. There appears to be no consistent relationship between the levels of expression of the mRNAs for aFGF and bFGF in human mammary carcinomas, suggesting that these two mRNAs are independently expressed (Anandappa *et al.*, 1994).

int-2 (Figure 10) is expressed by three metastatic and one non-metastatic sublines established from the MMTV-induced TPDMT-4 mouse mammary tumour, but *hst* (Figure 10) expression occurs only in the three metastatic sublines (Murakami *et al.*, 1990). The co-amplification and overexpression of the genes for *int*-2 and *hst* occur frequently in MMTV-induced tumours, and are likely to result from these genes being within 17 kb of each other on mouse chromosome 7 in the same transcriptional orientation (Peters *et al.*, 1989). Engineered overexpression of the *hst* gene in MCF-7 cells by transfection results in the generation of the metastatic phenotype in these cells, when they are injected into nude mice (Kurebayashi *et al.*, 1993; McLeskey *et al.*, 1993). The *int*-2 and *hst* genes are also amplified in human mammary carcinomas, but the concomitant overexpression of their respective gene products does not occur and hence a causal relationship between gene amplification and malignancy has not been demonstrated for these genes (Lidereau *et al.*, 1988; Theillet *et al.*, 1989). However, it has been suggested that the amplification of the *int*-2 and *hst* genes in human mammary carcinomas is the result of their belonging to an amplicon which is driven by the PRAD-1 gene that encodes a cyclin-like protein which may be crucial for malignancy/metastasis (Motokura *et al.*, 1991; Motokura and Arnold, 1993).

The unusual androgen-dependent malignant mouse Shionogi carcinoma tumour and its clonal cell lines such as SC-3 and CS-2 produce a variety of heparin-binding, growth-stimulatory activities, particularly in response to androgens (Akakura *et al.*, 1990; Tanaka *et al.*, 1990; Yamaguchi *et al.*, 1992). The CS-2 cells show the same integration pattern of MMTV as the parental tumour and in response to androgens they express mRNA for *int*-2 weakly and *hst* more strongly (Akakura *et al.*, 1990). In contrast, the activity of the heparin-binding growth factor produced by the SC-3 cells after stimulation by androgens or pharmacological doses of glucocorticoids can be neutralized with antibodies to bFGF (Tanaka *et al.*, 1990; Yamanishi *et al.*, 1991). Thus, stimulation of cell lines derived from the SC-115 tumour with androgens would appear to induce mRNA for *hst* and a bFGF-like growth factor. However, it has also been reported that the heparin-binding growth factor induced by androgens is not sensitive to anti-bFGF neutralizing antibodies (Kawamoto *et al.*, 1992), and sequencing of such a growth factor has revealed that SC-3 cells produce a novel member of the FGF family, FGF-8, in response to androgens (Tanaka *et al.*, 1992). It remains to be determined whether the production of bFGF, *int*-2, *hst*, and FGF-8 by Shionogi carcinoma cells is directly relevant to human mammary carcinomas, since the balance of evidence suggests that bFGF, and possibly aFGF, are the pertinent FGFs. However, the production of FGFs by the Shionogi carcinoma cells appears to be related to their metastatic potential, thus supporting the notion that bFGF, acting through its cellular receptors may be causally related to the malignant phenotype.

Potential autocrine loop for basic fibroblast growth factor

bFGF and receptors for bFGF are thus products of the myoepithelial and intermediate cells of the normal mammary gland and of benign mammary tumours (pages 180 and 175). The production of FGFs and high-affinity receptors for bFGF by murine, rat and human malignant mammary epithelial cell lines could be interpreted as reflecting differentiation towards the myoepithelial cell. Indeed some malignant mammary cells, such as the weakly metastasizing Rama 600 cells, also possess low-affinity HSPG receptors for bFGF similar to those found on myoepithelial-like cells. However, the overwhelming majority of the rat and human malignant cell lines all fail to produce low-affinity receptors for bFGF and, when this has been examined, other, more clearly recognizable, markers of the myoepithelial cells (page 192). The presence of low-affinity receptors for bFGF in the weakly metastasizing cells such as Rama 600 may therefore reflect in tumours the presence of the undifferentiated more-elongated cells which retain some properties of myoepithelial cells (page 165). Therefore any such differentiation process towards the myoepithelial cell is of a very incomplete nature, and thus the production of high-affinity receptors for bFGF and bFGF-like activity by malignant mammary epithelial cells is probably ectopic. At the cellular level, one possible mechanism for the ectopic expression of FGFs and high-affinity receptors for FGFs by malignant epithelial cells is the notion that the first carcinogenic insult occurs in the intermediate cells of the terminal structures of the ducts, particularly the TEBs, which normally express both bFGF and receptors for bFGF (page 177) consistent with the hypothesis advanced on page 199. Since bFGF is both mitogenic and angiogenic, and angiogenesis is an important feature of tumour growth and metastasis (Weidner *et al.*, 1991), the adaptive retention of some of the differentiation potential of the intermediate cells of the TEBs by the malignant epithelial cells, observed as, for example, the ectopic expression of FGFs and high-affinity receptors for FGFs, may thus favour the establishment of a more-successfully metastasizing breast cancer.

As well as being ectopic, the expression of FGFs and FGFRs by malignant epithelial cells shows some important differences with that observed in normal mammary cells. Thus, some malignant epithelial cells may splice FGFR-1 transcripts differently (Luqmani *et al.*, 1992) and they may express all the FGFRs (Leholta *et al.*, 1992). Moreover, the expression of high-affinity receptors for bFGF by the malignant epithelial cells is unusual in that these are expressed in cells, e.g. Rama 800, MCF-7, that do not appear to express low-affinity receptors for bFGF. The absence of detectable HSPG receptors for bFGF on the more malignant carcinoma cells may be caused by their failure to produce HS or HSPG core proteins. Alterations in HS structure, and in particular undersulphation in a syngeneic murine model of metastasis (Smith *et al.*, 1987) and a reduction in 3-O-sulphation and anti-thrombin III binding in NMuMG cells (Pejler and David, 1987) have been found to correlate closely with increasing metastatic potential. Only a low level of HS is required to enable the binding of bFGF to FGFRs (Yayon *et al.*, 1991). The bFGF-binding characteristics of the HSPG receptor are heterogeneous and the relationship between FGF-binding and the two functions of this receptor, activation and storage, has yet to be elucidated at the structural level. Thus, it remains an open question whether the more malignant cells produce only low levels of HS which may not be detected by the binding assays but which are sufficient to activate the binding of bFGF, or whether they produce only HS that contains oligosaccharide sequences which activate FGF binding to FGFRs but are unable to sequester FGFs. However, it is clear that the carcinoma cells lack the HSPG sink for the FGFs. Consequently bFGF released from either the malignant cells themselves or the surrounding normal tissue as a result, for example, of the digestion of extracellular matrix/basement membrane by enzymes released by the carcinoma cells (Briozzo *et al.*, 1991) during local invasion would not be sequestered into the HSPG sink. Instead the bFGF would be available to stimulate not only the growth of the malignant cells but also that of cells, in particular endothelial cells (Rifkin and Moscatelli, 1989), in the adjacent stroma. Thus the absence of low-affinity receptors for bFGF on the malignant rat and human mammary epithelial cells with the greatest malignant potential suggests that alterations to the HSPG receptors for bFGF may also play a key role in the contribution to the growth and dissemination of malignant mammary tumours by bFGF.

DEVELOPMENT OF THE MALIGNANT STATE IN BREAST CANCER

Generation of the malignant, metastatic phenotype by transfection of a benign stem cell system

In the mammary gland, both in rodents and particularly in humans it has proved difficult to identify unambiguously steps in a malignant pathway, since the vast majority of observations are of a comparative nature and, as a consequence, are indirect. However, it has proved possible to transfect genes and/or DNA fragments into non-transformed cells so as to identify those which may be important in neoplastic processes. Transfection of restriction enzyme-HindIII-fragmented DNA with the drug selectable plasmids pSV2*neo* or pSV2*gpt* into the Rama 37 epithelial cell line produces drug-resistant transformants with a frequency of 10^{-4} to 10^{-5}. The Rama 37 cells produce benign, non-metastasizing adenomatous tumours in syngeneic rats which contain epithelial, intermediate stem and myoepithelial-like cells (page 193). If the fragmented DNA is obtained from either malignant metastatic rat cell lines (e.g. Rama 800: Table 2) (Jamieson *et al.*, 1990a) or cell lines from primary human metastatic carcinomas (Ca2-83) (Table 3) or a metastatic pleural effusion (MCF-7) (Table 3) (Davies *et al.*, 1993a, 1994), then approximately 1–5% of the resultant colonies of cells produce malignant metastatic carcinomas when reintroduced into syngeneic rats. Similarly-fragmented DNA from the benign rat cell line, Rama 37 itself, from a normal human cell line Huma 7, from a cell line, HMT-3522 isolated from benign breast disease (Table 2) and from salmon sperm DNA fail completely in this respect and produce only benign, non-malignant adenomatous tumours (Table 8). The sites of metastases are predominantly in the lungs and lymph nodes draining the mammary glands, but other sites have been observed (Jamieson *et al.*, 1990a; Davies *et al.*, 1993a, 1994). The cell lines isolated from metastatic lesions in the injected rats invariably produce metastases when reinjected into other syngeneic rats.

Unlike the parental Rama 37 cells which produce predominantly adenomatous tumours containing epithelial, intermediate stem and myoepithelial-like cells, the malignant metastatic lines produce, in the rat, invasive ductal carcinomas containing only rather undifferentiated epithelial cells, similar to their naturally-arising counterparts in rats and humans (page 193). Induction of metastatic ability has been achieved in various other cell systems by transfection with drug-resistance plasmids alone or by the transfection process *per se* (van Roy *et al.*, 1986; Kerbel *et al.*, 1987). However, the specific requirement for DNA from malignant, metastatic cells in the Rama 37 cell system tends to negate this explanation for induction of metastatic ability. Thus the malignant, metastatic phenotype appears to be transferred in a genetically dominant manner in the Rama 37 stem cell system and, at the same time results in a markedly reduced capacity for differentiation to myoepithelial-like cells. Moreover, although standard oncogenes EJ-*ras*,

Polyoma Large T and Middle T Antigens (Jamieson *et al.*, 1990b, 1990c) decrease the latent period for the appearance of the tumours from injections of the relevant transfectants, these tumours are still benign adenomatous lesions and fail to metastasize (Table 8). These results suggest that although transforming oncogenes such as EJ-*ras* and Polyoma T Antigens can increase the rate of formation of benign tumours, presumably by increasing rates of cellular proliferation in this stem cell system, nevertheless they fail to induce the malignant/metastatic phenotype. Presumably a second class of oncogenes, the metastagenes, are required to enable the benign stem cell system to progress to its full malignant potential and become metastatic. One such gene is that for our calcium-binding regulatory protein, p9Ka, as described below.

p9Ka was first discovered as a polypeptide induced when cultured rat mammary epithelial cells convert in culture to myoepithelial-like cells (pages 183 and 185). However, the possible involvement of p9Ka in the progression of benign tumour cells towards metastatic cancer cells has arisen. This suggestion has come from the observation (Dunnington, 1984) that one of our metastatic rat mammary epithelial cell lines, Rama 800, expresses high levels of p9Ka, in contrast to related cell lines from benign tumours and normal mammary glands (Barraclough *et al.*, 1984a) (Figure 19). This finding,

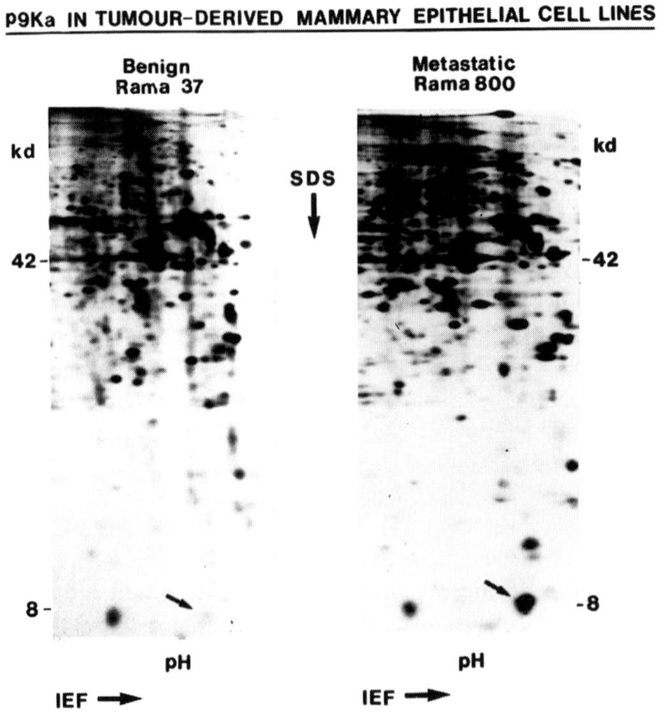

Figure 19 Polypeptide patterns of benign epithelial and metastatic epithelial cell lines on two-dimensional polyacrylamide gels. Extracts of benign epithelial cells, Rama 37 (left panel) or metastatic epithelial cells, Rama 800 (right panel), radioactively labelled overnight with [^{35}S]methionine, were subjected to isoelectric focusing in the first dimension, and SDS gel electrophoresis in the second dimension. The autoradiographs of the dried-down gels are shown. The arrow on each panel points to the expected position of migration of p9Ka.

which linked p9Ka expression with the metastatic phenotype, has been extended by a correlation between the level of elevated expression of the mRNA for p9Ka and the metastatic potential of a series of closely-related murine mammary adenocarcinoma cell lines (Ebralidze *et al.*, 1989). These cells show widely differing potential to spread to the lungs when examined 4–5 weeks after intramuscular injection into recipient mice (Ebralidze *et al.*, 1989). Such a correlation is not confined to cells derived from the mammary tumours of rodents. In the murine B16 melanoma cell line, induced variants of differing metastatic potential also show a correlation between metastatic (lung colonization) ability and the level of the mRNA for murine p9Ka (Parker *et al.*, 1991).

The direct role of p9Ka (or its mRNA or gene) in the process of metastasis has been examined by transfecting into our benign rat mammary epithelial stem cell system multiple additional (10–100) expressed copies of the p9Ka gene (Barraclough *et al.*, 1987b) linked to a plasmid containing the selectable marker, pSV2*neo* as described above. The levels of expression are higher than in the myoepithelial-like cell lines described earlier (pages 183 and 185). When such cells which are transfected with plasmid containing the entire p9Ka gene (Barraclough *et al.*, 1987b) are injected into the recipient rats, there is a two- to three-fold reduction in the latent period of tumours, and in over 50% of the animals, the tumours have acquired the metastatic phenotype (Davies *et al.*, 1993b) (Table 7). The primary tumours and metastases (Figure 15D) contain elevated levels of immunochemically detectable p9Ka (Davies *et al.*, 1993b). These results suggest that the metastatic phenotype in the p9Ka expressing cells observed previously (Dunnington, 1984; Ebralidze *et al.*, 1989) is a consequence of a direct effect of the increased levels of p9Ka mRNA/p9Ka protein in the cells. That the incidence of metastases does not increase when the cells cultured from metastases themselves are injected into the mammary fat pads of rats (Davies *et al.*, 1993b) suggests either that the metastatic phenotype can be inactivated in some cells either during the growth of the metastasis or during the subsequent cell culture, or that host factors limit the incidence of metastases. This loss of metastatic phenotype is possibly a manifestation of hypermethylation of the p9Ka gene, similar to that which is believed to lead to the altered levels of expression of p9Ka in the murine mammary tumour cells of differing metastatic potential (Tulchinsky *et al.*, 1993).

The cellular target for p9Ka in the malignant/metastatic transformants may once again be the cytoskeleton (Davies *et al.*, 1993b). Thus the Rama 37 cells transfected with the p9Ka gene possess strongly staining cytoplasmic filaments which are concentrated in the perinuclear region of the cell, in a very similar, but more intense and slightly more disorganized manner than that shown in Figure 15 for the myoepithelial-like Rama 29 cell line (Gibbs, 1993). As with the Rama 29 cells, the pattern of fluorescence is almost identical to that of the actin cytoskeleton obtained on the same section using rhodamine-conjugated phalloidin, suggesting that the metastasis-inducing activity of p9Ka may be mediated through the cytoskeleton of the metastatic epithelial cells. These malignant, metastatic transformants also fail to differentiate to myoepithelial-like cells, and one may speculate that perhaps supra-elevated levels of p9Ka cause dissemination and concomitant reduction in the differentiation potential of the Rama 37 stem cell system.

p9Ka contributes to the metastatic phenotype in the rat mammary cell lines by being over-expressed at either the level of mRNA or protein. Increased expression of genes in human tumours can arise through chromosomal abnormalities in the cancer cells. In the human genome, the gene for p9Ka has been mapped to chromosome 1, in the region q21 (Dorin *et al.*, 1990; Engelkamp *et al.*, 1993). From cytogenetic evidence, the commonest abnormalities in breast cancer are those that involve chromosome 1 (Chen *et al.*, 1989), particularly in the region q21, where a fragile site has been identified in the DNA (Yunis and Soreng, 1984). Changes in chromosome 1 include both over-representation of the entire chromosome and translocations of fragments of the chromosome, both of which may lead to increased expression of involved genes (Trent, 1985). It is accepted that the wide occurrence of abnormalities of chromosome 1 represent progressive rather than initiative events in the development of breast cancer. Such changes would be expected to be characteristic of a metastasis-inducing gene, such as p9Ka. Thus it is possible that gross changes in the distal region of chromosome 1 in human tumour cells may lead to the over-expression of p9Ka, which may in turn contribute to the induction of the metastatic phenotype in some human breast cancers. Preliminary experiments using probes, derived from the existing cloned rat p9Ka gene (Barraclough *et al.*, 1987b), and from the recently-isolated human p9Ka gene (Lloyd *et al.*, 1996), to screen RNA from human breast cancer cell lines, suggests that high levels of p9Ka are expressed in some such cell lines, raising the possibility that, as in rat cells, elevated expression of p9Ka contributes to the induction of the metastatic phenotype.

The stem cell model and mammary carcinoma

In normal development of the mammary gland the intermediate stem cells require both systemic and local growth factors to enable them to replicate and hence cause mammary growth. The presence of autocrine loops in normal development, however, pose a problem, in that they would provide for enhancement of growth rate of the tissue but no mechanism for its eventual cessation. The mechanism for limitation of glandular growth may be provided by the differentiation of the myoepithelial cell to a terminal, non-proliferating phenotype (page 153). Since invasive carcinomas have lost the ability to produce fully differentiated myoepithelial cells (page 193), the tendency towards continued growth may thus be an irreversible result.

However, normal growth is not purely autocrine, otherwise the growth of the mammary gland would not be restricted to the periods of puberty and pregnancy. External signals (e.g. PMGF) and the stroma (e.g. PGE_2) are also required. For a carcinoma to grow outside these normal confines it must provide a signal to replace the external signals, or enhance the effectiveness of those produced locally. Thus the TGFα/EGFR pathway may be enhanced by overproduction of either molecule. Furthermore, upwards of 20% of invasive carcinomas produce a closely-related receptor to that of EGFR, and this is the c-erbB-2 receptor (page 205). Although the ligand for this receptor is not well identified as yet, overexpression of c-erbB-2 can compensate for lack of ability to stimulate cell growth through EGFR (page 205) (Figure 20). What compensatory mechanisms occur in the remaining 80% of human invasive carcinomas is not clear at present, although many more carcinomas may use this compensatory mechanism at early, *in situ*, stages of the disease and have lost it on progression (B. B. Green, J. Winstanley and P. S. Rudland, unpublished results).

The intermediate stem cells may also require bFGF and a functional receptor to assist their growth (pages 180 and 175). It is not surprising then that some of the malignant

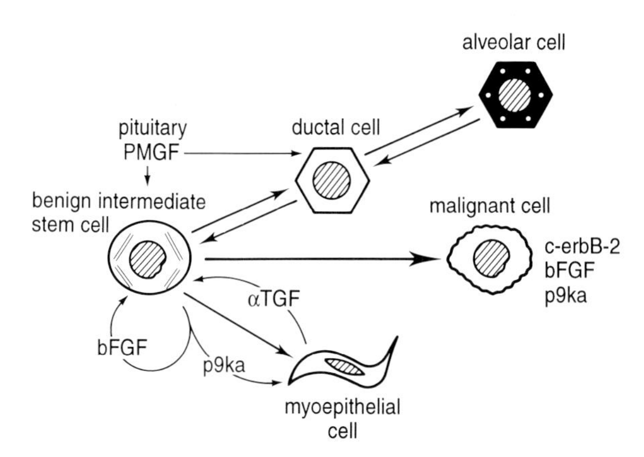

Figure 20 Possible molecular model for mammary cancer. The benign intermediate stem cells in terminal ductal structures and elsewhere in the mammary gland can still differentiate to ductal epithelial and then to alveolar-like cells (⇄) or to myoepithelial-like cells (→). This differentiating ability is largely lost in the malignant state. Some of the growth factors and other molecules produced by the myoepithelial cell are compensated in the malignant state by their ectopic production or production of related molecules in some rat and human mammary carcinomas.

cell lines and primary invasive carcinomas ectopically produce both bFGF and its high-affinity receptors (page 209). They have also lost much of their ability to produce the low-affinity receptors that serve to sequester this family of growth factors and so prevent them from activating the high-affinity receptors which trigger cell proliferation (page 212). Thus the growth-controlling element exerted by the FGF sink provided by the HSPGs present in the basement membrane/extracellular matrix adjacent to the myoepithelial cells has been lost in invasive carcinomas, at least as far as stimulation of cell growth by bFGF is concerned. There is thus a potentially viable autocrine loop provided by selective expression of some of the components of the bFGF system in the carcinoma cells. It is tempting to speculate that these components of the bFGF system are provided by malignant cells that bear some similarity to the intermediate stem cells found in normal mammary glands and in benign tumours (Figure 20), as suggested on page 199.

Although the requirement for circulating ovarian, adrenal and pituitary hormones is largely lost in three quarters of invasive carcinomas (page 201), the mechanism by which this autonomy is achieved in the malignant cells is largely unknown. If steroid hormones like oestrogen stimulate cell growth indirectly through intermediary growth factors such as TGFα and/or IGFs (page 172), then alterations in such growth-delivery systems that impart a degree of autonomy on the malignant cells would automatically impart a degree of hormone-independency as well (pages 204 and 206). In the case of pituitary-independent growth, differentiation of epithelial cells along the myoepithelial cell pathway yields an ever declining response to the growth-stimulatory ability of pituitary PMGF. Hence, retention of vestiges of differentiation along this pathway by intermediate stem cells may account for the refractory nature of most carcinoma cells to hormonal control of their growth (page 199).

Retention of other molecules involved more with cell differentiation than with cell proliferation has been reported in rat and more recently in human invasive carcinomas (page 194). Overexpression of p9Ka at higher levels than those encountered in the intermediate stem cells seems to be causally related to the induction of the malignant phenotype, at least when this is monitored by the ability to metastasize in rats (page 215). This result may be an example of where expression of a normal molecule in a system which no longer end differentiates may have a different biological effect on its target stem cells.

It can thus be seen that in addition to retaining some markers of myoepithelial cells, as described on page 194, carcinoma cells also selectively retain or adapt molecular systems involved in cell growth and perhaps differentiation. These results are again consistent with the target for malignant carcinogenic insult being the intermediate stem cells of the mammary gland. In addition to its much more limited differentiating ability, the insulted stem cell also has a more loose or uncoupled control over the expression of its own genes (Rudland, 1987a, 1987b). This state of affairs may well result from loss of function of suppressor oncogenes such as p53. Finally, the requirements of the local environment for epithelial/intermediate stem cell proliferation, be it TGFα, IGF, PGE_2, transferrin or any other agents may be provided by cell types in close proximity to the malignant cells in primary carcinomas, e.g. macrophages, lymphocytes, (myo)fibroblasts, etc. or by the parenchymal cells of tissues which support the growth of metastases.

In conclusion, the available evidence is that normal growth and differentiation, although signalled from outside the gland (e.g. oestrogen), rely to some extent on autocrine/paracrine loops. Such loops explain how growth may become self-perpetuating but not how it is controlled. Possibly control is exerted by the myoepithelial cells becoming terminally differentiated (page 153), whereupon the paracrine cycle is broken. Thus the end-differentiated myoepithelial cell contains TGFα, bFGF and p9Ka, yet does not divide. If the process of end-differentiation is prevented, however, as in the malignant conditions (pages 195 and 199), the means of perpetual growth and dissemination are immediately available.

ACKNOWLEDGEMENTS

We thank our colleages in the Cancer and Polio Research Fund Laboratories and the CANDIS Cancer Tissue Bank Research Centre at Liverpool University for helpful discussions and suggestions, Drs J. McAndrew, D. Dunnington, F. E. M. Gibbs and B. Lloyd for access to results before publication and the Cancer and Polio Research Fund, the North West Cancer Research Fund and the Cancer Research Campaign for financial support.

REFERENCES

Acland, P., Dixon, M., Peters, G. and Dickson, C. (1990). *Nature* **343**:662–665.
Adamo, M.L., Shao, Z.M., Lanau, F. *et al.* (1992). *Endocrinol.* **131**:1858–1866.
Adams, E.F., Rafferty, B. and White, M.C. (1991). *Int. J. Cancer* **49**:118–121.
Adnane, J., Gaudray, P., Dionne, C.A. *et al.* (1991). *Oncogene* **6**:659–663.

Ahmed, S.R., Badger, B., Wright, C. and Manni, A. (1991). *J. Steroid Biochem. Mol. Biol.* **38**:687–693.
Akakura, K., Furuya, Y., Sato, N. *et al.* (1990). *Jpn. J. Cancer Res.* **81**:554–556.
Albrechstein, R., Nielson, R., Wewer, U., Engvall, E. and Ruoslahti, E. (1981). *Cancer Res.* **41**:5076–5081.
Allen, R., Dulbecco, R., Syka, P., Bowman, M. and Armstrong, B. (1984). *Proc. Natl Acad. Sci. USA* **81**:1203–1207.
Anandappa, S.Y., Winstanley, J.H.R., Leinster, S., Green, B., Rudland, P.S. and Barraclough, R. (1994). *Br. J. Cancer* **69**:772–776.
Ariad, S., Seymour, L. and Bezwoda, W.R. (1991). *Breast Cancer Res. Treat.* **30**:11–17.
Arteaga, C.L., Coronado, E. and Osborne, C.K. (1988). *Mol. Endocrinol.* **2**:1064–1069.
Aung, W., Zakhour, H.D., Platt-Higgins, A. and Rudland, P.S. (1993). *The Breast* **2**:165–170.
Azzopardi, J.G. (1979). In *Major Problems in Pathology* (ed. J. L. Bennington), pp. 240–258. W. B. Saunders, London.
Bandyopadhyay, G.K., Imagawa, W., Wallace, D.R. and Nandi, S. (1988). *J. Biol. Chem.* **263**:7567–7573.
Barker, S., Panahy, C., Puddefoot, J.R., Goode, A.W. and Vinson, G.P. (1989). *Br. J. Cancer* **60**:673–677.
Barnes, D., Sirbasku, D.A. and Sato, G.H. (eds). (1984). *Cell Culture Methods for Molecular Cell Biology*, Vols 1–4. Alan Liss, New York.
Barnes, D.M., Lammie, G.A., Millis, R.R., Gullick, W.L., Allen, D.S. and Altman, D.G. (1988). *Br. J. Cancer* **58**:448–452.
Barraclough, R. and Rudland, P.S. (1991). In *Novel Calcium-Binding Proteins: Fundamental and Clinical Implications* (ed. C. W. Heizmann), pp. 105–123. Springer-Verlag, Berlin.
Barraclough, R., Dawson, K.J. and Rudland, P.S. (1982). *Eur. J. Biochem.* **129**:335–341.
Barraclough, R., Dawson, K.J. and Rudland, P.S. (1984a). *Biochem. Biophys. Res. Commun.* **120**:351–358.
Barraclough, R., Kimbell, R. and Rudland, P.S. (1984b). *Nucleic Acids Res.* **12**:8097–8114.
Barraclough, R., Kimbell, R. and Rudland, P.S. (1987a). *J. Cell. Physiol.* **131**:393–401.
Barraclough, R., Savin, J., Dube, S.K. and Rudland, P.S. (1987b). *J. Mol. Biol.* **198**:13–20.
Barraclough, R., Fernig, D.G., Rudland, P.S. and Smith, J.A. (1990a). *J. Cell. Physiol.* **144**:333–344.
Barraclough, R., Gibbs, F., Smith, J.A., Haynes, G.A. and Rudland, P.S. (1990b). *Biochem. Biophys. Res. Commun.* **169**:660–666.
Barsky, S.H., Siegal, G.P., Janotta, F. and Liotta, L.A. (1983). *Lab. Invest.* **49**:140–147.
Bartkova, J., Burchell, J., Bartek, J. *et al.* (1987). *Eur. J. Cancer Clin. Oncol.* **23**:1557–1563.
Bates, S.E., Valverius, E.M., Ennis, B.W., Bronzert, D.A., Sheridan, J.P. and Stampfer, M.R. (1990). *Endocrinol.* **126**:596–607.
Baudier, J., Glasser, N. and Gerard, D. (1986). *J. Biol. Chem.* **261**:8192–8203.
Baumrucker, C.R. and Stemberger, B.H. (1989). *J. Anim. Sci.* **67**:3503–3514.
Ben-David, M. (1968). *Endocrinol.* **83**:1217–1223.
Bennett, D.C. (1979). Ph.D. Thesis, University of London.
Bennett, D.C., Peachey, L.A., Durbin, H. and Rudland, P.S. (1978). *Cell* **15**:296–308.
Bernfield, M. and Sanderson, R.D. (1990). *Philos. Trans. R. Soc. Lond. (Biol.)* **327**:171–186.
Berns, E.M., Klijn, J.G., van Staveren, I.I., Portengen, H. and Foekens, J.A. (1992). *Cancer Res.* **52**:1036–1039.
Betta, P.G., Robutti, F., Pilato, F.P., Spinoglio, G. and Bottero, G. (1989). *Eur. J. Gynaecol. Oncol.* **10**:433–437.
Beuving, L.J., Bern, H.A. and DeOme, K.B. (1967). *J. Natl. Cancer Inst.* **39**:431–447.
Bjorge, J.D., Paterson, A.J. and Kudlow, J.E. (1989). *J. Biol. Chem.* **264**:4021–4027.
Bolufer, P., Miralles, F., Rodriguez, A. *et al.* (1990). *Eur. J. Cancer* **26**:283–290.
Bonneterre, J., Peyrat, J.P., Beuscart, R. and Demaille, A. (1990). *Cancer Res.* **50**:6931–6935.
Borg, A., Baldetorp, B., Ferno, M., Killander, D., Olsson, H. and Sigurdsson, H. (1991). *Oncogene* **6**:137–143.
Borsi, L., Carnemolla, B., Nicolo, G., Spina, B., Tanara, G. and Zardi, L. (1992). *Int. J. Cancer* **52**:688–92.
Boyer, B., Tucker, G.C., Valles, A.M., Franke, W.W. and Thiery, J.P. (1989). *J. Cell Biol.* **109**:1495–1509.

Breier, B.H., Gluckman, P.D., McCutcheon, S.N. and Davis, S.R. (1991). *J. Dairy Sci.* **2**:20–34.
Briozzo, P., Badet, J., Capony, F. *et al.* (1991). *Exp. Cell Res.* **194**:252–259.
Bronzert, D.A., Bates, S.E., Sheridan, J.P. *et al.* (1990). *Mol. Endocrinol.* **4**:981–989.
Brown, K.D. and Blakeley, D.M. (1983). *Biochem. J.* **212**:465–472.
Brown, K.D., Blakeley, D.M., Fleet, I.R., Hamon, M. and Heap, R.B. (1986). *J. Endocrinol.* **109**:325–332.
Brunner, N., Yee, D., Kern, F.G., Spangthomsen, M., Lippman, M.E. and Cullen, K.J. (1993). *Eur. J. Cancer* **29a**:562–569.
Burgess, W.H. and Maciag, T. (1989). *Ann. Rev. Biochem.* **58**:575–606.
Bussolati, G., Alfani, V., Weber, K. and Osborn, M. (1980). *J. Histochem. Cytochem.* **28**:169–173.
Cailleau, R., Young, R., Olive, M. and Reeves, W.J. (1974). *J. Natl Cancer Inst.* **53**:661–674.
Calabretta, B., Battini, R., Kaczmarek, L., de Riel, J.K. and Baserga, R. (1986). *J. Biol. Chem.* **261**:12628–12632.
Championarnaud, P., Ronsin, C., Gilberg, E., Gesnel, M.C., Houssaint, E. and Breathnach, R. (1991). *Oncogene* **6**:979–987.
Chen, L.-C., Dollbaum, C. and Smith, H.S. (1989). *Proc. Natl Acad. Sci. USA* **86**:7204–7207.
Chen, L., Shulman, L.M. and Revel, M. (1991). *J. Biol. Regul. Homeost. Agents* **5**:125–136.
Chiquet, E.R., Kalla, P. and Pearson, C.A. (1989). *Cancer Res.* **49**:4322–4325.
Coleman, S., Silberstein, G.B. and Daniel, C.W. (1988). *Dev. Biol.* **127**:304–315.
Colletti, R.B., Roberts, J.D., Devlin, J.T. and Copeland, K.C. (1989). *Cancer Res.* **49**:1882–1884.
Coombes, R.C., Barrett, L.P. and Luqmani, Y. (1990). *J. Steroid Biochem. Mol. Biol.* **37**:833–836.
Corps, A.N. and Brown, K.D. (1987), *J. Endocrinol.* **113**:285–290.
Cowley, G., Gusterson, B.A., Smith, J.A., Headler, F. and Ozanne, B. (1984). In *Cancer Cells, the Transformed Phenotype* (ed. J. A. Smith), pp. 5–10. Cold Spring Harbor Labs, Cold Spring Harbor, New York.
Crabb, J.W., Armes, L.G., Carr, S.A. *et al.* (1986). *Biochemistry (USA)* **25**:4988–4993.
Crouch, T.H. and Klee, C.L. (1980). *Biochemistry (USA),* **19**:3692–3698.
Cullen, K.J., Smith, H.S., Hill, S., Rosen, N. and Lippman, M.E. (1991). *Cancer Res.* **51**:4978–4985.
Cullen, K.J., Allison, A., Martire, I., Ellis, M. and Singer, C. (1992a). *Breast Cancer Res. Treat.* **22**:21–29.
Cullen, K.J., Lippman, M.E., Chow, D., Hill, S., Rosen, N. and Zwiebel, J.A. (1992b). *Mol. Endocrinol.* **6**:91–100.
Dai, S., Klagsbrun, M., Ogle, C.W. and Shing, Y.W. (1987). *Med. Sci. Res.* **15**:355–356.
Dairkee, S.H., Blayney, C.M., Smith, H.S. and Hackett, A.J. (1985). *Proc. Natl Acad. Sci. USA* **82**:7409–7413.
Dairkee, S.H., Puett, L. and Hackett, A.J. (1988). *J. Natl Cancer Inst.* **80**:691–695.
Daly, R.J., King, R.J.B. and Darbre, P.D. (1990). *J. Cell. Biochem.* **43**:199–211.
Daly, R.J., Harris, W.H., Wang, D.Y. and Darbre, P.D. (1991). *Cell Growth Differ.* **2**:457–464.
Danforth, D.J. and Sgagias, M.K. (1991). *Cancer Res.* **51**:1488–1493.
Danforth, D.J. and Sgagias, M.K. (1993). *Cancer Res.* **53**:1538–1545.
Daniel, C.W. and Silberstein, G.B. (1991). In *Breast Cancer: Cellular and Molecular Biology* (eds M. E. Lippman and R. Dickson), pp. 79–92. Kluwer Nijhoff, Norwell, MA.
Danielson, K.B., Knepper, J.E., Kittrell, F.S., Butel, J.S., Medina, D. and Durban, E.M. (1989). *In Vitro* **25**:535–543.
Dao, T.L., Bock, F.G. and Greiner, M.J. (1960). *J. Natl Cancer Inst.* **25**:991–1003.
Davies, B.R., Davies, M.P.A., Barraclough, R. and Rudland, P.S. (1993a). *Cell Biol. Int.* **17**:871–879.
Davies, B.R., Davies, M.P.A., Gibbs, F.E.M., Barraclough, R. and Rudland, P.S. (1993b). *Oncogene* **8**:999–1008.
Davies, B.R., Barraclough, R. and Rudland, P.S. (1994). *Cancer Res.* **54**:2785–2793.
Dawson, E.K. (1934). *Edinburgh Med. J.* **41**:653–682.
Delli Bovi, P., Curatola, A.M., Kern, F.G., Greco, A., Ittman, M. and Basilico, C. (1987). *Cell* 50:729–737.
Delvenne, C.G., Winkler, G.R., Piccart, M.J. *et al.* (1992). *Eur. J. Cancer* **28**:700–705.
Dembinski, T.C., Leung, C.K.H. and Shin, R.P.C. (1985). *Cancer Res.* **45**:3083–3089.
DeOme, K.B., Faulkin, L.J., Bern, H.A. and Blair, P.B. (1959). *Cancer Res.* **19**:515–520.
Dickson, C. (1990). *Int. J. Cancer* **S5**:51–54.
Dickson, R.B., Huff, K.K., Spencer, E.M. and Lippman, M.E. (1986). *Endocrinol.* **118**:138–142.

Dixon, M., Deed, R., Acland, P. *et al.* (1989). *Mol. Cell Biol.* **9**:4896–4902.
Dorin, J.R., Emslie, E. and van Heyningen, V. (1990). *Genomics* **8**:420–426.
Du Bois, M. and Elias, J.J. (1984). *Dev. Biol.* **106**:70–75.
Dulbecco, R., Henahan, M. and Armstrong, B. (1982). *Proc. Natl Acad. Sci. USA* **79**:7346–7350.
Dulbecco, R., Unger, M., Amstrong, B., Bowman, M. and Syka, P. (1983). *Proc. Natl Acad. Sci. USA* **80**:1033–1037.
Dunnington, D.J. (1984). Ph.D. Thesis, University of London.
Dunnington, D.J., Monaghan, P., Hughes, C.M. and Rudland, P.S. (1983). *J. Natl Cancer Inst.* **71**:1227–1240.
Dunnington, D.J., Kim, U., Hughes, C.M., Monaghan, P., Ormerod, E.J. and Rudland, P.S. (1984a). *J. Natl Cancer Inst.* **72**:455–466.
Dunnington, D.J., Kim, U., Hughes, C.M., Monaghan, P. and Rudland, P.S. (1984b). *Cancer Res.* **44**:5338–5346.
Earl, H. and McIlhinney, R.A. (1985). *J. Mol. Immunol.* **232**:981–991.
Earl, H.M., McIlhinney, R.A., Wilson, P., Gusterson, B.A. and Coombes, R.C. (1989). *Cancer Res.* **49**:6070–6076.
Easty, G.C., Easty, D.M., Monaghan, P., Ormerod, M.G. and Neville, A.M. (1980). *Int. J. Cancer* **26**:577–584.
Ebralidze, A., Tulchinsky, E., Grigorian, M. *et al.* (1989). *Genes and Develop.* **3**:1086–1093.
Edwards, P.A.W. and Brooks, I.M. (1984). *J. Histochem. Cytochem.* **32**:531–537.
Edwards, P.A.W., Brooks, I.M. and Monaghan, P. (1984). *Differentiation* **25**:247–258.
Edwards, P.A.W., Brooks, I.M., Bunnage, H.J., Foster, A.K., Ellison, M.C. and O'Hare, M.J. (1986). *J. Cell. Sci.* **86**:91–101.
Einspanier, R. and Schams, D. (1991). *J. Dairy Res.* **58**:171–178.
Elenius, K., Maatta, A., Salmivirta, M. and Jalkanen, M. (1992). *J. Biol. Chem.* **267**:6435–6441.
Elliott, B., Ostman, A., Westermark, B. and Rubin, K. (1992). *J. Cell. Physiol.* **152**:292–301.
Emerman, J.T., Enami, J., Pitelka, D.R. and Nandi, S. (1977). *Proc. Natl Acad. Sci. USA* **74**:4466–4470.
Engel, L.W., Young, N.A., Tralka, T.S., Lippman, M.E., O'Brien, S.J. and Joyce, M.J. (1978). *Cancer Res.* **38**:3352–3364.
Engelkamp, D., Schafer, B.W., Mattei, M.G., Erne, P. and Heizmann, C.W. (1993). *Proc. Natl Acad. Sci. USA* **90**:6547–6551.
Engleka, K.A. and Maciag, T. (1992). *J. Biol. Chem.* **267**:11307–11315.
Ennis, B.W., Valverius, E.M., Bates, S.E. *et al.* (1989). *Mol. Endocrinol.* **3**:1830–1838.
Eppstein, D.A., Marsh, Y.V., Schryver, B.B. and Bertics, P.J. (1989). *J. Cell. Physiol.* **141**:420–430.
Esch, F., Baird, A., Ling, N. *et al.* (1985). *Proc. Natl Acad. Sci. USA* **82**:6507–6511.
Ewing, T.M., Murphy, L.J., Ng, M.-L. *et al.* (1989). *Int. J. Cancer* **44**:744–752.
Fechner, R.E. (1987). In *Diagnostic Histopathology of the Breast* (eds D. L. Page and T. J. Anderson), pp. 72–85. Churchill Livingstone, Edinburgh.
Fenton, S.E. and Sheffield, L.G. (1991). *Biochem. Biophys. Res. Commun.* **181**:1063–1069.
Ferguson, J.E., Schor, A.M., Howell, A. and Ferguson, M.W. (1990). *Differentiation* **42**:199–207.
Fernandez, P. and Safe, S. (1992). *Toxicol. Lett.* **61**:185–197.
Fernandez-Pol, J.A., Klos, D.J. and Hamilton, P.D. (1989). *J. Cell. Biochem.* **41**:159–170.
Fernig, D.G., Smith, J.A. and Rudland, P.S. (1990). *J. Cell. Physiol.* **142**:108–116.
Fernig, D.G., Rudland, P.S. and Smith, J.A. (1992). *Growth Factors* **7**:27–39.
Fernig, D.G., Barraclough, R., Ke, Y.Q., Wilkinson, M.C., Rudland, P.S. and Smith, J.A. (1993). *Int. J. Cancer* **54**:629–635.
Finch, P.W., Rubin, J.S., Miki, T., Ron, T. and Aarsonson, S.A. (1989). *Science* 241:752–755.
Flaumenhaft, R., Moscatelli, D., Saksela, O. and Rifkin, D.B. (1989). *J. Cell. Physiol.* **140**:75–81.
Florkiewicz, R.Z. and Sommer, A. (1989). *Proc. Natl Acad. Sci. USA* **86**:3978–3981.
Foekens, J.A., Portengen, H., Janssen, M. and Klijn, J.G. (1989a). *Cancer* **63**:2139–2147.
Foekens, J.A., Portengen, H., Vanputten, W. *et al.* (1989b). *Cancer Res.* **49**:7002–7009.
Fontana, J.A., Burrows, M.A., Clemmons, D.R. and LeRoith, D. (1991). *Endocrinol.* **128**:1115–1122.
Fontana, J.A., Nervi, C., Shao, Z.-M. and Jetten, A.M. (1992). *Cancer Res.* **52**:3938–3945.
Foster, C.S., Edwards, P.A.W., Dinsdale, E.A. and Neville, A.M. (1982a). *Virchows Arch. (A)* **394**:279–293.
Foster, C.S., Dinsdale, E.A., Edwards, P.A.W. and Neville, A.M. (1982b). *Virchows Arch. (A)* **394**:295–305.

Freed, K.A. and Herington, A.C. (1989). *J. Mol. Endocrinol.* **3**:183–190.
Fullmer, C.S. and Wasserman, R.H. (1981). *J. Biol. Chem.* **256**:5669–5674.
Furuya, Y., Sato, N., Watabe, Y. and Shimazaki, J. (1991). *Jpn. J. Cancer Res.* **82**:1245–1251.
Gallagher, J.T. (1989). *Curr. Opinion Cell Biol.* **1**:1201–1218.
Gallagher, J.T., Turnbull, J.E. and Lyon, M. (1992). *Int. J. Biochem.* **24**:553–560.
Gerke, V. and Weber, K. (1985). *EMBO J.* **4**:2917–2920.
Ghosh, S., Roholt, O.A. and Kim, U. (1983). *In Vitro* **19**:919–928.
Gibbs, F.E.M. (1993). Ph.D. Thesis, University of Liverpool.
Gibbs, F.E.M., Wilkinson, M.C., Rudland, P.S. and Barraclough, R. (1994). *J. Biol. Chem.* **269**:18992–18994.
Gibbs, F.E.M., Barraclough, R., Platt-Higgins, A., Rudland, P.S., Wilkinson, M.C. and Parry, E.W. (1995). *J. Histochem. Cytochem.* **43**:169–180.
Ginsburg, E. and Vonderhaar, B.K. (1991). *Cancer Lett.* **58**:137–144.
Glimm, D.R., Baracos, V.E. and Kennelly, J.J. (1992). *J. Dairy Sci.* **75**:2687–2705.
Gomm, J.J., Smith, J., Ryall, G.K., Baillie, R., Turnbull, L. and Coombes, R.C. (1991). *Cancer Res.* **51**:4685–4692.
Gospodarowicz, D. (1974). *Nature* **249**:123–127.
Gospodarowicz, D., Greenburg, G. and Bialecki, H. (1978). *In Vitro* **14**:85–91.
Gospodarowicz, D., Neufeld, G. and Schweigerer, L. (1986). *Cell Diff.* 19:1–17.
Gould, M.N., Biel, W.F. and Clifton, K.H. (1977). *Exptl Cell Res.* 107:405–416.
Grey, A.M., Schor, A.M., Rushton, G., Ellis, I. and Schor, S.L. (1989). *Proc. Natl Acad. Sci. USA* **86**:2438–2442.
Grigor, M.R., McDonald, F.J., Latta, N., Richardson, C.L. and Tate, W.P. (1990). *Biochem. J.* **267**:815–819.
Grimaux, M., Romain, S., Remvikos, Y., Martin, P.M. and Magelenat, H. (1989). *Breast Cancer Res. Treat.* **14**:77–90.
Gullino, P.M., Pettigrew, H.W. and Grantham, F.H. (1975). *J. Natl Cancer Inst.* **54**:401–411.
Guner, G., Kirkali, G., Yenisey, C. and Tore, I.R. (1992). *Cancer Lett.* **67**:103–112.
Gusterson, B., Warburton, M.J., Mitchell, D., Ellison, M., Neville, A.M. and Rudland, P.S. (1982). *Cancer Res* **42**:4763–4770.
Gusterson, B.A., McIlhinney, R.A.J., Patel, S., Knight, J., Monaghan, P. and Ormerod, M.G. (1985). *Differentiation* **30**:102–110.
Gusterson, B.A., Monaghan, P., Mahendran, R., Ellis, J. and O'Hare, M.J. (1986). *J. Natl. Cancer Inst.* **77**:343–349.
Gusterson, B.A., Machin, L.G., Gullick, W.J. *et al.* (1988). *Br. J. Cancer* **58**:453–457.
Hackett, A.J., Smith, H.S., Springer, E.L. *et al.* (1977). *J. Natl Cancer Inst.* **58**:1795–1806.
Hadsell, D.L., Campbell, P.G. and Baumrucker, C.R. (1990). *Endocrinol.* 126:637–643.
Hagueneau, F. and Arnoult, J. (1959). *Bull. Ass. Franc. Cancer* 46:177–211.
Hallowes, R.C., Rudland, P.S., Hawkins, R.A., Lewis, D.J., Bennett, D.C. and Durbin, H. (1977). *Cancer Res.* **37**:2492–2504.
Hallowes, R.C., Peachey, L.A. and Cox, S. (1983). *In Vitro* **19**:286.
Hamamoto, S., Imagawa, W., Yang, J. and Nandi, S. (1988). *Cell Differ.* **22**:191–202.
Hamperl, H. (1970). *Curr. Top. Pathol.* **53**:161–210.
Harris, A.L., Nicholson, S., Sainsbury, J.R.C., Farndon, J. and Wright, C. (1989). *J. Steroid Biochem.* **34**:123–131.
Hauser, S.D., McGrath, M.F., Collier, R.J. and Krivi, G.G. (1990). *Mol. Cell. Endocrinol.* **72**:187–200.
Hayashi, K., Hayashi, M., Jalkanen, M., Firestone, J., Trelstad, J. and Bernfield, M. (1987). *J. Histochem. Cytochem.* **35**:1079–1088.
Herbert, D.C., Burke, R.E. and McGuire, W.L. (1978). *Cancer Res.* **38**:2221–2223.
Hilt, D.C. and Kligman, D. (1991). In *Novel Calcium-Binding Proteins* (ed. C. W. Heizmann), pp. 65–103. Springer Verlag, Berlin.
Hinton, C.P., Bishop, H.M., Holliday, H.W., Doyle, P.L. and Blamey, R.W. (1986). *Brit. J. Clin. Pract.* **40**:326–330.
Hiraiwa, N., Kida, H., Sakakura, T. and Kusakabe, M. (1993). *J. Cell Sci.* **104**:289–296.
Hollman, K.H. (1974). In *Lactation, A Comprehensive Treatise* (eds B. L. Larson and V. R. Smith), pp. 3–37. Academic Press, New York.

Holmes, W.E., Sliwkowski, M.X., Akita, R.W. *et al.* (1992). *Science* **256**:1205–1210.
Hoshino, K. (1963). *J. Natl Cancer Inst.* **30**:585–591.
Hoshino, K. (1964). *Anat. Record* **150**:221–236.
Hou, J.Z., Kan, M., McKeehan, K., McBride, G., Adams, P. and McKeehan, W.L. (1991). *Science* **251**:665–668.
Houssaint, E., Blanquet, P.R., Champion-Arnaud, P. *et al.* (1990). *Proc. Natl Acad. Sci. USA* **87**:8180–8184.
Howeedy, A.A., Virtanen, I., Laitinen, L., Gould, N.S., Koukoulis, G.K. and Gould, V.E. (1990). *Lab. Invest.* **63**:798–806.
Huggins, C., Grand, L.C. and Brillantes, F.P. (1961). *Nature* **189**:204–207.
Hughes, C.M. and Rudland, P.S. (1990a). *J. Histochem. Cytochem.* **38**:1632–1645.
Hughes, C.M. and Rudland, P.S. (1990b). *J. Histochem. Cytochem.* **38**:1647–1657.
Huynh, H.T., Tetenes, E., Wallace, L. and Pollak, M. (1993). *Cancer Res.* **53**:1727–1730.
Ignar-Trowbridge, D.M., Nelson, K.G., Bidwell, M.C. *et al.* (1992). *Proc. Natl Acad. Sci. USA* **89**:4658–4662.
Inaguma, Y., Kusakabe, M., Mackie, E.J., Pearson, C.A., Chiquet, E.R. and Sakakura, T. (1988). *Dev. Biol.* **128**:245–255.
Ishige, H., Komatsu, T., Kondo, Y., Sugano, I., Horinaka, E. and Okui, K. (1991). *Acta Pathol. Jpn.* **41**:227–232.
Ishihara, M., Tyrrell, D.J., Stauber, G.B., Brown, S., Cousens, L.S. and Stack, R.J. (1993). *J. Biol. Chem.* **268**:4675–4683.
Isobe, T. and Okuyama, T. (1978). *Eur. J. Biochem.* **89**:379–388.
Isobe, T. and Okuyama, T. (1981). *Eur. J. Biochem.* **116**:79–86.
Jackson-Grusby, L.L., Swiergiel, J. and Linzer, D.I.H. (1987). *Nucleic Acids Res.* **15**:6677–6690.
Jalkanen, M., Rapraeger, A., Saunders, S. and Bernfield, M. (1987). *J. Cell Biol.* **105**:3087–3096.
Jalkanen, M., Rapraeger, A. and Bernfield, M. (1988). *J. Cell Biol.* **106**:953–962.
Jamieson, S., Barraclough, R. and Rudland, P.S. (1990a). *Pathobiol.* **58**:329–342.
Jamieson, S., Barraclough, R. and Rudland, P.S. (1990b). *Int. J. Cancer* **46**:1071–1080.
Jamieson, S., Barraclough, R. and Rudland, P.S. (1990c). *Cell. Biol. Int. Rep.* **14**:717–725.
Jammes, H., Peyrat, J.P., Ban, E. *et al.* (1992). *Br. J. Cancer* **66**:248–253.
Jaye, M., Howk, R., Burgess, W. *et al.* (1986). *Science* **233**:541–545.
Jones, J.L., Critchley, D.R. and Walker, R.A. (1992). *J. Pathol.* **167**:399–406.
Joshi, K., Ellis, J.T.B., Hughes, C.M., Monaghan, P. and Neville, A.M. (1986a). *Lab. Invest.* **54**:52–61.
Joshi, K., Smith, J.A., Perusinghe, N. and Monaghan, P. (1986b). *Am. J. Pathol.* **124**:199–206.
Kanazawa, T. and Hosick, H.L. (1992). *J. Cell. Physiol.* **153**:381–391.
Kan, M.K., Wang, F., Xu, J.M., Crabb, J.W., Hou, J.Z. and McKeehan, W.L. (1993). *Science* **259**:1918–1921.
Kano-Sueoka, T., Cambell, G.R. and Gerber, M. (1977). *J. Cell. Physiol.* **93**:417–424.
Kano-Sueoka, T., Cohen, D.M. and Yamaizumi, Z. (1979). *Proc. Natl Acad. Sci. USA* **76**:5741–5744.
Karey, K.P. and Sirbasku, D.A. (1988). *Cancer Res.* **48**:4083–4092.
Kawamoto, K., Yamaguchi, T., Watanabe, S. and Uchida, K. (1992). *Biochim. Biophys. Acta* **1134**:183–188.
Ke, Y.Q., Fernig, D.G., Smith, J.A. *et al.* (1990). *Biochem. Biophys. Res. Comm.* **171**:963–971.
Ke, Y., Fernig, D.G., Wilkinson, M.C. *et al.* (1993). *J. Cell Sci.* **106**:135–143.
Keegan, K., Johnson, D.E., Williams, L.T. and Hayman, M.J. (1991). *Proc. Natl Acad. Sci. USA* **88**:1095–1099.
Keon, B.H. and Keenan, T.W. (1993). *Protoplasma* **172**:43–48.
Kerbel, R.S., Waghorne, C., Man, H.-S., Elliott, B. and Breitman, M.L. (1987). *Proc. Natl Acad. Sci. USA* **84**:1263–1267.
Keydar, I., Chen, L., Karby, S. *et al.* (1979). *Eur. J. Cancer* **15**:659–678.
Kidwell, W.R. (1986). In *Hormones, Oncogenes, Growth Factors*, p. 14. Proc. Inst. Sci. Roussel Symp., Paris.
Kiefer, M.C., Stephans, J.C., Crawford, K., Okino, K. and Barr, P.J. (1990). *Proc. Natl Acad. Sci. USA* **87**:6985–6989.
Kim, U. (1979). In *Breast Cancer* (ed. W. L. McGuire), pp. 1–39. Plenum Press, New York.
Kim, U. (1984). In *Cancer Invasion and Metastasis, Biologic and Therapeutic Aspects* (eds G. L. Nicolson

and L. Milas), pp. 337–351. Raven Press, New York.
Kimelman, D. and Kirschner, M.C. (1989). *Cell* **59**:687–696.
King, R.J.B., Wang, D.Y., Daly, R.J. and Darbre, P.D. (1989). *J. Steroid Biochem.* **34**:133–138.
Klagsbrun, M. and Baird, A. (1991). *Cell* **67**:229–231.
Klijn, J.G., Berns, P.M., Schmitz, P.I. and Foekens, J.A. (1992). *Endocr. Rev.* **13**:3–17.
Kosano, H., Yasutomo, Y., Kugai, N. *et al.* (1990). *Cancer Res.* **50**:3172–3175.
Kosano, M., Kubota, T., Ohsawa, N. *et al.* (1992). *Cancer Res.* **52**:1187–1191.
Kouhara, H., Kasayama, S., Saito, H., Matsumoto, K. and Sato, B. (1991). *Biochem. Biophys. Res. Comm.* **176**:31–37.
Kratchwil, K. (1969). *Dev. Biol.* 20:46–71.
Kraus, M.H., Issing, W., Miki, T., Popescu, N.C. and Aaronson, S.A. (1989). *Proc. Natl Acad. Sci. USA* 86:9193–9197.
Kudlow, J.E., Cheung, C.-Y.M. and Bjorge, J.D. (1986). *J. Biol. Chem.* **261**:4134–4138.
Kurebayashi, J., McLeskey, S.W., Johnson, M.D., Lippman, M.E., Dickson, R.B. and Kern, F.G. (1993). *Cancer Res.* **53**:2178–2187.
Kurokawa, T., Seno, M. and Igarashi, K. (1988). *Nucleic Acid. Res.* **16**:5201.
Kute, T.E. and Quadri, Y. (1991). *J. Histochem. Cytochem.* **39**:1125–1130.
Kuwano, R., Usui, H., Maeda, T. *et al.* (1984). *Nucleic Acids Res.* **12**:7455–7465.
Lane, E.B. (1982). *J. Cell Biol.* **92**:665–673.
Lasfargues, E.Y. and Ozzello, L. (1975). *J. Natl Cancer Inst.* **21**:1131–1147.
Lavandero, S., Santibanez, J.F., Ocaranza, M.P., Ferreira, A. and Sapag-Hagar, M. (1991). *Comp. Biochem. Physiol. A* **99**:507–511.
Lavialle, C., Modjtahedi, N., Lamonerie, T. *et al.* (1989). *Anticancer Res.* **9**:1265–1279.
Leal, J.A., Gangrade, B.K., Kiser, J.L., May, J.V. and Keel, B.A. (1991). *Steroids* **56**:247–251.
Lee, C.S.L., Koga, M. and Sutherland, R.L. (1989). *Biochem. Biophys. Res. Commun.* **162**:415–421.
Lee, Y.T. (1985). *Cancer Metastasis Rev.* 4:153–172.
Leholta, L., Partanen, J., Sistonen, L. *et al.* (1992). *Int. J. Cancer* 50:598–603.
Lemoine, N.R., Barnes, D.M., Hollywood, D.P. *et al.* (1992). *Br. J. Cancer* **66**:1116–1121.
Lengsfield, A., Low, I., Wieland, T., Dancker, P. and Hasselbach, W. (1974). *Proc. Natl Acad. Sci. USA* **71**:2803–2807.
Leppa, S., Harkonen, P. and Jalkanen, M. (1991). *Cell Reg.* **2**:1–11.
Leppa, S., Mali, M., Miettinen, H.M. and Jalkanen, M. (1992). *Proc. Natl Acad. Sci. USA* **89**:932–936.
Leung, B.S., Stout, L., Zhou, L., Ji, H.J., Zhang, Q.Q. and Leung, H.T. (1991). *J. Cell. Biochem.* **46**:125–133.
Levay-Young, B.K., Imagawa, W., Wallace, D.R. and Nandi, S. (1989). *Mol. Cell. Endocrinol.* **62**:327–336.
Li, M. and Bernard, O. (1992). *Proc. Natl Acad. Sci. USA* **89**:3315–3319.
Li, S. and Shipley, G.D. (1991). *Cell Growth Differ.* **2**:195–202.
Li, S.W., Plowman, G.D., Buckley, S.D. and Shipley, G.D. (1992). *J. Cell Physiol.* **153**:103–111.
Lidereau, R., Callahan, R., Dickson, C., Peters, G., Escot, C. and Ali, I.U. (1988). *Oncogene Res.* **3**:285–291.
Lien, E.A., Johannessen, D.C., Aakvaag, A. and Lonning, P.E. (1992). *J. Steroid Biochem. Mol. Biol.* **41**:541–543.
Liscia, D.S., Merlo, G., Ciardiello, F. *et al.* (1990). *Dev. Biol.* **140**:123–131.
Liu, S.C., Sanfilippo, B., Perroteau, I., Derynck, R., Salomon, D.S. and Kidwell, W.R. (1987). *Mol. Endocrinol.* **1**:683–692.
Llorens, M.A., Bermejo, M.J., Salcedo, M.C., Charro, A.L. and Puente, M. (1989). *J. Steroid Biochem.* **34**:505–509.
Lloyd, B.H., Platt-Higgins, A., Winstanley, J., Leinster, S., Rudland, P.S. and Barraclough, R. (1996). *Cancer Res.* (submitted).
Lofts, F.J. and Gullick, W.J. (1992). *Cancer Treat. Res.* **61**:161–179.
Lonning, P.E., Hall, K., Aakvaag, A. and Lien, E.A. (1992). *Cancer Res.* **52**:4719–4723.
Lories, V., Cassiman, J.J., Vandenberghe, H. and David, G. (1992). *J. Biol. Chem.* **267**:1116–1122.
Lundy, J., Schuss, A., Stanick, D., McCormack, E.S., Kramer, S. and Sorvillo, J.M. (1991). *Amer. J. Pathol.* **138**:1527–1534.

Lupu, R., Colomer, R., Zugmaier, G. *et al.* (1990). *Science* **249**:1552–1555.
Luqmani, Y.A., Graham, M. and Coombes, R.C. (1992). *Br. J. Cancer* **66**:273–280.
Lyons, R.W., Li, C.H. and Johnson, R.E. (1958). *Recent Prog. Horm. Res.* **94**:219–254.
McAndrew, J. (1993). Ph.D. Thesis, University of Liverpool.
McAndrew, J., Rudland, P.S., Platt-Higgins, A.M. and Smith, J.A. (1994a). *Histochem. J.* **26**:355–366.
McAndrew, J., Fernig, D.G., Rudland, P.S. and Smith, J.A. (1994b). *Growth Factors* **10**:281–287.
McCann, A., Johnston, P.A., Dervan, P.A., Gullick, W.J. and Carney, D.N. (1989). *Ir. J. Med. Sci.* **158**:137–140.
McCormick, D.L., Burns, F.J. and Albert, R.E. (1980). *Cancer Res.* **40**:1140–1143.
McGregor, D.H., Land, C.E., Choi, K., Tokuoka, S. and Liv, P.I. (1977). *J. Natl. Cancer Inst.* **59**:799–811.
McGuire, W.J., Jackson, J.G., Figueroa, J.A., Shimasaki, S., Powell, D.R. and Yee, D. (1992). *J. Natl Cancer Inst.* **84**:1336–1341.
McLeskey, S.W., Kurebayashi, J., Honig, S.F. *et al.* (1993). *Cancer Res.* **53**:2168–2177.
MacMahon, B., Cole, P. and Brown, J. (1973). *J. Natl Cancer Inst.* **50**:21–42.
Mackie, E.J., Chiquet, E.R., Pearson, C.A. *et al.* (1987). *Proc. Natl Acad. Sci. USA* **84**:4621–4625.
Madsen, M.W., Lykkesfeldt, A.E., Laursen, I., Nielsen, K.V. and Briand, P. (1992). *Cancer Res.* **52**:1210–1217.
Malaab, S.A., Pollak, M.N. and Goodyer, C.G. (1992). *Eur. J. Cancer* **4–5**:788–793.
Manni, A., Wright, C., Badger, B., Bartholomew, M. *et al.* (1990). *Breast Cancer Res. Treat.* **15**:73–83.
Manni, A., Wei, L., Badger, B. *et al.* (1992). *Endocrinology* **130**:1744–1746.
Mansel, R.E., Preece, P.E. and Hughes, L.E. (1978). *Brit. J. Surg.* **65**:724–727.
Mansukhani, A., Dellera, P., Moscatelli, D., Kornbluth, S., Hanafusa, H. and Basilico, C. (1992). *Proc. Natl Acad. Sci. USA* **89**:3305–3309.
Marics, I., Adelaide, J., Raybaud, F. *et al.* (1989). *Oncogene* 4:335–340.
Marx, D., Schauer, A., Reiche, C. *et al.* (1990). *J. Cancer Res. Clin. Oncol.* **116**:15–20.
Mira-y-Lopez, R., Joseph-Silverstein, J., Rifkin, D.B. and Ossowski, L. (1989). *Proc. Natl Acad. Sci. USA* **83**:7780–7784.
Minafra, S., Morello, V., Glorioso, F. *et al.* (1989). *Brit. J. Cancer* **60**:185–192.
Mittra, I. (1980). *Biochem. Biophys. Res. Commun.* **95**:1760–1767.
Moews, P.C. and Kretsinger, R.H. (1975). *J. Mol. Biol.* **91**:201–228.
Mori, M., Naito, R., Tsukitani, K., Okada, Y. and Tsujimura, T. (1989). *Acta Histochem. Cytochem.* **322**:15–34.
Moller, P., Mechtersheimer, G., Kaufmann, M. *et al.* (1989). *Virchows Archiv. (A) Pathol. Anat.* **414**:157–164.
Monaghan, P., Warburton, M.J., Perusinghe, N. and Rudland, P.S. (1983). *Proc. Natl Acad. Sci. USA* **80**:3344–3348.
Monti, M., Catania, S., Locatelli, E., Gandini, R., Reggiani, A. and Cunietti, E. (1990). *Breast Cancer Res. Treat.* **17**:77–82.
Moon, R.C. (1969). *Int. J. Cancer* **4**:312–317.
Moore, R., Casey, G., Brookes, S., Dixon, M., Peters, G. and Dickson, C. (1986). *EMBO J.* **5**:919–924.
Moroz, C., Kahn, M., Ron, E., Luria, H. and Chaimoff, C. (1989). *Cancer* **64**:691–697.
Motokura, T. and Arnold, A. (1993). *Genes Chromosomes Cancer* **7**:89–95.
Motokura, T., Bloom, T., Kim, H.G. *et al.* (1991). *Nature* **350**:512–515.
Muller, W.J., Lee, F.S., Dickson, C., Peters, G., Pattengale, P. and Leder, P. (1990). *EMBO J.* **9**:907–913.
Murakami, A., Tanaka, H. and Matsuzawa, A. (1990). *Cell Growth Differ.* **1**:225–231.
Murayama, Y. (1990). *Ann. Surg.* **211**:263–268.
Murphy, L.C. and Dotzlaw, H. (1989). *Cancer Res.* **49**:599–604.
Murphy, L.J., Sutherland, R.L., Stead, B., Murphy, L.C. and Lazarus, L. (1986). *Cancer Res.* **46**:728–734.
Myers, J.A. (1917). *Am. J. Anat.* **22**:195–223.
Myers, J.A. (1919). *Am. J. Anat.* **25**:394–435.
Nagle, R.B., Brocker, W., Davies, J.R. *et al.* (1986). *J. Histochem. Cytochem.* **34**:869–881.
Nakamura, N., Yamanishi, H., Lu, J. *et al.* (1989). *J. Steroid Biochem.* **33**:13–18.
Nandi, S.J. (1958). *J. Natl Cancer Inst.* **21**:1039–1055.

Neckers, L.M. and Trepel, J.B. (1986). *Cancer Invest.* **4**:461–470.
Nehri, A., Welch, D., Kawaguchi, T. and Nicolson, G.L. (1982). *J. Natl Cancer Inst.* **68**:507–517.
Newman, C.B., Crosby, H., Friesen, H.G. *et al.* (1987). *Proc. Natl Acad. Sci USA* **84**:8110–8114.
Newman, R.A., Klein, P.J. and Rudland, P.S. (1979). *J. Natl Cancer Inst.* **63**:1339–1346.
Nicholson, S., Sainsbury, J.R.C., Needham, G.K., Chambers, P., Farndon, J.R. and Harris, A.L. (1988a). *Int. J. Cancer* **42**:36–41.
Nicholson, S., Halcrow, P., Sainsbury, J.R.C. *et al.* (1988b). *Br. J. Cancer* **58**:810–814.
Nicholson, S., Halcrow, P., Farndon, J.R., Sainsbury, J., Chambers, P. and Harris, A.L. (1989). *Lancet* **1**:182–185.
Nickerson, S.C., Owens, W.E., Boddie, R.L. and Boddie, N.T. (1992). *J. Dairy Sci.* **75**:3339–3351.
Nicolson, G.L., Steck, P.A., Welch, D.R. and Lembo, T.M. (1983). In *Understanding Breast Cancer, Clinical and Laboratory Concepts* (eds M. A. Rich, J. C. Hager and P. Furmanski), pp. 145–166. Marcel Dekker Inc., New York.
Nielson, M., Christensen, L. and Albrechstein, R. (1983). *Acta Pathol. Microbiol. Immunol. Scand. (A)* **91**:257–264.
Noguchi, M., Koyasaki, N., Miyazaki, I. and Mizukami, Y. (1991). *Jpn. J. Cancer Res.* **82**:1199–1202.
Noguchi, M., Thomas, M., Koyasaki, N. *et al.* (1993). *Mol. Cell. Endocrinol.* **92**:69–76.
Nonomura, N., Lu, J., Tanaka, A. *et al.* (1990). *Cancer Res.* **50**:2316–2321.
Nordquist, R.E., Ishmael, D.R., Lovig, C.A., Hyder, D.M. and Hoge, A.F. (1975). *Cancer Res.* **35**:3100–3105.
Normanno, N., Qi, C.F., Gullick, W.J. *et al.* (1993). *Int. J. Oncol.* **2**:903–911.
Norris, H.J., Bahr, G.F. and Mickel, U.V. (1988). *Anal. Quant. Cytol.* **10**:1–9.
North, S.M. and Nicolson, G.L. (1985). *Brit. J. Cancer* **52**:747–755.
Nowell, P.C. (1976). *Science* **194**:23–28.
Nugent, M.A. and Edelman, E.R. (1992). *Biochemistry* **31**:8876–8883.
Odink, K., Cerletti, N., Brueggen, J. *et al.* (1987). *Nature* **330**:80–82.
O'Hare, M.J., Ormerod, M.G., Monaghan, P., Lane, E.B. and Gusterson, B.A. (1991). *Differentiation* **46**:209–221.
Okada, M., Ohmura, E., Kamiya, Y. *et al.* (1991). *Life Sci* **48**:1151–1156.
Oka, T. and Topper, Y.J. (1972). *Proc. Natl Acad. Sci. USA* **69**:1693–1696.
Oka, T., Kurachi, H., Yoshimura, M., Tsutsumi, O., Cossu, M.F. and Taga, M. (1987). *Int. J. Rad. Appl. Instrum.* **14**:353–360.
Ormerod, E.J. and Rudland, P.S. (1982). *Dev. Biol.* **91**:360–375.
Ormerod, E.J. and Rudland, P.S. (1984). *Am. J. Anat.* **170**:631–652.
Ormerod, E.J. and Rudland, P.S. (1985). *In Vitro* **21**:143–153.
Ormerod, E.J. and Rudland, P.S. (1986). *J. Embryol. Exptl Morphol.* **96**:229–243.
Ormerod, E.J. and Rudland, P.S. (1988). *In Vitro* **24**:17–27.
Ormerod, M.G., Steele, K., Edwards, P.A.W. and Taylor-Papadimitriou, J. (1984). *J. Exp. Pathol.* **1**:263–271.
Ormerod, M.G., McIlhinney, R.A.J., Steele, K. and Shimizu, M. (1985). *Mol. Immunol.* **22**:265–269.
Ornitz, D.M. and Leder, P. (1992). *J. Biol. Chem.* **267**:16305–16311.
Ornitz, D.M., Cardiff, R.D., Kuo, A. and Leder, P. (1992). *J. Natl Cancer Inst.* **84**:887–892.
Osborne, C.K., Ross, C.R., Coronado, E.B., Fuqua, S.A.W. and Kitten, L.J. (1988). *Breast Cancer Res. Treat.* **11**:211–219.
Osborne, C.K., Coronado, E.B., Kitten, L.J. *et al.* (1989). *Mol. Endocrinol.* **3**:1701–1709.
Ozello, L. (1971). *Pathol. Annu.* **6**:1–39.
Page, D.L., Anderson, T.J. and Rogers, L.W. (1987). In *Diagnostic Histopathology of the Breast* (eds D. L. Page and T. J. Anderson), pp. 120–192. Churchill Livingstone, Edinburgh.
Parham, D.M. and Jankowski, J. (1992). *J. Clin. Pathol.* **45**:513–516.
Parker, C., Whittaker, P.A., Weeks, R.J., Thody, A.J. and Sherbert, G.V. (1991). *Clin. Biotech.* **3**:217–222.
Parkes, H.C., Lillycrop, K., Howell, A. and Craig, R.K. (1990). *Br. J. Cancer* **61**:39–45.
Partanen, J., Makela, T.P., Eerola, E. *et al.* (1991). *EMBO J.* **10**:1347–1354.
Pasquale, E.B. and Singer, S.J. (1989). *Proc. Natl Acad. Sci. USA* **86**:5449–5453.
Paterson, F.C. and Rudland, P.S. (1985a). *J. Cell. Physiol.* **125**:135–150.
Paterson, F.C. and Rudland, P.S. (1985b). *J. Cell. Physiol.* **124**:525–538.

Paterson, F.C., Graham, J.M. and Rudland, P.S. (1985a). *J. Cell. Physiol.* **123**:89–100.
Paterson, F.C., Warburton, M.J. and Rudland, P.S. (1985b). *Dev. Biol.* **107**:301–313.
Pechere, J.F. (1977). In *Calcium Binding Proteins and Calcium Function* (eds R. H. Wasserman, R. Corradino, E. Carafoli, R. H. Kretsinger, D. MacLennan and F. Siegel), pp. 213. Elsevier, Amsterdam.
Pejler, G. and David, G. (1987). *Biochem. J.* **248**:69–77.
Pekonen, F., Nyman, T., Ilvesmaki, V. and Partanen, S. (1992). *Cancer Res.* **52**:5204–5207.
Perroteau, I., Kidwell, W.R., Pardue, R., DeBortoli, M. and Salomon, D.S. (1985). *Breast Cancer Res. Treat.* **6**:166.
Peters, G., Brookes, S., Smith, R., Placzek, M. and Dickson, C. (1989). *Proc. Natl Acad. Sci. USA* **86**:5678–5682.
Peterson, O.W. and van Deurs, B. (1987). *Cancer Res.* **47**:856–866.
Peterson, O.W., van Deurs, B., Nielsen, K.V. *et al.* (1990). *Cancer Res.* **50**:1257–1270.
Peto, R., Doll, R., Buckley, J.D. and Sporn, M.B. (1981). *Nature* **290**:201–208.
Petrides, P.E., Hosang, M., Shooter, E., Esch, F.S. and Bohlen, P. (1985). *FEBS Letts* **187**:89–95.
Peyrat, J.P., Bonneterre, J., Dusanter, F.I., Leroy, M.B., Djiane, J. and Demaille, A. (1989). *Bull. Cancer Paris* **76**:311–319.
Peyrat, J.P., Bonneterre, J., Vennin, P.H. *et al.* (1990). *J. Steroid Biochem. Mol. Biol.* **37**:823–827.
Peyrat, J.P., Hondermark, H., Louchez, M.M. and Boilly, B. (1991). *Cancer Comm.* **3**:323–329.
Peyrat, J.P., Bonneterre, J., Hondermarck, H. *et al.* (1992a). *J. Steroid Biochem. Mol. Biol.* **43**:87–94.
Peyrat, J.P., Hondermarck, H., Hecquet, B., Adenis, A. and Bonneterre, J. (1992b). *Bull. Cancer* **79**:251–260.
Peyrat, J.P., Bonneterre, J., Hecquet, B. *et al.* (1993). *Eur. J. Cancer* **29a**:492–497.
Picardo, M., Schor, S.L., Grey, A.M. *et al.* (1991). *Lancet* **337**:130–133.
Picardo, M., Grey, A.M., McGurk, M., Ellis, I. and Schor, S.L. (1992). *Exp. Mol. Pathol.* **57**:8–21.
Poste, G. (1983). In *Understanding Breast Cancer, Clinical and Laboratory Concepts* (eds M.A. Rich, J. C. Hager and P. Furmanski), pp. 119–144. Marcel Dekker, New York.
Prasad, K.A.N. and Church, J.G. (1991). *Exptl. Cell Res.* **195**:20–26.
Prats, H., Kaghad, M., Prats, A.C. *et al.* (1989). *Proc. Natl. Acad. Sci. USA* **86**:1836–1840.
Presta, M., Maier, J.A.M., Rusnati, M. and Ragnotti, G. (1989). *J. Cell. Physiol.* **40**:68–74.
Prestrelski, S.J., Fox, G.M. and Arakawa, T. (1992). *Arch. Biochem. Biophys.* **293**:314–319.
Raaberg, L., Nexoe, E., Tollund, L., Poulsen, S.S., Christensen, S.B. and Christensen, M.S. (1990). *Regul. Peptides* **30**:149–157.
Raaf, H.N., Jacobsen, D.W., Savon, S. and Green, R. (1993). *Am. J. Clin. Pathol.* **99**:232–237.
Radnor, C.J.P. (1971). MSc Thesis, University of Manchester.
Radnor, C.J.P. (1972). *J. Anat.* **111**:381–398.
Rapraeger, A.C., Jalkanen, M. and Bernfield, M. (1986). *J. Cell Biol.* **103**:2683–2696.
Rapraeger, A.C., Krufka, A. and Olwin, B.B. (1991). *Science* **252**:1705–1708.
Raynaud, A. (1961). In *Milk, The Mammary Gland and Its Secretions* (eds S. K. Kon and A. T. Cowie), pp. 3–46. Academic Press, New York.
Reed, M.J., Christodoulides, A., Koistinen, R., Seppala, M., Teale, J.D. and Ghilchik, M.W. (1992a). *Int. J. Cancer* **52**:208–212.
Reed, M.J., Coldham, N.G., Patel, S.R., Ghilchik, M.W. and James, V.H. (1992b). *J. Endocrinol.* 13:R5–R8.
Reed, M.J., Topping, L., Coldham, N.G., Purohit, A., Ghilchik, M.W. and James, V.H. (1993). *J. Steroid Biochem. Mol. Biol.* 44:589–596.
Reubi, J.C. and Torhorst, J. (1988). *Breast Cancer Res. Treat.* **12**:245–246.
Rifkin, D.B. and Moscatelli, D. (1989). *J. Cell Biol.* **109**:1–6.
Riss, T.L. and Sirbasku, D.A. (1987a). *Cancer Res.* **47**:3776–3782.
Riss, T.L. and Sirbasku, D.A. (1987b). *In Vitro* **23**:841–849.
Riss, T.L. and Sirbasku, D.A. (1989). *J. Cell. Physiol.* **138**:405–414.
Ro, J., Bresser, J., Ro, J.Y., Brasfield, F., Hortobagyi, G. and Blick, M. (1989). *Oncogene* **4**:351–354.
Rosen, H.R., Moroz, C., Reiner, A. *et al.* (1992a). *Breast Cancer Res. Treat.* **24**:17–26.
Rosen, H.R., Moroz, C., Reiner, A. *et al.* (1992b). *Cancer Lett.* **67**:35–45.
Rosen, H.R., Stierer, M., Gottlicher, J. *et al.* (1992c). *Int. J. Cancer* **52**:229–233.
Rosen, H.R., Stierer, M., Gottlicher, J., Wolf, H., Spoula, H. and Eibl, M. (1993). *Am. J. Surg.* **165**:213–217.

Rowe, J.M., Kasper, S., Shiu, R.C. and Friesen, H.G. (1986). *Cancer Res.* **46**:1408–1412.
Ruan, W., Newman, C.B. and Kleinberg, D.L. (1992). *Proc. Natl Acad. Sci. USA* **89**:10872–10876.
Rudland, P.S. (1987a). In *Cellular and Molecular Biology of Mammary Cancer* (eds D. Medina, W. Kidwell, G. Heppner and G. E. Anderson). pp. 9–28. Plenum Press, New York.
Rudland, P.S. (1987b). *Cancer Metastasis Rev.* **6**:55–83.
Rudland, P.S. (1991a). *J. Histochem. Cytochem.* **39**:1471–1484.
Rudland, P.S. (1991b). *J. Histochem. Cytochem.* **39**:1257–1266.
Rudland, P.S. (1992). *J. Cell. Physiol.* **153**:157–168.
Rudland, P.S. (1993). *Histol. Histopathol.* **8**:385–404.
Rudland, P.S. and Barraclough, R. (1990). *J. Cell. Physiol.* **142**:657–665.
Rudland, P.S. and Hughes, C.M. (1989). *J. Histochem. Cytochem.* **37**:1087–1100.
Rudland, P.S. and Hughes, C.M. (1991). *J. Cell. Physiol.* **146**:222–233.
Rudland, P.S. and Jimenez de Asua, L. (1979a). *Biochim. Biophys. Acta* **560**:91–133.
Rudland, P.S. and Jimenez de Asua, L. (1979b). *Brit. J. Cancer* **39**:464–465.
Rudland, P.S., Seifert, W.E. and Gospodarowicz, D. (1974). *Proc. Natl Acad. Sci. USA* **71**:2600–2604.
Rudland, P.S., Hallowes, R.C., Durbin, H. and Lewis, D. (1977a). *J. Cell Biol.* **73**:561–577.
Rudland, P.S., Durbin, H., Clingan, D. and Jiminez de Asua, L. (1977b). *Biochem. Biophys. Res. Commun.* **75**:556–562.
Rudland, P.S., Bennett, D.C. and Warburton, M.J. (1979). *Cold Spr. Hb. Symp. Cell Prolif.* 6:677–699.
Rudland, P.S., Gusterson, B.A., Hughes, C.M., Ormerod, E.J. and Warburton, M.J. (1982a). *Cancer Res.* **42**:5196–5208.
Rudland, P.S., Davies, A.T. and Warburton, M.J. (1982b). *J. Natl Cancer Inst.* **69**:1083–1093.
Rudland, P.S., Warburton, M.J., Monaghan, P. and Ritter, M.A. (1982c). *J. Natl Cancer Inst.* **68**:799–811.
Rudland, P.S., Paterson, F.C., Davies, A.T. and Warburton, M.J. (1983a). *J. Natl Cancer Inst.* **70**:949–958.
Rudland, P.S., Hughes, C.M., Twiston-Davies, A.C. and Warburton, M.J. (1983b). *Cancer Res.* **43**:3305–3309.
Rudland, P.S., Dunnington, D.J., Gusterson, B., Monaghan, P. and Hughes, C.M. (1984a). *Cancer Res.* **44**:2089–2101.
Rudland, P.S., Twiston Davis, A.C. and Tsao, S.-W. (1984b). *J. Cell. Physiol.* **120**:364–376.
Rudland, P.S., Hallowes, R.C., Cox, S.A., Ormerod, E.J. and Warburton, M.J. (1985). *Cancer Res.* **45**:3864–3877.
Rudland, P.S., Paterson, F.C., Monaghan, P., Twiston-Davies, A.C. and Warburton, M.J. (1986). *Dev. Biol.* **113**:388–405.
Rudland, P.S., Ollerhead, G. and Barraclough, R. (1989a). *Dev. Biol.* **136**:167–180.
Rudland, P.S., Hughes, C.M., Ferns, S.A. and Warburton, M.J. (1989b). *In Vitro* **25**:23–36.
Rudland, P.S., Dunnington, D.J., Kim, U., Gusterson, B.A., O'Hare, M.J. and Monaghan, P. (1989c). *Br. J. Cancer* **59**:854–864.
Rudland, P.S., Ollerhead, G.E. and Platt-Higgins, A.M. (1991). *In Vitro* **27A**:103–112.
Rudland, P.S., Platt-Higgins, A.M., Wilkinson, M.C. and Fernig, D.G. (1993a). *J. Histochem. Cytochem.* **41**:887–898.
Rudland, P.S., Leinster, S.J., Winstanley, J., Green, B., Atkinson, M. and Zakhour, H.D. (1993b). *J. Histochem. Cytochem.* **41**:543–553.
Rudloff, H.E., Schmalstieg, F.J., Palkowetz, K.H., Paszkiewicz, E.J. and Goldman, A.S. (1993). *J. Reprod. Immunol.* **23**:13–20.
Russo, I.H. and Russo, J. (1978). *J. Natl Cancer Inst.* **61**:1439–1449.
Russo, I.H., Ireland, M., Isenburg, W., Russo, J. and Russo, I.H. (1976). *Proc. Electron Microscop. Soc. Am.* **34**:146–147.
Russo, J., Saby, J., Isenburg, W.M. and Russo, I.H. (1977). *J. Natl Cancer Inst.* **59**:435–455.
Russo, J., Tay, L.K. and Russo, I.H. (1982). *Breast Cancer Res. Treat.* **2**:5–73.
Russo, J. and Russo, I.H. (1987). In *The Mammary Gland, Development, Regulation and Function* (eds M. C. Neville and C. W. Daniel), pp. 67–93. Plenum Press, New York.
Russo, J., Reina, D., Frederick, J. and Russo, I.H. (1988). *Cancer Res.* **48**:2837–2857.
Saeki, T., Cristiano, A., Lynch, M.J. *et al.* (1991). *Mol. Endocrinol.* **5**:1955–1963.
Saito, H., Kasayama, S., Kouhara, H., Matsumoto, K. and Sato, B. (1991a). *Biochem. Biophys. Res. Commun.* **174**:136–141.

Saito, S., Maruyama, M., Kato, Y., Moriyama, I. and Ichijo, M. (1991b). *J. Reprod. Immunol.* **20**:267–276.
Sakakura, T., Nishizuka, T. and Dawe, C.J. (1976). *Science* **194**:1439–1441.
Sakakura, T., Kusano, I., Kusakabe, M., Inaguma, Y. and Nishizuka, Y. (1987). *Development* **100**:421–430.
Salazar, J. and Tobon, H. (1974). In *Lactogenic Hormones, Fetal Nutrition and Lactation* (ed. J. B. Josimovich), pp. 221–227. Wiley and Sons, New York.
Salomon, D.S., Zwiebel, J.A., Bano, M., Losonczy, I., Fehnel, P. and Kidwell, W.R. (1984). *Cancer Res.* **44**:4069–4077.
Sanchez, L., Lujan, L., Oria, R. *et al.* (1992). *J. Dairy Sci.* **75**:1257–1262.
Schams, D. and Einspanier, R. (1991). *Endocr. Regul.* **25**:139–143.
Schor, S.L., Schor, A.M., Grey, A.M. and Rushton, G. (1988). *J. Cell Sci.* **90**:391–399.
Schor, S.L., Schor, A.M., Grey, A.M. *et al.* (1989). *In Vitro* **25**:737–746.
Schor, S.L., Grey, A.M., Picardo, M. *et al.* (1991). *Exs.* **59**:127–146.
Schulman, H.M., Ponka, P., Wilczynska, A., Gauthier, Y. and Shyamala, G. (1989). *Biochim. Biophys. Acta* **1010**:1–6.
Seddiki, T., Delpal, S. and Ollivier, B.M. (1992). *J. Histochem. Cytochem.* **40**:1501–1510.
Segaloff, A. (1978). In *Breast Cancer: Advances in Research and Treatment* (ed. W. L. McGuire), Vol. 2, pp. 1–22. Plenum Press, New York.
Sehgal, P.B. and Tamm, I. (1991). *Exs* **49**:178–193.
Sheikh, M.S., Shao, Z.M., Chen, J.C. *et al.* (1992a). *Biochem. Biophys. Res. Commun.* **188**:1122–1130.
Sheikh, M.S., Shao, Z.M., Clemmons, D.R., LeRoith, D., Roberts, C. and Fontana, J.A. (1992b). *Biochem. Biophys. Res. Commun.* **183**:1003–1010.
Shing, Y. and Klagsbrun, M. (1984). *Endocrinol.* **115**:273–282.
Shirasuna, K., Watatani, K., Sugiyama, M., Morioka, S. and Miyazaki, T. (1986). *Cancer Res.* **46**:1418–1426.
Shoji, T., Kamiya, T., Tsubura, A. *et al.* (1992). *Virchows Arch. A* **421**:53–56.
Shterman, N., Kupfer, B. and Moroz, C. (1991). *Pathobiology* **59**:19–25.
Sirbasku, D.A., Officer, J.B., Leland, F.E. and Iio, M. (1982). In *Growth of Cells in Hormonally Defined Media* (eds G. H. Sato, A. B. Pardee and D. A. Sirbasku), Vol. B, pp. 765–788. Cold Spr. Hb. Lab., New York.
Simpson, H.W., Mutch, F., Halberg, F., Griffiths, K. and Wilson, D. (1982). *Cancer* **50**:2417–2422.
Simpson, R.J., Smith, J.A., Moritz, L. *et al.* (1985). *Eur. J. Biochem.* **153**:629–637.
Skaar, T.C., Vega, J.R., Pyke, S.N. and Baumrucker, C.R. (1991). *J. Endocrinol.* **131**:127–133.
Skilton, R.A., Earl, H.A., Gore, M.E. *et al.* (1990). *Tumor Biol.* **11**:20–38.
Skoog, L., Macias, A., Azavedo, E., Lombardero, J. and Klintenberg, C. (1986). *Br. J. Cancer* **54**:271–276.
Slamon, D.J., Clark, G.M., Wong, S.G., Levin, W.J., Ullrich, A. and McGuire, W.L. (1987). *Science* **235**:177–182.
Slemmer, G. (1974). *J. Invest. Dermatol.* **63**:27–47.
Sloane, J.B. and Ormerod, M.G. (1981). *Cancer* **47**:1786–1795.
Smith, A.D., Winterbourne, D.J., McFarland, V.W. and Mora, P.T. (1987). *Oncogene Res.* **1**:325–341.
Smith, C.A., Monaghan, P. and Neville, A.M. (1984a). *Virchows Arch. (A)* **402**:319–329.
Smith, G.H. and Medina, D. (1988). *J. Cell. Sci.* **89**:173–183.
Smith, J.A., Ham, J., Winslow, D.P., O'Hare, M.J. and Rudland, P.S. (1984b). *J. Chromatog.* **305**:295–308.
Smith, J.A., Winslow, D., O'Hare, M.J. and Rudland, P.S. (1984c). *Biochem. Biophys. Res. Comm.* **119**:311–318.
Smith, J.A., Winslow, D.P. and Rudland, P.S. (1984d). *J. Cell. Physiol.* **119**:120–126.
Smith, J.A., Barraclough, R., Fernig, D.G. and Rudland, P.S. (1989). *J. Cell. Physiol.* **141**:362–370.
Snedeker, S.M., Brown, C.F. and DiAugustine, R.P. (1991). *Proc. Natl Aacd. Sci. USA* **88**:276–280.
Sonnenberg, A., Daams, H., Calafat, J. and Hilgers, J. (1986). *Cancer Res.* **46**:5913–5922.
Sordillo, L.M., Redmond, M.J., Campos, M., Warren, L. and Babiuk, L.A. (1991). *Can. J. Vet. Res.* **55**:298–301.
Soule, H.D., Vasquez, A., Long, A., Albert, S. and Brennan, M. (1973). *J. Natl. Cancer Inst.* **51**:1409–1416.

Spitzer, E., Grosse, R., Kunde, D. and Schmidt, H.E. (1987). *Int. J. Cancer* **39**:279–282.
Sporn, M.B. and Newton, D.L. (1979). *Fed. Proc.* **38**:2528–2534.
Stampfer, M., Hallowes, R.C. and Hackett, A.J. (1980). *In Vitro* **16**:415–425.
Stewart, A.J., Johnson, M.D., May, F.E. and Westley, B.R. (1990). *J. Biol. Chem.* **265**:21172–21178.
Stewart, A.J., Westley, B.R. and May, F.E. (1992). *Br. J. Cancer* **66**:640–648.
Stierer, M., Rosen, H.R., Forster, E. and Moroz, C. (1991). *Breast Cancer Res. Treat.* **19**:283–288.
Stirling, J.W. and Chandler, J.A. (1976). *Virchows Arch. (A)* **372**:205–226.
Sumitani, S., Kasayama, S. and Sato, B. (1993). *Endocrinology* **132**:1199–1206.
Supowit, S.C. and Rosen, J.M. (1982). *Cancer Res.* **42**:1355–1360.
Sutherland, R., Delia, D., Schneider, C., Newman, R., Kemshead, J. and Greaves, M. (1981). *Proc. Natl Acad. Sci. USA* **78**:4515–4519.
Szebenyi, D.M.E. and Moffat, K. (1986). *J. Biol. Chem.* **261**:8761–8777.
Szebenyi, D.M.E., Obendorf, S.K. and Moffat, K. (1981). *Nature* **294**:327–332.
Taga, M., Sakakura, T. and Oka, T. (1989). *Endocrinol. Jpn* **36**:559–568.
Takahashi, K., Suzuki, K., Kawahara, S. and Ono, T. (1989). *Int. J. Cancer* **43**:870–874.
Takahashi, K., Kawahara, S. and Ono, T. (1990). *Jpn. J. Cancer Res.* **81**:52–57.
Tamm, I., Cardinale, I. and Murphy, J.S. (1991). *Proc. Natl Acad. Sci. USA* **88**:4414–4418.
Tan, T.J., Schober, D.A. and Simmen, F.A. (1990). *Regul. Peptides* **37**:61–74.
Tanaka, A., Matsumoto, K., Nishizawa, Y. *et al.* (1990). *J. Steroid Biochem. Mol. Biol.* **37**:23–29.
Tanaka, A., Miyamoto, K., Minamino, N. *et al.* (1992). *Proc. Natl. Acad. Sci. USA* **89**:8928–8932.
Tauchi, K., Hori, S., Itoh, H., Osamura, R.Y., Tokuda, Y. and Tajima, T. (1989). *Virchows Archiv. (A)* **416**:65–73.
Taverna, D., Groner, B. and Hynes, N.E. (1991). *Cell Growth Differ.* **2**:145–154.
Taylor-Papadimitriou, J., Lane, E.J. and Chang, S.E. (1983). In *Understanding Breast Cancer* (eds M. A. Rich, J. C. Hager and P. Furmanski), pp. 215–246. Marcel Dekker, New York.
Taylor-Papadimitriou, J. and Lane, E.J. (1987). In *The Mammary Gland, Development, Regulation and Function* (eds M. C. Neville and C. W. Daniel), pp. 181–215. Plenum Press, New York.
Theillet, C., Le Roy, X., De Lapeyriere, O. *et al.* (1989). *Oncogene* **4**:915–922.
Thorsen, T., Lahooti, H., Rasmussen, M. and Aakvaag, A. (1992). *J. Steroid Biochem. Mol. Biol.* **41**:537–540.
Toi, M., Hamada, Y., Nakamura, H. *et al.* (1989). *Int. J. Cancer* **43**:220–225.
Toi, M., Bicknell, R. and Harris, A.L. (1992). *Cancer Res.* **52**:275–279.
Topper, Y.J. and Freeman, C.S. (1980). *Physiol. Rev.* **60**:1049–1105.
Travers, M.T., Barrett-Lee, P., Berger, U. *et al.* (1988). *Br. Med. J. Clin. Res.* **296**:1621–1624.
Trempe, G. and Fogh, J. (1973). *In Vitro* **8**:433.
Trent, J.M. (1985). *Breast Cancer Res. Treat.* **5**:221–229.
Trowbridge, I.S. and Omary, B. (1981). *Proc. Natl Acad. Sci. USA* **78**:3039–3043.
Tulchinsky, E., Kramerov, D., Ford, H.I., Reshetnyak, E., Lukanidin, E. and Zain, S. (1993). *Oncogene* **8**:79–86.
Turnbull, J.E., Fernig, D.G., Ke, Y., Wilkinson, M.C. and Gallagher, J.T. (1992). *J. Biol. Chem.* **267**:10337–10341.
Turner, J.D. and Huynh, H.T. (1991). *J. Dairy Sci.* **74**:2801–2807.
Tyrrell, D.J., Ishihara, M., Rao, N. *et al.* (1993). *J. Biol. Chem.* **268**:4684–4689.
Ullrich, A. and Schlessinger, J. (1990). *Cell* **61**:203–212.
Umekita, Y., Enokizono, N., Sagara, Y. *et al.* (1992). *Virchows Archiv. A., Pathol. Anat.* **420**:345–351.
Valverius, E.M., Walker, J.D., Bates, S.E. *et al.* (1989). *Cancer Res.* **49**:6269–6274.
Valverius, E.M., Ciardiello, F., Heldin, N.E. *et al.* (1990). *J. Cell. Physiol.* **145**:207–216.
van der Burg, B., Isbrucker, L., van Selm-Miltenburg, A.J., de Laat, S.W. and van Zoelen, E.J. (1990). *Cancer Res.* **50**:7770–7774.
Vandewalle, B., Hornez, L., Revillion, F. and Lefebvre, J. (1989). *Biochem. Biophys. Res. Commun.* **163**:149–154.
Vandewalle, B., Hornez, L., Revillion, F. and Lefebvre, J. (1991). *Biochem. Biophys. Res. Commun.* **177**:1041–1048.
van Eldik, L.J., Zendegui, J.G., Marshak, D.R. and Watterson, D.M. (1982). *Int. Rev. Cytol.* **77**:1–61.
van Roy, F.M., Messiaen, L., Liebout, G. *et al.* (1986). *Cancer Res.* **46**:4787–4795.
Venesio, T., Taverna, D., Hynes, N.E. *et al.* (1992). *Cell Growth Differ.* **3**:63–71.

Verhasselt, B., van Damme, J., van Larebeke, N. *et al.* (1992). *Eur. J. Cell Biol.* **59**:449–457.
Vlodavsky, I., Barshavit, R., Ishaimichaeli, R., Bashkin, P. and Fuks, Z. (1991). *Trends Biochem. Sci.* **16**:268–271.
Volk, R., Koster, M., Poting, A., Hartmann, L. and Knochel, W. (1989). *EMBO J.* **8**:2983–2988.
Vonderhaar, B.K. (1987). *J. Cell. Physiol.* **132**:581–584.
Vorherr, H. (1974). *The Breast: Morphology, Physiology and Lactation.* Academic Press, London.
Vorherr, H. (1980). *Breast Cancer: Epidemiology, Endocrinology, Biochemistry and Pathobiology,* Urban and Schwarzenberg, Baltimore.
Wakeling, A.E., Newboult, E. and Peters, S.W. (1989). *J. Mol. Endocrinol.* **2**:225–234.
Walker, A.W., Turnbull, J.E., Lyons, M. and Gallagher, J.T. (1994). *J. Biol. Chem.* **269**:931–935.
Walker, R.A. and Day, S.T. (1986). *J. Pathol.* **148**:217–224.
Warburton, M.J., Ormerod, E.J., Monaghan, P., Ferns, S. and Rudland, P.S. (1981). *J. Cell Biol.* **91**:827–836.
Warburton, M.J., Mitchell, D., Ormerod, E.J. and Rudland, P.S. (1982). *J. Histochem. Cytochem.* **30**:667–676.
Warburton, M.J., Head, L.P., Ferns, S.A. and Rudland, P.S. (1983). *Eur. J. Biochem.* **133**:707–715.
Warburton, M.J., Ferns, S.A., Hughes, C.M. and Rudland, P.S. (1985). *J. Cell Sci.* **79**:287–304.
Warburton, M.J., Kimbell, R., Rudland, P.S., Ferns, S.A. and Barraclough, R. (1986). *J. Cell. Physiol.* 128:76–84.
Warburton, M.J., Hughes, C.M., Ferns, S.A. and Rudland, P.S. (1989). *J. Histochem. Cytochem.* **21**:679–685.
Warri, A.M., Laine, A.M., Majasuo, K.E., Alitalo, K.K. and Harkonen, P.L. (1991). *Int. J. Cancer* **49**:616–623.
Watanabe, Y., Usuda, N., Tsugane, S., Kabayashi, R. and Hidaka, H. (1992). *J. Biol. Chem.* **267**:17136–17140.
Watanabe, Y., Usuda, N., Minami, H. *et al.* (1993). *FEBS Letts* **234**:51–55.
Weidner, N., Semple, J.P., Welch, W.R. and Folkman, J. (1991). *New Eng. J. Med.* **324**:1–8.
Weinstein, R.E., Bond, B.H., Silberberg, B.K., Vaughan, C.B., Subbaiah, P. and Pieper, D.R. (1989). *Breast Cancer Res. Treat.* **14**:349–353.
Welch, D.R. and Nicolson, G.L. (1983). *Clin. Exptl Metastasis* **1**:317–325.
Wellings, S.R. and Misdrop, W. (1985). *Eur. J. Cancer* **19**:1721–1723.
Wellings, S.R., Jensen, M.M. and Marcum, R.C. (1975). *J. Natl Cancer Inst.* **55**:231–275.
Wetzel, R.H.W., Holland, R. and van Haelst, U.J.G.M. (1989). *Lab. Invest.* **50**:552–560.
Whitford, P., Mallon, E.A., George, W.D. and Campbell, A.M. (1990). *Br. J. Cancer* **62**:971–975.
Wilkinson, M.C., Nunez de Croker, C.A., Rudland, P.S. and Smith, J.A. (1996). *In Vitro* (submitted).
Williams, J.C., Gusterson, B.A., Humphreys, J. *et al.* (1981). *J. Natl Cancer Inst.* **66**:147–155.
Williams, J.C., Gusterson, B.A. and Coombes, R.C. (1982). *Br. J. Cancer* **45**:588–597.
Williams, J.C., Gusterson, B.A., Monaghan, P., Coombes, R.C. and Rudland, P.S. (1985). *J. Natl Cancer Inst.* **74**:415–428.
Williams, J.M. and Daniel, C.W. (1983). *Dev. Biol.* **97**:274–290.
Williams, M.R., Turkes, A., Pearson, D., Griffiths, K. and Blamey, R.W. (1990a). *Br. J. Cancer* **61**:126–132.
Williams, M.R., Turkes, A., Pearson, D., Griffiths, K. and Blamey, R.W. (1990b). *Eur. J. Surg. Oncol.* **16**:22–27.
Winder, S.J., Wheatley, S.D. and Forsyth, I.A. (1993). *J. Endocrinol.* **136**:297–304.
Winstanley, J.H.R., Cooke, T., Murray, D.G. *et al.* (1991). *Br. J. Cancer* **63**:447–450.
Wolman, S.R., Pauley, R.J., Mohamed, A.N., Dawson, P.J., Visscher, D.W. and Sarkar, F.H. (1992). *Cancer* **70**:1765–1774.
Wosikowski, K., Kung, W., Hasmann, M., Loser, R. and Eppenberger, U. (1993). *Int. J. Cancer* **53**:290–297.
Wrba, F., Chott, A., Reiner, A., Reiner, G., Marks Ritzinger, E. and Holzner, J.H. (1989). *Oncology* **46**:255–259.
Yamaguchi, T., Kawamoto, K., Uchida, N., Uchida, K. and Watanabe, S. (1992). *In Vitro* **4**:245–254.
Yamanishi, H., Nonomura, N., Tanaka, A., Nishizawa, Y., Terada, N., Matsumoto, K. and Sato, B. (1991). *Cancer Res.* **51**:3006–3010.

Yayon, A., Klargsbrun, M., Esko, J.D., Leder, P. and Ornitz, D.M. (1991). *Cell* 64:841–848.
Yee, D., Paik, S., Lebovic, G.S. *et al* (1989). *Mol. Endocrinol.* 3:509–517.
Yoshida, T., Miyagawa, K., Odagiri, H. *et al.* (1987). *Proc. Natl Acad. Sci. USA,* 84:7305–7309.
Yunis, J.J. and Soreng, A.L. (1984). *Science* 226:1199–1204.
Zajchowski, D., and Band, V., Pauzie, N., Tager, A., Stampfer, M. and Sager, R. (1988). *Cancer Res.* 48:7041–7047.
Zhao, X., McBride, B.W., Politis, I. *et al.* (1992). *J. Endocrinol.* 134:304–312.
Zhan, X., Bates, B., Hu, X. and Goldfarb, M. (1988). *Mol. Cell. Biol.* 8:3487–3495.
Zoll, B., Kynast, B., Corell, B., Marx, D., Fischer, G. and Schauer, A. (1992). *J. Cancer Res. Clin. Oncol.* 118:468–473.
Zwiebel, J.A., Bano, M., Nexo, E., Salomon, D.S. and Kidwel, W.R. (1986). *Cancer Res.* 46:933–939.

8 Liver stem cells*

J. W. Grisham and Snorri S. Thorgeirsson*

*J. W. Grisham, Department of Pathology, University of North Carolina at Chapel Hill, Chapel Hill, NC, USA; *Laboratory of Experimental Carcinogenesis, National Cancer Institute/National Institutes of Health, Bethesda, MD, USA*

PREFACE

The idea that the liver contains stem cells for hepatocytes and biliary epithelial cells has intrigued investigators of hepatic physiopathology for many years, and several recent reviews of observations and hypotheses have appeared on the subject (Aterman, 1992; Fausto, 1990, 1994; Marceau, 1994; Marceau *et al.*, 1989; Sell, 1990, 1993, 1994; Sell and Pierce, 1994; Sigal *et al.*, 1992; Thorgeirsson, 1993; Zajicek, 1992). The evidence adduced for the presence of liver stem cells is derived mainly from studies on embryogenesis of the liver, on experimental hepatocarcinogenesis, and on the properties of non-hepatocytic (stem-like) epithelial cells isolated from the liver and examined in culture and after transplantation into the liver and other sites *in vivo*. Some of this evidence was recently provided in a book of reviews (Sirica, 1992) and at a symposium on identification of liver stem cells (Brill *et al.*, 1993; Dabeva *et al.*, 1993; Fausto *et al.*, 1993; Grisham *et al.*, 1993; Thorgeirsson *et al.*, 1993; Yang *et al.*, 1993b). The opinion that the liver contains epithelial cells that share some of the major properties of stem cells of the well-studied stem cell-fed lineages that comprise bone marrow, intestinal epithelium and epidermis is now strongly supported. Nevertheless, the population dynamics of the major types of liver epithelial cells – hepatocytes and bile duct epithelial cells – differ drastically from the population dynamics of the classic stem cell-fed lineages. Both hepatocytes and bile duct epithelial cells are long-lived, with very little turnover in the absence of cell loss induced by toxic damage or surgical resection. Even after induced loss of cells, replacement of hepatocytes and bile duct epithelial cells occurs by proliferation of residual differentiated cells under most circumstances, without involving the activation of stem cells. Liver stem cells are, therefore, usually quiescent, unlike the stem cells of rapidly turning over populations; activation of liver stem cells and re-establishment of hepatocyte or bile duct cell lineages appears to occur only under circumstances in which the residual differentiated cells cannot proliferate, and it is not certain that liver stem cells have any important functions in normal circumstances. In

* The colour plate section for this chapter appears between pages 274 and 275.

STEM CELLS
ISBN 0–12–563455–2

this chapter we review the evidence for the presence of epithelial stem (or stem-like or progenitor) cells in the liver, and the roles that these cells may play in development, growth, and repair of the liver. Most of the data reviewed here resulted from experimental studies in rodents, since spatial constraints prevent a comprehensive review of data from other species, including humans. As a necessary setting for this review on liver stem cells we first briefly summarize the parenchymal structure of the adult liver and the population dynamics and cell kinetics of the hepatocyte and bile duct epithelial cell populations during embryogenesis and post-natal growth of this organ.

STRUCTURE OF THE LIVER PARENCHYMA IN ADULT MAMMALS

The liver is composed mainly of epithelial cells, of which hepatocytes make up the major fraction, the minor fraction including epithelial cells that form bile ducts and ductules (Weibel *et al.*, 1969). In the adult rat, hepatocytes are organized into plates of cells that are one cell wide, and stacked more-or-less as bricks in a wall, as depicted schematically in Figure 1. Individual plates, which are 0.3–0.5 mm long and contain up to 20 hepatocytes, bifurcate and merge freely (for reviews see Jones and Schmucker, 1977; McCuskey, 1993). Epithelial cells of bile ducts and ductules and hepatocytes form a continuous cellular array; ductules are joined at the portal ends of plates to bile canaliculi

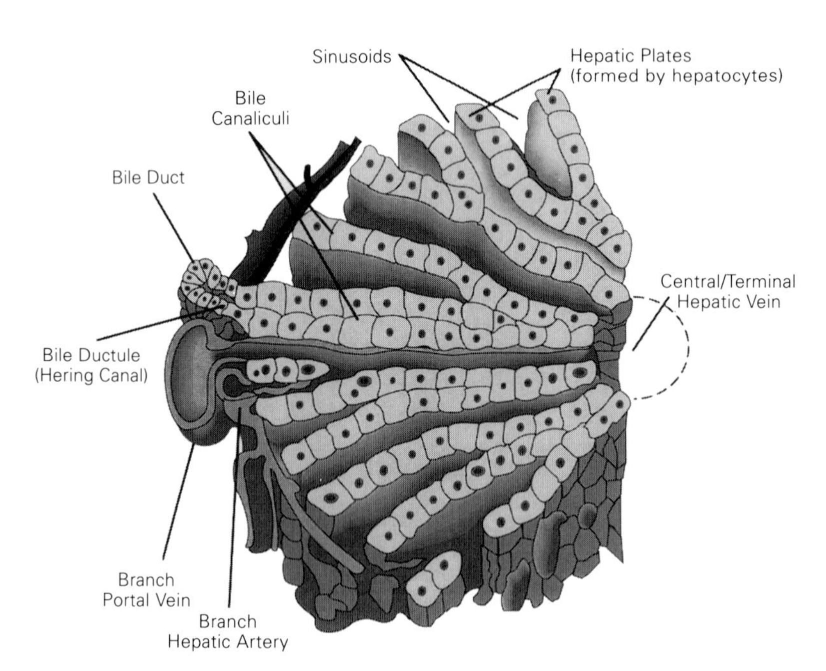

Figure 1 Schematic depiction of a microsegment of liver parenchyma, showing branches of afferent (portal vein and hepatic artery) and efferent (hepatic vein) vessels, the spaces for sinusoids that connect afferent and efferent vessels, bile duct/ductule, bile canaliculi, and hepatic plates. Hepatic plates, formed by stacks of hepatocytes, extend more-or-less directly from portal tracts (housing afferent vessels and bile ducts) toward adjacent terminal branches of hepatic veins.

that occupy the intercellular space between adjacent hepatocytes (Steiner and Carruthers, 1961). The continuous mass of hepatocytic plates is interpenetrated by the arborizing branches of vessels that supply blood to and remove blood from the liver (Figure 1). Intrahepatic branches of afferent vessels (portal vein and hepatic artery) occupy portal tracts or spaces (along with bile ducts), and the branches of the efferent vessels (hepatic veins) are generally located at a maximal distance from afferent vessels, together forming a grid-like pattern of regularly spaced afferent and efferent vessels. Terminal branches of the portal vein and the hepatic vein are connected only by capillary-sized sinusoids, which separate each adjacent, one-hepatocyte-wide hepatic plate. The resulting pattern of blood flow into and out of the smallest perfused units of parenchyma, gives to the parenchyma the appearance of lobulation, although true lobulation does not exist in most species. Microunits of liver structure are the smallest bits of parenchyma that are perfused by a terminal portal vein and/or that are drained by a terminal hepatic vein. Part of a microsegment of a liver parenchyma is shown schematically in Figure 1.

Hepatocytes vary in structure and phenotype in relation to their proximodistal location in hepatic plates (for reviews see Gebhardt, 1992; Jungermann, 1992). Epithelial cells of bile ducts and ductules of different sizes also differ structurally and enzymatically. The liver performs myriad functions, most of which are carried out by hepatocytes. Some metabolic functions, and the enzymes that are responsible for them, are most active in hepatocytes located in different parts of plates. Current evidence suggests that this zonation of metabolic function results mainly from local tissue microenvironmental conditions, which are determined by a complex combination of factors, including tissue matrix, direction of blood flow and efficiency of solute extraction (Gebhardt, 1992; Jungermann, 1992).

EMBRYONIC AND POSTNATAL GROWTH OF THE LIVER

Embryonic liver growth

Both hepatocytes and bile duct epithelial cells arise embryologically from a common founder cell, the hepatoblast, which has bipotential differentiation capabilities during liver development (LeDouarin, 1975; Houssaint, 1980). Hepatoblasts in rodents are derived from about 50 endodermal cells that are probabilistically assigned to the hepatic diverticulum of the foregut (Rabes *et al.*, 1982; Wareham and Williams, 1986). Proliferation of the endodermal cells of the presumptive liver is clonal or quasiclonal, since livers of chimeric mice and rats are composed of patches of genetically similar cells, separated by patches of genetically contrasting cells (West, 1976; Iannaccone *et al.*, 1987; Kusakabe *et al.*, 1988; Ng and Iannaccone, 1992a; Khokha *et al.*, 1994). In livers of adult mosaic animals, patches contain 35–100 cells (West, 1976; Rossant *et al.*, 1983), suggesting that clonal or quasiclonal growth during hepatic organogenesis occurs after a period of extensive cell mixing during earlier stages of embryogenesis (Gardner, 1985). Analysis of patch size and outline during rapid growth of the liver indicates that patches are formed by the quasiclonal iterative proliferation of cells, coupled with the random positioning of progeny in a fractal growth pattern (Iannaccone, 1990; Ng and Iannaccone, 1992b; Khokha *et al.*, 1994). Rather than continuous enlargement expected of clonal growth, the fractal growth of patches is characterized by oscillatory expansion and contraction, with the average patch size and complexity of outline remaining nearly constant even during periods of rapid hepatocyte proliferation and liver growth (Iannaccone, 1990; Ng and Iannaccone, 1992b; Khokha *et al.*, 1994). Oscillatory expansion and contraction of patches during fractal growth is explained by the probabilistic cycling of cells composing a patch, and by the random (non-directional) positioning of progeny cells in relation to the parents (Iannaccone, 1990; Ng and Iannaccone, 1992b; Khokha *et al.*, 1994). These characteristics allow individual patches to enlarge transiently when progeny cells occupy positions within the patch. At the interfaces between genetically distinct patches, random placement of progeny cells may result in their insertion into a contrasting patch, which may cause it to fragment. Alternatively, two genetically identical patches may coalesce. Since the growth centres are quasiclonal, this pattern of fractal growth is not compatible with centres of growth that place progeny directionally, which would be necessary to cause the streaming of presumptive hepatocytes from the hepatic diverticulum into the septum transversum in the form of coherent cords or plates. The development of the plate-like structure of the hepatic parenchyma appears to result secondarily from the cleavage of the hepatoblastic mass by capillary (sinusoidal) endothelium (Martinez-Hernandez and Amenta, 1993b; Stamatoglou and Hughes, 1994).

Hepatoblasts and differentiating hepatocytes increase rapidly in number during liver organogenesis, doubling approximately eight times between embryonic day 12 and 18 (Greengard *et al.*, 1972; Vassy *et al.*, 1988). Few studies have apparently been performed on the measurement of cell cycling by hepatoblasts or differentiating hepatocytes and bile duct epithelial cells of the embryonic liver. Between E12 and E18 in rats a decline from about 80% to about 20% in the fraction of hepatic cells labelled with a pulse dose of ^{3}H-thymidine has been described (Wright and Alison, 1984), but the details of this study are not available. On postnatal day 1 (P1) the fraction of hepatocytes labelled with a pulse of ^{3}H-thymidine is 5–25% (Stöcker and Butter, 1968; LeBouton and Marchand, 1970; Viola-Magni, 1972; Wright and Alison, 1984).

Postnatal liver growth

The kinetics of hepatocytic proliferation during postnatal growth and following induced cell loss (by partial hepatic resection and by toxic necrosis of cells) have been comprehensively reviewed by Wright and Alison (1984) and Alison (1986). Cell proliferation in the liver postnatally mirrors liver growth. A major feature of postnatal liver growth is the marked slowing of the proliferation of hepatocytes; in rats, the high pulse labelling rate of 5–25% at birth declines steadily to about 0.1% in the adult (Grisham, 1969; Wright and Alison, 1984), modulated only by a transient spurt of liver growth and hepatocyte proliferation that occurs about the time of weaning in rats (Grisham, 1969; Wright and Alison, 1984). Proliferation of bile duct epithelial cells appears to change similarly, although detailed studies are not available. The parenchymal localization of proliferating hepatocytes during the decelerating phase of liver growth postnatally and in adults is somewhat controversial. A shift in location of hepatocytes tagged with a pulse of ^{3}H-thymidine from a panlobular to a periportal-midlobular predominance has been found shortly after birth (LeBouton and Marchand, 1970), but the low rate of hepatocyte cycling makes it difficult to be statistically certain of the location of cycling cells in livers of adult animals (Grisham, 1962). Several authors have described S phase hepatocytes as being distributed throughout the lobular parenchyma with a slight (Messier and Leblond, 1960; Schultze and Oehlert, 1960; Edwards and Klein, 1961; Grisham, 1962; Bucher and Swaffield, 1964; Fabrikant, 1968a) or marked (Zajicek *et al.*, 1985; Arber *et al.*, 1988; Jezequel *et al.*, 1991) periportal predominance in livers of normal adult rats. The issue is of some importance, since it has implications for the 'streaming' liver theory (Zajicek *et al.*, 1985; Arber *et al.*, 1988; Sigal *et al.*, 1992; Zajicek, 1992), which holds that new hepatocytes arise from stem cells and are amplified in periportal zones to which cell formation is restricted, from which they migrate along hepatic plates toward the region of terminal hepatic veins.

Following 68% partial hepatic resection (partial hepatectomy) in the rat, hepatocyte proliferation is greatly accelerated for a relatively short period while the deficit in hepatocyte number is replaced (Grisham, 1962, 1969; Fabrikant, 1968b; Wright and Alison, 1984; Alison, 1986). Hepatocytes quickly begin to cycle at an increased rate, enter S-phase within 12–14 hours after surgery, reach a maximal rate of S-phase entry shortly thereafter, and continue to proliferate at an accelerated rate for the next several days until the liver mass is recuperated (Grisham, 1962; Fabrikant, 1968b). The first hepatocytes to begin S-phase are located in the periportal one-third of the parenchyma (Grisham, 1962; Fabrikant, 1968a; Rabes and Tuczek, 1970), because these cells have a shorter G_1 phase than do hepatocytes in other parts of the parenchyma (Rabes *et al.*, 1976). In order to replace the number of cells that are removed by 68% partial hepatic resection, the 'average' residual hepatocyte must cycle about 1.7 times during the period of accelerated growth that follows surgery, and virtually all hepatocytes proliferate at least once irrespective of their parenchymal location (Fabrikant, 1969; Stöcker and Heine, 1971). Many of the early replicating hepatocytes cycle at least twice in rapid succession, as shown by double labelling with ^{3}H- and ^{14}C-thymidine during the S-phases of two consecutive cell cycles (Grisham, 1969). Activation of stem cells to proliferate does not occur in livers in which new hepatocytes are being generated after 68% partial hepatectomy; newly formed hepatocytes arise from differentiated hepatocytes that proliferated (and were labelled) following a previous partial hepatic resection (Klinman and Erslev, 1963). In rats that are subjected to 5 consecutive partial hepatic resections at regular intervals, the 'average' residual hepatocyte is calculated to proliferate from 8 to 12 times during the prolonged process of residual liver growth (Simpson and Finckh, 1963). In a transgenic mouse in which the urokinase gene is expressed in liver under the control of an albumin promoter-enhancer, most hepatocytes are killed by the expressed gene product (Sandgren *et al.*, 1991). Some residual hepatocytes in this transgenic mouse are able to inactivate the toxic transgene and these cells proliferate consecutively for up to about 10–12 cycles to yield discrete nodular aggregates (clones) that repopulate the liver parenchyma (Rhim *et al.*, 1994). Transplanted mouse hepatocytes isolated from congenic animals whose cells are genetically tagged by the expression of bacterial β-galactosidase (encoded by a transfected *Escherischia coli* lac Z gene) can also replace the destroyed liver parenchyma of the albumin-urokinase transgenic mouse by clonal proliferation to form coherent nodules that coalesce (Rhim *et al.*, 1994). Based on the sizes of the resulting clonal nodules, the transplanted hepatocytes also appear to proliferate up to about 12 times. In each of these situations replacement appears to occur from the residual hepatocytes, and the ability of residual hepatocytes to proliferate does not appear to pose a limit to the growth and replacement of the liver parenchyma under the conditions employed.

Bile duct epithelial cells also undergo a self-limited burst of proliferation after partial hepatectomy, which follows the proliferation of hepatocytes by about 18–24 hours and is similar in magnitude (Grisham, 1962; Wright and Alison, 1984; Marucci *et al.*, 1993; Polimeno *et al.*, 1995). When biliary outflow is blocked, bile ducts proliferate for a prolonged period as the obstructed ducts enlarge (Gall and Bhathal, 1990a; Slott *et al.*, 1990; Marucci *et al.*, 1993; Polimeno *et al.*, 1995). As with hepatocytes following partial hepatectomy new bile duct epithelial cells appear to arise from the proliferation of pre-existing bile duct epithelial cells under these conditions (Grisham, 1962; Gall and Bhathal, 1990a; Slott *et al.*, 1990; Polimeno *et al.*, 1995).

Growth of liver parenchyma postnatally is also quasiclonal or fractal, as in the embryo. In adult chimeric rats and mice, genetically mosaic patches in the liver maintain their sizes and complexities of outline during growth, including the transient rapid growth that occurs after 68% partial hepatic resection (Ng and Iannaccone, 1992b). A clonal pattern of hepatocytic proliferation is also shown in studies in which hepatocytes proliferating *in vivo* after partial hepatectomy are tagged *in vivo* by infection with a replication-defective retroviral vector containing the *E. coli* lac Z (β-galactosidase) gene; progeny of the originally tagged hepatocyte are located in a group of adjacent tagged hepatocytes (Bralet *et al.*, 1994; Kennedy *et al.*, 1995). Although each originally tagged hepatocyte necessarily is located in a single hepatic plate, with time progeny of the tagged hepatocyte are distributed locally among several adjacent plates (Bralet *et al.*, 1994). This pattern results from the remodelling of the multicell-thick hepatic plates, produced by hepatocyte replication, by insertion of sinusoids between adjacent hepatocytes to cleave one-cell thick plates from micronodular aggregates (Martinez-Hernandez *et al.*, 1991; Martinez-Hernandez and Amenta, 1993b; Stamatoglou and Hughes, 1994). The occurrence of clonal or quasiclonal proliferation of hepatocytes is also supported by the pattern of hepatocyte repopulation in the albumin-urokinase transgenic mice in which residual hepatocytes proliferate clonally to form nodular aggregates (Rhim *et al.*, 1994), and by the clonal patterns of growth of preneoplastic lesions, adenomas and carcinomas of rodents undergoing hepatocarcinogenesis regimens (Rabes *et al.*, 1982; Howell *et al.*, 1985; Weinberg and Iannaccone, 1988; Lee *et al.*, 1991a). Even as patches do not change greatly in size during periods of rapid growth, neither does the average distance between portal tracts and hepatic veins, which reflects the sizes of hepatic lobules or acini and the lengths of hepatic plates. In fact, the distances between adjacent terminal portal and hepatic veins appear to oscillate following multiple partial hepatectomies, while the median measurements remain more-or-less stable (Simpson and Finckh, 1963). This evidence suggests that the sizes of lobules, acini, and hepatic plates are also determined by a fractal process; in this instance, the terminal portal and hepatic veins branch (McKellar, 1949) by a process that may be fractal to maintain the distance between the tips of these terminal afferent and efferent vessels.

HEPATIC PLATES AS STEM CELL-FED HEPATOCYTE LINEAGES

In contrast to the quasiclonal pattern of liver growth that is indicated by studies of genetically distinct patches in chimeric rats and mice and by the location of progeny of retrovirally tagged proliferating hepatocytes, other investigators hypothesize that the populations of hepatocytes and bile duct epithelial cells in livers of adult rats represent lineages derived from common, bipotential stem cells located in or near portal tracts (Zajicek *et al.*, 1985; Arber *et al.*, 1988; Arber and Zajicek, 1990; Brill *et al.*, 1993). Periportal hepatocytes newly formed from stem cells are hypothesized to undergo amplification in number in a restricted periportal zone of proliferation from which they migrate ('stream') slowly along the lengths of hepatic plates from the regions of portal tracts to the regions of hepatic veins (Zajicek *et al.*, 1985; Arber *et al.*, 1988; Brill *et al.*, 1993). During their hypothetical directional migration, hepatocytes are also posited to differentiate and mature (Zajicek *et al.*, 1985; Arber *et al.*, 1988; Zajicek, 1992; Brill *et al.*, 1993); age-dependent differentiation (rather than the microenvironment) is hypothesized to underlie the observed functional heterogeneity of hepatocytes located at different distances from terminal portal and hepatic veins (Brill *et al.*, 1993).

The proliferation kinetics and the dynamics of hepatocyte populations during liver growth and repair, as well as the pattern of liver growth in genetically mosaic, chimeric animals are incompatible with the major features of the 'streaming liver' hypothesis (Bralet *et al.*, 1994; Grisham, 1994; Kennedy *et al.*, 1995). Proliferative units in the liver in genetically mosaic, chimeric animals do not conform to any of the obvious hepatic landmarks, including portal tracts, lobules or acini and, of most importance to the 'streaming liver' hypothesis, hepatic plates (West, 1976; Iannaccone, 1987; Kusakabe *et al.*, 1988; Iannaccone, 1990; Ng and Iannaccone, 1992a, 1992b). Patches of genetically distinct hepatocytes are located randomly in the parenchyma, unrelated to the conventional landmarks. Each patch may be considered as a randomly sited growth centre, and each oscillates in size in a quasiclonal fractal pattern; hepatocytes within a patch are assigned to cycle probabilistically and the progeny cells are sited randomly (non-directionally) in relation to the parents (Iannaccone *et al.*, 1987; Iannaccone, 1990; Ng and Iannaccone, 1992a; Khokha *et al.*, 1994). Since patches are randomly located in the hepatic parenchyma, individual hepatic plates can be composed of two genetically distinct types of hepatocytes in livers of chimeric animals. If each plate were to represent a lineage derived from a single stem cell it would necessarily consist entirely of genetically identical hepatocytes, as do lineage columns of enterocytes migrating along the lengths of intestinal villi (Gordon *et al.*, 1992). Migration of hepatocytes along plates would require their formation periportally, as well as their regular placement and that of their progeny directionally along individual plates. This pattern occurs during neither normal embryogenesis and postnatal growth of the liver, nor during pathological growth responses exemplified by development of hepatocellular carcinoma (Lee *et al.*, 1991a).

Formation of new hepatocytes is not limited to a periportal zone of restricted proliferation, either during normal liver growth postnatally or during replacement growth following partial hepatectomy (Grisham, 1962; Fabrikant, 1968a; Rabes *et al.*, 1970, 1976). Furthermore, when hepatocytes located in selected parts of the parenchyma are killed by toxic chemicals, formation of new hepatocytes typically occurs by the proliferation of residual hepatocytes located adjacent to the areas of necrosis (Nostrant *et al.*, 1978). The appearance of a periportal zone of hepatocyte proliferation after partial hepatic resection results from a shorter G_1 phase in periportal hepatocytes as compared to hepatocytes in other parts of the parenchyma, which allows periportal hepatocytes to reach S-phase and (^{3}H-thymidine labelling) first (Rabes *et al.*, 1976), although virtually all hepatocytes eventually cycle at least once after partial hepatectomy (Fabrikant, 1969; Stöcker and Heine, 1971). A similar mechanism may underlie the predominance of replicating hepatocytes in periportal-midlobular areas of parenchyma during normal postnatal growth.

The appearance of time-dependent movement of ^{3}H-thymidine-tagged hepatocytes in relation to portal tracts and hepatic veins has been described many times (Grisham, 1962; Scherer and Friedrich-Freksa, 1970; Blikkendaal-Lieftinck *et al.*, 1977; Zajicek *et al.*, 1985; Arber *et al.*, 1988; Geisler *et al.*, 1994). The false appearance of directional movement of ^{3}H-thymidine-tagged hepatocytes from periportal locations can be caused by the labelling of late replicating cells, which are located closer to hepatic veins than are early replicating cells (Grisham, 1962; Fabrikant, 1968a; Rabes *et al.*, 1970), by reutilization of radiolabelled thymidine released into the bloodstream from the effete cells of rapidly turning over tissues (such as bone marrow and intestinal epithelium) (Bryant, 1962; Heiniger *et al.*, 1971). Radiolabelled nucleosides that may be incorporated into replicating DNA of S-phase cells (including hepatocytes) are present in the bloodstream for several hours after a 'pulse' dose of ^{3}H-thymidine (Heiniger *et al.*, 1971). Actual translocation of hepatocytes relative to portal tracts and hepatic veins may also occur by the displacement of residual and newly formed hepatocytes into focal areas of necrosis (Nostrant *et al.*, 1978; Rajvanshi and Gupta, 1994) and by the remodelling of lobules (and hepatic plates) by growth of the tips of terminal hepatic and portal veins (McKellar, 1949). However, a regular or continuous migration of hepatocytes along hepatic plates from periportal to perihepatic areas does not appear to occur.

EVIDENCE FOR LIVER STEM CELLS

The brief summary of proliferation kinetics and population dynamics during embryonic and postnatal growth of the liver indicates that the organization of hepatocytes into plate-like structures does not represent the anatomic location of differentiating lineages during either embryogenesis or normal postnatal growth of the liver. Furthermore, epithelial stem cells do not appear to be involved in normal liver growth postnatally. The evidence for the presence of inactive epithelial stem cells in the livers of normal animals during embryogenesis and postnatal life and for the participation of such cells in liver development, repair and carcinogenesis is given in this section. But first, a brief discussion of the availability and use of markers of liver epithelial cell differentiation in the tracing of lineages, and the use of cell-tags in determining the fate of cells is presented here, since much of the evidence for stem cells depends on the application and interpretation of such techniques.

Markers of differentiation and lineage of liver epithelial cells

Differentiation of hepatoblasts into hepatocytes or bile duct epithelial cells is correlated with the expression of structural and functional properties that distinguish the differentiating (differentiated) cell from its precursor. These distinguishing properties include the expression and specific cellular location of a variety of proteins and of associated structural features that are characteristic of the differentiated cell. Differentiated hepatocytes are characterized by the expression of a unique combination of liver-enriched (but not liver-unique) transcription factors (representing mainly the HNF1, HNF3, HNF4, and

C/EBP families) which, along with ubiquitous transcription factors, bind to multiple sites on promoter and enhancer elements of target genes to regulate the expression of hepatocyte-specific proteins (Johnson, 1990; Lai and Darnell, 1991; DeSimone and Cortese, 1992; Crabtree *et al.*, 1992). HNF1α and β are distantly related to homeobox proteins (Courtois *et al.*, 1988; Frain *et al.*, 1989; Lichsteiner and Schibler, 1989; Bamhueter *et al.*, 1990; Kuo *et al.*, 1990a; Mendel *et al.*, 1991). HNF3α, β and γ belong to the 'forkhead' protein gene family (Lai *et al.*, 1990, 1991). HNF4 is a member of the nuclear steroid-thyroid receptor family and contains a zinc-finger binding domain (Sladek *et al.*, 1990). C/EBP is the original leucine zipper protein (Landschulz *et al.*, 1988), and other members of this family, including C/EBPβ and γ (Descombes *et al.*, 1990; Poli *et al.*, 1990) and DBP (Mueller *et al.*, 1990) are also important regulators of hepatocyte gene expression. Hepatocyte-specific proteins, whose synthesis is regulated by these transcription factors acting in combinations, include, among the secreted hepatic proteins, α-fetoprotein (transient expression during lineage establishment) and albumin, as well as a large number of intracellular enzymes involved in hepatic intermediary metabolism, such as tyrosine aminotransferase. Extracellular proteins (laminins, collagens, fibronectins, etc.) are also specifically related to differentiated liver cells through their binding to membrane proteins (integrins and others) expressed by the cells (Ekblom *et al.*, 1986; Reid *et al.*, 1992; Stamatoglou *et al.*, 1992; Martinez-Hernandez and Amenta, 1993a; Maher and Bissel, 1993; Piredale and Arthur, 1994). The surfaces of differentiated hepatocytes that are involved in specific processes (absorption or secretion) are polarized by the location of specific membrane proteins, which are associated with identifiable structural modifications (Bartles *et al.*, 1985; Moreau *et al.*, 1988). Together all of these features allow the fully differentiated cell to be distinguished with ease and certainty from other types of differentiated epithelial cells. Embryogenesis of the liver is characterized by the unfolding of a programme of differentiation of primitive epithelial cells, during which a sequential and progressive pattern of acquisition of individual features of differentiation ultimately eventuates in the formation of fully differentiated hepatocytes and biliary epithelial cells. Disclosure of a similar differentiation programme can assist in the identification of lineage development under pathological conditions *in vivo* or *in vitro*.

Based on the phenotypic properties of differentiated hepatocytes and bile duct epithelial cells, several of these characteristic phenotypic properties are used as markers of differentiation and lineage *in vivo* and *in vitro*. Markers based on the expression of specific molecules include the localization of specific mRNAs by *in situ* hybridization and the decoration of specific antigens with antibodies (monoclonal and polyclonal) raised against a pure antigen molecule. Antibodies to cytokeratins appear to be particularly useful in evaluating liver epithelial cell lineage and differentiation. All types of epithelial cells express combinations of between two and ten cytokeratins (Moll *et al.*, 1982; Osborn and Weber, 1982), assembled as heteropolymeric pairs of one acidic (type I) and one basic (type II) subunit to form 10 nm filaments (Steinert and Roop, 1988). The particular combination of cytokeratins expressed by an epithelial cell reflects that cell's state of differentiation and perhaps its lineage (Moll *et al.*, 1982; Osborn and Weber, 1982). All epithelial cells in livers of fetal and adult rats express combinations of cytokeratins that are equivalent to human CKs 8, 18, 7, and 19 (Moll *et al.*, 1982; Marceau, 1990; Shiojiri *et al.*, 1991) and (rarely) CK 14 (Blouin *et al.*, 1992). In addition to cytokeratins, antibodies have been raised to antigens that are parts of the surfaces of whole cells or are components of cellular substructures isolated from developing, normal adult and pathological livers. Cells to which such antibodies have been raised include normal and pathologically altered hepatocytes and bile duct epithelial cells, oval cells, tumour cells, cultured cells of various types, etc. (see references to Table 1). Although some of these antibodies decorate a limited range of cell types in fetal and adult livers, the specific antigens to which they are directed are not known in many instances. Nevertheless, when the results of their use are interpreted judiciously, and when they are applied in conjunction with markers of cell functions known to be specific for either hepatocytes or biliary epithelial cells, these antibodies to unknown cellular epitopes can be helpful in tracing the development of liver epithelial cell lineages. When applied in groups these markers can give a liver epithelial cell a morphological and functional gestalt that identifies it, and establishes its stage of development along a differentiating programme of lineage development. However, the identification of partially (incompletely) differentiated cells is not a simple matter when only a single or a few such markers of unknown identity and uncertain specificity are applied. It is also not certain that some markers of known specificity are lineage-specific under all circumstances, or whether their expression may be illegitimately induced by non-physiological environmental conditions. A list of antibodies commonly used as phenotypic markers of liver epithelial lineage and differentiation is provided in Table 1; the types of liver epithelial cells which these antibodies have been found to mark are also shown.

Table 1 Antibody markers commonly used to assess differentiation and to trace lineage of liver epithelial cells.

Markers	Hepatoblasts	Oval cells	Hepatocytes	Bile duct cells	References
CK7	−	−	−	+	1,2,3
CK8	+	+	+	+	1,2
CK18	+	+	+	+	1,2
CK19	+	+	−	+	1,2,4
CK14	[+]	[+]	−	−	2,5
ALB	+	+/−	+	−	1,2,4
AFP	+	+	−	−	1,2,4
GGT	+	+	−	+	1,2
OV-6	(+)	+	−	+	6
OV-1	(+)	+	−	+	6
BDS7	+	+	−	+	4,7
BD1	−	−	−	+	8
BPC5	+	−	−	−	9
HES6	−	−	+	−	4,7
OC.2	+	+	−	+	10,11,12
OC.3	+	+	−	+	10,11,12
H.1	−	−	+	−	10,11,12
H.2	+ (transient)	−	−	−	10,11,12
HBD.1	+	−	+	+	10,11,12
A6	+/−	+	−	+	13,14

1. Shiojiri *et al.*, 1991; 2. Marceau, 1990; 3.VanEyken and Desmet, 1993; 4. Germain *et al.*, 1988a; 5. Bisgaard *et al.*, 1993, 1994a, 1994b; 6. Dunsford and Sell, 1989; 7. Marceau *et al.*, 1986; 8. Yang *et al.*, 1993a; 9. Marceau *et al.*, 1992; 10. Hixson *et al.*, 1990; 11. Hixson and Allison, 1985; 12. Faris *et al.*, 1991; 13. Engelhardt *et al.*, 1990; 14. Engelhardt *et al.*, 1993.

Lineage markers may be combined with cell labelling to attempt to trace the fates of cells, including the direct transition of a cell that expresses certain phenotypic characteristics into a cell that expresses other phenotypic features to suggest the progression of a differentiating lineage. Tagging of replicating DNA with ^{3}H-thymidine has been most often used for this purpose but such an application poses many problems that complicate interpretation, including the potential reutilization of label, the loss (dilution) of label as the tagged cells proliferate, and the simultaneous tagging of both potential progenitor and progeny populations by the initial pulse of thymidine. However, carefully performed and conservatively interpreted studies may provide useful information. Genetic tagging of cells with a uniquely expressed marker has been usefully applied to the analysis of lineage development *in vivo* in bone marrow (Lemischka *et al.*, 1986; Jordan and Lemischka, 1990), but it has not yet been used to study lineage development in the liver.

Another strategy for identifying cells with stem/progenitor cell properties is to isolate them based on some selected characteristics, culture them, if possible, tag them with a reliable marker (such as a distinct genetic marker that is expressed by all progeny), and evaluate the ability of the tagged cells to differentiate after transplanting them into appropriate sites in host animals. This general strategy was first used to define the presence of bone marrow stem cells in the spleen colony forming assay (Till and McCulloch, 1961). Subsequently, this method has enabled the isolation, on the basis of the pattern of expression of cell surface markers, of a small fraction of bone marrow cells that is highly enriched in multipotent stem cells (Uchida *et al.*, 1994).

Founding of hepatocytes and bile duct epithelial cells from hepatoblasts

The hepatic diverticulum is first visible in the mouse embryo on E9.5 and in the rat embryo on E10–10.5, as a thickening of the epithelium of the ventral foregut adjacent to the developing heart and projecting into the loose mesenchyme of the septum transversum (LeDouarin, 1975; Houssaint, 1980; Cascio and Zaret, 1991). Multiple inductive influences of surrounding tissues, especially of the cardiac area and of the mesenchyme of the septum transversum, are required for the development of the liver (LeDouarin, 1975; Houssaint, 1980; Cascio and Zaret, 1991; DiPersio *et al.*, 1991). During most of the period of liver ontogenesis in mammals, a large fraction of the fetal liver is composed of actively proliferating and differentiating haematopoietic cells (Houssaint, 1981), intimately mixed with the proliferating and differentiating epithelial cells. The first functional evidence of the impending development of the liver epithelium is seen at E9 in mice and E10 in rats with the weak expression in the hepatic diverticulum of the ventral foregut of α-fetoprotein (Cascio and Zaret, 1991; DiPersio *et al.*, 1991; Shiojiri *et al.*, 1991), as well as of several liver-enriched transcription factors (Ang *et al.*, 1993; Monaghan *et al.*, 1993).

The temporal sequence of embryonic differentiation of hepatocytes and biliary epithelial cells is depicted schematically in Figure 2, as a framework for the data presented below. Hepatoblasts initially bear little structural or functional resemblance to mature hepatocytes or bile duct epithelial cells, both of which morphologically differentiate during the last half of fetal development and the first few postnatal days (Wood, 1965; Herzfeld *et al.*, 1973; Luzzato, 1981; Feracci *et al.*, 1987; Vassy *et al.*, 1988). Morphological

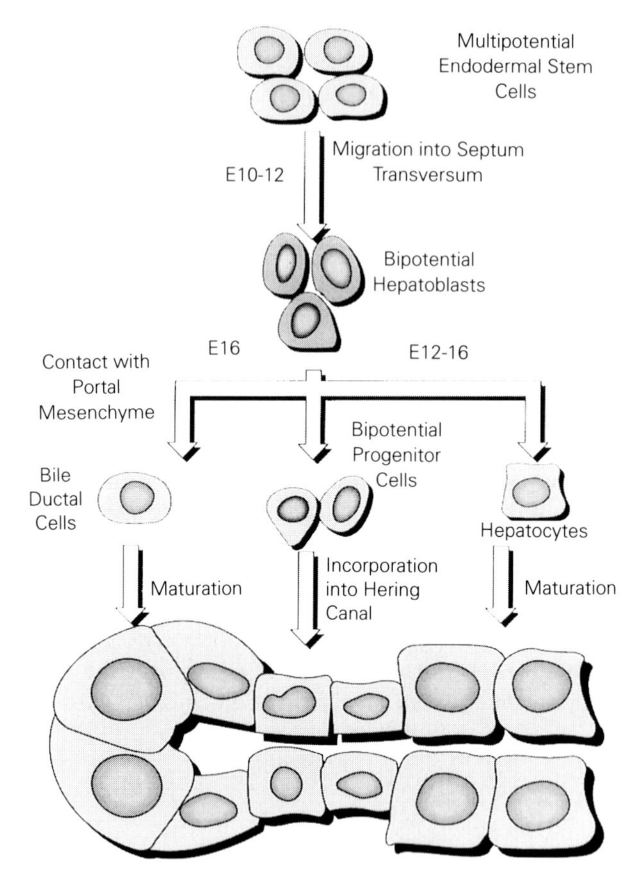

Figure 2 Schematic diagram showing the general temporal sequence of the differentiation of hepatocytes and biliary epithelial cells from multipotential endodermal cells of the hepatic diverticulum, through the development of hepatoblasts with bipotential differentiation options. Hepatoblasts throughout the mass of cells located in the septum transversum begin to differentiate into hepatocytes in an apparently stochastic fashion, while only those hepatoblasts that contact the mesenchyme of portal tracts differentiate into bile ducts. The drawing includes the hypothetical inclusion in the canal of Hering (bile ductule), which joins bile canaliculi and bile ducts, of incompletely differentiated cells that retain bipotential progenitor capablities to form both hepatocytes and biliary epithelial cells.

and functional changes in cells mirror the development of the extracellular matrix and adhesion molecules (Odin and Öbrink, 1988; Stamatoglou *et al.*, 1992; Martinez-Hernandez *et al.*, 1993b). Expression of the hepatocyte-enriched transcription factors are among the first evidences of hepatoblastic and hepatocellular differentiation (for reviews see Lai and Darnell, 1991; Kuo and Crabtree, 1992; Lai, 1992; Xanthopoulos and Mirkovitch, 1993; Zaret, 1993). HNF3β and α are highly expressed in the invaginating foregut at E8, and their expression becomes most intense at the site of the hepatic diverticulum at about E9 to E9.5 in mice (Ang *et al.*, 1993; Monaghan *et al.*, 1993). Heavy expression of HNF3 continues in the proliferating cells of the hepatic diverticulum, increasing further in intensity at about E10.5 to E11, and remaining high during the remainder of the fetal period, before declining (Ang *et al.*, 1993). Expression of HNF3α and β, the primordial endodermal transcription factors, appears to be essential, but not sufficient, for the differentiation of hepatocytes. Expression of HNF4 (Sladek *et al.*, 1990) and HNF1 (Courtois *et al.*, 1988; Frain *et al.*, 1989; Lichsteiner and Schibler, 1989; Bamheuter *et al.*, 1990; Kuo *et al.*, 1990a; Mendel *et al.*, 1991) is necessary for synthesis of significant levels of hepatocyte-specific proteins. These transcription factors are first detectable at low levels in the hepatic diverticulum of mice between E10 to E10.5, peaking around E12 to E16 (Blumenfeld *et al.*, 1991; Ott *et al.*, 1991; Duncan *et al.*, 1994). Sequential activation of expression of HNF4 and HNF1α appears to be part of a regulatory cascade (Tian and Schibler, 1991; Kuo *et al.*, 1992), which is initiated by an unidentified regulatory gene locus. In this hypothetical cascade HNF4 functions as a positive regulator of HNF1α, while expression of HNF3 and AP-1 are permissive and confer tissue specificity (Kuo *et al.*, 1992). Although insufficient for optimal hepatic function, expression of both HNF3 and AP-1 appears to be essential for the development and differentiation of the liver. The liver does not develop in mouse embryos in which the gene for c-jun, a component of AP-1, is genetically deleted (Hilberg *et al.*, 1993). By E16 in rats, hepatocytes express high steady-state levels of mRNAs for HNF1α, HNF3α, and HNF4, which gradually decrease during the remainder of liver development, while levels of mRNAs for HNF1β, HNF3β, C/EBPβ and DBP all increase, associated with increased expression of albumin (Nagy *et al.*, 1994). C/EBPα and β, and DBP, first expressed during the late fetal stages of hepatocyte differentiation (Kuo *et al.*, 1990b), do not reach peak expression until cell proliferation and liver growth decelerate postnatally (Diehl *et al.*, 1994), and their high postnatal expression is correlated with the onset of expression of tyrosine aminotransferase, and other enzymes of hepatic intermediary metabolism (Greengard, 1969). C/EBPα and β proteins are readily detectable in rat hepatocytes by P6, their levels escalating 10- to 22-fold by P35 (Diehl *et al.*, 1994).

Embryonic hepatoblasts in the rat express CK8 and CK18, both of which continue to be expressed in differentiating and differentiated hepatocytes (Moll *et al.*, 1982; Germain *et al.*, 1988a; Marceau, 1990; Shiojiri *et al.*, 1991). In addition to α-fetoprotein, the earliest affinities for antibody markers that are expressed in hepatoblasts committed to hepatocytic differentiation are HBD.1 and OC.2 (Shiojiri *et al.*, 1991). Further acquisition of hepatocytic differentiation is associated with rapid loss of OC.2 reactivity, transient acquisition of H.2 reactivity, and gradual loss of expression of the 2.1 kb mRNA for α-fetoprotein (Shiojiri *et al.*, 1991).

Differentiation of bile duct epithelial cells and formation of bile ducts is also a gradual and continuous process that begins in rats on about E15–E15.5, and continues postnatally. Little is known about specific changes in expression of transcription factors or synthesis of specific proteins during the differentiation of bile duct epithelium. Bile duct formation is restricted to two rows of hepatoblasts (the ductal plate) that are located immediately adjacent to the presumptive portal tracts (VanEyken *et al.*, 1988; Shiojiri *et al.*, 1991). Biliary differentiation in hepatoblasts that touch portal mesenchyme in rats is heralded by the expression of reactivity for BD.1 antibody, following which the ductal plate then develops into a series of tubules that bind BD.1, OC.2, HBD.1 and CK19; biliary duct epithelium gradually expresses CK7 as ducts develop further (VanEyken *et al.*, 1988; Shiojiri *et al.*, 1991; Hixson *et al.*, 1992). The smallest radicles of the intrahepatic bile duct system, the ducts of Hering, are posited by Hixson *et al.* (1992) to develop in a transitional zone containing BD.1-negative, OC.2- and HBD.1-positive cells that are ambiguously committed to either hepatocytic or biliary epithelial differentiation; Hixson *et al.* (1992) posit that these cells represent the most 'indifferent' population of epithelial cells in the adult liver.

Confirmation of the differentiation of bile duct epithelial cells (and ducts) from hepatoblasts, and the timing of differentiation, has been obtained by the transplantation of embryonic rat and mouse liver into other sites *in vivo*. When transplanted into the testis, E13 hepatoblasts located adjacent to connective tissue differentiate into epithelial cells that form bile ducts, while hepatoblasts that do not touch connective tissue form hepatocytes (Shiojiri, 1984; Shiojiri *et al.*, 1991). The capability of hepatoblasts from embryonic livers to undergo bipolar differentiation in tissue culture under the stimulus of selected differentiating agents has also been shown. There is little morphological evidence of differentiation along either hepatocytic or bile ductular lines in fragments of E12.5 mouse liver maintained in organ culture for 6 days unless dexamethasone is present in the medium, which stimulates development of both lineages; bile duct development is especially prominent when explants are cultured on Matrigel in the presence of dexamethasone (Shiojiri and Mizuno, 1993). Hepatocytes differentiate morphologically in fragments of E13, E16, or E19 liver maintained in organ culture for 6 days, with or without dexamethasone, but bile ducts form in only a small fraction of cultures (Gall and Bhathal, 1990b). Supplementation of culture medium with serum, insulin, dexamethasone, and dimethyl sulfoxide, induces epithelial cells isolated from E12 rat liver to express markers typical of hepatocytes, while exposure of similar cultures to sodium butyrate causes the cultured epithelial cells to express markers of bile duct epithelial cells (Germain *et al.*, 1988a). Using this same technique to estimate the fraction of liver epithelial cells that are able to differentiate into either hepatocytes or bile duct epithelial cells indicates that cells capable of bipolar differentiation decrease from 70% at E12, to 20% at E18, and to 5% at P5 (Germain *et al.*, 1988a). Based on these observations and on the paucity of cells that react with the monoclonal antibody BPC5, a putative marker of bipotential liver progenitor cells, Marceau *et al.* (1992) suggested that the livers of adult rats may lack bipolar progenitor cells. Nevertheless, other evidence indicates that hepatocyte and bile duct epithelial progenitor cells (stem-like epithelial cells) are present in the livers of adult rats and mice. Lemire and Fausto (1991) identified a few cells that express 2.1 kb mRNA for α-fetoprotein, a marker of primitive liver epithelial cells, in bile ductules of adult rats and as rare small cells in the lobular parenchyma. Furthermore, small epithelial cells that share some phenotypic properties with both hepatocytes and bile duct epithelial cells may be isolated from normal livers of adult rats and maintained for prolonged periods in culture as diploid cells. As discussed subsequently, these small liver epithelial cells possess some of the characteristics of stem cells, including the ability to differentiate into hepatocytes when they are transplanted into livers of syngeneic animals. Furthermore, cells with stem-like properties (oval cells) expand *in vivo* under certain pathological circumstances and some oval cells are able to differentiate into hepatocytes and other types of differentiated cells *in vivo*.

Re-establishment of hepatocyte lineages *in vivo* from stem-like (oval) cells

Rats have been used most extensively to generate experimental models with which to examine the development and outcome of oval cell proliferation, although a model

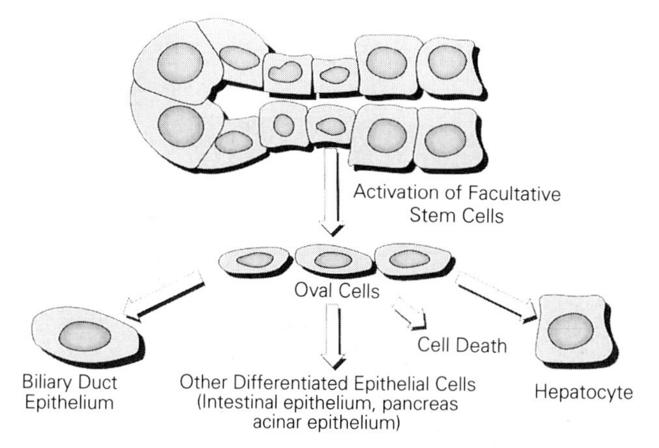

Figure 3 Schematic diagram illustrating the hypothesis that oval cells develop from facultative stem cells (bipotential progenitor cells) located in the canal of Hering (biliary ductule). The drawing shows the developmental options which oval cells possess, including differentiation into hepatocytes and biliary epithelium, formation of intestinal epithelium and pancreatic acinar epithelium by differentiation and/or metaplasia, and death.

employing Dipin-treated mice is also available (Factor *et al.*, 1994). The most commonly used models in rats are produced by: (a) treatment with azodyes (AZD) (Inaoka, 1967); (b) feeding of a choline-deficient diet, with or without supplements of ethionine (CDE) (Shinozuka *et al.*, 1978) or 2-acetylaminofluorene (CDA) (Sell *et al.*, 1981a), (c) treatment with 2-acetylaminofluorene and partial hepatectomy (AAF/PH) (Tatematsu *et al.*, 1984), and (d) treatment with D-galactosamine (GALN) (Lesch *et al.*, 1970). Central to all of these experimental models is the extensive destruction or compromised function of hepatocytes, coupled with the apparent inability of residual hepatocytes to proliferate. The common cellular response to the implementation of any of these experimental regimens is the proliferation of small cells with scant cytoplasm and ovoid nuclei that have been termed oval cells (Farber, 1956). A schematic outline of the emergence and evolution of oval cell populations is shown in Figure 3 as a guide for considering the data that follow.

Oval cells originate in portal zones in the regions of terminal bile ductules and then quickly emerge from portal regions and invade the entire lobular parenchyma (Figure 4). As they migrate through the parenchyma, oval cells proliferate rapidly, with labelling rates after pulse doses of ^{3}H-thymidine of 5–20% (Sell *et al.*, 1981b; Lemire *et al.*, 1991) during peak proliferation. As oval cells proliferate, individual and small groups of small, highly basophilic hepatocytes typically appear among them, after which the oval cells disappear and the parenchyma is gradually reconstructed. The timing of these events differs in the various models in rats. Appearance, evolution and resolution of oval cells is most rapid in GALN-treated rats, being largely completed within 8–10 days (Lesch *et al.*, 1970; Lemire *et al.*, 1991; Dabeva and Shafritz, 1993); in the AAF/PH model, the process is completed within 16–18 days (Evarts *et al.*, 1987, 1989), while both the AZD (Inaoka, 1967; Ogawa *et al.*, 1974; Dempo *et al.*, 1975) and CDE (Shinozuka *et al.*, 1978; Lenzi *et al.*, 1992) models produce more chronic reactions that evolve over several weeks.

Proliferating oval cells comprise a heterogeneous, expanding population of cells,

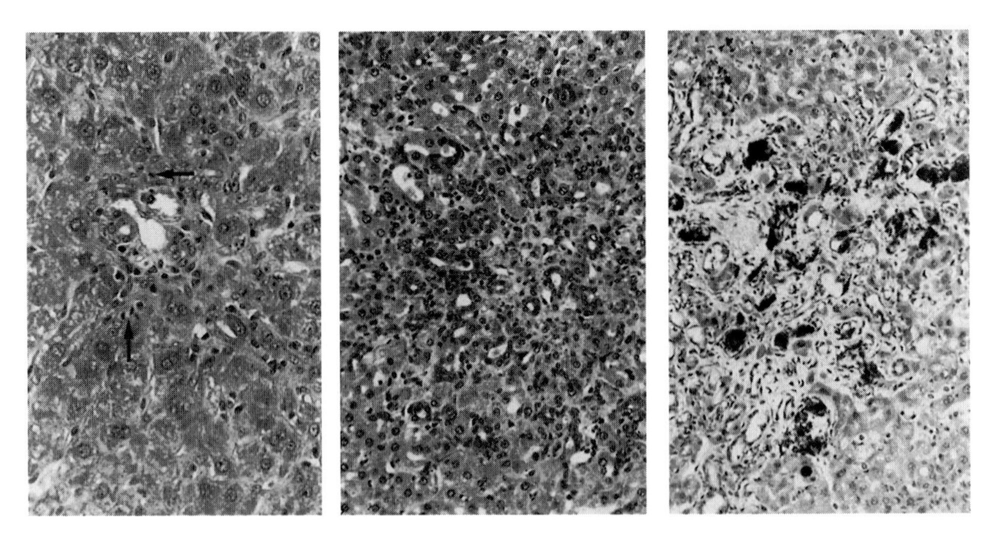

Figure 4 Development of oval cells in the liver of rats (AAF/PH model) shown in light micrographs. In the left panel, 5 days after PH, portal areas are hypercellular and some cells are proliferating (mitotic figure, arrow). By 8 days, middle panel, small oval cells forming duct-like structures penetrate into the parenchyma around a portal tract. These newly formed oval cell ductules contain pigmented gelatin (dark material) after injection of gelatin into the common bile duct (right panel), showing that the oval cell ductules communicate with bile ducts. Original magnifications, left ×400, middle ×200, right ×200. Reproduced from Sarraf *et al.* (1994) with permission.

which Fausto *et al.* (1992) termed an oval cell 'compartment', because of its cellular heterogenicity. Ultrastructurally individual oval cells closely resemble cells that form terminal bile ductules (Grisham and Hartroft, 1961; Lenzi *et al.*, 1992; Sarraf *et al.*, 1994), and, with other oval cells, they form irregular duct-like structures that enclose lumens (Figure 5) which connect to adjacent, pre-existing bile ducts (Dunsford *et al.*, 1985; Makino *et al.*, 1988; Sarraf *et al.*, 1994). In addition to the oval cells that ultrastructurally resemble the epithelium of small bile ductules, the oval cell 'compartment' contains transitional cells that are characterized by morphological features that are intermediate between oval cells and hepatocytes (Inaoka, 1967; Sarraf *et al.*, 1994) (Figure 5), as well as numerous mesenchymal cells (Popper *et al.*, 1957; Evarts *et al.*, 1993). Oval cells express some markers of normal intrahepatic biliary epithelium (Sirica *et al.*, 1990; Lemire *et al.*, 1991; Lenzi *et al.*, 1992), as well as phenotypic properties that distinguish them from biliary epithelial cells. A subset of oval cells expresses the 2.1 kb mRNA for α-fetoprotein (Lemire *et al.*, 1991; Fausto *et al.*, 1992), and reacts positively with antibodies to α-fetoprotein (Onoé *et al.*, 1973; Dempo *et al.*, 1975; Onda, 1976; Kuhlmann, 1978; Germain *et al.*, 1985; Marceau, 1990; Fausto *et al.*, 1992). Similarly, variable numbers of oval cells express albumin mRNA and bind antibodies to albumin (Dabeva and Shafritz, 1993), and some individual oval cells express both α-fetoprotein and albumin simultaneously (Alpini *et al.*, 1992; Dabeva and Shafritz, 1993). Other phenotypic properties expressed by some cells in oval cell populations include α_1-acid glycoprotein (Onda, 1976), both fetal and adult isozymes of aldolase and pyruvate kinase (Hayner *et al.*, 1984; Fausto *et al.*, 1992), peroxisomal catalase (Plenat *et al.*, 1988), glucose-6-phosphatase (Ogawa *et al.*, 1974; Yokoyama *et al.*, 1986; Plenat *et al.*, 1988; Steinberg *et al.*, 1991; Dabeva and Shafritz,

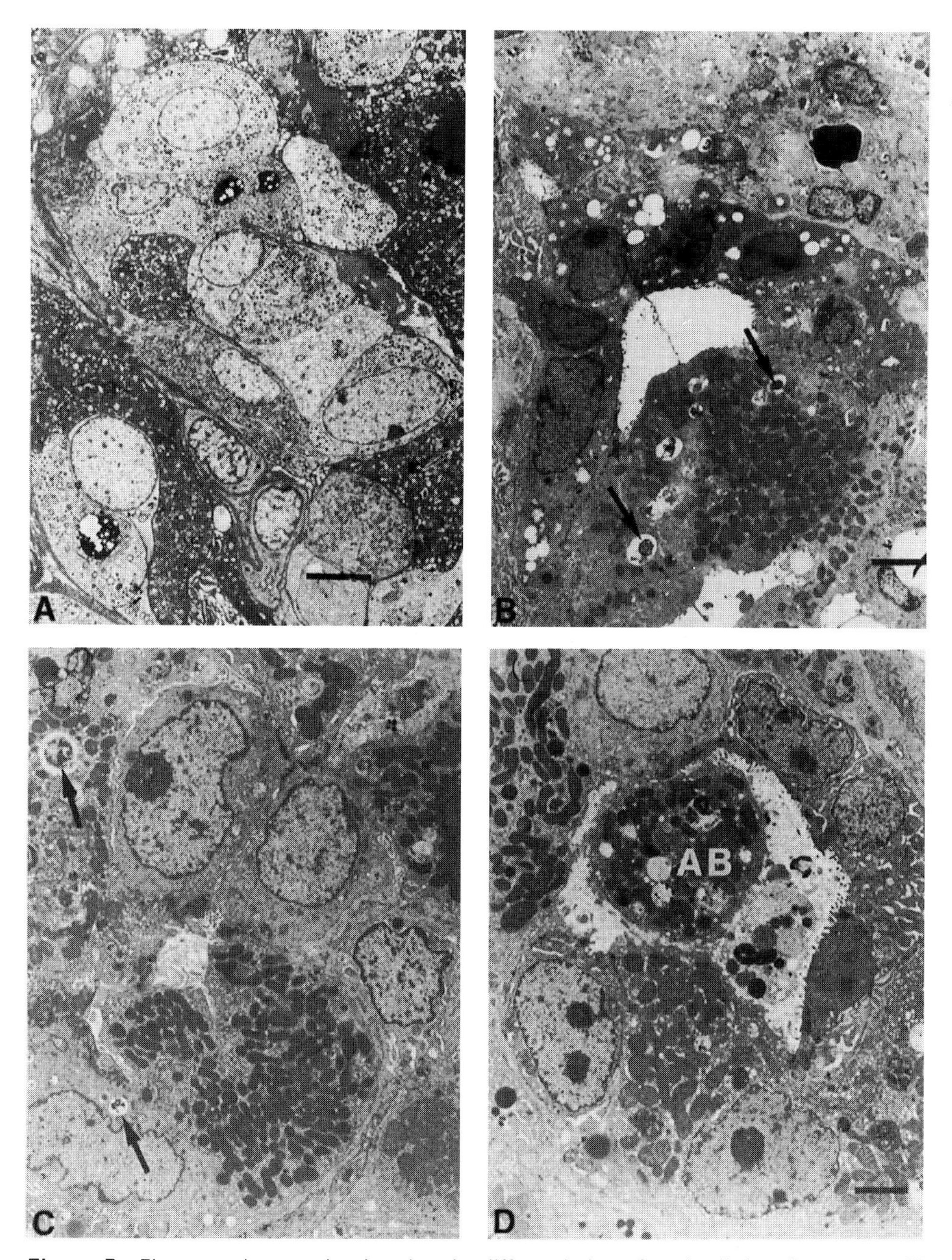

Figure 5 Electron micrographs showing the differentiation of oval cells into hepatocyte-like transitional cells at 7–10 days after PH in the AAF/PH model in rats. Panel A shows a tangentially sectioned longitudinal column of oval cells, which contain small mitochondria. Panels B, C, and D depict cross-sections of such columns, each of which contains small hepatocyte-like transitional cells (containing large mitochondria) in continuity with oval cells. Hepatocyte differentiation appears to be erratic and poorly organized. Many hypertrophied oval cells contain phagocytic vacuoles (arrows), suggesting enhanced organelle turnover in them. Apoptotic bodies (AB in panel D) are frequent. Scale bars: A.5.5μm, B.4.2 μm, C.3.1μm, D.3.1μm. Reproduced from Sarraf *et al.* (1994) with permission.

1993), and other enzymes of glycogen metabolism (Steinberg *et al.*, 1991). Oval cells also express the bile duct type cytokeratins, CK8, 18, and 19 (Germain *et al.*, 1985; Alpini *et al.*, 1992; Lemire *et al.*, 1991; Steinberg *et al.*, 1991; Marceau *et al.*, 1992; Lenzi *et al.*, 1992), γ-glutamyl transpeptidase (Sirica and Cihla, 1984; Yokoyama *et al.*, 1986; Sirica *et al.*, 1990; Lemire *et al.*, 1991; Steinberg *et al.*, 1991; Fausto *et al.*, 1992; Marceau *et al.*, 1992), and connexin 43 (Zhang and Thorgeirsson, 1994). Oval cells also may express several of the other antigenic markers that have been shown to react to either hepatocytes or biliary epithelial cells of adult rat livers (Table 1).

The hypothesis that oval cells may differentiate into hepatocytes is not new, having been proposed nearly 60 years ago by Kinosita (1937), and subsequently suggested as a possibility by other investigators (Price *et al.*, 1952; Farber, 1956; Wilson and Leduc, 1958). Wilson and Leduc postulated that 'cholangiole cells' of mouse liver proliferated to form oval cells which might differentiate to form both hepatocytes and biliary epithelial cells. An attempt to document the fate of oval cells was first made more than 30 years ago by Grisham and Porta (1964), who failed to find evidence for the direct transition of ^{3}H-thymidine-tagged oval cells into hepatocytes. Subsequently, Tatematsu *et al.* (1984) also could not demonstrate that ^{3}H-thymidine-tagged oval cells differentiate directly into hepatocytes, and Gerlyng *et al.* (1994) did not show the transition into hepatocytes of oval cells, tagged by incorporation of bromodeoxyuridine (BrdU) into DNA and subsequent reaction with an antibody to BrdU-containing DNA. In contrast, evidence that oval cells are precursors of hepatocytes, based on the transformation of ^{3}H-thymidine-tagged oval cells into hepatocytes, has been adduced by Onoé *et al.* (1973), Evarts *et al.* (1987, 1989), Lemire *et al.* (1991), and Dabeva and Shafritz (1993). The varying success in demonstrating the transfer of labelled DNA from oval cells to hepatocytes no doubt stems from the complex problems posed both by the heterogeneous cellular composition of the liver that is the site of oval cell proliferation and degeneration of residual hepatocytes, together with the kinetic complexities inherent in tracing tagged DNA, previously discussed. Additionally, defects in experimental design and limitations posed by the reagents that are used may have affected results of some studies. The experimental design of the early studies by Grisham and Porta (1964) could have prevented detection of a label-shift between oval cells and hepatocytes because of the long intervals between initial tagging and assessment of the transfer of tagged DNA. The tagging method used by Gerlyng *et al.* (1994) is particularly sensitive to label dilution during cell cycling; relative insensitivity of available anti-BrdU/DNA antibodies prevents detection of tagged progeny if potential progenitor or potential progeny cells have together proliferated more than about three times (Boswald *et al.*, 1990). Transitions between oval cells and hepatocytes could have been missed if either type of cell cycles many times during the process. To explain their observations that oval cells have a very brief life and that transfer of labelled DNA from oval cells to hepatocytes is not detectable, Rubin (1964) and Grisham and Porta (1964) suggested that many oval cells simply die and disappear, rather than differentiate. This prediction has been confirmed in the AAF/PH model in which substantial apoptotic death occurs in the ductal/oval cell population (Sarraf *et al.*, 1994).

Additional evidence for the differentiation of oval cells into hepatocytes comes from studies that demonstrate the unfolding of a hepatocytic differentiation programme in oval cells. Using the AAF/PH model, Evarts *et al.* (1989) combined the evaluation of oval cell fate, as assessed by tracing the morphological transformation of oval cells tagged with ^{3}H-thymidine, and the use of phenotypic markers to identify clearly oval cells and hepatocytes. The phenotypic properties expressed by oval cells gradually merge into properties that are typical of hepatocytes, and this change in differentiation is correlated with the transfer of label from oval cells to basophilic hepatocytes (Evarts *et al.*, 1989). Specifically, oval cells expressing several markers of fetal liver epithelial cells, including CK7 and CK19, reactivity to OV-6 antibody, α-fetoproten and, weakly, albumin, were heavily tagged with a pulse of ^{3}H-thymidine (Evarts *et al.*, 1989). Residual hepatocytes were not tagged by this pulse, but ^{3}H-thymidine-tagged small hepatocytes later appeared in basophilic foci; these ^{3}H-thymidine-tagged small hepatocytes expressed higher levels of albumin and much lower levels of α-fetoprotein than did oval cells and, as well, they expressed other markers of hepatocyte differentiation. Employing the GALN model, Lemire *et al.* (1991) also combined phenotypic marking of oval cells and hepatocytes with the assessment of time-dependent ^{3}H-thymidine labelling of both cell types. Shortly after administration of ^{3}H-thymidine, label was found in oval cells that expressed CK7 and CK19, γ-glutamyl transpeptidase, 2.1 kb mRNA for α-fetoprotein, and bound peanut agglutinin. ^{3}H-thymidine-labelled small hepatocytes expressing albumin and other hepatocytic markers appeared subsequently, in concert with the loss of the biliary epithelial phenotype (Lemire *et al.*, 1991). Dabeva and Shafritz (1993) analysed the simultaneous occurrence of a programme of hepatocyte differentiation in oval cells and the transfer of ^{3}H-thymidine between oval cells and hepatocytes in the GALN model in rats. Tagged oval cells initially expressed fetal α-fetoprotein mRNA and γ-glutamyl transpeptidase; during the next few days α-fetoprotein declined while the expression of albumin and glucose-6-phosphatase increased, as the tagged cells concurrently acquired the morphology of small hepatocytes. During late stages of recovery from GALN toxicity, morphologically and phenotypically identifiable hepatocytes also proliferated to augment the production of new hepatocytes. Dabeva and Shafritz (1993) concluded that in the GALN model the hepatocyte population is replaced by both differentiation of oval cells *and* by proliferation of the residual and newly formed hepatocytes.

The expression of hepatocyte-enriched transcription factors also develops sequentially in oval cells and basophilic hepatocytes during the peak of oval cell proliferation and formation of small foci of hepatocytes at 7–10 days after instituting the AAF/PH model (Nagy *et al.*, 1994). Measurement of steady-state levels of mRNA in the whole tissue shows that expression of C/EBPα and HNF4 does not vary greatly over the entire 16 days. In contrast, HNF3β and α reach maximal levels on day 9, and HNF3γ reaches a maximal level on about day 13, after which their expression gradually declines, while both C/EBPβ and DBP mRNAs reach their highest steady-state levels on day 16 (Nagy *et al.*, 1994). *In situ* hybridization shows that HNF1α and β, HNF3γ (the only HNF3 family member whose cellular localization was examined), and HNF4 are expressed at much higher levels by basophilic hepatocytes than by either oval cells or mature hepatocytes (Nagy *et al.*, 1994). Oval cells express HNF1α and β, and HNF3γ, more intensely than do mature hepatocytes, while HNF4 is most weakly expressed by oval cells. Biliary epithelial cells express HNF3γ at about the intensity of oval cells, but HNF1α and β are expressed weakly and HNF4 not at all by biliary epithelial cells. C/EBPα is expressed most intensely by oval cells and basophilic hepatocytes, while expression of C/EBPβ occurs in only a small subpopulation of oval cells that form ductular structures. DBP is expressed most intensely by oval cells, with lesser, but about equal, expression over both basophilic and mature hepatocytes (Nagy *et al.*, 1994).

With the exception of HNF4, oval cells express mRNAs for all of the liver-enriched transcription factors at higher intensity than do either basophilic or mature hepatocytes; the cellular expression and the pattern of the sequential changes in the expression of transcription factors in oval cells and hepatocytes indicates the unfolding of a hepatocytic differentiation programme in oval cells during the evolution of the complex cellular populations in the AAF/PH model (Nagy *et al.*, 1994). Most of the liver 'establishment' transcription factors including HNF1α and HNF3γ are upregulated in oval cells, as compared to mature hepatocytes, and high expression, or even further upregulation, of these liver-enriched transcription factors occurs in basophilic hepatocytes. The sudden upregulation of HNF4 in basophilic hepatocytes, from the basal level of expression in oval cells, may be the event that entrains the additional upregulation of the other transcription factors in an HNF4-regulated cascade (Tian and Schibler, 1991; Kuo *et al.*, 1992), and thus may play a triggering role in the development of hepatocyte lineages from oval cells. The formation of small foci of basophilic hepatocytes appears to represent a critical step in the differentiation of oval cells, possibly reflecting a point at which they are irreversibly committed to the hepatocyte lineage. Coincidental with the high levels of expression of liver-enriched transcription factors, albumin expression increases sharply in basophilic hepatocytes (Evarts *et al.*, 1987), expression of P-glycoprotein (Nakatsukasa *et al.*, 1992) and connexin 32 (Zhang and Thorgeirsson, 1994) (proteins produced only by mature hepatocytes) is initiated, and expression of α-fetoprotein (Evarts *et al.*, 1989; Bisgaard *et al.*, 1994a) and connexin 42 (Zhang and Thorgeirsson, 1994) decreases. The expression of a permissive combination of hepatocyte-enriched transcription factors (HNF3 and HNF1) appears to set the stage for the triggering of hepatocytic differentiation by the sudden upregulation of HNF4; expression of both HNF3 and HNF1 is known to be necessary but not sufficient for differentiation of hepatocytes during embryogenesis, which requires HNF4 expression also (Kuo *et al.*, 1992; Noda and Ichihara, 1993; Nagy *et al.*, 1994).

In this regard, the temporal unfolding of differentiation in embryonic hepatoblasts and in oval cells appears to differ slightly; during liver ontogenesis hepatoblasts seem to acquire hepatocytic differentiation gradually, whereas differentiation of oval cells into hepatocytes appears to occur in two discontinuous steps. In the first step putative stem cells are activated to give rise to oval cells which, when first identified, express an assortment of transcription factors that 'permit' the expression of hepatocyte-specific functions; the sudden upregulation of HNF4 is the critical second step that enables the optimal expression of the differentiated hepatocyte phenotype. A similar triggering (virtually all-or-none) role for HNF4 expression was also found by Griffo *et al.* (1993), who compared the extinction and re-expression of differentiated hepatocytic functions and simultaneous expression of transcription factors in rat hepatoma/human fibroblast hybrid cells. Expression of differentiated hepatocytic functions in rat hepatoma/human fibroblast cells that expressed HNF3, HNF1β, C/EBPα and β, and DBP *required* the simultaneous expression of HNF4 and HNF1α; since HNF1 could be induced by HNF4, HNF4 alone acted as a differentiation trigger in these cells, as it seems also to do in oval cells.

The proliferation of mesenchymal cells in concert with oval cells has long been known (Popper *et al.*, 1957). Evarts *et al.* (1993) showed that desmin-positive Ito cells represent an important category of mesenchymal cell intimately involved in the early stages of oval cell proliferation. The earliest cells that proliferate in the AAF/PH model are OV-6 antibody-reactive epithelial cells, as shown by combined ^{3}H-thymidine tagging and immunohistochemistry (Evarts *et al.*, 1993). Both OV-6-positive and desmin-positive cells are labelled with ^{3}H-thymidine within 4 hours after starting the AAF/PH regimen. ^{3}H-thymidine-tagged cells are identified as individual desmin-positive cells embedded in the portal connective tissue matrix, or as OV-6-positive cells in biliary ductules located in close proximity to branches of the portal vein (Plate 8.1). During the first 12 hours after PH, when increasing numbers of OV-6-positive ductular cells and desmin-positive Ito cells are labelled with ^{3}H-thymidine, cells in the large ducts in portal tracts are unlabelled. However, by 72 hours after PH, the majority of the biliary epithelial cells in portal tracts, including approximately 50% of the cells in large bile ducts are labelled, while hepatocytes remain unlabelled (Plate 1). In agreement with data obtained in other models that produce oval cell proliferation (Lesch *et al.*, 1970; Sell, 1990; Lenzi *et al.*, 1992), these observations indicate that both ductular and periductular cells proliferate coincidentally at early stages of oval cell accumulation.

These observations show that the earliest cycling epithelial cells that lead to oval cell proliferation are closely associated with mesenchymal cells, particularly Ito cells. Although the importance of epithelial cells located in terminal biliary ductules (possibly as stem cells) in generating oval cells is indicated by these studies, they do not exclude the possibility that stem cells may be located outside ductules, as proposed by Sell and Salman (1984). However, the early cycling periductal cells identified by Sell and Salman (1984) in electron microscopic autoradiographs may actually represent Ito cells. Interactions between stem cells/oval cells and Ito cells may be mediated by growth factors. Coincident with the initiation of DNA synthesis in OV-6-positive and desmin-positive cells in portal tracts, expression of transforming growth factor-α (TGF-α), hepatocyte growth factor (HGF) and α fibroblast growth factor (αFGF) is observed, and expression of transforming growth factor β1 (TGF-β1) begins within 24 hours (Evarts *et al.*, 1990; Nakatsukasa *et al.*, 1991; Marsden *et al.*, 1992; Hu *et al.*, 1993) (Figure 6a). This group of growth factors continues to be expressed at high levels throughout the period of expansion and differentiation of the oval cell population (Figure 6b). Transcripts for TGFα and αFGF are expressed by both oval cells and Ito cells (Evarts *et al.*, 1992; Marsden *et al.*, 1992), whereas transcripts for HGF are expressed only by Ito cells (Hu *et al.*, 1993). Transcripts for TGF-β1 are also highest in Ito cells, although the earliest population of oval cells also expresses lower levels of TGF-β1 mRNA (Nakatsukasa *et al.*, 1991). Oval cells express receptors for all of these growth factors (Lenzi *et al.*, 1992; Marsden *et al.*, 1992; Hu *et al.*, 1993), providing a molecular pathway by which Ito cells may influence the growth and development of oval cells. Production of matrix proteins by Ito cells may be another mechanism by which they interact with oval cells.

The stem cell factor/c-kit (SCF/c-kit) ligand/receptor system is also involved in the earliest stages of liver stem cell activation and oval cell proliferation (Fujio *et al.*, 1994). In the AAF/PH model, expression of SCF/c-kit occurs as early as does the expression of α-fetoprotein transcripts, and the levels of SCF/c-kit transcripts reach a peak and decline prior to that of the other growth factors (Figure 7). Individual oval cell precursors express both SCF and c-kit, providing the basis for autocrine stimulation. The SCF/c-kit signal transduction system is believed to play a fundamental role in the survival, proliferation and migration of stem cells during gametogenesis, melanogenesis and haematopoiesis (Morrison-Graham and Takahashi, 1993). In combination with selective multipotential colony stimulating factors, SCF can influence the relative frequency with

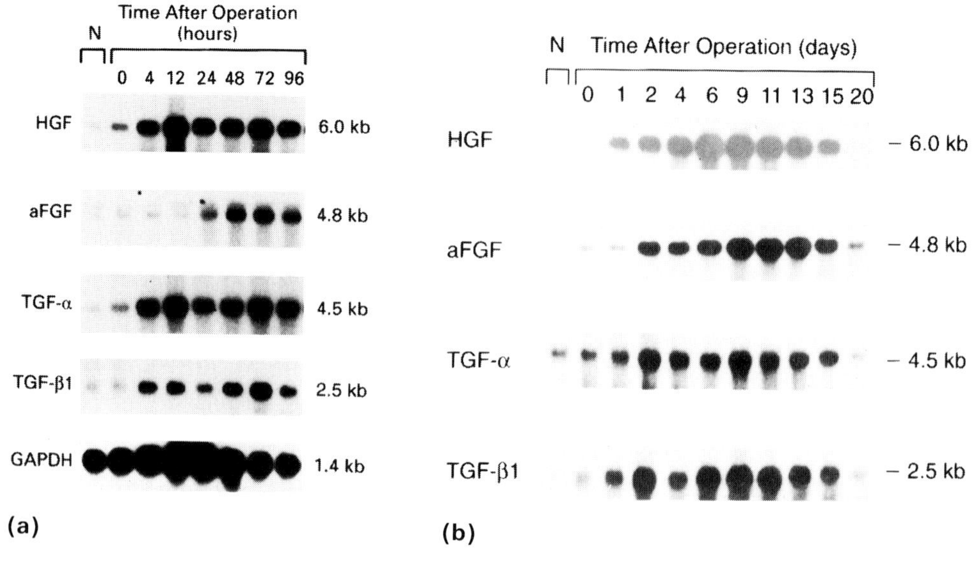

Figure 6 Analysis of expression of mRNAs (5 μg per lane) for HGF, αFGF, TGFα, and TGFβ1 by northern blotting during the activation and proliferation of oval cells in the AAF/PH model. N, normal liver; GADPH, glyceraldehyde phosphate dehydrogenase mRNA used as loading control. Time in panel A is in hours and in panel B in days after PH. Reproduced from Thorgeirsson *et al.* (1993) with permission.

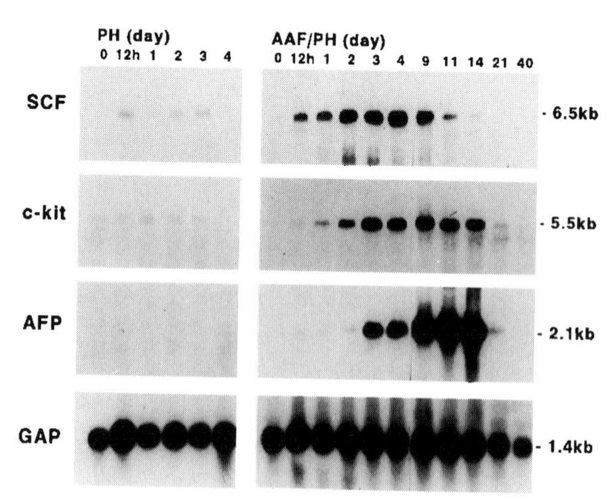

Figure 7 Analysis of expression of mRNAs (5 μg per lane) for stem cell factor (SCF), c-kit receptor (c-kit) and α-fetoprotein (AFP) following partial hepatectomy (PH) alone or after implementation of the AAF/PH regimen. Times indicated are hours or days after PH. Reproduced from Thorgeirsson *et al.* (1993) with permission.

which progeny of haematopoietic stem cells are committed to various differentiated lineages (Metcalf, 1991). Although the SCF/c-kit ligand/receptor pair appears to be involved in the early stages of liver stem cell activation, it is not yet known whether the SCF/c-kit system interacts with the other hepatic growth factors to influence the frequency of lineage commitment by stem/oval cells.

Together the studies reviewed suggest that oval cells are the immediate progeny of liver stem cells, and indicate that at least some of the oval cells differentiate into hepatocytes, as depicted schematically in Figure 3. In addition to hepatocytes, oval cells are also capable of differentiating into biliary (Steinberg *et al.*, 1991) and intestinal epithelium (Tatematsu *et al.*, 1985). These observations support the notion that oval cells have options for differentiation that are similar to the differentiation options of hepatoblasts during embryogenesis of the liver, and may be regarded as 'bipotential progenitors' for the two hepatic cell lineages. However, in some models many of the proliferated oval cells die (Sarraf *et al.*, 1994). The capacity of oval cells to differentiate into non-hepatic lineages, such as intestinal epithelium and, possibly, exocrine pancreatic epithelium (Rao *et al.*, 1986), may represent a pathological response which, nevertheless, reflects the relationship of oval cells to their more primitive endodermal origins (Thorgeirsson and Evarts, 1992).

The precise anatomical location of the oval cell precursors (stem cells) in normal liver is still unclear, although the data reviewed appear to narrow the possibilities. The most likely location for hepatic stem cells seems to be within the epithelium of the terminal biliary ductules (canals of Hering) as proposed originally by Wilson and Leduc (1958) or within a heterogeneous population of periductal cells (Sell, 1990).

Isolation, culture and characterization of liver epithelial cells with stem-like properties

Several types of epithelial cells can be isolated from the liver and established in primary or propagable culture, including fully differentiated hepatocytes and bile duct epithelial cells, and also poorly differentiated non-hepatocytic epithelial cells with a simplified phenotype (liver epithelial cells). Maintenance of isolated hepatocytes is limited to short-term primary cultures, and most biliary epithelial cell lines also have a limited life span (for reviews see Alpini *et al.*, 1994; Joplin, 1994; Strain, 1994). In contrast, liver epithelial cells can be readily established in propagable culture, some of which have maintained quasidiploid phenotypes during a hundred passages *in vitro* (for reviews see Grisham *et al.*, 1975a; Grisham, 1979, 1980).

The presence of phenotypically simple, non-hepatocytic epithelial cells in liver cultures was evident in explant cultures, and such cells were termed clear or simple liver epithelium to distinguish them from the more granular, structurally and functionally more complex hepatocytes (Alexander and Grisham, 1970). All of the early propagable cultures of liver epithelial cells were established from mass cultures of dispersed liver cells in which all of the various types of cells from the liver are represented (Grisham *et al.*, 1975a; Grisham, 1979, 1980). Under these conditions of culture establishment it is impossible to exclude any cell contained in the liver, including hepatocytes, as a possible source for the derivation of the cultured cells. However, epithelial cells can be isolated from livers and established in culture under conditions that exclude the reasonable possibility of their origin from differentiated cells, either hepatocytes or biliary epithelial cells. Coon (1968) first demonstrated that a propagable hepatic epithelial line (the BRL line) could be established from an individual cell by primary cloning directly from a cytologically complex enzymic suspension of liver. Grisham *et al.* (1975a) confirmed Coon's observation and demonstrated that neither mature hepatocytes nor bile duct epithelial cells are required for the establishment of liver epithelial cell cultures; individual cells from a collagenase suspension of liver (prepared so that portal tracts and bile ducts were not digested), were plated in individual wells of microtitre plates, following which they were examined microscopically and identified as either hepatocytes or non-hepatocytic cells. Identifiable hepatocytes never gave rise to colonies under the conditions of the study, whereas much smaller epithelial cells did so with an overall efficiency of over 2% (Grisham *et al.*, 1975a). Grisham *et al.* (1975a) also took advantage of the extreme sensitivity of differentiated hepatocytes to the lytic actions of strong proteases such as trypsin or Pronase, a technique used to isolate non-parenchymal liver cells from the more numerous hepatocytes (Joplin, 1994). Proteolytic removal of hepatocytes from a collagenase suspension does not reduce the relative number of colonies of small epithelial cells that can be grown from a suspension, indicating that highly clonogenic epithelial cells, distinct from hepatocytes or biliary epithelial cells, are present in the livers of normal adult rats and can be readily established in propagable cultures (Grisham *et al.*, 1975a). Using either differential centrifugation or strong proteases to remove hepatocytes from small epithelial cells, Furukawa *et al.* (1987) confirmed the observation that continuously propagable rat liver epithelial cell cultures can be established from small liver epithelial cells that are morphologically distinct from differentiated hepatocytes; they also developed a medium that improved the efficiency with which small epithelial cells can be clonally isolated from liver cell suspensions and established in culture. Herring *et al.* (1983) showed that propagable cultures of liver epithelial cells can be established from livers of rats of various ages. They posited that propagable liver epithelial cells were derived from hepatocytes that underwent culture-induced dedifferentiation, but their cell lines were established from mass (non-clonal) cultures, which prevented morphological identification of the cells from which cultured cells arise (Herring *et al.*. 1983). Although the ability to isolate small liver epithelial cells from collagenase suspensions of liver cells in which all hepatocytes have been removed (Grisham *et al.*, 1975a; Furukawa *et al.*, 1987) shows that mature hepatocytes are not a necessary source, it does not exclude hepatocytes or other cells as a possible source for the establishment of cultured liver epithelial cells. Based on their phenotypic similarities, Marceau *et al.* (1986, 1992) have proposed that propagable liver epithelial cultures may originate from mesothelial cells of Glisson's capsule. Cultured mesothelial cells share some phenotypic characteristics with propagable liver epithelial cells (Faris *et al.*, 1994).

Tsao *et al.* (1984) established a line of diploid liver epithelial cells (WB-F344) by primary cloning of small non-hepatocytic epithelial cells from the collagenase suspension of cells from the liver of a normal young adult male Fischer 344 rat. Tsao and Liu (1988) independently established other lines of cloned liver epithelial cells (RL-F344) from normal rats. Thorgeirsson's group (McMahon *et al.*, 1986; Hampton *et al.*, 1990; Huggett *et al.*, 1990, 1991; Bisgaard and Thorgeirsson, 1991) has also established several cloned lines of normal rat liver epithelial cells (called RLE lines). Cells of all of these propagable liver epithelial lines are morphologically similar, being small cells (9–12 μm in diameter) that grow in closely packed, regular monolayers, as shown in Figure 8. Some of the functional

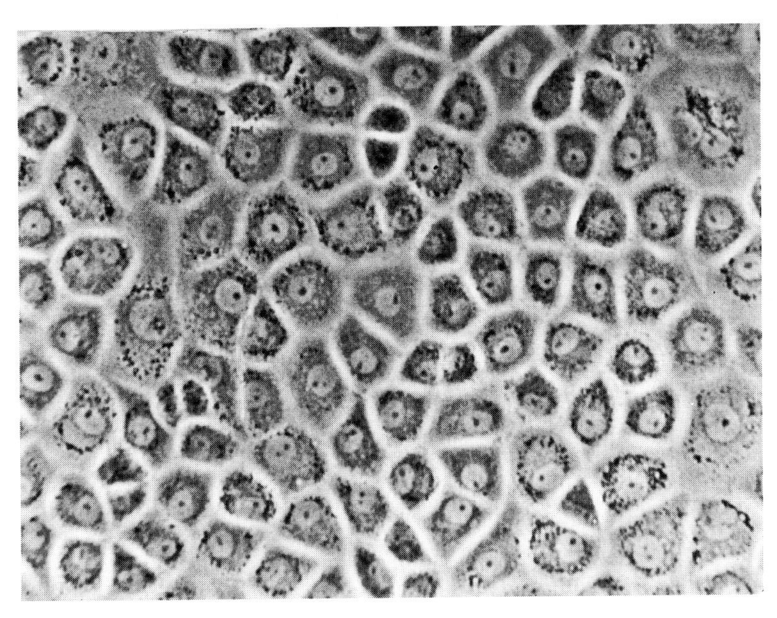

Figure 8 Morphology of WB-F344 rat liver epithelial cells growing in monolayer culture. Phase contrast, ×300.

and phenotypic properties of these propagable liver epithelial cell lines that were established from livers of normal adult rats are listed in Table 2.

Several investigators have isolated oval cells from the livers of rats in which these cells have proliferated in response to hepatocarcinogenic regimens. Germain *et al.* (1988b) isolated oval cells from rats fed 3-methyl-4-dimethylaminoazobenzene by panning (immunoabsorption) of liver cell suspensions in dishes coated with the HES6 antibody, followed by centrifugal sedimentation. Sells *et al.* (1981) and Yaswen *et al.* (1984) employed centrifugal sedimentation to isolate oval cells that proliferated after exposure of rats to a choline-deficient, ethionine-supplemented diet. Yoshimura *et al.* (1983) and Braun *et al.* (1987) established lines of propagable liver epithelial cells from isolated oval cells. The functional and phenotypic properties of two cultured oval cell lines established by Braun *et al.* (1987) from livers of rats fed a choline-deficient, ethionine-supplemented diet for either 2 or 6 weeks, named LE/2 and LE/6, respectively, are also listed in Table 2. Pack *et al.* (1993) also isolated oval cells by centrifugal elutriation from rats fed a choline-deficient, ethionine-supplemented diet for 6, 14, and 22 weeks, and they established three cell lines (named OC/CDE6, OC/CDE14, and OC/CDE22). These freshly isolated oval cells expressed γ-glutamyl transpeptidase; CKs 7, 8, 18 and 19; albumin; lactate dehydrogenase isoenzymes 1 to 5; and (weakly) glucose-6-phosphatase and alkaline phosphatase. The cultured lines lost the ability to express γ-glutamyl transpeptidase and albumin, but maintained expression of glucose-6-phosphatase and lactate dehydrogenase isoenzymes 2 to 5 (Pack *et al.*, 1993).

It is noteworthy that the phenotypic properties of some liver epithelial cells isolated from livers of normal rats (WB-F344, RL-F344, and RLE) resemble the properties of cell lines established from pathologically proliferated oval cells (LE/2 and LE/6); clearly the

Table 2 Phenotypic characteristics of normal rat liver epithelial cell lines and oval cell lines.

Marker[3]	Normal rat liver epithelial cell lines[1]			Oval cell lines[2]	
	WB-F344[4]	RLE[5]	RL-F344[6]	LE/2[7]	LE/6[8]
Antigenic markers for hepatocytes, bile ducts and oval cells					
OV-1	+				
OV-6	+ (weak)	+ (weak)			
OC.1					–
OC.2	–				–
OC.3	–				
BDS7	–				
BD.1	–				–
BD.2	+ (moderate)				
HES6	–				
H.1					–
H.2	–[9]				–
H.4	+				
Ep.1	+				
TuAg.1	–[9]				–
Intermediate filaments, extracellular matrix proteins					
CK5		–			
CK7	–	+			
CK8	–	+			
CK14	+ (moderate)	+/–			
CK18	–	–			
CK19	–	–			
Vimentin	+	+			
Desmoplakin I	+				–
β-Actin	+	+	+		
β-Tubulin		+			
Fibronectin	+	+			+
Collagen I		+			
Collagen IV		+			
Laminin		+			

Connexins, cell adhesion molecules					
Cx26	+ (−)	−			
Cx32	−	− (+)[10]			
Cx43	+	+			
cCAM 105	−				−
Integrin β1		+			
Enzymes and secreted proteins					
γ-GT	− (or weakly +)	−	+ (weak)	−	−
Albumin	− (or weakly +)	+ (weak)	+ (weak)	−	−
αFetoprotein	+	+ (weak)	+ (weak)	− (+ mRNA)	− (+ mRNA)
G6Pase	−			+	+
Alkaline Phosphatase	+			+	+
Acid Phosphatase				+	+
NADH Diaphorase	+ (low)				
Isozymes					
Aldolase A	+	+	+	+	+
Aldolase B	−	−	−	+ (weak)	−
Aldolase C	+	−	+		
Pyruvate Kinase K	+				
Hexokinase I	+				
Hexokinase II	+ (weak)				
LDH-1	− (weak)	−	−	−	−
LDH-2	− (weak)	+ (weak)	−	−	−
LDH-3	+ (moderate)	+	+	−	+
LDH-4	+	+	+	+ (weak)	+
LDH-5	+	+	+	+	+
GST Ya	−	+ (weak)			
GST Yb	+	+	+		
GST Yc		+			
GST Ye		+			
GST Yp	+	+	+		
PFK Liver Type	+				

Table 2 *Continued*

Marker[3]	Normal rat liver epithelial cell lines[1]			Oval cell lines[2]	
	WB-F344[4]	RLE[5]	RL-F344[6]	LE/2[7]	LE/6[8]
Growth factors/receptors					
TGFα	−	−	+ (weak)		
TGFβ	+	+	+	+	+
IGF-I		−			
IGF-II	+	−		+	+
FGF (α and β)		−			
Stem Cell Factor		+			
EGF Receptor	+	+	+		
Transferrin Receptor					+
c-kit Receptor		+			
Respond to TGFβ	+	+	+		+
Respond to TGFα	+	+			
Respond to EGF	+	+	+	−	+
Respond to IL-6		+			
Respond to Transferrin	+				
Oncogenes, tumour suppressor genes and other transformation-related genes					
mdr-1	+	+	+		
mdr-2	+		+		
WT1	+				+
p53	+	−		+	+
c-Ha-ras	+	+	+		
c-K-ras	+		+		
c-raf		+			
c-fos	+				
c-myc	+	−	+	+	+
Morphology, ultrastructure and other phenotypic traits					
Ploidy	Diploid	Diploid	Diploid	Diploid	Diploid
Glycogen	−				
Peroxisomes				+ (Scant)	+ (Scant)

Clofibrate-induced				+	+
Desmosomes	+				
Gap Junction Complexes	+	+			
Electrically Coupled	+				
Dye Coupled	+	+			
Soft Agar Growth	−	−	−	−	−
Soft Agar Growth (+EGF)	−	−	−	−	−
Tumorigenicity	−	−	−	−	−

[1] Normal rat liver epithelial cell lines were established from the livers of rats that had received no prior carcinogen treatment. WB-F344 cells were established from the liver of a young adult male Fischer 344 rat as described by Tsao *et al.* (1984). RLE refers to various cell lines established from the livers of neonatal or young adult rats by Thorgeirsson and colleagues (McMahon *et al.*, 1986; Huggett *et al.*, 1990, 1991; Hampton *et al.*, 1990; Bisgaard and Thorgeirsson, 1991). RL-F344 refers to various cell lines established from young adult Fischer 344 rats by Tsao and colleagues (Tsao and Liu, 1988; Tsao *et al.*, 1990; Tsao and Zhang, 1992).

[2] Cell lines were established from oval cell populations isolated from the livers of rats which were maintained on a choline-deficient diet containing 0.1% ethionine for 2 weeks (LE/2) or 6 weeks (LE/6) (Braun *et al.*, 1987; Fausto *et al.*, 1987).

[3] The monoclonal antibodies utilized to examine the expression of antigenic markers of hepatocytes, bile duct epithelium and oval cells have been extensively described (Hixson and Allison, 1985; Dunsford and Sell, 1989; Faris *et al.*, 1994). The abbreviations for the other markers included in this table are as follows: CK, cytokeratin; Cx, connexin; GGT, γ-glutamyl transpeptidase; αFP, α-fetoprotein; G6Pase, glucose-6-phosphatase; LDH, lactate dehydrogenase; GST, glutathione S-transferase; PFK, phosphofructokinase; TGF, transforming growth factor; FGF, fibroblast growth factor; EGF, epidermal growth factor; mdr, multi-drug resistance gene.

[4] Data on characterization of the WB-F344 rat liver epithelial cell line was taken from the following references: Tsao *et al.*, 1984, 1986, 1987, 1989, 1991; Earp *et al.*, 1986, 1988; Lin *et al.*, 1987; Liu *et al.*, 1988; Spray *et al.*, 1991; Batist *et al.*, 1991; Stutenkamper *et al.*, 1992; Marceau *et al.*, 1992; Woo *et al.*, 1992; Tsao, 1993; Lee *et al.*, 1991b; Faris *et al.*, 1994; Neveu *et al.*, 1994; Ren *et al.*, 1994; Zhang *et al.*, 1994. Additional data was from A. E. Wennerberg and J. W. Grisham (unpublished).

[5] Data on the characterization of the RLE rat liver epithelial cells was taken from the following references: Huggett *et al.*, 1989, 1990, 1991; Nagy *et al.*, 1989; Hampton *et al.*, 1990; Bisgaard and Thorgeirsson, 1991; Kalimi *et al.*, 1992; Bisgaard *et al.*, 1994a, 1994b; Fujio *et al.*, 1994; Zhang and Thorgeirsson, 1994; Tan *et al.*, 1994.

[6] Data on the characterization of the RL-F344 rat liver epithelial cells was taken from the following references: Woo *et al.*, 1992; Tsao and Liu, 1988; Tsao *et al.*, 1990; Tsao and Zhang, 1992.

[7] Data on the characterization of the LE/2 rat liver oval cell line was taken from the following references: Braun *et al.*, 1987; Fausto *et al.*, 1987, 1992; Plenat *et al.*, 1988.

[8] Data on the characterization of the LE/6 rat liver oval cell line was taken from the following references: Braun *et al.*, 1987, 1989; Fausto *et al.*, 1987, 1992; Plenat *et al.*, 1988; Goyette *et al.*, 1990; Faris *et al.*, 1994.

[9] WB-F344 cells in exponential growth are antigenically null for most markers examined (A. E. Wennerberg and J. W. Grisham, unpublished observations). At confluence WB-F344 cells are positive or weakly positive for OV-1, OV-6, BD.2, H.4, and Ep.1. Markers H.2 and TuAg.1 were examined in exponentially growing cells only.

[10] Cx32 is expressed in transformed RLE cells (Zhang and Thorgeirsson, 1994).

livers of normal rats contain cells that resemble oval cells (Tsao *et al.*, 1984). Cultured liver epithelial cells generally express phenotypic properties that are fetal or undifferentiated in character, although they may express a few differentiated properties that resemble those of either hepatocytes or biliary epithelial cells. The expression of genes such as c-kit and SCF (RLE lines), or the failure to express differentiation-specific cytokeratins (WB-F344), may suggest a general phenotypic resemblance of these liver epithelial cells to stem cells or immediate stem cell progeny of other tissues.

Germain *et al.* (1988a) and Marceau *et al.* (1992) isolated presumptive hepatoblasts from the livers of fetal rats and maintained them in primary culture for a few days. Rogler (1994) has reported the isolation of a propagable epithelial cell line from the hepatic diverticulum of mice, but details are not yet available. Propagable epithelial cell lines have been established from the epiblast of porcine blastocysts (the cell line is named PICM-19) (Talbot *et al.*, 1993, 1994a) and from the liver of an E27-28 fetal pig (the line is named PFL-2) (Talbot *et al.*, 1994b). Line PICM-19 has been passaged continuously for more than 100 population doublings (Talbot *et al.*, 1994a).

Reid and colleagues (Brill *et al.*, 1993; Sigal *et al.*, 1994) have pioneered studies on the use of monoclonal antibodies OC.2 and OC.3, which were raised against oval cells (Hixson and Allison, 1985; Hixson *et al.*, 1990), to isolate putative stem cells from livers of fetal rats. Haematopoietic cells (to which OC.2 and OC.3 also bind) are removed by panning (immunoadsorption) of the liver cell suspensions in dishes coated with antibodies to rat haematopoietic and endothelial cells (OX-43 and OX-44). Multiparametric fluorescence activated sorting of the panned suspension yields oval cell antigen-positive and oval cell antigen-negative populations, each of which are separated into granular and agranular subpopulations. OC.3-positive, granular cells express α-fetoprotein and γ-glutamyl transpeptidase, but not albumin, and are thought to represent committed progenitor cells for bile ducts (Sigal *et al.*, 1994). OC.3-negative, granular cells express both albumin and α-fetoprotein, but not γ-glutamyl transpeptidase, and are thought to represent committed progenitors for hepatocytes (Sigal *et al.*, 1994). Final proof that hepatocytic and bile duct epithelial progenitor cells have been isolated will await appropriate studies to determine the developmental potencies of each cell type.

Differentiation of cultured liver epithelial cells *in vitro* and *in vivo*

The capability of cultured liver epithelial cells to differentiate has been tested by exposing the cells to specific differentiation stimuli *in vitro* and by transplanting them into sites *in vivo*, such as the liver, in which liver-specific differentiation would be expected. Another method that has been used to assess the latent capacity of liver epithelial cells to differentiate involves the assessment of the differentiation of tumours produced *in vivo* after transplantation of liver epithelial cells that were neoplastically transformed *in vitro*.

Cultured liver epithelial cells can be induced to acquire some features of hepatocytic or biliary epithelial cell differentiation when exposed to appropriate conditions *in vitro*. When undifferentiated epithelial cells (hepatoblasts) from E12 fetal rat livers are exposed to sodium butyrate in primary culture they acquire features of biliary epithelial cell differentiation, as indicated by the expression of cytokeratins and the cellular antigen reacting to the monoclonal antibody BDS7, while the same types of cells express features of hepatocyte differentiation (cytokeratins and HES6) when they are cultured in serum thought to contain various growth factors (Germain *et al.*, 1988a). Marceau *et al.* (1992) subsequently showed that epithelial cells isolated from E18 fetal rat livers differentiate along the hepatocyte pathway when exposed to medium containing TGF-β and IGF-II, while supplementing the medium with tryptose phosphate broth is associated with acquisition of features of biliary epithelial differentiation. Studies by Thorgeirrson's group (McMahon *et al.*, 1986; Nagy *et al.*, 1989; Chapekar *et al.*, 1990) and Lin *et al.* (1987) showed the ability of TGF-β to induce differentiated hepatocyte-like properties in cultured rat liver epithelial cells. Exposure of rat liver epithelial cells to physiological levels of TGF-β quickly suppresses cell proliferation and causes a dramatic increase in the absolute and relative (to nuclei) increase in cytoplasmic area (McMahon *et al.*, 1986; Lin *et al.*, 1987; Nagy *et al.*, 1989). Furthermore, when exposed to TGF-β rat liver epithelial cells secrete albumin (Nagy *et al.*, 1989), and express other proteins typical of mature hepatocytes, including the multidrug resistance transporter protein (Chapekar *et al.*, 1990).

Exposure of the WB-F344 rat liver epithelial cell line to sodium butyrate *in vitro* simultaneously blocks their proliferation, causes a dramatic increase in absolute and relative cytoplasmic area, and greatly elevates the rate of protein synthesis (Coleman *et al.*, 1994). Dexamethasone treatment induces sodium butyrate-primed WB-F344 cells to express the hepatocyte-specific enzyme tyrosine aminotransferase at levels characteristic of hepatocytes *in vivo* or in primary culture (Coleman *et al.*, 1994). In keeping with these observations, growth of isolated oval cells for 5 days in the presence of sodium butyrate upregulates the synthesis of both albumin and tyrosine aminotransferase (Germain *et al.*, 1988b). Similarly exposure of cultured oval cells of the OC/CDE lines to dimethyl sulfoxide and sodium butyrate induces them to express albumin and γ-glutamyl transpeptidase, and enhances their expression of glucose-6-phosphatase and alkaline phosphatase (Pack *et al.*, 1993).

Exposure of two mouse liver epithelial cell lines (BNL CL.2 and NMuLi) to hepatocyte growth factor in culture induces their proliferation when growing as monolayers, but causes them to form hepatic plate-like trabeculae (BNL CL.2) or branching duct-like tubules (NMuLi) when grown on collagen gels (Johnson *et al.*, 1993). Although functional phenotypic markers were not examined, the morphological changes seen on collagen gels were considered to resemble hepatocytes and cells of bile ductules. In view of the influence of hepatocyte growth factor on the proliferation and differentiation of liver epithelial cells, it is of interest that this growth factor appears to be necessary for embryogenesis of the liver. When the gene for hepatocyte growth factor is deleted from the genome, the liver fails to develop completely (Schmidt *et al.*, 1995).

The fetal porcine liver epithelial cell lines PICM-19 and PFL-2 differentiate spontaneously along both hepatocytic and biliary epithelial pathways when cultured on feeder layers of STO cells (Talbot *et al.*, 1994a, 1994b). The STO cells probably condition the culture medium with several unidentified growth factors, which may have important effects on the differentiation of the cultured liver epithelial cells. However, an equally important condition that appears to stimulate the differentiation of PICM-19 and PFL-2 cells is the contact of adjacent cells in culture. As indicated by the expression of mRNAs for the liver-specific proteins α-fetoprotein, albumin, and β-fibrinogen, and the secretion of porcine albumin, α-fetoprotein and transferrin into serum-free medium, hepatocytic differentiation begins spontaneously in either PICM-19 or PFL-2 lines when the cells touch in culture (Talbot *et al.*, 1994a, 1994b). Biliary canaliculi form between adjacent hepatocytes, and some of the cultured cells subsequently differentiate into structures that are morphologically identical to bile ducts (Talbot *et al.*, 1994a, 1994b).

The results of these *in vitro* studies show that some hepatic epithelial cell lines are able to acquire differentiated properties of both hepatocytes and biliary epithelial cells. The results also suggest some of the probably complex group of conditions that directs an undifferentiated liver epithelial cell to enter a differentiation pathway, including various growth factors, extracellular tissue components, and contact between adjacent epithelial cells. Certain of these liver epithelial cell lines may provide useful resources with which to identify some of the elements that determine whether a bipotential liver stem/progenitor cell differentiates along hepatocyte or biliary epithelial cell pathways.

Transplantation of putative liver stem/progenitor cells into specific sites *in vivo* that may provide conditions that force their differentiation provides a well-tested technique to evaluate cell differentiation capacities and options. Transplantation of bone marrow cells into the spleens of lethally irradiated mice, the spleen colony-forming assay, first allowed the firm interpretation that some cells in bone marrow possessed multipotent capability to differentiate along all of the different bone marrow lineages (Till and McCulloch, 1961). Many studies have demonstrated the feasibility of transplanting epithelial cells isolated from livers. Differentiated hepatocytes from mice and rats have been successfully transplanted into several sites *in vivo*, including the spleen, pancreas, interscapular fat pad, renal subcapsular space and liver (see Gupta *et al.*, 1992 for review). Similarly, isolated bile duct epithelial cells have been transplanted into the interscapular fat pad (Sirica *et al.*, 1991). The liver is perhaps the most relevant transplantation site. When transplanted into livers of congenic or syngeneic hosts by either injection as single-cell suspensions into the spleen or into a tributary of the portal vein, isolated hepatocytes whose membranes are tagged with radioactive indium (Gupta *et al.*, 1991) or whose genomes are tagged with the *E. coli* lac Z gene (β-galactosidase) (Ponder *et al.*, 1991), incorporate into hepatic plates among host hepatocytes. Although the phenotypic features of transplanted hepatocytes have not been studied in great detail, the transplanted cells appear to function adequately, as indicated by the partial correction of certain genetically predicated metabolic abnormalities, when even a relatively small number of competent hepatocytes are transplanted into appropriate sites (usually the liver) of genetically defective animals (Gupta *et al.*, 1992). As already noted, when transplanted into the livers of transgenic mice whose hepatocytes have been killed by over expression of a urokinase transgene (Rhim *et al.*, 1994), lac Z gene-tagged normal hepatocytes can proliferate up to 12 times and replace the destroyed hepatocytes. These studies indicate that hepatocytes maintain key functional capabilities when transplanted into the liver.

Coleman *et al.* (1993) genetically tagged early passage WB-F344 cells with a construct containing the *E. coli* lac Z gene (β-galactosidase) and the Neo gene (neomycin resistance) from the Tn5 transposon, all under the regulation of LTR and SV40 promoters (Price *et al.*, 1987). Tagged cells (BAG2-WB) express a high activity of bacterial β-galactosidase, which enables their specific histochemical identification in the presence of other rat liver cells (Price *et al.*, 1987; Coleman *et al.*, 1993). When transplanted into the livers of syngeneic Fischer 344 rats, by either intrasplenic or intrahepatic transcapular injection of suspensions of $1–5 \times 10^6$ cells, tagged BAG2-WB cells accumulate in periportal sinusoids and in capillary vessels in portal tracts (Coleman *et al.*, 1993; McCullough *et al.*, 1994) (Plate 8.2a). Over a period of a few days, many of the BAG-2WB cells

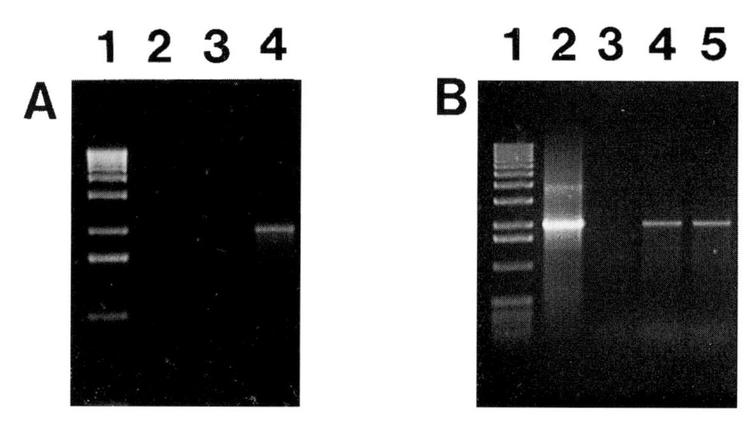

Figure 9 Detection of marker DNA sequences from the CRE BAG2 retroviral construct in genomic DNA from livers that received transplants of BAG2-WB cells, as well as in genomic DNA from liver epithelial cells recovered from transplanted cells. The Neo gene contained in the viral construct was amplified by polymerase chain reaction using primers directed against Neo sequences and against the adjacent SV40 polyadenylation signal (Coleman *et al.*, 1993). Panel A contains: Lane 1, DNA molecular size standards (1 kb ladder); Lane 2, no DNA (control); Lane 3, genomic DNA isolated from liver of rat not receiving transplanted cells; Lane 4, genomic DNA from liver of rat transplanted with BAG2-WB cells. Panel B contains: Lane 1, DNA molecular size standards (1 kb ladder); Lane 2, positive control template DNA (pSV2 neo plasmid DNA); Lane 3, no DNA (control); Lane 4, genomic DNA from cultured BAG2-WB cells used in transplants; Lane 5, genomic DNA from liver epithelial cells isolated from liver of rat receiving a transplant of BAG2-WB cells and re-established in culture under neomycin selection (the cell line recovered from the transplanted liver is termed BAG2-WB-H3). Genomic DNA from both BAG2-WB and BAG2-WB-H3 cells contain identical Neo sequences, indicating their identity.

incorporate into hepatic plates, where they are identified as individual, β-galactosidase-positive cells among host hepatocytes that are histochemically negative for β-galactosidase under conditions in which the bacterial enzyme is active (Plates 8.2a and 8.2b). DNA from livers containing the transplanted cells includes the Neo sequence from the tagging construct (Figure 9A), demonstrating that the hepatocytes are derived from the transplanted cells. BAG2-WB cells that are incorporated into hepatic plates enlarge to the size of host hepatocytes (Coleman *et al.*, 1993) and acquire the ability to synthesize and/or store the hepatocyte-specific proteins albumin, tyrosine aminotransferase, transferrin, and α-1 antitrypsin (Grisham *et al.*, 1993). These results provide strong evidence that the cells of the WB-F344 liver epithelial cell line express the functions of hepatocyte progenitor cells.

Hepatocytes and other epithelial cells that are derived from intrahepatic transplants of BAG2-WB cells can be recovered from livers of recipient rats by culturing liver cell suspensions in media containing neomycin (Grisham *et al.*, 1993). The liver epithelial cells that are re-established in culture (termed BAG2-WB-H3) are morphologically identical to the BAG2-WB cells and their genomes also contain the Neo gene (Figure 9B), showing that they are direct descendants of the original BAG2-WB cells (Grisham *et al.*, 1993). With the use of similar tagging, transplantation and recovery techniques it should be possible ultimately to determine whether some transplanted stem-like cells occupy a niche that protects them from the stimulus to differentiate that occupancy in hepatic plates seems to engender, or whether the tagged hepatocytes that differentiate from transplanted stem-like cells can modulate their phenotype in the culture environment and reacquire stem-like properties.

Transplantation of BAG2-WB cells into the liver parenchyma has not yet provided evidence for their ability to differentiate into bile duct epithelium, suggesting that the intraparenchymal microenvironment may force them along a pathway toward hepatocytic differentiation. However, the extensive range of differentiated morphological properties expressed by tumours formed by neoplastically transformed liver epithelial cells indicates that these cells possess multipotent differentiation options and may form both bile duct epithelium and hepatocytes. The first detailed studies that examined the morphology of tumours produced *in vivo* by subcutaneous or intraperitoneal transplantation of chemically transformed rat liver epithelial cells showed that the tumours were poorly differentiated neoplasms, possibly of mesenchymal origin, although some tumours expressed morphological features of both hepatocellular carcinomas and adenocarcinomas of biliary duct type (Montesano *et al.*, 1973; Williams *et al.*, 1973). Since the transformed rat liver epithelial cells used in these early studies were derived from uncloned mass cultures of liver cells, it could not be determined whether the tumours that expressed various epithelial or mesenchymal morphologies all arose from a common epithelial cell with multipotent properties. Nevertheless, it was clear that some cells in the cultures could express morphological evidence of specific liver cell differentiation. Tsao and Grisham (1987) first demonstrated that tumourigenic derivatives of the cloned WB-F344 rat liver epithelial cell line, neoplastically transformed by exposure *in vitro* to *N*-methyl-*N*-nitro-*N'*-nitrosoguanidine, yielded tumours that expressed a wide range of morphological differentiations, including hepatocellular carcinomas, adenocarcinomas of both biliary and intestinal (mucinous) types, hepatoblastomas (containing cartilage and osteoid), epidermoid (squamous) carcinomas, adenocarcinomas with sarcomatous elements, and poorly differentiated neoplasms of both epithelial and mesenchymal types (Table 3). Although the WB-derived tumours have not yet been examined for the expression of cytokeratin filaments, the results of this study indicate that transformed WB cells have a wide range of options for morphological differentiation, suggesting that the parental WB-F344 rat liver epithelial cells also have multipotential differentiation options.

Table 3 Classification of tumours produced by chemically transformed rat liver epithelial cells

	Number of tumours
Carcinomas (Epithelial):	54
Epidermoid	15
Adenocarcinoma	13
Hepatocellular	4
Poorly differentiated/anaplastic	22
'Sarcomas' (mesenchymal)	19
'Mixed epithelial-mesenchymal' tumours	30
Unclassified	22

(From Tsao and Grisham, 1987).

Garfield *et al.* (1988) showed that transplantation of RLE rat liver epithelial cells neoplastically transformed by transfection with either v-raf, v-raf + v-myc, or v-Ha-ras oncogenes yields tumours with poorly differentiated mesenchyme-like morphologies, as well as tumours with a variety of differentiated epithelial morphologies, including a few well-differentiated hepatocellular carcinomas and hepatoblastomas. They suggested that different oncogenes might 'switch' the ability of these liver epithelial cells to express differentiated properties of alternate lineages (Garfield *et al.*, 1988). Lee *et al.* (1989) showed that spontaneous transformation of the WB-F344 line of rat liver epithelial cells occurs predictably after the cells are maintained in confluent cultures. Acquisition of tumourigenicity is associated with the development of aneuploidy, and the tumours produced include well-differentiated hepatocellular carcinomas and adenocarcinomas. Huggett *et al.* (1991) confirmed these results with the RLE line of cloned rat liver epithelial cells, and demonstrated that the tumours produced in nude mice by spontaneously transformed RLE cells are mainly poorly or well-differentiated trabecular hepatocellular carcinomas (Huggett *et al.*, 1991; Williams *et al.*, 1992a). Tsao and Zhang (1992) and Zhang *et al.* (1994) assessed the differentiation of tumours produced in syngeneic rats by transplantation of cells of the RL-F344 line of cloned rat liver epithelial cells that spontaneously transformed *in vitro* in the model system of Lee *et al.* (1989), either in the presence or absence of the growth factors EGF or TGF-β. EGF-exposed transformants produced mainly intestinal-type adenocarcinomas, while the same cells that were transformed in the absence of EGF yield tumours that expressed evidence of hepatocytic differentiation (Tsao and Zhang, 1992); cells that transformed in the presence of TGF-β produced poorly differentiated adenocarcinomas (Zhang *et al.*, 1994). These studies suggest that the presence of growth factors during the process of transformation may influence the differentiation of the tumours produced by the transformed cells.

Bisgaard *et al.* (1994b) investigated alterations that occur in the expression of the cytokeratins CK8, 14 and 18 and of vimentin as RLE cells undergo spontaneous transformation during confluent growth *in vitro*. RLE cells normally express CK8 and vimentin, but not CK18 (Bisgaard *et al.*, 1994b), and expression of CK8 is maintained during confluent culture. RLE cells also express transcripts for CK14; after continued passaging and confluent growth RLE cells acquire the ability to express transcripts for CK18 and α-fetoprotein, while losing the ability to express CK14 and vimentin (Bisgaard *et al.*, 1994b); the fully tumorigenic variants continue to express CK8, CK18 and α-fetoprotein. Bisgaard *et al.* (1994b) concluded that RLE cells partially differentiate as they transform spontaneously, and that their ability to differentiate along the 'hepatoblastic' pathway may be responsible for the differentiated tumours that spontaneously transformed RLE derivatives produce. Neoplastic transformation of RLE cells spontaneously, by infection with retroviral constructs (v-raf or v-raf and v-myc oncogene), and by exposure to aflatoxin β_1, produced distinct patterns of cytokeratin expression and different extents of tumour differentiation, suggesting that expression of different intermediate filaments may determine the differentiation of a tumour (Bisgaard *et al.*, 1994b).

The powerful influence of the liver microenvironment on the differentiation of transformed derivatives of the WB-F344 line of rat liver epithelial cells has been demonstrated by Coleman *et al.* (1993) and McCullough *et al.* (1994), who compared the tumourigenicity of two neoplastically transformed WB lines when they were transplanted either subcutaneously or into the liver. The tumorigenicity of two transformed WB lines, GN6TF and GP7TB, each of which yields tumours when transplanted subcutaneously with an efficiency of nearly 100% and a latency period of 20–30 days, is unaffected when they are tagged with the *E. coli* lac Z (β-galactosidase) construct (to yield BAG2-GN6TF and BAG2-GP7TB) (Coleman *et al.*, 1993). However, BAG2-GN6TF cells are not tumorigenic and BAG2-GP7TB cells produce a reduced yield of tumours with greatly prolonged latency when transplanted into the livers of 3-month-old syngeneic rats (McCullough *et al.*, 1994). After intrahepatic transplantation β-galactosidase-positive hepatocytes are present in hepatic plates, even when tumours are also produced in the liver. β-galactosidase-positive cells recovered from livers in which tumours fail to develop after transplantation of BAG2-GN6TF again produce tumours with high efficiency and short latency when transplanted subcutaneously (Grisham *et al.*, 1993). The liver microenvironment thus has a strong regulatory influence on the differentiation of the diploid WB-F344 rat liver epithelial cells and on the differentiation and tumorigenicity of aneuploid tumorigenic cells derived from WB cells. The molecular nature of the intrahepatic regulation of hepatocyte differentiation is not known, but McCullough *et al.* (1994) have shown that the differentiation stimulus decays as host animals age. In rats older than 18 months, tumours are produced in the liver after transplantation of either BAG2-GN6TF or BAG2-GP7TB cells with approximately the same efficiency and latency as occurs after subcutaneous transplantation.

Studies by Faris and Hixson (1989) indicate that isolated, but untransformed, oval cells may differentiate into hepatocytes when transplanted into the liver, whereas Germain *et al.* (1988b) showed that they form bile duct-like structures when transplanted into the interscapular fat pad. Yoshimura *et al.* (1983) found that spontaneously transformed oval cell lines (uncloned) yield mainly anaplastic carcinomas when transplanted into rats or nude mice. However, many tumours contained hepatocyte-like cells, suggesting that some of the cultured cells could differentiate along the hepatocytic pathway. Braun *et al.* (1987) neoplastically transformed the LE/6 line of cultured oval cells by transfecting them with the activated H-ras (EJ) oncogene; these oncogene-transformed rat liver epithelial cells produce trabecular hepatocellular carcinomas and hepatoblastomas after transplantation to subcutaneous or intraperitoneal sites. Additional evidence that neoplastically transformed liver epithelial cells acquire differentiated properties is provided in a study of the tumours produced in nude mice by EJ ras-transfected LE/6 (Goyette *et al.*, 1990). Untransformed LE/6 cells are poorly differentiated and express neither oval cell nor hepatocyte antigens. Cells from the trabecular hepatocellular carcinomas produced by EJ ras-transfected LE/6 variants express a combination of oval cell antigens (react with OC.2 and OC.2 antibodies) and hepatocyte antigens (react with desmoplakin, cell CAM 105, and the H.1 antigen) (Goyette *et al.*, 1990). In contrast, cultured oval cells of the OC/CDE6 or OC/CDE22 lines produce cholangiocellular carcinomas of solid or adenoid types when transformed *in vitro* spontaneously or after exposure to *N*-methyl-*N*-nitro-*N*′-nitrosoguanidine (Steinberg *et al.*, 1994). Most of the tumour cells express CKs 7, 8, 18 and 19, characteristic of biliary epithelial cells, although a few islands of cells express only CKs 8 and 18, as well as peroxisome-like organelles and biliary canaliculus-like intercellular spaces, all features of hepatocytes. However, neither type of tumour cell expresses albumin or α-fetoprotein (Steinberg *et al.*, 1994).

Together these studies provide decisive evidence that both normal (diploid) and transformed (aneuploid) variants of cultured liver epithelial cells possess options to express multiple differentiations that are equivalent to the range of options of hepatic stem cells. Cultured liver epithelial cells, either normal or transformed, represent useful substrates with which to examine the mechanisms that regulate cellular differentiation.

DISCUSSION

This review clarifies some major features of the cell renewal processes in the epithelial populations of the liver and the involvement of stem cells in these renewal processes, but major questions about the nature of liver stem cells and their role in hepatic physiopathology also remain to be answered. We attempt to highlight both aspects in the discussion which follows.

Many of the properties of differentiated hepatocytes and of more poorly differentiated oval cells indicate that both types of cells can participate in the renewal of the hepatic parenchyma in rats. The material reviewed here shows that bipotential stem cells participate in the embryogenesis of the liver, but not in the postnatal growth of the liver and not in the maintenance of hepatocyte or biliary epithelial cell populations in adult animals under normal conditons or after most types of induced cell loss. Turnover of hepatocytes and biliary epithelial cells is practically nil in the livers of adult animals; both hepatocyte birth (Wright and Alison, 1984) and death (Benedetti *et al.*, 1988) occur at very low levels. Reparative renewal of the hepatocyte and biliary epithelial cell populations is effected by the proliferation of the residual differentiated cells of each type, without the need to activate stem cells and establish new lineages of differentiated cells. Certain extreme types of liver injury, however, can lead to the proliferation and accumulation of large populations of poorly differentiated cells, which appear to be produced by the activation of stem cells, and which can differentiate into hepatocytes, biliary epithelial cells, and other types of differentiated cells in favourable experimental circumstances. Additionally, small, poorly differentiated cells that possess some major properties of liver epithelial stem cells can be isolated from both normal and pathological livers of rats and established in culture; when transplanted into livers of syngeneic animals, cells of some of these lines differentiate into hepatocytes. Furthermore, the tumours that result when transformed variants of these stem-like cells are transplanted into different sites of host animals express various morphological differentiations, among them hepatocytic and biliary epithelial. These observations indicate that, rather than being a static system of cells whose only function is a metabolic factory for the body, the liver is, in fact, a kinetically dynamic tissue in which parenchymal cells and their precursors have wide and varied, but tightly regulated, options to proliferate and/or to differentiate. Although the mechanisms that regulate the proliferation and differentiation of liver epithelial cells are not clear as yet, the better understanding of the kinetic status and structural organization of cells that make up the liver provides a rational basis for the design and implementation of studies that may eventually lead to this knowledge. In fact, the liver and its component cells afford excellent tissue and cellular 'reagents' with which to investigate further the controls of cell proliferation and differentiation, and the interaction of these two important cellular processes during normal and pathological replacement of cells.

The liver contains both poorly differentiated epithelial cells with stem-like properties and highly differentiated epithelial cells that carry out complex functions, as do conventional stem cell-fed lineages, such as intestinal mucosa, epidermis, and bone marrow. Nevertheless, it is not immediately apparent that the liver epithelial cell population is structurally organized into a series of kinetic and functional compartments that resemble the compartmental structures of the conventional stem cell-fed lineages. Hall and Watt (1989) and Potten and Loeffler (1990) recently analysed some of the major functional and structural features of the lineages of intestinal mucosa, epidermis, and bone marrow. According to Hall and Watt (1989), the most basic property of stem cells is their potential to generate new functioning cells. Potten and Loeffler (1990) classify cells on the basis of replicative potential as *actual stem cells, potential stem cells* and, *committed stem cells*; actual stem cells are defined as 'undifferentiated cells capable of: (a) proliferation; (b) self-maintenance; (c) production of large numbers of differentiated progeny; (d) regeneration of the tissue after injury; and (e) flexibility in the use of their options'. Potten and Loeffler (1990) consider that potential stem cells are latent or quiescent or reserve counterparts of actual stem cells, which may be reactivated to become functioning stem cells. They propose that some of the essential properties of stemness may be retained by proliferating cells located distally in a lineage, such as the enterocytes in the dividing-transit compartment of the small bowel mucosa; these stem-like properties include the abilities to proliferate, self-maintain partially, and regenerate. Dividing-transit enterocytes also share with terminally differentiated cells in the functional compartment the ability to carry out tissue-specific functions. Potten and Loeffler (1990) suggest that the cells previously termed 'committed stem cells' or 'progenitor cells' may actually be dividing transit cells.

As shown by this review, mature, differentiated hepatocytes possess some of the major characteristics of stemness defined by Potten and Loeffler (1990): hepatocytes are able to: (a) proliferate; (b) produce large numbers of differentiated progeny; and (c) regenerate liver tissue. However, hepatocytes do not meet Potten and Loeffler's stem cell criterion of being undifferentiated; in fact, mature hepatocytes are able to proliferate without losing either structural (Grisham *et al.*, 1975b) or functional (Mohn *et al.*, 1990; Haber *et al.*, 1993) characteristics of differentiation. Whether hepatocytes can continue to renew the hepatocyte population in the face of cell loss sustained throughout life has not been determined. Stem cells of rapidly renewing tissues may need to cycle more than a thousand times during the life of an animal; although hepatocytes of rats can cycle at least 10–15 times, their ability to continue cycling beyond this number of cycles before senescing or neoplastically transforming has not been determined. Individual hepatocytes appear to be 'committed stem cells' or 'progenitor cells', which are normally quiescent, but can be activated to produce progeny whose only differentiation option is hepatocytic.

Evidence presented in this review suggests that oval cells are more primitive and more poorly differentiated progeny of liver epithelial stem cells than are hepatocytes or biliary epithelial cells, and the multiple differentiation options of oval cells are firmly established. Oval cells meet more of the stem cell criteria of Potten and Loeffler (1990) than do hepatocytes: (a) they are poorly differentiated; (b) they can proliferate; and (c) they have multiple differentiation options, including hepatocytic and bile ductular epithelial. Although they appear capable of producing 'large numbers' of differentiated progeny, this capacity has not yet been quantified. Whether they possess the ability to proliferate and differentiate throughout the life span of an animal is uncertain, since appropriate studies have not been (perhaps cannot be) done. The immediate precursors of oval cells

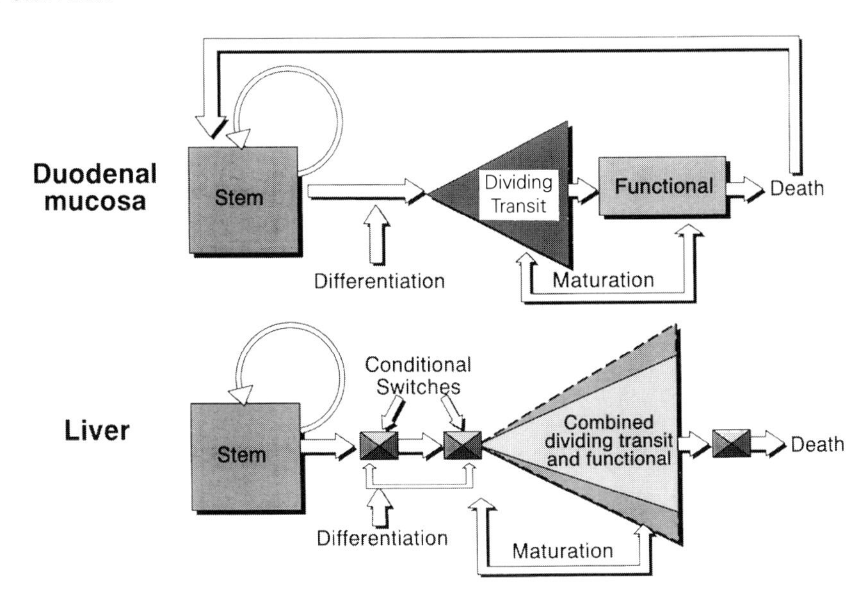

Figure 10 Schematic diagram comparing the hypothetical compartmental structure of the enterocyte lineage of the ileal mucosa (Potten and Loeffler, 1990) and the hepatocyte lineage of the liver. As discussed in the text the major visible compartment in the hepatocyte lineage of a normal liver is a functional compartment of hepatocytes which can acquire the combined properties of dividing-transit and functional compartments after simple cell loss. However, after more severe injuries the quiescent liver stem cell compartment can be activated to produce oval cells, temporarily entraining a complete lineage. For the structure of the normally fluxless hepatocyte lineage to resemble that of the enterocyte lineages requires stringent regulation (conditional switches) on both proliferation of stem cells and hepatocytes, as well as on the death of hepatocytes.

appear reasonably to approximate Potten and Loeffler's (1990) definition of a potential stem cell, that is a stem cell whose functions of stemness are suppressed or latent, but which can be reactivated by appropriate conditions. This definition of a potential stem cell echoes Grisham's (1980) hypothesis of 'facultative' liver stem cells, whose activation he posited to lead to the proliferation of oval cells.

A lineage structure that meets most of the stem-like and functional properties reviewed here for the various epithelial cells that are contained in or can be cultured from the liver, including oval cells and hepatocytes, can be patterned on the compartment model of the enterocyte lineage, as outlined by Potten and Loeffler (1990). In this model (depicted schematically in Figure 10), enterocytes of the small intestine are linked in three kinetic and functional compartments: stem cell, dividing-transit, and functional. The flux of cells through these compartments in the enterocyte lineage is continuous, new cells are supplied from the stem cell compartment, their number is amplified by proliferation in the dividing-transit compartment, where they begin to function; they reach their optimal differentiation and functional competence in the functional compartment; and they are eliminated from the population by cell death at the terminus of the lineage. In the enterocyte lineage cell generation and amplification in stem and dividing transit compartments balances the loss of cells from the lineage by death. The kinetic status of the hepatocytic lineage differs markedly from that of the small intestinal epithelium. Unlike the enterocyte lineage, through which the flux of cells occurs at a continuing high rate, flux through the hepatocyte lineage is episodic and dependent on induced loss of cells; turnover of hepatocytes under normal conditions is practically nil. The hepatic parenchyma contains highly functional, but proliferationally inactive, hepatocytes; in these characteristics the hepatocyte population in the unperturbed liver of a normal adult animal most closely resembles the functional compartment of the enterocyte lineage, but without cell flux (either input or output). However, after the hepatocyte population is reduced, residual hepatocytes proliferate promptly, they cycle repeatedly until the deficit is repaired, and they continue to function while proliferating; in these characteristics the population of hepatocytes proliferating after cell loss, such as is caused by 68% partial hepatectomy, most closely resembles enterocytes in the dividing-transit compartment. The population of adult hepatocytes can be conceptualized as a functional compartment that is usually static, but which can revert on demand (created by cell loss) to a compartment in which hepatocytes combine division and function. The residual hepatocytes remaining in the functional compartment of the liver at the time it reverts to a dividing-transit-like status take on the characteristics of monopotent progenitor cells, and they rapidly proliferate to amplify the population, following which they again become reproductively quiescent. Activation of oval cell proliferation and differentiation by injury that is more severe and/or qualitatively different from the simple loss of hepatocytes that triggers only hepatocyte proliferation, results in the temporary re-establishment of a hepatocyte lineage in which all of the kinetic and functional compartments of the enterocyte lineage can be delineated (Figure 10). Cells in a (normally quiescent) stem cell compartment (potential or facultative stem cells) are activated to produce poorly differentiated oval cell progeny. Oval cells proliferate extensively to yield a large population of cells that migrates through the parenchyma, some of which differentiate as they migrate; in these features, the oval cell population meets the essential attributes of the dividing-transit compartment of the enterocyte lineage, complete (temporarily) with cell flux. Hepatocytic progeny of oval cells merge into the functioning compartment of mature hepatocytes, and help to restore the parenchyma. As with the generation of new hepatocytes following their simple loss, the production of hepatocytes by the oval cell mechanism is also episodic and transient.

Thus the characteristics of these two distinct mechanisms of hepatocyte formation transiently reflect the reactivation of a part or all of the compartments in the structure proposed by Potten and Loeffler (1990) for the enterocyte lineage. The hepatocyte lineage appears, therefore, to include each of the compartments contained in the enterocyte lineage (Figure 10). However, the hepatocyte lineage normally lacks any flow of cells through it; the only highly visible compartment under normal conditions is the functioning compartment containing mature hepatocytes. Several points of stringent control must be necessary to enable the suppression and reactivation of cell flux through the hepatocyte lineage. Controls are required to regulate the reinitiation of hepatocyte formation is the normally quiescent, functional compartment, as well as to regulate the activation of potential stem cells and the proliferation of oval cells which energizes cell flow through the entire lineage (Figure 10). Death of hepatocytes also appears to be well-controlled, as indicated by their normally long life-span.

The status of the biliary epithelial lineage appears to resemble closely that of the hepatocyte lineage, although detailed data on the kinetics and population dynamics in the biliary epithelial cell lineage is less complete than is that for hepatocytes. Both biliary epithelial cells and hepatocytes appear to share a common stem cell that links these lineages.

Despite the strong evidence that the liver contains undifferentiated cells with stem-like properties, virtually nothing is known about their specific features which would allow their firm identification either in the liver structure or after their isolation. As a reflection of this situation, the location of stem cells in the liver structure can be suspected only by evidence that is indirect, such as the detection of small epithelial cells that express certain genes thought to be characteristic of stem cells or that first synthesize DNA following the implementation of a regimen that leads to oval cell proliferation. Based on these bits of indirect evidence, stem cells are thought to be located in or near portal tracts, more precisely within the cellular lining of small biliary ductules or in the mesenchyme of portal tracts immediately outside such small biliary ductules. Obviously, little is known about the phenotypic properties that liver stem cells may express *in vivo*, other than what may be inferred from the analysis of the phenotypic properties that are expressed *in vitro* by cultured liver epithelial cells that have the ability to differentiate into hepatocytes when transplanted into the liver *in vivo*. Of course, it is not certain that the phenotypic properties expressed by stem-like cells *in vitro* are equivalent to the phenotypic properties expressed by authentic stem cells *in vivo*.

Conditions that prevent the identification and handicap the functional analysis of liver stem cells are not unique to the liver system, and they differ insignificantly from similar constraints that also prevent the certain identification and precise characterization of the stem cells for any tissue, including bone marrow and epithelium of the small intestine for which the physiological necessity of constant stem cell activity is evident. Individual bone marrow stem cells cannot yet be identified by their morphological structure or their phenotypic properties, even though they can be highly concentrated as a result of their expression of specific surface antigens (Uchida *et al.*, 1994). The presence of both bone marrow stem cells and liver epithelial stem cells must be inferred from their ability to establish differentiated lineages. Potten and Loeffler (1990) have pointed out the major problem that this situation poses for investigators of stem cells: isolating stem cells and testing their properties inevitably may alter those properties. Furthermore, because the precise characterization and identification of stem cells (including liver epithelial stem cells) is uncertain, the definition of stem cells, as outlined previously, is loose and ambiguous (Potten and Loeffler, 1990). Despite these handicaps, progress is being made toward a more fundamental understanding of stem cell biology, including the biology of liver stem cells.

Hepatocytes and biliary epithelial cells can be activated in adult animals to function as committed progenitor cells with a potency limited to the generation of only hepatocytes or biliary epithelial cells, respectively. Whether hepatocytes and biliary epithelial cells have the plasticity to modulate their phenotype properties (Leduc, 1959), and if so, over what range of differentiation and under what physiopathological conditions, are intriguing questions. The development of hepatocyte-like cells within biliary ducts and the expression of ductal/ductular phenotypes by hepatocytes (for review see Sirica, 1995) suggests the possibility of such modulation or metaplasia. Marceau (1994) has hypothesized that such plasticity of differentiation might occur through the operation of a 'differentiation window' that allows cells of either hepatocyte or biliary epithelial lineages, when located at a certain point in their differentiation, to switch differentiated phenotypes. This hypothesis merits additional study in terms of its implications for understanding hepatic pathophysiology. Differentiated hepatocytes rapidly lose many of their differentiated properties when they are isolated and maintained in culture, and hepatocytes that have lost functional properties in culture can regain some of them when they are transplanted into the spleen (Nishikawa *et al.*, 1994). Whether some (or all) hepatocytes can dedifferentiate in culture sufficiently to assume the properties of liver epithelial cells, including propagability, needs to be ascertained.

Liver epithelial stem cells have no known physiological role, and their potential roles in pathological processes are still not entirely clear. The mere presence of stem cells in tissues, such as liver and brain (Marvin and McKay, 1992; Reynolds and Weiss, 1992), in which they do not normally function to generate new cells for a lineage, suggests the possibility that they may serve another, unknown, function. The capacity of cells with stem-like properties from various tissues to produce numerous substances that can regulate not only their own growth and differentiation, but also affect the expression of those properties by other types of cells (Baldwin, 1992; Graham and Pragnell, 1992; Lee, 1992; Smith *et al.*, 1992; Williams *et al.*, 1992b), suggests the possibility that stem cells may serve an important paracrine function in tissues, possibly to control the balance between proliferation and differentiation in the parenchyma (Smith *et al.*, 1992). Liver epithelial cells with stem-like properties also produce many of these growth and differentiation factors (Smith and Hooper, 1987; Zsebo *et al.*, 1990; Hata *et al.*, 1993; Fujio *et al.*, 1994; Nanno *et al.*, 1994); further studies are needed to examine the paracrine role of liver stem cells.

The process of stem cell activation to generate new hepatocytes and biliary epithelial cells has potential importance for two major pathological topics: development of liver cancer and repair of massively destroyed parenchyma. Evidence supporting the participation of oval cell progeny in the development of hepatocellular carcinoma in rodents has been mustered (reviewed by Thorgeirsson, 1995). The evidence that hepatocellular carcinomas originate from liver stem cells through proliferation and partial differentiation of oval cells is disputed (Farber, 1992). However, since new hepatocytes can be generated from both mature hepatocytes and from oval cells that result from stem cell activation, it seems likely that hepatic neoplasms could arise from cells produced through either pathway. It must be admitted that the demonstration of the cellular pathway to the development of hepatocellular carcinoma from either mature hepatocytes that dedifferentiate (Farber and Sarma, 1987) or oval cells that differentiate (Sell and Dunsford, 1989), is based on evidence that is indirect, using markers to trace cells through presumed developmental pathways. Before this important issue can be settled, more definitive studies must be designed and carried out. The possibility that new hepatocyte lineages generated by liver epithelial stem cells can repopulate parenchyma totally destroyed by hepatic necrosis is intriguing and potentially most important. In rats the parenchyma is efficiently repopulated following its ablation by galactosamine toxicity, and in this model many of the new hepatocytes appear to arise by differentiation of oval cells. However, the overall efficiency (or effectiveness) with which replacement hepatocytes are generated from stem cells by differentiation of oval cells has not yet been precisely determined under any experimental circumstance, and merits further study.

Both of these pathological processes – hepatocellular cancer and massive hepatic necrosis – represent important causes of morbidity and mortality in humans worldwide, and a more detailed understanding of the potential role of stem cells in the initiation or

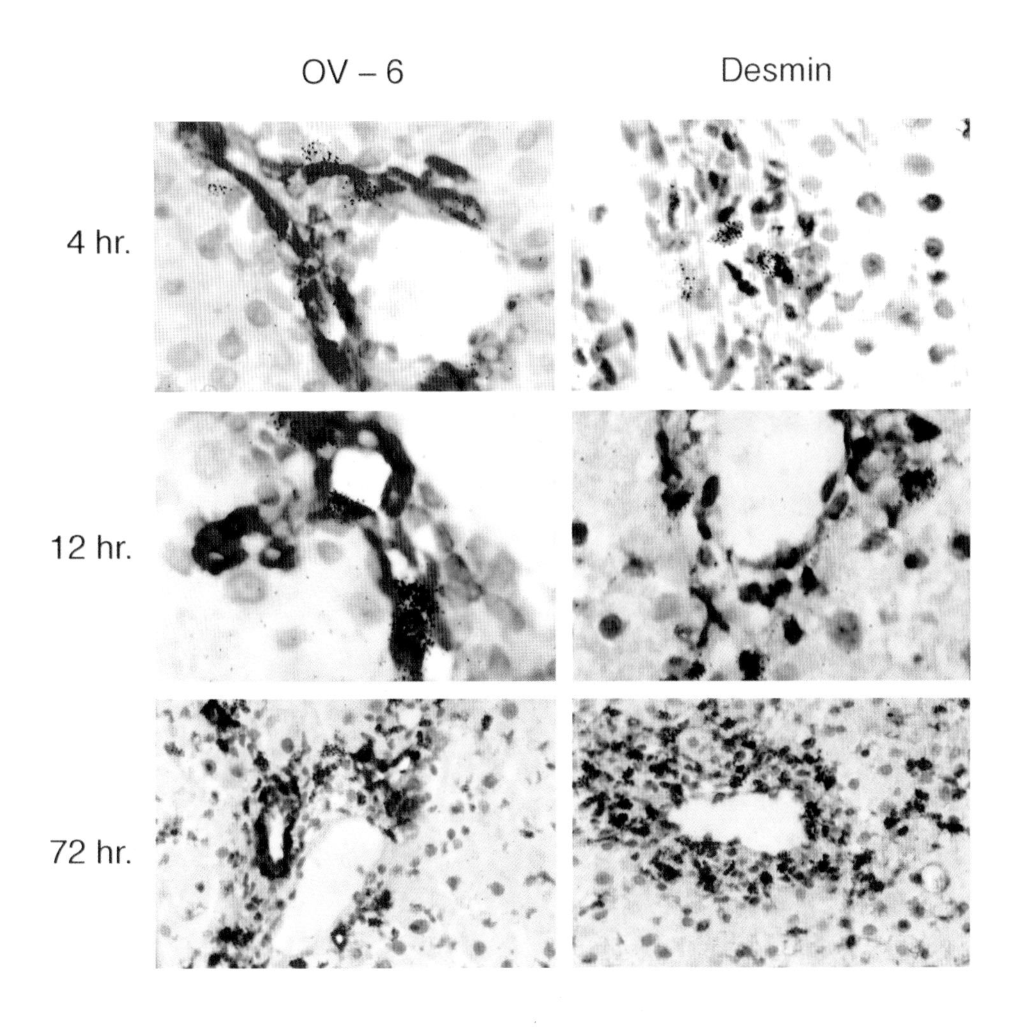

Plate 8.1 Initially proliferating cells in portal tracts of rats subjected to the AAF/PH model. ^{3}H-thymidine-labelled cells at early time points include both OV-6-positive (biliary-type epithelial cells) and desmin-positive cells (Ito cells). ^{3}H-thymidine 1μCi or 37kBq gm body weight) was administered to all animals 2 hours before the time of sacrifice shown. Reproduced from Thorgeirsson *et al.* (1993) with permission.

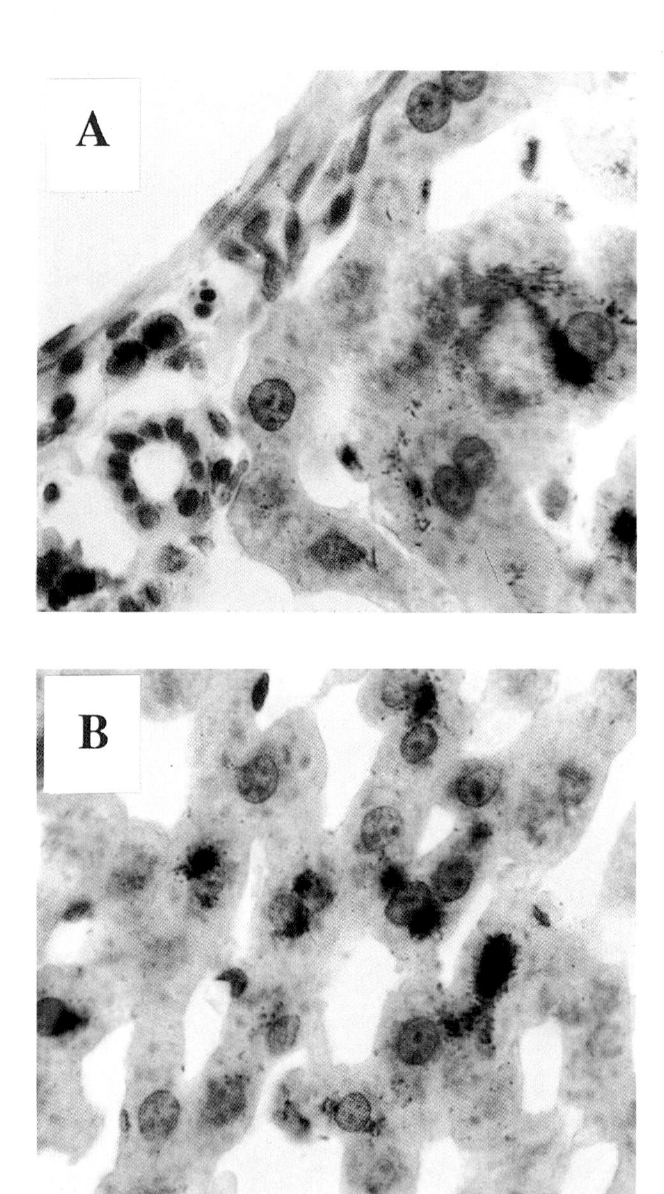

Plate 8.2 Location of progeny of BAG2-WB cells two weeks after transplatation into the liver of a syngeneic Fischer 344 rat. Cryosections were prepared and processed for β-galactosidase and counterstained with Mayer's hematoxylin as described (Coleman *et al.*, 1993; Grisham *et al.*, 1994). Panel A shows small β-galactosidase-positive cells in the tissue of the portal tract, as well as β-galactosidase-positive hepatocytes integrated into hepatic plates among β-galactosidase-negative host hepatocytes. Panel B illustrates several β-galactosidase hepatocytes in hepatic plates of a similar recipient animal. These β-galactosidase-negative hepatocytes differentiated from the small BAG2-WB cells that were transplanted into the liver.

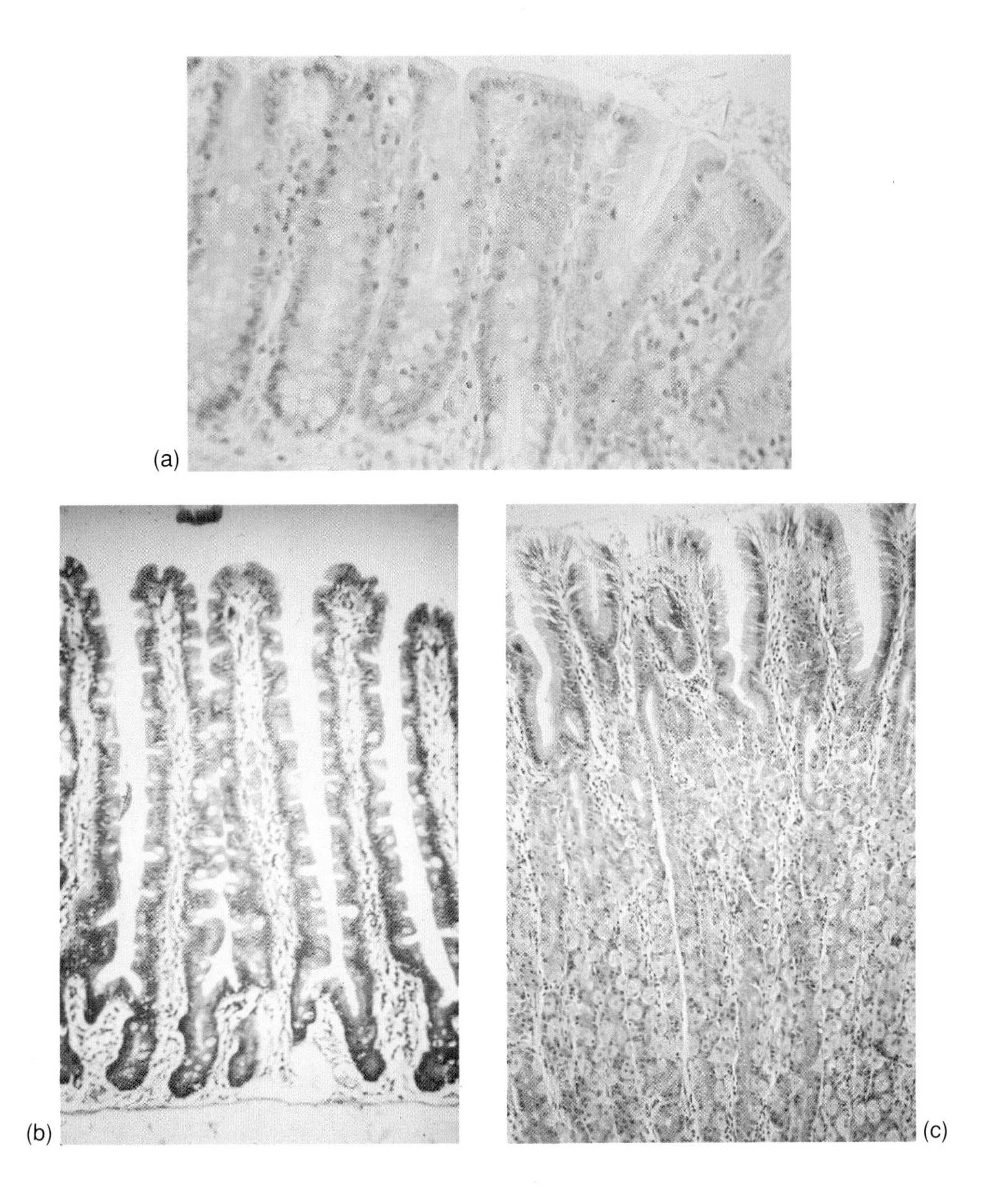

Plate 10.1 The histological appearances of (a) the colon, (b) the small intestine and (c) the acid-secreting gastric mucosa.

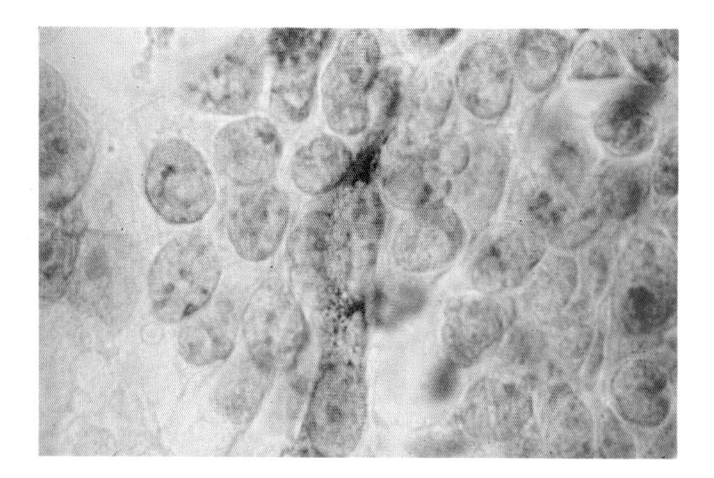

Plate 10.2 An endocrine cell, immunostained with chromogranin A, growing in a culture of HRA-19 cells. Note the typical bipolar morphology. (Courtesy of Dr Susan Kirkland.)

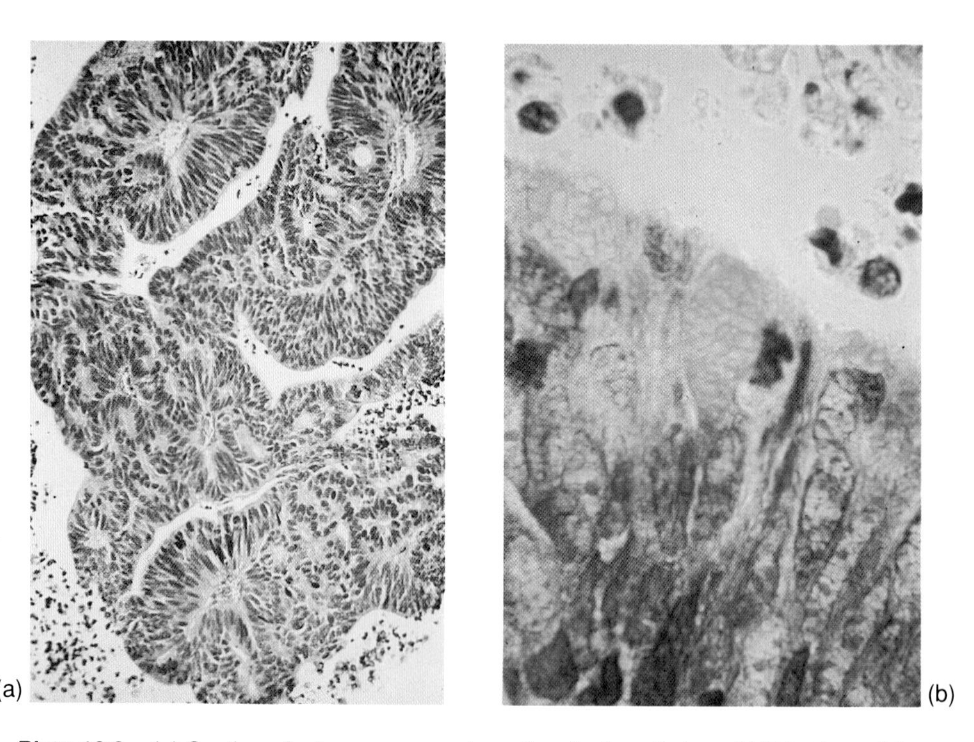

Plate 10.3 (a) Section of a tumour grown from the single-cell cloned HRA-19a cell line in the flank of a nude mouse; (b) from a xenograft of the HRA-19a single cell cloned cell line growing in a nude mouse, demonstrating the presence of goblet cells stained with Alcian Blue. (Courtesy of Dr Susan Kirkland.)

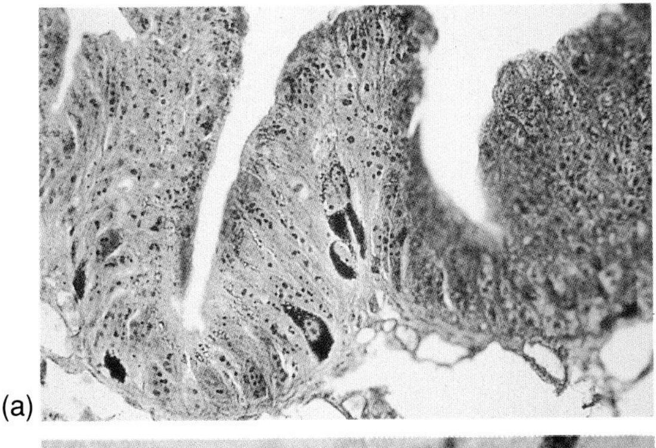

(a)

(b)

Plate 10.4 Endocrine cells in the HRA-19a xenograft demonstrated by (a) the Grimelius technique and (b) immunostaining with chromogranin A. (Courtesy of Dr Susan Kirkland.)

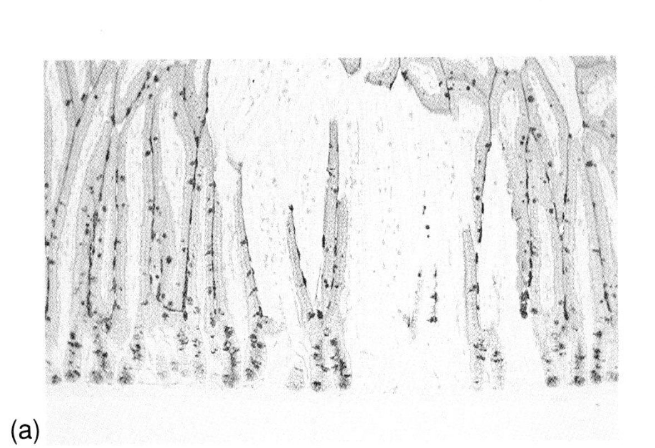

(a)

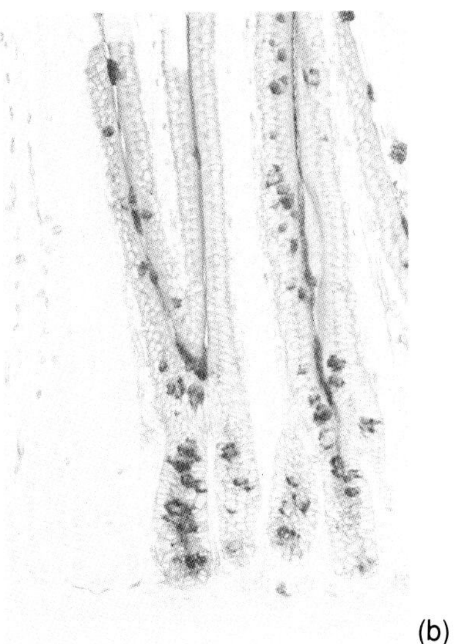

(b)

Plate 10.5 A section from the jejunal mucosa of a tetraparental allophenic mouse stained with the lectin *Dolichos biflorus* agglutinin, which differentiates between the two types of cell present in the chimaera. Note in (a) and confirm in (b) (at high power) that crypts are either totally positive or totally negative, indicating that each crypt was originally the progeny of a single cell, or group of cells, of the same type, and supporting the concept of a clonal origin of cells. The polymorphism is also shown in the endothelial cells. (Photographed from a preparation made available through the courtesy of Professor B. Ponder.)

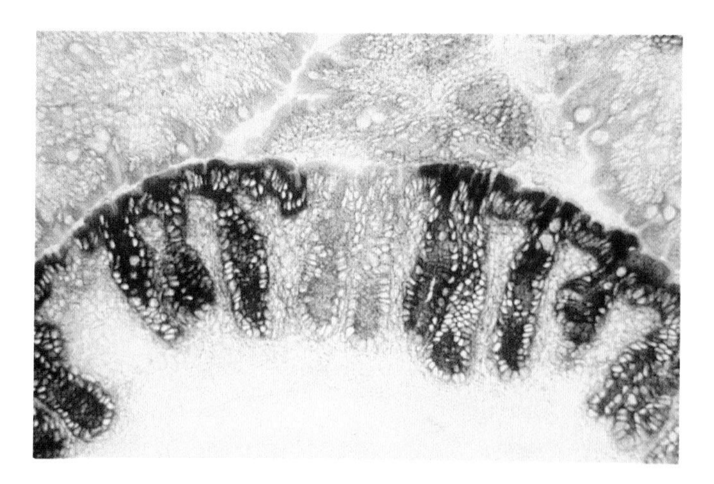

Plate 10.6 A section from the colon of a mouse heterozygous for a defective glucose-6-phosphate dehydrogenase (G6PD) gene, stained for G6PD activity. Most of the mucosa is positive, but single crypts are wholly negative. (Courtesy of Professor Geraint Williams.)

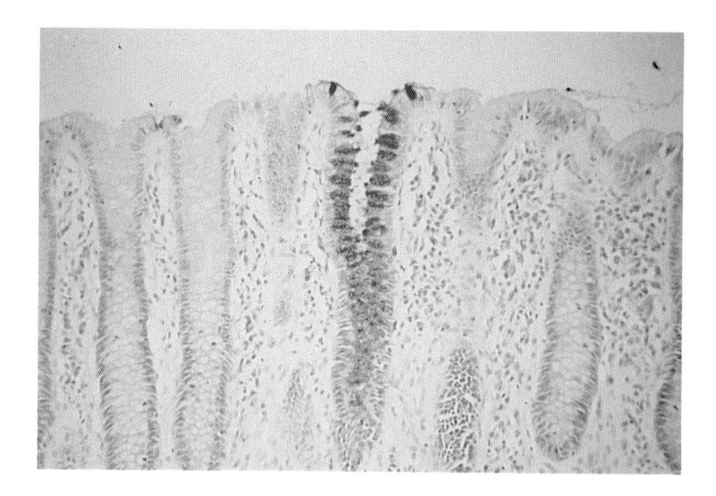

Plate 10.7 A section from the human colon stained with the mild periodic acid-Schiff method for non-acetylated sialomucin; note that the loss of acetylation is confined to single-crypt – crypt restricted, and that all cells in the crypt are negatively stained. (Courtesy of Dr Fiona Campbell.)

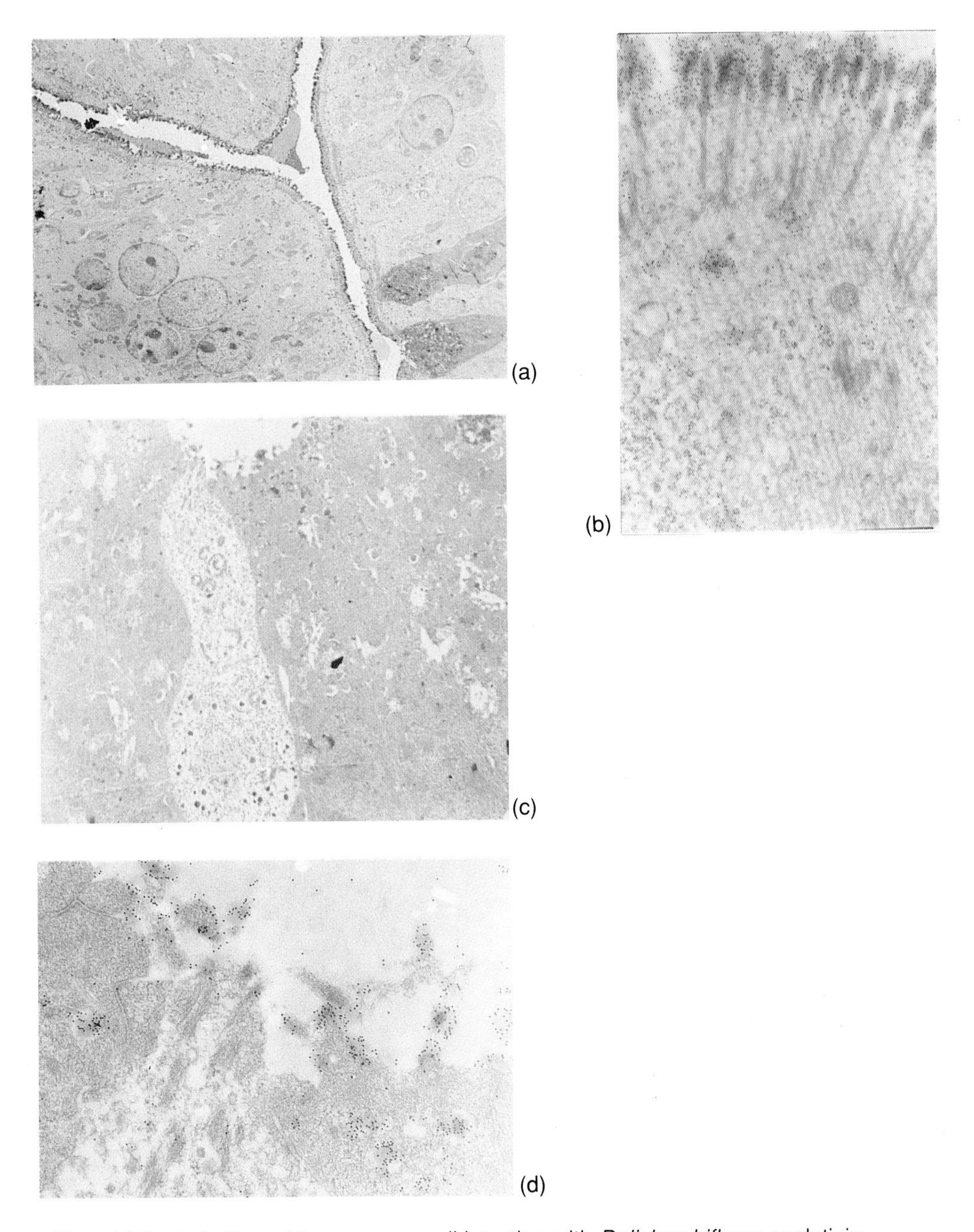

Plate 10.8 Labelling of the mouse small intestine with *Dolichos biflorus* agglutinin (DBA)-gold conjugate, viewed at the ultrastructural level: (a) a low-power view to show definitive surface labelling in a control C57/BL mouse; (b) a high-power view showing that indeed labelling is associated with the glycocalyx; (c) an endocrine cell in the crypt in a control C57/BL mouse, showing the characteristic neuroendocrine granules; (d) the same endocrine cell, viewed at high-power, showing that even in control crypts, endocrine cells do not label with the lectin. (Courtesy of Dr Mary Thompson.)

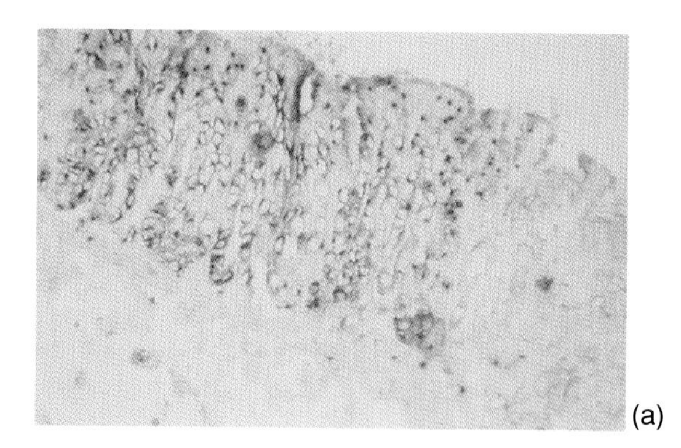

(a)

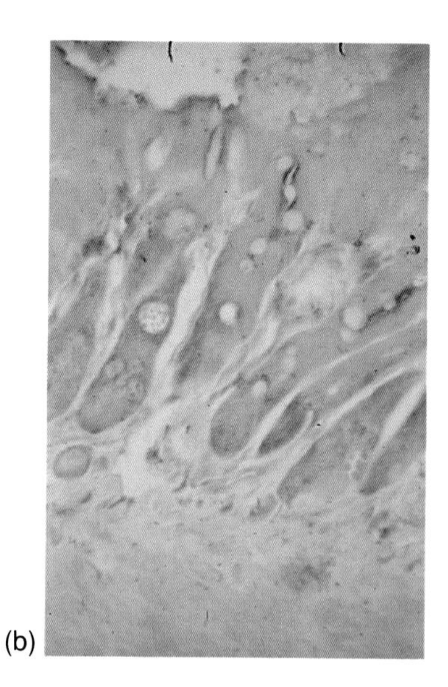

(b)

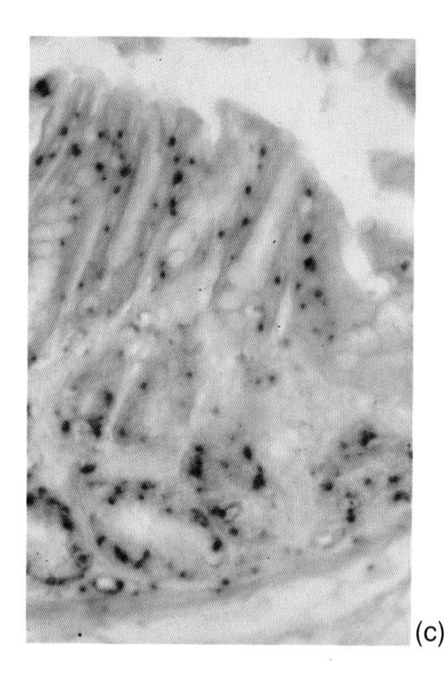

(c)

Plate 10.9 A section of the stomach of an XX/XY chimaeric mouse, contemporaneously subjected to *in situ* hybridization to demonstrate the highly-repetitive sequences in the mouse Y chromosome using digoxigenin, and to show gastrin cells using an antigastrin antibody: (a) a low power view showing contiguous male and female areas; (b) a female area showing endocrine cells which do not bear a blue spot, indicating their female nature; (c) a male area containing definitive male endocrine cells. (Courtesy of Dr Mary Thompson.)

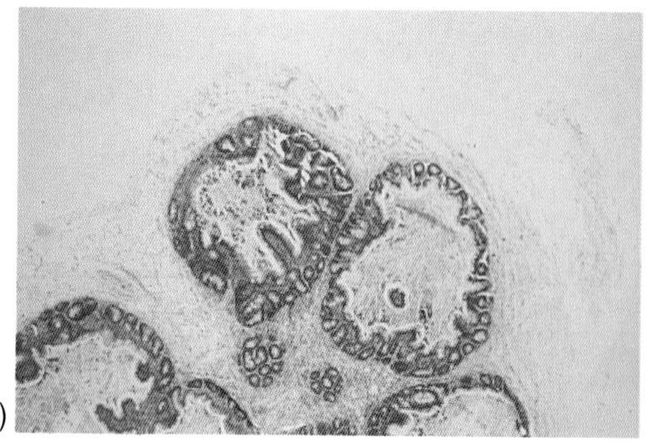

(a)

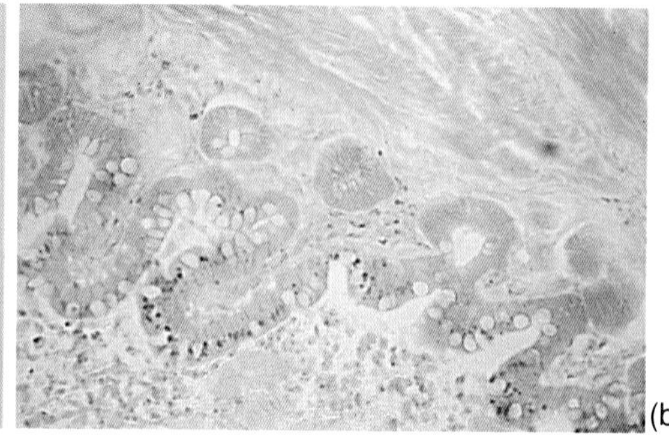

(b)

Plate 10.10 (a) Rat small intestine which has been transplanted subcutaneously into nude mice as a piece of developing gut endoderm embedded in collagen, and sectioned some 14 days later, stained with diastase PAS/Alcian Blue; (b) endoderm which has been transfected with the BAG retrovirus, showing expression of β-galactosidase. (Courtesy of Dr R. Del Buono.)

repair of these processes could hold important benefits for their prevention and/or treatment. Although not discussed in this review, evidence from studies in humans indicates that human livers contain stem-like cells that are also involved in embryogenesis of the liver and possibly in the repair of the liver after massive hepatic necrosis and development of liver cancer (for reviews see Gerber and Thung, 1992; VanEyken and Desmet, 1992). Further study of these and other pathological processes in the human liver are needed to understand the potential roles of liver epithelial stem-like cells in the development and evolution of liver diseases.

Whatever is the eventual outcome of future experiments and debates on liver epithelial stem cells, their investigation has already provided useful new insights about the cytological composition of the liver, and about the repair of the liver by replacement of parenchymal cells. Ultimately this knowledge will aid the determination of the mechanism(s) that regulate the proliferation and differentiation of hepatocytes and other liver epithelial cells and may lead to new insights into control of hepatic functions. The future continuation of experiment and debate also can be expected to enhance our knowledge of hepatic pathology and our ability to understand, and to cope with, hepatic diseases.

ACKNOWLEDGEMENTS

The writing of a portion of this review by J. W. Grisham, and the experimental studies from his laboratory that are reported in it, were supported by grant CA29323 from the National Institutes of Health. He thanks his colleague William B. Coleman for preparing Table 2 and, along with Karen D. McCullough, for reading and perceptively commenting on the text. Mrs Lynn S. Koehneke and Ms Fumi Wells expertly organized and produced several drafts of the text.

REFERENCES

Alexander, R.W. and Grisham, J.W. (1970). *Lab. Invest.* **22**:50–62.
Alison, M. (1986). *Physiol. Rev.* **66**:499–541.
Alpini, G., Aragona, E., Dabeva, M., Salvi, R., Shafritz, D.A. and Tavoloni, N. (1992). *Am. J. Pathol.* **141:**623–632.
Alpini, G., Phillips, J.O., Vroman, B. and LaRusso, N. (1994). *Hepatology* **20**:494–514.
Ang, S.-L., Wierda, A., Wong, D., Stevens, K.A., Cascio, J. and Zaret, K.S. (1993). *Development* **119**:1301–1315.
Arber, N. and Zajicek, G. (1990). *Liver* **10**:205–208.
Arber, N., Zajicek, G. and Ariel, I. (1988). *Liver* **8**:80–87.
Aterman, K. (1992). *J. Cancer Res. Clin. Oncol.* **118**:87–115.
Baldwin, G.C. (1992). *Develop. Biol.* **151**:352–362.
Bamheuter, S., Mendel, D.B., Conley, P.B. *et al.* (1990). *Genes Develop.* 4:372–379.
Bartles, J.R., Braiterman, L.T. and Hubbard, A.L. (1985). *J. Cell Biol.* 100:1126–1138.
Batist, G., Woo, A. and Tsao, M.-S. (1991). *Carcinogenesis* **12**:2031–2034.
Benedetti, A., Jezequel, A.M. and Orlandi, F. (1988). *Liver* **8**:172–177.
Bisgaard, H.C. and Thorgeirsson, S.S. (1991). *J. Cell. Physiol.* **147**:333–343.
Bisgaard, H.C., Parmelee, D.C., Dunsford, H.C., Sechi, S. and Thorgeirsson, S.S. (1993). *Mol. Carcinog.* **7**:60–66.

Bisgaard, H.C., Nagy, P., Ton, P.T., Hu, Z. and Thorgeirsson, S.S. (1994a). *J. Cell. Physiol.* **159**:476–484.
Bisgaard, H.C., Ton, P.T., Nagy, P. and Thorgeirsson, S.S. (1994b). *J. Cell. Physiol.* **159**:485–494.
Blikkendaal-Lieftinck, L.F., Kooij, M., Kramer, M.F. and Den Otter, W. (1977). *Exp. Mol. Pathol.* **26**:184–192.
Blouin, R., Blouin, M., Royal, I., Grenier, A., Roop, D.R., Loranger, A. and Marceau, N. (1992). *Differentiation* **52**:45–54.
Blumenfeld, M., Maury, M., Chouard, T., Yaniv, M. and Condamine, H. (1991). *Development* **113**:589–599.
Boswald, M., Harasim, S. and Maurer-Schultze, B. (1990). *Cell Tissue Kinetics* **23**:169–181.
Bralet, M.-P., Branchereau, S., Brechot, C. and Ferry, N. (1994). *Am. J. Pathol.* **144**:896–905.
Braun, L., Goyette, M., Yaswen, P., Thompson, N.L. and Fausto, N. (1987). *Cancer Res.* **47**:4116–4124.
Braun, L., Mikuno, R. and Fausto, N. (1989). *Cancer Res.* **49**:1554–1561.
Brill, S., Holst, P., Sigal, S. *et al.* (1993). *Proc. Soc. Exp. Biol. Med.* **204**:261–269.
Bryant, B.J. (1962). *Exp. Cell. Res.* **27**:70–79.
Bucher, N.L.R. and Swaffield, M.N. (1964). *Cancer Res.* **24**:1611–1626.
Cascio, S. and Zaret, K.S. (1991). *Development* **113**:217–225.
Chapekar, M.S., Huggett, A.C., Cheng, C.C., Hampton, L.L., Lin, K.-H. and Thorgeirsson, S.S. (1990). *Cancer Res.* **50**:3600–3604.
Coleman, W.B., Wennerberg, A.E., Smith, G.J. and Grisham, J.W. (1993). *Am. J. Pathol.* **142**:1373–1382.
Coleman, W.B., Smith, G.J. and Grisham, J.W. (1994). *J. Cell. Physiol.* **161**:463–469.
Coon, H.G. (1968). *Carnegie Inst. Washington Yearbook* **68**:419–421.
Courtois, G., Bamhueter, S. and Crabtree, G.R. (1988). *Proc. Natl. Acad. Sci. USA* **85**:7937–7941.
Crabtree, G.R., Schibler, V. and Scott, M.P. (1992). In *Transcriptional Regulation* (eds S. S. McKnight and K. R. Yamamoto), pp. 1063–1102. Cold Springer Harbor Laboratory, Cold Spring Harbor, New York.
Dabeva, M.D. and Shafritz, D.A. (1993). *Am. J. Pathol.* **143**:1606–1620.
Dabeva, M.D., Alpini, G., Hurston, E. and Shafritz, D.A. (1993). *Proc. Soc. Exp. Biol. Med.* **204**:242–252.
Dempo, K., Chisaka, N., Yoshida, Y., Kaneko, A. and Onoe, T. (1975). *Cancer Res.* **35**:1282–1287.
Descombes, P., Chojkier, M., Lichsteiner, S., Falvey, E. and Schibler, U. (1990). *Genes Develop.* **4**:1541–1551.
DeSimone, V. and Cortese, R. (1992). *Biochim. Biophys. Acta* **1132**:119–126.
Diehl, A.M., Michaelson, P. and Yang, S.Q. (1994). *Gastroenterology* **106**:1625–1627.
DiPersio, C.M., Jackson, D.A. and Zaret, K.S. (1991). *Mol. Cell. Biol.* **11**:4405–4414.
Duncan, S.A., Manova, K., Chen, W.S. *et al.* (1994). *Proc. Natl Acad. Sci. USA* **91**:7598–7602.
Dunsford, H.A. and Sell, S. (1989). *Cancer Res.* **49**:4887–4893.
Dunsford, H.A., Maset, R., Salman, J. and Sell, S. (1985). *Am. J. Pathol.* **118**:218–224.
Earp, H.S., Austin, K.S., Blaisdell, J. *et al.* (1986). *J. Biol. Chem.* **261**:4777–4780.
Earp, H.S., Hepler, J.R., Petch, L.A. *et al.* (1988). *J. Biol. Chem.* **263**:13868–13874.
Edwards, J.L. and Klein, R.E. (1961). *Am. J. Pathol.* **38**:437–455.
Ekblom, P., Vestweber, D. and Kemler, R. (1986). *Annu. Rev. Cell. Biol.* **2**:27–47.
Engelhardt, N.V., Factor, V.M., Yasova, A.K. *et al.* (1990). *Differentiation* **45**:29–37.
Engelhardt, N.V., Factor, V.M., Medvinsky, A.L., Baranov, V., Lazareva, M.N. and Poltoranina, V.S. (1993). *Differentiation* **55**:19–26.
Evarts, R.P., Nagy, P., Marsden, E. and Thorgeirsson, S.S. (1987). *Carcinogenesis* **8**:1737–1740.
Evarts, R.P., Nagy, P., Nakatsukasa, H., Marsden, E. and Thorgeirsson, S.S. (1989). *Cancer Res.* **49**:1541–1547.
Evarts, R.P., Nakatsukasa, H., Marsden, E.R., Hsia, C.-C., Dunsford, H.A. and Thorgeirsson, S.S. (1990). *Cancer Res.* **50**:3439–3444.
Evarts, R.P., Nakatsukasa, H., Marsden, E.R., Hu, Z. and Thorgeirsson, S.S. (1992). *Mol. Carcinog.* **5**:25–31.
Evarts, R.P., Hu, Z., Fujio, K., Marsden, E.R. and Thorgeirsson, S.S. (1993). *Cell Growth Differ.* **4**:555–561.

Fabrikant, J.I. (1968a). *Johns Hopkins Med. Bull.* **120**:137–147.
Fabrikant, J.I. (1968b). *J. Cell Biol.* **36**:551–565.
Fabrikant, J.I. (1969). *Exp. Cell. Res.* **55**:277–279.
Factor, V.M., Radaeva, S.A. and Thorgeirsson, S.S. (1994). *Am. J. Pathol.* **145**:409–422.
Farber, E. (1956). *Cancer Res.* **16**:142–148.
Farber, E. (1992). In *The Role of Cell Types in Hepatocarcinogenesis* (ed. A. E. Sirica), pp. 1–28. CRC Press, Boca Raton, FL.
Farber, E. and Sarma, D.S.R. (1987). *Lab. Invest.* **56**:4–22.
Faris, R.A. and Hixson, D.C. (1989). *Transplantation* **48**:87–92.
Faris, R.A., Monfils, B.A., Dunsford, H.A. and Hixson, D.C. (1991). *Cancer Res.* **51**:1308–1317.
Faris, R.A., McBride, A., Yang, L., Affigne, S., Walker, C. and Cha, C.-J. (1994). *Am. J. Pathol.* **145**:1432–1443.
Fausto, N. (1990). *Current Opinion Cell. Biol.* **2**:1036–1042.
Fausto, N. (1994). In *The Liver: Biology and Pathobiology* (eds I. M. Arias, J. L. Boyer, N. Fausto, W. B. Jakoby, D. A. Schachter and D. A. Shafritz), pp. 1501–1518. Raven Press, New York.
Fausto, N., Thompson, H.L. and Braun, L. (1987). In *Cell Separation Methods and Selected Applications*, Vol. 4 (eds T. G. Pretlow, II and T. R. Pretlow), pp. 45–77. Academic Press, Orlando, FL.
Fausto, N., Lemire, J.M. and Shiojiri, N. (1992). In *The Role of Cell Types in Carcinogenesis* (ed. A. E. Sirica), pp. 89–108. CRC Press, Boca Raton, FL.
Fausto, N., Lemire, J.M. and Shiojiri, N. (1993). *Proc. Soc. Exp. Biol. Med.* **204**:237–241.
Feracci, H., Connolly, T.P., Margolis, R.N. and Hubbard, A.L. (1987). *Develop. Biol.* **123**:73–84.
Frain, M., Swart, G., Monaci, P. *et al.* (1989). *Cell* **59**:145–157.
Fujio, K., Evarts, R.P., Hu, Z., Marsden, E.R. and Thorgeirsson, S.S. (1994). *Lab. Invest.* **70**:511–516.
Furukawa, K., Shimada, T., England, P., Mochizuki, Y. and Williams, G.M. (1987). *In Vitro. Cell. Develop. Biol.* **23**:339–348.
Gall, J.A.M. and Bhathal, P.S. (1990a). *Liver* **10**:106–115.
Gall, J.A.M. and Bhathal, P.S. (1990b). *J. Exp. Pathol.* **71**:41–50.
Gardner, R.L. (1985). *Phil. Trans. R. Soc. Lond. B* **312**:163–178.
Garfield, S., Huber, B.E., Nagy, P., Cordingley, M.G. and Thorgeirsson, S.S. (1988). *Mol. Carcinog.* **1**:189–195.
Gebhardt, R. (1992). *Pharmac. Therap.* **53**:275–353.
Geisler, A., Stiller, K. and Machnik, G. (1994). *Exp. Toxicol. Pathol.* **46**:247–250.
Gerber, M.A. and Thung, S.N. (1992). In *The Role of Cell Types in Hepatocarcinogenesis* (ed. A. D. Sirica), pp. 209–226. CRC Press, Boca Raton, FL.
Gerlyng, P., Grotmol, T., Stokke, T., Erikstein, B. and Seglen, P.O. (1994). *Carcinogenesis* **15**:53–59.
Germain, L., Goyette, R. and Marceau, N. (1985). *Cancer Res.* **45**:673–681.
Germain, L., Blouin, M.-J. and Marceau, N. (1988a). *Cancer Res.* **48**:4909–4918.
Germain, L., Noel, M., Gourdeau, H. and Marceau, N. (1988b). *Cancer Res.* **48**:363–378.
Gordon, J.I., Schmidt, G.H. and Roth, K.A. (1992). *FASEB J.* **6**:3039–3050.
Goyette, M., Faris, R., Braun, L., Hixson, D. and Fausto, N. (1990). *Cancer Res.* **50**:4809–4817.
Graham, J.G. and Pragnell, I.B. (1992). *Develop. Biol.* **151**:377–381.
Greengard, O. (1969). *Science* **163**:891–895.
Greengard, O., Federman, M. and Knox, W.E. (1972). *J. Cell. Biol.* **52**:261–272.
Griffo, G., Hamon-Benais, C., Angrand, P.-O. *et al.* (1993). *J. Cell Biol.* **121**:887–898.
Grisham, J.W. (1962). *Cancer Res.* **26**:842–849.
Grisham, J.W. (1969). *Recent Results Cancer Res.* **17**:28–43.
Grisham, J.W. (1979). *Int. Rev. Exp. Pathol.* **20**:123–210.
Grisham, J.W. (1980). *Ann. NY Acad. Sci.* **349**:128–137.
Grisham, J.W. (1994). *Am. J. Pathol.* **144**:849–854.
Grisham, J.W. and Hartroft, W.S. (1961). *Lab. Invest.* **10**:317–332.
Grisham, J.W. and Porta, E.A. (1964). *Exp. Mol. Pathol.* **3**:242–261.
Grisham, J.W., Thal, S.B. and Nagel, A. (1975a). In *Gene Expression and Carcinogenesis in Cultured Liver* (eds L. E. Gerschenson and E. B. Thompson), pp. 1–23. Academic Press, New York.
Grisham, J.W., Tillman, R.L., Nägel, A.E.H. and Compagno, J. (1975b). In *Liver Regeneration after Experimental Injury* (eds R. Lesch and W. Reutter), pp. 6–23. Stratton Intercontinental Medical Book Corp., New York.

Grisham, J.W., Coleman, W.B. and Smith, G.J. (1993). *Proc. Soc. Exp. Biol. Med.* **204**:270–279.
Gupta, S., Aragona, E., Vemuru, R.P., Bhargava, K.K., Burk, R.D. and Chowdhury, J.R. (1991). *Hepatology* **14**:144–149.
Gupta, S., Wilson, J.M. and Chowdhury, J.R. (1992). *Sem. Liver Dis.* **12**:321–331.
Haber, B.A., Mohn, K.L., Diamond, R.H. and Taub, R. (1993). *J. Clin. Invest.* **91**:1319–1326.
Hall, P.A. and Watt, F.M. (1989). *Development* **106**:619–633.
Hampton, L.L., Worland, P.J., Yu, B., Thorgeirsson, S.S. and Huggett, A.L. (1990). *Cancer Res.* **50**:7468–7475.
Hata, M., Nanno, M., Doi, H. *et al.* (1993). *J. Cell. Physiol.* **154**:381–392.
Hayner, N.T., Braun, L., Yaswen, P., Brooks, M. and Fausto, N. (1984). *Cancer Res.* **44**:332–338.
Heiniger, H.J., Friedrich, G., Feinendegen, L.E. and Cantelmo, F. (1971). *Proc. Soc. Exp. Biol. Med.* **137**:1381–1384.
Herring, A.S., Raychaudhuri, R., Kelley, S.P. and Iype, P.T. (1983). *In Vitro* **19**:576–588.
Herzfeld, A., Federman, M. and Greengard, O. (1973). *J. Cell. Biol.* **57**:475–483.
Hilberg, F., Aguzzi, A., Howells, N. and Wagner, E.F. (1993). *Nature* **365**:179–181.
Hixson, D.C. and Allison, J.P. (1985). *Cancer Res.* **45**:3750–3760.
Hixson, D.C., Faris, R.A. and Thompson, N.L. (1990). *Pathobiology* **58**:65–77.
Hixson, D.C., Faris, R.A., Yang, L. and Novikoff, P. (1992). In *The Role of Cell Types in Hepatocarcinogenesis* (ed. A. E. Sirica), pp. 151–182. CRC Press, Boca Raton, FL.
Houssaint, E. (1980). *Cell. Differ.* **9**:269–279.
Houssaint, E. (1981). *Cell Differ.* **10**:243–252.
Howell, S., Wareham, K.A. and Williams, E.D. (1985). *Am. J. Pathol.* **121**:426–432.
Hu, Z., Evarts, R.P., Fujio, K., Marsden, E.R. and Thorgeirsson, S.S. (1993). *Am. J. Pathol.* **142**:1823–1830.
Huggett, A.C., Ford, C.P. and Thorgeirsson, S.S. (1989). *Growth Factors* **2**:83–89.
Huggett, A.C., Hampton, L.L., Ford, C.P., Wirth, P.J. and Thorgeirsson, S.S. (1990). *Cancer Res.* **50**:7468–7475.
Huggett, A.C., Ellis, P.A., Ford, C.P., Hampton, L.L., Rimoldi, D. and Thorgeirsson, S.S. (1991). *Cancer Res.* **51**:5929–5936.
Iannaccone, P.M. (1987). *Cell. Differ.* **21**:79–91.
Iannaccone, P.M. (1990). *FASEB J.* **4**:1508–1512.
Iannaccone, P.M., Weinberg, W.C. and Berkwits, L. (1987). *Development* **99**:187–196.
Inaoka, Y. (1967). *Gann.* **58**:355–366.
Jezequel, A.M., Paolucci, F., Benedetti, A., Mancini, R. and Orlandi, F. (1991). *Dig. Dis. Sci.* **36**:482–484.
Johnson, P.F. (1990). *Cell Growth Differ.* **1**:47–51.
Johnson, M., Koukoulis, G., Matsumoto, K., Nakamura, T. and Iyer, A. (1993). *Hepatology* **17**:1052–1061.
Jones, A.L. and Schmucker, D.L. (1977). *Gastroenterology* **73**:833–851.
Joplin, R. (1994). *Gut* **35**:875–878.
Jordan, C.T. and Lemischhka, I.R. (1990). *Genes Develop.* **4**:220–232.
Jungermann, K. (ed.) (1992). *Enzyme* **46**:5–168.
Kalimi, G.H., Hampton, L.L., Trosko, J.E., Thorgeirsson, S.S. and Huggett, A.C. (1992). *Mol. Carcinog.* **5**:301–310.
Kennedy, S., Rettinger, S., Flye, M.W. and Ponder, K.P. (1995). *Hepatology* **22**:160–168.
Khokha, M.K., Landini, G. and Iannaccone, P.M. (1994). *Develop. Biol.* **165**:545–555.
Kinosita, R. (1937). *Trans. Soc. Pathol. Jpn.* **27**:665–727.
Klinman, N.R. and Erslev, A.J. (1963). *Proc. Soc. Exp. Biol. Med.* **112**:338–340.
Kuhlmann, W.D. (1978). *Int. J. Cancer* **21**:368–380.
Kuo, C.J. and Crabtree, G.R. (1992). In *Development the Molecular Genetic Approach* (eds V. E. A. Russo, S. Brody, D. Core and S. Ottolenghi), pp. 379–498. Springer-Verlag, Berlin.
Kuo, C.J., Conley, P.B., Hsieh, C.-L., Francke, U. and Crabtree, G.R. (1990a). *Proc. Natl Acad. Sci. USA* **87**:9838–9842.
Kuo, C.J., Xanthopoulos, K.G. and Darnell, J.R. (1990b). *Development* **109**:473–489.
Kuo, C.J., Conley, P.B., Chen, L., Sladek, F.M., Darnell, J.E. and Crabtree, G.R. (1992). *Nature* **355**:457–461.

Kusakabe, M., Yokoyama, M., Sakakura, T., Nomura, T., Hosick, H.L. and Nishizuka, Y. (1988). *J. Cell. Biol.* **107**:257–251.
Lai, E. (1992). *Sem. Liver Dis.* **12**:246–251.
Lai, E. and Darnell, J.E. (1991). *Trends Biochem. Sci.* **16**:427–430.
Lai, E., Prezioso, V.R., Smith, E., Litvin, O., Costa, R.H. and Darnell, J.E. (1990). *Genes Develop.* **4**:1427–1436.
Lai, E., Prezioso, V.R., Tao, W., Chen, W.S. and Darnell, J.E. (1991). *Genes Develop.* **5**:416–427.
Landschulz, W.H., Johnson, P.F., Adashi, E.Y., Graves, B.J. and McKnight, S.L. (1988). *Genes Develop.* **2**:786–800.
LeBouton, A.V. and Marchand, R. (1970). *Develop. Biol.* **23**:524–533.
LeDouarin, N.M. (1975). *Med. Biol.* **53**:427–455.
Leduc, E.H. (1959). *J. Histochem. Cytochem.* **7**:253–255.
Lee, F.D. (1992). *Develop. Biol.* **151**:331–338.
Lee, L.W., Tsao, M.-S., Grisham, J.W. and Smith, G.J. (1989). *Am. J. Pathol.* **135**:63–71.
Lee, G.-H., Nomura, K., Kanda, H. *et al.* (1991a). *Cancer Res.* **51**:3257–3260.
Lee, L.W., Raymond, V.W., Tsao, M.-S., Lee, D.C., Earp, H.S. and Grisham, J.W. (1991b). *Cancer Res.* **51**:5238–5244.
Lemire, J.M. and Fausto, N. (1991). *Cancer Res.* **51**:4656–4664.
Lemire, J.M., Shiojiri, N. and Fausto, N. (1991). *Am. J. Pathol.* **139**:535–552.
Lemischka, I.R., Raulet, D.H. and Mulligan, R.C. (1986). *Cell* **45**:917–927.
Lenzi, R., Liu, M.H., Tarsetti, F. *et al.* (1992). *Lab. Invest.* **66**:390–402.
Lesch, R., Reutter, W., Keppler, D. and Decker, K. (1970). *Exp. Mol. Pathol.* **13**:58–69.
Lichsteiner, S. and Schibler, U. (1989). *Cell* **57**:1179–1187.
Lin, P., Liu, C., Tsao, M.-S. and Grisham, J.W. (1987). *Biochem. Biophys. Res. Commun.* **143**:26–30.
Liu, C., Tsao, M.-S. and Grisham, J.W. (1988). *Cancer Res.* **48**:850–855.
Luzzatto, A.C. (1981). *Cell. Tissue Res.* **215**:133–142.
McCullough, K.D., Coleman, W.B., Smith, G.J. and Grisham, J.W. (1994). *Cancer Res.* **54**:3668–3671.
McCuskey, R.S. (1993). In *Hepatic Transport and Bile Secretion: Physiology and Pathophysiology* (eds N. Tavoloni and P. D. Berk), pp. 1–10. Raven Press, New York.
McKellar, M. (1949). *Am. J. Anat.* **85**:263–307.
McMahon, J.B., Richards, W.L., del Campo, A.A., Song, M.K. and Thorgeirsson, S.S. (1986). *Cancer Res.* **46**:4665–4671.
Maher, J.J. and Bissell, D.M. (1993). *Sem. Cell Biol.* **4**:189–201.
Makino, Y., Yamamoto, K. and Tsuji, T. (1988). *Acta. Med. Okayama* **42**:143–150.
Marceau, N. (1990). *Lab. Invest.* **63**:4–20.
Marceau, N. (1994). *Gut* **35**:294–296.
Marceau, N., Germain, L., Goyette, R., Nel, M. and Gourdeau, H. (1986). *Biochem. Cell Biol.* **64**:788–802.
Marceau, N., Blouin, M.J., Germain, L. and Noël, M. (1989). *In Vitro Cell. Develop. Biol.* **25**:336–341.
Marceau, N., Blouin, M.J., Noël, M., Török, N. and Loranger, A. (1992). In *The Role of Cell Types in Hepatocarcinogenesis* (ed. A. E. Sirica), pp. 121–149. CRC Press, Boca Raton.
Marsden, E.R., Hu, Z., Fujio, K., Nakatsukasa, H., Thorgeirsson, S.S. and Evarts, R.P. (1992). *Lab. Invest.* **67**:427–433.
Martinez-Hernandez, A. and Amenta, P.S. (1993a). *Virchows Arch. A* **423**:1–11.
Martinez-Hernandez, A. and Amenta, P.S. (1993b). *Virchows Arch. A* **423**:77–84.
Martinez-Hernandez, A., Delgado, F.M. and Amenta, P.S. (1991). *Lab. Invest.* **64**:157–166.
Marucci, L., Sregliati Baroni, G., Mancini, R., Benedetti, A., Jezequel, A.M. and Orlandi, F. (1993). *J. Hepatol.* **17**:163–169.
Marvin, M. and McKay, R. (1992). *Sem. Cell Biol.* **3**:401–411.
Mendel, D.B., Hanse, L.P., Graves, M.K., Conley, P.B. and Crabtree, G.R. (1991). *Genes Develop.* **5**:1042–1056.
Messier, B. and Leblond, C.P. (1960). *Am. J. Anat.* **106**:247–285.
Metcalf, D. (1991). *Proc. Natl Acad. Sci. USA* **88**:11310–11314.
Mohn, K.L., Laz, T.M., Melby, A.E. and Taub, R. (1990). *J. Biol. Chem.* **265**:21914–21920.
Moll, R., Franke, W.W., Schiller, D.L., Geiger, B. and Krepler, R. (1982). *Cell* **31**:11–24.
Monaghan, A.P., Kaestner, K.H., Grau, E. and Schütz, G. (1993). *Development* **119**:567–578.

Montesano, R., Saint Vincent, L. and Tomatis, L. (1973). *Br. J. Cancer* **28**:215–220.
Moreau, A., Maurice, M. and Feldmann, G. (1988). *J. Histochem. Cytochem.* **36**:87–94.
Morrison-Graham, K. and Takahashi, Y. (1993). *BioEssays* **15**:77–83.
Mueller, C.R., Maire, P. and Schibler, U. (1990). *Cell* **61**:279–291.
Nagy, P., Evarts, R.P., McMahon, J.B. and Thorgeirsson, S.S. (1989). *Mol. Carcinog.* **2**:345–354.
Nagy, P., Bisgaard, H.C. and Thorgeirsson, S.S. (1994). *J. Cell Biol.* **126**:223–233.
Nakatsukasa, H., Evarts, R.P., Hsia, C.-C., Marsden, E.R. and Thorgeirsson, S.S. (1991). *Lab. Invest.* **65**:511–517.
Nakatsukasa, H., Evarts, R.P., Burt, R.K., Nagy, P. and Thorgeirsson, S.S. (1992). *Mol. Carcinog.* **6**:190–198.
Nanno, M., Hata, M., Doi, H., Satomi, S. *et al.* (1994). *J. Cell. Physiol.* **160**:445–454.
Neveu, M.J., Sattler, C.A., Sattler, G.L. *et al.* (1994). *Mol. Carcinog.* **11**:145–154.
Ng, Y.-K. and Iannaccone, P.M. (1992a). *Current Topics Develop. Biol.* **27**:235–274.
Ng, Y.-K. and Iannaccone, P.M. (1992b). *Develop. Biol.* **151**:419–430.
Nishikawa, Y., Ohta, T., Ogawa, K. and Nagase, S. (1994). *Lab. Invest.* **70**:925–932.
Noda, C. and Ichihara, A. (1993). *Cell Struct. Funct.* **18**:189–194.
Nostrant, T.T., Miller, D.L., Appelman, H.D. and Gumucio, J.J. (1978). *Gastrenterology* **75**:181–186.
Odin, P. and Öbrink, B. (1988). *Exp. Cell Res.* **179**:89–103.
Ogawa, H., Minase, T. and Onoe, T. (1974). *Cancer Res.* **34**:3379–3386.
Onda, H. (1976). *Gann.* **67**:253–262.
Onoé, T., Dempo, K., Kaneko, A. and Watabe, H. (1973). *Gann Monogr. Cancer Res.* **14**:233–247.
Osborn, M. and Weber, K. (1982). *Cell* **31**:303–306.
Ott, M.-O., Rey-Campos, J., Cereghini, S. and Yaniv, M. (1991). *Mech. Develop.* **36**:47–58.
Pack, R., Heck, R., Dienes, H.P., Oesch, F. and Steinberg, P. (1993). *Exp. Cell Res.* **204**:198–209.
Piredale, J.P. and Arthur, M.J.P. (1994). *Gut* **35**:729–732.
Plenat, F., Braun, L. and Fausto, N. (1988). *Am. J. Pathol.* **130**:91–102.
Poli, V., Mancini, F.P. and Cortese, R. (1990). *Cell* **63**:643–653.
Polimeno, L., Azzarone, A., Zeng, Q.H. *et al.* (1995). *Hepatology* **21**:1070–1078.
Ponder, K.P., Gupta, S., Leland, F. *et al.* (1991). *Proc. Natl Acad. Sci. USA* **88**:1217–1221.
Popper, H., Kent, G. and Stein, R. (1957). *J. Mt Sinai Hosp.* **24**:551–556.
Potten, C.S. and Loeffler, M. (1990). *Development* **110**:1101–1020.
Price, J.M., Harman, J.W., Miller, E.C. and Miller, J.M. (1952). *Cancer Res.* **12**:192–200.
Price, J., Turner, D. and Cepko, C. (1987). *Proc. Natl Acad. Sci. USA* **84**:156–160.
Rabes, H. and Tuczek, H.-V. (1970). *Virchows Arch. B Cell Pathol.* **6**:302–312.
Rabes, H.M., Wirsching, R., Tuczek, H.-V. and Iseler, G. (1976). *Cell Tissue Kinet.* **9**:517–532.
Rabes, H.M., Bücher, T., Hartmann, A., Linke, I. and Dünnwald, M. (1982). *Cancer Res.* **42**:3220–3227.
Rajvanshi, P. and Gupta, S. (1994). *Hepatology* **20**:266A (abstract 678).
Rao, M.S., Bendayan, R.D., Kimbrough, R.D. and Reddy, J.K. (1986). *J. Histochem. Cytochem.* **34**:197–201.
Reid, L.M., Fiorino, A.S., Sigal, S.H., Brill, S. and Holst, P.A. (1992). *Hepatology* **15**:1198–1203.
Ren, D., deFeijter, A.W., Paul, D.L. and Ruch, R.J. (1994). *Carcinogenesis* **15**:1801–1813.
Reynolds, B.A. and Weiss, S. (1992). *Science* **255**:1707–1710.
Rhim, J.A., Sandgren, E.P., Degen, J.L., Palmiter, R.D. and Brinster, R.L. (1994). *Science* **263**:1149–1152.
Rogler, L.E. (1994). *Hepatology* **20**:211A (abstract 457).
Rossant, J., Vijh, M., Siracusa, L.D. and Chapman, V.M. (1983). *J. Embryol. Exp. Morphol.* **73**:179–191.
Rubin, E. (1964). *Exp. Mol. Pathol.* **3**:279–336.
Sandgren, E.P., Palmiter, R.D., Heckel, J.L., Daugherty, C.C., Brinster, R.L. and Degen, J.L. (1991). *Cell* **66**:245–256.
Sarraf, C., Lalani, E., Golding, M., Anilkumar, T.V., Poulsom, R. and Alison, M. (1994). *Am. J. Pathol.* **145**:1114–1126.
Scherer, E. and Friedrich-Freksa, H. (1970). *Z. Naturforsch. B* **256**:637–642.
Schmidt, C., Bladt, F., Goedecke, S. *et al.* (1995). *Nature* **373**:699–702.
Schultze, B. and Oehlert, W. (1960). *Science* **131**:737–738.

Sell, S. (1990). *Cancer Res.* **50**:3811–3815.
Sell, S. (1993). *Int. J. Develop. Biol.* **37**:189–201.
Sell, S. (1994). *Mod. Pathol.* **7**:105–112.
Sell, S. and Dunsford, H.A. (1989). *Am. J. Pathol.* **134**:1347–1363.
Sell, S. and Pierce, G.B. (1994). *Lab. Invest.* **70**:6–22.
Sell, S. and Salman, J. (1984). *Am. J. Pathol.* **114**:287–300.
Sell, S., Leffert, H.L., Shinozuka, H., Lombardi, B. and Goochman, N. (1981a). *Gann.* **72**:479–487.
Sell, S., Osborn, K. and Leffert, H.L. (1981b). *Carcinogenesis* **2**:7–14.
Sells, M., Katyal, S.L., Shinozuka, H., Estes, L.W., Sell, S. and Lombardi, B. (1981). *J. Natl Cancer Inst.* **66**:355–362.
Shinozuka, H., Lombardi, B., Sell, S. and Iammarino, R.M. (1978). *Cancer Res.* **38**:1092–1098.
Shiojiri, N. (1984). *J. Embryol. Exp. Morphol.* **79**:23–39.
Shiojiri, N. and Mizuno, T. (1993). *Anat. Embryol.* **187**:221–229.
Shiojiri, N., Lemire, J.M. and Fausto, N. (1991). *Cancer Res.* **51**:2611–2620.
Sigal, S.H., Brill, S., Fiorino, A.S. and Reid, L.M. (1992). *Am. J. Physiol.* **263**:G139–G148.
Sigal, S.H., Brill, S., Reid, L.M. *et al.* (1994). *Hepatology* **19**:999–1006.
Simpson, G.E.C. and Finckh, E.S. (1963). *J. Pathol. Bacteriol.* **86**:361–370.
Sirica, A.E. (ed.) (1992). *The Role of Cell Types in Hepatocarcinogenesis.* CRC Press, Boca Raton, FL.
Sirica, A.E. (1995). *Histol. Histopathol.* **10**:433–456.
Sirica, A.E. and Cihla, H.P. (1984). *Cancer Res.* **44**:3454–3466.
Sirica, A.E., Mathis, G.A., Sano, N. and Elmore, L.W. (1990). *Pathobiology* **58**:44–64.
Sirica, A.E., Elmore, L.W. and Sano, N. (1991). *Digest. Dis. Sci.* **36**:494–501.
Sladek, F.M., Zhong, W., Lai, E. and Darnell, J.E. (1990). *Genes Develop.* **4**:2353–2365.
Slott, P.A., Liu, M.H. and Tavaloni, N. (1990). *Gastroenterology* **99**:466–477.
Smith, A.G. and Hooper, M.L. (1987). *Develop. Biol.* **121**:1–9.
Smith, A.G., Nichols, J., Robertson, M. and Rathjen, P.D. (1992). *Develop. Biol.* **151**:339–351.
Spray, D.C., Chanson, M., Moreno, A.P., Dermietzel, R. and Meda, P. (1991). *Am. J. Physiol.* **260**:C513–C527.
Stamatoglou, S.C. and Hughes, R.C. (1994). *FASEB J.* **8**:420–427.
Stamatoglou, S.C., Enrich, C., Manson, M.M. and Hughes, R.C. (1992). *J. Cell. Biol.* **116**:1507–1515.
Steinberg, P., Hacker, H.J., Dienes, H.P., Oesch, F. and Bannasch, P. (1991). *Carcinogenesis* **12**:225–231.
Steinberg, P., Steinbrecher, R., Radaeva, S. *et al.* (1994). *Lab. Invest.* **71**:700–709.
Steiner, J.W. and Carruthers, J.S. (1961). *Am. J. Pathol.* **38**:639–661.
Steinert, P.M. and Roop, D.R. (1988). *Am. Rev. Biochem.* **37**:593–625.
Stöcker, E. and Butter, D. (1968). *Experientia* **24**:704–705.
Stöcker, E. and Heine, W.-D. (1971). *Beitr. Pathol.* **144**:400–408.
Strain, A.J. (1994). *Gut* **35**:433–436.
Stutenkamper, R., Geisse, S., Schwarz, H.J. *et al.* (1992). *Exp. Cell Res.* **201**:43–54.
Talbot, N.C., Rexroad, C.E., Pursel, V.G., Powell, A.M. and Nel, N.D. (1993). *In Vitro Cell. Dev. Biol.* **29A**:543–554.
Talbot, N.C., Rexroad, C.E., Powell, A.M. *et al.* (1994a). *In Vitro Cell Dev. Biol.* **30A**:843–850.
Talbot, N.C., Pursel, V.G., Rexroad, C.E., Caperna, T.J., Powell, A.M. and Stone, R.T. (1994b). *In Vitro Cell Dev. Biol.* **30A**:851–858.
Tan, T.B., Marino, P.A., Padmanabhan, R., Hampton, L.L., Hanley-Hyde, J.M. and Thorgeirsson, S.S. (1994). *In Vitro Cell. Dev. Biol.* **30A**:615–621.
Tatematsu, M., Ho, R.H., Kaku, T., Ekem, J.K. and Farber, E. (1984). *Am. J. Pathol.* **114**:418–430.
Tatematsu, M., Kaku, T., Medline, A. and Farber, E. (1985). *Lab. Invest.* **52**:354–362.
Thorgeirsson, S.S. (1993). *Am. J. Pathol.* **142**:1331–1333.
Thorgeirsson, S.S. (1995). In *Liver Regeneration and Carcinogenesis: Molecular and Cellular Mechanisms* (ed. R. L. Jirtle), pp. 99–112. Academic Press, Orlando, FL.
Thorgeirsson, S.S. and Evarts, R.P. (1992). In *The Role of Cell Types in Hepatocarcinogenesis* (ed. A. E. Sirica), pp. 109–120. CRC Press, Boca Raton, FL.
Thorgeirsson, S.S., Evarts, R.P., Bisgaard, H.C., Fujio, K. and Huz, Z. (1993). *Proc. Soc. Exp. Biol. Med.* **204**:253–260.
Tian, J.-M. and Schibler, U. (1991). *Genes Develop.* **5**:2225–2234.

Till, J.E. and McCulloch, E.A. (1961). *Radiat. Res.* **14**:213–222.
Tsao, M.-S. (1993). *Pathobiology* **61**:25–30.
Tsao, M.-S. and Grisham, J.W. (1987). *Am. J. Pathol.* **127**:168–181.
Tsao, M.-S. and Liu, C. (1988). *Lab. Invest.* **58**:636–642.
Tsao, M.-S. and Zhang, X.-Y. (1992). *Am. J. Pathol.* **140**:85–94.
Tsao, M.-S., Smith, J.D., Nelson, K.G. and Grisham, J.W. (1984). *Exp. Cell Res.* **154**:38–52.
Tsao, M.-S., Earp, H.S. and Grisham, J.W. (1986). *J. Cell. Physiol.* **126**:167–173.
Tsao, M.-S., Duong, M. and Batist, C. (1989). *Mol. Carcinog.* **2**:144–149.
Tsao, M.-S., Shepherd, J. and Batist G. (1990). *Cancer Res.* **50**:1941–1947.
Tsao, M.-S., Zhang, X.-Y., Liu, C. and Grisham, J.W. (1991). *Exp. Cell Res.* **195**:214–217.
Uchida, N., Agulla, H.L., Fleming, W.H., Jerabek, L. and Weissman, J.L. (1994). *Blood* **83**:3758–3779.
VanEyken, P. and Desmet, V.J. (1993). *Liver* **13**:113–122.
VanEyken, P., Sciot, R. and Desmet, V. (1988). *Lab. Invest.* **59**:52–59.
VanEyken, P. and Desmet, V.J. (1992). In *The Role of Cell Types in Hepatocarcinogenesis* (ed. A. E. Sirica), pp. 227–263. CRC Press, Boca Raton, FL.
Vassy, J., Kraemer, M., Chalumeau, M.T. and Foucrier, J. (1988). *Cell. Differ.* **24**:9–24.
Viola-Magni, M.P. (1972). *J. Micros.* **96**:191–203.
Wareham, K.A. and Williams, E.D. (1986). *J. Embryol. Exp. Morphol.* **95**:239–246.
Weinberg, W.C. and Iannaccone, P.M. (1988). *J. Cell. Sci.* **80**:423–431.
Weibel, E.R., Stäubli, W., Gnägi, H.R. and Hess, F.A. (1969). *J. Cell Biol.* **42**:68–91.
West, J.D. (1976). *J. Embryol. Exp. Morphol.* **36**:151–161.
Williams, G.M., Elliott, J.M. and Weisburger, J.H. (1973). *Cancer Res.* **33**:606–612.
Williams, A.O., Huggett, A.C. and Thorgeirsson, S.S. (1992a). *Int. J. Exp. Pathol.* **73**:99–114.
Williams, D.E., DeVries, P., Namen, A.E., Widmer, M.B. and Lyman, S.D. (1992b). *Develop. Biol.* **151**:368–376.
Wilson, J.W. and Leduc, E.H. (1958). *J. Pathol. Bacteriol.* **76**:441–449.
Woo, A., Tsao, M.-S. and Batist, G. (1992). *Carcinogenesis* **13**:1675–1677.
Wood, R.L. (1965). *Anat. Rec.* **151**:507–530.
Wright, N. and Alison, M. (1984). In *The Biology of Epithelial Cell Populations*, Vol. 2, pp. 880–980. Clarendon Press, Oxford.
Xanthopoulos, K.G. and Mirkovitch, J. (1993). *Eur. J. Biochem.* **216**:353–360.
Yang, L., Faris, R.A. and Hixson, D.C. (1993a). *Hepatology* **18**:357–366.
Yang, L., Faris, R.A. and Hixson, D.C. (1993b). *Proc. Soc. Exp. Biol. Med.* **204**:280–288.
Yaswen, P., Hayner, N.T. and Fausto, N. (1984). *Cancer Res.* **44**:324–331.
Yokoyama, S., Satoh, M. and Lombardi, B. (1986). *Carcinogenesis* **7**:1215–1219.
Yoshimura, H., Harris, R., Yokoyama, S. *et al.* (1983). *Am. J. Pathol.* **110**:322–332.
Zajicek, G. (1992). *Path. Res. Pract.* **188**:410–412.
Zajicek, G., Oren, R. and Weinreb, W. (1985). *Liver* **5**:293–300.
Zaret, K.S. (1993). In *Hepatic Transport and Bile Secretion Physiology and Pathophysiology* (eds N. Tavoloni and P. D. Berk), pp. 135–143. Raven Press, New York.
Zhang, M. and Thorgeirsson, S.S. (1994). *Exp. Cell Res.* **213**:37–43.
Zhang, X., Wang, T., Batist, G. and Tsao, M.-S. (1994). *Cancer Res* **54**:6122–6128.
Zsebo, D.M., Wypych, J., McNiece, I.K. *et al.* (1990). *Cell* **63**:195–201.

9 Regulation of proliferation and differentiation of stem cells in the male germ line

Dirk G. de Rooij*† and Federica M. F. van Dissel-Emiliani†

**Department of Cell Biology, Medical School; †Department of Functional Morphology, Veterinary School, Utrecht University, The Netherlands*

INTRODUCTION

In the spermatogenic lineage, from the primordial germ cells in the fetus up to the spermatozoa that are released in the adult seminiferous epithelium, regulatory mechanisms induce or inhibit the proliferation, differentiation and further development of the various cell types. Only a few of the (growth) factors involved in these regulatory mechanisms are known. Some progress has been made in understanding the regulation of the development of fetal and neonatal germ cells but, as yet, little information is available for the spermatogenic process in the adult.

In the first part of this review the subsequent types of cells in the germ cell lineage, from those in the fetus to those in the adult, will be described. Then the focus will be on the cells which are at the base of this lineage, primordial germ cells (PGCs), gonocytes and undifferentiated spermatogonia, i.e. the cells with stem-cell properties and their direct descendants. These cell types will be described as well as the methods which are used to study the regulation of the kinetics of these cells. Finally, the present data on the role of various (growth) factors on proliferation and differentiation of PGCs, gonocytes and undifferentiated spermatogonia will be reviewed. In addition, some factors will be mentioned that are available to the undifferentiated spermatogonia in the adult testis but that have not been studied as yet with respect to their effect on these cells.

CELL TYPES IN THE SPERMATOGENIC LINEAGE

Primordial germ cells

PGCs, the origin of the germ cell lineage, derive from a small population of epiblast (embryonal ectoderm) cells, that are set aside at the egg cylinder stage prior to

STEM CELLS
ISBN 0-12-563455-2

gastrulation (Lawson and Pederson, 1992). By 7 days *post coitum* (pc) in the mouse embryo, about 100 alkaline phosphatase positive PGCs can be detected in the extraembryonal mesoderm posterior to the definitive primitive streak (Ginsberg *et al.*, 1990). Later in development the PGCs migrate from the base of the allantois, along the hindgut to finally reach the genital ridges (Figure 1A). The PGCs divide during migration and by day 13 of fetal life in the mouse, when the PGCs have reached the genital ridges, their numbers have increased to about 10,000 in each gonad (Tam and Snow, 1981).

Primordial germ cells are single cells that under certain culture conditions can form colonies of cells which morphologically resemble undifferentiated embryonic stem cells (ES cells) (Resnick *et al.*, 1992). These cells can be maintained on feeder layers for extended periods of time and can give rise to embryoid bodies and to multiple differentiated cell phenotypes in monolayer culture and in tumours in nude mice. PGC-derived ES cells can also contribute to chimeras when injected into host blastocysts (Resnick *et al.*, 1992). Clearly, PGCs are stem cells, still having the capacity to renew themselves and to differentiate in various directions.

Gonocytes

Once arrived in the genital ridges, the PGCs become enclosed by the differentiating Sertoli cells, and seminiferous cords are formed. The germ cells present within the seminiferous cords differ morphologically from PGCs and are called prospermatogonia (Hilscher *et al.*, 1974) or, as in this review, gonocytes (Clermont and Perey, 1957; Sapsford, 1962; Huckins and Clermont, 1968) (Figures 1B and C). In rats and mice, after formation the gonocytes proliferate for a few days and then become arrested in the G_0/G_1 phase of the cell cycle (Clermont and Perey, 1957; Franchi and Mandl, 1964; Huckins and Clermont, 1968; Hilscher *et al.*, 1974; Vergouwen *et al.*, 1991). Within a few days after birth the gonocytes resume proliferation to give rise to A spermatogonia (Sapsford, 1962; Novi and Saba, 1968; Huckins and Clermont, 1968; Bellvé *et al.*, 1977; Vergouwen *et al.*, 1991). This event marks the start of spermatogenesis (Figure 2).

Up to now, it has not been possible to study the gonocytes in a way comparable to the PGCs. *In vitro* studies have demonstrated some differences between gonocytes and PGCs. For example, gonocytes can only survive *in vitro* in the presence of Sertoli cells, while PGCs can be co-cultured with other types of somatic cells. This suggests that the gonocytes are restricted in their differentiation potential in comparison to the PGCs. Furthermore, at the mitoses of the gonocytes, cytokinesis is often not complete and many gonocytes remain interconnected by intercellular bridges (Zamboni and Merchant, 1973). As discussed below, in the adult testis the first visible sign of differentiation of the stem cells is the formation of a pair of cells interconnected by an intercellular bridge. These paired spermatogonia are destined ultimately to become spermatozoa. Conceivably, the gonocytes may well be a heterogeneous population of cells of which only some have spermatogonial stem cell properties, while the rest may be destined to become differentiating cells directly at the start of spermatogenesis after birth.

Undifferentiated and differentiating spermatogonia

As reviewed by De Rooij (1983) and Meistrich and Van Beek (1993a), the experimental evidence favours the model of spermatogonial multiplication and stem cell renewal in

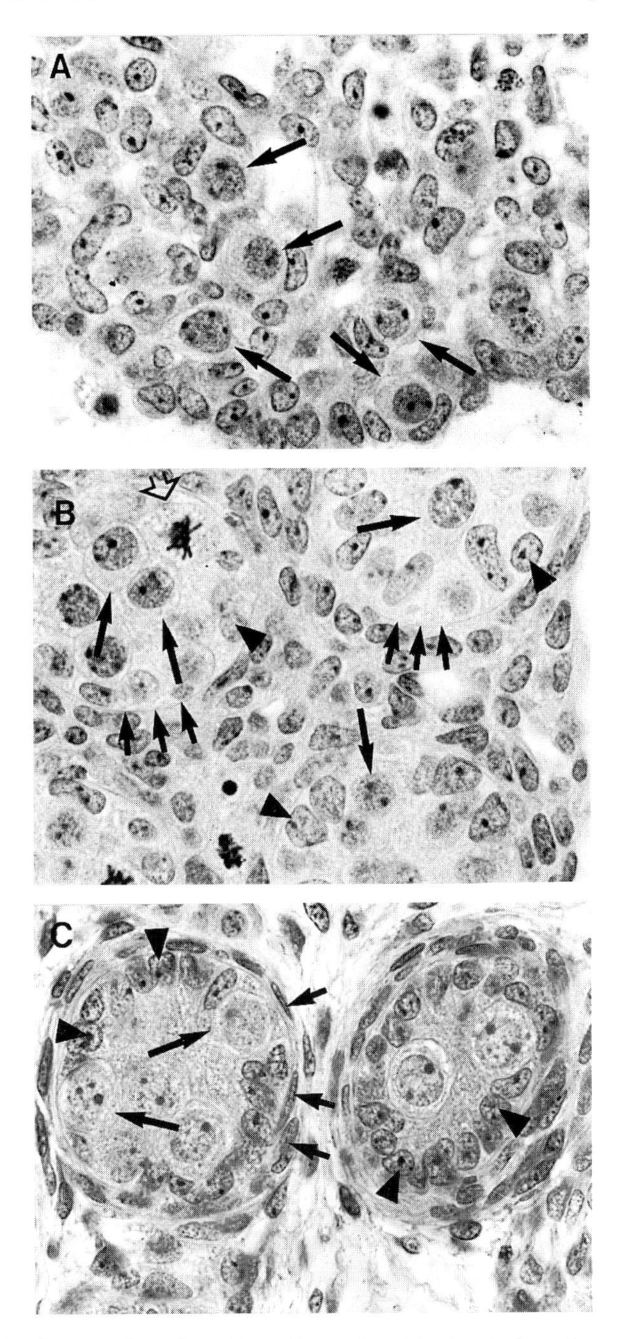

Figure 1 Photomicrographs of sections through rat testes before the onset of testicular differentiation and thereafter. At 13 days pc (A), the PGCs (arrows) are randomly distributed in the gonadal primordium, which consists mostly of undifferentiated mesenchymal-like cells. By 15 days pc (B) seminiferous cords, consisting of a layer of Sertoli cells (arrowheads) and a clear basement membrane (small arrows), can be observed. The large round gonocytes (large arrows) are located in the centre of the cords. A dividing gonocyte is indicated by the hollow arrow. By 1 day pp (C), the seminiferous cords have become more conspicuous and are surrounded by a layer of flattened peritubular myoid cells (small arrows). Some gonocytes have migrated to the basement membrane. ($\times$ 450.)

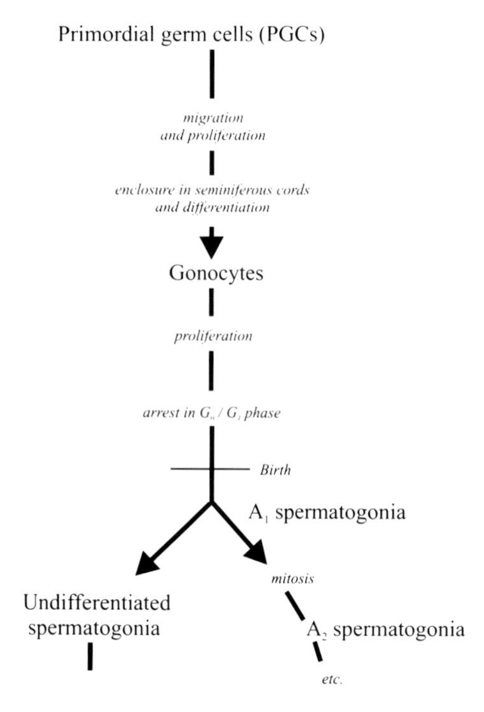

Figure 2 Development of the spermatogenic lineage up to the start of spermatogenesis after birth.

the adult testis, as originally proposed by Huckins (1971a) and Oakberg (1971). Hence, in this review this model will be used. According to this model, a compartment of undifferentiated A spermatogonia stands at the beginning of spermatogenesis. Using whole mounts of seminiferous tubules, these cells can be subdivided into A_{single} (A_s), A_{paired} (A_{pr}) or $A_{aligned}$ (A_{al}) spermatogonia according to their topographical arrangement on the basement membrane. The A_s spermatogonia are considered the stem cells of spermatogenesis. Upon division of the A_s spermatogonia the daughter cells can either migrate away from each other and become two new stem cells or they can stay together connected by an intercellular bridge and become A_{pr} spermatogonia. In the normal situation about half of the stem cell population will divide to form A_{pr} spermatogonia, while the other half will go through a self-renewing division, thus maintaining stem cell numbers. The A_{pr} spermatogonia divide further to form chains of 4, 8 or 16 A_{al} spermatogonia (Figure 3).

The A_{al} spermatogonia are able to differentiate into A_1 spermatogonia that are the first generation of the differentiating spermatogonia. These A_1 spermatogonia synchronously go through a series of six divisions and via A_2, A_3, A_4, intermediate (In) and B spermatogonia become primary spermatocytes. In whole mounts of seminiferous tubules undifferentiated spermatogonia can be distinguished from differentiating type A spermatogonia (A_1–A_4)) in areas in which the latter cells are in late G_2 or M phase, by the fact that the undifferentiated spermatogonia do not cycle synchronously with the A_1–A_4 spermatogonia. However, there is no clear morphological difference between

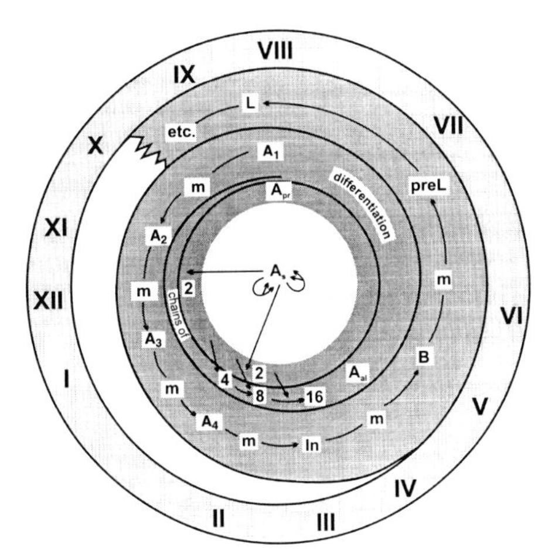

Figure 3 Scheme of spermatogonial multiplication and stem cell renewal in mice. The roman numerals in the outer ring indicate the stages of the cycle of the seminiferous epithelium. preL, preleptotene spermatocytes; L, leptotene spermatocytes.

undifferentiated and differentiating type A spermatogonia to distinguish these cells in all cell cycle phases.

According to the model of Huckins (1971a) and Oakberg (1971), in the adult testis differentiation of the stem cells, the A_s spermatogonia, occurs when following a division the daughter cells remain connected by an intercellular bridge. However, it is not known whether the presence of intercellular bridges among A_{pr} and A_{al} spermatogonia reflects a functional differentiation of the A_s spermatogonia and whether these cells still have stem cell properties when the intercellular bridges become severed, as occurs after irradiation (Van Beek *et al.*, 1984). The A_s, A_{pr} and A_{al} spermatogonia can only be distinguished by their topographical arrangement as there are no further morphological differences between these cells. Also, the A_s spermatogonia more or less follow the proliferation pattern of the total population of undifferentiated spermatogonia. Hence, the A_s, A_{pr} and A_{al} spermatogonia and the regulation of the proliferation of these cells, may be rather similar.

The model of Huckins (1971a) and Oakberg (1971) opposes that of Clermont and Bustos-Obregon (1968) and Clermont and Hermo (1975) in which the A_s and A_{pr} spermatogonia are supposed to be quiescent in the normal testis and together are called A_0 spermatogonia. The A_0 spermatogonia would only come into action after cell loss, for example after irradiation. Furthermore, in this model the A_1, A_2, A_3 and A_4 spermatogonia still retain stem cell properties, the A_4 spermatogonia at their division rendering both A_1 and In spermatogonia. The A_{al} spermatogonia in this model are considered to be out of phase A_1–A_4 spermatogonia as these cells do not participate in the synchronous waves of division characterizing the A_1–A_4 spermatogonia. The A_0 model is in conflict with several experimental findings. First, the A_s and A_{pr} spermatogonia are not quiescent in the normal testis but have a distinct pattern of proliferative activity during the epithelial cycle (Huckins, 1971a; Lok *et al.*, 1983b). Second, the A_{al} spermatogonia have cell cycle properties similar to those of the A_s and A_{pr} spermatogonia and different from those of the A_1–A_4 spermatogonia (Huckins, 1971a;

Lok and De Rooij, 1983a, 1983b), explaining the asynchronous behaviour of the A_{al} spermatogonia. Third, when locally the A_1 spermatogonia are specifically removed, leaving the undifferentiated spermatogonia unharmed, in these areas supranormal numbers of A_{al} but no In spermatogonia are formed during the ensuing epithelial cycle (De Rooij *et al.*, 1985). The latter argues against the idea that the A_{al} spermatogonia are merely out of phase A_1–A_4 spermatogonia. Fourth, there is a considerable difference in radiosensitivity between proliferating A_{al} spermatogonia and A_1–A_4 spermatogonia (Van der Meer *et al.*, 1992a, 1992b) also suggesting that A_{al} and A_1–A_4 spermatogonia are different types of cells.

The transition from gonocytes to adult type spermatogonia

The nature of the gonocytes and of their daughter cells at the start of spermatogenesis after birth, is still a matter of debate. The gonocytes may precede the undifferentiated spermatogonia in the spermatogenic line and may consist of one cell type only, or of several subsequent types of cells (Hilscher *et al.*, 1974), or they may be similar to the undifferentiated spermatogonia in the adult (Kluin and De Rooij, 1981). The latter view was based on cell kinetics and the morphology which are rather similar for gonocytes and undifferentiated spermatogonia. However, it was also found that fetal gonocytes bear a specific antigen, as demonstrated by a monoclonal antibody, which is not present in spermatogonia of the adult testis (Van Dissel-Emiliani *et al.*, 1993a).

In addition, it is still unclear what happens at the start of spermatogenesis. The daughter cells of the gonocytes after birth may be a special type of prespermatogonia that after one or more divisions give rise to adult-type spermatogonia (Novi and Saba, 1968; Huckins and Clermont, 1968; Bellvé *et al.*, 1977; Hilscher *et al.*, 1974). However, the morphological evidence for the existence of a special kind of prespermatogonia is not conclusive and the gonocytes might also directly give rise to adult type spermatogonia (Figure 2) (Kluin and De Rooij, 1981). Van Haaster and De Rooij (1993) recently studied the appearance of the subsequent spermatogenic cell types after the start of spermatogenesis in the Djungarian and the Chinese hamster and in the rat. From the rate of appearance of these cell types with age, extrapolation to the moment of the start of spermatogenesis in each of these species invariably pointed to the formation of A_2 spermatogonia at the first division of the gonocytes (Van Haaster, 1993). In theory, generations of differentiating spermatogonia could be skipped during the first spermatogenic wave, but in the adult testis in a variety of abnormal hormonal situations or during recovery after cell loss inflicted by irradiation or cytotoxic agents, the latter has never been seen to occur. Hence, it is likely that the gonocytes directly generate A_2 spermatogonia.

If gonocytes give rise to A_2 spermatogonia these cells function as A_1 spermatogonia. Furthermore, since these A_2 spermatogonia can only differentiate further, some undifferentiated spermatogonia, including stem cells, must also be formed. Indeed, during the first spermatogenic wave in some tubular cross-sections only A spermatogonia were found, indicating that in these tubules the gonocytes had only developed into undifferentiated spermatogonia and that locally no differentiating spermatogonia were produced. Hence, at the start of spermatogenesis while some of the gonocytes divide into A_2 spermatogonia the rest become undifferentiated spermatogonia.

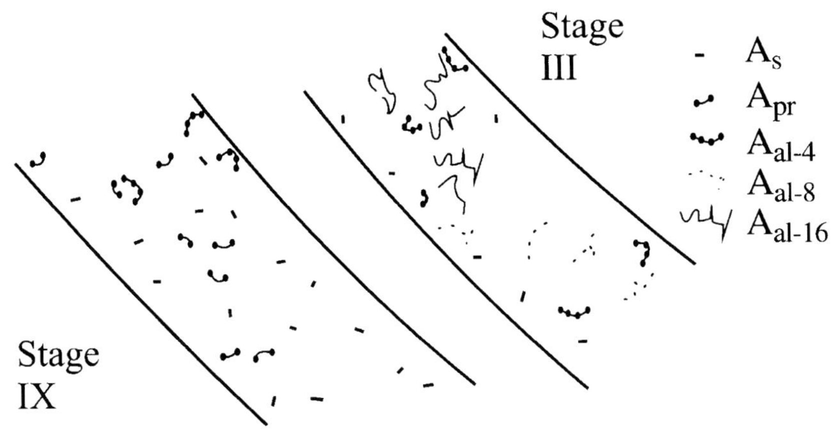

Figure 4 Topographical arrangement of undifferentiated spermatogonia on the basement membrane of two seminiferous tubules. The undifferentiated spermatogonia lie intermingled with differentiating type spermatogonia, Sertoli cells and in some stages with early spermatocytes. In epithelial stage IX only A_s, A_{pr} and a few chains of A_{al-4} are present. In epithelial stage III, after the period of proliferation many more A_{al} spermatogonia in longer chains are present.

THE CYCLE OF THE SEMINIFEROUS EPITHELIUM

Spermatogenesis is a cyclic process. During part of that cycle the undifferentiated spermatogonia are mostly quiescent and their number is low, only few A_{al} spermatogonia being present. Then the cells are stimulated and for a certain period of time, active proliferation results in the formation of more and more A_{al} spermatogonia (Figures 3 and 4) (De Rooij, 1983). The total number of A_s and A_{pr} spermatogonia does not change much during the epithelial cycle. After a period of active proliferation, most of the cells become arrested in the G_1 phase of the cell cycle. This period of relative quiescence ends when almost all of the A_{al} spermatogonia formed during the period of proliferation, differentiate into the first generation of the differentiating spermatogonia. In most animals these cells are called A_1 spermatogonia. These A_1 spermatogonia then go through the first of a series of six divisions and ultimately become spermatocytes. Following the division of the A_1 spermatogonia the undifferentiated spermatogonia start to proliferate again to form a new cohort of A_1 spermatogonia for the next epithelial cycle.

The interval of time between subsequent cohorts of new A_1 spermatogonia is always similar in a particular species and is the duration of the epithelial cycle. Also the duration of the various steps in the development of the spermatogenic cells is always the same within a particular species. As a consequence the same associations of generations of cells in particular developmental stages are always found together along the tubular wall. Thus, for example, spermatids at a particular step in their transformation into spermatozoa are always found with spermatogonia and spermatocytes of a particular stage in their respective development. This makes it possible to divide the epithelial cycle into stages according to steps in the development of spermatogenic cells (Figure 5). In most species 12 stages are recognized based on the development of the spermatids (review by Courot *et al.*, 1970).

I	II	III	IV	V	VI	VII	VIII	IX	X	XI	XII
13	14	14	15	15	15	16	16				
1	2	3	4	5	6	7	8	9	10	11	12
P	P	P	P	P	P	P	P	P	P	D	m_I m_{II}
A_3A_{4m}	A_{4m}	In	In_m	B	B_m	preL	preL	L	Z	Z	P
						→ A_1	A_1	A_{1m}	A_2	A_{2m}	A_3
Undifferentiated spermatogonia											

Figure 5 Diagram of the cellular composition of the 12 stages of the epithelial cycle in the mouse. The columns numbered with a roman numeral show the various cell types within each cellular association that are encountered in the various cross-sections of seminiferous tubules. A_1, A_2, A_3, A_4, In and B: subsequent types of differentiating spermatogonia; preL: preleptotene spermatocytes; m: mitosis; L, Z, P, D: respectively spermatocytes in leptotene, zygotene, pachytene and diplotene phase of the prophase of the first meiotic division; m_I, m_{II}: first and second meiotic division; 1–16 steps in spermatid development.

METHODS TO STUDY THE REGULATION OF PROLIFERATION AND DIFFERENTIATION IN THE GERM LINE

With increasing maturity of the animals it becomes more difficult to study the regulation of the germ line stem cells. Primordial germ cells can be isolated and cultured, and consequently the effects of growth factors on these stem cells can be studied (Table 1). Gonocytes too, can be isolated from fetal testes, when they are in G_1 arrest (Van Dissel-Emiliani *et al.*, 1989a). It is also possible to subsequently culture these cells on a feeder layer of mature Sertoli cells (Van Dissel-Emiliani *et al.*, 1993b). In this system the gonocytes survive for extended periods of time, remaining quiescent. This system is very suitable to look for the factors that regulate the start of spermatogenesis itself. After birth, when the gonocytes resume proliferation, the isolation of these cells is much more difficult and no method rendering highly purified postnatal gonocytes has yet been developed. However, it is possible to isolate seminiferous tubules from newborn-rat testes, stripped of most peritubular myoid cells which can then be digested into a cell suspension containing merely Sertoli cells and gonocytes, and to culture these cells (Orth and Boehm, 1990). In such a co-culture system of Sertoli cells and gonocytes, the gonocytes start to proliferate following a time-course similar to that *in vivo*. Hence, this system is suitable to characterize (growth) factors and hormones that modulate gonocyte proliferation and differentiation.

After the start of spermatogenesis the situation becomes very complex as many differentiating types of spermatogenic cells appear in the seminiferous tubules. Nevertheless, there are a few ways to study regulatory effects on stem cells (see for review Meistrich and Van Beek, 1993b). First, in whole mounts of seminiferous tubules it is possible to distinguish the stem cells, the A_s spermatogonia, by their topographical arrangement. This makes it possible to count the stem cells and to study their behaviour by light microscopy in the normal situation and after disturbances by for instance irradiation, (intratesticular) injection of hormones or growth factors. Unfortunately, this is a

Table 1 Methods for *in vitro* studies of spermatogenic cells.

Cell type	Methods	References
PGC	Purification/co-culture with feeder cells	DeFelici and Siracusa, 1985 Donovan and Scott, 1986
Gonocytes day 18 pc rat	Organ culture Purification/co-culture with mature Sertoli cells	Van Dissel-Emiliani *et al.*, 1989a, 1993b
neonatal	Co-culture gonocytes/ Sertoli cells	Orth and Boehm, 1990
Undifferentiated and possibly differentiating A spermatogonia	Culture of testicular fragments of cryptorchid mouse testes	Aizawa and Nishimune, 1979 Haneji and Nishimune, 1982
Mixture of undifferentiated and differentiating A spermatogonia	Culture of fragments of 9-day-old rat testes	Boitani *et al.*, 1993
Differentiating spermatogonia	Culture of (selected) tubular segments of immature or adult rats	Mather *et al.*, 1990 Parvinen *et al.*, 1991

PGC, Primordial germ cell.

very laborious way of studying these problems, also requiring extensive experience. Furthermore, only *in vivo* studies can be carried out as cultured whole mounts of seminiferous tubules quickly lose their integrity. Second, when stem cell killing is involved, the numbers of surviving stem cells can be determined by estimating the size of the progeny of the surviving stem cells after a certain time has elapsed (in the mouse at least 1 week). The degree of repopulation of the seminiferous epithelium can then be taken as a measure of the number of surviving stem cells. At best, this method can give an indirect measure of the effects of (growth) factors on spermatogonial stem cells.

All other systems to evaluate the effects of factors on proliferation and differentiation of spermatogonia do not detect effects on stem cells but rather effects on a mixture of undifferentiated and differentiating spermatogonia. Aizawa and Nishimune (1979) and Haneji and Nishimune (1982) have developed an experimental system in which C57/Bl mice are made cryptorchid. Two months after the operation the seminiferous epithelium of these mice only contains A spermatogonia. Testicular fragments of these mice can then be used for *in vitro* studies into the regulation of proliferation and differentiation of these A spermatogonia. The type of A spermatogonia present in the cryptorchid testis, undifferentiated with or without differentiating type A spermatogonia, has not been determined as yet. Boitani *et al.* (1993) have developed a system in which fragments of 9-day-old rat testes are cultured for extended periods of time. In view of the age of the donor rats, these cultures will contain undifferentiated as well as differentiating type A spermatogonia.

Finally, there are two systems in which primarily effects on differentiating spermatogonia can be detected. Mather *et al.* (1990) use a system in which pieces of seminiferous tubules of 20–25 day old rats, stripped of peritubular cells and the basement membrane, are cultured. This system represents a Sertoli cell/germ cell co-culture. Parvinen *et al.* (1991) have developed an experimental system in which segments of seminiferous tubules in particular stages of the cycle of the seminiferous epithelium are isolated under the microscope by way of the transillumination technique. These tubular segments can then be cultured. These last two systems can only be used to study the effects of factors on differentiating type spermatogonia, as in both cases these cells by far outnumber the undifferentiated spermatogonia.

REGULATION OF PROLIFERATION AND DIFFERENTIATION IN THE GERM LINE

Regulatory mechanisms involved in germ line proliferation and differentiation

A specific regulation of the proliferation and differentiation of the germ cells has to take place at many developmental steps along the spermatogenic lineage. The PGCs have to be stimulated to proliferate on their way to the genital ridge and the necessary growth factors will have to be available to these cells on their migratory pathway. When the PGCs become enclosed in the seminiferous cords they differentiate into gonocytes. Subsequently, the gonocytes are stimulated to proliferate for a few days and then these cells become quiescent. After birth, at the start of spermatogenesis, differentiation of the gonocytes into A_1 spermatogonia occurs and undifferentiated spermatogonia are formed. During the normal adult cyclic pattern of proliferation and differentiation of undifferentiated spermatogonia again, stimulation of the proliferative activity of the undifferentiated spermatogonia (epithelial stages IX–XI) and inhibition (stages II/III) takes place. Then there is an induction of the differentiation of most A_{al} spermatogonia formed into A_1 spermatogonia (stages VII/early VIII). Finally, during the time spermatogenesis proceeds in the animal the number of spermatogonial stem cells remains constant. All of these processes, whether they are actively or passively initiated will have to be regulated by possibly stimulatory or inhibitory factors.

In the adult testis the factors that regulate these processes are most likely locally produced. This is necessary because in the testis all epithelial stages are present simultaneously. Hence, in some tubular areas the proliferation of the undifferentiated spermatogonia will have to be inhibited while in others, sometimes even areas of neighbouring seminiferous tubules, these cells will have to be stimulated (Figure 6). Hormones and growth factors in the serum or produced by the interstitial tissue can only have a permissive function and except for retinoic acid will not be discussed. Furthermore, Sertoli cells form the Sertoli cell barrier which by way of tight junctions separates the tubular wall into an outer basal and an inner adluminal compartment (Dym and Fawcett, 1970; Russell, 1978). All types of spermatogonia and early spermatocytes are in the basal compartment. At the leptotene step of the meiotic prophase the spermatocytes move to the adluminal compartment and hence, the rest of the spermatocytes and the spermatids reside in this compartment. The Sertoli cell barrier is such that only very small molecules can pass from one compartment to the other without having to traverse the cytoplasm of the Sertoli cells (Figure 7). As a result, growth factors secreted by spermatocytes beyond zygotene stage and spermatids can reach, or influence,

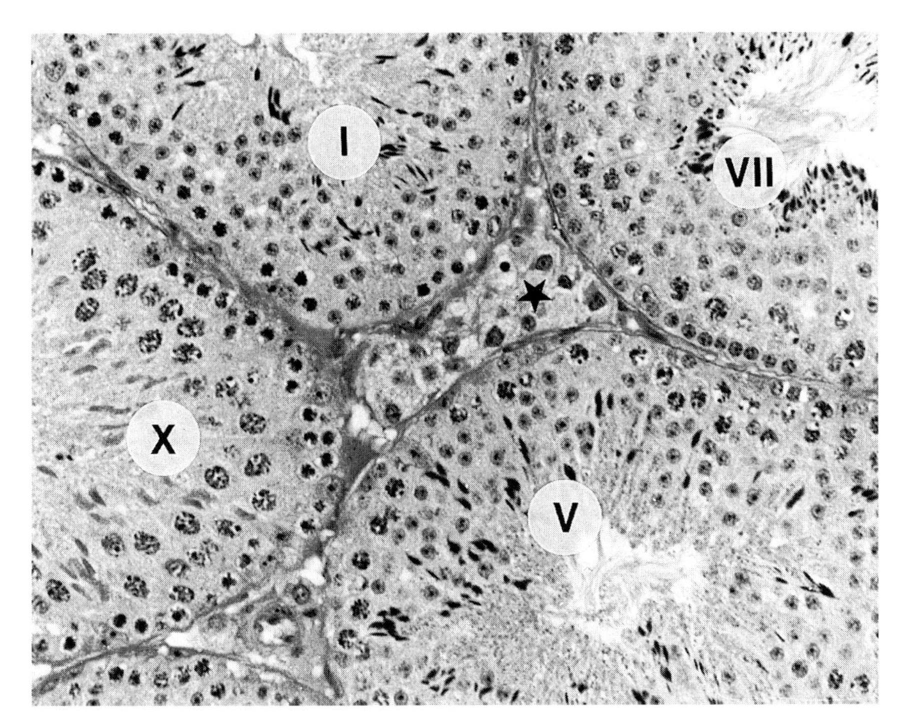

Figure 6 Testis section showing interstitial tissue (asterisk) surrounded by tubules in epithelial stages I, V, VII and X. In stage I the undifferentiated spermatogonia are actively proliferating, in stages V and VII these cells are quiescent and in stage X they are about to start to proliferate again (70 ×).

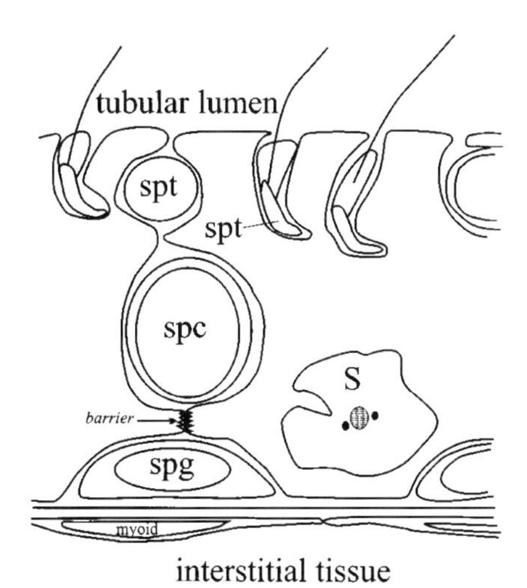

Figure 7 Schematic drawing of the seminiferous epithelium showing the intimate contact of Sertoli cells with the germ cells. The Sertoli cell barrier splits the epithelium into two parts, the adluminal compartment containing spermatocytes (spc) and spermatids (spt) and the basal compartment containing spermatogonia (spg) and very early spermatocytes. Factors secreted by spermatocytes and spermatids cannot reach the spermatogonia. S, Sertoli cell nucleus.

Table 2 Growth factors produced by Sertoli cells and/or peritubular myoid cells.

	Reference
Known growth factors	
inhibin	Review: De Jong (1988)
AMH	Review: Josso and Picard (1986)
IGF-1	Review: Sharpe (1990)
TGFα	Skinner *et al.* (1989)
TGFβ	Skinner and Moses (1989); Ailenberg *et al.* (1990)
Il-1α	Syed *et al.* (1988); Pöllänen *et al.* (1989) Gérard *et al.* (1991); Parvinen *et al.* (1991)
Il-6	Syed *et al.* (1993)
Stem cell factor (SCF)	Rossi *et al.* (1991); Manova *et al.* (1993)
bFGF	Smith *et al.* (1989); Mullaney and Skinner (1992b)
activin	Kaipia *et al.* (1993)
Unknown growth factors	
Sertoli cell secreted growth factor	Buch *et al.* (1988)
Seminiferous growth factor	Feig *et al.* (1980, 1983)

the spermatogonial compartment only indirectly via the Sertoli cells (review Jégou, 1992). Hence, in this chapter only those factors will be considered potentially important for the direct and specific regulation of spermatogonial proliferation and differentiation that are produced by Sertoli cells and peritubular myoid cells, both of which are very close to the spermatogonia (Table 2).

While in the adult testis generally occurring factors cannot induce a specific regulation, the situation is different during development. PGCs proliferation and differentiation, gonocyte proliferation and inhibition, and the start of spermatogenesis (Van Haaster and De Rooij, 1994) take place in a more or less synchronized way. In these situations hormone levels and systemic growth factors may play a role.

Cell–cell interaction

In several situations a clear reaction of the seminiferous epithelium is seen after the loss of certain cell types (review De Rooij, 1988). While in the normal testis spermatogonial stem cell renewal and differentiation is balanced in such a way that the number of stem cells remains the same, this is not the case after cell loss. During the first few divisions after irradiation when the surviving stem cells start to repopulate the testis, differentiation of stem cells decreases to very low percentages (Van Beek *et al.*, 1990). The mechanism of this decreased differentiation of the stem cells is not known.

In the normal seminiferous epithelium the proliferative activity of the undifferentiated spermatogonia is inhibited around stages III–IV in the mouse, rat and Chinese hamster. However, in the Chinese hamster this inhibition does not take place when the differentiating spermatogonia are specifically removed from the epithelium (De Rooij *et al.*, 1985). Apparently, the In and B spermatogonia inhibit the proliferation of the undifferentiated spermatogonia that are present among them by way of a negative feedback system, possibly by the production of an inhibiting factor (see below). Further evidence for the existence of a feedback system between differentiating and undifferentiated spermatogonia was obtained by Tiba and co-workers (Tiba and Kita, 1991; Tiba *et al.*, 1992) in seasonal breeders such as mink and the Syrian hamster. These authors found increased numbers of undifferentiated spermatogonia in those tubular areas in which the differentiating spermatogonia were lacking because of testicular atrophy induced by the non-breeding season or light deprivation.

REGULATION OF PROLIFERATION AND DIFFERENTIATION OF PRIMORDIAL GERM CELLS

First of all, it is worthwhile to look for mutants in which the regulation of PGC migration, proliferation and/or differentiation is changed. Mice homozygous for mutations in the dominant white spotting (W) locus or the Steel (Sl) locus are infertile and show hypoplastic anaemia and depletion of mast cells and melanocytes (Mintz and Russell, 1957; Russell, 1979; Kitamura *et al.*, 1985). The W and Sl loci were found to encode the c-kit receptor and its ligand, stem cell factor (SCF), respectively (Yarden *et al.*, 1987; Qiu *et al.*, 1988; Anderson *et al.*, 1990; Huang *et al.*, 1990; Zsebo *et al.*, 1990). The c-kit proto-oncogene encodes a receptor tyrosine kinase, which is a member of the Platelet Derived Growth Factor (PDGF) receptor/Colony Stimulating Factor-1 (CSF-1) receptor family. It recently became clear that SCF and its receptor c-kit are important both for PGC migration and proliferation (Buehr *et al.*, 1993). Another mutation that affects proliferation and/or migration of PGC is called 'germ cell deficient' (*gcd*) (Pellas *et al.*, 1991). In mice homozygous for this mutation by gestational day 11.5 a severe depletion of PGC was observed both in male and female genital ridges. Furthermore, the inbred strain of 129/Sv-*ter* mice has a high incidence of congenital teratomas which are tumours that derive from PGCs (Stevens 1967, 1973; Noguchi and Stevens, 1982). When genital ridges from these mice of up to day 13 pc are grafted into testes of normal mice, teratomas are formed with an even higher frequency than *in vivo* in 129/Sv-*ter* mice (Stevens, 1970), suggesting that the defect is in the germ cells themselves. Mice homozygous for the recessive autosomal mutation *ter* are germ cell deficient (Noguchi and Noguchi, 1985). Probably, the expression of the normal gene that encodes the *ter* mutation is necessary for the differentiation of primordial germ cells into gonocytes.

The possibility to isolate and purify mouse PGCs and to maintain these cells in culture has offered the opportunity to characterize some of the regulatory mechanisms involved in their development. PGCs can survive and even differentiate *in vitro* when entire fetal gonads are cultured (Taketo *et al.*, 1986; McLaren and Buehr, 1990), suggesting that factors from within the gonad are sufficient for the normal development of PGCs.

Although isolated PGCs can survive *in vitro*, conditions are not optimal and a high rate of degeneration occurs (DeFelici and McLaren, 1983), unless suitable cell feeder layers are employed (DeFelici and Siracusa, 1985; Donovan and Scott, 1986; DeFelici *et al.*, 1989). By using feeder dependent cultures, putative molecules involved in the survival and proliferation of PGCs have been identified. The c-kit ligand, SCF, as well as leukaemia inhibitory factor (LIF), a cytokine with a broad range of effects on different cell types including embryonic stem cells (for a review see Metcalf, 1991) were shown to increase the survival and in some cases the proliferation of mouse PGCs (Dolci *et al.*, 1991; DeFelici and Dolci, 1991; Matsui *et al.*, 1990, 1991; Godin *et al.*, 1991). Similar results were obtained with TNFα, alone or in combination with SCF or LIF (Kawase *et al.*, 1994). However, these factors, alone or in combination with conditioned media from feeder layers, cannot substitute for the feeder support, indicating that PGC survival and proliferation are regulated by both contact-dependent mechanisms and diffusible factors (for a review see DeFelici *et al.*, 1992). None the less, SCF or LIF markedly reduce the occurrence of apoptosis in isolated PGCs during the first hours of culture (Pesce *et al.*, 1993). Interestingly, the c-kit and SCF genes are expressed in PGCs and in cells associated with both the migratory and homing sites of these cells, respectively (Manova and Bachvarova, 1991). It has been suggested that LIF directly acts on PGCs (Dolci *et al.*, 1993). However, as yet no information is available on the presence of LIF receptors on PGCs or if LIF is produced by somatic cells surrounding PGCs *in vivo*. The same holds true for TNFα, although TNFα is expressed in 8.5–9.5 days pc embryos, which coincides with the stages when PGCs are reactive to TNFα in culture. While SCF, LIF and TNFα have a stimulatory effect, TGFβ1 inhibits the proliferation of PGCs of mice at 8.5 days pc (Godin and Wylie, 1991).

In the presence of SCF, LIF and TNFα, PGCs have a finite proliferative capacity *in vitro* that correlates with their cessation of division *in vivo*. However, if basic fibroblast growth factor (bFGF) is also added, PGCs continue to proliferate in culture and give rise to colonies of cells resembling undifferentiated embryonic stem (ES) cells in morphology and also in being alkaline phosphatase and SSEA-1 antigen positive (Matsui *et al.*, 1992; Resnick *et al.*, 1992). The long-term culture of PGCs and their transition into pluripotential ES cells has important implications for germ cell biology and the induction of teratocarcinomas.

Recently, DeFelici *et al.* (1993) demonstrated that the *in vitro* proliferation of mouse PGCs is markedly enhanced by agents known to enhance intracellular cAMP, provided PGC survival is sustained by SCF and/or LIF (Dolci *et al.*, 1993). This indicates that PGC survival and proliferation *in vitro* is regulated at least by two distinct mechanisms: cytokine-mediated tyrosine phosphorylation and cAMP-dependent protein kinase activation. The intracellular targets for cAMP, SCF and LIF and the interrelationships between their molecular pathways are currently being investigated.

REGULATION OF PROLIFERATION AND DIFFERENTIATION OF GONOCYTES

When the PGCs become enclosed in the seminiferous cords, these cells differentiate into gonocytes that proliferate for several days and then become quiescent until after birth. Not much is known yet about the factors involved in the regulation of the proliferative activity of the gonocytes. Possibly, the proliferation of gonocytes is affected by TGFβ1 since TGFβ1 is able to inhibit PGCs proliferation (Godin and Wylie, 1991). Furthermore, in the rat Vigier *et al.* (1987) found that anti-Mullerian hormone (AMH), another member of the TGFβ family, drastically reduces the number of oogonia in the fetal ovary. Both TGFβ1 and AMH were produced by Sertoli cells at this stage of development (Ao *et al.*, 1993; Teerds and Dorrington, 1993; Tran and Josso, 1982; Münsterberg and Lovell-Badge, 1991). Hence, AMH may be involved in the regulation of the proliferative activity of gonocytes in the developing testis. Also, extratesticular factors may play a role. For instance, it has been demonstrated that when co-cultured with fetal testis, thymus tissue can stimulate the proliferation of gonocytes from 13.5 days pc rats (Prépin, 1993). The factor involved was named thymuline and is secreted in the medium by cultured thymus tissue.

Like PGCs, gonocytes cannot be cultured without feeder cells. However, in contrast to PGCs, in the case of gonocytes, the feeder layer has to consist of Sertoli cells (Figure 8). In the absence of Sertoli cells, even when co-cultured with other somatic cells or even in the presence of Sertoli cells but without being in actual contact with these cells, gonocytes will not survive for more than a few days *in vitro* (Van Dissel-Emiliani *et al.*, 1993b). Hence, Sertoli cells are likely to trigger the start of spermatogenesis both through secreted factor(s) and through contact-dependent mechanisms. The nature of such interactions, however, is still virtually unexplored.

It has been demonstrated that gonocytes can survive and resume proliferation in organ culture systems (Steinberger, 1967; McGuinness and Orth, 1989, 1992a; Kancheva *et al.*, 1990) or in co-culture with neonatal Sertoli cells (Orth and Boehm, 1990; Kancheva *et al.*, 1990; Orth and McGuinness, 1991) independently of the addition of specific factors or hormones to the culture medium. In organ-cultured neonatal testes, gonocytes behave as *in vivo*. They relocate from the more central part of the seminiferous cord to its periphery, where they are in contact with the basement membrane and located between the Sertoli cells (McGuinness and Orth, 1992a, 1992b). In co-culture with Sertoli cells, many gonocytes also become elongated and move from an apical to a more basal position relative to the somatic cells (Orth and Boehm, 1990; McGuinness and Orth, 1992b; Van Dissel-Emiliani *et al.*, 1993c). These observations argue against involvement of extratesticular factors in stimulating gonocytes division and relocation, and point to the Sertoli cells as the source of factor(s) regulating the postnatal maturation of these cells.

The capacity of Sertoli cells to trigger the gonocytes into division is dependent on the age of the Sertoli cells. *In vitro* studies have demonstrated that in co-culture with Sertoli cells isolated from 21-day-old rat testes, the gonocytes remain quiescent (Van Dissel-Emiliani *et al.*, 1993b) while in co-culture with neonatal Sertoli cells, the gonocytes resume proliferation with a timing apparently similar to that *in vivo* (Orth and Boehm, 1990; Orth and McGuinness, 1991; Maekawa and Nishimune, 1991; McGuinness and Orth, 1992a). Possibly, Sertoli cells secrete mitogenic factors specific for gonocytes in an age-dependent manner. This has also been suggested by Kancheva *et al.* (1990) who demonstrated that Sertoli-cell conditioned medium (SCCM) from 6-day-old rats can stimulate gonocyte proliferation while SCCM from 12-day-old rats did not stimulate gonocytes to the same extent. Comparing known factors from testicular origin, Kancheva *et al.* (1990) suggested that their putative growth factor(s) was most similar to SGF in its characteristic properties. The concentration of this factor is highest in seminiferous cords of prepubertal mice and newborn calves (Feig *et al.*, 1980, 1983).

Recently, we have used gonocytes – Sertoli cell co-cultures from newborn testes, as described by Orth and Boehm (1990) – to study the effect of specific growth factors on the proliferation of the gonocytes (Figure 9). The basic fibroblast growth factor (bFGF) gene is expressed during early prepubertal testicular development (Mullaney and Skinner, 1992b) and bFGF is produced by cultured Sertoli cells (Smith *et al.*, 1989; Mullaney and Skinner, 1992b). bFGF appeared to be a potentially important factor at the start of the spermatogenesis as it was found to significantly increase the survival of gonocytes and to stimulate their proliferation (Van Dissel-Emiliani *et al.*, 1993c, 1996). Since immature and adult Sertoli cells possess bFGF receptors (Han *et al.*,

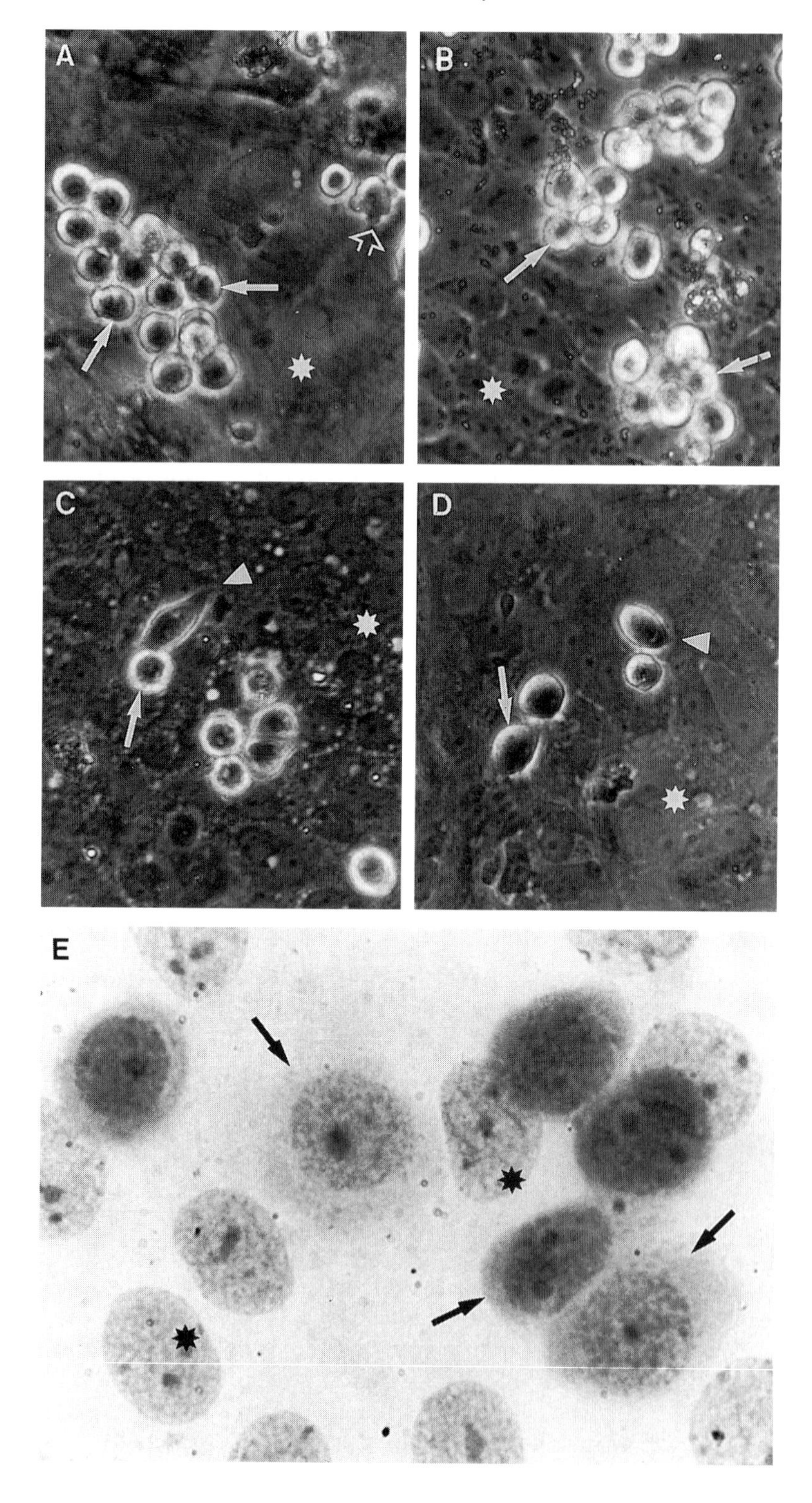
A
B
C
D
E

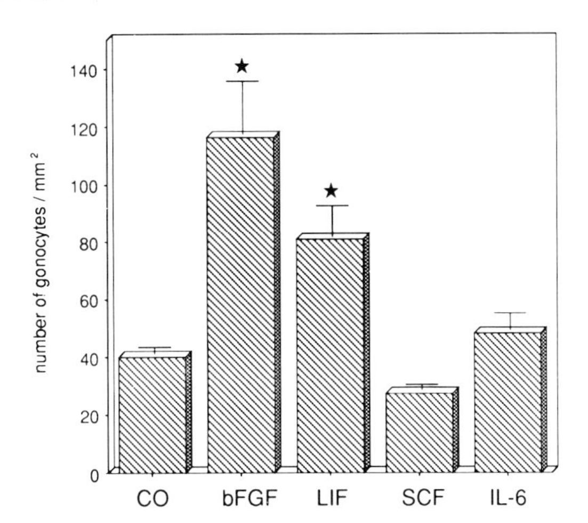

Figure 9 The effect of various growth factors (concentration range 1–60 ng ml^{-1}) on the number of gonocytes in co-culture with Sertoli cells, both isolated from newborn rat testes: Leukaemia Inhibitory Factor (LIF) and basic Fibroblast Growth Factor (bFGF) only caused a marked increase in the gonocyte number. Bar – mean ± SEM.

1993) this effect may be mediated by the Sertoli cells and may not be a direct effect of bFGF on the gonoctyes. The recent finding that LIF and the c-kit ligand may play an important role in PGCs development prompted us to investigate the effect of these factors on gonocyte survival and proliferation in co-culture with neonatal Sertoli cells. LIF was found to significantly increase the survival and proliferation of the gonocytes whereas the c-kit ligand did not show any effect (De Miguel *et al.*, 1996). Expression of the LIF gene or protein has been demonstrated in the adult testis (Robertson *et al.*, 1993) and at the start of spermatogenesis (De Miguel *et al.*, 1996). However, as yet no information is available concerning the presence of the LIF receptor during testicular ontogeny. This and other questions are currently being investigated in our laboratory. The fact that the c-kit ligand does not affect gonocytes *in vitro* may result from down-regulation of c-kit in the gonocytes or the uncoupling of this receptor from the intra-cellular signalling pathway (Manova and Bachvarova, 1991; Matsui *et al.*, 1991).

In co-culture with Sertoli cells and in the presence of either bFGF, LIF or follicle stimulating hormone (FSH), or a combination of these factors, the gonocytes do not develop into differentiating spermatogonia (Van Dissel-Emiliani *et al.*, 1996). Apparently in these cultures, a factor necessary for further development is missing. Zhou *et al.* (1993) demonstrated that in an organ-culture system, AMH induces newborn rat gonocytes to mature up to primary spermatocytes. Hence, AMH may be this missing factor.

Figure 8 Photomicrographs of gonocytes in co-culture with STO- (A), BRL-cells (B), or Sertoli cells isolated from 21-day-old rats after 1 day (C) or 9 days (D, E), after plating the gonocytes. On somatic cells other than Sertoli cells, the gonocytes form aggregates with each other that are removed when the medium is replenished. Note the presence of blebs on the gonocytes in this case (hollow arrow). In co-culture with Sertoli cells, the gonocytes become attached to the monolayer and some of them even relocate between the Sertoli cells through the production of cellular processes. Arrows, gonocytes or aggregates of gonocytes; Asterisks, somatic cell monolayer; Arrowheads, cellular processes of gonocytes (A–D 360 ×; E 1100 ×).

FACTORS (POSSIBLY) INVOLVED IN THE REGULATION OF PROLIFERATION AND DIFFERENTIATION OF UNDIFFERENTIATED SPERMATOGONIA

Role of vitamin A

Vitamin A is essential for the maintenance of the spermatogenic process. Vitamin A deficiency (VAD) causes extensive loss of germ cells and eventually only Sertoli cells, spermatogonia and some spermatocytes are left in the seminiferous tubules (Mitranond *et al.*, 1979; Unni *et al.*, 1983). The damage caused by VAD is reversible as replacement of vitamin A in the diet restores normal spermatogenesis (Huang and Hembree, 1979; Huang *et al.*, 1983). In the *in vitro* system using fragments of mouse cryptorchid testes it was found that both vitamin A and retinoic acid stimulate the proliferation and differentiation of the A spermatogonia present in the cryptorchid testis (Haneji *et al.*, 1982, 1986).

In recent years it has been established that during VAD the production of A_2 spermatogonia becomes arrested, while there is also an arrest at the level of preleptotene spermatocytes (De Rooij *et al.*, 1989, 1994; Griswold *et al.*, 1989; Van Pelt and De Rooij, 1990a; Ismail *et al.*, 1990). Administration of either vitamin A (De Rooij *et al.*, 1989; Griswold *et al.*, 1989; Van Pelt and De Rooij, 1990a; Ismail *et al.*, 1990) or retinoic acid (RA) (Van Pelt and De Rooij, 1991) causes a mass production of A_1 spermatogonia throughout the testis that synchronously develop along the spermatogenic line. In this way synchronized testes of rats and mice can be obtained in which a restricted number of epithelial stages is present, enabling the study of epithelial stage related processes in rats and mice (Morales and Griswold, 1987; Van Pelt and De Rooij, 1990b).

The precise nature of the A spermatogonia in the VAD testis is still a matter of debate (Figure 10). According to Van Pelt and De Rooij (1990a) the A spermatogonia in the severe VAD state are comparable to the A spermatogonia normally present in stage VII, i.e. a mixture of largely quiescent A_s, A_{pr} and A_{al} spermatogonia. In this model RA is necessary to induce the differentiation of the A_{al} spermatogonia into A_1 spermatogonia that subsequently start their cell cycle to divide into A_2 spermatogonia. In contrast, Griswold *et al.* (1989) and Ismail *et al.* (1990) suppose that in the VAD testis the arrest of spermatogenesis is in the G_2 phase of the A_1 spermatogonia, these spermatogonia not being able to start their mitosis into A_2 spermatogonia. The difference in opinion arose from the fact that Griswold *et al.* (1989) and Ismail *et al.* (1990) reported an increased

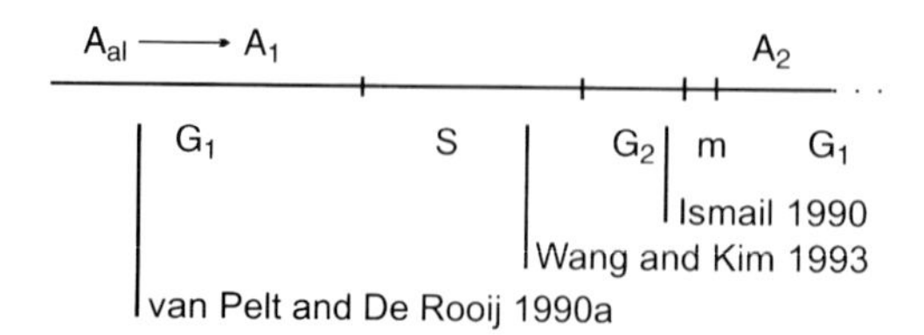

Figure 10 The various points of view with respect to the nature of the A spermatogonia left in the vitamin A deficient testis.

number of spermatogonial mitoses within the first few hours after administration of vitamin A while Van Pelt and De Rooij (1990a) did not observe such an increase. Furthermore, Wang and Kim (1993) found a highly increased level of H1 histone kinase activity associated with the cdc2 kinase/cyclin B complex at 12 hours after administration of retinol suggesting that an increased number of cells, presumably A_1 spermatogonia, pass from G2 to mitosis at that time. These authors suggested that the A spermatogonia present in the VAD testis are arrested in S-phase. However, in more extensive studies we have recently been able to show that there is no peak in mitotic activity of the A spermatogonia during the first 24 hours after retinol administration and that in the VAD testis the great majority of these cells have a DNA content of cells in G_1 phase (Van Pelt *et al.*, 1995). Preleptotene spermatocytes remaining in the VAD testis have an S-phase DNA content, i.e. between 2n and 4n. So, it is likely that the increased levels of H1 histone kinase activity associated with the cdc2 kinase/cyclin B complex at 12 hours after administration of retinol is caused by the transition of spermatocytes from preleptotene to the leptotene phase. It is intriguing that the A spermatogonia in stage VIII are the only spermatogonia really affected by the onset of the VAD. All other spermatogonia more or less proliferate and differentiate normally into preleptotene spermatocytes (De Rooij *et al.*, 1994). This suggests that vitamin A/RA has a highly specific effect on the development of the A spermatogonia in stage VIII and that this effect is not likely to be due to the inhibition of the expression of a general cell cycle related gene.

Therefore, the VAD testis offers an unique opportunity to study the induction of differentiation of the undifferentiated spermatogonia into A_1 spermatogonia. In addition, because of the arrest in the differentiation of the undifferentiated spermatogonia, the VAD testis can serve as a source for the purification of these cells (Van Pelt *et al.*, 1996).

The c-kit receptor and its ligand stem cell factor (SCF)

As described above SCF is involved in the development of the PGCs during fetal life. In addition, in the adult testis c-kit and its ligand also seem to play an important role. Rossi *et al.* (1991) found SCF to be produced by Sertoli cells of 18-day-old mice and showed that FSH increases SCF-mRNA levels in these cells. This was also found by Manova *et al.* (1993), who in addition demonstrated that a membrane-bound form of SCF predominates over the soluble form, from day 5 *post partum* (pp) onwards. Furthermore, the c-kit-mRNA was found to be present in mouse spermatogonia from day 6 pp onwards (Manova *et al.*, 1990; Sorrentino *et al.*, 1991). Yoshinaga *et al.* (1991) gave intraperitoneal injections of a monoclonal antibody against c-kit to mice and found that these injections did not kill gonocytes and undifferentiated spermatogonia. However, all differentiating type A spermatogonia were killed by the c-kit antibodies.

Recent data indicate that c-kit and its ligand may play a role in the differentiation of spermatogonia. First, as described above in the postnatal mouse testis c-kit gene expression in germ cells starts at day 6 pp, i.e. when differentiating type A spermatogonia appear (Manova *et al.*, 1990; Sorrentino *et al.*, 1991). Second, Koshimizu *et al.* (1991) studied the effect of W mutations on the differentiation of mouse testicular germ cells using experimental cryptorchidism and its surgical reversal. The mutant mice studied were $W^v/+$, $W^{sh}/+$, $W^f/+$ and W^f/W^f. Like in $+/+$ mice all these mice showed a normal seminiferous epithelium, and when made cryptorchid, in all mice spermatogenesis became arrested at the level of A spermatogonia. However, upon surgical reversal of cryptorchidism, spermatogenesis recovered in +/+ mice but not or incompletely in the mice carrying W mutations, suggesting that the W mutation affects the differentiation of the A spermatogonia. Finally, SCF was found to increase the number of differentiating A spermatogonia in S-phase but not of B spermatogonia *in vitro* (Rossi *et al.*, 1993).

Taken together, these data indicate that the expression of the c-kit receptor on A spermatogonia and the concomitant secretion of SCF by Sertoli cells is necessary for the differentiation of the A spermatogonia and also stimulates the proliferative activity of the differentiating type A spermatogonia. However, c-kit/SCF does not seem to play an important role in the compartment of undifferentiated spermatogonia, except that it may be necessary for their differentiation into differentiating type A spermatogonia.

Interleukin 1 alpha (Il-1α)

Sertoli cells produce Il-1α (Gérard *et al.*, 1991) in an epithelial stage dependent way, correlating with the numbers of spermatogonia in DNA synthesis in the various epithelial stages (Söder *et al.*, 1991). Pöllänen *et al.* (1989) and Parvinen *et al.* (1991) found a significant stimulation by Il-1α of the ^{3}H-thymidine incorporation by differentiating type spermatogonia both *in vivo* and *in vitro* in the rat. The effect of Il-1α on undifferentiated spermatogonia was not studied specifically.

Interleukin 6 (Il-6) and Leukaemia Inhibitory Factor (LIF)

Two members of the Il-6 family have been identified in the testis so far, Il-6 and LIF. Il-6 has been shown to be produced by Sertoli cells in an epithelial-stage-dependent way (Syed *et al.*, 1993), the levels being highest during stages II–VI. To date no effects of IL-6 on spermatogonial proliferation and differentiation have been described.

In view of the effects of LIF on PGCs and gonocytes it is interesting that LIF mRNA transcripts are present in the adult testis (Robertson *et al.*, 1993). It is not known yet which cell type(s) produce(s) the mRNA and whether the LIF protein is available to spermatogonia. However, the role of LIF in spermatogenesis does not seem to be crucial, as male mice homozygous for a mutated LIF gene are fertile (Stewart *et al.*, 1992).

Anti-Mullerian Hormone (AMH)/Mullerian Inhibiting Substance

AMH is a member of the TGFβ family of growth factors and is produced by Sertoli cells (review Josso and Picard, 1986). No specific effect on spermatogonia in the adult testis is known, although it has been suggested that like in the ovary (Takahashi *et al.*, 1986; Ueno *et al.*, 1988) AMH in the testis may prevent germ cells from entering the meiotic prophase prematurely (Jost, 1972; review Lee and Donahoe, 1993). As mentioned above, *in vitro* it induces differentiation of gonocytes.

Inhibin and activin

Inhibin is secreted by Sertoli cells (review De Jong, 1988) and its production is regulated by FSH (Morris *et al.*, 1988). Inhibin exerts a negative feedback on the secretion of FSH by the pituitary. Intratesticular injections of inhibin were found to cause a decrease in the numbers of the last four generations of differentiating spermatogonia in the Chinese hamster (Van Dissel-Emiliani *et al.*, 1989b). No effect was seen on the undifferentiated spermatogonia. In accordance with this, Hakovirta *et al.* (1993) found an inhibition of DNA synthesis of In spermatogonia by inhibin in cultured rat seminiferous tubular segments.

Activin is also produced by Sertoli cells (De Winter *et al.*, 1993) and has effects opposite to those of inhibin on the FSH secretion by the pituitary. Mather *et al.* (1990) found that in germ cell–Sertoli cell co-cultures of 20–25 day old rats, activin increased spermatogonial numbers while inhibin had no effect in this system. No distinction was made between spermatogonial cell types. However, in view of the numbers of spermatogonia present in the cultures, these were presumably mostly differentiating type spermatogonia. Recently, Moore and Mather (1994) found that activin in these co-cultures promotes cell survival rather than cell proliferation. Hakovirta *et al.* (1993) found a stimulation of ^{3}H-thymidine incorporation by In spermatogonia and preleptotene spermatocytes by activin-A in cultured rat seminiferous tubular segments. Kaipia *et al.* (1993) using *in situ* hybridization with activin receptor (ActR-IIb2) mRNA, found that at the base of the seminiferous tubules the mRNA was expressed in Sertoli cells and in differentiating type A1 and A2 spermatogonia. However, due to the kind of vector used, the labelling might in fact be localized in the residual bodies left in the epithelium after spermiation in stage VIII (Millar *et al.*, 1994).

In conclusion, both inhibin and activin may play a role in spermatogonial proliferation, and may have opposing effects on the differentiating type spermatogonia. Inhibin does not seem to have an effect on undifferentiated spermatogonia. For activin this has not been studied as yet.

Insulin-like growth factors 1 and 2 (IGF-1, IGF-2)

IGF-1 is produced by Sertoli cells (Hall *et al.*, 1983; Benahmed *et al.*, 1987) and Sertoli cells also have receptors for IGF-1 (Chatelain *et al.*, 1987; Oonk and Grootegoed, 1988). So far no IGF-1 or IGF-1 receptors have been found in spermatogonia (Forti *et al.*, 1989). Nevertheless, Söder *et al.* (1992) found a significant stimulation of ^{3}H-thymidine uptake by A_4 and B spermatogonia *in vitro* after 48 hours. In view of the long time interval needed to see an effect *in vitro*, the effects of IGF-1 on spermatogonial proliferation may be mediated by the Sertoli cells. It is not known yet whether the proliferation of the undifferentiated spermatogonia is affected by IGF-1.

IGF-2 mRNA and protein have been detected in all testicular cells (Dombrowicz *et al.*, 1992). In the experimental system of Söder *et al.* (1992), IGF-2 only had a slight effect on ^{3}H-thymidine incorporation by B spermatogonia.

Transforming growth factor-alpha (TGFα)/Epidermal growth factor (EGF)

Sertoli cells and peritubular cells have been found to contain both the mRNA for TGFα and the protein itself (Skinner *et al.*, 1989; Teerds *et al.*, 1990; Mullaney and Skinner, 1992a; Radhakrishnan *et al.*, 1992). EGF receptors were found on Sertoli cells, peritubular cells, pachytene spermatocytes and round spermatids (Suarez-Quian *et al.*, 1989; Radhakrishnan and Suarez-Quian, 1992; Mullaney and Skinner, 1992a). However, the EGF receptors on Sertoli cells do not seem to be functional (Mullaney and Skinner, 1992a). Although as yet these receptors have not been detected on spermatogonia, EGF has been found to stimulate the proliferation and differentiation of the A spermatogonia in cryptorchid testes *in vitro* (Haneji *et al.*, 1991).

Transforming growth factor-beta (TGFβ)

TGFβ1, TGFβ2, and TGFβ3 are produced by Sertoli cells and peritubular myoid cells in a developmentally specific pattern (Skinner and Moses, 1989; Mullaney and Skinner, 1993; Teerds and Dorrington, 1993). Hakovirta *et al.* (1993) found that TGFβ1 slightly stimulated the DNA synthetic activity of preleptotene spermatocytes in cultured segments of seminiferous tubules, but had no effect on differentiating spermatogonia. As yet there are no data on possible direct or indirect effects of TGFβs on proliferation and differentiation of undifferentiated spermatogonia.

Basic fibroblast growth factor (bFGF)

Immunocytochemical studies have revealed the presence of bFGF in all types of germ cells including A spermatogonia in the 5-day-old rat and in the adult rat testis (Mayerhofer *et al.*, 1991; Suzuki *et al.*, 1991; Han *et al.*, 1993). Sertoli cells and peritubular cells have been shown to produce bFGF in culture (Smith *et al.*, 1989; Mullaney and Skinner, 1992b), but Han *et al.* (1993) did not find bFGF-like proteins in secreted medium from cultured Sertoli cells. Receptors for bFGF have been detected on Sertoli cells and germ cells (Han *et al.*, 1993, Le Magueresse-Battistoni *et al.*, 1995) but not specifically on spermatogonia. Taken together, in the immature and adult testis bFGF seems to be a germ cell product that may influence Sertoli cell or germ cell function. Nevertheless, the expression patterns of bFGF and its receptors may change during development and not all issues have been resolved unequivocally. For the adult testis no indications of a direct or indirect role for bFGF in the regulation of spermatogonial proliferation have been reported so far.

Growth factors of unknown identity

A mitogenic activity produced by Sertoli cells has been called seminiferous growth factor (SGF) (Feig *et al.*, 1980, 1983; Bellvé and Feig, 1984). SGF is a 15.7 kDa protein which is able to stimulate the proliferation of prepuberal Sertoli cells (Bellvé and Feig, 1984) and is not identical to acidic or basic FGF (Zheng *et al.*, 1990). The effects of SGF on spermatogonial proliferation has not been studied as yet.

Sertoli cells have also been shown to produce a unique mitogenic activity which has been called Sertoli cell secreted growth factor (SCSGF) (Holmes *et al.*, 1986; Buch *et al.*, 1988). It appears to be a heat- and acid-stable, protease sensitive 14 kDa protein which is mitogenic for the A431 cell line (Lamb *et al.*, 1991). The secretion of SCSGF by Sertoli cells seems to be regulated by FSH and testosterone (Shubhada *et al.*, 1993). However, the effect of SCSGF on spermatogonial proliferation and differentiation has not been studied as yet.

Finally, SCCM has been shown to contain a factor which affects the numbers of undifferentiated spermatogonia. A partially purified inhibin preparation from rat SCCM was also found to contain a factor which, upon intratesticular injection in Chinese hamsters induced a 50–70% increase in the numbers of undifferentiated spermatogonia in stages IV–VII (Van Dissel-Emiliani *et al.*, 1989b). Apparently, this factor, which was not further characterized, partially counteracts the decrease in the proliferative activity of the undifferentiated spermatogonia normally taking place in these stages.

Inhibitory substances

It has been shown by various groups that the proliferation of the undifferentiated spermatogonia can be partly inhibited by testicular extracts (Clermont and Mauger, 1974; Thumann and Bustos-Obregon, 1978; Irons and Clermont, 1979; De Rooij, 1980). The inhibiting factor which was found to be tissue- but not species-specific was thought to be a chalone and has not been purified as yet.

In the normal seminiferous epithelium the proliferative activity of the undifferentiated spermatogonia is inhibited around stages III–IV in the mouse, rat and Chinese hamster. In the Chinese hamster it was found that when the differentiating spermatogonia were specifically removed from the epithelium this inhibition did not take place (Figure 7) (De Rooij *et al.*, 1985). It was concluded that from stage III onwards the proliferation of the undifferentiated spermatogonia is inhibited by the In and B spermatogonia present in these stages by way of a negative feedback system (Figure 4). This feedback regulation might work via the production of a chalone by the In and B spermatogonia. In this respect it is interesting that the inhibiting activity of the testicular extracts was absent when they were made of testes from which all spermatogonia were removed (De Rooij, 1980).

An important question is whether or not the stem cells are also inhibited in their proliferative activity by this feedback system. In the Chinese hamster the LI of the A_{pr} and A_{al} spermatogonia drops sharply in stage III while the LI of the A_s spermatogonia drops in early stage VII (Lok and De Rooij, 1983b). In the rat the LI of the undifferentiated spermatogonia drops in stage II and that of the A_s spermatogonia in stage V (Huckins, 1971b). Hence it can be concluded that the A_s spermatogonia are much less sensitive or perhaps completely insensitive to the feedback regulation. In case these cells are insensitive we have to assume that the decrease in their proliferative activity later during the epithelial cycle is caused by a lack of stimulation. This would be consistent with the notion that the secretory activity of the Sertoli cells varies during the epithelial cycle (Parvinen, 1982).

Mutations affecting spermatogonial proliferation

Although many mutations have been found that affect the spermatogenic process, only few affect the regulation of spermatogonial multiplication and stem cell renewal. Of course, as discussed above, much work has been done on mutations in the dominant white spotting (W) and Steel (Sl) loci.

Male mice homozygous for the mutation juvenile spermatogonial depletion (jsd), experience one wave of spermatogenesis, but fail to continue proliferation of type A spermatogonia (Beamer *et al.*, 1988). This suggests that in these mice the regulation of either the ratio between self-renewal and differentiation of the spermatogonial stem cells is disturbed or the differentiation of undifferentiated spermatogonia into A_1 spermatogonia. Recently, Mizunuma *et al.* (1992) found that in seminiferous tubules of a cryptorchid normal testis transplanted into the testis of a jsd/jsd mouse, the A spermatogonia could develop into spermatids. Thus, the jsd mutation does not affect the extratubular environment. Also, as initially fully development of spermatogenic cells can take place the defect is probably not in the spermatogonia themselves but possibly in the Sertoli cells that might fail to produce some factor(s) crucial for spermatogonial differentiation.

Sutcliffe and Burgoyne (1989) found that testes of $XOSxr^b$ mice, which are males because of the sex-reversed factor, almost totally lack spermatocytes. Apparently, a Y chromosomal gene called Spy affects the proliferation of differentiating A spermatogonia, while no effect was seen on undifferentiated spermatogonia. Later this gene appeared to encode the ubiquitin-activating enzyme E1 (Mitchell *et al.*, 1991).

Another gene that may well be important in spermatogonial proliferation in the human is the Fmr-1 gene which, when mutated, is responsible for the fragile X syndrome (De Vries *et al.*, 1994). Using *in situ* hybridization Bächner *et al.* (1993a, 1993b) found expression of Fmr-1 mRNA in mouse gonocytes and early spermatogonia. As patients with the fragile X syndrome have hypospermatogenesis and as spermatozoa with the full mutation are not formed these authors suppose that Fmr-1 gene expression is necessary for normal spermatogonial proliferation, the spermatogonia carrying the full mutation being selected out. Interestingly, most of the patients also show enlarged testes.

CONCLUSIONS AND PROSPECTS

From the above it will have become clear that as yet we are far from understanding the factors involved in the regulation of the proliferation and differentiation of PCG, gonocytes and undifferentiated spermatogonia. The present knowledge is summarized in Tables 3 and 4.

With respect to PGCs, growth factors have been found that affect these cells *in vitro* (Table 3). However, the only factor known to be indispensable to these cells is SCF as mutants in which either the SCF protein (Sl) or the c-kit receptor protein (W) is severely affected have much less PGCs and are infertile. Also, SCF is present in cells in the PGCs migratory pathway and in the genital ridges (Matsui *et al.*, 1990). Unfortunately, as yet no information is available concerning the site of production of the other growth factors known to affect PGCs proliferation *in vitro*, and whether or not PGCs have receptors

Table 3 Effects of various genes/growth factors on proliferation and differentiation of PGCs, gonocytes and undifferentiated spermatogonia

Cell type	(Growth) factor	Effect	Reference
Primordial germ cell (PGC)	SCF and LIF	enhanced survival and proliferation *in vitro*	Dolci *et al.*, 1991, 1993; DeFelici and Dolci, 1991; Matsui *et al.*, 1991; Godin *et al.*, 1991; Pesce *et al.*, 1993; Manova and Bachvarova, 1991
	SCF	stimulation of both migration and proliferation *in vivo*	Buehr *et al.*, 1993
	gcd gene product	stimulation of migration and/or proliferation *in vivo*	Pellas *et al.*, 1991
	TGFβ1	inhibition of proliferation	Godin and Wylie, 1991
	bFGF	sustained proliferation *in vitro*	Matsui *et al.*, 1992 Resnick *et al.*, 1992
	TNFα	enhanced survival and proliferation *in vitro*	Kawase *et al.*, 1994
	ter gene product	differentiation into gonocytes	Stevens, 1967, 1970
Gonocytes	thymuline	stimulation of proliferation in organ culture	Prépin, 1993
	AMH	induction of differentiation in organ culture	Zhou *et al.*, 1993
	LIF, CNTF and bFGF	enhanced survival *in vitro*	De Miguel *et al.*, 1996; Van Dissel-Emiliani *et al.*, 1996
Undifferentiated spermatogonia	retinoic acid, SCF?, *jsd* gene product? *jsd* gene product?	differentiation into A_1 spermatogonia	Van Pelt and De Rooij, 1990a; Manova *et al.*, 1990; Sorrentino *et al.*, 1991; Koshimizu *et al.*, 1991; Beamer, 1988; Mizunuma, 1992;
	jsd gene product	k-regulation of self-renewal of stem cells	Beamer, 1988

for such factors. It is hazardous to extrapolate data obtained *in vitro* to the *in vivo* situation. For example, LIF has been shown to be important for the culture of PGCs (Dolci *et al.*, 1991; DeFelici and Dolci, 1991; Matsui *et al.*, 1991; Godin *et al.*, 1991). However, male mice homozygous for a mutated LIF gene are fertile (Stewart *et al.*, 1992). This indicates that there is no absolute requirement for LIF *in vivo*. Since LIF is a member

Table 4 Effects of various genes/growth factors on proliferation and differentiation of differentiating type spermatogonia.

Differentiating spermatogonial type	Factor	Action	References
A	SCF	stimulation of proliferative activity	Rossi *et al.*, 1993
	Il-1α	stimulation of ^{3}H-thymidine incorporation *in vitro*	Pöllänen *et al.*, 1989; Parvinen *et al.*, 1991
In and B In	inhibin	decreases cell numbers inhibition DNA synthesis	Van Dissel-Emiliani *et al.*, 1989b; Hakovirta *et al.*, 1993
In	activin	stimulation of ^{3}H-thymidine incorporation	Hakovirta *et al.*, 1993
A_4 and B	IGF-1	stimulation of ^{3}H-thymidine incorporation	Söder *et al.*, 1992

of a family of growth factors (review Patterson, 1992) an absolute dependence on LIF for PGCs development might either be circumvented by other members of the family using the same receptors or in reverse, one of the other family members might be the crucial factor. The nature of such factors and their expression at this early stage of testicular development remains, however, to be elucidated.

With regard to the gonocytes and the undifferentiated spermatogonia, we are still at the early stages of understanding the mechanisms involved in their development. The ability to culture the gonocytes either with Sertoli cells or in organ culture has allowed us to gain some insight into the factors regulating their proliferation and differentiation. However, in such cultures, it is difficult to establish whether the effects found on the germ cells are direct, or indirect and mediated by as yet uncharacterized factors produced by the Sertoli cells or other somatic cells of the testis. Studies on the regulation of proliferation and differentiation of the undifferentiated spermatogonia has been hampered by the lack of methods to purify these cells from the adult testis and to culture them. Although *in vivo* studies have been valuable to understand some regulatory mechanisms involved in their development, the nature of the testicular architecture and the multiple interactions occurring at the cellular level make interpretation of *in vivo* data often difficult. More attention should be paid to the possibilities to purify these cells using testes in which the undifferentiated spermatogonia are relatively enriched as is the case in the VAD testis and likely also in the cryptorchid testis model of the group of Nishimune (Aizawa and Nishimune, 1979; Haneji and Nishimune, 1982). Also, as discussed above, the gonocytes in the newborn testis may well represent a population of undifferentiated spermatogonia. Understanding the regulation of the start of spermatogenesis may therefore also help elucidate some regulatory mechanisms involved in spermatogonial development in the adult testis. In view of the simplicity with which gonocytes can be cultured, such a system may turn out to be of great importance.

The number of factors isolated from testicular tissue, reported to possibly affect spermatogonial proliferation and differentiation is constantly increasing. A majority of these substances was initially discovered in Sertoli cell culture media or in isolated Sertoli cells (mainly from immature rats, but also from piglets, lambs and humans). This highlights the unique contribution of cell separation and culture techniques to the investigation of testicular function. However, a prerequisite for the identification of growth factors important for the regulation of spermatogonial development should be the demonstration of the synthesis and secretion of the factor at the right time and place, demonstration of an effect *in vitro* and *in vivo* at physiological concentrations, and the localization of receptors on the target cells. The present analysis of the literature reveals that this information is lacking in most cases. It is likely that the use of recently developed models in which the undifferentiated spermatogonia are enriched, together with the increasing application of molecular biology technology, including the production of transgenic mice in which the function of specific factors and receptors is abolished, will generate more precise data about the paracrine system of the regulation of the proliferation and differentiation of germ line stem cells.

ACKNOWLEDGEMENTS

The authors are grateful to Drs K. J. Teerds and M. E. A. B. van Beek for critical reading of the manuscript and Mr Van Rijn for preparing the illustrations.

REFERENCES

Ailenberg, M., Tung, P.S. and Fritz, I.B. (1990). *Biol. Reprod.* **42**:499–509.
Aizawa, S. and Nishimune, Y. (1979). *J. Reprod. Fert.* **56**:99–104.
Anderson, D.M., Lyman, S.D., Baird, A. *et al.* (1990). *Cell* **63**:235–243.
Ao, A., Erichson, R.P. and Stalvey, J.R.D. (1993). *Mol. Reprod. Dev.* **35**:159–164.
Bächner, D., Steinbach, P., Wöhrle, D. *et al.* (1993a). *Nature Genetics* **4**:115–116.
Bächner, D., Manca, A., Steinbach, P. *et al.* (1993b). *Human Molec. Genetics* **2**:2043–2050.
Beamer, W.G., Cunliffe-Beamer, T.L., Shultz, K.L., Langley, S.H. and Roderick, T.H. (1988). *Biol. Reprod.* **38**:899–908.
Bellvé, A.R. and Feig, L.A. (1984). *Recent Progr. Hormone Res.* **40**:531–567.
Bellvé, A.R., Cavicchia, J.C., Millette, C.F., O'Brien, D.A., Bhatnagar, Y.M. and Dym, M. (1977). *J. Cell Biol.* **74**:68–85.
Benahmed, M., Morera, A.M., Chauvin, M.C. and De Peretti, E. (1987). *Mol. Cell. Endocrinol.* **50**:69–77.
Boitani, C., Politi, M.G. and Menna, T. (1993). *Biol. Reprod.* **48**:761–767.
Buch, J.P., Lamb, D.J., Lipshultz, L.I. and Smith, R.G. (1988). *Fertil. Steril.* **49**:658–665.
Buehr, M., McLaren, A., Bartley, A. and Darling, S. (1993). *Develop. Dynam.* **198**:182–189.
Chatelain, P., Perrard-Sapori, M.H., Jaillard, C., Naville, D., Ruitton, A. and Saez, J. (1987). *Int. J. Rad. Appl. Instrum. B.* **14**:617–622.
Clermont, Y. and Bustos-Obregon, E. (1968). *Am. J. Anat.* **122**:237–248.
Clermont, Y. and Hermo, L. (1975). *Am. J. Anat.* **142**:159–176.
Clermont, Y. and Mauger, A. (1974). *Cell Tissue Kinet.* **7**:165–172.
Clermont, Y. and Perey, B. (1957). *Am. J. Anat.* **100**:241–268.
Courot, M., Houchereau-de Reviers, M.T. and Ortavant, R. (1970). In *The Testis* (eds A. D. Johnson, W. R. Gomes and N. L. Vandemark), pp. 339–432. Academic Press, New York.
DeFelici, M. and McLaren, A. (1983). *Exp. Cell Res.* **144**:417–423.

DeFelici, M. and Dolci, S. (1991). *Dev. Biol.* **147**:281–284.
DeFelici, M. and Siracusa, G. (1985). *J. Embryol. Exp. Morphol.* **87**:87–97.
DeFelici, M., Dolci, S. and Siracusa, G. (1989). *Cell Differ. Dev.* **28**:65–70.
DeFelici, M., Dolci, S. and Pesce, M. (1992). *Int. J. Dev. Biol.* **36**:205–213.
DeFelici, M., Dolci, S. and Pesce, M. (1993). *Dev. Biol.* **157**:277–280.
De Miguel, M.P., De Boer-Brouwer, M., Paniagua, R., Van den Hurk, R., De Rooij, D.G. and Van Dissel-Emiliani, F.M.F. (1996). *Endocrinol.* **137**:1885–1893.
De Jong, F.H. (1988). *Inhibin. Physiol. Rev.* **68**:555–607.
De Rooij, D.G. (1980). *Virchows Arch. B. Cell Path.* **33**:67–75.
De Rooij, D.G. (1983). In *Stem Cells. Their Identification and Characterization* (ed. C. S. Potten), pp. 89–117. Churchill Livingstone, Edinburgh.
De Rooij, D.G. (1988). *J. Cell Sci. Suppl.* **10**:181–194.
De Rooij, D.G., Lok, D. and Weenk, D. (1985). *Cell Tissue Kinet.* **18**:71–81.
De Rooij, D.G., Van Dissel-Emiliani, F.M.F. and Van Pelt, A.M.M. (1989). In *Regulation of Testicular Function: Signaling Molecules and Cell–Cell Communication* (eds L. L. Ewing and B. Robaire) *Ann. NY Acad. Sci.* **564**:140–153.
De Rooij, D.G., Van Pelt, A.M.M., Van de Kant, H.J.G. *et al.* (1994). In *Function of Somatic Cells in the Testis* (ed. A. Bartke), pp. 345–361. Springer Verlag, New York.
De Vries, L.B.A., Halley, D.J.J., Oostra, B.A. and Niermeijer, M.F. (1994). *J. Intellectual Disability Res.* **38**:1–8.
De Winter, J.P., Van der Stichele, H.M., Timmerman, M.A., Blok, L.J., Themmen, A.P. and De Jong, F.H. (1993). *Endocrinology* **132**:975–982.
Dolci, S., Williams, D.E., Ernst, M.K. *et al.* (1991). *Nature* **352**:809–811.
Dolci, S., Pesce, M. and De Felici, M. (1993). *Mol. Reprod. Dev.* **35**:134–139.
Dombrowicz, D., Hooghe-Peters, E.L., Gothot, A. *et al.* (1992). *Arch. Int. Physiol. Biochim. Biophys.* **100**:303–308.
Donovan, P.J. and Scott, D. (1986). *Cell* **44**:831–838.
Dym, M. and Fawcett, D.W. (1970). *Biol. Reprod.* **3**:308–326.
Feig, L.A., Bellvé, A.R., Horbach-Erickson, N. and Klagsbrun, M. (1980). *Proc. Natl Acad. Sci. USA* **77**:4774–4778.
Feig, L.A., Klagsbrun, M. and Bellvé, A.R. (1983). *J. Cell Biol.* **97**:1435–1443.
Forti, G., Barni, T., Vanelli, B.G., Balboni, G.C., Orlando, C. and Serio, M. (1989). *J. Steroid. Biochem.* **32**:135–144.
Franchi, L.L. and Mandl, A.M. (1964). *J. Embryol. Exp. Morph.* **12**:289–308.
Gérard, N., Syed, V. and Bardin, C.W. (1991). *Molec. Cell. Endocrinol.* **82**:R13–R16.
Ginsburg, M., Snow, M.H.L. and McLaren, A. (1990). *Development* **110**:521–528.
Godin, I. and Wylie, C.C. (1991). *Development* **113**:1451–1457.
Godin, I., Deed, R., Cooke, J., Zsebo, K., Dexter, M. and Wylie, C.C. (1991). *Nature* **352**:807–809.
Griswold, M.D., Bishop, P.D., Kim, K.H., Ping, R., Siiteri, J.E. and Morales, C. (1989). In *Regulation of Testicular Function: Signaling Molecules and Cell–Cell Communication* (eds. L. L. Ewing and B. Robaire). *Ann. NY Acad. Sci.* **564**:154–172.
Hakovirta, H., Kaipia, A., Söder, O. and Parvinen, M. (1993). *Endocrinology* **133**:1664–1668.
Hall, K., Ritzen, E.M., Johnsonbaugh, R.E. and Parvinen, M. (1983). In *Insulin-like Growth Factors/Somatomedins* (ed. E. M. Spencer), pp. 611–614. Walter de Gruyter, New York.
Han, I.S., Sylvester, S.R., Kim, K.H. *et al.* (1993). *Mol. Endocrinol.* **7**:889–897.
Haneji, T. and Nishimune, Y. (1982). *J. Endocr.* **94**:43–50.
Haneji, T., Maekawa, M. and Nishimune, Y. (1982). *Biochem. Biophys. Res. Comm.* **108**:1320–1324.
Haneji, T., Koide, S.S., Nishimune, Y. and Oota, Y. (1986). *Endocrinology* **119**:2490–2496.
Haneji, T., Koide, S.S., Tajima, Y. and Nishimune, Y. (1991). *J. Endocrinol.* **128**:383–388.
Hilscher, B., Hilscher, W., Bülthoff-Ohnolz, B. *et al.* (1974). *Cell Tissue Res.* **154**:443–470.
Holmes, S.D., Spotts, G. and Smith, R.G. (1986). *J. Biol. Chem.* **261**:4076–4080.
Huang, H.F.S. and Hembree, W.C. (1979). *Biol. Reprod.* **21**:891–904.
Huang, H.F.S., Durenfurth, I. and Hembree, W.C. (1983). *Endocrinology* **112**:1163–1171.
Huang, E., Nocka, K., Beier, D.R. *et al.* (1990). *Cell* **63**:225–233.
Huckins, C. (1971a). *Anat. Rec.* **169**:533–558.
Huckins, C. (1971b). *Cell Tissue Kinet.* **4**:313–334.

Huckins, C. and Clermont, Y. (1968). *Arch. Anat. Hist. Embryol.* **51**:343–354.
Irons, M.I. and Clermont, Y. (1979). *Cell Tissue Kinet.* **12**:425–433.
Ismail, N., Morales, C. and Clermont, Y. (1990). *Am. J. Anat.* **188**:57–63.
Jégou, B. (1992). *Ballière Clin. Endocrinol. Metab.* **6**:273–311.
Jost, A. (1972). *Johns Hopkins Med. J.* **130**:38–53.
Josso, N. and Picard, J.Y. (1986). *Physiol. Rev.* **66**:1038–1090.
Kaipia, A., Parvinen, M. and Toppari, J. (1993). *Endocrinology* **132**:477–479.
Kancheva, L.S., Martinova, Y.S. and Georgiev, V.D. (1990). *Mol. Cell. Endocrinol.* **69**:121–127.
Kawase, E., Yamamoto, H., Hashimoto, K. and Nakatsuji, N. (1994). *Dev. Biol.* **161**:91–99.
Kitamura, Y., Sonoda, T., Nakano, T., Hayashi, C. and Asai, H. (1985). *Int. Arch. Allergy Appl. Immunol.* **77**:144–150.
Kluin, Ph.M. and De Rooij, D.G. (1981). *Int. J. Andrology* **4**:475–493.
Koshimizu, U., Sawada, K., Tajima, Y., Watanabe, D. and Nishimune, Y. (1991). *Biol. Reprod.* **45**:642–648.
Lamb, D.J., Spotts, G.S., Shubhada, S. and Baker, K.R. (1991). *Mol. Cell. Endocrinol.* **79**:1–12.
Lawson, K.A. and Pederson, R.A. (1992). In CIBA foundation symposium 165 *Post Implantation Development in the Mouse,* **165**:3–26. John Wiley & Sons, New York.
Lee, M.M. and Donahoe, P.K. (1993). *Endocr. Rev.* **14**:152–164.
Le Margueresse-Battistoni, B., Wolff, J., Morera, A.M. and Benahmed, M. (1995). *Endocrinology* **135**: 2404–2411.
Lok, D. and De Rooij, D.G. (1983a). *Cell Tissue Kinet.* **16**:7–18.
Lok, D. and De Rooij, D.G. (1983b). *Cell Tissue Kinet.* **16**:31–40.
Lok, D., Jansen, M.T. and De Rooij, D.G. (1983). *Cell Tissue Kinet.* **16**:19–29.
Maekawa, M. and Nishimune, Y. (1991). *Cell Tissue Res.* **265**:551–554.
McGuinness, M. and Orth, J. (1989). *Anat. Rec.* **223**:72A (abstract).
McGuinness, M.P. and Orth, J.M. (1992a). *Anat. Rec.* **233**:527–537.
McGuinness, M.P. and Orth, J.M. (1992b). *Eur. J. Cell Biol.* **59**:196–210.
McLaren, A. and Buehr, M. (1990). *Cell Differ. Dev.* **31**:185–195.
Manova, K. and Bachvarova, R.F. (1991). *Dev. Biol.* **146**:312–324.
Manova, K., Nocka, K., Besmer, P. and Bachvarova, R.F. (1990). *Development* **110**:1057–1069.
Manova, K., Huang, E.J., Angeles, M. *et al.* (1993). *Dev. Biol.* **157**:85–99.
Mather, J.P., Attie, K.M., Woodruff, T.K., Rice, G.C. and Phillips, D.M. (1990). *Endocrinology* **127**:3206–3214.
Matsui, Y., Zsebo, K. and Hogan, B.L.M. (1990). *Nature* **347**:667–669.
Matsui, Y., Toksoz, D., Nishikawa, S. *et al.* (1991). *Nature* **353**:750–752.
Matsui, Y., Zsebo, K. and Hogan, B.L.M. (1992). *Cell* **70**:841–847.
Mayerhofer, A., Russell, L.D., Grothe, C., Rudolf, M. and Gratzl, M. (1991). *Endocrinology* **129**:921–924.
Meistrich, M.L. and Van Beek, M.E.A.B. (1993a). In *Cell and Molecular Biology of the Testis* (eds C. Desjardins and L. L. Ewing), pp. 266–295. Oxford University Press, New York.
Meistrich, M.L. and Van Beek, M.E.A.B. (1993b). In *Methods in Reproductive Toxicology* (eds R. E. Chapin and J. J. Heindel), **3A**:106–123. Academic Press, New York.
Metcalf, D. (1991). *Int. J. Cell Cloning* **9**:95–108.
Millar, M.R., Sharpe, R.M., Maguire, S.M., Gaughan, J., West, A.P. and Saunders, P.T.K. (1994). *Int. J. Androl.* **17**:149–160.
Mintz, B. and Russell, E.S. (1957). *J. Exp. Zool.* **134**:207–237.
Mitchell, M.J., Woods, D.R., Tucker, P.K., Opp, J.S. and Bishop, C.E. (1991). *Nature* **354**:483–486.
Mitranond, V., Sobhon, P., Tosukhowong, P. and Chindaduangrat W. (1979). *Acta Anat.* **103**:159–168.
Mizunuma, M., Dohmae, K., Tajima, Y., Koshimizu, U., Watanabe, D. and Nishimune, Y. (1992). *J. Cell Physiol.* **150**:188–193.
Moore, A. and Mather, J.P. (1994). *Biol. Reprod.* **50 (Suppl. 1)**:79.
Morales, C. and Griswold, M.D. (1987). *Endocrinology* **121**:432–434.
Morris, P.L., Vale, W.W., Cappel, S. and Bardin, C.W. (1988). *Endocrinology* **122**:717–725.
Mullaney, B.P. and Skinner, M.K. (1992a). *Mol. Endocrinol.* **6**:2103–2113.
Mullaney, B.P. and Skinner, M.K. (1992b). *Endocrinology* **131**:2928–2934.

Mullaney, B.P. and Skinner, M.K. (1993). *Mol. Endocrinol.* **7**:67–76.
Münsterberg, A. and Lovell-Badge, R. (1991). *Development* **113**:613–624.
Noguchi, T. and Noguchi, M. (1985). *J. Natl Cancer Inst.* **75**:385–392.
Noguchi, T. and Stevens, L.C. (1982). *J. Natl Cancer Inst.* **69**:907–913.
Novi, A.M. and Saba, P. (1968). *Z. Zellforsch.* **86**:313–326.
Oakberg, E.F. (1971). *Anat. Rec.* **169**:515–532.
Oonk, R.B. and Grootegoed, J.A. (1988). *Mol. Cell. Endocrinol.* **55**:33–43.
Orth, J.M. and Boehm, R. (1990). *Endocrinology* **127**:2812–2820.
Orth, J.M. and McGuinness, M.P. (1991). *Endocrinology* **129**:1119–1121.
Parvinen, M. (1982). *Endocrine Rev.* **3**:404–417.
Parvinen, M., Söder, O., Mali, P., Fröysa, B. and Ritzen, E.M. (1991). *Endocrinology* **129**:1614–1620.
Patterson, D.M. (1992). *Curr. Opp. Neurobiol.* **2**:94–97.
Pellas, T.C., Ramachandran, B., Duncan, M., Pan, S.S., Marone, M. and Chada, K. (1991). *Proc. Natl Acad. Sci. USA* **88**:8787–8791.
Pesce, M., Farrace, M.G., Piacentini, M., Dolic, S. and DeFelici, M. (1993). *Development* **118**:1089–1094.
Pöllänen, P., Söder, O. and Parvinen, M. (1989). *Reprod. Fertil. Dev.* **1**:85–87.
Prépin, J. (1993). *C.R. Acad. Sci. Paris, Life Sciences* **316**:451–454.
Qiu, F., Ray, P., Brown, K. *et al.* (1988). *EMBO J.* **7**:1003–1011.
Radhakrishnan, B. and Suarez-Quian, C.A. (1992). *J. Reprod. Fertil.* **96**:13–23.
Radhakrishnan, B., Oke, B.O., Papadopoulos, V., DiAugustine, R.P. and Suarez-Quian, C.A. (1992). *Endocrinology* **131**:3091–3099.
Resnick, J.L., Bixler, L.S., Cheng, L. and Donovan, P.J. (1992). *Nature* **359**:550–551.
Robertson, M., Chambers, I., Rathjen, P., Nichols, J. and Smith, A. (1993). *Dev. Genetics* **14**:165–173.
Rossi, P., Albanesi, C., Grimaldi, P. and Geremia, R. (1991). *Biochem. Biophys. Res. Commun.* **176**:910–914.
Rossi, P., Dolci, S., Albanesi, C., Grimaldi, P., Ricca, R. and Geremia, R. (1993). *Dev. Biol.* **155**:68–74.
Russell, E.S. (1979). *Adv. Genetics* **20**:357–459.
Russell, L.D. (1978). *Anat. Rec.* **190**:99–112.
Sapsford, C.S. (1962). *Austr. J. Zool.* **10**:178–192.
Sharpe, R.M. (1990). *Clin. Endocrinol.* **33**:787–807.
Shubhada, S., Glinz, M. and Lamb, D.J. (1993). *J. Androl.* **14**:99–109.
Skinner, M.K. and Moses, H.L. (1989). *Mol. Endocrinol.* **3**:625–634.
Skinner, M.K., Takacs, K. and Coffey, R.J. (1989). *Endocrinology* **124**:845–854.
Smith, E.P., Hall, S.H., Monaco, L., French, F.S., Wilson, E.M. and Conti, M. (1989). *Mol. Endocrinol.* **3**:954–961.
Söder, O., Syed, V., Callard, G.V. *et al.* (1991). *Int. J. Androl.* **14**:223–231.
Söder, O., Bang, P., Wahab, A. and Parvinen, M. (1992). *Endocrinology* **131**:2344–2350.
Sorrentino, V., Giorgi, M., Geremia, R., Besmer, P. and Rossi, P. (1991). *Oncogene* **6**:149–151.
Steinberger, A. (1967). *Anat. Rec.* **157**:327 (abstract).
Stevens, L.C. (1967). *J. Natl Cancer Inst.* **38**:549–552.
Stevens, L.C. (1970). *J. Natl Cancer Inst.* **44**:923–929.
Stevens, L.C. (1973). *J. Natl Cancer Inst.* **50**:235–242.
Stewart, C.L., Kaspar, P., Brunet, L.J. *et al.* (1992). *Nature* **359**:76–79.
Suarez-Quian, C.A., Dai, M.Z., Onoda, M., Kriss, R.M. and Dym, M. (1989). *Biol. Reprod.* **41**:921–932.
Sutcliffe, M.J. and Burgoyne, P.S. (1989). *Development* **107**:373–380.
Suzuki, K., Kamel, T., Hakamata, Y., Kikukawa, K., Shiota, K. and Takahashi, M. (1991). *Proc. Soc. Exp. Biol. Med.* **198**:728–731.
Syed, V., Soder, Ö., Arver, S., Lindh, M., Khan, S. and Ritzen, E.M. (1988). *Int. J. Androl.* **11**:437–448.
Syed, V., Gérard, N., Kaipia, A., Bardin, C.W., Parvinen, M. and Jégou, B. (1993). *Endocrinology* **132**:293–299.
Takahashi, M., Koide, S.S. and Donahoe, P.K. (1986). *Mol. Cell. Endocrinol.* **47**:225–234.

Taketo, T., Seen, C.D. and Koide, S.S. (1986). *Biol. Reprod.* **34**:919–924.
Tam, P. and Snow, M.H.L. (1981). *J. Embryol. Exp. Morphol.* **64**:133–147.
Teerds, K.J. and Dorrington, J.H. (1993). *Biol. Reprod.* **48**:40–45.
Teerds, K.J., Rommerts, F.F.G. and Dorrington, J.H. (1990). *Molec. Cell. Endocr.* **69**:R1–R6.
Tiba, T. and Kita, I. (1991). *Anat. Histol. Embryol.* **20**:118–128.
Tiba, T., Takahashi, M., Igura, M. and Kita, I. (1992). *Anat. Histol. Embryol.* **21**:9–22.
Thumann, A. and Bustos-Obregon, E. (1978). *Andrologia* **10**:22–25.
Tran, D. and Josso, N. (1982). *Endocrinology* **111**:1562–1567.
Ueno, S., Manganaro, T.F. and Donahoe, P.K. (1988). *Endocrinology* **123**:1652–1659.
Unni, E., Rao, M.R.S. and Ganguly, J. (1983). *Indian J. Exp. Biol.* **21**:180–192.
Van Beek, M.E.A.B., Davids, J.A.G., Van de Kant, H.J.G. and De Rooij, D.G. (1984). *Radiat. Res.* **97**:556–569.
Van Beek, M.E.A.B., Meistrich, M.L. and De Rooij, D.G. (1990). *Cell Tissue Kinet.* **23**:1–16.
Van der Meer, Y., Huiskamp, R., Davids, J.A.G., Van der Tweel, I. and De Rooij, D.G. (1992a). *Radiat. Res.* **130**:289–295.
Van der Meer, Y., Huiskamp, R., Davids, J.A.G., Van der Tweel, I. and De Rooij, D.G. (1992b). *Radiat. Res.* **130**:296–302.
Van Dissel-Emiliani, F.M.F., De Rooij, D.G. and Meistrich, M.L. (1989a). *J. Reprod. Fertil.* **86**:759–766.
Van Dissel-Emiliani, F.M.F., Grootenhuis, A.J., De Jong, F.H. and De Rooij, D.G. (1989b). *Endocrinology* **125**:1898–1903.
Van Dissel-Emiliani, F.M.F., Van Kooten, P.J.S., De Boer-Brouwer, M., Van Eden, W., De Rooij, D.G. and Van der Donk, J.A. (1993a). *J. Reprod. Immunol.* **23**:93–108.
Van Dissel-Emiliani, F.M.F., De Boer-Brouwer, M., Van der Donk, J.A. and De Rooij, D.G. (1993b). *Cell Tissue Res.* **273**:141–147.
Van Dissel-Emiliani, F.M.F., De Boer-Brouwer, M., van der Donk, J.A. and de Rooij, D.G. (1993c). Abstract book, XIIth North American Testis Workshop, Tampa, Florida, USA, A 54.
Van Dissel-Emiliani, F.M.F., De Boer-Brouwer, M. and De Rooij, D.G. (1996). *Endocrinology* **137**:647–654.
Van Haaster, L.H. (1993). *Regulation of postnatal testicular development in rodents.* Thesis, Utrecht.
Van Haaster, L.H. and De Rooij, D.G. (1993). *Biol. Reprod.* **49**:1229–1235.
Van Haaster, L.H. and De Rooij, D.G. (1994). *J. Reprod. Fertil.* **101**:321–326.
Van Pelt, A.M.M. and De Rooij, D.G. (1990a). *Biol. Reprod.* **42**:677–682.
Van Pelt, A.M.M. and De Rooij, D.G. (1990b). *Biol. Reprod.* **43**:363–367.
Van Pelt, A.M.M. and De Rooij, D.G. (1991). *Endocrinology* **128**:697–704.
Van Pelt, A.M.M., Van Dissel-Emiliani, F.M.F., Gremers, I.C., Van de Burg, M., Tanke, H.J. and De Rooij, D.G. (1995). *Biol. Reprod.* **53**:568–576.
Van Pelt, A.M.M., Morena, A.R., Van Dissel-Emiliani, F.M.F., Boitani, C., Gaemers, I.C., De Rooij, D.G. and Stefanini, M. (1996). *Biol. Reprod.* In press.
Vergouwen, R.P.F.A., Jacobs, S.G.P.M., Huiskamp, R., Davids, J.A.G. and De Rooij, D.G. (1991). *J. Reprod. Fert.* **93**:233–243.
Vigier, B., Watrin, F., Magre, S., Tran, D. and Josso, N. (1987). *Development* **100**:43–55.
Wang, Z. and Kim, K.H. (1993). *Biol. Reprod.* **48**:1157–1165.
Yarden, Y., Kuang, W.J., Yang-Feng, T. *et al.* (1987). *EMBO J.* **6**:3341–3351.
Yoshinaga, K., Nishikawa, S., Ogawa, M. *et al.* (1991). *Development* **113**:689–699.
Zamboni, L. and Merchant, H. (1973). *Am. J. Anat.* **137**:299–336.
Zheng, W., Butwell, T.J., Heckert, L., Griswold, M.D. and Bellvé, A.R. (1990). *Growth Factors* **3**:73–82.
Zhou, B., Watts, L.M. and Hutson, J.M. (1993). *J. Urol.* **150**:613–616.
Zsebo, K.M., Williams, D.A., Geissler, E.N. *et al.* (1990). *Cell* **63**:213–224.

10 Stem cell repertoire in the intestine*

Nicholas A. Wright

ICRF Histopathology Unit, London and Department of Histopathology, RPMS, Hammersmith Hospital, London

INTRODUCTION

Many characteristics have been claimed for intestinal stem cells. One of the more important is *pluripotentiality*, or the ability to give rise to cells of more than one lineage. However, this is a stem cell property which, in epithelia such as that lining the gastrointestinal tract, has received much less attention than other putative properties. This is singular when we recall that, until comparatively recently, *direct* evidence for pluripotent stem cells in the gut was lacking, and that even now, we have little or no idea of the mechanisms which decide fate in gut epithelia; this is maybe a reflection of the lack of tractable models upon which such studies can be based.

The renewal systems of the gastrointestinal tract are based upon a series of tubules of different type. Conceptually the simplest tubule is seen in the colon, whose tubule – called a *crypt* – is closed at one end while the other opens out onto the luminal surface (Plate 10.3a). The arrangement in the small intestine is similar, but here the surface is adorned with projections called *villi*, themselves covered with a single layered sheet of epithelium continuous with the underlying crypts (Plate 10.3b). In the stomach, longer tubules called gastric glands are seen, again closed at one end and opening onto the surfaces as *foveolae* at the other.

Stem cells in the gut are specifically-located, and in epithelia, a realization has emerged that, notionally at any rate, stem cells should be found at the *origin of the cell flux*. In the colonic and small intestinal crypt, cell migration apparently begins at the base of the crypt, and cells migrate from here, emerging onto the surface in the colon or onto the villi in the small intestine; consequently, the basal crypt cells, or a subset thereof, are candidate stem cells. In the gastric gland the kinetic arrangements are quite different: cell proliferation is confined to the middle portion of the tubule, and cells are thought to migrate *bidirectionally* to supply the gastric surface and the base of the tubule. Consequently, the origin of the flux is in the upper middle of the gland, and this is where the candidate stem cells may be housed.

A further important proposal is that stem cells form a minority population within the proliferating cell complement of the tubule. There has been a lengthy debate on the

* The colour plate section for this chapter appears between pages 274 and 275.

STEM CELLS
ISBN 0–12–563455–2

actual number of stem cells in each tubule (Potten and Loeffler, 1990), but suffice it to say that exhaustive clonal regeneration studies have demonstrated that the clonogenic population, or the number of cells capable of regenerating a crypt after severe irradiation damage, is small compared with the number of proliferating cells in the crypt (Hendry *et al.*, 1992). Moreover, as we shall see, several experiments with mutagens, which induce phenotypic changes in colonic and small intestinal crypts, can best be explained by mutations induced in a single stem cell, or a very small number of stem cells.

Thus the concept of a few basally-sited crypt stem cells which supply progeny to the proliferative compartment of *dividing transit cells* has emerged (Figure 1); these cells have long cell cycles compared with the other proliferating cells, divide (at least statistically) asymmetrically, and the daughter cells migrate up the crypt, undergoing a further series of divisions before leaving the cell cycle to begin differentiation towards the top of the crypt. While less defined, we might imagine the migration kinetics in the gastric gland to be similar – cells migrating away from the stem cell zone in the neck of the gland, but in two directions (Figure 1). Stem cells have other kinetic and operational properties which need not concern us here.

Most writers on gut stem cells are coy when it comes to suggesting how the stem cells are actually organized and the implications of this organization for the control of stem cell proliferation and differentiation; however, among the proposals are (i) *the stem cell selection hypothesis* (Bjerknes and Cheng, 1990), in which a mechanism exists which actively selects a stem cell, or one of the daughters of a recent stem cell division, for differentiation while allowing other stem cells to remain as stem cells; (ii) *the stem cell zone hypothesis* (Bjerknes and Cheng, 1982), which constrains stem cells to a defined region of the crypt in which no differentiation signals are present – stem cell progeny only differentiate if they migrate out of this zone, and (iii) *the stem cell niche hypothesis,*

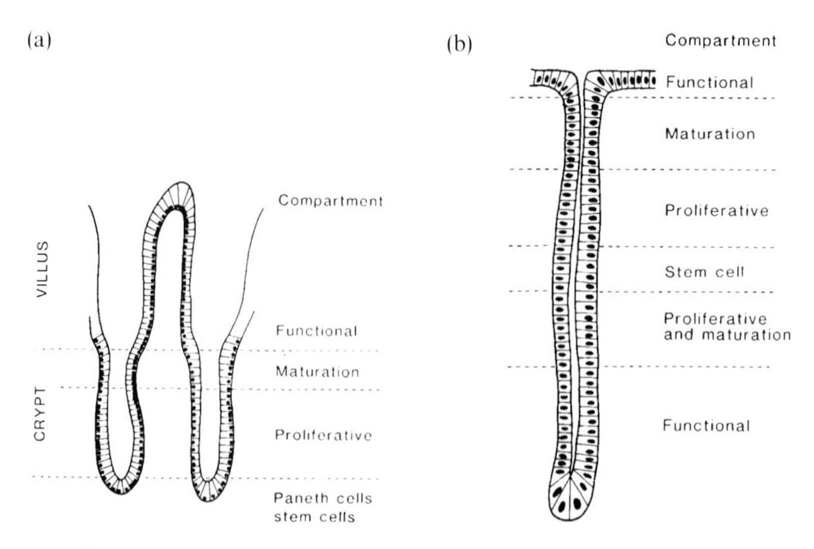

Figure 1 Classical ideas about the organization of cell proliferation in (a) the intestinal crypt, and (b) the gastric gland.

in which stem cells are again confined within anatomical or conceptual 'niches', in its original form where cell divisions are in some way polarized so that defined daughters of a stem cell division are selected for differentiation (Potten *et al.*, 1979), or in its resurrected form in which commitment to differentiation is random (Williams *et al.*, 1992).

The gastrointestinal epithelium contains a series of cells of different type; these are usually dignified by the term *cell lineage*; there is no general agreement what this term means. The word *lineage* is redolent of descent or ancestry. Thus we speak of a family of cells derived from a stem cell as a lineage. But in usage the term has come to imply more than a group of cells derived from a common progenitor. For example, in the intestine we talk freely about the goblet or (mucin-secreting) cell lineage; we do not necessarily imply by this that we believe that goblet cells have their own stem cell which gives rise exclusively to goblet cells and to no other, although this is of course a possible hypothesis. What we usually mean by the term is a discrete group of differentiated cells of defined phenotype which have the same progenitor or stem cell; but we make no presumption of exclusivity of production of that group from this particular progenitor. Within this argument lies the crux of this chapter – do gut stem cells give rise to multiple cell types, i.e. are they pluripotential, or does each cell type have its own stem cell?

CELL LINEAGES IN THE INTESTINAL EPITHELIUM

There are four main cell lineages in the intestinal epithelium; *columnar cells, mucin-secreting cells, endocrine cells* and *Paneth cells*. There are other less abundant lineages such as *caveolated cells* and *M cells*, which we shall ignore for the purposes of this discussion. The columnar cells are the most numerous population in the intestine, so much so that they are sometimes called *enterocytes* in the small bowel and *colonocytes* in the colon. They are polarized cells with a basal nucleus and an apical brush border bearing the glycocalyx (Figure 2). These cells have tight junctions (Figure 2a) and functionally are responsible for the secretory and absorptive properties, on the villus and in the crypt, which are central to the gut's existence. The mucin-secreting cells are colloquially termed 'goblet cells', because of the obvious resemblance (Figure 2a); the goblet cell theca contains mucigen granules, which are discharged onto the surface as intestinal mucus, protecting and lubricating the mucosa. The endocrine cells of the gut, sometimes called 'neuroendocrine' or 'enteroendocrine' cells, form a deceptively large population; these cells sometimes, but not always communicate with the lumen (Figure 2b), and contain numerous basally-sited dense core or neurosecretory granules which contain the secreted peptide hormones, now a burgeoning family, which are secreted basally in an endocrine or paracrine manner. Endocrine cells are found throughout the epithelium, unlike the Paneth cell population, which is found more or less exclusively at the crypt base (Figure 2c). These cells contain large secretory granules, and express lysozyme, tumour necrosis factor, and small molecular weight peptides called cryptins (related to defensins), with anti-bacterial properties. Paneth cells also have a phagocytic properties; these characteristics indicate a role in the maintenance of a sterile environment in the crypt. While Paneth cells are usually confined to the small intestine, they are seen in the proximal colon, and in some inflammatory conditions are prominent in the colon as *Paneth cell metaplasia*.

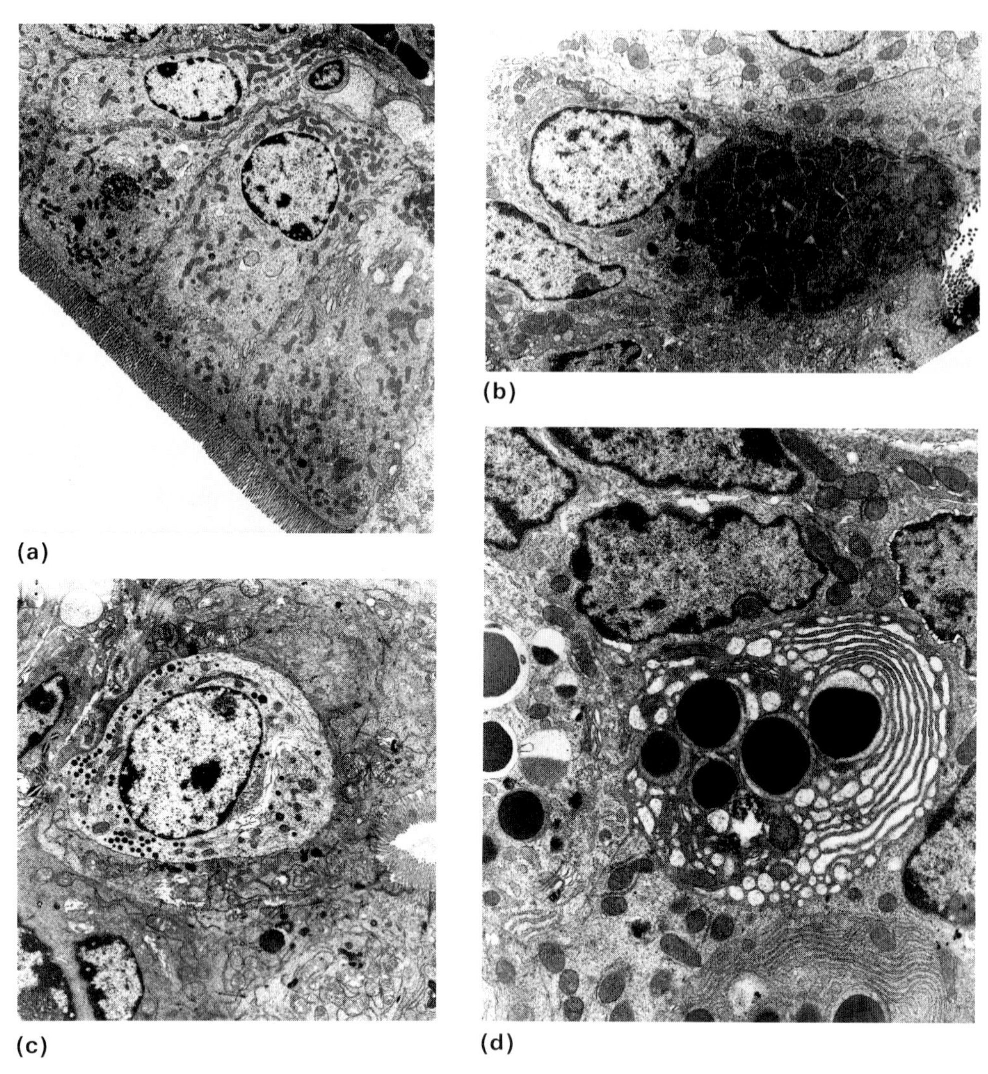

Figure 2 The main cell lineages in the intestinal epithelium: (a) columnar cells; (b) mucin-secreting goblet cells; (c) gut endocrine cells; (d) the basal crypt Paneth cells. (Courtesy of Dr Catherine Sarraf.)

CONCEPTS OF STEM CELL PLURIPOTENTIALITY

The most popular concept which has been proposed is the unitarian hypothesis, in which a single stem cell gives rise to all cell lineages in the epithelium (Figure 3) (Cheng and Leblond, 1974). Developing from the concept of Holzer (1978) that pluripotentiality can only be the property of a *group of cells* rather than a single cell, the concept of 'committed progenitor cells' arose (Figure 3b); but note that, ultimately, at any rate, all cell lineages

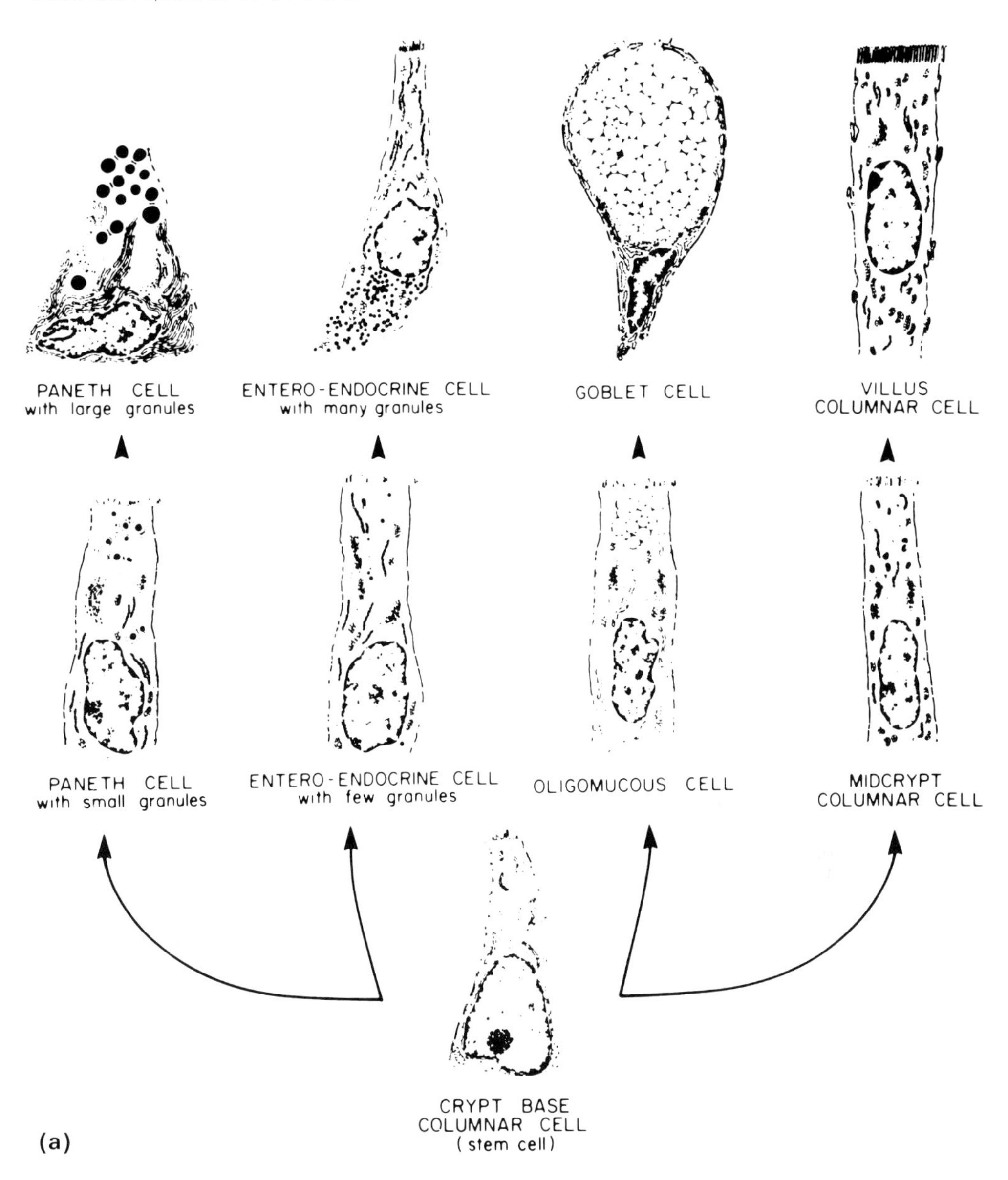

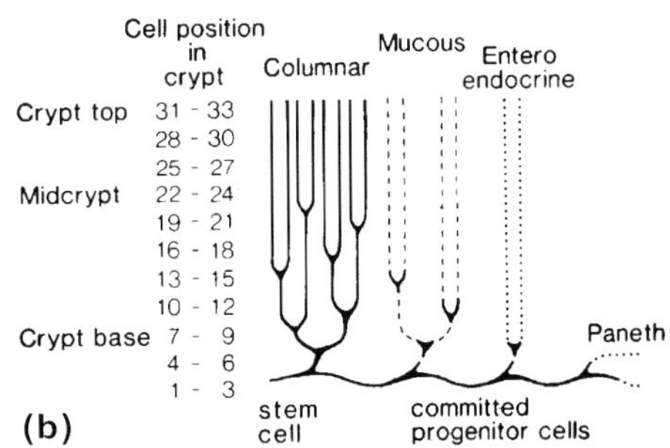

Figure 3 (a) A diagrammatic representation of the unitarian theory of cytogenesis in the intestine. The crypt base columnar stem cells gives rise to the cel lineages shown. (Courtesy of Professor C. P. Leblond.) (b) This diagram, redrawn from Cheng and Leblond (1974), introduces the concept of the committed progenitor cell, which arises from the stem cell and establishes and maintains the respective cell lineages.

take origin from a single cell. At the time this proposal was made, the main evidence proferred was as follows:

(i) the finding of radiolabelled debris in all cell lineages in the small intestinal epithelium after a dose of isotope (here triatiated thymidine) that was claimed only to kill basal crypt cells (Cheng and Leblond, 1974); however, this experiment is certainly not without its critics (Wright and Alison, 1984), who question the selective phagocytic ability of only crypt base cells, and point to the finding of non-radioactive debris in a *single* endocrine cell as a scant base for such a definitive conclusion;

(ii) the finding that all cell lineages were present in colonies of intestinal cells regenerating after irradiation, assuming that each colony came from a single cell (Withers and Elkind, 1970), or the observation that fundic stomach mucosa, transplanted to the subcutaneous tissue, following necrosis is colonized by a layer of simple immature mucous cells, which with time give rise to parietal, mucous neck and argyrophil cells, with eventual reconstitution of the gastric mucosa (Matsuyama and Suzuki, 1970). Neither experiment definitively shows that all developed cell lineages are derivative from a single (stem) cell;

(iii) the existence of so-called *amphicrine cells*, which contain both endocrine and mucous granules, indicating the possibility that these cells had the potential to give rise to at least two types of cell (Cheng and Leblond, 1974); parietal-endocrine cells have also been described in the stomach. These amphicrine cells have been identified in many species, more commonly in pathological states such as tumours, but are also found in normal tissues. On the other hand, membranous partitions in these cells may suggest the possibility of *in vivo* cell fusion, a concept definitely advocated by Pearse (1984);

(iv) there is the well-known existence of tumours with both mucous and endocrine differentiation patterns, the so-called 'goblet cell carcinoid' (Isaacson, 1981), which can include Paneth cell elements (McDonald and Hourihane, 1977). Many tumours have been shown to be clonal in origin – derived from a single cell – and this would predict that such mixed tumours were derived from a single multipotential stem cell, albeit malignant (Gatter *et al.*, 1985); however, this is a considerable assumption (Wright, 1990).

Perhaps the most vociferous argument against the concept of pluripotentiality in gut stem cells, encompassing all cell lineages, has come from Pearse and his disciples (Pearse and Takor Takor, 1976); Pearse originally maintained that gut endocrine cells are derived from the neural crest, presumably by neuroendocrine stem cells during migration, in the same way that the ultimo-branchial body is colonized by migrating neuroectoderm, eventually to produce the C cell lineage in the thyroid (Le Douarin and Le Lievre, 1970). Faced with data in which the neural crest had been transplanted from quail to chick embryo (Le Douarin and Teillet, 1973), and neural crest extirpation experiments (Andrew, 1973), all of which indicated that gut endocrine cells were of local endodermal origin, Pearse (1984) modified the hypothesis to suggest that the gut endoderm was colonized by 'neuroendocrine-programmed stem cells' which migrated from the primitive epiblast, to establish the gut endocrine and pancreatic cell endocrine lineages (see Figure 4); the quail/chick chimera experiments do not of course exclude this concept, and we must look to other data from other models.

PLURIPOTENTIAL COLORECTAL CARCINOMA CELL LINES

The microscopical study of tumours from both humans and animals has shown that neoplastic proliferations rarely consist of one cell type. This has led to an unfortunate and totally unwarranted extrapolation by several authors that, since tumours are held to be monoclonal in origin, this observation, *ipso facto*, indicates that the contained cell lineages are perforce progeny of a single stem cell (Gatter *et al.*, 1985); this extravagant speculation has even been dignified by the sobriquet 'the Gatter conjecture' (Wright, 1990), after a particularly blatant transgression of this type (Gatter *et al.*, 1985). On the other hand there have been several genuine *experimental* attempts to show that the repertoire of malignant cells emanating from usually colonic carcinomas have the ability to differentiate into the several cell lineages found in normal colonic crypt (Goldenberg and Fisher, 1970; Kobori and Oota, 1979), but perhaps more convincing was the report of Cox and Pierce (1982) that putatively single cells from a rat colonic adenocarcinoma injected subcutaneously into recipient mice occasionally gave rise to tumours which showed all colonic cell lineages. For the human situation, this matter was settled by the work of Kirkland (1988), working on the HRA19 cell line, derived from a moderately well-differentiated colorectal carcinoma; in culture HRA19 cells form a polarized monolayer of epithelial cells, in which it was noticed that neuroendocrine differentiation was present: Plate 10.4 shows a typical bipolar neuroendocrine cell stained with a specific marker for endocrine cells, a monoclonal antibody against chromogranin A. Consequently, single cells were selected with a Pasteur pipette, and the single epithelial cell was confirmed by observation in a Petri dish. These single cells were transferred, in a drop of medium, to 96-well plates; epithelial cells are notoriously difficult to clone by themselves, but feeder cells have been shown to be particularly helpful; 5.5×10^3 lethally-irradiated fibroblasts were added to each well. The epithelial cells were grown up, and to ensure the single cell origin of the cloned cell line, the cloning procedure was repeated.

Perhaps the most powerful differentiating environment for colorectal cell lines is the

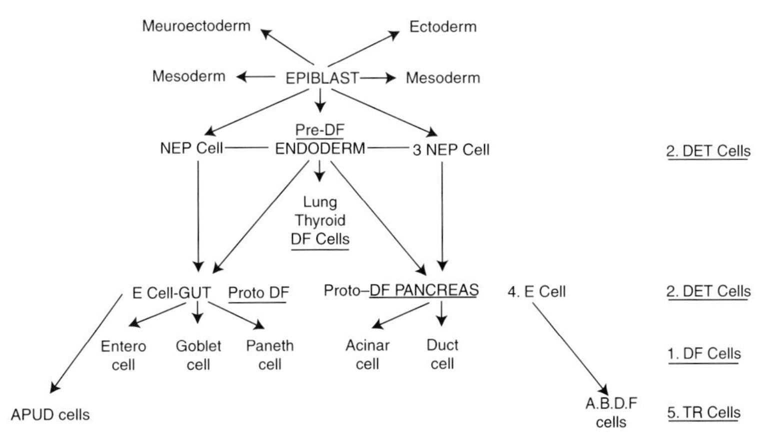

Figure 4 The modified Pearse hypothesis, in which neuroendocrine-programmed stem cells from the primitive epiblast colonize the gut and pancreatic primordia. (Courtesy of Professor A. G. E. Pearse and Churchill Livingstone.)

subcutaneous tissues of the nude mouse, so 10^6 cloned epithelial cells were grafted into this site. Within 4 months a sizeable tumour forms (Plate 10.5a; electron microscopy shows the presence of both columnar and goblet cells – Plate 10.5b). A Grimelius stain, which recognizes so-called argentophil neuroendocrine cells, also shows the typical triangular morphology of endocrine cells (Plate 10.6a); a chromogranin A immunostain confirms the presence of neuroendocrine differentiation (Plate 10.6b). The human origin of these endocrine cells has been shown using double-labelling with Grimelius and the Hoechst dye 33258 (Kirkland, 1988).

These observations make it possible to conclude that a single epithelial cell, albeit malignant, can give rise to all cell types seen in the colorectal epithelium.

ALLOPHENIC TETRAPARENTERAL CHIMERIC MICE

Application of the technology for producing chimeric mice has given considerable insight into the way intestinal crypts are organized. The technique is well-known (Mintz, 1965), but briefly, consists of selecting two strains of mice, having remembered to develop a marker specific for the tissue of interest of one of the strains. Fertilized zygotes from each strain are treated with pronase and on incubation, form a single zygote; these are introduced into the uterus of a pseudopregnant female (previously mated with a vasectomized male) and, in the fullness of time, pups are born which are chimeras of the two strains selected. Where the intestine is concerned, several markers are available: use has been made of the H-2 locus, applying monoclonal antibodies to H2b and H-2k; however, these work mainly in frozen sections, with resulting limited resolution at the morphological level. But a polymorphism in lectin-binding capacity also exists in the intestine in the Dlb1 locus, carried on chromosome 1 (Ponder *et al.*, 1985); thus mice of the C57Bl strains bind the lectin *Dolichos biflorus* agglutinin, which selectively binds to the *N*-acetylgalactoseamine residues present on blood group markers present on the cell surfaces. Mouse strains such as RIII/Lac-*ro* and DDK do not bind the lectin in the intestinal epithelium.

Thus Ponder *et al.* (1985) prepared such chimeras, and staining the small intestine with DBA shows a fascinating pattern (Plate 10.7a). Note that the crypts are either positive or negative – there are no mixed crypts. Ponder and his colleagues could not detect mixed crypts in *adult* mice after observing tens of thousands of crypts. The important conclusion from this seminal study is that each crypt forms a clonal population, i.e. crypts are each derived, ultimately at any rate, from a single cell. High power examination (Plate 10.7b) shows that goblet cells and Paneth cells share in this clonality. In neonatal animals, up to about 2 weeks, mixed crypts are found, but these quickly disappear (Schmidt *et al.*, 1988). Thus it appears that, in development, crypts are pleoclonal – thus multiple stem cells form individual crypts, but, by an as yet ill-understood process, these crypts sort themselves out to become monoclonal. The development of crypts is interesting; Figure 5 shows a schematic of cryptogenesis. Secondary lumina form in the stratified gut endoderm, which coalesce and form the precursors of crypts, but definitive crypts themselves are not formed until after birth. Presumably in excess of one stem cell is incorporated into these early crypts. Mechanisms of 'cleansing' of these crypts include overgrowth and extrusion of one stem cell lineage by the other, or, since there is

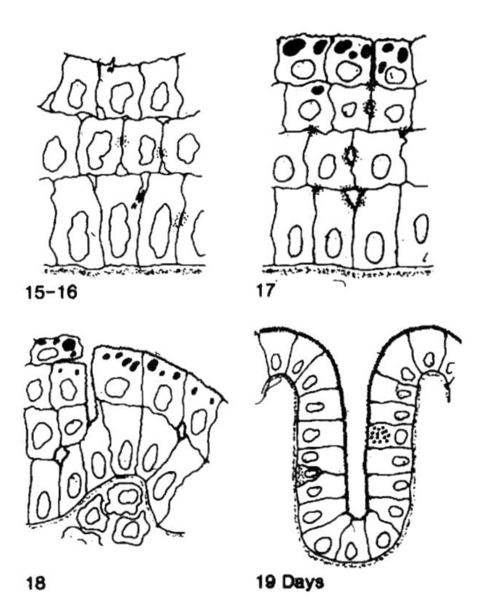

Figure 5 The histogenesis of crypts in the small intestine: 15–16 days, note the junctional complexes and cilia; 17 days, note the secondary lumina and cytolysosome bodies in the superficial epithelium; 18 days, with mesenchymal cells invaginating the epithelium from below. Note the secondary lumina, cleft-like extension of the main lumen and exfoliating superficial cells; 19 days, showing the sides and tips of two villi, with columnar epithelium and endocrine and goblet cells (redrawn from Mathan *et al.*, 1976).

extremely active replication of crypts by fission at this time, it is conceivable that segregation of the lineages could occur at this point.

Taken at face value, the concept of a monoclonal crypt evokes the immediate image of a single multipotential stem cell which naturally gives rise, ultimately at any rate, to all contained crypt cell lineages. There are however, several caveats to this proposal: there is the possibility of some sort of 'chimeric artefact' in which cells from each partner in the chimera could, at least theoretically, segregate independently in development, so that apparently monoclonal crypts are nothing more than monophenotypic. Admittedly the finding that crypts are not monophenotypic *ab initio*, but only acquire this characteristic during development, would of course be against such a concept. Nevertheless, it is reassuring to note that mice who are heterozygous for a defective glucose-6-phosphate (G6PD) gene, a gene carried on the X chromosome, show crypts which, as a consequence of Lyonization, have a crypt-restricted pattern of G6PD histochemical staining (Plate 10.8) (Griffiths *et al.*, 1988). In these animals, the crypts are also monophenotypic, and again no examples of mixed crypts are seen; this confirms the conclusion that crypts are derived from a single stem cell, in this case either showing normal G6PD expression or lacking it.

STEM CELLS MUTATIONS, CRYPT CLONALITY AND IMPLICATIONS FOR STEM CELL REPERTOIRE

A very interesting model in which stem cell behaviour can be studied has been exploited by Winton and Ponder (1990), Williams *et al.* (1992), Loeffler *et al.* (1993) and Park

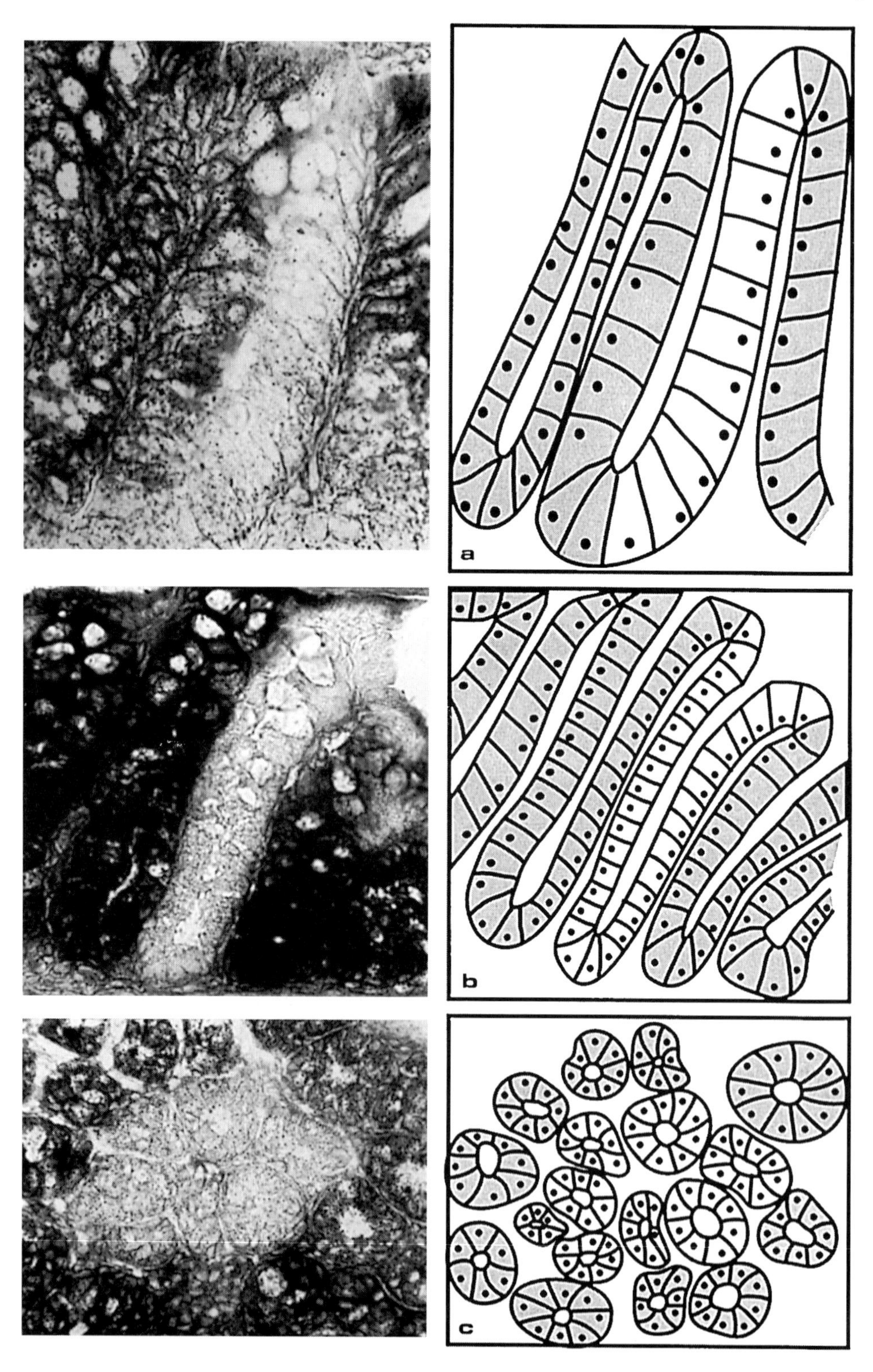
a
b
c

et al. (1995). If mice showing uniform staining for DBA or for G6PD are given a single dose of mutagen, in the weeks that follow, crypts appear which are apparently composed of cells with a different, mutated phenotype. In these experiments there is an induction of a rapid, but transient, increase in crypts which show a partial, or segmented, mutated phenotype (Figure 6a); later, there is an increase in the frequency of crypts showing a completely or wholly mutated phenotype, an increase which plateaus contemporaneously with the disappearance of partially or segmented crypts (Figure 6b). It is interesting that the small intestine and colon show a major difference in the rate and timing of these events (Figure 7): the plateau is reached at between 5 and 7 weeks in the colon (Park *et al.*, 1995), but not until some 12 weeks in the small intestine. The mechanism behind the emergence of the partially-mutated crypts, and their supercedence by wholly-mutated crypts is possibly best explained by a mutation at the Dlb-1 or G6PD locus in the single stem cell from which all lineages derive; on this hypothesis, the partially-mutated crypts are either crypts in the process of being colonized by progeny from the mutated stem cell, a crypt which would ultimately develop into a wholly-mutated crypt; alternatively, some of these partially-mutated crypts could derive from mutations in non-stem proliferative cells, and these would of course disappear as the mutated clone was lost through migration out of the crypt.

These experiments again indicate that a single stem cell can give rise to all crypt lineages, in both colon and small intestine. The reasons for the disparity in the timing of this process between the small intestine and the colon is perhaps beyond this discussion; possible hypotheses include different durations of the stem cell cycle time (Winton and Ponder, 1990), the presence of a stem cell 'niche' with differences in the number of stem cells between the two tissues (Williams *et al.*, 1992) and the possibility that crypt fission plays an important part in the genesis of the wholly-mutated phenotype (Park *et al.*, 1995). Loeffler *et al.* (1993) have also formulated a model which explains this phenomenon on the basis of several indistinguishable stem cells per crypt, which can replace each other.

Extension of these concepts to the human sphere has proved possible through the observation that approximately 9% of the Caucasian population secrete sialic acid lacking in *O*-acetyl substituents; this is manifested by their colonic goblet cells staining with the mild periodic acid Schiff (mPAS) technique and negatively, or weakly, with techniques which show *O*-acetyl sialic acid, such as the periodate-borohydride/potassium hydroxide saponification/PAS (PB/KOH/PAS) method; in turn this is explained by genetic variability in the expression of the enzyme *O*-acetyl transferase (OAT) (Jass and Robertson, 1994). Consequently, this 9% of the Caucasian population is homozygous for inactive OAT genes – OAT−/OAT−. The Hardy-Weinberg equilibrium then predicts that some 42% of the population are heterozygous – OAT−/OAT+; however, *O*-acetylation proceeds since there is one active OAT gene. However, loss of this gene would convert the genotype to OAT−/OAT−. In heterozygotes this shows up as

---

Figure 6 Sections from the colon of a mouse treated with a single injection of a mutagen (ethyl nitrosourea, ENU), and histochemically-stained for glucose-6-phosphatase activity: (a) a partially-negative crypt, where only a portion of the crypt is positive, and (b) a wholly-negative crypt without any positive cells; (c) a patch of negative-staining cells. (Courtesy of Hyun Sook Park.)

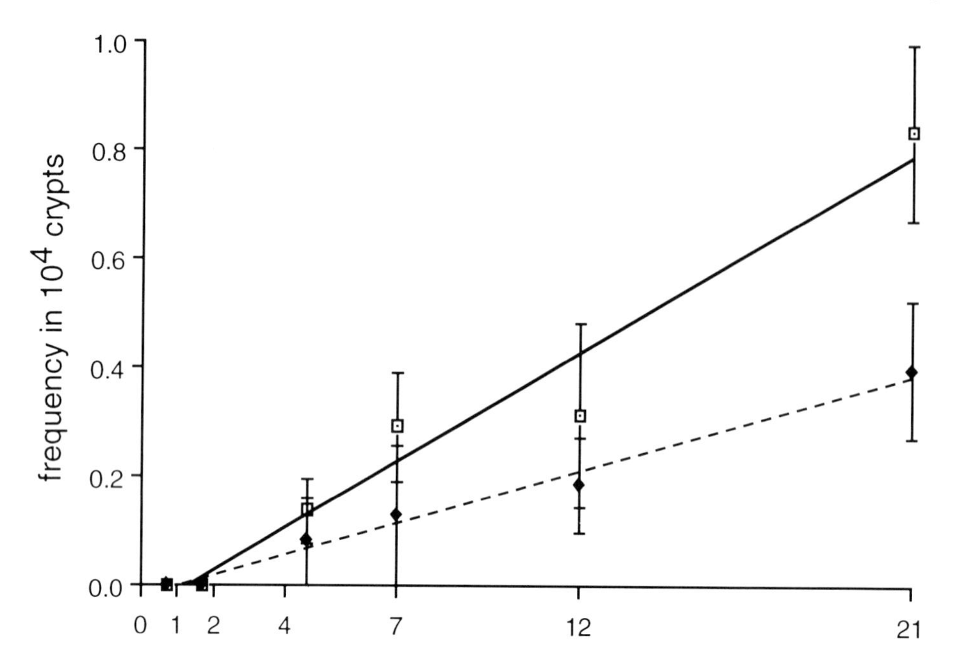

Figure 7 The rate of appearance of partially and wholly-negative crypts in mice treated with ethyl nitrosourea, expressed per 10^4 crypts; animals were injected with ENU at 6 weeks of age. (Courtesy of Hyun Sook Park.) Small intestine: $y = -2.015e^{-2} + 1.9338e^{-2x}$, $R^2 = 0.988$. Colon: $y = -5.0854e^{-2} + 3.9981e^{-2x}$, $R^2 = 0.98$.

crypt-restricted mPAS staining in a negative background (Plate 10.9). This phenomenon is indeed seen in about 42% of the population, and of course is most simply explained by a mutation or loss of the gene by non-dysjunction in the single crypt stem cell and the colonization of this crypt by its clonal progeny. This unicryptal loss of heterozygosity occurs randomly, is increased by age, as would be expected (Fuller *et al.*, 1990), and also increased in individuals who have received pelvic irradiation (Campbell *et al.*, 1994). These observations point to the same conclusion about the repertoire of human colonic stem cells – that they ultimately generate all crypt cell lineages.

However, it must be conceded that all these observations have been made at the level of the light microscope, and without taking particular pains to differentiate the several cell lineages; despite this, it is relatively easy to descry enterocytes, goblet cells and Paneth cells in Plate 10.7. The G6PD marker is not amenable to such exactitude, since frozen sections are perforce used. The lineage which cannot be routinely identified in sections is of course the enigmatic endocrine cell.

ARE GUT ENDOCRINE CELLS CLONAL WITH OTHER CRYPT CELL LINEAGES?

The last forty years has seen a spirited debate on the origin of gut endocrine cells; the discussion above would indicate that, at least in neoplastic tissues, endocrine cells derive from single epithelial, and here endodermally-derived, stem cells. In fact, classical embryological observations supported a local endodermal origin, as did detailed observations of carcinoid tumours (neoplastic proliferations of gut endocrine cells) (Masson, 1928, 1932). The counter-proposal came from Danisch (1924), an origin from migrating neural cells. This was the concept advocated by Pearse (1966a, 1966b), as part of his all-embracing APUD cell concept, which took cognizance of the remarkable cytochemical similarities of apparently unrelated cells, viz. fluorogenic *A*mine content, amine *P*recursor *U*ptake and amino acid *D*ecarboxylase (plus α-glycerophosphate dehydrogenase, non-specific esterase and/or cholinesterase), side-chain carboxyl groups, the presence of endocrine granules and specific immunocytochemistry, including chromogranin A positivity (Pearse, 1969). Cells with these properties included not only gastrointestinal and pancreatic endocrine cells, but also some of proven origin in migrating cells from the neural crest, such as phaechromocytes from the adrenal medulla, cells from sympathetic ganglia and cutaneous melanocytes; this neat and attractive hypothesis regarded these cells with common properties as having a common embryological origin: neuroendocrine stem cells from the neural crest colonize primordia, such as the developing gut, and maintain the endocrine cell lineage separate from the other crypt cell lineages.

Evidence from embryonic extirpation experiments of the neural crest (Andrew, 1973), and from quail/chick chimeras (Le Douarin and Teillet, 1973), have shown fairly conclusively that gut neuroendocrine cells are not of neural crest origin; however, the concept that gut, and for that matter pancreatic neuroendocrine cells were derivative from 'neuroendocrine-programmed' cells, themselves derived from the primitive epiblast prior to migration through Hensen's node, and again seed the developing gut with neuroendocrine stem cells, was proposed by Pearse (1984) (Figure 4). This concept was supported by Scothorne (1988), who suggested that there may be experimental support for this concept, at least at the level of organogenesis, from the work of Svajger *et al.* (1981), who transplanted pre-primitive streak ectoderm from mouse embryos beneath the kidney capsule of recipient mice and observed the formation of benign teratomas which frequently showed organotypical differentiation, such as stomach or small bowel. This would imply predetermination of primary ectodermal cells prior to their migration.

None of the preceding evidence with the single potential exception of the chimeric mouse system, can disprove this modified Pearse neural origin theory: if it were possible to show definitively that the gut endocrine cells showed the same phenotype as the cells in the monophenotypic crypts, such as those portrayed in Plate 10.7, then it would be very reasonable to conclude that endocrine cells are indeed of local endodermal origin. Unfortunately, the Ponder system, using the Dlb-1 polymorphism, cannot be used to answer this question. Conjugating the DBA lectin with protein A gold allowed the possibility of labelling endocrine cells with the lectin and using the contained neuroendocrine granules as a cell lineage marker (Thompson, 1993). There was no difficulty in labelling the positive control intestine of C57/BL mice with the lectin-gold complex (Plate 10.10a and b), or indeed the C57/BL-derived crypts of the chimeras; the endocrine cells in chimeric crypts from the RIII/ro component were also negative, but so were endocrine cells in the putatively positive C57/BL component; crypt endocrine cells in the intestines of control C57/BL mice were negatively stained (Plate 10.10c), even though villus endocrine cells in these C57/BL animals were positive. We are therefore forced to the conclusion that DBA staining of gut endocrine cells, even in those animals whose intestines are uniformly positive, is acquired during migration from the crypts to the villus; thus this system is non-informative about the clonal origin of gut endocrine cells.

A different chimeric system was therefore used by Thompson *et al.* (1990); making male:female chimeras, the contributing tissues were differentiated by using *in situ* hybridization with a digoxigenin-labelled probe which identified the highly repetitive sequences of the mouse Y chromosome, pY 353, to identify the male component of the chimera (Plate 10.11a). Endocrine cells in the antral stomach of these animals were recognized by staining the gastrin cells (G cells) with a polyclonal anti-gastrin antibody. There are easily recognizable male and female areas of the chimeric stomach (Plate 10.11a), and within these areas, it was clear that male gastric glands contained G cells which were Y-spot positive (Plate 10.11b), while female areas contained G cells which were Y-spot negative (Plate 10.9c). It is therefore safe to conclude that gut endocrine cells share the same clonal origin as the other cell types.

GENE TRANSFER TO GUT EPITHELIUM

One possibility which has remained elusive up until now is the possibility of marking gut stem cells during development and watching their development into the several contained cell lineages of the intestinal crypt. Such a possibility was indicated by the work of Del Buono *et al.* (1992) on a model of intestinal development: rat or mouse developing intestine is removed and the endodermal component is separated from the mesenchyme by proteinase treatment. This endoderm is then embedded in rat type I collagen and transplanted subcutaneously into nude mice. Within about 12 days a fully-developed small intestine is seen (Plate 10.12a) in which all cell lineages can be identified (Del Buono *et al.*, 1992). Using the BAG retrovirus (B-gal At GAG), a retroviral vector in which the LTR drives the *Lac Z* gene as a reporter gene, these endodermal cells can be labelled, apparently permanently. The endoderm is either injected or incubated with the retrovirus, and again transplanted subcutaneously in nude mice. It is possible to show that cell lineages in the transplant express β-galactosidase (Plate 10.12b). This experimental tool not only gives information about stem cell repertoire in development, but also allows the possibility of examining the rate at which gut stem cells are committed during development; not to mention the scope for transferring genes to the developing gut endoderm.

THE CONTROL OF GUT CELL LINEAGE DEVELOPMENT

It is one thing to definitively show that intestinal stem cells have a repertoire which includes all the main cell lineages found in the gut; it is quite another to refine the detailed arrangement whereby the mature representative of each lineage is produced. How this single, multipotential stem cell produces up to four differentiated progeny, whether directly by its own division potential or by producing other cells with perhaps limited 'stemness', which produce differentiating progeny favouring a particular lineage is, as yet, unknown. We are far from the rather happy situation in the bone marrow, where individual growth factors which induce differentiation along a defined pathway are now identified. However, it is clear that we are beginning to identify systems which are tractable in terms of identifying putative factors which induce differentiation along a particular pathway; for example, TGFβ-1 has been shown to inhibit endocrine differentiation in the HRA.19 cell line (Kirkland and Henderson, 1994).

CONCLUSIONS

We have perhaps come rather a long way in recent years in defining the intestinal stem cel repertoire; as has been observed before, it is quite one thing to speculate, often on scant and inconclusive data, that 'stem cells' are responsible for generating all manner of cell lineages in epithelial and other tissues: it is quite another thing to show that this is so. This albeit brief survey of the available evidence has, I believe, shown quite definitively that the intestinal stem cell has a repertoire which includes all the main cell lineages observed within intestinal crypt systems; the challenge of the next few years is to define and isolate this elusive stem cell, identify its properties and to define those mechanisms which induce it to produce individual cell lineages at particular times.

ACKNOWLEDGEMENTS

I am especially grateful to Dr Susan Kirkland, Dr Mary Thompson, Professor Bruce Ponder, Dr K. C. Lui, Dr Raff Del Buono, Professor Julia Polak and Dr Hyun Sook Park for donating material, biological, photographic and intellectual, which have assisted in the preparation of this chapter. I also, believe it or not, owe a considerable debt to Professor Tony Everson Pearse, whose energy and resourcefulness deemed it always worthwhile to find out where indeed gut endocrine cells came from; and to Professor R. J. Scothorne, who, no doubt somewhat to his surprise, first stimulated my interest in histology; and to the Imperial Cancer Research Fund.

REFERENCES

Andrew, A. (1973). *J. Embryol. Exp. Morphol.* **31**:581–598.
Bjerknes, M. and Cheng, H. (1990). *Am. J. Anat.* **166**:76–92.
Campbell, F., Fuller, C.E., Williams, G.T. and Williams, E.D. (1994). *J. Pathol.* **174**:175–182.
Cheng, H. and Leblond, C.P. (1974). *Am. J. Anat.* **141**:537–562.
Cox, W.F. and Pierce G.B. (1982). *Cancer* **50**:1530–1538.
Danisch, F. (1924). *Beitr. Pathol. Anat.* **72**:687–709.
Del Buono, R., Fleming, K.A., Morey, A.L., Hall, P.A. and Wright, N.A. (1992). *Development* **114**:67–73.
Fuller, C.E., Davies, R.P., Williams, G.T. and Williams, E.D. (1990). *Br. J. Cancer* **61**:382–384.
Gatter, K.C., Dunnill, M.S., Pulford, K.A., Duyret, A. and Mason, D.Y. (1985). *Histopathology* **9**:805–823.
Goldenberg, D.M. and Fisher E.R. (1970). *Br. J. Cancer* **24**:610–614.

Griffiths, D.F., Davies, S.J., Williams, D., Williams, G.T. and Williams, E.D. (1988). *Nature* **333**:461–463.
Hendry, J.H., Roberts, S.A. and Potten, C.S. (1992). *Radiat. Res.* **132**:115–119.
Holzer, H. (1978). In *Stem Cells and Tissue Homeostasis* (eds C. S. Potten, B. I. Lord and R. Schofield), pp. 1–27. Cambridge University Press, Cambridge.
Isaacson, P. (1981). *Am. J. Surg. Pathol.* **5**:213–224.
Jass, J.R. and Robertson, A.M. (1994). *Pathol. Int.* **44**:487–504.
Kirkland, S.C. (1988). *Cancer* **61**:1359–1363.
Kirkland, S.C. and Henderson, K. (1994). *J. Cell Sci.* **107**:1041–1046.
Kobori, O. and Oota, K. (1979). *Int. J. Cancer* **23**:536–541.
Le Douarin, N. and Le Lievre, C. (1970). *C.R. Seances Acad. Sci.* **270**:2857–2860.
Le Douarin, N. and Teillet, M.A. (1973). *J. Embryol. Exp. Morphol.* **30**:31–38.
Loeffler, M., Birke, A., Winton, D. and Potten, C.S. (1993). *J. Theor. Biol.* **162**:471–491.
McDonald, G.S.A. and Hourihane D.O'B. (1977). *Ir. J. Med. Sci.* **146**:386–394.
Masson, P. (1928). *Am. J. Pathol.* **4**:181–212.
Masson, P. (1932). In *Cytology and Pathology of the Nervous System*, Vol. 3, (ed. W. Penfield), pp. 1095–1130. Harper and Row.
Mathan, M., Moxey, P.C. and Trier, J.S. (1976). *Am. J. Anat.* **146**:73–92.
Matsuyama, M. and Suzuki, H. (1970). *Science* **169**:385–387.
Mintz, B. (1965). In *Preimplantation Stages of Pregnancy*, pp. 194–216. CIBA Foundation Symposium. J and A Churchill, London.
Park, H.S., Goodlad, R.G. and Wright, N.A. (1995). *Am. J. Pathol.* 147:1416–1421.
Pearse, A.G.E. (1966a). *Nature* 211:598–600.
Pearse, A.G.E. (1966b). *Vet. Rec.* **79**:587–590.
Pearse, A.G.E. (1969). *J. Histochem. Cytochem.* **17**:303–313.
Pearse, A.G.E. (1984). In *Endocrine Tumours* (eds J. M. Polak and S. L. Bloom), pp. 82–94. Churchill Livingstone, Edinburgh.
Pearse, A.G.E. and Takor Takor, T. (1976). *Clin. Endocrinol. Suppl.* **5**:299–344.
Ponder, B.A.J., Schmidt, G.H., Wilkinson, M.M., Wood, M.J., Monk, M. and Reid, A. (1985). *Nature* **313**:689–691.
Potten, C.S. and Loeffler, M. (1990). *Development* **110**:1001–1020.
Potten, C.S., Schofield, R. and Lajtha, L.G. (1979). *Biochem. Biophys. Acta* **560**:281–299.
Schmidt, G.H., Winton, D.J. and Ponder, B.A.J. (1988). *Development* **103**:785–790.
Scothorne, R.J. (1988). *Histopathology* **13**:355–359.
Svajger, A., Levak-Scajger, B., Kostovic-Knezevic, L. and Brademante, Z. (1981). *J. Embryol. Exp. Morphol. Suppl.* **65**:243–267.
Thompson, M. (1993). Enteroendocrine cells; an investigation into their origin, differentiation and renewal. MD Thesis, University of Cambridge.
Thompson, M., Fleming, K., Evans, D.J. and Wright, N.A. (1990). *Development* **110**:477–481.
Williams, E.D., Lowes, A.P., Williams, D. and Williams, G.T. (1992). *Am. J. Pathol.* **141**:773–776.
Winton, D.J. and Ponder, B.A.J. (1990). *Proc. R. Soc. B* **241**:13–18.
Withers, H.R. and Elkind, M.M. (1970). *Int. J. Rad. Biol.* **17**:261–267.
Wright, N.A. (1990). *J. Pathol.* **161**:85–87.
Wright, N.A. and Alison, M.R. (1984). *The Biology of Epithelial Cell Populations*, Vol. 2. Oxford University Press.

11 Keratinocyte stem cells of cornea, skin and hair follicles

Stanley J. Miller, Robert M. Lavker* and Tung-Tien Sun†

*Department of Dermatology, Johns Hopkins University Medical School, Baltimore, MD, USA; *Department of Dermatology, University of Pennsylvania School of Medicine, Philadelphia, PA, USA; †Epithelial Biology Unit, Ronald O. Perelman Department of Dermatology and Department of Pharmacology, Kaplan Comprehensive Cancer Center, New York University Medical School, New York, NY, USA*

INTRODUCTION

All external surfaces of the body, including the skin and cornea, are covered by a stratified squamous epithelia composed of keratinocytes. These cells synthesize tissue-restricted keratin intermediate filament proteins, and form a specialized submembrane structure known as the cornified envelope in the final stages of differentiation that is composed of covalently crosslinked proteins, including involucrin and loricrin (Eckert, 1989; Fuchs, 1990; Goldsmith, 1991). Stem cells in these epithelia, as in all self-renewing tissues, are the progenitor cells that give rise to the terminally-differentiated cell population.

In this chapter, we will review our present knowledge of keratinocyte stem cells of the cornea, epidermis, and hair follicle. Nine characteristics of stem cells (see Table 1), derived initially from studies of the haematopoietic system (for reviews, see Keller, 1992; Spangrude, 1991; Baum *et al.*, 1992; Uchida *et al.*, 1993), will be used as a framework to discuss keratinocyte populations.

LOCATION

Corneal epithelium

The stem cells of the corneal epithelium are located in a narrow section of the peripheral cornea, bordering the bulbar conjunctiva, termed the limbus (Figure 1). Many early studies

STEM CELLS
ISBN 0–12–563455–2

Table 1 General stem cell characteristics.

A	Stem cells exist in specific locations within a given tissue.
B	Stem cells comprise a small percentage of the total cell population.
C	Stem cells are ultrastructurally unspecialized, with a large nuclear-to-cytoplasmic ratio and few organelles.
D	Stem cells are pluripotent.
E	Stem cells are slow-cycling, but may be induced to proliferate more rapidly in response to certain stimuli.
F	Stem cells have a proliferative potential that exceeds an individual's lifetime.
G	Because stem cells cycle slowly, and represent only a small percentage of a cellular population, an intermediate group of more rapidly proliferating cells exists, that form clonal expansions resulting in the final, differentiated cell population.
H	A stem cell's microenvironment plays an important role in its homeostasis and in the differentiation of its progeny.
I	Many cancers arise from stem cells or early progenitor cells.

*Modified from Miller *et al.* (1993c).

documented centripetal migration of surface epithelial cells onto the cornea during normal homeostasis (Buck, 1985; Huang and Tseng, 1991) and following damage of central corneal epithelium (Maumenee and Scholz, 1948; Buschke, 1949; Mann, 1944; Davanson and Evensen, 1971; Bron, 1973; Kuwabara *et al.*, 1976; Buck, 1979, 1982; Kaye, 1980; Kinoshita *et al.*, 1981; Alldridge and Krachmer, 1981), but the origin of the cells was never determined. The possibility that the limbus was the source of these migrating epithelial cells was first raised by Davanger and Evensen (1971). They observed that a centripetal movement of pigment from the region of the limbus onto the cornea of guinea pigs occurred following corneal epithelial wounding, accompanied by a marked increase in mitotic figures within limbal basal epithelial cells. These early observations were, for the most part, ignored and little attention was directed towards the limbal epithelium as the source of corneal epithelial cells following a wound. Instead, the issue of conjunctival transdifferentiation came into fashion. In this process, conjunctival epithelium assumes some, but not all, of the characteristics of corneal epithelium upon migration onto the corneal stroma. However, there is now ample evidence that this conversion or 'transdifferentiation' is both incomplete and readily reversible. Therefore, it reflects environmental modulation of cells rather than transdifferentiation (see page 346 for more details). The idea that the limbal epithelium was the site of origin of corneal epithelial cells re-emerged when Schermer *et al.* (1986) showed that a 64 kDa keratin [K3] marker of advanced corneal epithelial differentiation was present in all epithelial layers of the cornea and all suprabasal limbal layers, but not in the basal layer of the limbus. This observation suggested that the basal layer of the limbus contained a 'less-differentiated' population of cells than existed in the basal layer of the cornea. In addition, the K3 keratin was expressed in significantly larger quantities in the limbal/corneal epithelium than in conjunctival epithelium, providing the much needed evidence that conjunctival epithelium was not a source of corneal epithelium. These and other significant data, to be presented in further sections in this chapter, led to the concept that corneal epithelial stem cells are located in the basal layer of the limbal zone.

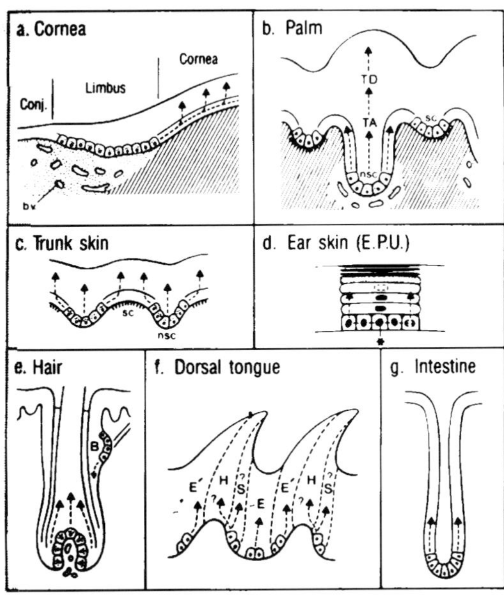

Figure 1 Localization of keratinocyte stem cells in (a) cornea, (b) palm, (c) trunk, (d) ear (the epidermal proliferative unit or EPU), (e) hair follicle, (f) dorsal tongue and (g) small intestine epithelium. (a) This panel summarizes the existing knowledge of corneal epithelial stem cells. They are located in the basal layer of the limbal region. These cells are slow-cycling (Cotsarelis *et al.*, 1989), relatively undifferentiated (K3 keratin$^-$ and enolase$^+$) (Schermer *et al.*, 1986; Zieske *et al.*, 1992b), heavily pigmented (schematically represented by perinuclear dots), close to blood vessels (b.v.), and they undergo centripetal as well as vertical migration. Hatched and dotted areas denote the dense and loose connective tissues of the cornea and limbus, respectively. Conj = conjunctiva. (b, c) Stem cells of palm (b) and trunk epidermis (c) are thought to be located at the tips of rete ridges where they have the morphological appearance of non-serrated cells (nsc) (Lavker and Sun, 1982, 1983). They are slow-cycling, heavily pigmented, and are believed to give rise to mitotically active, transient amplifying cells (TA), which, in turn, become terminally differentiated (TD). SC = serrated basal cells. (d) Mouse ear skin is believed to be composed of 'epidermal proliferative units' (EPUs) as proposed by Potten (1974). In this unusually thin and flat epidermis, stem cells appear to be located at the centre of each EPU in the basal layer. These putative stem cells are slow-cycling, and evidence exists that they are not Langerhans' cells (Morris *et al.*, 1985). Their progenitor (TA) offspring appear to migrate sideways, undergo additional divisions, and then finally move upward into the suprabasal compartment (sequentially becoming spinous, granular and finally cornified cells). (e) Hair follicular stem cells were classically thought to reside in the matrix area of the hair bulb. Matrix cells are heavily pigmented, relatively undifferentiated, and apparently pluripotent, but they are also rapidly cycling. A growing body of evidence suggests that these matrix cells are progenitor (TA) cells, and that the stem cells actually reside in the bulge of the outer root sheath (Cotsarelis *et al.*, 1990), a portion of the upper follicle that is the follicular attachment site of the arrector pili muscle. For evidence supporting the stem cell nature of bulge cells, see the text. (f) Dorsal tongue epithelium can be divided into several compartments, each expressing different major keratins that are characteristic of hair-type (H), esophageal-type (E), and skin-type (S) differentiation (Dhouailly *et al.*, 1989). Label-retaining cells, indicative of stem cell-like behaviour, have been located at the bottom of the E and S compartments (Bickenbach, 1981; Hume and Potten, 1976). Dotted lines depict presumed directions of cell movement. (g) Intestinal epithelium has pluripotent stem cells located at the bottom of the crypts (Leblond, 1981). Adapted from Cotsarelis *et al.* (1989).

Epidermis

In discussing epidermal stem cells, we want to emphasize that 'epidermis' of different body sites – e.g., trunk, palm/sole and foreskin – are not identical. Although this may seem obvious, it is an assessment that has not been widely noted in the literature. It is well-known that epidermis of different body sites is significantly different in its thickness and in certain morphological details. For example, palm and sole epidermis is much thicker than normal trunk epidermis, and bovine snout epidermis has been shown to be morphologically distinct (Steinert *et al.*, 1980). Nevertheless, these differences have usually been attributed to local environmental modulation. We believe this notion is incorrect, and that rather, the epidermis from different body sites is intrinsically divergent. Several lines of evidence support such a distinction. (1) palm and sole (volar) epithelium is much thicker than epidermis that is found at all other body sites. This increased thickness is present at the time of birth (K. Holbrook, personal communication), and thus is not the result of repeated trauma, as is commonly assumed. (2) Volar epithelium contains large quantities of a unique keratin, K9, which is found only in minute amounts at other epidermal sites (Knapp *et al.*, 1986). (3) Transplantation of volar skin to the chest in a guinea pig results in a graft that maintains the unique histologic features of volar skin for the remainder of the animal's life (Billingham and Silvers, 1963). (4) In a mouse, volar skin has a cell replacement rate higher than that of any other cutaneous location (Lavker and Sun, 1983; Allen and Potten, 1976). (5) Foreskin epithelium produces trichohyalin granules, which are not found in any other non-follicular epidermal sites (O'Guin and Manabe, 1991). Together, these observations suggest that keratinocytes of the epidermis of different body sites may be genetically distinct, much like keratinocytes of skin versus other organs such as the cornea or oesophagus (Doran *et al.*, 1980).

Regional variation also appears to exist in the spatial organization of epidermal stem, transient-amplifying (TA) and differentiated cells from different body sites. Such variation may relate to the volar-glabrous skin differences noted above. Historically, two distinct theories have developed to describe epidermal stem cell organization. The first arose from ultrastructural studies of murine and hamster ear and trunk skin (Mackenzie, 1969, 1970; Christophers, 1971; Menton and Eisen, 1971; Potten, 1974), and was termed the 'epidermal proliferative unit' (EPU) by Potten (1974). According to the EPU theory, the epidermis is composed of multiple hexagonal-shaped, microscopically-sized regions, termed EPUs (Figure 2). Each EPU contains a stem cell, located in the centre of the basal layer, 8–10 TA cells in the surrounding basal and overlying suprabasal region, and multiple, more superficially located terminally-committed cells. Recent kinetic and morphological data support the EPU model (Morris *et al.*, 1985). Another investigation showed that injection of fluorescein dye into epidermal cells resulted in vertical patterns of spread that resemble the cellular columns of EPUs (Kam *et al.*, 1986). Most recently, Mackenzie transduced murine keratinocytes using a replication-deficient retroviral vector carrying the β-galactosidase gene. Transplantation of these transduced keratinocytes back to *in vivo* sites resulted in clusters of cells that developed epidermal columns with the β-galactosidase-positive cells at the centre of an EPU (Mackenzie, 1996). However, one chimeric mouse study failed to reveal the clonal organization of EPUs in trunk epidermis (Schmidt *et al.*, 1987), and EPUs have not been found in human volar skin (Christophers, 1971; Menton and Eisen, 1971).

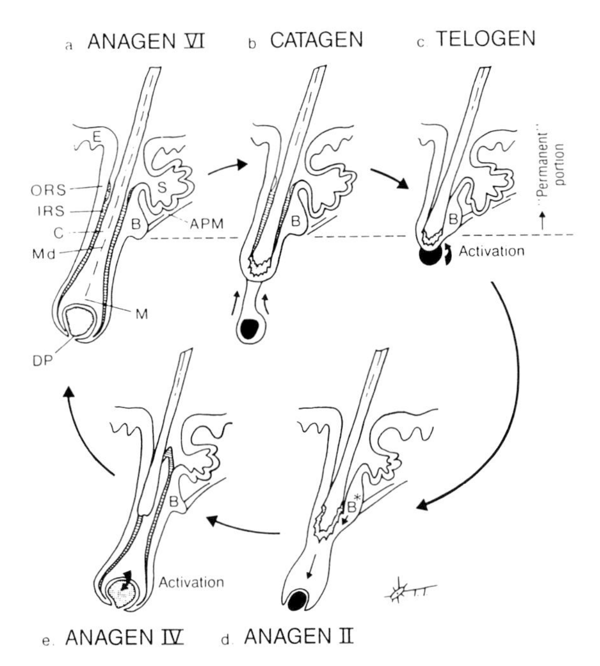

Figure 2 Hair Cycle: The Bulge Activation Theory. Illustrated are the sequential phases of the hair cycle. Abbreviations include epidermis (E), sebaceous gland (S), bulge (B), arrector pili muscle (APM), outer root sheath (ORS), inner root sheath (IRS), hair cortex (C), hair medulla (Md), matrix (M) and dermal papilla (DP). Structures above the dashed line constitute the permanent portion of the hair follicle; keratinocytes below this line degenerate during catagen and telogen and are therefore dispensable. Four important tenets of the bulge activation theory include (1) upward movement of the dermal papilla during catagen that will lead to its close proximity to the bulge during telogen, (2) activation of the bulge cells by the adjacent dermal papilla, (3) downward movement of the dermal papilla as anagen begins, followed shortly thereafter by a 'turning off' of the now proliferating (B*) bulge cells, and (4) activation of the mesenchymal papilla by the new matrix. This theory predicts that a hair cycle will have a duration based on the finite proliferative potential of the matrix (transient amplifying) cells. Adapted from Cotsarelis *et al.* (1989).

A completely different spatial organization (Figure 1) was noted by Lavker and Sun in monkey palm epithelium, which is composed of regular and alternating 'deep' and 'shallow' rete ridges (Lavker and Sun, 1982, 1983). Non-serrated cells located in the basal layer of the 'deep' rete ridges are morphologically primitive and slow-cycling, and apparently give rise to a population of rapidly cycling cells located immediately above them in the suprabasal layers. These non-serrated basal cells are believed to represent palmar epithelial stem cells. Serrated basal cells are located in the basal layer of the 'shallow' rete ridges, and contain abundant keratin filament bundles, few melanosomes and many cellular processes which extend deeply into the dermis, suggesting an anchorage function. These cells are believed to be TA cells that have ascended up the shoulder of the deep rete ridges. Epidermal growth factor (GF) receptors have recently been shown to be localized at the base of the deep rete ridges (Misumi and Akiyoshi, 1990, 1991), where the presumptive stem cells are located. The lichenoid tissue reaction occurring in graft-versus-host disease has also been show to originate at the base of the deep epidermal rete ridges (Sale *et al.*, 1985, 1991), again coinciding with stem cell location. Interestingly, the location of serrated and non-serrated basal cells in the shallow and deep rete ridges has been observed in other human epidermal regions as well, including the arm, leg, back, abdomen, and face (Lavker and Sun, 1983). For a discussion of a current related controversy, see page 353. How this stem cell organization pattern relates to the EPU concept is uncertain. It may be that each model describes the organizational pattern of a distinct epithelium, i.e. glabrous versus volar epithelium, or the two models may reflect the innate differences in human as opposed to mouse skin.

Follicle

Two competing theories regarding the location of follicular stem cells exist. For many years it was presumed that follicular stem cells are located in the germinative matrix of the hair bulb (Van Scott *et al.*, 1963; Kligman, 1959). This theory is based on: (1) the high proliferative rates of germinative cells during anagen phase of the hair cycle, (2) the known interactions between the matrix and the dermal papilla that regulate hair growth and (3) the differentiation of the matrix cells into several different terminal pathways which form the hair medulla, the hair cortex and the inner root sheath. As recently as 1991, it has been suggested that matrix germinative cells constitute a form of stem cell, based on their ultrastructurally undifferentiated appearance, their retention following hair-plucking, their ability to undergo *in vitro* growth, and their unique ability to form complex organoid structures when combined with cultured dermal papilla cells in *in vitro* settings (Reynolds and Jahoda, 1991a).

However, several lines of evidence point against the presence of stem cells in the hair matrix, and suggest instead a residence in the upper portion of the outer root sheath known as the bulge. Some early evidence comes from clinical experiments by Butcher (1964), Oliver (1966a, 1966b), Inaba *et al.* (1979) and Ibrahim and Wright (1982). Similar work has also recently been presented by Kim and Choi (1994, 1995) and by Kappenman *et al.* (1994). The results indicate that a significant portion of the lower hair follicle can be removed, either the lower one third of the follicle (Oliver, 1966b; Kim and Choi, 1994, 1995; Kappenman *et al.*, 1994), or a portion purportedly up to the follicular entrance of the sebaceous canal (Inaba *et al.*, 1979), and follicular regrowth with hair production will occur. In a similar fashion, Fukuda and Ezaki (1975) observed that hairs will occasionally grow out of transplanted split thickness skin grafts, i.e. in a situation in which the vast majority of the lower follicle is gone. Finally, in a very early study, Montagna and Chase (1956) noticed a similar phenomenon in the setting of radiation damage. They observed that following complete, radiation-induced degeneration of the lower, transient portion of a hair follicle, new hair would grow following reformation of the follicle from the upper, permanent section. They suggested that this permanent outer root sheath is the source of follicular regeneration.

Cell-cycle studies have provided strong experimental evidence suggesting that follicular stem cells are located in the outer root sheath. Al-Barwari and Potten (1976) noted 3H-TdR-labelling of many cells around the upper follicle and follicular canal following epidermal damage, and on the basis of this, Potten postulated the existence of 'a reserve of potent stem cells within the hair follicle' (1981). Recent cell culture studies have shown that bulge cells have a proliferative potential in culture that is superior to bulb cells (Yang *et al.*, 1993). Finally, Cotsarelis and co-workers found that label-retaining cells, indicative of slow-cycling (stem) cells, were not found in the bulb portion of either pelage, eyelash or vibrissae follicles of mice (Cotsarelis *et al.*, 1990; Lavker *et al.*, 1991a). They were noted only in the upper outer root sheath in the bulge. However, single pulsing with 3H-TdR, which marks rapidly cycling (TA) cells, revealed prominent labelling throughout the hair matrix, but none in the region of the bulge. Because an integral aspect of our current understanding of stem cells involves their slow-cycling nature, this latter work argues strongly against the presence of stem cells in the matrix region.

The exact location of the outer root sheath stem cells has recently been called into question. However, the discussion may be more semantic than factual. In examining keratinocyte colony-forming cells from different, discrete horizontal sections of rat vibrissae, Kobayashi *et al.* (1993) found that 95% of all colonies arose from follicle fragments that contained the bulge. When identical experiments were performed by the same group using human scalp hair follicles (Rochat *et al.*, 1994), however, the majority of colony-forming cells appeared to reside in a middle portion of outer root sheath, approximately half way between the arrector pili muscle attachment and the bulb. Similar results have recently been obtained by Moll (1995). In all three studies, very few colony-forming cells were present in the hair matrix region, and therefore, the work is consistent with the theory that follicular stem cells reside in the outer root sheath region of the upper follicle and not in the matrix. Because visualization of the bulge in adult terminal human hairs is difficult, the apparent discrepancy between this recent study and earlier work may be explained. Alternatively, the functional 'bulge' in adult hair follicles, unlike rodents, may exist below the level of the arrector pili muscle attachment. In addition, cell culture studies may not be capable of differentiating between stem and TA cells (see discussion on page 345).

That it is upper outer root sheath cells that are the source of new follicular downgrowth during anagen is further supported by two early observations. Lyne (1957, 1970) noted that follicular degeneration and regeneration in certain marsupials does not occur with each hair cycle. Rather, a new hair follicle simply buds down, and it originates from the outer root sheath of an existing follicle at the level of the sebaceous duct. Evidence that the origin is specifically the bulge comes from Taylor (1949) and Butcher (1951), who noted that in rats at the onset of second anagen 'cells of the germ are proliferated from the base of the resting hair . . . [and] the cells of the resting bulb had oval nuclei whose surfaces were thrown into folds' (Butcher, 1951). These cells with nuclear indentations correspond histologically with the irregularly-shaped label-retaining cell nuclei observed by Cotsarelis *et al.* (1990) in the region of the mouse bulge, a morphological appearance not seen in any other follicular cells. Finally, recent work suggests that graft-versus-host disease, which has been previously shown to develop at the bottom of the deep rete ridges of epidermis (Sale *et al.*, 1985, 1991), coinciding with the location of stem cells, also develops in a highly focal fashion, at the level of the bulge, in hair follicles (Sale and Beauchamp, 1993).

Two, more theoretical, lines of reasoning support specifically a bulge location for follicular stem cells. One is the degradation of the lower follicle and the simultaneous upward movement of the dermal papilla that occurs during the catagen phase of every hair cycle. To survive a complete hair cycle, a stem cell located anywhere below the permanent portion of the follicle would have to rise upwards with the papilla. Studies in the haematopoietic system (Schofield, 1983) suggest that stem cells occupy relatively 'fixed' positions within tissue. In addition, if stem cells are located in the bulge, which occupy the lowermost region of the 'permanent' follicle (Pinkus, 1958), at the end of catagen they will be juxtaposed against the uplifted dermal papilla, and thus be ideally located to participate in the epidermal–dermal interactions that will initiate the next anagen phase.

A second line of reasoning has to do with the aetiology of cyclical hair production. For many years, this phenomenon has not been well explained. One theory has invoked the notion of a tissue-specific inhibitor or 'chalone' (Chase, 1954). Chalones were originally studied in mouse epidermis (Bullough and Laurence, 1960), and a telogen-specific, epidermally-derived molecule (Paus *et al.*, 1990, 1991) has recently been shown to delay the onset of anagen after hair plucking. Whether this molecule plays a regulatory role in the natural, non-plucked setting is unknown. Other researchers have not found a relationship between plucking during different phases of the hair cycle and the time interval to subsequent new hair growth, and thus dispute the existence of a follicular chalone (Johnson and Ebling, 1964). However, if the hair matrix is composed of TA cells which are generated by the transient replication of bulge stem cells at the onset of

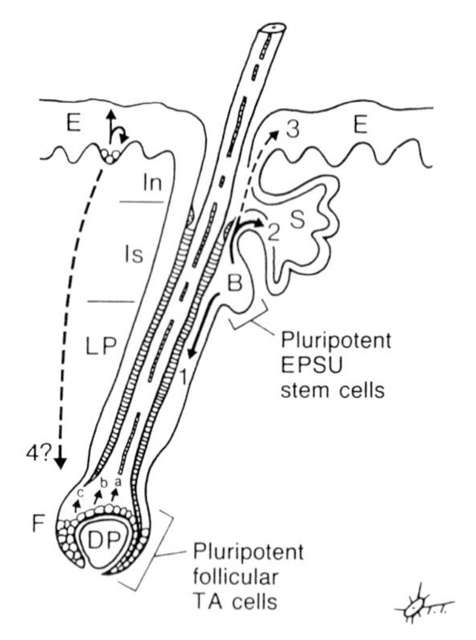

Figure 3 A schematic diagram illustrating the possible relationship between putative pluripotent stem cells of the bulge and other cell types located within the epidermis-pilosebaceous unit. Abbreviations include interfollicular epidermis (E), sebaceous gland (S), bulge (B), epidermis-pilosebaceous unit (EPSU), infundibulum (In), isthmus (Is), lower portion of the follicle (Lp), follicular matrix (F), dermal papilla (DP), and transient amplifying (TA) cells. (1) Bulge cells may give rise to follicular matrix cells which in turn differentiate into the various components of hair. (2) No slow-cycling cells have been observed within the sebaceous gland; nearby bulge cells may serve as the stem cells for this structure. (3) Following complete interfollicular epidermal loss (e.g. a burn patient), follicular cells migrate upwards to cover the wound, suggesting that bulge cells can serve as a source for the interfollicular epidermis. (4) Interfollicular epidermal cells may be able to form a hair follicle in response to inductive signals from the dermal papilla. Adapted from Lavker *et al.* (1993).

anagen, then the hair cycle will, by definition, be self-limited, and continue only for the number of replications that occur in the committed, non-renewing matrix TA cells.

The above evidence and considerations led Cotsarelis and co-workers to propose the 'bulge-activation hypothesis' of follicular cycling (Figure 2) (Cotsarelis *et al.*, 1990; Sun *et al.*, 1991). According to this hypothesis, the normally slow-cycling bulge stem cells are signalled to replicate during early anagen by the adjacent dermal papilla cells. Recent work has verified this discrete, early-anagen cycling of bulge cells (Lavker *et al.*, 1992; Wilson *et al.*, 1993). This leads to the formation of a downgrowth of TA cells which forms the new hair bulb, in conjunction with the dermal papilla. As the papilla moves away from the bulge during this downgrowth, the bulge stem cells return to their normal, quiescent, slow-cycling state. The return of bulge cells to quiescence by mid-anagen has also been documented (Lavker *et al.*, 1992; Wilson *et al.*, 1993). Finally, when the proliferative potential of the matrix TA cells is exhausted, the follicular cycle completes itself. There is degeneration of the lower follicle, coupled with upward movement of the dermal papilla which comes to rest against the bulge. This last aspect of the hair cycle, the upward movement of the dermal papilla in catagen that then allows future bulge-papilla interactions, has been shown to be crucial to the development of the next hair cycle. In hairless HR/Ch mice (Montagna *et al.*, 1952), the first anagen growth phase is normal, but at catagen, a normal club strand fails to form, and the papilla is thus not drawn up to the base of the permanent follicle. Subsequently, no further hair growth occurs. This strongly suggests that without papilla-bulge apposition, the next anagen phase cannot begin.

SUBPOPULATION

Corneal epithelium

Stem cell presence in the corneal epithelium is small relative to the entire corneal epithelial population, because the stem cells are localized to the basal layer of the limbal epithelium. However, because specific markers for corneal epithelial stem cells do not yet exist, exact quantitation of stem cell numbers is not yet possible. The limbus has been defined as a narrow band of tissue, located between the cornea and the conjunctiva, with an anterior border in humans and rabbits coinciding with the termination of Bowman's membrane, and a posterior border approximately 1.5 mm posterior to this (Lavker *et al.*, 1991b, Wiley *et al.*, 1991). Histologically, one notes a change from dense, avascular dermal stroma under the cornea to loose, vascular stroma under the limbus, and a decrease in pigment and onset of goblet cells at the transition from limbus to conjunctiva. Zieske *et al.* (1992a) have proposed a possible candidate for a corneal epithelial stem cell marker. They described a monoclonal antibody that reacts with a 50 kDa cytoplasmic protein present only in the 64 kDa keratin (K3)-negative basal cells of the limbal epithelium. This protein has been identified as alpha-enolase (Zieske *et al.*, 1992b). The number of $K3^-$, enolase$^+$ limbal basal cells increases following corneal epithelial injury and then rapidly returns to normal. Studies on developing cornea have shown that in embryonic cornea all epithelial basal cells are $K3^-$, enolase$^+$ (Chung *et al.*, 1992). What causes these central basal cells to lose their stem cell features and become $K3^+$, enolase$^-$ is unknown. One possibility is that the basement membrane zone may undergo developmentally-related changes so that only limbal basal cells retain their $K3^-$, $enolase^+$ features in the adult eye. Although these studies of enolase have provided useful information regarding the developmental changes of corneal basal cells, it should be noted that this enzyme is detected in all basal cells of several stratified epithelia (Zieske *et al.*, 1992b), so it seems unlikely to be an exclusive stem cell marker.

Epidermis

Several different approaches have been used to determine the fraction of epidermal cells that are clonogenic. Barrandon and Green (1987) found that cultured epidermal colonies can be divided into three groups on the basis of their replicative capacity. The group showing the greatest proliferative potential, termed holoclones, made up about 28% of the growing colonies. However, the relationship between these *in vitro* growing cells and *in vivo* keratinoctye stem cells is uncertain (see discussion on page 345). In his original article defining the EPU, Potten (1974) found an average of 10.6 basal cells per EPU, with one central stem cell in its midst. Radiation repopulation studies (Potten and Hendry, 1973) have also suggested that approximately 10% of basal cells are clonogenic. However, Morris *et al.* (1985), using 3H-TdR labelling techniques, found that only 1% of epidermal basal cells showed label-retaining features. In addition, Pavlovitch *et al.* (1991) isolated a population of possible epidermal stem cells from actively proliferating basal keratinocytes by unit gravity centrifugation, and found that they constituted 1% of the total basal cell population. Taken together, these studies suggest that stem cells represent between 1% and 10% of the epidermal basal cell population.

Follicle

The bulge cells constitute a small portion of the follicular structure, but the exact percentage is unknown. Keratin 19 may represent a marker specific for bulge cells (Lane *et al.*, 1991), and it has recently been shown that keratin 19-positive cells are also label-retaining cells, as determined by tritiated thymidine studies, and are present in the bulge region of hair-bearing areas and at the tips of the rete ridges of sole skin in mice (Germain *et al.*, 1994). However, the specificity of this potential keratinocyte stem cell marker has not yet been proven.

DIFFERENTIATION FEATURES

Corneal epithelium

By studying the expression of keratin in the cornea and conjunctiva, Schermer *et al.* (1986) showed that a 64 kDa (K3) keratin associated with differentiated epithelial cells is present in all layers of the corneal epithelium and all suprabasal layers of the limbus, but not in the basal layer of the limbus. This finding strongly suggests that limbal epithelial basal cells are in a less differentiated state than the corneal epithelial basal cells. In keeping with this concept of limbal cells as phenotypically more primitive, limbal epithelial basal cells are small and round or cuboidal, as opposed to corneal epithelial basal cells which are more columnar (Lavker *et al.*, 1991b).

Epidermis

Potten (1974) stated that the central (stem) basal cell of an EPU showed no morphological characteristics to distinguish it from other basal cells by light microscopy. Lavker and Sun (1982, 1983) found that the non-serrated basal cells at the base of deep rete ridges of palm epithelium were small and cuboidal, with a large nuclear-to-cytoplasmic ratio, a diffuse distribution of nuclear chromatin, a paucity of keratin filaments and an abundance of ribosomes. These histologic characteristics are all consistent with a primitive cell. Finally, existing data have shown that increasing keratinocyte size correlates with a decreased proliferative capacity and, at some point, with a commitment to terminal differentiation (Barrandon and Green, 1985; Sun and Green, 1976). Several investigators (Pavlovitch *et al.*, 1989, 1991; Staiano-Coico *et al.*, 1986; Furstenberger *et al.*, 1986) have noted a relationship between small cell size (in the range of 7 μm) and a number of physical (undifferentiated phenotype, low RNA content) and kinetic (long generation time, presence of label-retaining cells, proliferative response to phorbol esters) features of stem cells. Increased cell density has also been shown to correlate with these stem cell features (Morris *et al.*, 1990).

Follicle

In mouse pelage hair follicles, bulge cells have a large nuclear-to-cytoplasmic ratio, a cytoplasm filled with ribosomes but lacking in keratin fibres, and a surface covered with microvilli (Cotsarelis *et al.*, 1990). These features were also observed in the non-serrated cells located at the tips of the rete ridges (putative stem cells; see above), and are quite consistent with an undifferentiated or 'primitive' cell. Interestingly, the nuclei of the bulge cells have many convolutions (Cotsarelis *et al.*, 1990; Butcher, 1951). Morphologically, the bulge cells are smaller and quite distinct from the surrounding outer root sheath cells, which contain numerous keratin filaments, are smooth-surfaced, and have multiple desmosomal connections to their neighbours.

PLURIPOTENCY

Corneal epithelium

Limbal stem cells do not appear to be pluripotent; their only known differentiation pathway is to form corneal epithelium. Although the possibility that limbal epithelial stem cells could also produce conjunctival epithelium has been suggested (Thoft *et al.*, 1989), a recent study found that when limbal and corneal epithelial cells are grown in an identical *in vitro* environment, they showed remarkably similar keratin profiles, which were distinct and separate from the keratin profiles of various conjunctival epithelia (Wei *et al.*, 1993). In a more recent study, Wei *et al.* analysed the *in vivo* differentiation of cultured corneal, limbal and conjunctival epithelial cells when placed subcutaneously in the flanks of athymic mice. In this *in vivo* model, the athymic mouse stroma provides a permissive environment that allows cells to faithfully reproduce their original tissue phenotype. Corneal, limbal and conjunctival epithelial cells all recapitulated their *in vivo* phenotype when placed in this setting. These findings provide the strongest evidence to-date that the corneal/limbal lineage is distinct from the conjunctival lineage (Wei *et al.*, 1996).

The issue of pluripotency has most recently been addressed with respect to the conjunctival epithelium. Conjunctival epithelium is a self-renewing tissue whose role in maintaining the ocular surface is critical. Conjunctival epithelium is comprised of two cell types: keratinocytes, which form the outermost barrier and are the predominant cell type, and goblet cells, that elaborate mucins which are critical for maintaining the tear film. Both of these morphologically and functionally distinct cell types have subpopulations of cells with stem cell (slow-cycling) and transit amplifying cell (normal-cycling) characteristics. This has raised the question of whether the conjunctival keratinocyte and goblet cell originate from two distinct populations or from a single population of bipotent conjunctival epithelial cells. To answer this question of lineage, a method was developed that enabled the analysis of the differentiation potential of single cells in an *in vivo* setting. Utilizing this method, we have determined that keratinocytes and goblet cells of the rabbit conjunctival epithelium share a common progenitor cell, and this progenitor cell undergoes numerous divisions prior to acquiring a specific phenotype. Furthermore, following commitment, both keratinocytes and goblet cells have proliferative capacity.

Epidermis

As mentioned in the discussion of corneal stem cells, there are no known pluripotent stem cells in the epidermis (however, see discussion of follicular stem cell pluripotency below). Like corneal epithelial stem cells, interfollicular 'stem' cells more closely approximate haematopoietic 'progenitor' cells (for example see Pavlovitch *et al.*, 1991), because they appear to be relatively committed to the development of only a single cell lineage, the epidermis.

Bulge

It is possible that the bulge stem cells are a pluripotent population capable of regenerating follicular, epidermal and sebaceous structures (see Figure 3). However, this hypothesis is not yet proven. The idea of pluripotent stem cells was first put forth hypothetically by Chase (1954). He postulated that these cells were located in the 'upper outer [root] sheath' of the hair follicle. The notion was raised again by Pinkus and Mehregan (1981) to explain epidermal and follicular tumour formation. Several lines of evidence have developed to support the concept of follicular stem cell pluripotency.

The first line of evidence is the role follicular stem cells appear to play in epidermal repair and homeostasis. A number of early investigators (Al-Barwari and Potten, 1976; Bishop, 1945; Eisen *et al.*, 1956; Sanford *et al.*, 1964; Argyris, 1976) noted follicular cell proliferation in response to epidermal injury or loss. Several reported a generalized change in the follicles, including the entry into the anagen phase (Argyris, 1976), in the area of the epidermal damage, but some observed a more discrete mitotic response, specifically within the region of the upper follicle (Al-Barwari and Potten, 1976) or at the junction of the follicle and sebaceous duct (Eisen *et al.*, 1956). This region corresponds, roughly, to the vicinity of the bulge. Lavker *et al.* (1993) noted, in both naturally developing and chemically induced anagen follicles, 3H-TdR-labelling of upper outer root sheath cells above the level of the bulge in early anagen, following single pulse 3H-TdR administration. The location of this labelling suggested that a TA population of rapidly-cycling cells was moving upwards, towards the interfollicular epidermis. In another series of experiments, Regauer and Compton (1990) cultured sequentially obtained, 20/1000 inch horizontal sheets of dermis for epidermal growth, and observed that the greatest proliferative potential was obtained from mid-dermal explants that contained follicular infundibula. The keratinocyte proliferative potential of this region is likely due to the presence of bulge follicular stem cells. Similar results were obtained by Yang *et al.* (1993), who found that upper follicular keratinocytes have a proliferative potential exceeding even that of epidermal keratinocytes. Potten (1981) noted that in irradiation experiments, some follicular clonogenic cells were more radio-resistant than those from interfollicular regions. This suggested that they could 'outlive' interfollicular germinative cells in the event of a radiation insult. Finally, several groups (Lenoir *et al.*, 1988; Limat *et al.*, 1991) have shown that outer root sheath cells, although not clearly bulge cells, can be cultured on a dermal equivalent at an air–liquid interface to produce a tissue with morphological, immunological and biological features of normal interfollicular epidermis. In summary, a number of studies suggest that follicular stem cells may play a role in the normal and reparative maintenance of the epidermis.

Several studies have also suggested a role for follicular stem cells in sebaceous gland reproduction. Montagna and Chase (1950) destroyed sebaceous glands with 3-methylcholanthrene and observed gland regeneration from external root sheath cells. Eisen *et al.* (1956) noted localized collections of sebaceous cells within regions of new, follicular-derived epidermis formed on the backs of human subjects following epidermal injury. This suggested a pluripotent capacity of the repopulating cells. Finally, Lavker *et al.* (1993) were unable to detect any label-retaining cells in sebaceous glands or ducts, suggesting that the sebaceous gland germinative population must reside in another, nearby, location such as the bulge.

Until marker studies can be done to trace the putative stem cells through the regeneration of follicular, interfollicular, and sebaceous gland regions, the pluripotent nature of follicular stem cells must remain speculative.

SLOW-CYCLING NATURE

Corneal epithelium

The slow-cycling nature of limbal epithelial basal cells in a steady-state setting has been documented in several ways. Cotsarelis *et al.* (1989) made use of the fact that continuous 3H-TdR labelling followed by a prolonged 'chase' period results in the labelling of only those cells that cycle slowly. The slow-cycling cells have been termed 'label-retaining cells' (Morris *et al.*, 1985; Bickenbach, 1981). Cotsarelis *et al.* (1989) found that label-retaining cells were present in the basal layer of limbal epithelium, where corneal epithelial stem cells are believed to exist, and not in the corneal epithelium itself. Interestingly, label-retaining cells of the conjunctival epithelium have been found in the fornix (Wei *et al.*, 1993).

A number of studies examining cell mitoses in an *in vivo* setting have found that the mitotic index of limbus and cornea is approximately equal (Buschke *et al.*, 1943; Friedenwald, 1951; Kaufmann *et al.*, 1944; Haskjold *et al.*, 1988, 1989). However, Lavker *et al.* (1991b) found that the labelling index of rabbit and mouse limbal epithelium was slightly lower than that of corneal epithelium, suggesting a more slowly-cycling nature of the limbal cells. Finally, Tseng and Zhang (1992) showed that the anti-metabolite 5-fluorouracil, which inhibits cell division, suppressed explant cultures from cornea to a much greater degree than limbal explants, again supporting the slow-cycling nature of limbal cells.

Epidermis

Several lines of *in vivo* evidence in both the mouse ear epidermal EPU and the deep rete ridges of monkey palm epithelium point to the slow-cycling nature of the presumed stem cells. In one of the original EPU studies, Mackenzie (1970) showed by colcemid arrest techniques that mitoses occurred more frequently in the peripheral basal layer of an EPU compared with the central region. This suggested a more slow-cycling nature of the central cells. Potten (1974) also noted, using labelling index studies, that central basal cells of an EPU appear to cycle more slowly than those in the periphery. Finally, Morris *et al.* (1985) showed that slow-cycling label-retaining cells were primarily located in the centre of EPUs.

In the monkey palm epithelial model, Lavker and Sun (1982, 1983) observed that following a single ^{3}HTdR injection, over 80% of labelled nuclei were present in the deep rete ridges, but over 75% of these were in suprabasal layers. Basal cells at the lowermost rete tips, i.e. the site of presumed stem cells, were not labelled, thus suggesting their slow-cycling nature.

Finally, several *in vitro* studies have also been performed. Clausen *et al.* (1983, 1984) showed that a subset of epidermal cells in culture are slow-cycling, and traverse slowly through S-phase. Pavlovitch *et al.* (1991) isolated a small keratinocyte sub-population from intact epidermis, that constituted 1% of the basal cell population; it contained all the label-retaining cells present and formed colonies consisting of homogeneously small keratinocytes. Several investigators (Staiano-Coico *et al.*, 1986; Morris *et al.*, 1990; Albers *et al.*, 1986; Jensen *et al.*, 1985) have documented *in vitro* subpopulations of epidermal cells with different cell cycle times. The slowest cycling sub-populations exhibit features, such as low RNA content, small size, and increased density, consistent with their being stem cells.

Proliferative response of presumed stem cells to stimuli has also been documented. Potten (1974) found that the central cells of an EPU showed a greater proliferative response to hair plucking than did the cells of the periphery. A number of investigators (Furstenberger *et al.*, 1986; Yuspa *et al.*, 1982; Parkinson *et al.*, 1983) have documented a sub-population of cultured epidermal cells that do not differentiate in the presence of 12-*O*-tetradecanoylphorbol-13-acetate (TPA) or phorbol,12-myristate,13-acetate (PMA); instead, they proliferate. Morris *et al.* (1985) showed that this paradoxical proliferative response to phorbol esters is exhibited by central but not by peripheral basal cells of the EPU. In another study, Morris *et al.* (1990) noted that the sub-population of small, dense epidermal cells containing the label-retaining cells (the presumptive stem cells) also exhibited a proliferative response to TPA and an ability to proliferate in high calcium medium, which were features not seen in the other epidermal subpopulations. Taken together, these studies suggest that epidermal stem cells are able to proliferate in response to stimuli that would cause more committed cells to differentiate.

Follicle

Cotsarelis *et al.* (1990) showed that the only label-retaining cells present in mouse pelage follicles are located in the region of the bulge. Further, these cells proliferated in response to topically applied tumour promoter. Thus, bulge cells appear to be both slow-cycling, and capable of proliferating in response to stimuli. Al-Barwari and Potten (1976) also noted ^{3}H-TdR-labelling of cells in the upper follicle and follicular canal following epidermal damage. This again suggests a proliferation in the region of the bulge in response to stimuli. Temporally, bulge cell cycling has been shown to occur, in a naturally occurring hair cycle, only in early anagen (Lavker *et al.*, 1992; Wilson *et al.*, 1993). This is consistent with the bulge-activation hypothesis, which predicts that papilla-bulge interactions stimulate the onset of each new anagen hair cycle.

PROLIFERATIVE POTENTIAL

Proliferative ability of keratinocytes is usually measured *in vivo* as a labelling index indicative of the proliferative rate. In the *in vitro* setting, colony-forming ability, doubling time, mitotic index and total *in vitro* life span, as measured by number of cell passages, are gauges of proliferation. One of the experimental difficulties, which has been a hindrance both in defining the precise location of keratinocyte stem cell populations and in comparing stem and TA cell populations, has been the proper interpretation of *in vitro* proliferation study results. In the *in vivo* setting, single pulse labelling experiments label rapidly proliferating cells, which represent TA cell populations. Long-term continuous labelling studies followed by a washout detect label-retaining cells, which are slow-cycling stem cells. However, stem cells are slow-cycling, and the growth-intensive environment of cell culture systems selectively enhance the presence of rapidly proliferating cells that constitute TA cell populations, so after a period of time it is probable that no stem cells remain within the culture system, as Kruse and Tseng (1991, 1993a, 1993b), Yang *et al.* (1993) and Moll (1995) have noted. While stem cells that are initially present may be recruited in the cell culturing process to form additional rapidly proliferating colonies by converting to TA cells, it is essentially impossible to determine on the basis of these *in vitro* studies the percentage of stem versus TA cells present in a given population of cells at the time of plating. This caveat must be kept in mind whenever *in vitro* keratinocyte proliferative studies are assessed.

Corneal epithelium

The ability of limbal versus conjunctival epithelia to cover corneal wounds has been evaluated clinically in a number of studies. Kruse *et al.* (1989) surgically removed the entire limbus prior to wounding cornea epithelium in 54 rabbits. They found that these limbal-deficient corneas displayed profound neovascularization and conjunctivalization (the covering of the corneal surface by a conjunctival-like epithelium). Huang and Tseng (1991) showed that repetitive wounding of rabbit corneal epithelium following the surgical removal of limbus resulted in the coverage of the cornea by conjunctival epithelium, with delayed healing time, development of recurrent erosions and vascularization following each wounding episode. Finally, Tsai *et al.* (1990) and Chen and Tseng (1990) showed that limbal transplants result in much better corneal epithelial repair than conjunctival transplants. Similar results were obtained in human patients (Kenyon and Tseng, 1989; Dua and Forrester, 1990). Thus, limbal but not conjunctival grafting was able to provide an appropriate tissue source for repairing corneal epithelium. A discussion refuting the long-standing concept that conjunctival epithelium can convert into a bona fide corneal epithelium through 'transdifferentiation' can be found in Wei *et al.* (1996).

The proliferative potential of limbal versus corneal epithelial populations in response to the removal of central corneal epithelium has also been compared. Cotsarelis *et al.* (1989) showed that limbal epithelial basal cells underwent an 8–9-fold increase in their labelling index in response to central corneal epithelial injury versus only a twofold increase in central corneal epithelial basal cells. Similar results were obtained in response to the topical application of a tumour promoter, *O*-tetradecanoylphorbol 13-acetate.

The *in vitro* proliferative activity of limbal and corneal epithelial cells appears to parallel the results obtained from *in vivo* studies. Two reports (Ebato *et al.*, 1987, 1988) found that limbal epithelial explants exhibited a greater plating efficiency, a shorter doubling time, and a higher mitotic activity than corneal epithelial explants. A third study (Wei *et al.*, 1993) noted similar results in cell culture (instead of explant culture). A fourth (Kruse and Tseng, 1991), using a particular serum-free medium found that limbal cells had a lower colony-forming efficiency than corneal epithelial cells. More recent work with that serum-free medium has shown that addition of fetal bovine serum (Kruse and Tseng, 1993a) or retinoic acid (Kruse and Tseng, 1994) stimulates the colony-forming ability of limbal but not peripheral corneal cultures, and that a larger percentage of limbal as opposed to peripheral corneal culture cells are resistant to the terminal differentiation-inducing effects of phorbol 12-myristate 13-acetate (Kruse and Tseng, 1993b). The results of all three latter studies suggest that a higher percentage of stem cells are present in the limbus than in peripheral cornea.

Although both the *in vivo* and *in vitro* proliferative potential of limbal cells seems to be greater than that of central corneal epithelial cells, their proliferative reserve relative to an animal's or a human's lifetime is unknown.

Epidermis

Cultured epidermal cells are capable of reconstituting a long-lasting *in vivo* epidermis in humans (Gallico *et al.*, 1984; Compton *et al.*, 1989). However, serial transplantations have not been performed. It has been shown that cultures of newborn foreskin epithelial cells have a significantly greater replicative capacity than keratinocytes from trunk epidermal cells of older donors (Rheinwald and Green, 1975). This is consistent with the finding that newborn foreskin contains a much greater proportion of holoclones (colonies with high proliferative capacities) than trunk skin of aged individuals (Barrandon and Green, 1987), and that the proliferative response of newborn keratinocytes to GFs is greater than that of cultured adult keratinocytes (Gilchrest, 1983).

Follicle

Allograft hair transplantation has been a clinical treatment of male-pattern baldness for over 30 years (Orentreich and Orentreich, 1988). Transplants appear to remain viable and produce hair for the life of an individual. Within the follicle itself, bulge cells in culture have been shown to have a proliferative potential that exceeds that of bulb cells (Yang *et al.*, 1993).

TRANSIENT AMPLIFYING CELLS

The haematopoietic system contains multiple lineages, each containing multiple stages of development. Thus it is already clear that this replicative system has many sub-divisions. For instance, subdivisions are known to exist between pluripotent 'stem' or 'CFU-S' cells, and 'stem' cells committed to a single lineage, perhaps better called 'progenitor' cells, which are known to exist in the lymphoid, megakaryocyte, erythroid, and monocyte/granulocyte lines (Quesenberry and Levitt, 1979a, 1979b).

Among the keratinocyte stem cell populations we have examined, only the follicular stem cells of the bulge show evidence of pluripotency. This has raised the question of whether other, committed keratinocyte 'stem' cells also exist. On the basis of experimental radiation recovery data, Potten (1981) suggested that the epidermis might contain two 'classes' of stem cells, one located in the EPU, responsible for day-to-day epidermal integrity and a second, more primitive 'reserve' population, located in the upper follicle. In a similar fashion, comparing the cells of the follicular bulge with those of the matrix, Reynolds and Jahoda (1991a) have suggested that while the bulge appears to contain pluripotent stem cells, the bulb germinative cells, with their undifferentiated ultrastructural features, their capacity for significant proliferation and their apparently unique ability to form organoid structures when combined with dermal papilla cells, may also represent a class of different 'stem' cells. Finally, Lavker *et al.* (1993) have recently proposed a different scheme (Figure 3). Based on histologic observations showing that in early anagen rapidly proliferating cells are located both above and below the bulge, Lavker *et al.* (1993) suggest that only the bulge contains pluripotent stem cells, while the EPU and matrix regions contain TA cells with varying degrees of proliferative potential and commitment to differentiation. While further experimentation will be required to differentiate between the specifics of these three theories, the conceptual framework is similar. These diverse viewpoints can be unified if one considers that a stage in follicular differentiation exists, i.e. the matrix cells, in which cells exhibit a relatively long proliferative life span, a rapid cell turnover time, and some degree of commitment to differentiation. It would perhaps be best, as it is done in the haematopoietic literature, to refer to these cells as 'progenitor cells', instead of 'stem cells'. To do this would eliminate much of the confusion currently existing in this area.

Corneal epithelium

Because markers for corneal epithelial cells at different stages of differentiation do not yet exist, the precise location of the TA population in the cornea is unknown. Presumably, these cells reside along the physical route from limbal basal cell layer to corneal basal and then finally corneal suprabasal layers.

Epidermis

A number of *in vivo* studies suggest the presence of a multi-compartment proliferative system within the epidermis. Morris *et al.* (1985) showed that while central basal cells of the EPU are slow-cycling, the peripheral cells contain the majority of label after a 1-hour incubation with ^{3}H-TdR, indicating their highly proliferative nature. Further, these peripheral cells are displaced to the suprabasal layers in response to TPA application rather than proliferating as the central cells do, indicating their commitment to terminal differentiation. These peripheral EPU cells appear to represent a population of TA cells. In a similar fashion, Lavker and Sun (1982, 1983) used ^{3}H-TdR labelling to document the existence of three proliferative basal cell populations within volar epithelium: slow-cycling cells at the base of rete ridges, a rapidly proliferating population situated suprabasally, and a mitotically quiescent, differentiating population located more superficially.

Several studies have suggested the presence of cells within the basal layer already committed to differentiation. Using a double-labelling technique, in which pulse-labelling with ^{3}H-TdR represented a commitment to terminal differentiation, Albers *et al.* (1987) showed that in confluent keratinocyte cultures, withdrawal from the replicative cycle occurs while cells are still in the basal position. In other studies utilizing confluent cell cultures as well as neonatal mouse epidermis, markers of terminal differentiation, including a 67 kDa keratin (K1) (Regnier *et al.*, 1986), and a 60 kDa keratin (K10) (Schweizer *et al.*, 1984), have been observed in about 5–10% of basal cells. This sub-population of basal keratinocytes are probably part of the TA population. Finally, several *in vitro* studies (Pavlovitch *et al.*, 1989; Staiano-Coico *et al.*, 1986; Albers *et al.*, 1986) have separated cultured epidermal cells into three populations on the basis of their replicative abilities: (1) slow-cycling, small cells that have undifferentiated features and low RNA levels, (2) larger cells with greater RNA content and a more rapid cell-cycle time and (3) still-larger cells with low RNA levels, a low or non-existent proliferation rate, expressing markers of differentiation. These three epidermal populations appear to represent stem, TA and committed cells.

Follicle

In their cell kinetic studies, Cotsarelis *et al.* (1990) found that essentially all hair matrix cells are rapidly proliferating, a capacity of the matrix that has been known for some time

(Malkinson and Keane, 1978). Interestingly, no slow-cycling cells were detected in the matrix. On this basis, they hypothesized that the matrix constitutes a collection of TA cells. In a temporal analysis of the same process (Wilson *et al.*, 1993), this group of investigators observed that 3 days after the onset of anagen, rapidly cycling cells were present both in, below and above the bulge region, suggesting an outward spread of new TA cells, both down to the newly forming follicular matrix, and possibly upwards, to the epidermis or sebaceous gland, as well. TA cells thus appear to originate by bulge stem cell replication, and they go on to produce possibly several populations of more committed, terminally differentiated cells.

One additional point to make in this discussion of the follicular matrix TA cell population, is the role that dermal papilla cells appear to play in inducing and, indeed, specifying their terminally differentiated end-product, the hair. Oliver (1966b) showed that the presence of dermal papillae is necessary for hair growth to occur; without it, hair growth ceases. Van Scott and Ekel (1958) and Ibrahim and Wright (1982) demonstrated that hair size is proportional to papilla volume. In transplantation studies, Oliver (1967) stimulated hair growth from the upper portion of a transected vibrissae follicle that would not normally be capable of producing hair, by surgically implanting a dermal papilla adjacent to it. Jahoda and co-workers (Jahoda *et al.*, 1984; Horne *et al.*, 1986) obtained similar results using cultured dermal papilla cells. Oliver (1970) also stimulated the development of follicle-like formations and hairs in rats, through the transplantation of dermal papillae into ear epidermis, afollicular scrotal skin, and keratinizing (afollicular) oral epithelium. Reynolds and Jahoda (1990, 1991b) obtained similar results by transplanting cultured dermal papilla cells into afollicular rat footpads. Finally, it appears that dermal papilla cells are not only necessary for hair formation, and can possibly stimulate hair production by epithelium that is normally non-hair bearing, but they also carry the genetic information which specifies the characteristics of the hair that is produced. Thus, when vibrissa papilla are transplanted into a rat ear, a vibrissa and not an ear hair results (Reynolds and Jahoda, 1991b), and when pelage papilla are transplanted into afollicular rat footpads, a pelage hair grows (Reynolds and Jahoda, 1990, 1991b). One final note, however, is that the complex organoid structures observed by Reynolds and Jahoda (1991a), which occurred when cultured germinative matrix cells but not outer root sheath or interfollicular epidermal cells were combined with dermal papilla cells, suggest that matrix germinative TA cells are intrinsically divergent from other epidermal cells.

STEM CELL 'NICHE'

The keratinocyte stem cell microenvironment or 'niche' probably functions in several important ways: (1) to protect a stem cell from damage by the external, physical environment, (2) to maintain or help to maintain a stem cell's slow proliferative rate and continuous 'non-commitment' to terminal differentiation, and (3) to assist in the signalling process that calls for stem cell division, and possibly for the differentiation of early progeny cells. All three functions probably arise from regional organizational structure, because the location of stem cells in keratinocyte populations is so localized (in the limbus, in the centre of an EPU or at the bottom of deep rete ridges and in the bulge).

Keratinocyte stem cell niches exist in sites that specifically minimize environmental insults, such as ultraviolet light-mediated DNA damage, or stem cell loss through epidermal shearing or hair plucking, and they are always adjacent to a good blood supply. Three-dimensional spatial organization is probably maintained through the presence of specific extracellular molecules to which the stem cells have membrane receptors, and control of stem cell proliferation and differentiation is probably mediated through cytokines, perhaps secreted by local accessory cells, localized in the extracellular milieu, or contained within the molecular structure of the matrix itself. The role of inherent genetic programming versus microenvironment signalling in maintaining 'stem cellness' is not well-understood. It is still an open question whether any of a number of cells that have not progressed beyond a certain point in the differentiation pathway can assume the functional role of a stem cell, given the right microenvironment.

Corneal epithelium

In terms of minimizing environmental damage, the limbal location of corneal epithelial cells makes sense in several ways. First, unlike the cornea proper, which has a smooth epithelial-stromal junction that maintains corneal transparency but is vulnerable to shear damage, the limbus has an undulating epithelial-stromal junction that is extremely resistant to traumatic shearing (Cotsarelis *et al.*, 1989). This helps to ensure the presence of residual stem cells in the event of corneal epithelial tearing or loss. Secondly, the transparent corneal basal cells are devoid of pigment, while the limbal epithelial basal cells, with no such functional constraint, are heavily pigmented (Lavker *et al.*, 1991b; Wiley *et al.*, 1991; Cotsarelis *et al.*, 1989). This limbal pigmentation protects the cells from the carcinogenic effects of ultraviolet light. Finally, the corneal epithelium rests on a largely avascular stroma, again to maintain the necessary transparency for vision, while the limbus lies on a vessel-rich loose connective tissue base (Davanson and Evensen *et al.*, 1971; Lavker *et al.*, 1991b; Cotsarelis *et al.*, 1989) which provides a constant and rich supply of nutrients.

A second function of the extracellular microenvironment is to maintain the resting state of stem cells. In terms of the cornea, one may ask why limbal basal cells do not express K3 keratin (a marker for an advanced stage of corneal epithelial differentiation) while central corneal epithelial basal cells do. Schermer *et al.* (1986) speculated that the basement membrane (BM) of these two zones may be different. It has been shown that significant molecular heterogeneity exists among BMs obtained from different body sites, in terms of laminin and type IV collagen presence (Sanes *et al.*, 1990). It has also been suggested that BMs may be heterogeneous within a given tissue (Sengel, 1986). Consistent with Schermer *et al.*'s hypothesis, Kolega *et al.* (1989) found that a BM antigen, defined by monoclonal antibody AE27, is present strongly in corneal epithelial BM, but only weakly in conjunctival BM (Kolega *et al.*, 1989). Interestingly, in some limbal regions where AE27 antigen is present in a heterogeneous fashion, its presence in BM appears to be related to K3 expression in the overlying keratinocyte basal cells. Unfortunately, the biochemical identity of this AE27 antigen is unknown, due to the fact that the antibody does not work in immunoprecipitation or immunoblotting. Another BM component, collagen type IV, is readily detectable throughout the BM of embryonic conjunctival, limbal and corneal epithelia, but it is difficult to detect in the corneal BM of adults

((Kolega *et al.*, 1989; Fitch *et al.*, 1982; Fujikawa *et al.*, 1984); however, see Newsome *et al.*, 1981). Thus, it is possible that unique, localized regions of keratinocyte BMs could provide molecular ligands required for the maintenance of a keratinocyte stem cell niche. However, other well-known BM components, including laminin, entactin, heparin sulfate proteoglycan, KF-1 (Kolega *et al.*, 1989) and bullous pemphigoid antigen (Ben-Zvi *et al.*, 1986) appear to be present uniformly in BMs of conjunctiva, limbus and cornea.

In haematopoietic repopulation studies using stem cells 'marked' by retrovirus-mediated gene transfer, it appears that at a given time, the proliferation of only one or two clones account for the majority of the resulting mature haematopoietic cell population (Lemischka *et al.*, 1986). Thus, some form of 'overall' stem cell regulation appears to be occurring in the haematopoietic system. Wiley *et al.* (1991) has suggested that there may be regional differences in corneal epithelial stem cell presence in different quadrants of the eye; whether this relates to any type of coordinated regulation of these keratinocyte stem cells as a population is unknown.

Finally, there is some evidence to support the notion that intrinsic genetic differences exist in corneal stem cells. Recent work by Kruse and Tseng suggests that, in the *in vitro* setting, corneal and limbal cells have certain innate differences: their responses to serum (in the serum-free setting) (Kruse and Tseng, 1993a), to phorbol esters (Kruse and Tseng, 1993b) and to retinoic acid (Kruse and Tseng, 1994) are distinct.

Epidermis

Similar to the cornea situation, the physical location of epidermal stem cells appears to offer them maximal protection from environmental damage. In their studies of volar epithelium, Lavker and Sun (1982, 1983) showed that the presumptive stem cells reside at the bottom of deep rete ridges. In addition, these cells are heavily melanized. With the melanosomes forming a 'cap' at the apical end of the cell nuclei, protection of the stem cells from ultraviolet radiation is maximized. Histologic examination of the deep and shallow rete ridges also reveal different local dermal environments (Lavker and Sun, 1983). The dermis surrounding the deeper rete ridges, where the presumptive stem cells reside, contains a high density of fibroblasts and blood vessels. In contrast, the shallow rete ridges have a surrounding dermis that is quite acellular. This suggests that the deep rete ridge basal cells have a richer blood supply and greater interaction with mesenchymal cells than the basal cells of the shallow rete ridges.

Three primarily clinical observations suggest a relative importance of the micro-environment over intrinsic genetic programming in the establishment and maintenance of epidermal stem cells: (1) As discussed on page 345, it appears that slow-cycling stem cells are absent from cell cultures and explants (Yang *et al.*, 1993; Moll, 1995; Kruse and Tseng, 1991, 1993a, 1993b, 1994). Instead, an early stage of TA cells appears to proliferate. Yet, cultured epidermal cells are capable of re-constituting a long-lasting epidermis in humans (Gallico *et al.*, 1984; Compton *et al.*, 1989). Therefore, the transplanted cells must in some way be capable of re-establishing stem cell presence. (2) Similarly, epidermal wounds are healed by an inward movement of surrounding keratinocytes that develop a specialized 'migratory' phenotype prior to migrating (Clark, 1993; Kirsner and Ealgstein, 1993). These cells show specific morphologic changes similar to those seen in cultured keratinocytes and psoriatic skin cells, including retraction of intracellular keratin filaments, dissolution of intracellular desmosomes and hemidesmosomes, and formation of peripheral cytoplasmic actin filaments. Immunohistochemically, the migrating cells lose certain markers of stratified squamous differentiation, such as keratins 1 and 10 and fillaggrin, but retain others such as involucrin and transglutaminase. The cells are thus 'differentiated', in a specific fashion, at least temporarily. However, after the process of migration and re-epithelialization is completed, a durable stratified squamous epithelium remains, that apparently contains a normal cadre of stem cells, even in the absence of residual appendageal structures. (3) In keratinocyte cultures, raising cells to an air–liquid interface, in and of itself, precipitates the development of a vertically-organized, stratified squamous epidermis. In the process, does a horizontal patterning of stem cells also develop within the basal layer (a patterning that presumably initiates the subsequent, vertically-organized pattern of differentiation)?

Cairns (1975) proposed the 'immortal strand hypothesis', which suggested that stem cells replicate asymmetrically. It is not known whether asymmetric epidermal stem cell division takes place; one study has provided evidence for (Potten *et al.*, 1978) and one against (Kuroki and Murakami, 1989) its occurrence. Such a non-random replication of genetic material would undoubtably require stem cell orientation at the time of DNA division, and this in turn would require a degree of organization of the stem cell micro-environment.

Cell–cell communication may be important in the early replicative stages of the epidermis, since it has been shown that gap junctions are present only in the lower epidermal layers, but they disappear as markers of differentiation appear (Kam *et al.*, 1987). Kam *et al.* (1986) found that cells within an EPU appear to form a unit of cell–cell communication. Whether these potential avenues for intercellular communication are specifically important at the stem cell level is unknown.

The epidermal Merkel cell population appears to be concentrated at the bottom of the deep rete ridges (Moll *et al.*, 1984, 1995), as well as in the vicinity of the bulge region of the vibrissae follicle (Narisawa *et al.*, 1993a, 1994a, 1994b). This raises the interesting question of whether neural factors may play a role in the maintenance of the epidermal stem cell micro-environment.

Haematopoietic stem cells appear to have receptors for many haematopoietic GFs (Heyworth *et al.*, 1988), including myeloid GFs, interleukin 3, granulocyte-macrophage colony stimulating factor (CSF), granulocyte CSF and macrophage CSF. One or several of these may mediate signals for stem cell proliferation. Differentiation inhibiting factor/leukaemia inhibiting factor has been shown to prevent differentiation of embryonic stem cells (Heath *et al.*, 1990; Rathjen *et al.*, 1990), and a molecule with a similar function, termed haematopoietic stem cell inhibitor or macrophage inflammatory protein 1-alpha, appears to maintain haematopoietic stem cells in the undifferentiated state (Graham *et al.*, 1990). Haematopoietic stem cell inhibitor has recently been shown to inhibit keratinocyte proliferation as well (Parkinson *et al.*, 1993), and messenger RNA for its production is present exclusively in Langerhans' cells, and not keratinocytes, in the epidermis. Interestingly, the Langerhans' cell was originally described by Potten (Potten, 1981; Potten and Allen, 1976) as being immediately adjacent to the stem cell in every epidermal proliferative unit (EPU) and Potten hypothesized that the Langerhans' granule 'contains an inhibitor for keratinocytes' (Potten and Allen, 1976). This theory has not been widely accepted, but it is possible that the Langerhans' cell may play a role in epidermal stem cell regulation. In the haematopoietic system, known molecular ligands for stem cell binding exist. In transplant studies, haematopoietic stem cells have receptors which recognize specific galactosyl and manosyl moieties in the haematopoietic stroma (Aizawa and Tavassoli, 1987, 1988; Matsuoka *et al.*, 1989). Not as much is known about keratinocyte populations. Fibronectin has been shown to inhibit terminal differentiation of keratinocytes in culture (Adams and Watt, 1989), and keratinocyte stem cells have been reported to be more 'adhesive' than transient-amplifying cells in an unpublished *in vitro* assay (Hall, 1989). Jones and co-workers have recently observed, in both an *in vitro* (Jones and Watt, 1993) and an *in vivo* model (Jones *et al.*, 1995), a direct relationship between beta$_1$ integrin expression in keratinocytes and subsequent colony-forming ability. Rapid keratinocyte adherence to type IV collagen, fibronectin, or a mixture of extracellular matrix proteins was also observed to correlate with increased colony formation. It is proposed that these integrin-rich, rapidly-adhering keratinocytes represent stem cells. However, in these experiments the putative stem cells constituted 10–30% of the total cell population in an *in vitro* setting, and 25–43% of the basal layer in an *in vivo* setting, and this is a significantly greater stem cell presence than has been previously calculated to exist in skin (1–10%; see page 340). In addition, integrin-bright cultured cells were shown to be capable of undergoing terminal differentiation without first dividing (Jones and Watt, 1993), and the rapidly-adherent populations of keratinocytes obtained from whole skin did contain transient-amplifying and committed cells (Jones *et al.*, 1995). Thus, while high beta$_1$ integrin expression and rapid adherence to type IV collagen constitutes a valuable inroad in our ability to purify early progenitor keratinocytes, at the moment one cannot exclude the possibility that these cells represent not just stem cells, but perhaps also an early form of transient-amplifying cell, which others have suggested is the earliest progenitor stage that regularly exists in cell culture (see page 345) (Yang *et al.*, 1993; Kruse and Tseng, 1993a, 1993b, 1994). One additional, interesting observation relates to the location of stem cells in epidermis of different body sites. Jones and Watt reported that integrin-rich cells occurred in groups or 'patches' within the basal layer of foreskin, scalp, and palm epidermis. Each patch contained an average of 9–14 cells. These integrin beta$_1$-enriched keratinocytes were suggested to represent the stem cells. Indeed, in the palm epithelium, these integrin-positive cells were located at the bottom of the deep rete ridges, an assignment in agreement with prior labelling studies (Lavker and Sun, 1982, 1983). In the epidermis of the foreskin and scalp, keratinocytes located at the bottom of the rete ridges were also integrin-positive; however, those located at the top of the rete ridges were labelled even more strongly. Based on this, it was suggested that while stem cells reside at the *bottom* of the deep rete ridges in palm epithelium, they reside at the *top* of the rete ridges in non-volar epidermis. As discussed earlier, the bottom of the rete ridges offers several advantages as a stem cell niche, including maximal protection from ultraviolet light and other environmental insults, and maximal potential for stem cell retention in the advent of epidermal loss due to mechanical shearing forces. These general advantages are equally applicable to volar and non-volar epithelium. That non-volar epidermis might have a fundamentally different architecture regarding stem cell location is both provocative and interesting. However, at this point it is important to ascertain whether the 'integrin-strong' keratinocytes observed after the epidermis was trypsin-dissociated and used for cell plating experiments indeed correspond to the 'integrin-strong' cells of the top rete ridges (the 'top cells') in intact epidermis. This is important because although the top cells have the strongest integrin staining, most of the integrins appear to be actually associated with the lateral cell surface. These top cells appear to have less basal cell surface-associated integrin than the cells located at the bottom of the rete ridges (the 'bottom cells'). If for any reason the integrins of the lateral cell surface are more susceptible to trypsinization than the basal cell surface-associated ones, then the 'integrin-strong', trypsin-dissociated keratinocytes may correspond to the bottom cells instead of the top cells. This is an exciting possibility, because if this is the explanation, then all the existing data fits nicely together pointing to the location of stem cells at the bottom of rete ridges in both volar and non-volar epithelium. In summary, although at present the location of stem cells in palm epithelium being at the bottom of the deep rete ridges seems to be generally agreed upon, the location of the stem cells in the non-volar epidermis is an interesting and important issue that clearly deserves continued experimentation.

Finally, it is clear that many GFs can bind tightly to extracellular matrix molecules, and thus potentially provide localized effects. Heparan sulfate has been shown to bind granulocyte-macrophage CSF, interleukin 3 (Roberts *et al.*, 1988) and possibly fibroblast GF (Gospodarowicz *et al.*, 1987), and fibronectin appears to bind a transforming GF-like substance (Fava and McClure, 1987). Further, laminin has been shown to contain epidermal GF-like sequences (Panayotou *et al.*, 1989), suggesting a possible direct role for extracellular matrix molecules in regulation of cellular proliferation.

Follicle

Several aspects of the bulge microenvironment are conducive to the protection and maintenance of a stem cell population. This region forms an outgrowth which points away from the hair shaft, and thus it remains with the skin following the trauma of hair plucking (Cotsarelis *et al.*, 1990; Bassukas and Hornstein, 1989). It is a site with a rich vascular (Montagna and Ellis, 1957) and neural (Ishibashi and Tsuru, 1976; Pasche *et al.*, 1990; Halata, 1988) supply. The suggestion has, in fact, been made that follicular sensory innervation is greatest in the region of the bulge (Halata, 1980). Environmental heterogeneity could also come from the arrector pili muscle, which attaches to the bulge region of the follicle. Merkel cells appear to be concentrated in the skin specifically in the vicinity of the bulge (Narisawa *et al.*, 1993a, 1994a, 1994b), where they may interact with and/or arise from follicular stem cells. Merkel cells have also been observed at the sites of developing hair pegs in early fetal skin, and in the vicinity of the bulge at later stages of development (Kim and Holbrook, 1995; Narisawa *et al.*, 1993b; Moll, 1994). Although little is known about extracellular structures in the bulge region, Couchman *et al.* (1991) have noted that a chondroitin sulfate proteoglycan is present in anagen follicles only in two regions: near the hair bulb matrix and dermal papilla, and at the upper outer root sheath in the vicinity of the bulge. Antibodies specific for the low affinity NGF (p75) receptor, EGF receptor and PDGF alpha and beta receptors did not find any receptor presence specific for the hair follicle bulge of human fetal skin (Haake, 1993). Finally, it has been observed (Reynolds and Jahoda, 1991b) that in the rat vibrissa follicle, which undergoes little shortening during catagen phase, there is an asymmetric build up of outer root sheath cells in the lower follicle during the pre-anagen period, involving only the portion of the follicle directly opposite the ascended club hair. This asymmetry suggests a highly coordinated reproductive process, the spatial organization of which undoubtably involves extracellular molecules. However, the nature and location of these putative molecules in relation to bulge stem cells is unknown.

Involution of the lower 'non-permanent' portion of the hair follicle during catagen phase involves extensive programmed cell death or apoptosis (Weedon and Strutton, 1981). As this process occurs, one presumes there are mechanisms to protect the stem cells of the bulge, which lie at the lowermost portion of the 'permanent' portion of the hair follicle, from this involutional process. One candidate molecule to perform such a protective function is the known potent inhibitor of apoptosis, bcl-2 (Hockenberry *et al.*, 1990, 1993). However, recent studies have shown that while the bcl-2 protein is present within the dermal papilla throughout the hair cycle (Stenn *et al.*, 1994; Polakowska *et al.*, 1994), it is observed in the region of the bulge only during anagen (Stenn *et al.*, 1994). Although confusing, these findings coupled with the observation that hair growth is essentially normal in bcl-2-deficient mice (Veis *et al.*, 1993), suggest that the bcl-2 protein may not have a significant functional role in protecting bulge stem cells from apoptosis during the catagen and telogen phases of the hair cycle. In accord with this, levels of p53, a mediator of apoptosis, did not change in whole skin samples throughout the hair cycle (Seiberg *et al.*, 1995).

Finally, the control of TA populations, like the control of stem cells, is probably mediated to a significant extent by local events, although in this larger population regional effects may play a role as well. Stem cells themselves may play a role in regulating the proliferation of TA cells. For instance, mouse embryonic stem cells have been shown to secrete several members of the fibroblast GF family (Heath *et al.*, 1989) which are capable of regulating the proliferation of their differentiated offspring (Heath and Rees, 1985). However, it is as yet unclear how keratinocyte TA populations, in the process of forming a new hair matrix, are able to differentiate into multiple differentiated lines, i.e. inner root sheath, hair cortex and hair medulla. One possibility is the existence of a diffusion gradient of a factor or growth factors. Such a gradient has been shown to exist in a very localized fashion during vulval induction in *Caenorhabditis elegans* (Sternberg, 1988), and in the initiation of primary hair follicles in sheep (Nagoreka and Mooney, 1985).

TUMOUR FORMATION

Corneal epithelium

The vast majority of corneal epithelial dysplasias and neoplasms arise in the region of the limbus (Russell *et al.*, 1956; Pizzarello and Jakobiec, 1978; Waring *et al.*, 1984), which coincides precisely with the location of corneal epithelial stem cells. This finding is in complete agreement with the concept that stem cells are the source of many cancers.

Epidermis

A number of studies have suggested that epidermal stem cells may be the target of chemical carcinogens. In two-stage carcinogenesis protocols (Stenback *et al.*, 1981; Loehrke *et al.*, 1983), a delay of months to years between initiation and promotion results in very little decrease in subsequent tumour development. In a similar fashion, polycyclic aromatic hydrocarbon-DNA adducts have been shown to persist for 5 months in mouse skin (Randerath *et al.*, 1983). Because only stem cells will be retained for this length of time, this result strongly suggests that they are the target of carcinogenic changes. In addition, Morris *et al.* (1986) showed, using double-labelling experiments in mouse skin, that label-retaining cells are the cells that retain labelled carcinogen 1 month after its application. In two other *in vitro* studies, the same group observed that a sub-population of cultured basal epidermal cells exhibiting stem cell characteristics retained labelled carcinogen (Staiano-Coico *et al.*, 1986), and exhibited the highest levels of carcinogen-DNA adducts (Baer-Dubowska *et al.*, 1990).

Barrandon *et al.* (1989), however, have shown that epidermal cells already committed to terminal differentiation, forming so-called 'paraclone' colonies, as discussed earlier, can be transformed by retroviral transduction of an adenovirus early 1A gene, forming immortalized lines. It would thus appear that while epidermal stem cells may be a major target of carcinogens *in vivo*, one cannot exclude the possibility at the present time that TA cells may also, at least under certain experimental conditions, give rise to tumours.

Follicle

It has also been known for some time that hair follicles are an important source of skin cancers (reviewed in (Miller *et al.*, 1993a)). This has been shown in: (1) experiments applying carcinogen to haired versus hairless animals (Giovanella *et al.*, 1970), and to animals in which areas of their hair-bearing skin have been replaced by a scar (Lacassagne and Latarjet, 1946), (2) skin grafting experiments involving initiation of only interfollicular or follicular components (Billingham *et al.*, 1951; Steinmuller, 1971), (3) studies irradiating animals to differing depths, involving interfollicular or interfollicular plus follicular components (Albert *et al.*, 1967a, 1967b, 1969) and (4) hair cycle studies, in which carcinogen is applied to experimental animals during different phases of the hair cycle (Andreasen and Engelbreth-Holm, 1953; Borum, 1954; Berenblum *et al.*, 1958; Whiteley, 1956, 1957; Miller *et al.*, 1993b). As mentioned in the section on interfollicular epidermis, several lines of evidence suggest that stem cells specifically, because of their longevity, are an important origin of epidermal carcinomas (Staiano-Coico *et al.*, 1986; Stenback *et al.*, 1981; Loehrke *et al.*, 1983; Randerath *et al.*, 1983; Morris *et al.*, 1986; Baer-Dubowska *et al.*, 1990). Thus, it is important to consider the follicular stem cells as a possible source of skin cancers. In the category of hair cycle experiments, Miller *et al.* (1993b) have shown that application of carcinogen to mice at the onset of anagen, in either a complete or an initiating dose, produces greater numbers of tumours than application in telogen. Because proliferating cells have been shown to be more susceptible to tumour initiation (Hennings *et al.*, 1968, 1978, 1981; Bowden and Boutwell, 1974; Pound and Withers, 1963), and it has recently been observed that bulge cells cycle only during early anagen (Wilson *et al.*, 1993), Miller *et al.*'s results are consistent with the hypothesis that follicular stem cells reside in the bulge.

CONCLUSIONS

Keratinocyte populations must continuously regenerate, as their terminally-differentiated cells are shed. A highly organized proliferative process has developed in these tissues, involving three subpopulations of cells with differing capacities for proliferation and differentiation: stem cells, TA cells, and terminally differentiated cells. Keratinocyte stem cells, although exhibiting some intrinsically distinct and specialized features among different tissues (Doran *et al.*, 1980), appear to have a number of common characteristics. In this chapter, we have reviewed general stem cell characteristics, as they are known to exist in the haematopoietic population, and then examined their presence in three keratinocyte tissues: the corneal epithelium, the interfollicular epidermis, and the follicle. Although much is known about the reproductive processes in these three tissues, many interesting and incompletely answered questions remain.

One major question is, 'What role do keratinocyte stem cells and TA cells play in diseases of the cornea, skin and hair?' Because corneal epithelial stem cells exist within a discrete region of macroscopic size that may be surgically removed, a clinical model of complete corneal epithelial stem cell absence has been developed (Huang and Tseng, 1991). In this model, delayed wound healing with recurrent epithelial breakdowns, corneal vascularization, and conjunctival epithelial ingrowth result, following removal of the limbal (stem cell-rich) epithelium. Tseng (1989) has theorized, on the basis of these clinical features, that corneal epithelial stem cell deficiency may be a component of a number of ocular surface disorders, including chemical injuries, Stevens-Johnson syndrome, aniridia and some contact-lens-induced keratopathies.

Because of their more dispersed and microscopic locations, little is known with certainty about the role of interfollicular and follicular progenitor cells in disease. The development of molecular markers for stem and TA cells would greatly expand our understanding of the roles these different classes of progenitor cells play in keratinocyte diseases. Our current methods of stem cell assessment depend on either label-retaining capacity (indicating low proliferative rate), or relative cell size and density. The first is a physiological marker, requiring time and the occurrence of cellular reproduction; the second produces only a gradient of separation between keratinocytes of differing proliferative and differentiative potentials. The lack of cellular markers has inhibited experimental ability to 'follow' stem cells through proliferative cycles, or to separate them from their immediate offspring, the TA cells, in healthy or diseased populations.

In summary, the search for keratinocyte stem cells has led to their identification in unexpected sites: the periphery of the cornea, the bottom of epidermal rete ridges, and the upper, outer root sheath of the hair follicle. In each instance, the discovery has provided explanations for many previously unexplained or paradoxical observations, often made over a span of many years by different investigators. This, in turn, has validated the usefulness of the stem cell concept. Much remains to be learned about the progenitor cells of keratinocyte populations: the structural and biochemical basis of the stem cell niche; and the regulatory processes involved in stem cell, TA cell and committed cell functioning. The development of cellular markers, specific for distinct stages in this differentiation process, will greatly improve our understanding of the role these cell populations play in normal and disease states of cornea, skin and hair follicle.

REFERENCES

Adams, J.C. and Watt, F.M. (1989). *Nature* **340**:307–309.
Aizawa, S. and Tavassoli, M. (1987). *Proc. Natl Acad. Sci. USA* **84**:4485–4489.

Aizawa, S. and Tavassoli, M. (1988). *Proc. Natl Acad. Sci. USA* **85**:3180–3183.
Al-Barwari, S.E. and Potten, C.S. (1976). *Int. J. Radiat. Biol.* **30**:201–216.
Albers, K.M., Setzer, R.W. and Taichman, L.B. (1986). *Different.* **31**:134–140.
Albers, K.M., Greif, F., Setzer, R.W. and Taichman, L.B. (1987). *Different.* **43**:236–240.
Albert, R.E., Burns, F.J. and Heimbach, R.D. (1967a). *Radiat. Res.* **30**:515–524.
Albert, R.E., Burns, F.J. and Heimbach, R.D. (1967b). *Radiat. Res.* **30**:590–599.
Albert, R.E., Phillips, M.E., Bennett, P. and Heimbach, R. (1969). *Cancer Res.* **29**:658–668.
Alldridge, O. and Krachmer, J.H. (1981). *Arch. Ophthalmol.* **99**:599–604.
Allen, T.D. and Potten, C.S. (1976). *J. Cell Sci.* **21**:341–359.
Andreasen, E. and Engelbreth-Holm, J. (1953). *Acta Pathol. Microbiol. Scand.* **32**:165–169.
Argyris, T.S. (1976). *Am. J. Pathol.* **83**:329–337.
Baer-Dubowska, W., Morris, R.J., Gill, R.D. and Digiovanni, J. (1990). *Cancer Res.* **50**:3048–3054.
Barrandon, Y. and Green, H. (1985). *Proc. Natl Acad. Sci. USA* **82**:5390–5394.
Barrandon, Y. and Green, H. (1987). *Proc. Natl Acad. Sci. USA* **84**:2302–2306.
Barrandon, Y., Morgan, J.R., Mulligan, R.C. and Green, H. (1989). *Proc. Natl Acad. Sci. USA* **86**:4102–4106.
Bassukas, I.D. and Hornstein, O.P. (1989). *Arch. Dermatol. Res.* **281**:188–192.
Baum, C.M., Weissman, I.L., Tsukamoto, A.S., Buckle, A. and Peault, B. (1992). *Proc. Natl Acad. Sci. USA* **89**:2804–2808.
Ben-Zvi, A., Rodrigues, M.M., Krachmer, J.H. and Fujikawa, L.S. (1986). *Curr. Eye Res.* **5**:105–117.
Berenblum, I., Haran-Ghera, N. and Trainin, N. (1958). *Br. J. Cancer* **12**:402–413.
Bickenbach, J.R. (1981). *J. Dent. Res.* 60:1611–1620.
Billingham, R.E. and Silvers, W.K. (1963). *New Engl. J. Med.* 268:539–545.
Billingham, R.E., Orr, J.W., Woodhouse, D.L. (1951). *Br. J. Cancer* **5**:417–432.
Bishop, G.H. (1945). *Am. J. Anat.* **76**:153–181.
Borum, K. (1954). *Acta Pathol. Microbiol. Scand.* **34**:542–553.
Bowden, G.T. and Boutwell, R.K. (1974). *Cancer Res.* **34**:1552–1563.
Bron, A.J. (1973). *Trans. Ophthalmol. Soc. UK* **93**:455–472.
Buck, R.C. (1979). *Invest. Opthalmol.Vis. Sci.* **18**:767–784.
Buck, R.C. (1982). *Virchows Arch. (Cell Pathol.)* **41**:1–16.
Buck, R.C. (1985). *Invest. Opthalmol. Vis. Sci.* **26**:1296–1299.
Bullough, W.S. and Laurence, E.B. (1960). *Proc. R. Soc. B.* **151**:517–536.
Buschke, W. (1949). *Arch. Ophthalmol.* **41**:306–316.
Buschke, W., Friedenwal, J.S. and Fleischmann, W. (1943). *Bull. Johns Hopkins Hosp.* **73**:143–162.
Butcher, E.O. (1951). *Ann. NY Acad. Sci.* **53**:508–515.
Butcher, E.O. (1964). *Anat. Rec.* **151**:231–238.
Cairns, J. (1975). *Nature* **255**:197–200.
Chase, H.B. (1954). *Physiol. Rev.* **34**:112–126.
Chen, J.J.Y. and Tseng, S.C.G. (1990). *Invest. Opthalmol. Vis. Sci.* **31**:1301–1314.
Christophers, E. (1971). *J. Invest. Dermatol.* **56**:165–169.
Chung, E.H., Bukusoglu, G. and Zieske, J.D. (1992). *Invest. Opthalmol. Vis. Sci.* **33**:121–128.
Clark, R.A.F. (1993). *J. Dermatol. Surg. Oncol.* **19**:693–706.
Clausen, O.P.F., Elgjo, K., Kirkhus, B., Pedersen, S. and Bolund, L. (1983). *J. Invest. Dermatol.* **81**:545–549.
Clausen, O.P.F., Aarnaes, E., Kirkhus, B., Pedersen, S., Thorud, E. and Bolund, L. (1984). *Cell Tissue Kinet.* **17**:351–365.
Compton, C.C., Gill, J.M., Bradford, D.A., Regauer, S., Gallico, G.G. and O'Connor, N.E. (1989). *Lab. Invest.* **60**:600–612.
Cotsarelis, G., Cheng, S.Z., Dong, G., Sun, T.T. and Lavker, R.M. (1989). *Cell* **57**:201–209.
Cotsarelis, G., Sun, T.T. and Lavker, R.M. (1990). *Cell* **61**:1329–1337.
Couchman, J.R., McCarthy, K.J. and Woods, A. (1991). *Ann. NY Acad. Sci.* **642**:243–252.
Davanson, M. and Evensen, A. (1971). *Nature* **229**:560–561.
Dhouailly, D., Xu, C., Manabe, M. and Sun, T.T. (1989). *Exp. Cell Res.* **181**:141–198.
Doran, T.I., Vidrich, A. and Sun, T.T. (1980). *Cell* **22**:17–25.
Dua, H.S. and Forrester, J.V. (1990). *Am. J. Ophthalmol.* **110**:646–656.
Ebato, B., Friend, J. and Thoft, R.A. (1987). *Invest. Opthalmol. Vis. Sci.* **28**:1450–1456.
Ebato, B., Friend, J. and Thoft, R.A. (1988). *Invest. Opthalmol. Vis. Sci.* **29**:1533–1537.

Eckert, R.L. (1989). *Physiol. Rev.* **69**:1316–1341.
Eisen, A.Z., Holyoke, J.B. and Lobitz, W.C. (1956). *J. Invest. Dermatol.* **25**:145–156.
Fava, R.A. and McClure, D.B. (1987). *J. Cell Physiol.* **131**:184–189.
Fitch, J.M., Gibney, E., Sanderson, R.D., Mayne, R. and Linsenmayer, T.F. (1982). *J. Cell Biol.* **95**:641–647.
Friedenwald, J.S. (1951). *Doc. Ophthalmol.* **5–6**:184–187.
Fuchs, E. (1990). *J. Cell Biol.* **111**:2807–2814.
Fujikawa, L.S., Foster, C.S., Gipson, I.K. and Calvin, R.B. (1984). *J. Cell Biol.* **98**:128–138.
Fukuda, O. and Ezaki, T. (1975). *Jpn Plast. Reconstr. Surg.* **18**:109–114.
Furstenberger, G., Gross, M., Schweizer, J., Vogt, I. and Marks, F. (1986). *Carcinogen.* **7**:1745–1753.
Gallico, G.G., O'Connon, N.E., Compton, C.C., Kehind, O. and Green, H. (1984). *New Engl. J. Med.* **311**:448–451.
Germain, L., Torok, N., Lussier, M., Gaudreau, P. and Royal, A. (1994). *J. Invest. Dermatol.* **102**:545–545.
Gilchrest, B.A. (1983). *J. Invest. Dermatol.* **81 (suppl)**:184s–189s.
Giovanella, B.C., Liegel, J. and Heidelberger, C. (1970). *Cancer Res.* **30**:2590–2597.
Goldsmith, L. (1991). *Physiology, Biochemistry, and Molecular Biology of the Skin.* Oxford University Press, New York.
Gospodarowicz, D., Neufeld, G., Schweigerer, L. (1987). *J. Cell Physiol.* **suppl. 5**:15–26.
Graham, G.J., Wright, E.G., Hewick, R. *et al.* (1990). *Nature* **344**:442–444.
Haake, A.R. and Paolakowska R.R. (1993). *J. Invest. Dermatol.* **101**:107–112.
Halata, Z. (1980). In *The Skin of Vertebrates* (eds R. I. C. Spearman and P. A. Riley), pp. 303–307. Adademic Press, London.
Halata, Z. (1988). In *Hair and Hair Diseases* (eds C. E. Orfanos and R. Happle), pp. 149–164. Springer-Verlag, Berlin.
Hall, P.A. (1989). *J. Pathol.* **158**:275–277.
Haskjold, E., Refsum, S.B. and Bjerkness, R. (1988). *Acta Ophthalmol. (Copenh.)* **66**:533–537.
Haskjold, E., Bjerkness, R. and Refsum, S.B. (1989). *Acta Ophthalmol. (Copenh.)* **67**:174–180.
Heath, J.K. and Rees, A.R. (1985). In *Growth Factors in Biology and Medicine* (eds D. Evered and M. Stoker), pp. 3–22. Ciba Symposium (Pitman), London.
Heath, J.K., Paterno, G.D., Lindon, A.C. and Edwards, D.R. (1989). *Different.* **107**:113–122.
Heath, J.K., Smith, A.G., Hsu, L.W. and Rathjen, P.D. (1990). *J. Cell Sci.* **13 (suppl)**:75–85.
Hennings, H., Bowden, G.T. and Boutwell, R.K. (1968). *Cancer Res.* **29**:1773–1780.
Hennings, H., Michael, D. and Patterson, E. (1978). *Proc. Soc. Exp. Biol.* **158**:1–4.
Hennings, H., Devor, D. and Wenk, M.L. *et al.* (1981). *Cancer Res.* **41**:773–779.
Heyworth, C.M., Ponting, I.L.O. and Dexter, T.M. (1988). *J. Cell Sci.* **91**:239–247.
Hockenberry, D., Nunez, G., Milliman, C., Schreiber, R.D. and Korsmeyer, S.J. (1990). *Nature* **348**:334–336.
Hockenberry, D.M., Oltval, Z.N., Yin, X.M., Milliman, C.L. and Korsmeyer, S.J. (1993). *Cell* **75**:241–251.
Horne, K.A., Jahoda, C.A.B. and Oliver, R.F. (1986). *J. Embryol. Exp. Morph.* **97**:111–124.
Huang, A.J.W. and Tseng, S.C.G. (1991). *Invest. Opthalmol. Vis. Sci.* **32**:96–105.
Hume, W.J. and Potten, C.S. (1976). *J. Cell Sci.* **22**:140–160.
Ibrahim, L. and Wright, E.A. (1982). *J. Embryol. Exp. Morph.* **72**:209–224.
Inaba, M., Anthony, J. and McKinstry, C. (1979). *J. Invest. Dermatol.* **72**:224–231.
Ishibashi, Y. and Tsuru, N. (1976). In *Biology and Disease of the Hair.* (eds T. Kobori, and W. Montagna), pp. 73–85. University Park Press, Baltimore.
Jahoda, C.A.B., Horne, K.A. and Oliver, R.F. (1984). *Nature* **311**:560–562.
Jensen, P.K.A., Pedersen, S. and Bolund, L. (1985). *Cell Tissue Kinet.* **18**:201–215.
Johnson, E. and Ebling, F.J. (1964). *J. Embryol. Exp. Morph.* **12**:465–474.
Jones, P.H. and Watt, F.M. (1993). *Cell* **73**:713–724.
Jones P.H., Harper, S. and Watt, F.M. (1995). *Cell* **80**:83–93.
Kam, E., Melville, L. and Pitts, J.D. (1986). *J. Invest. Dermatol.* **87**:748–753.
Kam, E., Watt, F.M. and Pitts, J.D. (1987). *Exp. Cell Res.* **173**:431–438.
Kappenman, K.E., Kawabe, T.T., Waldon, D.J. and Buhl, A.E. (1994). *J. Dermatol. Surg. Oncol.* **102**:533.
Kaufmann, B.P., Cay, H. and Hollaender, A. (1944). *Anat. Rec.* **90**:161–178.

Kaye, D.B. (1980). *Am. J. Ophthalmol.* **89**:381–387.
Keller, G. (1992). *Curr. Opinion Immunol.* **4**:133–139.
Kenyon, K.R. and Tseng, S.C.G. (1989). *Ophthalmol.* **96**:709–723.
Kim, D.K. and Holbrook, K.A. (1995). *J. Invest. Dermatol.* **104**:411–416.
Kim, J.C., Choi, Y.C. (1994). *Stem cells in human hair follicle.* Paper presented at Meeting of American Academy of Cosmetic Surgery. January: Palm Springs, CA.
Kim, J.C. and Choi, Y.C. (1995). *J. Dermatol. Surg. Oncol.* **21**:312–313.
Kinoshita, S., Friend, J. and Thoft, R.A. (1981). *Invest. Opthalmol. Vis. Sci.* **21**:434–441.
Kirsner, R.S. and Ealgstein, W.H. (1993). *Derm. Clin.* **11**:629–640.
Kligman, A.M. (1959). *J. Invest. Dermatol.* **33**:307–316.
Knapp, A.C., Franke, W.W., Heid, H., Hatzfeld, M., Jorcano, J.L. and Moll, R. (1986). *J. Cell Biol.* **103**:657–667.
Kobayashi, K., Rochat, A. and Barrandon, Y. (1993). *Proc. Natl Acad. Sci.* **90**:7391–7395.
Kolega, J., Manabe, M. and Sun, T.T. (1989). *Different.* **42**:54–63.
Kruse, F.E. and Tseng, S.C.G. (1991). *Invest. Opthalmol. Vis. Sci.* **32**:2086–2095.
Kruse, F.E. and Tseng, S.C.G. (1993a). *Invest. Ophthalmol. Vis. Sci.* **34**:2976–2989.
Kruse, F.E. and Tseng, S.C.G. (1993b). *Invest. Ophthalmol. Vis. Sci.* **34**:2501–2511.
Kruse, F.E. and Tseng, S.C.G. (1994). *Invest. Ophthalmol. Vis. Sci.* **35**:2405–2420.
Kruse, F.E., Chen, J.J.Y., Tsai, R.J.F. and Tseng, S.C.G. (1989). *Invest. Opthalmol. Vis. Sci.* **30** (suppl):520.
Kuroki, T. and Murakami, Y. (1989). *Jpn J. Cancer Res.* **80**:637–642.
Kuwabara, T., Perkins, D.G. and Cogan, D.G. (1976). *Invest. Opthalmol.* **15**:4–14.
Lacassagne, A. and Latarjet, R. (1946). *Cancer Res.* **6**:183–188.
Lane, E.B., Wilson, C.A., Hughes, B.R. and Leigh, I.M. (1991). *Ann. NY Acad. Sci.* **642**:197–213.
Lavker, R.M. and Sun, T.T. (1982). *Science* **215**:1239–1241.
Lavker, R.M. and Sun, T.T. (1983). *J. Invest. Dermatol.* **81**:121s–127s.
Lavker, R.M., Cotsarelis, G., Wei, Z.G.and Sun, T.T. (1991a). *Ann. NY Acad. Sci.* **642**:214–225.
Lavker, R.M., Dong, G., Cheng, S.Z., Kudoh, K., Cotsarelis, G. and Sun, TT. (1991b). *Invest. Opthalmol. Vis. Sci.* **32**:1864–1875.
Lavker, R.M., Margolis-Fryer, E., Ostad, M., Cotsarelis, G. and Sun, T.T. (1992). *J. Invest. Dermatol.* **98**:581a.
Lavker, R.M., Miller, S., Wilson, C. *et al.* (1993). *J. Invest. Dermatol.* **101**:16s–26s.
Leblond, C.P. (1981). *Am. J. Pathol.* **160**:114–158.
Lemischka, I.R., Raulet, D.H. and Mulligan, R.C. (1986). *Cell* **45**:917–927.
Lenoir, M.C., Bernard, B.A., Pautrat, G., Darmon, M. and Shroot, B. (1988). *Dev. Biol.* **130**:610–620.
Limat, A., Breitkreutz, D., Hunziker, T. *et al.* (1991). *Exp. Cell Res.* **194**:218–227.
Loehrke, H., Schweizer, J., Dederer, E., Hesse, B., Rosenkranz, G. and Goerttler, K. (1983). *Cell* **41**:771–775.
Lyne, A.G. (1957). *Aust. J. Biol. Sci.* **10**:197–216.
Lyne, A.G. (1970). *Aust. J. Biol. Sci.* **23**:1241–1253.
Mackenzie, I.C. (1969). *Nature* **222**:881–882.
Mackenzie, I.C. (1970). *Nature* **226**:653–655.
Mackenzie, I.C. (1996). *J. Invest. Dermatol.* In press.
Malkinson, F.D. and Keane, J.T. (1978). *Int. J. Dermatol.* **7**:536–551.
Mann, I. (1944). *Br. J. Ophthalmol.* **28**:26–40.
Matsuoka, T., Hardy, C. and Tavassoli, M. (1989). *J. Clin. Invest.* **83**:904–911.
Maumenee, A.E. and Scholz, R.O. (1948). *Bull. Johns Hopkins Hosp.* **82**:121–147.
Menton, D.N. and Eisen, A.Z. (1971). *J. Ultrastruc. Res.* **35**:247–264.
Miller, S.J., Lavker, R.M. and Sun, TT. (1993a). *J. Invest. Dermatol.* **100**:288s–294s.
Miller, S.J., Wei, Z.G., Wilson, C., Dzubow, L., Sun, T.T. and Lavker, R.M. (1993b). *J. Invest. Dermatol.* **101**:591–594.
Miller, S.J., Lavker, R.M. and Sun, T.T. (1993c). *Sem. Dev. Biol.* **4**:217–240.
Misumi, Y. and Akiyoshi, T. (1990). *Acta Anat.* **137**:202–207.
Misumi, Y. and Akiyoshi, T. (1991). *Am. J. Anat.* **191**:419–428.
Moll, I. (1994). *Cell Tissue Res.* **277**:131–138.
Moll, I. (1995). *J. Invest. Dermatol.* **105**:14–21.

Moll, I., Kuhn, C. and Moll, R. (1995). **104**:910–915.
Moll, R., Moll, I. and Franke, W. (1984). *Different.* **28**:136–154.
Montagna, W. and Chase, H.B. (1950). *Anat. Rec.* **107**:83–91.
Montagna, W. and Chase, H.B. (1956). *Am. J. Anat.* **99**:415–435.
Montagna, W. and Ellis, R.A. (1957). *J. Natl Can. Inst.* **19**:451–463.
Montagna, W., Chase, H.B. and Melaragno, H.P. (1952). *J. Invest. Dermatol.* **19**:83–94.
Morris, R.J., Fischer, S.M. and Slaga, T.J. (1985). *J. Invest. Dermatol.* **84**:277–281.
Morris, R.J., Fischer, S.M. and Slaga, T.J. (1986). *Cancer Res.* **46**:3061–3066.
Morris, R.J., Fischer, S.M., Klein-Szanto, A.J.P. and Slaga, T.J. (1990). *Cell Tissue Kinet.* **23**:587–602.
Nagoreka, B.N. and Mooney, J.R. (1985). *J. Theor. Biol.* **114**:243–272.
Narisawa, Y., Hashimoto, K. and Kohda, H. (1993a). *Arch. Dermatol. Res.* **285**:269–277.
Narisawa, Y., Hashimoto, K., Nakamura, Y., Kohda, H. (1993b). *Arch. Dermatol. Res.* **285**:261–268.
Narisawa, Y., Hashimoto, K. and Kohda, H. (1994a). *J. Invest. Dermatol.* **102**:506–510.
Narisawa, Y., Hashimoto, K. and Kohda, H. (1994b). *J. Invest. Dermatol.* **103**:191–195.
Newsome, P.A., Foidart, J.M., Hassell, J.H., Rodrigues, M.M. and Katz, S.I. (1981). *Invest. Opthalmol. Vis. Sci.* **20**:738–750.
O'Guin, W.M. and Manabe, M. (1991). *Ann. NY Acad. Sci.* **642**:51–62.
Oliver, R.F. (1966a). *J. Embryol. Exp. Morph.* **15**:331–347.
Oliver, R.F. (1966b). *J. Embryol. Exp. Morph.* **16**:231–244.
Oliver, R.F. (1967). *J. Embryol. Exp. Morph.* **18**:43–51.
Oliver, R.F. (1970). *J. Embryol. Exp. Morph.* **23**:219–236.
Orentreich, D. and Orentreich, N. (1988). In *Hair Transplantation* (eds W. P. Unger and R. E. A. Nordstrom), pp. 1–36. Marcel Dekker, New York.
Panayotou, G., End, P., Aumailley, M., Timpl, R. and Engel, J. (1989). *Cell* **56**:93–101.
Parkinson, E.K., Grabham, P. and Emmerson, A. (1983). *Cell* **4**:857–861.
Parkinson, E.K., Graham, G.J., Daubersies, P. *et al.* (1993). *J. Invest. Dermatol.* **101**:113–117.
Pasche, F., Merot, Y., Cevraux, P. and Saurat, J.H. (1990). *J. Invest. Dermatol.* **95**:247–254.
Paus, R., Stenn, K.S. and Link, R.E. (1990). *Brit. J. Dermatol.* **122**:777–784.
Paus, R., Stenn, K.S. and Elgjo, K. (1991). *Dermatologica* **183**:173–178.
Pavlovitch, J.H., Rizk-Rabin, M., Gervaise, M., Metezeau, P. and Grunwald, D. (1989). *Am. J. Physiol.* **256**:C977–C986.
Pavlovitch, J.H., Rizk-Rabin, M., Jaffray, P., Hoehn, H. and Poot, M. (1991). *Am. J. Physiol.* **261**:C964–C972.
Pinkus, H. (1958). In *The Biology of Hair Growth* (eds W. Montagna and W. Ellis), pp. 1–32. Academic Press, New York.
Pinkus, H. and Mehregan, A.H. (1981). *A Guide to Dermatohistopathology*, pp. 458–459. New York, Appelton-Century-Crofts.
Pizzarello, L.D. and Jakobiec, F.A. (1978). In *Ocular and Adenexal Tumors* (ed. F. A. Jakobiec), pp. 553–571. Aesculappius, Birmingham.
Polakowska, R.R., Piacentini, M., Bartlett, R. and Goldsmith, L.A. (1994). *Dev. Dynamic* **199**:176–188.
Potten, C.S. (1974). *Cell Tissue Kinet.* **7**:77–88.
Potten, C.S. (1981). *Int. Rev. Cytol.* **69**:271–318.
Potten, C.S. and Allen, T.D. (1976). *Different.* **5**:43–47.
Potten, C.S. and Hendry, J.H. (1973). *Int. J. Radiat. Biol.* **24**:537–540.
Potten, C.S., Hume, W.J., Reid, P. and Cairns, J. (1978). *Cell* **15**:899–906.
Pound, A.W. and Withers, H.R. (1963). *Br. J. Cancer* **17**:460–470.
Quesenberry, P. and Levitt, L. (1979a). *New Engl. J. Med.* **301**:755–760.
Quesenberry, P. and Levitt, L. (1979b). *New Engl. J. Med.* **301**:819–823.
Randerath, E., Agrawal, H.P., Reddy, M.V. and Randerath, K. (1983). *Cancer Lett.* **20**:109–114.
Rathjen, P.D., Toth, S., Edwards, P.R., Heath, J.K. and Smith, A.G. (1990). *Cell* **62**:1105–1114.
Regauer, S. and Compton, C.C. (1990). *J. Invest. Dermatol.* **95**:341–346.
Regnier, M., Vaigot, P., Darmon, M. and Prunieras, M. (1986). *J. Invest. Dermatol.* **87**:472–476.
Reynolds, A.J. and Jahoda, C.A.B. (1990). *J. Invest. Dermatol.* **95**:485.
Reynolds, A.J. and Jahoda, C.A.B. (1991a). *J. Cell Sci.* **99**:373–385.
Reynolds, A.J. and Jahoda, C.A.B. (1991b). *Ann. NY Acad. Sci.* **642**:226–244.

Rheinwald, J. and Green, H. (1975). *Cell* **6**:331–334.
Roberts, R., Gallagher, J., Spooncer, E., Allen, T.D., Bloomfield, F. and Dexter, T.M. (1988). *Nature* **332**:376–378.
Rochat, A., Kobayashi, K. and Barrandon, Y. (1994). *Cell* **76**:1063–1073.
Russell, W.O., Wynne, E.S., Loquvam, G.S. (1956). *Cancer* **9**:1–52.
Sale, G.E. and Beauchamp, M. (1993). *B M Transplant.* **11**:223–225.
Sale, G.E., Shulman, H.M., Gallucci, B.B. and Thomas, E.D. (1985). *Am. J. Pathol.* **118**:278–287.
Sale, G.E., Farr, A. and Hamilton, B.L. (1991). *Bone Marrow Transplant.* **7**:263–267.
Sanes, J.R., Engvall, E., Butkowski, R. and Hunter, D.D. (1990). *J. Cell Biol.* **111**:1685–1699.
Sanford, B.H., Chase, H.B., Carroll, S.B. and Arsenault, C.T. (1964). *Anat. Rec.* **152**:17–24.
Schermer, A., Galvin, S. and Sun, T.T. (1986). *J. Cell Biol.* **103**:49–62.
Schmidt, G.H., Blount, M.A. and Ponder, B.A.J. (1987). *Develop.* **100**:535–541.
Schofield, R. (1983). *Biomed. Pharm.* **37**:375–380.
Schweizer, J., Kinjo, M., Furstenberger, G. and Winter, H. (1984). *Cell* **37**:159–170.
Seiberg, M., Marthinuss, J. and Stenn, K.S. (1995). *J. Invest. Dermatol.* **104**:78–82.
Sengel, P. (1986). In *Biology of the Integument. Volume II. Vertebrates* (eds J. Bereiter-Hahn, A. G. Maltoltsy and K. S. Richards), pp. 374–408. Springer-Verlag, Berlin.
Spangrude, G.J. (1991). *Curr. Opinion Immunol.* **3**:171–178.
Staiano-Coico, L., Higgins, P.J., Darzynkiewicz, Z. *et al.* (1986). *J. Clin. Invest.* **77**:396–404.
Steinert, P.M., Idler, W.W. and Wantz, M.L. (1980). *Biochem. J.* **187**:913–916.
Steinmuller, D. (1971). *Cancer Res.* **31**:2080–2084.
Stenback, F., Peto, R. and Shubik, P. (1981). *Br. J. Cancer* **44**:1–14.
Stenn, K.S., Lawrence, L., Veis, D., Korsmeyer, S. and Seiberg, M. (1994). *J. Invest. Dermatol.* **103**:107–111.
Sternberg, P.W. (1988). *Nature* **335**:551–554.
Sun, T.T. and Green, H. (1976). *Cell* **179**:217–221.
Sun, T.T., Cotsarelis, G. and Lavker, R.M. (1991). *J. Invest. Dermatol.* **96**:77s–78s.
Taylor, A.C. (1949). *J. Exp. Zool.* **110**:77–112.
Thoft, R.A., Wiley, L.A. and SundarRaj, N. (1989). *Eye* **3**:109–113.
Tsai, R.J.F., Sun, T.T. and Tseng, S.C.G. (1990). *Ophthalmol.* **97**:446–455.
Tseng, S.C.G. (1989). *Eye* **3**: 141–157.
Tseng, S.C.G. and Zhang, S.H. (1992). *Invest. Opthalmol. Vis. Sci.* **33**:1177.
Uchida, N., Fleming, W.H., Alpern, E.J. and Weissman, I.L. (1993). *Curr. Opinion Immunol.* **5**:177–184.
Van Scott, E.J. and Ekel, T.M. (1958). *J. Invest. Dermatol.* **31**:281–294.
Van Scott, E.J., Ekel, T.M. and Auerbach, R. (1963). *J. Invest. Dermatol.* **4**:269–273.
Veis, D.J., Sorenson, C.M., Shutter, J.R. and Korsmeyer, S.J. (1993). *Cell* **75**:229–240.
Waring, G.O., Roth, A.M. and Ekins, M.B. (1984). *Am. J. Ophthalmol.* **97**:547–559.
Weedon, D. and Strutton, G. (1981). *Acta Derm. Venereol. (Stockh)* **61**:335–339.
Wei, Z.G., Wu, R.L., Lavker, R.M. and Sun, T.T. (1993). *Invest. Opthalmol. Vis. Sci.* **34**:1814–1828.
Wei, Z.G., Sun, T.T. and Lavker, R.M. (1996). *Invest. Ophthal. Vis. Sci.* In press.
Whiteley, H.J. (1956). *J. Pathol. Bacteriol.* **72**:1–13.
Whiteley, H.J. (1957). *Br. J. Cancer* **11**:196–205.
Wiley, L., SundarRaj, N., Sun, T.T. and Thoft, R.A. (1991). *Invest. Opthalmol. Vis. Sci.* **32**:594–602.
Wilson, C.A., Cotsarelis, G., Wei, Z.G. *et al.* (1994). *Different.* **55**:127–135.
Yang, J.S., Lavker, R.M. and Sun, T.T. (1993). *J. Invest. Dermatol.* **101**:652–659.
Yuspa, S.H., Ben, T., Hennings, H. and Lichti, U. (1982). *Cancer Res.* **42**:2344–2349.
Zieske, J.D., Bukusoglu, G. and Yankauckas, M.A. (1992a). *Invest. Opthalmol. Vis. Sci.* **33**:143–152.
Zieske, J.D., Bukusoglu, G., Yankauckas, M.A., Wasson, M.E. and Keutmann, H. (1992b). *Different.* **151**:18–26.

12 Tumour stem cells

Johann Kummermehr and Klaus-Rüdiger Trott*

*GSF – Forschungszentrum für Umwelt und Gesundheit, Institut für Strahlenbiologie, Neuherberg, Germany; *Department of Radiation Biology, St. Bartholomew's Medical College, London, UK*

THE PROLIFERATIVE STRUCTURE OF MALIGNANT TUMOURS

There is evidence that malignant tumours arise by malignant transformation from a single cell which has the capacity for unlimited proliferation, i.e. a stem cell (Buick and Tannock, 1992). The persistent growth of cancers, clonal diversification and evolution, tumour metastasis and tumour recurrence after therapy, are all indicators that some tumour cells in the clinically manifest cancer have retained the proliferative potential of the stem cell from which the tumour originated in the first place. Those cells which have the capacity of starting a distant metastasis or a local recurrence after nearly curative treatment by clonal expansion fulfil all criteria of a stem cell.

There is a characteristic difference between the concept of stem cells in tumours and stem cells in normal tissues. In *tumours*, stem cells may be defined as the biological units which have the potential of giving rise to a new tumour, either by metastasis or by tumour recurrence after therapy. This definition provides a much more rigorous test for identifying stem cells than is available in most normal tissues except for the bone marrow stem cells CFU-S. In *normal tissue* stem cells are defined on the basis of a compartimentalized system of constant cell renewal with cells progressing through a series of differentiation and maturation steps, and by their capacity to maintain the steady state of cell production, cell differentiation and cell loss although it is only by severe perturbation of the steady state that they can be tested.

In order to clarify the concept of stem cells with regard to the proliferative organization of tumour cell populations it may be useful to contrast two extreme models.

Model A, the *all stem cell model*, argues that stemness is conveyed in all stem cell divisions and hence is a feature of all viable tumour cells. The reason that in cell kinetic measurements the growth fraction usually is less than 1, and that cell loss may be high is thought to occur through micro-environmental factors, i.e. nutritional constraint. The most simple and straightforward morphological example is the tumour cord. In many transplantable animal tumours, cuffs of parenchymal cells can be seen that are apparently supported by a central capillary (Tannock, 1968). This geometrical arrangement and the thickness of the viable rim of tumour cells is in agreement with the critical diffusion range of oxygen in metabolizing tissues. As a regular finding, the labelling index is highest in the most central, pericapillary layer, often compatible with a growth fraction

STEM CELLS
ISBN 0-12-563455-2

of one. The decrease in labelling index as one moves outward is interpreted as enforced arrest, due mostly though not exclusively to the fall of oxygen concentration (Tannock, 1968). At the periphery of the cuff, a hypoxic zone must be assumed through which cells pass before they eventually undergo necrosis. Cells that leave the proliferative state may become resting stem cells or also may lose their clonogenicity, as a consequence of the environmental situation. The population of a cuff is virtually in steady state, with cell production in the inner layers being the driving force.

This concept may, indeed, reflect the true situation in some anaplastic tumours. It explains the observed existence of subpopulations, in terms of proliferating and non-proliferating cells, even of clonogens and non-clonogens and of hypoxic cells. In some transplantable rodent tumours all viable looking cells have the capacity for unlimited proliferation (Hewitt, 1958). However, this may not apply to all tumours, in particular not to tumours with a differentiated morphology.

Model B is based on the concept of *hierarchical proliferation* which is characteristic for all steady state normal tissues (Chapter 1). Normal renewal tissues, e.g. the stratified epithelia, maintain their steady state by a balance between cell production and cell loss. Whatever the details of further compartmentalization, to keep up this equilibrium and to maintain the tissue at all requires a constant supply of cells that are produced by stem cell divisions. On average, a dividing stem cell will give rise to one stem cell and one differentiating daughter cell which may perform several transit divisions. Whether this asymmetry is reproduced in every single stem cell division or must be regarded as an average is of no concern for the overall result. Under steady state conditions, the partitioning factor p or probability of self-maintenance must be 0.5; it may temporarily expand when trauma or cytotoxic insult have depleted the stem cell pool but it will fall back to 0.5 once the damage has been repaired. Likewise, it follows that at the stem cell level, a p value of 0.5 is synonymous with a cell loss factor of 100% which is a necessary condition for all tissues in steady state. In certain conditions when demand exceeds self-replication the cell loss factor may even rise to $>100\%$, temporarily.

The hierarchical stem cell model of tumour growth proposes that neoplasms might be regarded as stem cell systems in which a minority of cells have the proliferative capacity to maintain the tumour whereas the majority of cells demonstrate differentiation features and have limited proliferative potential (Bush and Hill, 1975; Tepper, 1981). Obviously, the tumour stem cell concept derives primarily from studies of normal renewing cell populations. It is consistent with the view that tumours arise primarily in the stem cells of the respective tissue by a series of transformation steps which leave the tumour with some sort of memory of its original proliferative organization, presenting the tumour as a caricature of its tissue of origin.

This model has been developed further mostly by Mackillop *et al.* (1983) who proposed that the crucial event in the development of cancer is an irreversible, heritable change that confers on a stem cell a probability of self-maintenance of greater than 0.5. This will result in an expanding clone of stem cells and this may constitute the neoplastic growth. The concept that within neoplasms the cell renewal hierarchy of normal tissues continues to function predicts the existence of three categories of cells (Figure 1):

(i) stem cells (S);
(ii) non-stem transitional cells with limited proliferative capacity (T);
(iii) end cells (EC).

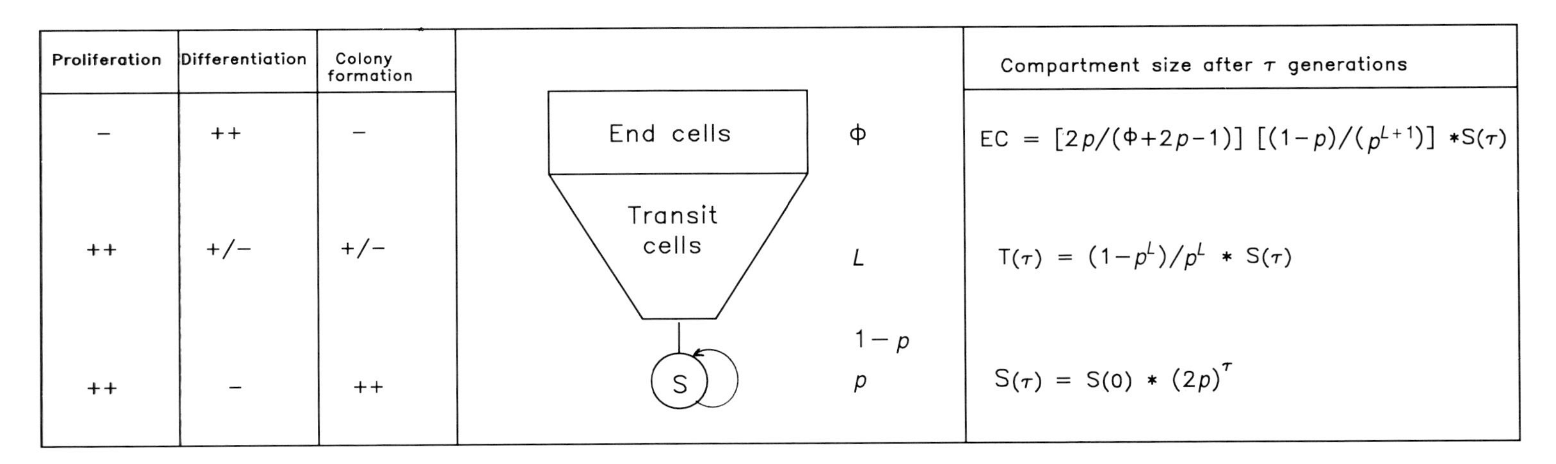

Figure 1 Schematic representation of the hierarchical organization of tumour cell proliferation, modified after Mackillop *et al.* (1983). S, stem cell.

The relative size of these populations is mostly determined by two factors:

(i) The probability of self-maintenance of stem cells (p).
(ii) The potential of the non-stem, transitional cells for further cell division as defined by the clonal expansion number L (i.e. the number of generations between the first generation non-stem cells and the end cells).

Figure 2 demonstrates how the proportion of tumour stem cells in an established tumour depends on both factors provided no specific, preferential cell loss occurs from any of the various subpopulations. For the most simple case in which all non-stem cells do not proliferate any more and are end cells, the proportion of stem cells increases linearly with increasing probability of self-maintenance. The more transitional cell divisions a non-stem cell performs before becoming an end cell the smaller is the proportion of stem cells in the total tumour cell population. On the other hand, the proportion of end cells depends primarily on the probability of self-maintenance, the closer p is to 0.5, the greater is the proportion of end cells (Table 1). For a value of 0.55, 90% of the tumour cell population are end cells; for a value of 0.75, 50% of the tumour cell population are end cells. Since the formation of organ-specific structures which is the main criterion of the histopathological grading of tumours requires the presence of a sufficient number of differentiated end cells, tumour grade may be related to the probability of self-maintenance in this way. The growth rate of a tumour is independent of the number of transit cell divisions and only related to p (Figure 3). Any random loss of cells by necrosis, migration or differentiation would not affect the relative proportion of tumour stem cells, tumour transit cells and tumour end cells as they are

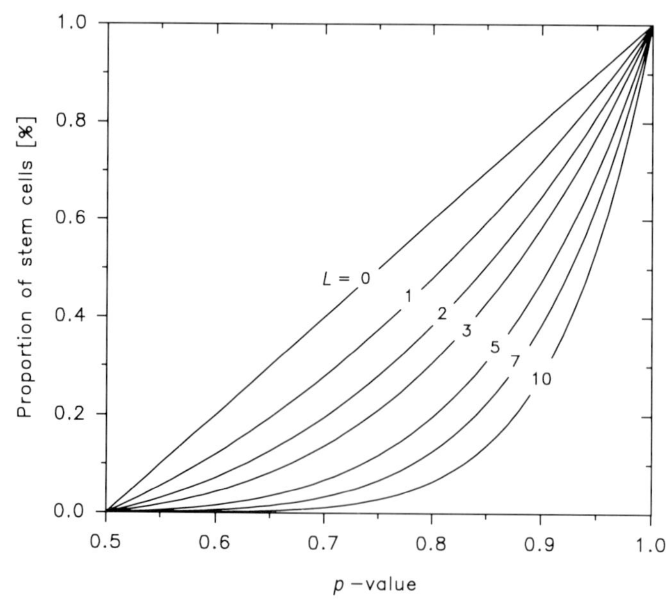

Figure 2 The proportion of stem cells in an established tumour population as a function of stem cell self-renewal probability (p) and number of transit generations (L) assuming no selective cell loss from any of the compartments.

defined by p and L. Whereas a selective loss from the end cell compartment, e.g. necrosis of post-mitotic cells has little influence on the proportions of the different tumour cell categories, selective loss from the stem cell compartment – although this appears an unlikely event – would decrease the proportion of stem cells in the total tumour cell population dramatically.

Table 1 The relationship of tumour cell category to p and L assuming no preferential cell loss from any compartment; modified after Mackillop *et al.* (1983).

p	L	Stem cells, %	Transit cells, %	End cells, %
0.55	0	10	0	90
	1	5.5	4.5	90
	2	3.0	7	90
	3	1.7	8.3	90
	4	0.9	9.1	90
0.75	0	50	0	50
	1	38	12	50
	2	28	22	50
	3	21	29	50
	4	16	34	50

p, probability of self-maintenance.
L, number of transit generations.

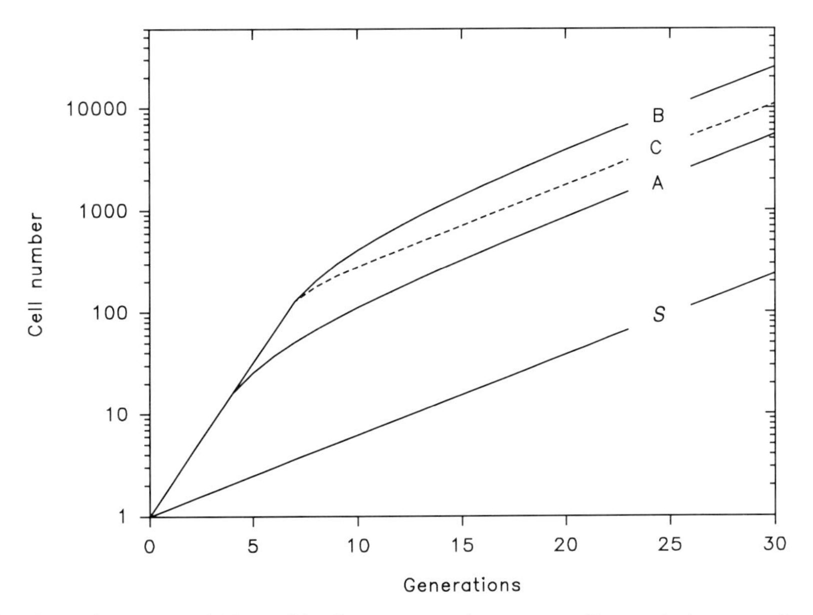

Figure 3 Growth curves of hierarchically structured tumour cell populations starting from single stem cells and gradually reaching equilibrium with non-stem cells. Curve labels denote the stem cell number (S), the total cell number for numbers of transit generations $L=3$ (A) of $L=6$ (B) and total cell number for $L=6$ but loss of end cells with a cell loss factor $\Phi=0.5$ (C).

These predictions of the simple stem cell model of tumour growth may also become distorted if the generation times are different for stem cells and transit cells which they may well be as has been observed in the bone marrow.

In normal tissues, the crucially important property of stem cells is self-maintenance. This cannot be measured directly but is implied by the fact that a steady state of cell production and loss of terminally differentiated cells exists throughout adult life. In modelling this steady state it has to be assumed that 50% of the progeny of stem cell divisions remain stem cells and 50% progress through the transit compartment into terminal differentiation. This could be achieved by either asymmetrical stem cell divisions leading to one stem cell and one non-stem cell at each division or as a mean value of a random process, or by symmetrical divisions leading always to two stem cells with removal of 50% of the new stem cells per generation by differentiation-inducing signals before the next stem cell division. Neither in normal tissues nor in tumours can it be decided which of the potential mechanisms is actually operating. Yet because of the observed flexibility of stem cell responses to any disturbance of the steady state we favour the second option. The probability of self-maintenance as defined above does not imply any particular biological mechanism but is a purely mathematical description of the behaviour of cell populations.

One particular feature of stem cell models of proliferation and differentiation is the definition of a particular cell which is not yet a fully differentiated cell but no longer a stem cell but in transit from one to the other, progressing on an essentially one-way street to terminal differentiation. As extensively discussed by Potten and Loeffler (1990) *dividing transit cells* would thus be expected to share some properties with stem cells and some properties with mature cells, with stem-like properties declining as the mature-like properties increase. The fundamental question is, however, to which extent the acquisition of more mature-like properties is necessarily associated with the loss of stem-like properties and whether any of these steps in the phenotypical progression is reversible or not. Although this problem applies to all renewal tissues in view of the disorganized structure of a typical tumour, this problem may be particularly pertinent to tumour transit cells.

Most published models of normal tissue turnover kinetics assume irreversible progression along a one-way street into terminal differentiation. Though this may be true for the normal steady state, Potten and Loeffler (1990) suggested that some experimental results on stem cell responses of the intestinal mucosa to a cytotoxic insult might be better explained by a more flexible model of gradual loss of stemness with progressive differentiation, i.e. that transit cells possess a progressively declining *spectrum* of stemness (which may be unrelated to their proliferative activity) and an increasing spectrum of differentiation and maturation.

The proliferative capacity of an immature transit cell may be considerable such as the early transit cells in the bone marrow which have the potential for more than ten divisions to produce more than 1,000 progeny, although they are clearly committed to differentiation and have lost their stemness. They may even satisfy some of the criteria used to define stemness such as self-maintenance over a short period of several cell cycles or colony-forming ability *in vitro*. Their limits become apparent if they are called into extensive renewal, such as tissue regeneration after severe cytotoxic injury or tumour recurrence after potentially curative therapy. There is evidence that in some human and rodent tumours, transit cells possess a similarly extensive proliferative capacity.

The parameters commonly used to describe the proliferation kinetics of tumours such as tumour doubling time and growth fraction (Steel, 1977) can be related to the parameters which govern the hierarchical organization of tumour growth in the unperturbed state. In the simplified model described above, the tumour growth rate (1/tumour doubling time) depends on the cycle time of tumour stem cells (but not on the cycle time of tumour transit cells) and on p (Figure 3). The tumour volume doubling time is defined as

$$T_d = \ln 2 / \ln 2p \times T_{c(\text{stem cell})}$$

The *tumour growth fraction* is the sum of the proportion of tumour stem cells and tumour transit cells which, in turn, is defined by p alone as

$$GF = 1 - 2(1 - p)$$

The cell loss factor Φ was defined by Steel (1977) by the discrepancy of the tumour volume doubling time and the potential doubling time. It is, however, not well defined in the hierarchical model of tumour cell proliferation. Since the model predicts that most proliferating cells are transit cells the potential doubling time is mostly defined by the generation times of the transit cells and the growth fraction. If the generation time of transit cells is shorter than the generation time of stem cells (which appears very likely) a cell loss factor Φ results even if not a single cell gets lost from the entire tumour cell population. If no cell loss occurs, the proportion of stem cells in the total cell population is

$$(2p - 1)p^L$$

In a real tumour in which cell differentiation as well as nutritional insufficiency may play a role and in which cells in different compartments may have different generation times and different cell loss factors, these simple predictions may be substantially modified (although little mathematical modelling has been done on this problem yet). The principles of the hierarchical model of tumour cell proliferation remains, however, valid.

In the following discussion we want to examine the evidence that, indeed, malignant tumours in rodents and in man contain a population of cells which meet the criterion of a tumour stem cell, which is the ability of a single cell to form a metastasis or to form a local recurrent tumour after treatment. Furthermore we shall examine which proportion of tumour cells have the functional potential of stem cells in various rodent and human tumours and the implication this has for the proliferative organization of the tumour. This will be followed by a discussion of the clonal evolution and differentiation of a tumour recurrence from a single surviving stem cell. Finally, the implications of a stem cell model of tumour growth on the response of tumours to various types of treatment and its importance for research into cancer biology and cancer treatment will be addressed.

THE IDENTIFICATION AND MEASUREMENT OF TUMOUR STEM CELLS

Experimental demonstration of stemness in individual cells and quantification of stem cells requires severe disruption of the unperturbed population. Such disruption can be induced by massive cytotoxic insult and the response then assessed *in situ*, in terms of tumour recurrence rate. The ability to regenerate a tumour is the ultimate proof of unlimited proliferative capacity. The *in situ* assessment necessarily falls back to an all or none response but has the attractive advantage that the performance of the functional stem cell is tested under conditions that are immediately relevant from a therapeutic point of view.

The criterion of unlimited regenerative capacity is also tested if the disaggregated tumour cell population is transplanted into a host animal. In a mathematical sense, the response obtained, take or failure, is similar to that after *in situ* treatment, i.e. cure or recurrence: the graft will either succeed or fail, and the response hence is an all or none, or quantal response. It is obvious, however, that otherwise the two assays provide very different settings of the experimental procedures and environments to which the potential stem cells are exposed and which may greatly influence their ability or readiness to act as functional stem cells.

An even more artificial setting is created when the disaggregated cells are seeded *in vitro* and their unlimited proliferative potential tested by a colony formation assay. The term 'unlimited' has a relative meaning, here. A successful colony is commonly defined by a progeny of some 50 cells, a number that corresponds to no more than 5.5 effective doublings. In principle the observation period could be extended and the threshold be raised to deliberately high numbers of doublings. However, there are some inherent limitations, e.g. if colony formation is tested in soft agar which is the most suitable technique to grow cells from solid tumours, colony growth usually ceases at a certain size of a few hundred cells. What is in fact demonstrated by these assays, therefore, is clonogenicity rather than stemness (Steel and Stephens, 1983). The colony forming cells are more aptly defined as 'clonogenic cells' or 'clonogens'. There are many possibilities how the colony forming efficiency (CFE) *in vitro* may under- or overestimate the fraction of functional stem cells *in vivo*. Overestimation may result from the inclusion of transit cells with extensive but not unlimited proliferative capacity. On the other hand, the CFE may be an underestimate due to suboptimal growth conditions offered *in vitro*. With increasingly improved culture conditions and enriched media, however, it is conceivable that potential stem cells (e.g. early, proliferating transit cells) may become functional stem cells that would not have done so *in situ*. Moreover, disaggregation and *in vitro* plating of treated or untreated tumours may rescue cells that would have perished *in situ* for nutritional or other micro-environmental reasons. The prevailing opinion is that apart from well adapted tumours, i.e. systems frequently passaged from *in vivo* to *in vitro* and back, stem cell numbers tend to be underestimated by *in vitro* plating, due to these factors (Denekamp, 1994).

In vitro colony formation

To prepare a single cell suspension from a solid tumour following aseptic excision, the tissue is minced using scalpels, scissors or razor blades and further disaggregated in buffered enzyme solutions. A great variety of enzyme cocktails and incubation conditions have been used dependent on tumour type and personal experience. The cell yield is usually below 10% and often below 1% which raises serious questions about whether the recovered cell population is, indeed, representative of the original tumour cell population *in vivo* (Pallavicini, 1987). Useful reviews of the different methods can be found in textbooks (e.g. Masters, 1991). To get rid of the enzymes which are aggressive to cell membranes, the resulting suspension is spun down by centrifugation, the pellet is resuspended in saline and, to remove cell clumps, the resuspended pellet is then filtered through an appropriate mesh or forced through injection cannulae of decreasing gauge. The proportion of viable cells is counted by a dye exclusion test. Microscopic examination of the final suspension is imperative to verify the presence of a proper single cell suspension if the results are intended to be strictly quantitative. The concentration of cells in the suspension is counted and serial dilutions prepared to seed a defined cell number into culture vessels. The addition of an excess number of feeder cells, i.e. heavily irradiated cells, is often helpful to increase the CFE. During the subsequent incubation period which may run from 8 days to 5 weeks, a fraction of the seeded cells will give rise to colonies. Inevitably, colony sizes will vary even within untreated cells. The usual cut-off is a threshold of 50 cells but this may have to be adjusted to the tumour line used.

Every single step in this procedure may have a greater influence on the outcome than the actual proportion of stem cells in the investigated tumour. These include the method of disaggregation, type of culture vessel and its surface coating, the addition of feeder cells, incubation conditions such as temperature, oxygen and carbon dioxide tension, and medium constituents. Ideally, the CFE from *in vitro* plating should reflect the number of cells that would have also been performing as stem cells *in situ*. In addition to the technical problems mentioned above, there may also be biological reasons such as clone extinction that would make the relationship between the proportion of stem cells and CFE less straightforward.

Some transplantable rodent tumours have CFE values as high as established cell lines *in vitro*. Examples are the mouse tumours RIF, EMT6, B16, SSK, Lewis lung and the rat tumour R1 all of which have been developed over many cycles of alternating *in vivo/in vitro* passaging in order to optimize their colony forming efficiency (Steel and Stephens, 1983). While for the study of many basic biological questions they may be desirable, they are not suitable experimental models for examining the proliferative structure of spontaneous rodent or human cancers or their responses which are closely related to their proliferative structure.

There are quite a number of published data on the colony forming efficiency of primary human tumour cell cultures. West *et al.* (1989) reported on the CFE values of 54 primary explants of carcinoma of the cervix (Figure 4). Their mean CFE was 0.22($\pm$0.08)% but they found a wide range of values from 0.003% to 4.3%. There was no apparent correlation between radiosensitivity and colony forming efficiency of these cancer cells. The largest collection of CFE values from human cancer biopsies has been reported in ovarian cancers. Whelan *et al.* (1991) summarized the results of 575 investigations listed in 11 publications in which the range of obtained CFE values were given. The highest CFE value was 1%, all studies observed CFE values as low as 0.01% or even lower. Table 2 lists the range of CFE values recorded in some of the bigger studies on primary human cell cultures, based on different methods: mostly the Hamburger–Salmon assay, but also more refined assays such as the Courtenay–Mills assay were employed, while other authors used simple monolayer techniques. All available CFE data from primary human cancer cell cultures indicate that, almost independent of the refinement which went into the tissue culture techniques, the colony forming efficiency of primary explants of human tumour biopsies of different histology is very low at around 0.1% and only rarely exceeds 1%.

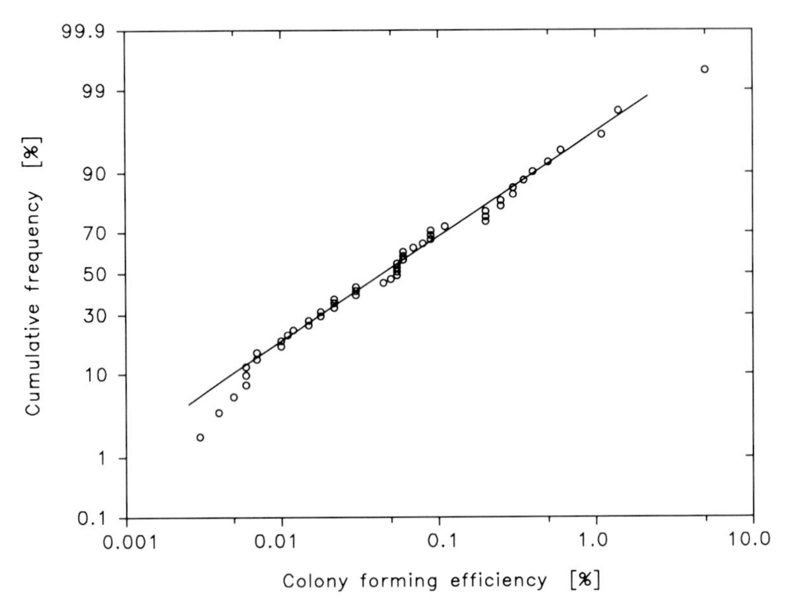

Figure 4 The cumulative frequency distribution of CFE values measured by West *et al.* (1989) for carcinoma of the cervix.

Table 2 CFE values of human cancers in primary culture.

Tumour type	Number of values	Assay type *)	Range, %	References
Ca cervix	54	C+M	0.003–4.3	West *et al.*, 1989
Ca ovary	575	H+S	0.001–1	Whelan *et al.*, 1991
SCLC lung	45	H+S	0.05–1.5	Carney *et al.*, 1980
	3	C+M	0.0002–2	Walls and Twentyman, 1985
Ca breast	99	H+S	0.001–0.3	Sandbach *et al.*, 1982
	12	improved H+S	0.39±0.05	Besch *et al.*, 1986
Ca kidney	8	Mono	0.001–0.3	von Lieven *et al.*, 1978
Glioblastoma	13	Mono	0.06±0.03	Rosenblum *et al.*, 1978
Melanoma	285	C+M	0.1–20	Tveit *et al.*, 1988

CFE, colony forming efficiency.
*C+M, Courtenay–Mills test.
H+S, Hamburger–Salmon test.
Mono, direct monolayer plating.

More information on the proportion of cells with stem-like properties can be derived from re-plating experiments such as those reported by Buick and Mackillop (1984) and Bizzari and Mackillop (1985). Colonies grown in soft agar from 25 primary human tumours were isolated and dispersed into single cells and their colony forming ability determined (re-plating efficiency or PE2). Although there is evidence that each colony, in particular the larger colonies are the clonal offspring of one clonogenic cell the vast majority of the cells in these colonies are not clonogenic but a small proportion of them retain their clonogenic potential. This indicates that the clonal expansion *in vitro* in some way reflects the hierarchical organization of tumour cell proliferation *in vivo* as studied in more detail on page 386.

Multicellular spheroids

Many tumour cells including those from human cancers can readily be grown *in vitro* as multicellular spheroids. Growth can be initiated and sustained in spinner culture, but a technique frequently used nowadays is cultivation in vessels coated by agar or agarose to prevent attachment. This enables the response of individual spheroids to treatment to be studied. In addition, the spheroids can be disaggregated and both number and radiosensitivity of the constituting clonogens can be measured in a conventional colony formation assay. In parallel, the relationship between gross therapeutic response of the spheroid and cellular parameters have been studied including the number of effective spheroid rescuing units (SRUs). This endpoint is analogous to cure experiments in tumours with spheroid survival or eradication tested by observing successful outgrowth of cells from treated spheroids that have been placed into uncoated vessels to allow attachment (Carlsson and Nederman, 1989). The actual criterion adopted to define spheroid survival varies considerably, e.g. from 100 cells in 3 weeks (Kuwashima *et al.*, 1988) to $>1{,}000$ cells in 5 weeks (Rofstad *et al.*, 1986).

In an early study using V79 Chineses hamster cells good evidence was provided that every viable cell surviving radiation was also capable of rescuing the spheroid. The fraction of SRUs was virtually 1 (Durand, 1975) which was further demonstrated by the coincidence of the clonogen survival curve and the dose response curve of spheroid survival after irradiation of spheroids over a considerable range of sizes (Figure 5).

While these conclusions were reached by a direct comparison of surviving clonogen number and gross spheroid response, more complex analyses have sometimes introduced ambiguity. For example it has been argued that the cellular D_0 derived from the cure curve itself bears greater relevance as it more immediately reflects radiosensitivity of the last surviving cells *in situ* (Moore and Hendry, 1984). However, on closer examination such measurements may be critically affected by heterogeneity. Any variability in the number of initial cells per spheroid is translated as a greater variance of the individual dose point of the clonogen survival curve but with no systematic bearing on the curve slope. The same heterogeneity would, however, in the spheroid control experiment result in a shallower slope of the cure curve which, in turn would result in overestimation of the true cellular D_0 and underestimation of the number of stem cells.

Two large sets of spheroid control data have been published together with the parameters of cellular radiosensitivity. From these, the fraction of SRUs in small spheroids of a few thousand cells can be calculated. For four malignant melanomas, studied by Rofstad *et al.* (1986), the estimated fraction of SRUs ranged from 1% to 100%. For 6 soft tissue sarcomas and 11 glioblastomas studied by Stuschke *et al.* (1995), the estimated fraction of SRUs ranged from 3% to 100%.

These studies suggest that the fraction of cells with stem-like properties (i.e. SRU) in these *in vitro* spheroid systems is very much higher than found in direct tumour biopsies tested for colony formation *in vitro*. This reflects the fact that all of these spheroids were

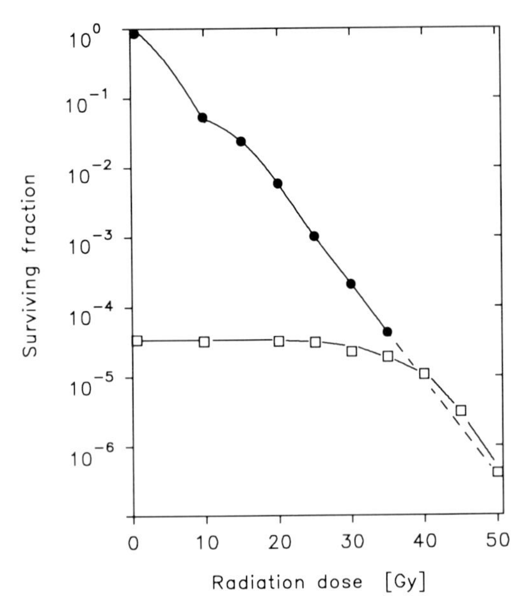

Figure 5 The cellular survival curve for large spheroids (21 d) established from V-79 Chinese hamster cells determined either directly by single cell plating and colony formation (●) or by spheroid survival (□) (Redrawn using data from Durand, 1975.)

established using cells which had been grown over several passages and selected for growth *in vitro*. Therefore it is not surprising that with regard to SRU fraction they are more like the CFE of established cell lines than the CFE of primary human cancers.

Quantitative transplantation

With the assay of quantitative transplantation, graded numbers of cells derived from a cell suspension as described above are injected into a host animal to generate a tumour graft. The site has to be readily accessible to inspection and palpation. The usual choice is the subcutaneous space in the flank, back or leg, or the gastrocnemius muscle. Depending on the number of cells injected and on their growth rate, a tumour may become palpable after a variable lag period. Formation of a tumour will be an all or nothing event, i.e. the graft will either succeed or fail. The rationale of the assay is that whenever the graft succeeds, one or more stem cells must have been present in the inoculum while failure indicates that not a single stem cell was present or was grafted. Apparently, the frequency of stem cells in an inoculum is a function of the total number of injected cells, along with the stem cell fraction.

Since the dilution process is stochastic, the actual number of stem cells in any inoculum must be randomly distributed around the mean, according to binomial statistics, or in the practical case of high dilution factors, to Poisson statistics. The simple prediction from this is that the frequency of no stem cell being present (p_0) is defined by the term

$$p_0 = \exp^{-\mu}$$

with μ being the arithmetic mean of the number of stem cells per inoculum. Thus, when on average the inoculum contains 1 stem cell, the probability that a particular inoculum contains no stem cell at all, will be e^{-1} or 37%. Similarly no stem cell is to be expected in 50% of the cases when the mean value is ln 2 (=0.69) while the remaining 50% will contain one or more stem cells to make up for the arithmetic mean.

When the stem cells constitute only a fraction of all inoculated cells the probability p_0 is given by the equation

$$p_0 = \exp^{-Nz}$$

where N is the total cell number and z denotes the stem cell fraction, the quantity we are considering here. In practice, the mean take dose (TD_{50}) is obtained from a mathematical fit to the actual take rates as a function of injected cell numbers. From these considerations, a theoretical take curve is expected with a steepness predicted by the Poisson distribution in the case of one cell take kinetics. The probability of 0 stem cells in an inoculum stays small as long as the mean number is greater than 5 and rises to assume the maximum steepness when the mean stem cell number of 1 per inoculum is reached and then asymptotically approaches 100% (Figure 6).

A large number of data is available on the TD_{50} values of experimental tumours and to some extent also on the actual take kinetics. The range observed for the TD_{50} is extremely wide (Figure 7). Hewitt *et al.* (1976) reported on the TD_{50} values of 27 spontaneous tumours which arose in their mouse colony, all of which were proven to be non-immunogenic. The TD_{50} values ranged from 1 to 17,000. Hill (1987) showed that there was close correlation between the TD_{50} values determined by transplantation and the CFE value determined by the soft agar method from the same cell suspensions prepared from 12 autochthonous C3H mouse tumours. In a compilation of the data

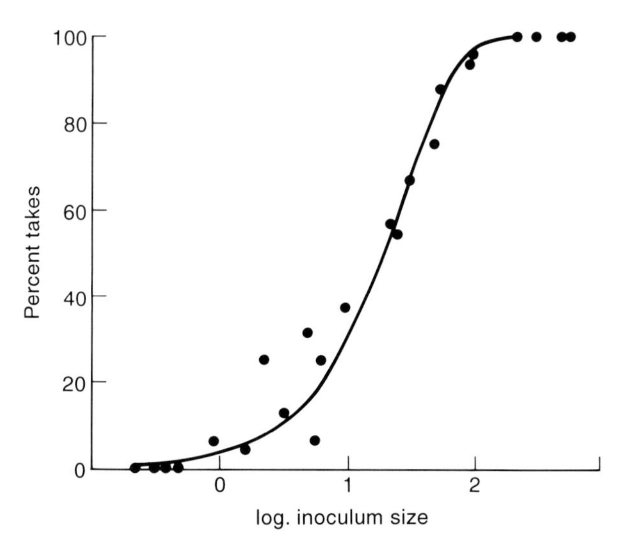

Figure 6 The transplantation kinetics of single cell suspensions prepared from the squamous cell carcinoma 'D'. The fitted curve is a cumulative Poisson distribution, the TD_{50} is 20 cells. (Redrawn using data from Porter *et al.*, 1973.)

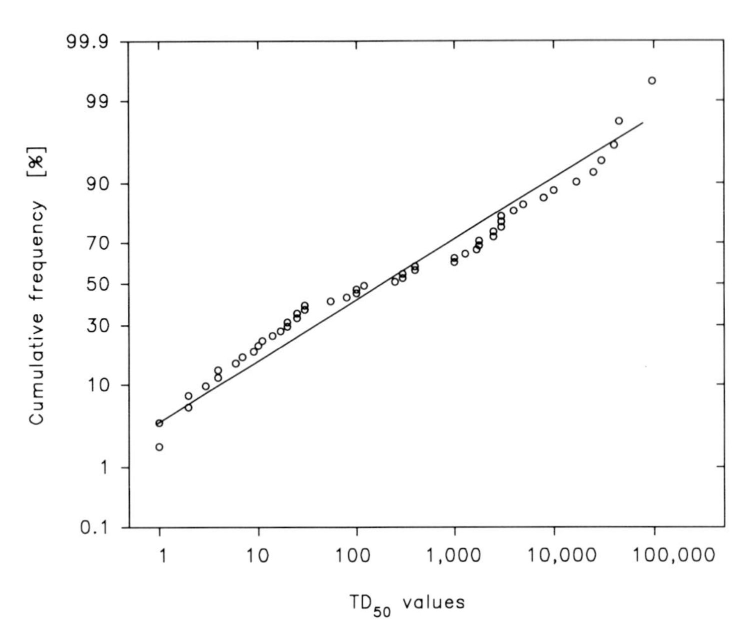

Figure 7 The cumulative frequency distribution of TD_{50} values determined in 52 different mouse tumours pooled from data of three laboratories (Hewitt *et al.*, 1976; Hill and Milas, 1989).

for 25 different mouse tumours from two institutions Hill and Milas (1989) found the TD_{50} values ranging from 2 cells to nearly 100,000 cells. In addition to the significant negative correlation between TD_{50} and CFE values they also reported a significant negative correlation between TD_{50} and the mean curative radiation dose TCD_{50} (see page 378) (Figure 8). They interpreted these observations as a reflection of pronounced differences in the stem cell content of the different tumours even though the actual proportion of stem cells in the tumours was not known. Although tumours which had been passaged several times tended to have lower TD_{50} values, i.e. a higher stem cell content, there were several primary tumours (indicated by solid symbols in Figure 8) which had TD_{50} values <10, i.e. they had more than 10% functional stem cells.

A less universally applicable transplantation assay uses injection of a tumour cell suspension into the venous bloodstream to let the tumour cells be trapped in the lung where they form visible nodular metastases. To facilitate lodging in the capillaries and to augment the yield, feeder cells and microspheres may be added to the inoculum, the tolerable burden being around 10^6 feeder cells and microspheres each. At autopsy, the superficial nodules can be counted readily. In some tumours, also the spleen is an organ where artificial metastases can be induced by injection into the venous bloodstream (Hill, 1987). The information these techniques yields are not all or nothing but rather comparable to the CFE *in vitro*. Using well standardized systems, this method yields clean data provided the effective range of inoculum sizes is not exceeded when saturation effects such as confluence of nodules become apparent. In general, the results of the lung or the spleen colony assay yield stem cell fractions which are smaller by a factor of 10 compared to subcutaneous injection in the quantitative transplantation assay.

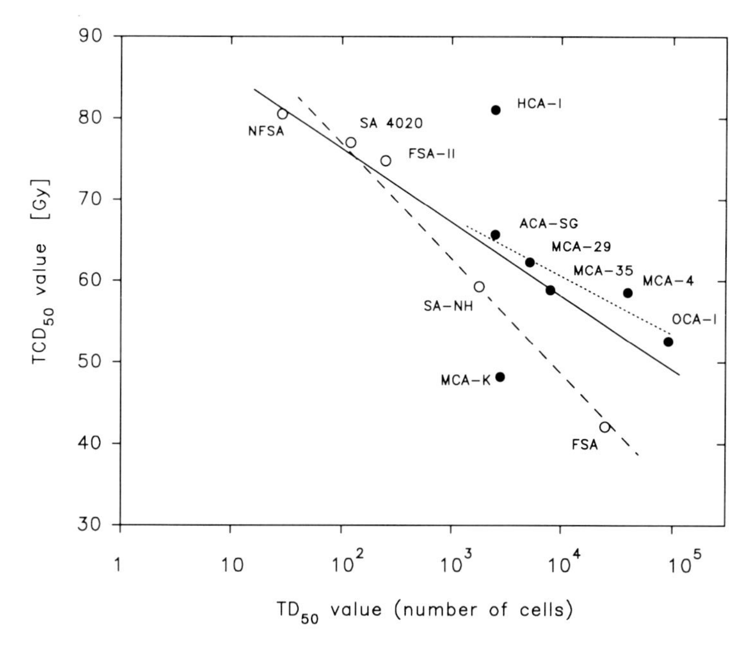

Figure 8 The mean curative radiation doses (TCD_{50}) for 7 carcinomas (●) and 5 sarcomas (○) plotted versus the TD_{50} values of the same tumours. Separate regression lines (dotted) are drawn for the two tumour types as well as one solid line for all tumours. This showed a significant correlation ($p<0.0005$) (Redrawn using data from Hill and Milas, 1989.)

Porter *et al.* (1973) discussed extensively the take kinetics in the quantitative transplantation experiments. The steepest curves which were actually obtained were compatible with the expectation of 1-cell Poisson take kinetics (Figure 6). This rules out the possibility that a tumour take required the presence of two or more interacting stem cells, the mathematical reason for this being that the probability of failure would then become

$$\Sigma P_0 + P_1 \cdots + P_{N-1}$$

From this equation an increasingly steeper curve follows than for p_0 alone. In reality, such curves have not been observed. Also, biologically the interaction of two or more stem cells in a inoculum would be difficult to explain given their spatial distance that would be inevitable when true single cell suspensions were injected.

Conversely, in roughly 50% of all cases, the take curves had slopes that were significantly less steep than predicted by one-cell kinetics. This finding was not correlated with the immunogenicity of the tumour (Porter *et al.*, 1973). While a biological explanation is not immediately at hand, a satisfactory mathematical fit to the data is obtained by attributing a power function to z:

$$p_0 = \exp(-Nz^k).$$

In many cases, k assumes a value around 0.7 or even 0.5. The latter exponent would imply that stem cells graft with a potential which is proportional to the square root of their number. A simple reduction of the grafting probability of any transplanted stem cells fails to explain this finding. As long as the take probability does *not differ between* individual hosts, a grafting probability <1 would simply become expressed as a smaller value of z, i.e. a smaller stem cell fraction. In other words, the Poisson distribution itself would be retained and only its mean would be altered as has been demonstrated by Hewitt *et al.* (1976). The situation changes, however, if there is heterogeneity in the recipient animals. This could be the case when the hosts raise an immunological response that varies between individuals. The survey of transplantation data shows indeed that shallow transplantation kinetics prevail with chemically induced systems. A stringent test of such interindividual variability is provided by the use of multiple sites in each host. Four sites can be used in the mouse model without difficulty although this may require excision of early takes or at least censoring of the data (Hill, 1987).

Of special interest is a study using quantitative autologous transplantation of human cancer cell suspensions into the identical, terminally ill cancer patient (Southam and Brunschwig, 1961). A total of 59 cancer patients with incurable disease was involved, including 36 cases of ovarian adenocarcinoma, 19 squamous cell carcinomas, mostly of the uterine cervix and 4 sarcomas. Tumour material was obtained at surgery or extensive biopsy and brought into cell suspension by techniques that can be regarded as adequate. Multiple inocula were injected subcutaneously into the frontal thigh ranging from 10^4 to 10^8 tumour cells. The overall take rate was only 16/59, in spite of the large number of injected cells, i.e. 12/36 in ovarian cancer, 4/10 in squamous cell carcinomas and 0/4 in sarcomas. There was a definite dependence on inoculum size, although the take kinetics were quite shallow both inter- or intra-individually. The latent period between injection and palpability was usually brief, with a median of around 2–4 weeks. Undeniably, many objections can be raised against these data, e.g. the fact that the observation period was too limited in many patients and that the statistical methods did not meet modern standards. The results do, however, indicate that large cell numbers are required for autologous transplantation of human tumours if methods are used that are comparable to those used in well-controlled animal experiments.

In situ stem cell assay: local tumour control

A functional assay information on tumour stem cells which appears immediately relevant is based on the analysis of the dependence of local tumour control on radiation dose. There is a direct analogy to the quantitative transplantation assay: a tumour can regrow into a recurrence after cytotoxic treatment only if at least one stem cell has survived or, conversely, cure must result when not a single surviving stem cell is left. The crucial difference, however, is that in this test stem cells are left to regrow in their natural environment. This, in turn, undergoes major post-treatment alterations that apart from disintegration of dead tumour cells also include damage to the vascular and connective tissues of the tumour, the tumour 'bed'. The assay therefore addresses the *functional* stem cell, responsible for tumour regrowth, in the conceivably adverse local growth conditions and possibly also subject to host defences raised in the post-irradiation period. Thus, the setting in which stem cell numbers assessed differs profoundly from those discussed on pages 370–378 but, unlike those, it addresses the complex therapeutic situation more directly.

The required complete inactivation of millions of stem cells in rodent cancers is readily achieved by X-rays but hardly by cytotoxic drugs. The experimental procedure therefore involves local irradiation using an adequate range of doses and the observation of the irradiated animals over an extended period of time to ascertain the proportion of tumours that are permanently controlled. Tumour systems with a propensity to loco-regional or distant metastasis are less suitable for the assay as dissemination may have occurred prior to treatment. This would lead to a loss of animals which is difficult to correct for by censoring if the frequency of dissemination is substantial because the required assay time is in the order of months. In many fast growing tumours the observation period ranges from 3–5 months, but it has to be defined for the individual system from the distribution of latent times to recurrence. Design and procedures of tumour control experiments have been described in detail by Suit *et al.* (1987a, 1987b).

The biological interpretation and mathematical evaluation of tumour recurrence rate is identical to that of the take rate in a transplantation experiment. Treatment failure or positive tumour graft are both evidence of the same event, i.e. survival of one or more stem cells that subsequently express their growth potential by giving rise to a huge progeny. Given the stochastic nature of cell inactivation by ionizing radiation, the probability of cell survival, or of no cell surviving should follow Poisson statistics as it does in physical serial dilution.

A factor that is often assumed to influence the number of stem cells assessed from cure experiments is the histological damage occurring in irradiated tumours. Massive cell death of tumour cells may lead to overt necrosis. In addition, radiation injury to the vascular and connective tissue components of the tumour produces additional alterations which have been shown to decrease the growth rate of tumour recurrences after therapy (i.e. the 'tumour bed effect', Milas, 1987). Although impressive, such changes are not effective in sterilizing tumour stem cells that survived the radiation insult in the first place. Experimental evidence comes from a study by Kummermehr and Buschmann (1987) in which graded numbers of untreated sarcoma cells were injected into autologous sarcomas or carcinomas that one week earlier had received a sterilizing radiation dose. Although the time of injection coincided with the onset of massive coagulative necrosis, the injected cells grafted as well as cells injected in parallel into the unirradiated subcutaneous site.

In experimental studies where the heterogeneity of the tumour population can be relatively well-controlled, the dose relationship of local control follows a typical sigmoidal curve which rises after a large threshold dose due to the large number of tumour stem cells that are initially present (Figure 9). Using the above definition that local control reflects the probability of no single stem cell surviving, a cure rate of 37% or e^{-1} should be accomplished when, on average, one stem cell survives. Similarly, a control rate of 50% should be associated with an average number of ln 2 (=0.69) cells. To increase the cure rate from 10% (2.3 stem cells on average) to 90% (0.1 stem cell on average) accordingly requires a further decrease of cell survival by a factor of 20, i.e. e^{-3}. As cellular radiosensitivity is commonly defined by D_0, i.e. the dose that decreases the surviving fraction to e^{-1}, the dose increment needed to increase the cure rate from 10% to 90% should be 3 D_0. Provided the tumour population is very homogenous, the shape of the dose response relationship should permit the derivation of both, the parameter of cellular radiosensitivity D_0 and the number of functional tumour stem cells.

In reality, however, the slope of even experimental cure curves are often found to be considerably shallower than expected from the D_0 values that can reasonably be inferred

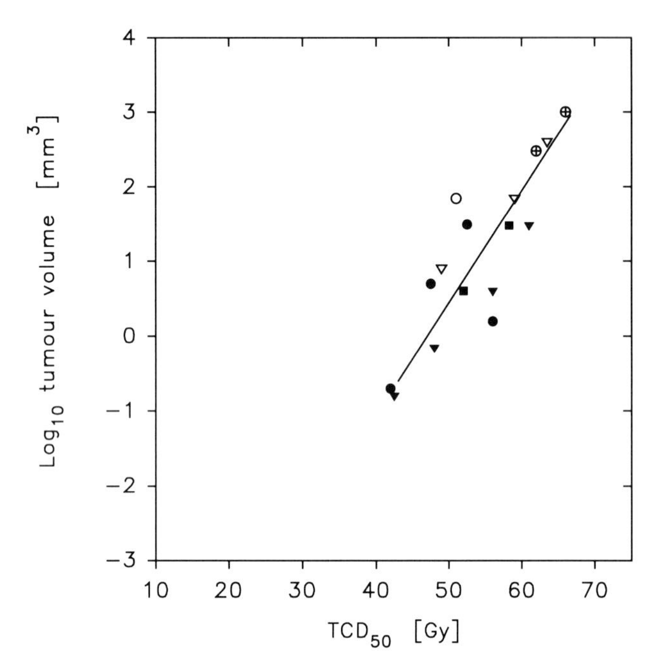

Figure 9 The dependence of the mean curative radiation dose on the tumour volume in a mouse mammary carcinoma (data of Suit *et al.*, 1965). Using the multitarget-single hit model with the extrapolation number ranging from 1 to 6, a value of D_0 of 3.0–3.2 Gy was calculated.

or which have been obtained by independent methods, e.g. by *in vitro* plating. The reason for this phenomenon is pronounced heterogeneity either in the initial number of stem cells or in their radiosensitivity. Variable oxygenation is a major factor in this regard and a broad distribution of hypoxic stem cell fractions can render quite shallow dose response curves. As a consequence, the number of stem cells at risk will be underestimated from such dose response curves. This is particularly evident if clinical data are subjected to this sort of analysis. Hendry (1992) reported from such straightforward analysis apparent stem cell numbers in human cancers as low as 20 stem cells in a typical advanced carcinoma of the head and neck.

Thus, due to inter-tumour heterogeneity of stem cell numbers and radiosensitivity, the stem cell fraction cannot be estimated from the slope of dose response curves of local tumour control. However, in any given transplantable tumour line, stem cell numbers wll be roughly proportional to total tumour mass. It is for this reason that larger tumours are more difficult to cure and require a higher radiation dose. If other confounding factors of radioresistance are excluded (such as variable oxygenation) a comparison of the mean curative doses (TCD_{50}) in tumours of different sizes can be used to estimate the absolute stem cell numbers.

In a classical experimental study, Suit *et al.* (1965) submitted mouse mammary carcinomas which had been transplanted to the ear lobe or to a subcutaneous site in the flank at a mass ranging from 0.2 mm^3 to 1,000 mm^3 to single radiation doses given under local clamp hypoxia. The results (Figure 10) demonstrate a linear increase in the

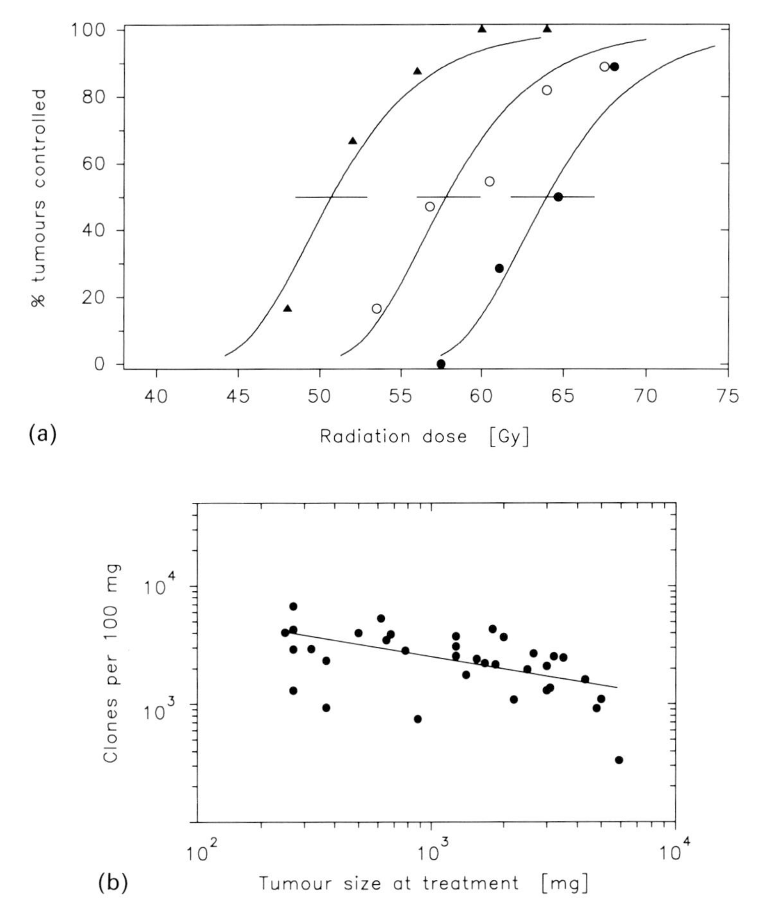

Figure 10 Local control curves of the adenocarcinoma AT17 irradiated at three different sizes (▲, 35 mg; ○, 330 mg; ●, 4200 mg). The lower panel shows the dependence of the stem cell density (stem cells per mg) in this tumour. Data from Guttenberger *et al.* (1990).

TCD_{50} from 42 to 65 Gy. From the measured differences in volume a D_0 of 3.2 Gy for the hypoxic stem cells was calculated. From the same regression analysis, the mean stem cell concentration was estimated to be approximately 10^6 functional stem cells per mm^3 tumour mass, which is 100% of the viable tumour cell number.

The uncertainty in this calculation is twofold. Over a change in tumour volume that covers more than three decades, the proportion of stem cells and the cellularity may well change systematically, in particular it may decrease in those large tumours in which necrosis develops. As a consequence, the stem cell radiosensitivity would be underestimated. Another problem is the mathematical back-extrapolation to zero dose as this requires assumptions on the shape of the dose effect relationship. Suit *et al.* (1965) used the single-hit multi-target model which graphically presents as a shouldered exponential curve. The linear-quadratic curve which is more commonly used nowadays would render considerably smaller back-extrapolates of stem cell numbers, the precise number depending on the α/β value used for these calculations.

Guttenberger *et al.* (1990) performed a similar study in the mouse carcinoma AT17 which was designed to overcome some of the problems encountered in the study of Suit *et al.* (1965). Tumour cell radiosensitivity was determined both from the tumour size dependent increase of TCD_{50} values and from cell survival curves of *in situ* tumour clonogens (page 389). Using the *in situ* assay, it was found that the stem cell concentration in this tumour fell slowly over a tumour volume range of two orders of magnitude (Figure 10). Also shown are the dose cure curves for tumours of 35, 330 or 4300 mg mass including histologically verified subclinical recurrences. From these data, a D_0 of 2.7 Gy was calculated whereas the *in situ* cell survival curve (see Figure 14) had a D_0 of 2.8 Gy. By back-extrapolation according to different dose response models approximately 3×10^6 tumour stem cells are calculated for the 100 mg tumour.

The principal advantage of analysing the impact of tumour size on radiocurability is its wide applicability that extends also to clinical material whenever sufficiently well-defined isoeffective doses of local control for tumours of different sizes are available. However, such data which should, preferentially, come from one single institution, are surprisingly rare. One data set comprises human skin cancers, mostly basaliomas, which have been analysed by Trott *et al.* (1984). A considerable proportion of the tumours had been treated by single dose irradiation which has an advantage with regard to stem cell estimation; as for fractionated radiotherapy, stem cell repopulation during the treatment course can confound the results. On the other hand, if curability by single dose radiation were dominated by the presence of radioresistant hypoxic cells, the derived stem cell number would denote a particular sub-population and not the larger critical target cell population to fractionated radiotherapy. The derived dose response curves for local control after single doses are shown in Figure 11. The two curves represent small tumours with a mean diameter of 8 mm and large tumours with a mean diameter of 32 mm. Taking into account the discoid shape of these primary skin cancers, the volume ratio was estimated to be 1:16. The dose response curves demonstrate greater radioresistance of the larger tumours which also demonstrate greater heterogeneity as reflected in the

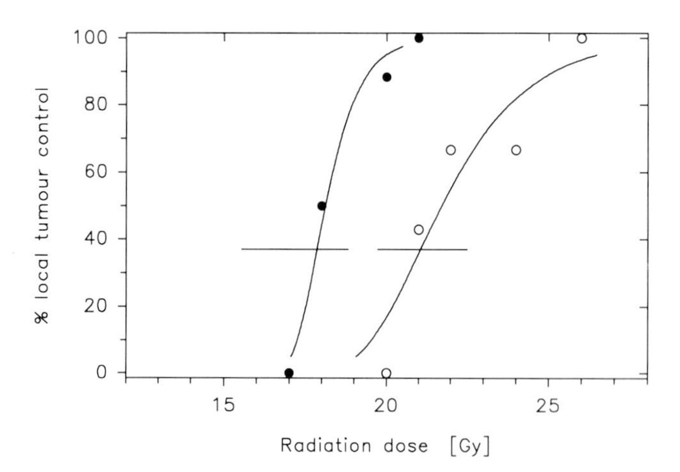

Figure 11 The dependence of local tumour control on single radiation dose given to skin cancers of different size, i.e. mean diameters of 8 mm (●) or 32 mm (○). Data from Trott *et al.* (1984).

shallower curve slope. Using the argument that tumour sterilization depends on the inactivation of all stem cells and assuming a Poisson distribution as mentioned earlier, the TCD_{37} must represent the dose where on average one stem cell survives. The incremental dose needed for the larger tumours to accomplish this cure rate of 37% is 3.5 Gy. As the difference in stem cells initially at risk should be proportional to the volume ratio, i.e. 1:16, this incremental dose should provide an additional cell kill by a factor of 16. From these data, a D_0 of 3.5 Gy/ln 16 = 1.2 Gy is calculated, denoting the cellular sensitivity of the very last surviving tumour stem cells. Since their D_0 is typical for oxic cells, the authors concluded that the last surviving stem cells which gave rise to a local recurrence after curative radiotherapy, were not hypoxic.

From these data, the number of stem cells in these skin cancers can be estimated, yet the results depend on the dose response model that is assumed to adequately describe their survival curve *in vivo*. Two alternative models have been used: the single-hit multi-target model with an extrapolation number of 4 calculated an apparent number of tumour stem cells of $1/Ne^{TCD37/D_0}$, i.e. $1/4e^{14}$ or 10^6. Using the linear-quadratic model, and an α/β ratio of 8 Gy as was derived from cure data of these tumours treated with multifraction radiotherapy, the critical cell number was 10^5. Since the total number of parenchymal tumour cells is approximately 10^9 per ml in these cancers, the fraction of functional stem cells has been estimated to be between 0.1 and 1%.

A similar estimate can be made from radiotherapy cure data on carcinomas of the larynx of different diameters published by Hjelm-Hansen *et al.* (1979). The calculated number of tumour stem cells for these squamous cell carcinomas of 20 mm diameter is approximately 10^6.

Conclusion: the dependence of the fraction of clonogenic tumour cells or tumour stem cells on the assay method

Very different values of clonogenic fractions have been observed with the different techniques described on pages 370–383. However, it is obvious that these differences depend less on the method of assessment than on the tumour which is investigated. Established tumour cell lines, whether tested as cultures, spheroids or by transplantation or by tumour cure, tend to show high clonogenic/stem cell fractions which may reflect the effort put into the establishment of a tumour cell line with optimal CFE and growth rate. Primary human tumours usually have much lower values in all stem cell/clonogenic cell assays. Primary rodent tumours are very variable with regard to their stem cell fraction, some apparently having 100% stem cells, whereas others look more like human tumours. The good correlations found between CFE and stem cell fractions determined by quantitative transplantation (Hill, 1987) and between stem cell fractions and tumour control doses in a number of tumours (Hill and Milas, 1989) suggests that the large differences in clonogenic fractions *in vitro* are not only attributable to the technical problems of culturing human cancer cells *in vitro*. We conclude, therefore, that the large variation of clonogenic fractions and of tumour stem cell fractions in rodent and human tumours reflects a genuine heterogeneity of stem cell fractions in these tumours.

Since human cancers yields CFE values *in vitro* of approximately 0.1%, TD_{50} values of $>1{,}000$ tumour cells and stem cell fractions as determined by the tumour size dependence of the tumour control dose of <1%, we suggest that human cancers characteristically have tumour stem cell fractions of less than 1%.

CLONAL EXPANSION OF THE TUMOUR STEM CELL

Most measurements on tumour stem cells and population structure are carried out in populations that have completed many cell cycles after growth of the malignant population has started. Over this time, an equilibrium has formed between the various sub-populations, the term equilibrium here simply denoting that over a limited period of observation the proportion of stem cells and non-stem cells will not vary very markedly.

On pages 364–369 this situation was explained for a hierarchically structured tumour cell population, with a relatively low fraction of stem cells forming the basis of all population growth. The data reviews on pages 371–383 have given evidence that in human tumours but also in differentiated animal tumours the stem cell fraction is indeed low. Along with the substantial growth fractions that are commonly measured in such tumours this finding supports the view that hierarchical organization is a realistic concept for tumour cell populations.

A 'mature' tumour cell population will be the result of clonal expansion of one parental malignant stem cell. Such a cell may arise *de novo* from transformation, or may have been left behind after major cytotoxic insult or may be a metastatic cell. In the case of successful further growth it will expand into a structured population through a dynamic process defined as *clonal expansion*. This particularly interesting process has been modelled by Mackillop *et al.* (1983) for a larger population of individual stem cells. If one looks at the fate of an individual stem cell, an additional aspect has to be taken into consideration. If proliferation occurs in a probabilistic way, there is a finite chance of a dividing stem cell producing two non-stem cell daughters and thus becoming extinct. This factor is of major importance for the fate of the individual stem cell and may also impact on the quantitation of stem cell numbers in any of the assays mentioned.

These theoretical considerations will form the background to an analysis of experimental data which give support to the concept of hierarchical organization and clonal expansion. The evidence is derived from a particular transplantable tumour system in which survival and expansion of stem cells can be quantitated by an unique *in situ* assay of intratumoural colony formation. Although constituting another assy of stem cell measurement (and therefore belonging in the previous section) the details and results of this experimental approach will be given here for greater consistency.

Hierarchical proliferation and clonal expansion

As has been shown on pages 365–369, the proportions of stem cells, transit cells and end cells in a mature hierarchical cell population can be defined by relatively simple equations in which the probability of stem cell self-maintenance p, plays a central role. These relationships hold when the number of generations elapsed is large. We will now view the early steps of clonal expansion, following the model of Mackillop *et al.* (1983) and, as was done by the authors, we will disregard clonal extinction at present. This is justifiable when the model is thought to describe the average course of evolution in many units of expanding clones. In this situation stem cell growth will be defined by

$$S(\tau) = (2p)^{\tau}$$

where τ represents the number of cell cycles. Apparently, growth is exponential, with the rate depending on the base $2p$ ($\leqslant 2$) and is constant as long as p retains a fixed value. In parallel, non-stem cells will be generated with the probability $1-p$ or q in the average stem cell division. As they will double in each subsequent transit divisions the proportion of transit cells will grow relative to the stem cells present, depending on q and the number of transit generations (L). Mathematically, the number of first generation transit cells must be

$$T_1 = 2(1-p)S_{(\tau-1)}$$

and that of second generation transit cells

$$T_2 = 2T_{1(\tau-1)}$$

and so forth until

$$T_L = 2T_{L-1(\tau-1)}$$

The total number of transit cells, when $\tau \gg T_c$ thus will become

$$T_{\tau} = (1-p^{L})/(1-p) \cdot (1-p)/p^{L} \cdot S_{\tau}$$

The evolution of end cells (EC) is readily calculated, as they derive from the T_L class of transit cells. After $\tau \gg L$ they will have become

$$EC(\tau) = S(\Phi/2p)^{\tau}(1-p)/p^{L+1}S(\tau)$$

The increase in stem cell and total cell number is depicted in Figure 3. While the stem cells multiply exponentially with a rate defined by p, the total cell number doubles over the first cycles, regardless whether or how many transit cells will form, i.e. regardless of p. This changes when the time elapsed (measured in cycles) exceeds the maximum number of transit generations. At this time end cells begin to emerge and to be produced continuously. The overall growth curve now bends over and gradually a balance betwee stem cells, transit cells and end cells will be reached. The overall growth rate will now become the same as that of the stem cell compartment. Mathematically this is reflected in the presence of the factor $S(\tau)$ in all the equations defining the number of T and EC cells. The actual proportion of end cells will also depend on their cell loss factor, defined as the fraction of end cells present that become lost per cycle. However, one must keep in mind that this regular time course, in particular the exponential growth of stem cells over the first steps of expansion, is an average taken over many individual stem cells. The fate of the individual stem cell can be described only in a probabilistic way.

Stem cell extinction

Stem cell replication in the above model apparently proceeds in a probabilistic way. As just stated, for larger starting stem cell numbers the equation therefore gives a correct approximation of their numerical increase, but for the individual stem cell the stochastic nature of proliferation has different implications. Whenever the probability that two stem cell daughters are produced is smaller than 1.0, a finite chance must exist that the division results in two non-stem cell daughters. In this case the parental cell becomes extinct as a stem cell, with the probability of this event being defined by

$$\Omega = (1-p)/p$$

There are also consequences for the early growth of stem cells and non-stem cells. In *surviving* clones growth is somewhat different from what is shown in Figure 3. To make up for the loss of stem cells in clones that become extinct and to maintain an overall partitioning factor of p, in *surviving* clones initially more stem cell and less differentiating daughters (T1 transit cells) must be produced than predicted by p or $1-p$, respectively. As a consequence, stem cell growth in a surviving clone is not strictly exponential initially but will soon become so after a few rounds of stem cell divisions. The 'hump' in the overall growth curve due to the emergence of transit and end cells will develop somewhat more gradually and it will be superimposed to a baseline of stem cells that remains permanently elevated.

Experimental evidence of clonal expansion

Clonal expansion of single surviving stem cells must occur in every tumour that has been treated by near curative doses of radiation. Yet, neither stem cell survival nor clonal expansion can readily be studied *in situ*, for reasons that are technical or rather reside in the histological pattern of tumour regression and regeneration. In order to enable recognition of viable clonal progeny as discrete colonies several conditions must be fulfilled. The dead but morphologically intact cells must clear away sufficiently fast to make the progeny of survivors visible, while proliferation of the survivors has to proceed in a reasonably demarcated way (and not diffusely). Moreover, to allow quantitative measurement the clonal outgrowth should proceed in a sufficiently regular temporal pattern. The vast experience of radiobiology with transplantable tumours suggests that these criteria are usually not met. A rare exception has been reported by Marsden *et al.* (1980) concerning a rat chondrosarcoma. However, the observable dose range was quite limited and the response not reproducible enough to make this a convincing system where clonal regrowth or clonal survival could be studied quantitatively.

To date, quantitative studies on clonal regrowth *in situ* have been successful only in one transplantable system (Kummermehr, 1985). This tumour, mammary carcinoma AT17, arose in our Neuherberg C3H mouse colony and is probably radiation-induced. The volume growth rate is relatively slow, with a volume doubling time of 5–7 days at a tumour mass of 100 mg and necrosis is remarkably absent up to tumour masses of several grams. Histologically, the structure is consistent with an adenoacanthoma, a relatively rare form of differentiated mammary adenocarcinoma. The fact that clonal regrowth can unambiguously be measured in this tumour is due to a particular pattern of damage expression and regeneration after major cytotoxic insult. This is demonstrated in Figure 12. Following a large radiation dose, the epithelial tumour nodules undergo rapid clearance or transformation into acellular debris/keratin, leaving a tumour mass that is largely depleted of parenchymal cells within 7–9 days; interstitial cells clear away

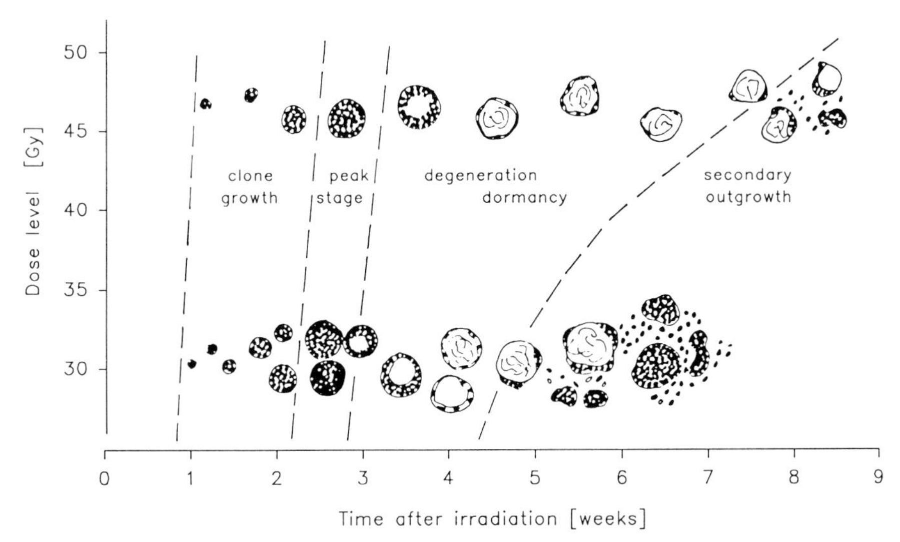

Figure 12 Schematic time course of clone growth and involution or transient degeneration in carcinoma AT17, shown at two different radiation dose levels (Kummermehr, 1985).

definitely slower but have mostly gone by 2–3 weeks. Within this mass, small buds of basophilic epithelial cells appear around day 8 and subsequently increase rapidly in size and number. There is no tendency to coalesce as even nodules in close vicinity stay well demarcated. As a result, after 21/2 weeks postirradiation, discrete spheroids of densely packed epithelial cells have formed (Figure 13A). The distribution of colony profiles in histological sections suggests no dependence of colony size on dose, and similarly the time course of their evolution is almost identical after all doses tested. In contrast, the frequency of these colonies, as related to treatment volume, decreases in a stringent dose-dependent manner. Given the vast dilution of surviving cells by the high radiation doses given the majority of regrowing colonies must have arisen from single cells.

Remarkably, after this phase of rapid regeneration further regrowth is interrupted. From day 20 onwards the clones undergo central differentiation and cell death in a perfectly synchronous wave, with differentiation spreading from the core to the periphery, until by day 29 only a narrow viable rim of flattened epithelial cells is left (Figure 13B). This is followed by a phase of dormancy that is dose-dependent and after high (but still subcurative) doses may last many weeks. Regrowth starts from these structures by outgrowth of epithelial cells into the vicinity with deposition of irregular satellites and eventually the original histology is restored.

This transient involution is insufficiently understood at present, but from retreatment experiments a nutritional crisis can be excluded. As long as the emerging clones are re-irradiated within the first 14 or 18 days, i.e. before degeneration sets in, a second wave of regrowth is started in quite the same time pattern, resulting in perfectly compact clones after 2.5 weeks and a similar profile distribution at that time as was originally seen in the primary clones. Notwithstanding this complex pattern the initial wave of clone evolution must reflect clonal expansion of single surviving tumour clonogens. Clone frequency as a function of radiation dose can therefore be used to establish *in vivo* cell survival curves from histological sections.

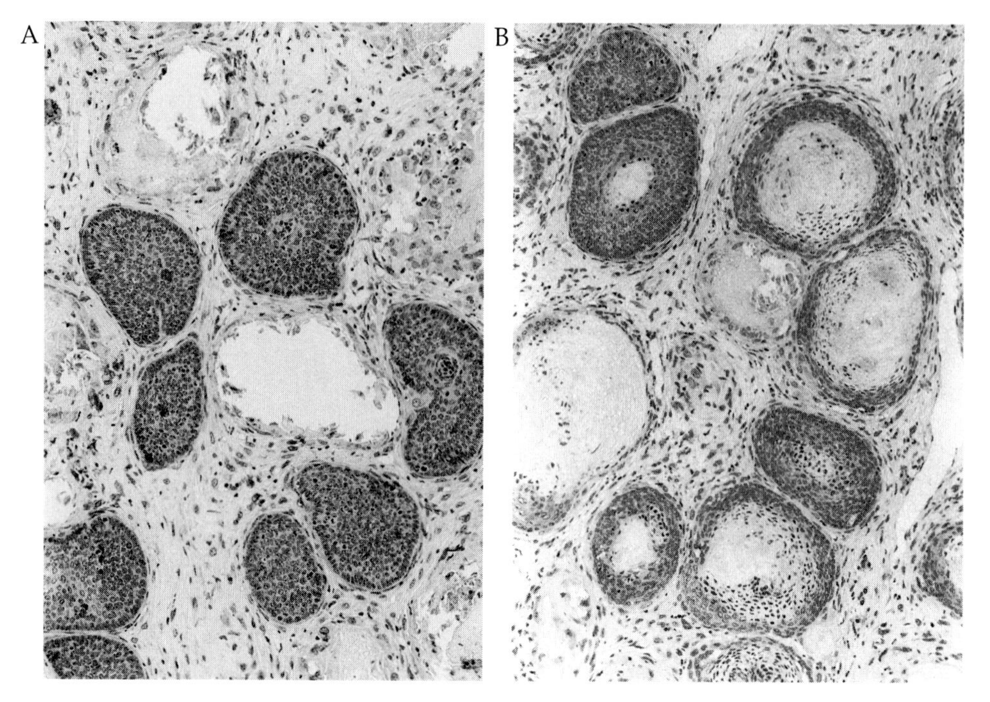

Figure 13 Clonal regeneration pattern in mouse mammary carcinoma AT17 after a radiation dose of 34 Gy. (A) Early regrowth of single surviving cells into discrete spheroidal clones observed at day 16. (B) Stage of synchronous degeneration at day 24. This stage is transient and followed by secondary outgrowth with eventual tumour regeneration.

The resulting survival curves for single dose and fractionated irradiation are shown in Figure 14. As with normal tissue clonogen survival assays, the dose threshold required to induce tissue breakdown and clone resolution is quite large, e.g. 30 Gy single dose given under local ischaemia (clamp hypoxia). With increasing dose levels the clone frequency decreases exponentially. As the probability of finding a clone in a histological section will depend both on the clone size distribution and the distance between sequential histological sections, the calculation of absolute clone numbers requires some calibration. In Figure 14 all clone numbers have been normalized to a section distance of 150 μm, which is about the median clone diameter. Fitting all normalized data to a linear-quadratic dose relationship using non-linear regression and a maximum likelihood procedure resulted in well defined survival parameters. Furthermore, the number of clonogens at risk (z) could be estimated by this *in situ* assay to be about 2.4×10^6 per 100 mg tumour mass. This can be compared to the total number of epithelial tumour cells, which in a 100 mg tumour is about 10^8, or to the number of proliferating epithelial cells, which is about 4×10^7. The apparent percentage of functional stem cells, defined as clone forming cells after irradiation, therefore is less than 3% of all parenchymal tumour cells or 6% of the growth fraction.

The question of whether these *in situ* colony-forming cells represent stem cells in the more stringent definition that they can form a recurrence has been studied in a parallel cure experiment where local control was assessed after an extended observation period

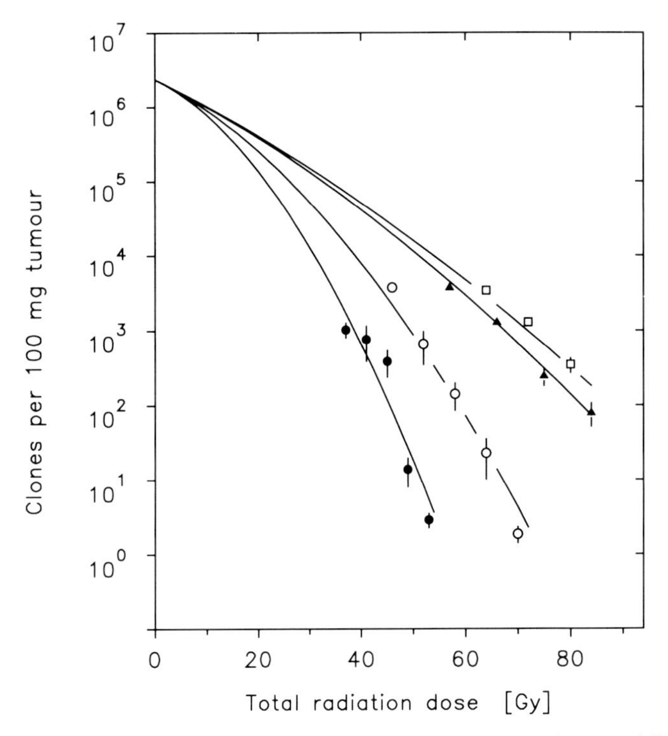

Figure 14 *In situ* survival curves of tumour clonogens in carcinoma AT17 for single dose and fractionated irradiation given under clamp hypoxia. Curve and parameter estimates were obtained by fitting all data simultaneously to the linear-quadratic model. (Best parameter estimates were $\alpha = 0.080\ \text{Gy}^{-1}$; $\beta = 0.0031\ \text{Gy}^{-2}$; $Z = 2.4 \times 10^6$.)

(Figure 15). In Figure 15A the number of clones per tumour for different radiation doses and the frequency of tumours in the *in situ* clonogen assay that contained no single cone is plotted. The theoretical cure curves to be expected from these data are shown as the dashed curves in Figure 15B, in contrast to the actual cure curves observed. There is a systematical trend of underestimating permanent control rate from the clonogen survival data, although the discrepancy is not large. The discrepancy was further reduced when the histology of tumours with no progressive growth was taken into account. In a small proportion there was viable tumour tissue, actively proliferating but apparently unable to grow into a clinical recurrence.

Considering the involution and subsequent dormancy of clones described above the nearest explanation is that not all clones will recover from the state of dormancy. Mathematically this may be simulated by allotting a finite probability of resuscitation to the individual clone. Assuming a probability of 0.5 is already sufficient to give a good match of the binomially 'corrected' clone survival and clinical tumour cure rates as shown in panel C of Figure 15. The quantitative interpretation of these results therefore is quite straightforward. All clones measured early (i.e. after 2.5 weeks) derive from surviving stem cells and have the potential to regrow the tumour. However, after clonal expansion

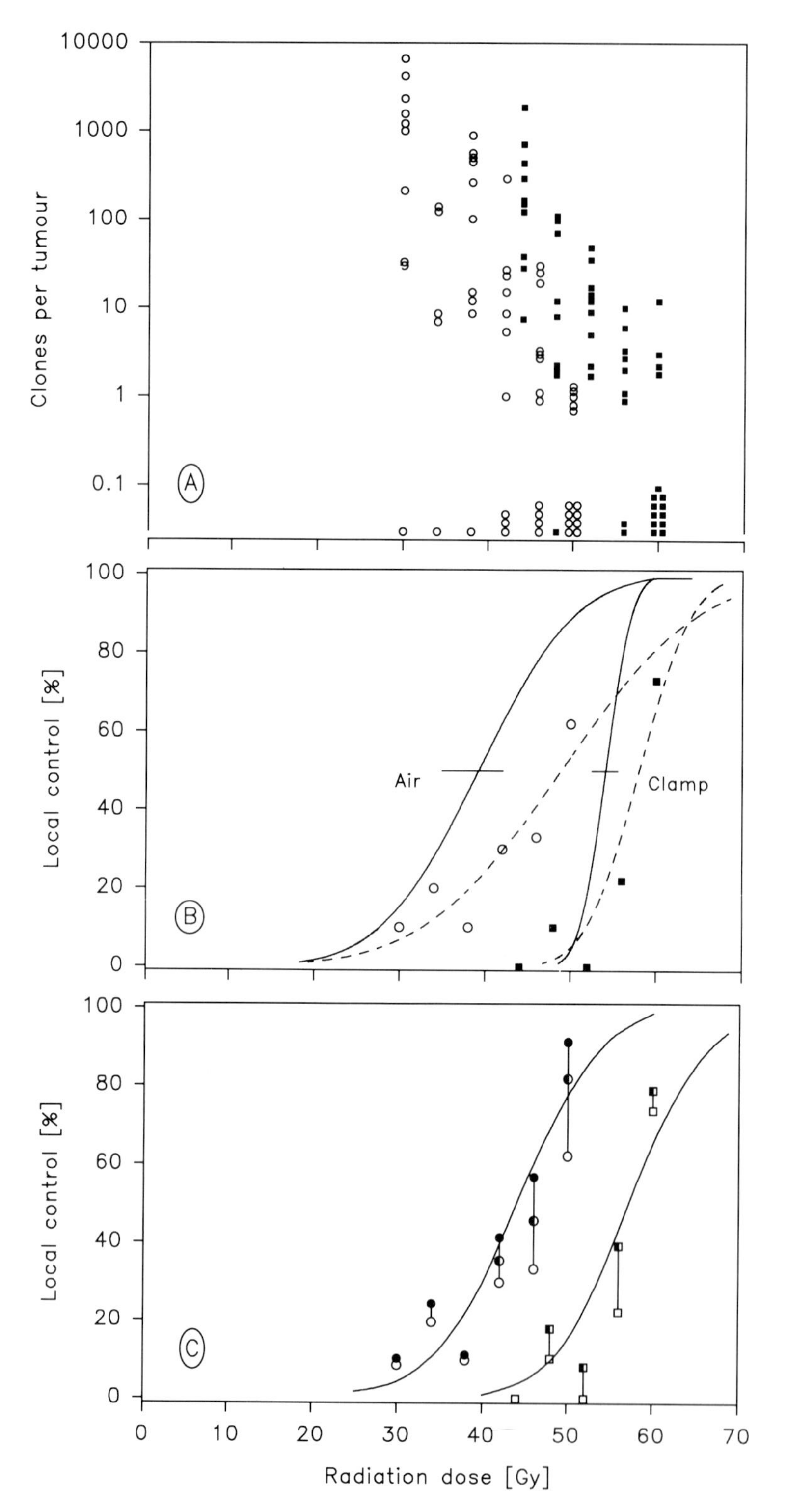

Clones per tumour
10000
1000
100
10
1
0.1
A
Local control [%]
100
80
60
40
20
0
Air
Clamp
B
Local control [%]
C
0
10
20
30
40
50
60
70
Radiation dose [Gy]

they undergo a stage of involution and dormancy from which only 25–50% of the clones eventually awake to initiate progressive tumour growth and clinical recurrence.

Modelling clonal expansion

During the phase of rapid clone evolution, the cell population of a clone undergoes marked changes in its proliferative structure. It must start from a single surviving stem cell which during the ensuing 2 weeks will multiply but at the same time will also give rise to non-stem cells, i.e. transit cells and end cells. In a series of experiments the characteristics of such clone populations were studied and an attempt was made to measure the proportion of stem cells, proliferating cells and non-proliferating cells and the total cell number in individual clones. Total cell number and its increase with time was assessed from morphometric measurements. As the original data were obtained from histological sections this step required a transformation of profile area distributions into volume distributions, done by a Saltykov transformation. Knowing the average cell size a given clone volume could then be recalculated into total cell number. It must be kept in mind, however, that clone growth data obtained are meaningful only when the number of clones has stabilized as otherwise the actual growth rate will be underestimated by the inclusion of later (smaller) clones that were undetectable at earlier stages. The available period of useful total cell number measurements is thus restricted to days 14 to 18. Median clone sizes derived from the volume distribution on these days are about 60 and 80 mm (diameter), corresponding to about 800 and 2,000 cells, respectively, yielding an effective doubling time of 3.5 days.

The proportion of proliferating cells can readily be measured at all stages from the 3H-thymidine labelling index and a previously measured S-phase duration of 8 hours. The growth fraction thus established decreases from about 5% on day 10 to 30% on day 18 during clone growth.

The estimation of stem cell numbers per clone requires a functional assay. The approach used was to re-irradiate clones in a given stage and to determine clone survival as a function of the test dose. The underlying assumption was that only clones that retain one or more surviving stem cells will successfully regrow while all others will perish. Depending on the number of stem cells per clone the resulting curves will show a distinct multi-target shoulder that is absent in the single cell survival curve (see above). The data in Figure 16 present the results of an experiment in which 3 different priming doses were given and retreatment was carried out on day 14 with a range of single doses. All

---

Figure 15 Clonogen survival and local control rate in carcinoma AT17 after single doses in air (circles) or clamp hypoxia (squares). (Panel A) Numbers of clones for individual tumours measured at day 19. Points at the 0.1 level represent tumours with no clone. (Panel B) Tumour cure curves calculated either from the frequency of local tumour control at 18 months follow-up (solid curves) or from the frequency of histological absence of tumour clonogens on day 19 (data from panel A) (data points and dashed lines). (Panel C) The same data on histological tumour control but corrected for probabilities of individual clones to regrow after the dormancy period of 1.0 (open symbols), 0.5 (half-closed symbols) or 0.25 (closed symbols). The results are compatible with the hypothesis that 1 out of 3 or 4 clones is resuscitated after transient degeneration.

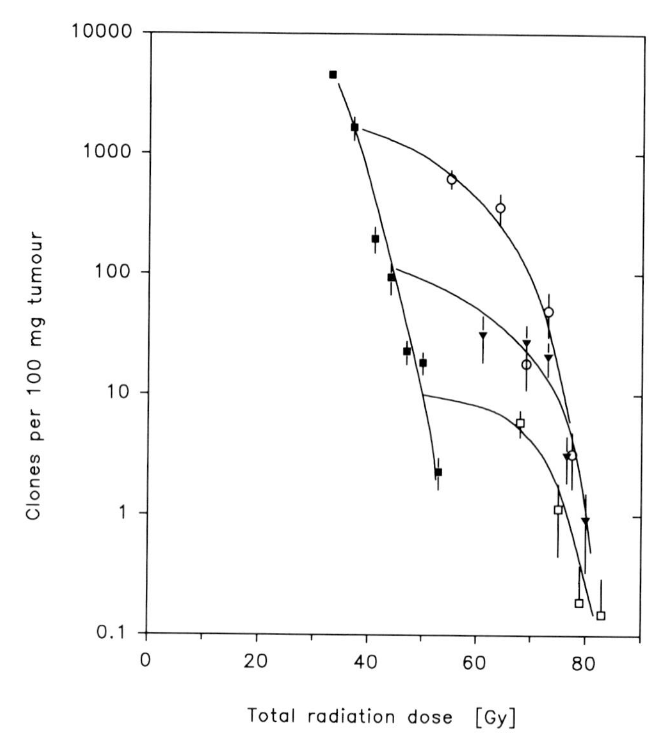

Figure 16 Results of a re-treatment experiment to measure stem cell numbers in growing tumour clones. Following priming doses of 37 (○), 44 (▼) or 50 (□) Gy, tumours were re-irradiated at day 14 with graded test doses and eventual clone survival was measured 19 days later. The wide shoulders in the eventual clone survival curves reflect the multiplicity of clonogens in the individual clone.

curves do indeed initially display a pronounced shoulder as expected and subsequently follow a steep course that is compatible with the single dose curvival curve, suggesting that beyond the shoulder regrowing clones arise from a single surviving stem cell.

This situation can readily be modelled by the Poisson statistics underlying clone eradication. If the number of stem cells in a clone at the time of test treatment is S_1, the test dose D will reduce the number to the mean value S2 according to the equation

$$S_2 = S_1 \exp(-\alpha D - \beta D^2)$$

The frequency of surviving stem cells will be distributed according to Poisson statistics, with the frequency of zero cells being $p(0) = \exp(-S_2)$. As this equals the percentage of eradicated clones it follows that with K_1 clones initially present the number K_2 of clones that eventually survive will be

$$K_2 = K_1(1 - p(0)) = K_1[1 - \exp(-S_2)]$$

or

$$K_2 = K_1\{1 - \exp[-S_1 \exp(-\alpha D - \beta D^2)]\}$$

Using non-linear regression and a maximum likelihood procedure this model was fitted to the data. The stem cell numbers estimated this way were quite low, ranging from 15 to 45.

One obvious drawback of this mathematical approach is that it deals with average numbers of stem cells per clone at the time of re-irradiation. Assuming stochastic growth of stem cells as outlined above in the clonal expansion model, considerable variability between clones must be expected. Although this might be approximated by assuming standard deviations to K_1 and S_1 a simulation procedure seems more straightforward.

A Monte Carlo simulation was therefore set up, parametrized according to the hierarchical model. Stem cell expansion per clone was thought to proceed with the same partioning factor p both after the priming dose *and* after the test doses, including a finite risk of extinction for the individual stem cell. While p is the only parameter that decides on eventual clone survival, simulation of total proliferating and non-proliferating cells in the primary clones (K_1) further requires the growth parameters L (number of transit divisions) and Φ (proportion of end cells lost per cell cycle). Radiation sensitivity parameters were taken from previous experiments, as very consistent figures had been obtained in a series of previous fractionation experiments.

In a first run the parameter p was estimated from iterations that gave the best fit to the re-irradiation data. Assuming that the stem cell number in the primary clones at the time of re-irradiation was independent of the priming dose, the clone numbers K_2 were normalized as shown in Figure 17. The best estimate obtained was $p=0.6$. The growth

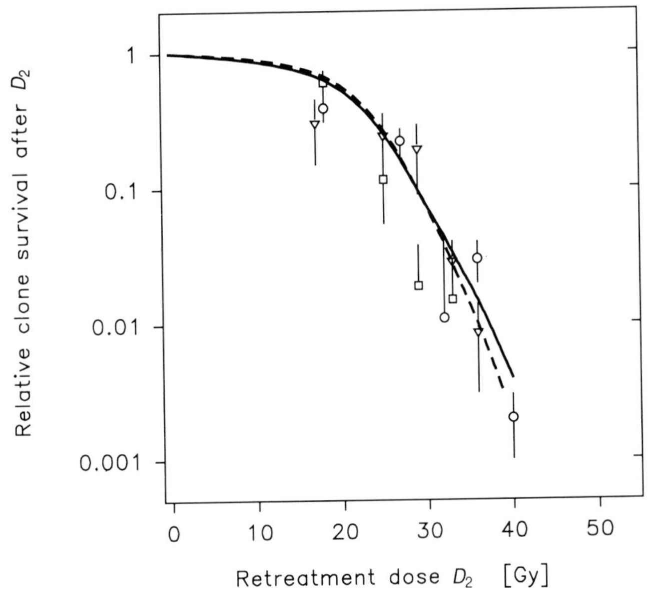

Figure 17 Clone survival curve derived from the data of Figure 16 but normalized to a relative survival of 1.0 after the various priming doses. The curves represent the results of Monte Carlo calculations assuming that stem cell expansion after the priming and test doses proceeds either with a constant p-value of 0.6 (dashed curve) or with an exponential decrease of the p-value from 0.73 to 0.5 (solid curve).

parameters L and Φ were subsequently optimized to give the best fit to the total cell number and growth fraction in the median clone at days 14 and 18; these parameter estimates were $L=3$ and $\Phi=0.4$.

However, the natural history of clone growth demonstrates clone involution soon after day 18, possibly heralding a dramatic change in stem cell expansion. This was supported by an additional re-irradiation experiment with series of test doses delivered on days 10, 14 and 18; the resulting clone survival curves showed a considerable increase in radioresistance from day 10 to day 14 but no further change between day 14 and day 18, indicating that stem cell expansion declined and came to a halt. To model such a dynamic change it was assumed that from a higher initial value p decreased exponentially with each cell cycle, to eventually reach a constant value of 0.5 by day 14; a similar decay was assumed to take place after the retreatment dose. The best initial p in this situation was 0.70–0.73 (see solid curve in Figure 17). In order to explain the actual increase in proliferating cell number that is observed over the entire growth period including the time between day 14 and day 18, the parameter L must run counter to p. This was done by assuming a stepwise increase in L with every 3 cycles, from a starting value of $L=1$. The eventual fit, including a Φ of 0.4 gave a satisfactory fit to the experimental data of proliferating and total cell number as shown in Figure 18.

Importantly, the median stem cell number per clone at the time of re-irradiation was quite similar in both simulations, regardless whether a constant or dynamically changing p was assumed. The estimate was about 60 stem cells, which out of the 800 total cells

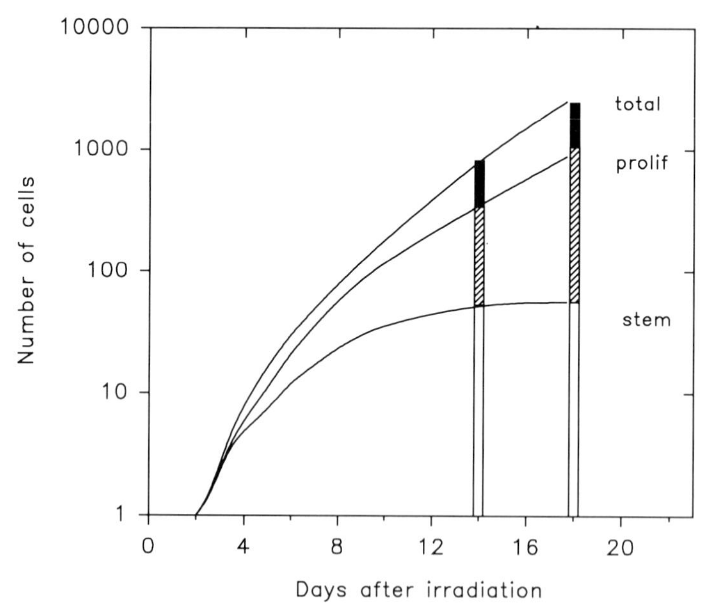

Figure 18 Simulated curves of clonal expansion based on the dynamic model with an exponentially decreasing p-value ($0.73 \rightarrow 0.5$), a gradual increase in the number of transit cell divisions from 1 to 5 and an end cell loss factor Φ of 0.4. Bars represent the proportion of cell populations in 14 and 18 day old clones.

present in the median clone at day 14 would constitute a stem cell fraction of less than 8%. The dynamic scenario seems more appropriate, however, to describe the entire growth history of the clone. The opposing changes in p and L suggest that surviving stem cells start with a greater degree of self-replication which then declines as the clone population grows, while concurrently a greater number of transit generations evolves. The collapse in the overall clone population after day 18 is best explained in this model by assuming that the stem cells stop to proliferate (supported by the drop in labelling index) and the transit tree consequently runs dry after L generations.

THE IMPLICATIONS OF THE HIERARCHICAL PROLIFERATIVE STRUCTURE OF TUMOURS ON THE INTERPRETATION OF EXPERIMENTAL RESULTS IN CANCER RESEARCH

Great progress has been made in recent years in the identification of genetic and molecular alterations in cancer by studying cell cultures, primary cultures as well as cancer cell lines, derived from human cancers. Present concepts of the processes of dysregulation of cell proliferation in carcinogenesis and cancer progression are based mostly on such cell culture studies. The clonal nature of cancer growth is some guarantee that stable genetic and genomic alterations whether incidental or causal, are transmitted to all progeny and thus can be studied in transit cells equally well as in stem cells. Problems may, however, arise in the interpretation of *in vitro* data if effects are to be studied which are supposed to be specific to stem cells and their proliferative regulation.

Although cancer cells *in vitro* share many similarities with cancer stem cells *in vivo*, e.g. their great proliferative potential and clonogenicity, they usually lack the propensity for spontaneously progressing through a series of differentiation steps into cell loss through postmitotic cell death. Cancer cells *in vitro* usually are selected for their ability to perform symmetrical divisions, although under certain experimental conditions asymmetric divisions and differentiation can be induced, e.g. by retinoids (e.g. Meyskens and Fuller, 1980). We can only speculate whether these experimental *in vitro* models bear any similarity to the structural proliferative organization which appears to be typical for differentiated cancers *in vivo*.

Whereas in various normal tissues specific stem cell markers have been developed such as CD34 in bone marrow stem cells (Watt and Visser, 1992) and bright integrin staining in epidermal stem cells (Jones *et al.*, 1995) which may also stain the respective stem cells *in vitro*, there is little prospect that specific stem cell markers for tumour stem cells *in vivo* or *in vitro* can be developed. Therefore, experiments with cancer cells *in vitro* are dealing with a cell population which is not well characterized with regard to its position in a hierarchical proliferative structure. Moreover, it may be very inhomogeneous with regard to the stemness of the cells and therefore, may show very different proliferative responses to the signals which are involved in the regulation of cancer growth *in vivo*. Even though, today, we cannot offer any better experimental models, it is important to bear in mind this fundamental difference between the cancer stem cells *in vivo* which are the driving force of the malignant process and cancer cells *in vitro* which in the optimal case may be a cancer cell population enriched in cancer stem cells. *In vitro*, they usually are prevented by the 'optimized' culture conditions to behave like genuine stem cells as they do not receive the pertinent signals or are unable to respond to those signals which regulate the asymmetry of stem cell divisions.

These problems are particularly relevant if we consider research into the response of cancer to therapeutic intervention (Denekamp, 1994). Since cure of cancer is only a question of tumour stem cell response to treatment and is not affected in any way by the response of the other sub-populations of a tumour (Trott, 1994) any experimental method which is not able to address tumour stem cells specifically can, at best, give only indirect answers (Selby *et al.*, 1978). The most obvious example is the sensitivity testing in cancer therapy research. Most systems for chemotherapy testing (Weisenthal, 1991) do not even attempt to look specifically at tumour stem cells. The aim of most tests is to identify metabolic features within the cells of individual cancers which may affect the cytotoxicity of the tested chemotherapeutic agent. As there is no good reason to assume that these intracellular metabolic pathways differ in a systematic manner between the stem cells and the transit cells, stem cell specificity may therefore not be required for chemotherapy sensitivity testing.

Great efforts have been made to develop clonogenic assays for radiotherapy sensitivity testing. The CFE values of these tests are generally low (Table 2), low enough to be consistent with the estimated stem cell fraction of those tumours – however, it is impossible to decide whether the clonogens which are tested for radiosensitivity, are indeed stem cells or early transit cells. Although this difference is important for understanding the growth pattern of tumours it may not be so important for radiosensitivity testing as this would only matter if we had evidence that the radiosensitivity of stem cells and transit cells as measured by colony formation differed in a systematic manner. In some normal tissues such as bone marrow (Nothdurft, 1991) and testes (Oakberg and Lorenz, 1972), such differences have, indeed, been described. We doubt, however, that in tumours such a difference has to be assumed. Unless proven otherwise, we suggest that the radiosensitivity testing of *tumour clonogens* should be considered as a practicable surrogate for the radiosensitivity of the only tumour cells that are relevant for cancer cure, i.e. the *tumour stem cells* (Selby *et al.*, 1978). This may, however, be different if new and faster techniques are developed which are no longer based on the small subpopulation of clonogenic cells in a human tumour (which conceptually at least may be close to the tumour stem cells) but on radiation effects in the DNA or in chromosomes of all proliferating (or even of non-proliferating) cancer cells.

Another important consequence of the tumour stem cell concept is its impact on the interpretation of cell loss in tumours and the mechanism of accelerated repopulation during a course of radiotherapy. The overall cell production rate in any human or rodent tumour generally exceeds the rate of net tumour volume growth, often by a factor of 10 or more (Steel, 1977). This discrepancy is numerically corrected for by the cell loss factor which describes the proportion of cells which are produced but which do not contribute to overall cell population expansion. Following the observations of Thomlinson and Gray (1955) that cell death in tumours occurs at a critical distance from the nutritive capillary, the cause of cell loss in tumours is commonly assumed to be cell necrosis due to nutritional constraint. Yet this mechanism may not be the only one, particularly not in differentiated carcinomas. A decision on cell loss is made at the stem cell level when a stem cell enters the progenitor cell pool and is irrevocably 'lost' although only after having gone through several more transit cell divisions. In tumours with a hierarchical proliferative organization the cell loss factor is largely determined by the inherent tendency of tumour stem cells to shed a certain proportion of their progeny through the transit compartment into terminal differentiation.

These considerations on the nature of cell loss are critical for our understanding of the mechanisms leading to accelerated repopulation in tumours (Trott and Kummermehr, 1991). In the mouse skin and oral mucosa, acceleration of repopulation during fractionated radiotherapy is mainly due to the shift from asymmetrical to more symmetrical stem cell divisions to meet the increased demand for stem cell regeneration during ongoing cytotoxic insult highlighting the enormous flexibility of the proliferative organization in replacing normal tissues.

Mathematical modelling predicts that accelerated repopulation in tumours requires a marked decrease in the cell loss factor (Fowler, 1991) which specifically affects the small sub-population of surviving tumour stem cells. Yet, experimental studies in mouse squamous cell carcinomas demonstrated that accelerated repopulation is actually associated with a pronounced *decrease* of overall cell proliferation (Begg *et al.*, 1991). Therefore, we do not consider it likely that the decreased cell loss factor during a course of radiotherapy results from improvement of the overall nutritive situation, maybe related to reoxygenation, as this would relate to the total proliferative population rather than the small subpopulation of surviving stem cells. We favour the concept that after tissue injury has been recognized by normal tissues as well as by some tumours, chemical signals such as cytokines are produced which reduce the transition rate of stem cell daughters into the transit cell compartment, and increase the value of p. Any increase in the value of p would mathematically result in a decrease in the cell loss factor (see page 364) whereas any proliferative change in the transit tumour cell population, be it an increase in growth fraction, or a shortening of cell cycle time, would have no influence on the repopulation rate which is specific to tumour stem cells. If, however, after accumulation of sufficient cytotoxic insult in the tumour, the probability of self-maintenance of the surviving tumour stem cells increases from say 0.6 to 1, the repopulation rate would take place at the rate determined by the generation time of stem cells which mathematically would be equivalent to disappearance of the cell loss factor.

The hierarchical organization model of tumour cell proliferation which is based on an increased probability of self-maintenance (p) as the crucial determinant of malignant growth has a major impact on the concept of differentiation therapy of cancers (Beere and Hickman, 1993). The therapeutic strategy is based on a perception of the malignant cell as one whose differentiation has been partly or totally blocked. By removing this block and re-establishing homeostatically determined population dynamics rather than by killing cancer cells the proponents of this strategy hope that it may offer the opportunity for the use of novel, relatively non-toxic agents, maybe combined with low doses of cytotoxic drugs. The best studied agent which selectively engages the process of terminal differentiation is retinoic acid. Retinoic acid has already been applied clinically with some success in some rare malignancies such as acute non-lymphoblastic leukaemias (Huang *et al.*, 1988). Whatever the precise mechanism of action may be, in order to lead to complete remission as described by Huang *et al.* (1988), a reduction in the probability of self-maintenance in the leukaemic stem cells to a value close to 0.5 must have taken place. The success of these studies gives further support for the concept of hierarchical organization of tumour cell proliferation as set out here and underline its importance for the further development of cancer therapy research.

ACKNOWLEDGEMENT

The authors are most indebted to Dr Hennitte Honoré, Department of Medical Physics, Community Hospital, Aarhus, Denmark, for implementing and performing the Monte Carlo simulations.

REFERENCES

Beere, H.M. and Hickman, J.A. (1993). *Anti-Cancer Drug Design* **8**:299–322.
Begg, A.C., Hofland, I. and Kummermehr, J. (1991). *Eur. J. Cancer* **27**:537–543.
Besch, G.D., Tanner, M.A., Howard, S.P., Wolberg, W.H. and Gould, M.N. (1986). *Cancer Res* **46**:2306–2313.
Bizzari, J.P. and Mackillop, W.J. (1985). *Br. J. Cancer* **52**:189–195.
Buick, R.N. and Mackillop, W.J. (1981). *Br. J. Cancer* **44**:349–355.
Buick, R.N. and Tannock, I.F. (1987). In *The Basic Science of Oncology* (eds I. F. Tannock and R. P. Hill), pp. 139–153, New York.
Bush, R.S. and Hill, R.P. (1975). *Laryngoscope* **85**:1119–1133.
Carlsson, J. and Nederman, T. (1989). *Eur. J. Cancer Clin. Oncol.* **25**:1125–1133.
Carney, D.N., Gazdar, A.F. and Minna, J.D. (1980). *Cancer Res.* **40**:1820–1823.
Denekamp, J. (1994). *Radiotherapy and Oncology* **30**:6–10.
Durand, R.E. (1975). *Br. J. Radiol.* **48**:556–571.
Fowler, J.F. (1991). *Radiotherapy and Oncology* **22**:156–158.
Guttenberger, R., Kummermehr, J., Wang, J. and Willich, N. (1990). 38th Annual Meeting of the Radiation Research Society, New Orleans, 1990, abstr. Cn-5. Radiation Research Society, Philadelphia.
Hendry, J.H. (1992). *Radiotherapy and Oncology* 25:308–312.
Hewitt, H.B. (1958). *Br. J. Cancer* 12:378–401.
Hewitt, H.B., Blake, E. and Walder, A.S. (1976). *Br. J. Cancer* **33**:241–259.
Hill, R.P. (1987). In *Rodent Tumor Models in Experimental Cancer Therapy* (ed. R. F. Kallman), pp. 67–75. Pergamon Press, New York.
Hill, R.P. and Milas, L. (1989). *Int. J. Radiat. Oncol. Biol. Phys.* **16**:513–518.
Hjelm-Hansen, M., Jorgenson, K., Anderson, A.P. and Lund, C. (1979). *Acta Radiol. Oncol.* **18**:385–407.
Huang, M.E., Ye, Y.I., Chen, S.R. *et al.* (1988). *Blood* **72**:567–572.
Jones, P.H., Harper, S. and Watt, F.M. (1995). *Cell* **80**:83–93.
Kummermehr, J. (1985). In *Cell Clones: Manual of Mammalian Cell Techniques* (eds C. S. Potten and J. H. Hendry), pp. 215–222. Churchill Livingstone, Edinburgh.
Kummermehr, J. and Buschmann, J. (1987). In *Rodent Tumor Models in Experimental Cancer Therapy* (ed. R. F. Kallman), pp. 101–102. Pergamon Press, New York.
Kuwashima Y., Majima, H. and Okada, S. (1988). *Int. J. Radiat. Biol.* **54**:91–104.
Mackillop, W.J., Ciampi, A., Till, J.E. and Buick, R.N. (1983). *J. Natl Cancer Inst.* **70**:9–16.
Marsden, J.J., Kember, N.F. and Shaw, J.E.H. (1980). *Br. J. Cancer* **41** (suppl. IV):91–94.
Masters, J.R.W. (ed.) (1991). *Human Cancer in Primary Culture.* Kluwer Academic Publishers, Dordrecht.
Meyskens, F. and Fuller, B. (1980). *Cancer Res.* **40**:2194–2199.
Milas, L. (1987). In *Rodent Tumour Models in Experimental Cancer Therapy* (ed. R. F. Kallman), pp. 174–178. Pergamon Press, New York.
Moore, J.V. and Hendry, J.H. (1984). *Br. J. Radiol.* **57**:935–937.
Nothdurft, W. (1991). In *Radiopathology of Organs and Tissues* (eds E. Scherer, C. Streffer and K. R. Trott), pp. 113–170. Springer Verlag, Berlin.
Oakberg, E.F. and Lorenz, E.C. (1972). In *Encyclopedia of Medical Radiology*, Vol II/3, Radiation Biology (eds O. Hug and A. Zuppinger), pp. 217–234. Springer Verlag, Berlin.

Pallavicini, M.G. (1987). In *Rodent Tumour Models in Experimental Cancer Therapy* (ed. R. F. Kallman), pp. 76–81. Pergamon Press, New York.
Porter, E.H., Hewitt, H.B. and Blake, E.R. (1973). *Br. J. Cancer* **27**:55–62.
Potten, C.S. and Loeffler, M. (1990). *Development* **110**:1101–1120.
Rofstad, E.K., Wahl, A. and Brustad, T. (1986). *Br. J. Radiol.* **59**:1023–1029.
Rosenblum, J.H., Vasquez, D.A., Hoshino, T. and Wilson, C.B. (1978). *Cancer* **41**:2305–2314.
Sandbach, J., von Hoff, D.D., Clark, G., Cruz, A.B. and O'Brien, M. (1982). *Cancer* **50**:1315–1321.
Selby, P., Buick, R.N. and Tannock, I. (1978). *New Engl. J. Med.* **298**:1321–1327.
Southam, C.M. and Brunschwig, A. (1961). *Cancer* **14**:971–978.
Steel, G.G. (1977). *The Growth Kinetics of Tumours.* Oxford University Press, Oxford.
Steel, G.G. and Stephens, T.C. (1983). In *Stem Cells: Their Identification and Characterisation* (ed. C. S. Potten), pp. 271–293. Churchill Livingstone, Edinburgh.
Stuschke, M., Budach, V., Stüben, G., Streffer, C. and Sack, H. (1995). *Int. J. Radiat. Oncol. Biol. Phys.* **32**:395–408.
Suit, H.D., Shalek, R.J. and Wette, R. (1965). In *Cellular Radiation Biology*, pp. 514–530. Williams and Wilkins, Baltimore.
Suit, H.D., Hwang, T.Y., Hsieh, C.C. and Thames, H.D. (1987a). In *Rodent Tumor Models in Experimental Cancer Therapy* (ed. R. F. Kallman), pp. 154–164. Pergamon Press, New York.
Suit, H.D., Sedlacek, R. and Thames, H.D. (1987b). In *Rodent Tumor Models in Experimental Cancer Therapy* (ed. R. F. Kallman), pp. 138–148. Pergamon Press, New York.
Tannock, I.F. (1968). *Br. J. Cancer* **22**:258–273.
Tepper, J. (1981). *Acta Radiol. Oncol.* **20**:283–288.
Thomlinson, R.H. and Gray, L.H. (1955). *Br. J. Cancer* **9**:539–549.
Trott, K.R. (1994). *Radiotherapy and Oncology* **30**:1–5.
Trott, K.R. and Kummermehr, J. (1991). *Radiotherapy and Oncology* **22**:159–160.
Trott, K.R., Maciejewski, B., Preuss-Bayer, G. and Skolyszewski, J. (1984). *Radiotherapy and Oncology* **2**:123–129.
Tveit, M., Gunderson, S., Hoie, J. and Pihl, A. (1988). *Br. J. Cancer* **58**:734–737.
von Lieven, H., Trott, K.R. and Sintermann, R. (1978). *Strahlentherapie* **154**:299–304.
Walls, G.A. and Twentyman, P.R. (1985). *Br. J. Cancer* 52:505–513.
Watt, S.M. and Visser, J.W.M. (1992). *Cell Proliferation* 25:263–297.
Weisenthal, L.M. (1991). In *Human Cancer in Primary Culture.* (ed. J. R. W. Masters), pp. 103–147. Kluwer Academic Publisher, Dordrecht.
West, C.M.L., Davidson, S.E. and Hunter, R.D. (1989). *J. Radiat. Biol.* **56**:761–765.
Whelan, R.D.H., Hosking, L.H., Shallard, S.A. and Hill, B.T. (1991). In *Human Cancer in Primary Culture* (ed. J. R. W. Masters), pp. 253–260. Kluwer Academic Publishers, Dordrecht.

13 Biology of the haemopoietic stem cell

Brian I. Lord

CRC Department of Experimental Haematology, Paterson Institute for Cancer Research, Christie Hospital NHS Trust, Manchester, UK

The term haemopoietic stem cell is often used in the literature very loosely, embracing a wide range of cells and including many that are directly descended from the true stem cells. From 1970, Lajtha (1970, 1983) defined stem cells as cells which maintain their own number, in spite of removal of cells for differentiation into all the haemopoietic cell lineages, throughout the lifetime of the animal. At that time, the spleen colony-forming cell or unit (CFU-S) (Till and McCulloch, 1961) was considered to fulfil these requirements. These cells were assayed by bone marrow transplantation into a lethally irradiated recipient, whereby stem cells in the inoculum lodged in the spleen and grew as colonies that could readily be counted 7–12 days later. By definition, the lineage restricted colony-forming cells, normally grown in soft agar and descended from the CFU-S, were excluded.

Doubt was thrown on the candidature of the CFU-S when it was shown that colonies seen at 8 days after the bone marrow transplant were being replaced by later developing colonies, appearing in the spleen at 12 days (Magli *et al.*, 1982). It was suggested that those 8-day colony-forming cells were merely transit cells without any capacity for self-maintenance. Heterogeneity within the CFU-S population had already been recognized (Worton *et al.*, 1969; Schofield and Lajtha, 1973); a pre-CFU-S had been described (Hodgson and Bradley, 1979) and an age-structure to the stem cell population had been hypothesized (Schofield, 1978; Rosendaal *et al.*, 1979). Thus, there were increasing doubts concerning the long-term repopulating capacity of the CFU-S population. As a result, many experimentalists now insist on a definition, for the stem cell, which precludes any cell other than those which specifically persist in an ablated animal, and repopulate its lymphohaemopoietic tissues for the rest of its lifespan. Since they exclude CFU-S as transit cells, destined only to differentiate and amplify as mature functional cells, their number is extremely small – perhaps 1 in 10^5 bone marrow cells – and what is known of their biology could probably be wound up in a couple of short paragraphs. However, I shall argue that limiting the stem cell population to this very restricted definition may not be warranted and that any cell which still retains the capacity to differentiate into all of the lympho-haemopoietic cell lineages, also retains some degree

STEM CELLS
ISBN 0–12–563455–2

of 'stem cell' capacity. Accordingly, I shall explore the characteristics and properties of haemopoietic progenitor cells, up to the point of commitment into one of the lineage restricted lines of development.

STEM CELL ASSAYS

At least 99% of haemopoietic cells in bone marrow can be ruled out as stem cells. These are the morphologically recognizable cell lineages, classically assessed by the clinician for diagnostic and prognostic purposes (*c.* 95%) and the lineage restricted progenitor cells which form colonies in soft agar gels when presented with specific haemopoietic growth factors (*c.* 4%). Functional assays are essential for the multi- or pluripotent cell populations which have no specific recognizable morphological characteristics.

The spleen colony-forming unit assay, CFU-S

This, the first of the modern stem cell assays, was introduced in 1961 (Till and McCulloch, 1961). Mice, lethally irradiated to destroy endogenous haemopoiesis and prevent its recovery, are transplanted intravenously with bone marrow cells (or any haemopoietic tissue) from syngeneic donors (Figure 1). The technique is described

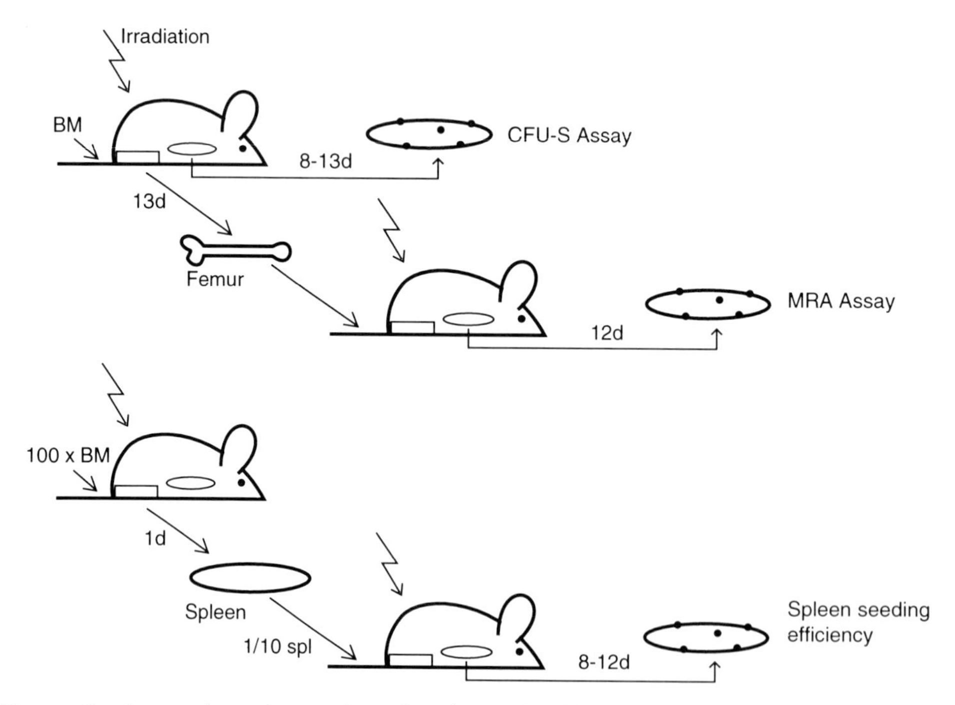

Figure 1 Assays for spleen colony forming units (CFU-S) marrow repopulating ability (MRA) and CFU-S spleen seeding efficiency.

in detail elsewhere (Lord, 1993a). A proportion of the colony forming cells (CFC-S) injected, lodges in the spleen to develop as macroscopic colonies of differentiated cells which are counted as colony forming units, CFU-S, in the fixed spleen 8–13 days later.

Note that the clonality of these colonies was demonstrated using unique chromosome markers, induced in the donor material by sublethal irradiation (Becker *et al.*, 1963; Wu *et al.*, 1967), thus justifying the equation that one colony equals one CFU-S. Furthermore, the majority of colonies contained a mixture of the maturing cell types and on secondary transplantation of a single colony, a further set of colonies was generated (Siminovitch *et al.*, 1963). Self-renewal and differentiation thus justified the identification of CFU-S as stem cells.

Since only a proportion of the inoculated CFC-S seed and develop in the spleen as CFU-S, a spleen seeding factor, *f*, is measured by a secondary transplantation and

Table 1 Stem cells in the mouse.

Fetal liver	13d	17d		1 wk postnatal
Number of $CFU\text{-}S_8$	400	3800		300
Number of $CFU\text{-}S_{12}$	200	2600		250
Spleen seeding efficiency		0.02		
Total $CFU\text{-}S_8$	2×10^4	1.9×10^5		1.5×10^4
Total $CFU\text{-}S_{12}$	10^4	1.3×10^5		1.3×10^4
Proportion of CFU-S in DNA-synthesis		~40		
Bone marrow (adult)	**MRA**	**$CFU\text{-}S_{12}$**		**$CFU\text{-}S_8$**
Number per 10^5 bone marrow cells	500	10–40		10–40
Number per femur	10^5	10^4		10^4
Spleen seeding efficiency			0.1	
Total CFC-S per femur		10^5		10^5
Probability of CFU-S self-renewal			0.68	
Proportion in DNA-synthesis	<5%	<5%		<10%
Spleen				
Number per 10^6 spleen cells		15		20
Number per spleen		2.3×10^3		3×10^3
Spleen seeding efficiency			0.05	
Total CFC-S per spleen		4.5×10^4		6×10^4
Probability of self-renewal			0.68	
Proportion in DNA synthesis		<10%		<10%
Peripheral Blood				
Number per 10^6 leucocytes	16	1.2		2
Number per blood volume	160	12		20
Spleen seeding efficiency			0.2	
Total CFC-S per blood volume		60		100
Probability of self-renewal			low	
Proportion in DNA-synthesis			~30%	
Whole mouse				
Total CFC-S per mouse			2×10^6	
CFU-S produced per day			4×10^5	
CFU-S cell cycle time			6–10 hrs	

spleen colony assay (Figure 1) (Siminovitch *et al.*, 1963; Fred and Smith, 1968; Lord, 1971, 1993a). Thus the relationship between spleen colony forming cells and units is given by CFU-S $= f \times$ CFC-S. Commonly, the value of f for bone marrow CFC-S is about 0.1 but it may vary with the method of recipient irradiation (Lord *et al.*, 1986) or with any treatment that may have altered the biological or biophysical state of the CFC-S in the donor haemopoietic tissues (Smith *et al.*, 1968; Monette and DeMello, 1979). By contrast, only about 1% of the injected colony forming cells seed into the femur (Lord *et al.*, 1975).

Typically, there are approximately 10^5 CFC-S per femur and 2×10^6 per mouse (Table 1).

Marrow repopulating ability (MRA)

Pre-CFU-S are recognized by their ability to repopulate the marrow and the usual short-term assay is to measure the production of more mature progenitors in the femur. This is illustrated in Figure 1 and described in detail by Lord (1993a). Marrow from the femur of a mouse transplanted 13 days earlier with bone marrow is retransplanted into secondary irradiated recipients from which spleen colonies are counted 12 days later. The MRA is then calculated as the number of $CFU\text{-}S_{12}$ generated in thirteen days' growth in the femur, per 10^5 cells (or per femur) originally inoculated.

Typically the normal murine marrow has an MRA of 10^5 per femur or 500 per 10^5 bone marrow cells (Table 1). It must be clear, however, that this is not a direct count of the marrow repopulating cells – only a measure of the ability to repopulate, i.e. the amount of production from those cells – one femur has sufficient MR cells to generate 10^5 $CFU\text{-}S_{12}$ in 13 days' growth in the femur. Assuming 9 days of actual growth – it takes 4 days for MR cells to enter cell cycle (Lord and Woolford, 1993) – and a population doubling time of say 24 hours, this would represent approximately 1 MR cell per 10^5 marrow cells, a figure in line with reported values for the most primitive stem cells (Harrison *et al.*, 1988).

Long-term culture initiating cells (LTC-IC)

Long-term culture of bone marrow is achieved by allowing the stromal elements to form an adherent layer in which stem cells lodge and develop haemopoiesis (Dexter *et al.*, 1977). Endogenous haemopoiesis can be ablated with radiation leaving a competent adherent layer into which exogenous marrow can lodge – but it needs primitive cells. $CFU\text{-}S_{12}$ generate only transient bursts of haemopoiesis (Spooncer *et al.*, 1985). Limiting dilution of mononuclear cells from bone marrow has defined a LTC-IC that gives rise to unipotential and multipotential clonogenic cells for 5–8 weeks at a frequency of 1–2 per 10^4 bone marrow cells (Szilvassy *et al.*, 1989; Sutherland *et al.*, 1990). Each LTC-IC can produce several clonogenic cells and therefore possesses some of the characteristics of primitive stem cells. Although these cells from LTC do not compete with fresh bone marrow cells for long-term repopulation, they are a useful measure for human primitive progenitors where direct experimentation is not practicable.

High proliferation potential colony forming cells (HPP-CFC) and blast-CFC

Two additional *in vitro* techniques detect cells with very high proliferative potential. HPP-CFC, cells resistant to the cytotoxic effects of 5-fluorouracil (5-FU) and therefore considered to be a very primitive population, generate macrophage colonies containing 5×10^4 cells (Bradley and Hodgson, 1979; McNiece *et al.*, 1990). Blast cell colonies (Leary and Ogawa, 1987) consist of 50–500 immature haemopoietic cells which, on replating, form a variety of secondary lineage restricted colonies. These blast cell CFC also are resistant to 5-FU and, therefore, considered very primitive.

Long-term repopulation (LTR)

Ultimately, the only test that defines a stem cell is one that demonstrates its capacity for full lymphohaemopoietic regeneration. This is conventionally carried out by transplanting test cells from a male donor into ablated female recipients and identifying the Y-chromosome in myeloid and lymphoid cells of the marrow, spleen and thymus, six or more months later. The various subsets of the stem cell compartment must demonstrate persistence of their Y-chromosome before they can be considered as haemopoietic stem cells.

THE STEM CELL POPULATION

Having defined the cells we can assay, more or less quantitatively, they can now be fitted into some sort of biological sequence. The primitive haemopoietic stem cell (PHSC)

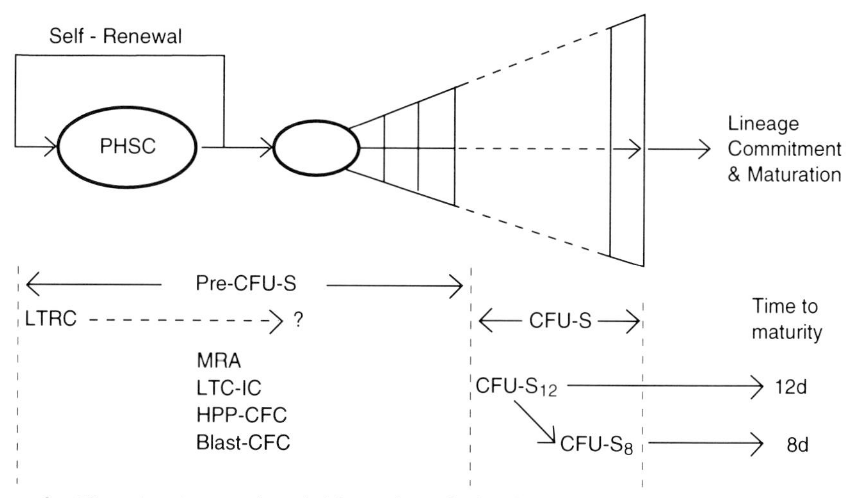

Figure 2 The time/age related hierarchy of the haemopoietic stem cell populations. PHSC, primitive haemopoietic stem cell; LTRC, long-term repopulating cell; MRA, marrow repopulating ability; LTC-IC, long-term culture initiating cell; HPP-CFC, high proliferative potential-colony forming cell; CFU-S, spleen colony forming units.

is, by definition, both self-renewing and capable of long-term repopulation (LTR). It resides at the base of progenitor cell development (Figure 2). How far along the development sequence the cells retain these two properties is still a matter of debate but pre-CFU-S, including MRA, LTC-IC, HPP-CFC and Blast-CFC express them, at least in part. CFU-S can be measured from 7 days to about 14 days following transplantation but assays are generally limited to 8–12 days for practical reasons of small size before

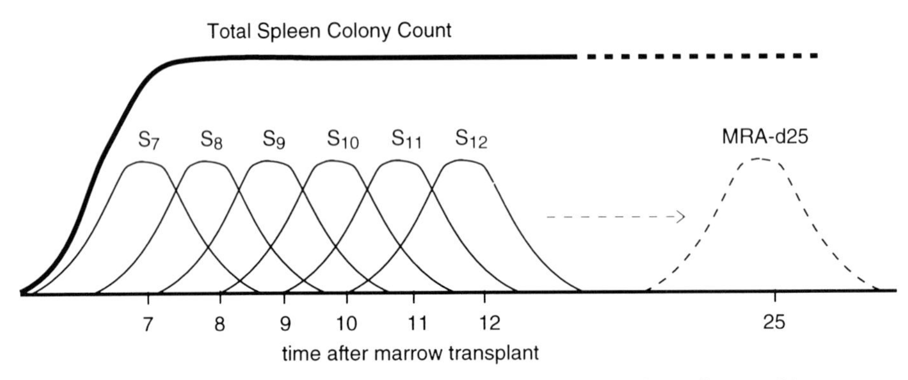

Figure 3 Observed spleen colony count following transplantation of normal bone marrow. The appearance and disappearance CFU-S_7; CFU-S_8 ... CFU-S_{12} ... CFU-S_{25} (MRA) as they arrive at maturity result in a constant count. The broken line anticipates the count should it be practicably possible to continue observing colonies beyond about 13 days.

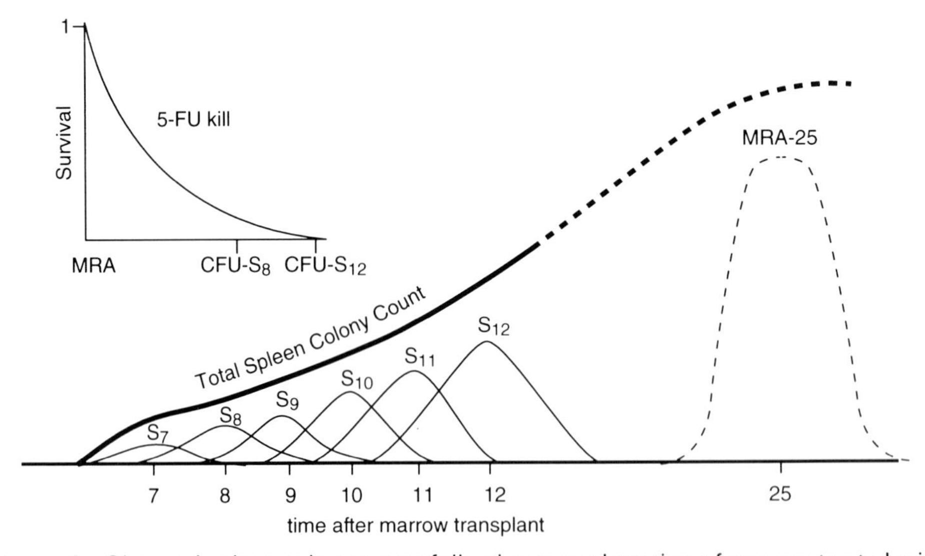

Figure 4 Observed spleen colony count following transplantation of marrow treated with 5-Fluorouracil. The superior sensitivity of CFU-S_8 and CFU-S_{12} to 5-FU (inset) means that the early colony-formers will be absent and the colony count will increase when the surviving MRA cells arrive at maturity as colony forming cells.

8 days and confluence beyond 12 days. Over this range, the colony count from normal bone marrow is constant so although CFU-S_8 and CFU-S_{12} counts nominally measure cells of different maturity, numerically a count represents either (Figure 3). Indeed, if the animals could be kept alive and counts were practical one might expect to see colonies from cells with MRA arising at the same level after 25 days.

Should the marrow be perturbed, as by some cytotoxic drug, for example 5-FU, the colony count would not remain constant. 5-FU kills proliferating but not non-proliferating cells. Cells with MRA will survive but many CFU-S, particularly the more mature ones, will not (see below). As a result, colony numbers will rise with time as cells from the earlier stages re-establish, through proliferation and maturation, the normal complement of CFC-S (Figure 4).

Currently, it is considered that CFU-S are transit cells with little or no self-renewal capacity and no long-term repopulating ability. Orlic and Bodine (1994) conclude that 'the long-term *in vivo* studies are the only measure of PHSC activity'.

PROLIFERATIVE STATUS OF CFU-S

The requirement for cell proliferation in the stem cell compartment is low. As illustrated by Lajtha (1983), if the committed and maturing cell populations go through 10 cell cycles (i.e. amplification is 1,000-fold) then for any 1,000 end cells produced, only 1 pluripotent progenitor cell needs to be mobilized into differentiation; and this requires only 1 cell division to replace it. Lajtha went on to show that a stem cell population of less than 0.5% of the total cells with 10% turnover per day and a 24-hour cycle time would service a production of 1024 mature cells per progenitor in 10 days.

The proliferative status of the cells is measured by the tritiated thymidine [^{3}H]TdR suicide test (Becker *et al.*, 1965; Lord *et al.*, 1974; Lord, 1993a). DNA-synthesizing cells are allowed to incorporate large amounts of tritium (via high specific activity [^{3}H]TdR) which then kills the S-phase cells by intranuclear irradiation. The loss of functional colony-forming competence of the population when assayed thus gives a direct measure of the proportion of cells in S-phase. For normal adult haemopoietic tissue, a non-significant proportion ($<10\%$) of the CFU-S_8, CFU-S_{13} and MR cells is killed (Lord and Woolford, 1993). This means that all the pluripotent progenitor cell populations either have a very long cell cycle time or that most of them are not normally in an active cell cycle. From studies on regulation (see below) and from considerations requiring maintenance of the integrity of the stem cell genome (Lajtha, 1983), the out-of-cycle G_0-phase stem cell model is probably to be preferred.

If the stem cell population(s) should suffer serious depletion, then during recovery it will be required to regenerate its own population(s) at the same time as filling the pressing need for functional differentiated cells. Under these circumstances, CFU-S_8 and CFU-S_{12} are rapidly triggered into cycle (76% and 41% in S-phase respectively). Cells with MRA, however, are not triggered until their numbers reach a nadir at about 4 days (Lord and Woolford, 1993) a property appropriate to the need for the most primitive cells to be most protected from the potential hazards of cell proliferation.

PROPERTIES OF HAEMOPOIETIC STEM CELLS

Number and kinetic properties of murine pluripotent progenitor cells

Haemopoiesis develops from the embryonic yolk sac and pluripotent cells are first reliably measured as CFU-S in the fetal liver. Appearing at about 11 days' gestation, they rise to a peak of about 4,000 at 17 days' gestation but are virtually extinct in the liver by 1 week postnatally (Mason, 1989; Wolf *et al.*, 1995; Table 1). Their low spleen seeding efficiency means that, at maximum, CFC-S numbers are about twice those in an adult femur which, along with the rest of the skeleton, ultimately becomes a major site of haemopoiesis. The spleen retains a few CFU-S but its net contribution to haemopoiesis is less than that of one femur which typically contains about 5,000 CFU-S – 10^4 if one includes the epiphyseal and diaphyseal ends of the bone (Carsten and Bond, 1969). MRA is about 10^5 per femur which contains *c.* 2×10^7 cells. At a concentration of 1 MR cell per 10^5 marrow cells (see section on MRA) then a femur contains about 200 marrow repopulating cells per femur compared to 10^5 CFC-S. Assuming the distribution of CFU-S in the mouse is comparable to the overall distribution of haemopoietic tissue, the total CFC-S in the mouse is about 2×10^6 and by the same token, MR cells should be about 4,000. Schofield (1974) made a series of calculations concerning CFU-S kinetics and found that about 4×10^5 CFC-S are produced per day.

This calculation depended on the cell cycle time of marrow CFU-S and its determination is not straightforward. It relies on information deduced from experiments with perturbed CFU-S conditions. Vassort *et al.* (1973) partially synchronized CFU-S with hydroxyurea treatment, followed this by plotting a 'fraction suicided curve' (cf. labelled mitosis curve) and derived a value of about 8 hours for the cycle time. A value of 6–7 hours was deduced from the kinetics of growing spleen colonies (Schofield and Lajtha, 1969; Guzman and Lajtha, 1970; Lajtha *et al.*, 1971) and an estimate based on measurements of CFU-S renewal gave a value of 8.6 hours assayed over 11 days of colony growth (Schofield *et al.*, 1980).

Peripheral blood contains a few MR cells and CFU-S but numerically they are of little significance. These CFU-S have a high cycling rate and spleen seeding efficiency (Table 1) (Gidali *et al.*, 1974) and few if any re-enter the marrow to contribute to haemopoiesis (Micklem *et al.*, 1975a). Their self-renewal quality is low (Micklem *et al.*, 1975b). Thus, their properties are more characteristic of an old, effete stem cell population and they are often regarded as an overflow of old stem cells from the bone marrow and, perhaps, *en route* to a stem cell graveyard.

Marrow stem cells can be mobilized into the peripheral blood using cytotoxic drugs and/or haemopoietic growth regulators giving CFU-S concentrations more comparable with bone marrow (Molineux *et al.*, 1990b) and MRA enhanced some 120-fold (Lord *et al.*, 1995). Mobilization is, in fact, so efficient that human peripheral blood is now commonly being used for transplantation in place of bone marrow.

Division potential of CFU-S

Since the majority of CFU-S are in an out-of-cycle G_0-phase ($<5\%$ in DNA-synthesis) Schofield (1978) calculated that about 20% of CFU-S turn over each day. Thus, during a mouse lifespan of about 1000 days, each CFU-S completes about 200 doublings. In spite of this, the overall quality of these CFU-S remains unchanged throughout the mouse's lifetime. The growth rate of CFU-S transplanted from an old mouse into an irradiated recipient is equal to that of CFU-S from a young mouse (Lajtha and Schofield, 1971). A single lifetime is not the limit, however. The stem cell deficiency of W/W^v mice can be cured by transplantation of normal marrow. Subsequent serial transplantation of this 'cured' marrow to young W/W^v mice, through at least five lifespans, showed no loss of ability to repopulate the deficient animals (Harrison, 1979). The CFU-S have thus undergone some 1,000 divisions without apparent loss of still further division potential. Is the division potential of those pluripotent CFU-S unlimited then? In some circumstances, it does seem to be limited; for example, the so-called 'decline phenomenon' on serial retransplantation (Siminovitch *et al.*, 1964) when recovery of the CFU-S population progressively falls. Even on increasing the recovery interval and then taking care to retransplant exactly the same number of CFU-S each time, this loss of repopulating ability was still observed (Lajtha and Schofield, 1971; Pozzi *et al.*, 1973). It showed up also in measurements of CFU-S per colony following serial transplantation (Schofield, 1978) and in CFU-S surviving a variety of cytotoxic treatments (Schofield *et al.*, 1980). The net capacity of the stem cell compartment for self-maintenance, sometimes referred to as the quality of the stem cells, declined. This was a paradox that needed explanation. Is the CFU-S population really a mere transit population, retaining its pluripotentiality but maintained by a real stem cell, way back in the pre-CFU-S zones?

An analysis of proliferation in the various components of the stem cell population following hydroxyurea showed that the most mature cells, CFU-S_8 were most easily triggered into cell cycle while the most immature MR cells, which can be selected on the basis of their extreme proliferative quiescence (Ploemacher and Brons, 1989) did not enter DNA-synthesis until 3–4 days had elapsed (Lord and Woolford, 1993). Similarly, Bodine *et al.* (1991) found the optimal time for retroviral transfection of immature cells was 4–5 days following 5-FU treatment. This delay was explained by a feedback mechanism whereby the MR cells continued to feed the CFU-S population and only responded themselves when their own population fell to a critical level (Lord and Woolford, 1993).

Thus CFU-S are indeed transit cells, fed from earlier, more primitive cell compartments which have a greater propensity for self-renewal. As such, the real primitive haemopoietic stem cell must be pre-CFU-S. Does this justify excluding the still pluripotent CFU-S as a cell with some stem cell properties?

Self-maintenance capacity and age structure

Spleen colonies, developing from single CFU-S, clearly contain increasing numbers of CFC-S. It was on this property that CFC-S were considered as stem cells and since their other property – differentiation – is linked inversely to the degree of self-renewal it became important to assess the probability, p, of self-renewal during regeneration: (probability of differentiation, $d = 1 - p$). The average probability that a CFU-S will renew itself under these conditions was determined theoretically (Vogel *et al.*, 1968; Lajtha *et al.*, 1971) and experimentally (Till *et al.*, 1964; Vogel *et al.*, 1968) giving an average value of $p=0.63$. A later measurement (Schofield *et al.*, 1980) gave $p=0.68$ (Table 1). In fact, the measured value depends to a degree on the experimental conditions. Thus, during CFC-S growth in a colony, 63% (or 68%) will on average produce another CFC-S while 37% (32%) will be lost to differentiation or death at, or during, each generation cycle.

The technique for measuring this probability requires a CFU-S assay to be made on individual retransplanted colonies and is described in detail by Lord (1993a). It should be realized, of course, that this does not give a measure of the physiological self-renewal, *in situ,* under normal steady state conditions. In these circumstances it must be assumed that stem cell self-renewal probability is 0.5. Instead, it is a measure of the *capacity* of the CFU-S population to grow under standardized, albeit extreme, conditions of elevated proliferation and/or differentiation pressures. Under a variety of cytotoxic insults, the capacity of marrow CFU-S for self-maintenance was often reduced – occasionally increased (Schofield *et al.*, 1980) and this has an important bearing on the kinetic structure of the stem cell compartment.

Early evidence for individual CFU-S with different capacities for self-renewal came from cells, separated on a linear density gradient (Worton *et al.*, 1969). On transplantation, some fractions containing CFU-S gave in excess of 100 new CFU-S per colony; others, the majority, gave about 20. The idea of different qualities of CFU-S was first mooted, however, when investigating the effects of an alkylating agent – isopropyl methane sulfonate (IMS) – on CFU-S kinetics (Schofield and Lajtha, 1973). The prolonged run-down of CFU-S numbers before eventual recovery suggested that an older population with limited proliferative potential, and largely unaffected by IMS, was slowly running out while the earlier and younger CFU-S were initially killed by the drug. Although defining two classes of CFU-S, the one with high and the other with low self-renewal capacity, this was merely a simplification of the continuum illustrated in Figures 2–4. Rosendaal *et al.* (1979) defined a category of CFU-S that is resistant to 5-FU and with extremely high self-renewal capacity. Subsequently, they classified a series of treatments giving a sequence of CFU-S with reducing capacities for self-renewal. Thence the principle of the age-structured stem cell compartment was born (Schofield, 1978; Rosendaal *et al.*, 1979).

Other groups have also reported CFU-S populations with different self-renewal properties. Micklem and Ogden (1976) demonstrated a sequential reduction in the number of serial transfers that could successfully be effected by CFU-S, respectively, from the 9-day yolk-sac, through the fetal and neonatal livers to the young and finally the old adult marrow. They showed that peripheral blood CFU-S have a very low self-renewal capacity. Late marrow failure after busulfan (myleran) when CFU-S never recovered their capacity for high self-renewal was described (Morley and Blake, 1974). Early brain-associated; antigen-independent CFU-S renewed themselves 3–4-fold better than older CFU-S (Monette and Stockel, 1981).

In addition, the self-renewal probability of a CFU-S was found to depend on its location in the marrow. Spatial structure was established in the marrow cell populations (Lord and Hendry, 1972; Lord, 1993b) whereby it was proposed that early stem cells reside close to the central venous sinus of the marrow space (Table 2, Figure 5) and 'mature' towards the bone surfaces where they interact with the differentiation signals that commit them to lineage development. This conclusion was based on three properties: (i) proliferative activity increases towards the bone surfaces (Lord *et al.*, 1975) (Table 2)

Table 2 Spatial distribution of stem cells in femur.

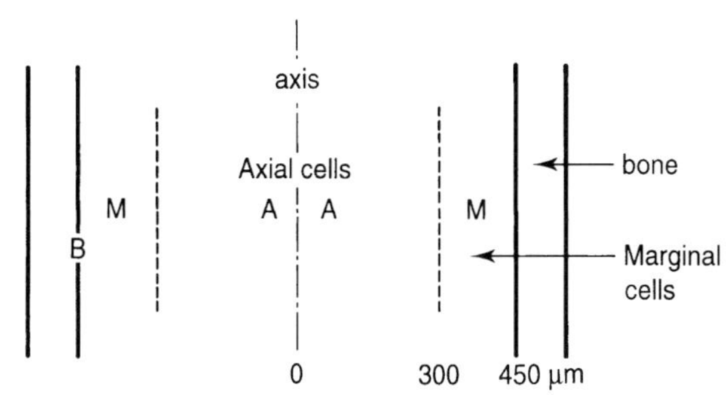

	Axial core (A) (0–300 μm)	Marginal (M) (300–450 μm)	Bone (B) associated
CFU-S per 10^5 cells	22	38	43
Proportion in DNA synthesis	2%	15%	50%
Probability of self-renewal	0.73	0.65	0.57
MRA per 10^5 cells	733	268	–
Mean cell cycle time (hours) (stimulated)	11.4	8.7	6.4

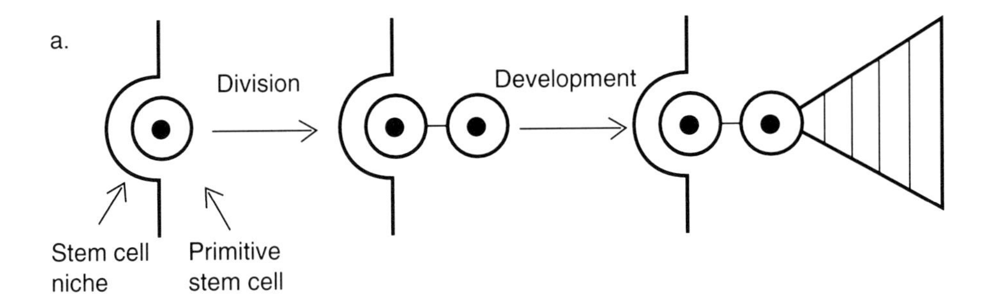

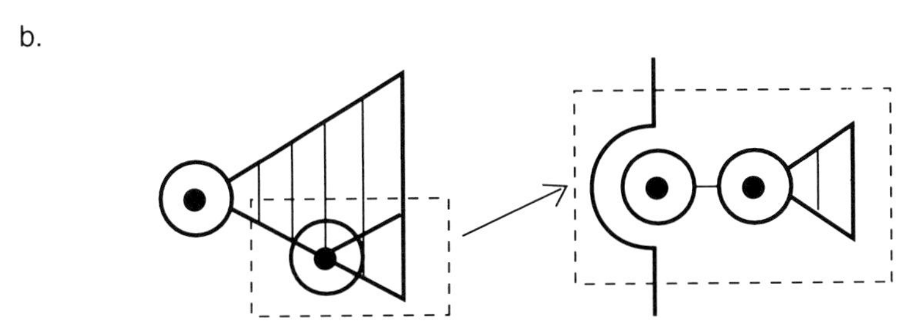

Figure 5 (a) A primitive stem cell in a micro-environmental 'niche' is induced to divide. One daughter remains in the niche as a primitive stem cell. The other daughter moves out of the sphere of influence of the niche where it is now free to undergo development as a stem or pluripotent progenitor cell population through a series of amplification divisions. (b) Stem cells from different age stages of their development sequence, on transplantation, are free to lodge in empty environmental niches where they behave as in (a) above but with only their remaining developmental potential.

– consistent with the principle that 'real' stem cells have a low turnover rate; (ii) the probability of self-renewal is highest in the axial zones (Lord and Schofield, 1980) and (iii) the pre-CFU-S, measured as MRA, predominate in the axial region (Lord, 1993b) (Table 2).

The principle of age-structure, whereby a CFU-S has a limited amount of potential amplification (self-renewal as a cell capable of generating further spleen colonies) and each time it divides, its potential for further division is reduced, is attractive. Furthermore, inclusion of the pre-CFU-S into this programme lends credence to the modern view that the CFU-S is simply a transit cell which is continuously ageing and which could, on its own, run out in a matter of days, let alone a lifetime. It could also explain the retransplantation decline phenomenon (Siminovitch *et al.*, 1964) which is based on CFU-S measurements. At the same time, however, there is no evidence for ageing of the stem cell population and this brings us back to the question of what is a real stem cell.

Based on the principles stated above, i.e. that a stem cell must be capable of generating long-term lymphohaemopoietic repopulation, there are now numerous publications describing stem cell isolation studies which exclude most of the more mature progenitors, including CFU-S, from this fundamental property. For example, van der Loo *et al.* (1994) separated CFU-S_{12} and pre-CFU-S using a combination of density differences and affinity for wheat germ agglutinin and rhodamine-123. They then demonstrated that only the pre-CFU-S had long-term repopulating capacity. On the other hand, Spangrude *et al.* (1988) claimed the property for CFU-S_{12} using the Sca-1 antibody for sorting. Orlic and Bodine (1994) however, maintain that 'cell size or a $CD34^+$, Sca-1^+ or c-kit^+ phenotype cannot define the PHSC population because each of those markers is shared with CFU-S and other short-term repopulating cells' and that therefore 'there is a need for strict adherence to that requirement for long-term repopulation'.

It is possible, however, that interpretation of these experiments is limited by the assay model. From the time of Till and McCulloch (1961) it has been customary to prepare (ablate haemopoiesis in) recipient animals using a radiation dose of about 8 Gy X-rays or 9–10 Gy γ-rays given at a dose rate of at least 15 cGy min^{-1}. Using standard radiobiological data (Hendry and Lord, 1983) one should expect endogenous survival of CFU-S in the femur to be about 4 for X-rays and 10 for γ-rays. To compete against these, it can be estimated that CFU-S giving a spleen-colony count of 10 will contribute only one additional exogenous CFC-S in the femur (Testa *et al.*, 1972) which, together with any potential transplantation trauma means it is unlikely to contribute much to long-term haemopoiesis. Our own experiments (unpublished results) based on low dose rate recipient conditioning, where a total dose of 15.25 Gy γ-rays is given at 0.84 Gy hr (Lord *et al.*, 1984) give a different picture. This irradiation regimen leaves a very low endogenous CFU-S survival against which exogenous CFC-S can compete more efficiently. Under these circumstances, we have been able to record the required long-term repopulation using small inocula of sorted CFU-S_{12} or of whole bone marrow such that conventional marrow repopulating (MRA) cells would be injected into only a small proportion of the recipients. It is the opinion of this worker, therefore, that under appropriate conditions of recipient preparation, the more mature, multipotent progenitor cells can act also as 'real stem cells'. It must be appreciated, however, that this does not negate the 'age structure' principle. On a level playing field it can be anticipated that the cells with MRA will always compete better than will CFC-S. To understand how the CFC-S can fulfil the 'stem cell' role it is necessary to consider the micro-environment.

ROLE OF THE HAEMOPOIETIC MICROENVIRONMENT

The role of the micro-environment is clear when growing bone marrow in long-term cultures (Dexter *et al.*, 1977). A successful culture requires an adherent layer of stromal cells including endothelial cells, fat cells and macrophages to be formed before stem cells can lodge and generate haemopoietic lineages. This implies spatial cellular organization such as was seen for haemopoietic cells and the fibroblastoid colony-forming unit, CFU-F, recognized (Friedenstein *et al.*, 1967, 1970) as a major component of the micro-environment, was found to be most concentrated in the axial zones of the marrow where the most primitive stem cells can be found (Xu and Hendry, 1981). Macrophages synthesizing a stem cell proliferation inhibitor (Wright *et al.*, 1980) later recognized as macrophage inflammatory protein-1α (MIP-1α) (Graham *et al.*, 1990) are also most concentrated in this region (Lord and Wright, 1984). Such a cellular organization was described as a stem cell niche (Schofield, 1978) a regulatory unit designed to maintain the stem cell's integrity. On the rare occasion that its resident stem cell may divide, one daughter remains in the niche as a stem cell while the other emerges to develop fully (Figure 5a).

This model helps us also to explain 'decline' and the ability of a more mature CFU-S to function as a stem cell. On transplantation (Figure 5b) into a haematologically ablated marrow, progenitors with less residual self-renewal capacity compete for the empty niches. Once lodged, the micro-environment stabilizes them in that state. Output is less per stem cell and each has to work harder (divide more frequently) to maintain the ultimate output of functional cells. A secondary transplantation will similarly be further restricted and the self-renewal quality of the population will similarly decline.

A major role of this micro-environment then is to maintain the integrity of the stem and multipotent progenitors and it does this by providing a network of regulatory cytokines or growth factors which ensure the correct balance of proliferation (see Chapter 9).

CONTROL OF STEM CELL PROLIFERATION

Under normal steady state conditions the early stem cells and most of the CFU-S population are relatively quiescent, less than 10% of even the more mature cells being in DNA-synthesis at any one time. It is generally considered that the majority of the cells are in an out-of-cycle G_0-state, a phase which Lajtha (1979) proposed may be important as a time for genetic housekeeping: a period during which any errors developed during self-replication may be repaired. On regeneration following drug or radiation-depletion, or during repopulation of an irradiated host, the stem cells undergo a long series of proliferation cycles in order both to re-establish their own population and to generate mature functional cells.

There is much literature on the cytokine regulation of the stem cell population including many books devoted to the effects of haemopoietic growth factors (e.g. Lord and Dexter, 1992) and full coverage is given in this Chapter 14. Three factors at least must, however, be given some thought. These are stem cell factor (SCF) or kit-Ligand, transforming growth factor-β (TGF-β) and MIP-1α. First, however, it is useful to consider the nature of stem cell control.

Local proliferation control

A series of early experiments illustrated a most important basic principle of stem cell proliferation control. The basis of these experiments was to probe the balance of haemopoietic populations in different parts of the body. One hind limb, shielded during irradiation, acts as a source of stem cells to initiate repopulation in the irradiation-depleted marrow (Croizat *et al.*, 1970; Gidali and Lajtha, 1972). The small disturbance to the shielded limb was quickly restored: CFU-S proliferation which was triggered by their depopulation, replenished the stem cells and they again became quiescent, all in a time which left the rest of the animal struggling with still grossly depressed haemopoiesis. It was thus concluded that CFU-S proliferation control is a local phenomenon, effected by a negative feedback process which operated in close proximity to the stem cells. The same effect was observed (Rencricca *et al.*, 1970; Wright and Lord, 1977) using a drug-induced CFU-S imbalance between bone marrow and spleen. Migration of CFU-S from bone marrow (proliferation induced in the remaining cells due to CFU-S depletion) to the spleen (proliferation suppressed due to CFU-S excess) ensued.

The localized nature of all these events meant that regulatory factors should be synthesized and utilized in the marrow itself.

Endogenous proliferation inhibitors

The best known of the endogenous proliferation inhibitors, specifically in relation to haemopoiesis, is MIP-1α which was discovered as a derivative of a macrophage cell line (Graham *et al.*, 1990). It appears to be identical to the inhibitor extracted from fresh bone marrow (Lord *et al.*, 1976) and indeed from a subset of bone marrow macrophages (Wright *et al.*, 1980; Simmons and Lord, 1985). It certainly bears all the properties of the crude preparation which was neutralized by antibody to MIP-1α (Graham *et al.*, 1990). Initially described by its ability to prevent CFU-S entering DNA-synthesis and thereby protect them from cytotoxic, cycle specific drugs, its *in vivo* activity has since been demonstrated (Lord *et al.*, 1992; Dunlop *et al.*, 1992) in models which exploit this property and suggest its potential as a chemotherapeutic protection agent.

MIP-1α is specific for the stem cell population (Lord *et al.*, 1976; Tejero *et al.*, 1984) with CFU-S_{12} showing greater sensitivity to the inhibitory effects than CFU-S_8 (Wright *et al.*, 1985). However, there are suggestions that the pre-CFU-S have not acquired sensitivity to MIP-1α (Quesniaux *et al.*, 1993). At the same time, there are indications, largely anecdotal, that MIP-1α may have pandemic inhibitory activity for a wide variety of stem cell systems. One report indicates inhibition of clonogenic epidermal keratinocyte proliferation (Parkinson *et al.*, 1993).

Two other molecules display similar properties. Derived from the original granulocyte chalone, Paukovits *et al.* (1990) synthesized a pentapeptide, pEEDKC and from fetal calf bone marrow extract (Frindel and Guigon, 1977), Lenfant *et al.* (1989) purified a tetrapeptide AcSDKP. This tetrapeptide probably acts through intermediary molecules while MIP-1α and the pentapeptide probably have more direct action. All of them are currently under trial as physiological proliferation inhibitors capable of blocking stem cell proliferation, though MIP-1α still appears to retain the highest specificity for cells of the stem cell populations.

TGF-β should be considered separately. It is ubiquitous in the body and may have inhibitory or stimulatory effects depending on the tissue. In its activated form, it is inhibitory to haemopoietic stem cells including the more primitive, 5-FU resistant cells with marrow repopulating ability (Keller *et al.*, 1990). The time scale of its inhibitory effects, however, is considerably longer than that of MIP-1α (Migdalska *et al.*, 1991) and it may require an intermediary in the form of a locally produced activator of the latent molecule to give it any kind of target specificity. It is interesting that TGF-β, like MIP-1α, is inhibitory to early normal human progenitor cells but only TGF-β is inhibitory to stem cells of chronic myeloid leukaemia (CML) (Eaves *et al.*, 1993; Holyoake *et al.*, 1993). Attempts to exploit this differential property of MIP-1α in CML treatment are currently in progress.

Stimulation of stem cell proliferation

It is well recognized that the normally quiescent stem cell population is readily triggered into proliferative activity as a result of drug or radiation-induced depopulation. It was quite quickly appreciated that simply removing endogenous inhibitor was not sufficient to bring about this change. It had to be counterbalanced and superseded by a stimulatory molecule(s) which could override the inhibitor to operate a $G_0 \rightarrow$cycle switch (Lord *et al.*, 1977). Conversely, superimposing inhibitor on a proliferating population was required to turn off the switch, back to quiescence. Necessary proliferation of the stem cell population thus appears to require a balanced production of stimulatory and inhibitory molecules.

The stimulator extracted from regenerating bone marrow (Lord *et al.*, 1977; Wright and Lord, 1977) has not yet been characterized other than as a product of another localized subset of macrophages (Wright et al., 1982). Two of the major growth factors, SCF and IL-3 (see Chapter 14) are appropriate to the stem cell populations. IL-3 can stimulate CFU-S proliferation and promote growth (Garland and Crompton, 1983; Lord *et al.*, 1986). SCF may not, on its own, promote proliferation and growth but in concert with a variety of other growth factors its effects are more than additive.

Inhibitor/stimulator interaction for stem cell regulation

Assuming stem cells generate the regulatory cells found in their micro-environment, namely the macrophages which are known to synthesize most of the haemopoietic growth factors, the inhibitor and stimulator producing macrophages (Ip and Sp) (Figure 6) feed back on to the stem cell population. There appears to be production

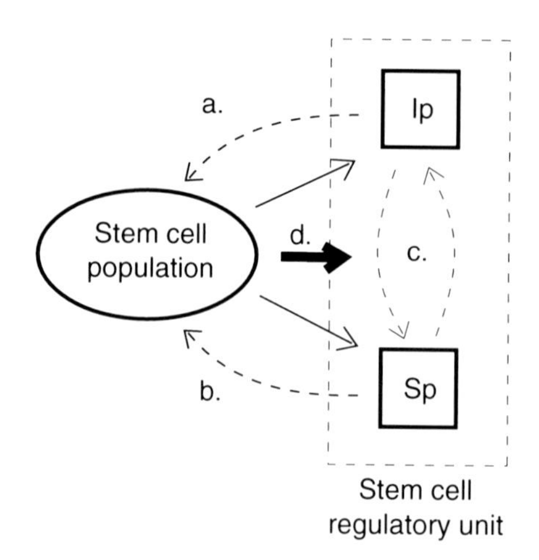

Figure 6 Regulation of stem cell proliferation by negative feedback control. Ip, inhibitor producing cell; Sp, stimulator producing cell; a, stem cell inhibitor; b, stem cell stimulator; c, interaction between inhibitor and stimulator production; d, feedback signal from stem cell population.

interaction between the Ip and Sp (Lord and Wright, 1982) and in order for the regulatory unit to provide the right balance it is necessary to hypothesize a feedback signal from the stem cell population to the regulatory unit (Figure 6). Such a signal has been demonstrated (though not molecularly characterized). It appears to work, however, via blocking stimulator production and thereby permitting passive inhibitor production (Lord, 1986).

DIFFERENTIATION OF THE STEM CELL

Pluripotentiality

It is well recognized that the stem cell compartment of the haemopoietic system generates a wide variety of mature cell types. The CFU-S are clearly demonstrated in spleen colonies to give rise to erythroid, granulocytic and megakaryocytic cells. By association with other clonal techniques, macrophages can be included and osteoclasts, often associated with macrophages, are also derived from the pluripotent haemopoietic progenitor cells (Ash *et al.*, 1980; Marks and Walker, 1981). Lymphocytes are generally thought to diverge at the pre-CFU-S phase.

Langerhans' cells represent about 5% of mammalian epidermal cells. These are not generally recognized in the marrow but, using transplantation techniques it was shown that 80% are derived from the marrow of donor animals (Katz *et al.*, 1979). Since they have surface markers characteristic of monocyte/macrophages and play a role in the initiation of the immune response (Stingl *et al.*, 1978a, 1978b) it is probable that the Langerhans' cells do originate in the haemopoietic stem cell complex, developing via the CFU-S stage.

Owing to their widespread distribution in gut mucosa, peritoneal cavity, subcutaneous tissue lymph nodes and thymic capsule; their responses to T-lymphocyte mitogens, parasitic infections and ease of generation in lymphoid cultures, it was thought that the origin of tissue mast cells was as a post-mitotic derivative of T-lymphocytes. Using a double transplant of bg^j/bg^j mouse marrow, first into wild type mice and the resulting spleen colonies secondarily into anaemic W/W^v recipients; then examining granulocytes and mast cells for the bg^j granules, Kitamura *et al.* (1977, 1981) demonstrated conclusively the haemopoietic stem cell origin of the tissue mast cells. Since the crucial aspect of this experiment demonstrated the development of mast cells from the intermediate spleen colonies, it is clear that this differentiation process developed also via the CFU-S.

Bone cells (osteoblasts) and other stromal elements (e.g. fibroblasts), although of marrow origin, appear not to be directly associated with the haemopoietic stem cell.

Diversity

Since the pluripotent haemopoietic progenitor cell population is clearly the source of at least eight functional cell types, the mechanisms and modes of differentiation become the next important question. In an earlier version of this book (Lord, 1983) various models involving stochastic or microenvironmentally direct differentiation were contrasted with those requiring quantal changes at cell division whereby a cell at any stage of its development has a programmed potential for only two specified further differentiation stages; then two more different ones at the next cycle (Figure 7a). This quantal development model (Holtzer *et al.*, 1975; Holtzer, 1978) has never found much favour among experimental haematologists although bipotentiality in development can be apparent. Humphries *et al.* (1979) demonstrated that some erythroid-burst forming units, early committed erythroid progenitors, led to the development of both erythroid and megakaryocytic cells. Metcalf (1981) replated single cells from first and second generation daughters of GM-CFC and showed that after amplifying their population as GM-CFC for 2 or 3 days, the two individual and non-transmutable lineages of granulocytic and macrophagic cells emerged.

Contrary to this principle of diversity being generated through successive division cycles, there was no evidence that proliferation is a prerequisite of differentiation. Rather, the balance of granulocyte and macrophage development is variable and can be regulated by the balance of prostaglandin E and colony-stimulating factor production (Motomura and Dexter, 1980).

This brings us back to the more common principle that the development of differentiated cell lineages depends on external differentiation stimuli. Furthermore, the more general acceptance that haemopoiesis depends on its regulatory micro-environment suggests that a stem cell is likely to be influenced by the growth factor(s) generated in the particular cellular environment in which a cell finds itself. At a rather gross level, for example, the erythroid:granulocytic-cell ratio in spleen colonies is approximately 3:1 while that in the bone marrow of the same animal is about 1:3.

This still does not shed any light on the relationship between proliferation and differentiation, and an alternative model (Figure 7b) is worth consideration. Here, in much the same way that cell division and maturation in the differentiated cell

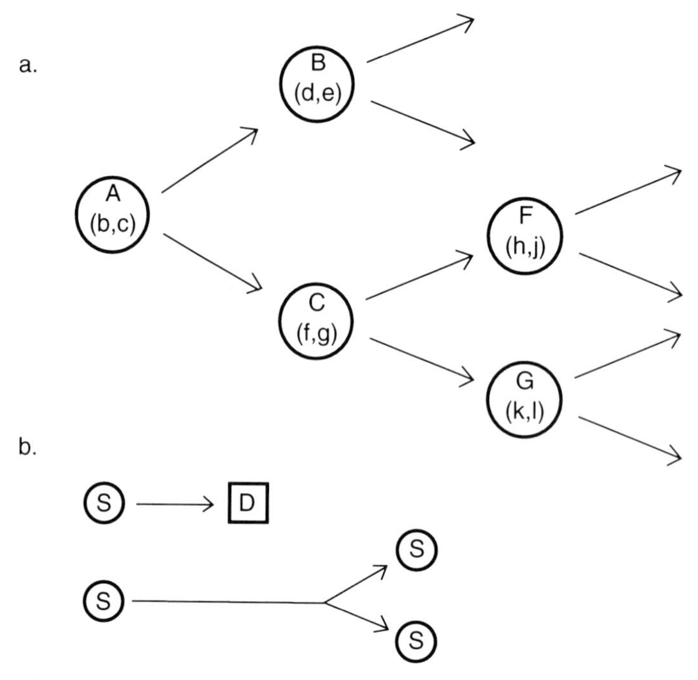

Figure 7 (a) Generation of diversity in a cell population by bipotential, quantal cell divisions. (b) Independent differentiation (D) and stem cell (S) proliferation which results from negative feedback from the depleted S population.

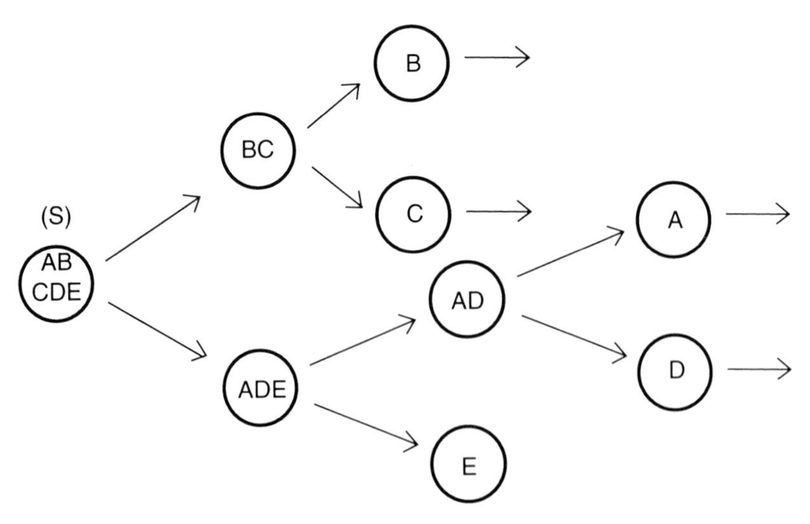

Figure 8 Generation of diversity in a stem cell population(s) with phase related loss of multipotentiality (A–E).

compartments are independent processes (Stohlman, 1959) differentiation (S→D) and proliferation (S→2S) also are independent processes whereby, according to demand, a stem cell differentiates and only then, by negative feedback regulation, another stem cell divides to make up the compartment's shortfall.

Two pieces of evidence support this principle. First, within the confines of the 'aging' stem cell population, it was found that following cytotoxic treatment which severely damaged the CFU-S populations, the earlier, undamaged marrow repopulating stem cells initially became more concentrated. They matured into the more developed CFU-S which diluted their concentration and only at a later stage when that concentration fell below normal did feedback trigger them into proliferation to replenish their own numbers (Lord and Woolford, 1993).

The second series of experiments utilized the factor-dependent stem cell line, FDCP-mix, which can be manipulated to differentiate into either erythroid or granulocyte cell development. Using the modern, elegant molecular technology, Fairbairn *et al.* (1993) incorporated the Bcl-2 gene into the FDCP-mix cells to suppress apoptosis and found that in the absence of IL-3, which is normally required for their survival and self-renewal, they would survive but not proliferate. After 4–6 days' incubation, without growth factors, cells which had been seen not to have divided were found to have undergone differentiation. They therefore concluded that proliferation is not a necessary prerequisite for differentiation. Furthermore, since this occurred in the absence of growth factors, they also concluded that differentiation of haemopoietic stem and progenitor cells is intrinsically determined, with growth factors supplying support via effects on survival and proliferation of their target cells.

This is clearly not the whole story. In *in vivo* experiments and in clinical trials there is no doubt that G-CSF can, for example, dictate granulopoiesis at the expense of erythropoiesis (Lord *et al.*, 1989, 1991; Molineux *et al.*, 1990a). This may indicate that as the stem cell population progresses through its development phases, it may present changing probabilities of susceptibility to the various growth factors (Figure 8). Johnson (1981), for example, re-plated mixed colonies and demonstrated a sequential loss of capacity to generate further colonies containing specific cell types. It appeared that in the absence of other differentiative signals, erythropoiesis remained a constant outcome. Those signals, however, applied before commitment to erythropoiesis, redirected their commitment programme. This is not incompatible with the outcome of the Bcl-2 experiments (Fairbairn *et al.*, 1993). Although differentiation was clearly taking place in the absence of growth factors (62% developed as erythroid cells) it was clear that exogenous growth factors were at least able to modulate lineage choice.

Undoubtedly, the differentiation capacity (and the ability to manipulate it) of the FDCP-mix cells is a very powerful experimental system and it will be capable of giving enormous insights into the intracellular signalling processes involved in proliferation and differentiation processes. Nevertheless, it will remain important not to forget the whole physiological being in developing a full understanding of stem cell behaviour and function.

ACKNOWLEDGEMENT

I wish to acknowledge the Cancer Research Campaign of Great Britain whose continued support is greatly appreciated.

REFERENCES

Ash, P., Loutit, J.F. and Townsend, K.M.S. (1980). *Nature* 283:669–670.
Becker, A.J., McCulloch, E.A., Siminovitch, L. and Till, J.E. (1965). *Blood* 26:296–308.
Bodine, D.M., McDonagh, K.T., Seidel, N.E. and Nienhuis, A.W. (1991). *Exp. Hematol.* **19**:206–212.
Bradley, T.R. and Hodgson, G.S. (1979). *Blood* **54**:1446–1450.
Carsten, A.L. and Bond, V.P. (1969). *Nature* **219**:1082–1084.
Croizat, H., Frindel, E. and Tubiana, M. (1970). *Int. J. Radiat. Biol.* **18**:347–358.
Dexter, T.M., Allen, T.D. and Lajtha, L.G. (1977). *J. Cell Physiol.* **91**:335–344.
Dunlop, D.J., Wright, E.G., Lorimore, S. *et al.* (1992). *Blood* **79**:2221–2225.
Eaves, C.J., Cashman, J.D., Wolpe, S.D. and Eaves, A.C. (1993). *Proc. Natl Acad. Sci. USA* **90**:12015–12019.
Fairbairn, L.J., Cowling, G.J., Reipert, B.M. and Dexter, T.M. (1993). *Cell* **74**:823–832.
Fred, S.S. and Smith, W.W. (1968). *Proc. Soc. Exp. Biol. Med.* **128**:364–366.
Friedenstein, A.J., Chailakhjan, R.K., Latsinik, N.V., Panasynk, A.F. and Keiliss-Borok, I.V. (1967). *Transplantation* **5**:74–80.
Friedenstein, A.J., Chailakhjan, R.K. and Lalykina, K.S. (1970). *Cell Tissue Kinet.***3**:393–403.
Frindel, E. and Guigon, M. (1977). *Exp. Hematol.* **5**:74–76.
Garland, J.M. and Crompton, S. (1983). *Exp. Hematol.* **11**:757–761.
Gidali, J. and Lajtha, L.G. (1972). *Cell Tissue Kinet.* **5**:147–157.
Gidali, J., Feher, I. and Antal, S. (1974). *Blood* **43**:573–580.
Graham, G.J., Wright, E.G., Hewick, R. *et al.* (1990). *Nature* **344**:442–444.
Guzman, E. and Lajtha, L.G. (1970). *Cell Tissue Kinet.* **3**:91–98.
Harrison, D.E. (1979). *Mech. Age Devel.* **9**:409–426.
Harrison, D.E., Astle, C.M. and Lerner, C. (1988). *Proc. Natl Acad. Sci. USA* **85**:822–826.
Hendry, J.H. and Lord, B.I. (1983). In *Cytotoxic Insult to Tissues* (eds C. S. Potten and J. H. Hendry), pp. 1–66. Churchill Livingstone, Edinburgh.
Hodgson, G.S. and Bradley, T.R. (1979). *Nature* **281**:381–382.
Holtzer, H. (1978). In *Stem Cells and Tissue Homeostasis* (eds B. I. Lord, C. S. Potten and R. J. Cole), pp. 1–27. Cambridge University Press, Cambridge.
Holtzer, H., Rubinstein, N., Fellini, S. *et al.* (1975). *Quart. Rev. Biophys.* **8**:523–557.
Holyoake, T.L., Freshney, M.G., Sproul, A.M. *et al.* (1993). *Stem Cells* **11 (Suppl. 3)**:122–128.
Humphries, R.K., Eaves, A.C. and Eaves, C.J. (1979). *Blood* **53**:746–763.
Johnson, G.R. (1981). In *Experimental Hematology Today* (eds S. S. Baum, G. D. Ledney and A. Khan), pp. 13–20. S. Karger, Basel.
Katz, S.I., Tamaki, K. and Sachs, D.H. (1979). *Nature* **282**:324–326.
Keller, J.R., McNiece, I.K., Sill, K.T. *et al.* (1990). *Blood* **75**:596–602.
Kitamura, Y., Shimada, M., Hatanaka, K. and Miyano, Y. (1977). *Nature* **268**:442–444.
Kitamura, Y., Yokoyama, M., Matsuda, H., Ohno, T. and Mori, K.J. (1981). *Nature* **291**:159–161.
Lajtha, L.G. (1970). In *Regulation of Hematopoiesis*, Vol. 1, (ed. A. S. Gordon), pp. 111. Appleton, New York.
Lajtha, L.G. (1979). *Differentiation* **14**:23–34.
Lajtha, L.G. (1983). In *Stem Cells: Their Identification and Characterisation* (ed. C. S. Potten), pp. 1–11. Churchill Livingstone, Edinburgh.
Lajtha, L.G. and Schofield, R. (1971). In *Advances in Gerontological Research*, Vol. 3 (ed. B. L. Strehler), pp. 131–146. Academia Press, New York, London.
Lajtha, L.G., Gilbert, C.W. and Guzman, E.E. (1971). *Brit. J. Haematol.* **20**:343–354.
Leary, A.G. and Ogawa, M. (1987). *Blood* **69**:953–956.
Lenfant, M., Wdzieczak-Bakala, J., Guittet, E., Prome, J.C., Sotty, D. and Frindel, E. (1989). *Proc. Natl Acad. Sci. USA.* **86**:779–782.
van der Loo, J.C.M., van der Bos, C., Baert, M.R.M., Wagemaker, G. and Ploemacher, R.E. (1994). *Blood* **83**:1769–1777.
Lord, B.I. (1971). *Cell Tissue Kinet.* **4**:211–216.
Lord, B.I. (1983). In *Stem Cells: Their Identification and Characterisation* (ed. C. S. Potten), pp. 118–154. Churchill Livingstone, Edinburgh, London, Melbourne, New York.

Lord, B.I. (1986). In *Biological Regeneration of Cell Proliferation* (eds R. Baserga, P. Foa, D. Metcalf and E. E. Polli), pp. 161–171. Serono Symposia Publications, **34**.
Lord, B.I. (1993a). In *Haemopoiesis: A Practical Approach* (eds N. G. Testa and G. Molineux), pp. 1–20. IRL Press, OUP, Oxford.
Lord, B.I. (1993b). In *Concise Reviews in Clinical and Experimental Hematology* (ed. M. J. Murphy Jr.), pp. 225–234. AlphaMed Press, Dayton, Ohio.
Lord, B.I. and Dexter, T.M. (eds.). (1992). *Growth Factors in Haemopoiesis. Baillière's Clinical Haematology*, Vol. 5, pp. 499–789. Baillière Tindall, London.
Lord, B.I. and Hendry, J.H. (1972). *Brit. J. Radiol.* **45**:110–115.
Lord, B.I. and Schofield, R. (1980). In *Biological Growth and Spread* (eds W. Jäger, H. Rost and P. Tautu), pp. 9–22. Lect. Notes in Biomaths. **38**, Springer Verlag: Berlin, Heidelberg, New York.
Lord, B.I. and Woolford, L.B. (1993). *Stem Cells* **11**:212–217.
Lord, B.I. and Wright, E.G. (1982). *Leuk. Res.* **6**:541–551.
Lord, B.I. and Wright, E.G. (1984). *Leuk. Res.* **8**:1073–1083.
Lord, B.I., Lajtha, L.G. and Gidali, J. (1974). *Cell Tissue Kinet.* 7:507–515.
Lord, B.I., Testa, N.G. and Hendry, J.H. (1975). *Blood* **46**:65–72.
Lord, B.I., Mori, K.J., Wright, E.G. and Lajtha, L.G. (1976). *Brit. J. Haematol.* **34**:441–445.
Lord, B.I., Mori, K.J. and Wright, E.G. (1977). *Biomed. Exp.* **27**:223–226.
Lord, B.I., Hendry, J.H., Keene, J.P. *et al.* (1984). *Cell Tissue Kinet.* **17**:323–334.
Lord, B.I., Molineux, G., Testa, N.G., Kelly, M., Spooncer, E. and Dexter, T.M. (1986). *Lymphokine Res.* **5**:97–104.
Lord, B.I., Bronchud, M.H., Owens, S. *et al.* (1989). *Proc. Natl Acad. Sci. USA* **86**:9499–9503.
Lord, B.I., Molineux, G., Pojda, Z., Souza, L.M., Mermod, J.J. and Dexter, T.M. (1991). *Blood* **77**:2154–2159.
Lord, B.I., Dexter, T.M., Clements, J.M., Hunter, M.G. and Gearing, A.J.H. (1992). *Blood* **79**:2605–2609.
Lord, B.I., Woolford, L.B., Wood, L.M. *et al.* (1995). *Blood* **85**:3412–3415.
McNiece, I.K., Bertoncello, I., Kriegler, A.B. and Quesenberry, P.J. (1990). *Int. J. Cell Cloning* **8**:146–160.
Magli, M.C., Iscove, N.N. and Odartchenko, N. (1982). *Nature* **295**:527–529.
Marks, S.C. Jr. and Walker, D.G. (1981). *Am. J. Anat.* **161**:1–10.
Mason, T.M. (1989). A study of the effects of perinatal plutonium contamination on the development of haemopoietic tissue. PhD Thesis, University of Manchester.
Metcalf, D. and Lord, B.I. (1981). *Stem Cells* **1**:140–147.
Micklem, H.S. and Ogden, D.A. (1976). In *Stem Cells of Renewing Cell Populations* (eds A. B. Cairnie, P. K. Lala and D. G. Osmond), pp. 331–341. Academic Press, New York.
Micklem, H.S., Anderson, N. and Ross, E. (1975a). *Nature* **257**:41–43.
Micklem, H.S., Ogden, D.A., Evans, E.P., Ford, C.E. and Gray, J.G. (1975b). *Cell Tissue Kinet.* **8**:233–248.
Migdalska, A., Molineux, G., Demuynck, H., Evans, G.S., Ruscetti, F. and Dexter, T.M. (1991). *Growth Factors* **4**:239–245.
Molineux, G., Pojda, Z. and Dexter, T.M. (1990a). *Blood* **75**:563–569.
Molineux, G., Pojda, Z., Hampson, I.N., Lord, B.I. and Dexter, T.M. (1990b). *Blood* **76**:2153–2158.
Monette, F.C. and DeMello, J.B. (1979). *Cell Tissue Kinet.* **12**:161–176.
Monette, F.C. and Stockel, J.B. (1981). *Stem Cells* **1**:38–52.
Morley, A. and Blake, J. (1974). *Blood* **44**:49–56.
Motomura, S. and Dexter, T.M. (1980). *Exp. Hematol.* **8**:298–303.
Orlic, D. and Bodine, D.M. (1994). *Blood* **84**:3991–3994.
Parkinson, E.K., Graham, G.J., Daubersies, P. *et al.* (1993). *J. Invest. Dermatol.* **101**:113–117.
Paukovits, W.R., Hergl, A. and Schulke-Hermann, R. (1990). *Molec. Pharmacol.* **38**:401–409.
Ploemacher, R.E. and Brons, R.H.C. (1989). *Exp. Hematol.* **17**:263–266.
Pozzi, L.V., Andreozzi, U. and Silini, G. (1973). *Curr. Top. Radiat. Res.* **8**:259–302.
Quesniaux, V.F.J., Graham, G.J., Pragnell, I. *et al.* (1993). *Blood* **81**:1497–1504.
Rencricca, N.J., Rizzola, V., Howard, D., Duffy, P. and Stohlman, F. Jr. (1970). *Blood* 36: 764–771.
Rosendaal, M., Hodgson, G.S. and Bradley, T.R. (1979). *Cell Tissue Kinet.* **12**:17–29.

Schofield, R. (1974). In *Proc. IX Int. Cancer Congress, Florence. Vol. 1: Cell Biology and Tumour Immunology*, pp. 18–23. Exerpta Medica, Amsterdam.
Schofield, R. (1978). *Blood Cells* **4**:7–25.
Schofield and Lajtha, L.G. (1969). *Cell Tissue Kinet.* **2**:147–155.
Schofield, R. and Lajtha, L.G. (1973). *Br. J. Haematol.* **25**:195–202.
Schofield, R., Lord, B.I., Kyffin, S. and Gilbert, C.W. (1980). *J. Cell Physiol.* **103**:355–362.
Siminovitch, L., McCulloch, E.A. and Till, J.E. (1963). *J. Cell Comp. Physiol.* **62**:327–336.
Siminovitch, L., Till, J.E. and McCulloch, E.A. (1964). *J. Cell Comp. Physiol.* **64**:23–31.
Simmons, P.J. and Lord, B.I. (1985). *J. Cell Sci.* **78**:117–131.
Smith, W.W., Wilson, S.M. and Fred, S. (1968). *J. Nat. Cancer Inst.* **40**:847–852.
Spangrude, G.J., Heimfeld, S. and Weissman, I.L. (1988). *Science* **241**:58–62.
Spooncer, E., Lord, B.I. and Dexter, T.M. (1985). *Nature* **316**:62–64.
Stingl, G., Katz, S.I., Clement, L., Green, I. and Shevach, E.M. (1978a). *J. Immunol.* **121**:2005–2013.
Stingl, G., Katz, S.I., Shevach, E.M., Rosenthal, A.S. and Green, I. (1978b). *J. Invest. Dermatol.* **71**:59–66.
Stohlman, F. Jr. (1959). In *The Kinetics of Cellular Proliferation* (ed. F. Stohlman Jr.), pp. 318–324. Grune and Stratton, New York, London.
Sutherland, H.J., Lansdorp, P.M., Henkelman, D.H., Eaves, A.C. and Eaves, C.J. (1990). *Proc. Natl Acad. Sci. USA* **87**:3584–3588.
Szilvassy, S.J., Fraser, C.C., Eaves, C.J., Lansdorp, P.M., Eaves, A.C. and Humphries, R.K. (1989). *Proc. Nat. Acad. Sci. USA* **86**:8798–8802.
Tejero, C., Testa, N.G. and Lord, B.I. (1984). *Br. J. Cancer* **50**:335–341.
Testa, N.G., Lord, B.I. and Shore, W.A. (1972). *Blood* **40**:654–661.
Till, J.E. and McCulloch, E.A. (1961). *Radiat. Res*, **14**:213–222.
Till, J.E., McCulloch, E.A. and Siminovitch, L. (1964). *Proc. Nat. Acad. Sci. USA* **51**:29–36.
Vassort, F., Winterholer, M., Frindel, E. and Tubiana, M. (1973). *Blood* **41**:789–796.
Vogel, H., Niewisch, H. and Matioli, G. (1968). *J. Cell Physiol.* **72**:221–228.
Wolf, N.S., Bertoncello, I., Jiang, D.Z. and Priestley, G. (1995). *Exp. Hematol.* **23**:142–146.
Worton, R.G., McCulloch, E.A. and Till, J.E. (1969). *J. Exp. Med.* **130**:91–101.
Wright, E.G. and Lord, B.I. (1977). *Biomed. Exp.* **27**:215–218.
Wright, E.G., Garland, J.M. and Lord, B.I. (1980). *Leuk. Res.* **4**:537–545.
Wright, E.G., Ali, A.M., Riches, A.C. and Lord, B.I. (1982). *Leuk. Res.* **6**:531–539.
Wright, E.G., Lorimore, S.A. and Lord, B.I. (1985). *Cell Tissue Kinet.* **18**:193–199.
Wu, A.M., Till, J.E., Siminovitch, L. and McCulloch, E.A. (1967). *J. Cell Physiol.* **69**:177–184.
Xu, C.X. and Hendry, J.H. (1981). *Biomed. Exp.* **35**:119–122.

14 Growth factors and the regulation of haemopoietic stem cells

Clare M. Heyworth, Nydia G. Testa, Anne-Marie Buckle* and Anthony D. Whetton*

*Cancer Research Campaign Department of Experimental Haematology, Paterson Institute for Cancer Research, Christie Hospital NHS Trust, Manchester, UK; *Leukaemia Research Fund, Cellular Development Unit, Department of Biochemistry and Applied Molecular Biology, UMIST, Manchester, UK*

INTRODUCTION

The proliferation and development of haemopoietic stem cells *in vivo* is promoted by contact with bone marrow stromal cells and the surrounding extracellular matrix (ECM). Whilst there is some ability of soluble cytokines or growth factors to promote survival and proliferation of stem cells and their progeny in the absence of a stromal cell matrix, the primitive haemopoietic stem cell can only be maintained, in the long term, when co-cultured with the appropriate stromal cell environment (Dexter *et al.*, 1990). Data on the specific molecular interactions that are involved in the association between haemopoietic cells, stromal cells and ECM have been obtained using a number of different approaches. Purification procedures to isolate populations of primitive haemopoietic cells (see below) have been devised such that the short term response to specific cytokines can be gauged. Identifying the capability of cytokines to promote, *in vitro*, purified primitive cells to give long term reconstitution of haemopoiesis capability, *in vivo*, is an important goal. The molecular cloning of the cDNA for many of the cytokine receptors means that expression of the mRNA for these proteins can be detected in primitive haemopoietic cells, and the use of receptor antibodies or fluorescently tagged cytokines can also tell us to which growth factors or integrins stem cells may respond. The study of haemopoiesis in gene knockout mice or mutant mice also gives a terrific insight into the physiological role of the cytokines in stem and progenitor cell biology. For example the roles *in vivo* of Stem Cell Factor (SCF) and Granulocyte Macrophage Colony Stimulating Factor (GM-CSF) have been elucidated using these approaches.

STEM CELLS
ISBN 0–12–563455–2

Furthermore, work on stromal cell/haemopoietic interactions *in vitro* has also characterized molecular interactions which support haemopoiesis. In this chapter we describe some of the recent work describing cell purification, molecular characterization of cytokines and their receptors and *in vivo* studies with growth factors which have provided the background for informed debate on the response of stem cells to growth factors.

PURIFICATION OF HAEMOPOIETIC STEM CELLS

The rarity of primitive haemopoietic cells has provided a major challenge to their isolation and characterization. Recent innovations in techniques employed to separate these cells from their more mature counterparts have enabled researchers to understand more about the stromal cell- and growth factor-mediated control of the early stages of haemopoiesis, and allowed clinicians to take new approaches to bone marrow transplantation.

Early primitive cell isolation procedures were mainly based on the physical characteristics of the different cells in the bone marrow, such as size and density (reviewed by Spangrude, 1989). We now know that early progenitors are small, lymphoid like cells which, by flow cytometry, have medium to high side scatter (Sutherland *et al.*, 1989, 1994). Counterflow centrifugal elutriation (CCE), which sorts cells on the basis of size and density, has been successfully used to enrich for primitive cells (Schwartz *et al.*, 1986; Wagner *et al.*, 1988; Jones *et al.*, 1990). In particular Jones and colleagues have used CCE to define sub-populations of mouse bone marrow, and were able to identify a primitive murine pre-colony forming unit–spleen (CFU-S) fraction. Clinically this has also proven a useful approach since there is a 2–3 log reduction in T cell content of bone marrow progenitors isolated by CCE (Schattenberg *et al.*, 1990; Wagner *et al.*, 1990). Unfortunately, this approach is hampered by the fact that this technique is not easy to establish and there may be a significant loss of stem cells because of their similarity in size and density with lymphocytes (Jones *et al.*, 1990).

Currently the most widely used technique for separating early progenitor cells is based on the use of monoclonal antibodies. Once labelled with antibody, cells can be separated using a range of methods including solid phase immunological methods, immunomagnetic particles or flow cytometry. The characterization of the CD34 antigen, which is expressed on only 0.5–5% of human bone marrow, has been central to the development of this approach for use on clinical samples. CD34 is expressed on early progenitor cells but not on their more mature counterparts (Civin *et al.*, 1990) and there are now numerous commercially available antibodies and devices for separating CD34 positive cells from bone marrow and cytokine mobilized peripheral blood progenitor and stem cells.

Solid phase immunological methods such as panning are a useful technique for a preliminary enriching of $CD34^+$ cells, but do not have the high levels of reproducibility, purity and yield that are required for clinical use. However, an avidin–biotin immuno-absorption device has recently been developed for the selection of $CD34^+$ cells from large scale clinical bone marrow and peripheral blood samples (Berenson *et al.*, 1992). This system has now been used to achieve clinical engraftment of $CD34^+$ cells in cancer patients (Brugger *et al.*, 1994) using both bone marrow and peripheral blood as the source of $CD34^+$ cells.

For research purposes paramagnetic beads that can be directly conjugated to antibody have been widely used (Kannourakis *et al.*, 1988; Umemura *et al.*, 1989). Cells labelled with $CD34^+$ conjugated beads can be easily separated from unlabelled cells using a magnet; however, the use of such beads for positive selection of $CD34^+$ cells results in the enriched cells being coated with beads, which may compromise cell behaviour patterns. Several approaches have been taken to remove the beads after selection, including various enzyme treatments (Strauss *et al.*, 1991; Sutherland *et al.*, 1992). These do remove the beads but may result in cleavage of other functionally important cell surface markers at the same time. Other approaches such as linkage via a cleavable bond (Grimsley *et al.*, 1993) have also been used.

More recently a superparamagnetic microparticle has been developed in which small (< 100 nm) beads are used to label cells (Miltenyi *et al.*, 1990). The resulting magnetic cell separation system (MACS) can be used for enrich for $CD34^+$ cells, and has the advantage that the beads do not significantly affect the light scatter or fluorescent properties of the cells, so that selection for CD34 can be followed by flow cytometry.

The techniques described above are primarily enrichment steps which are able to select populations of progenitor cells from large numbers of whole bone marrow or peripheral blood samples. Flow cytometry can take this a step further and purify the most primitive cells away from the more committed cells in the bone marrow. The CD34 compartment is now known to be heterogeneous, and contains cells committed to the various lymphoid and myeloerythoid lineages. Several cell surface markers have been used to identify the more primitive population of cells found within the CD34 compartment, such as CD38 (Visser and Van, 1990; Terstappen *et al.*, 1991; Baum *et al.*, 1992), Thy-1 (Spangrude *et al.*, 1988b) and CD33 (Brandt *et al.*, 1992). These subpopulations of CD34 can be identified and separated using flow cytometry, which has the advantages of being able to detect and discriminate between low and high levels of antigen expression as well as being able to separate cells with a very high degree of purity ($> 90\%$). In fact, flow cytometric analysis of human bone marrow cells has revealed subpopulations based on the degree of CD34 expression, the most primitive cells appear to express high levels of CD34 (Sutherland *et al.*, 1989; DiGiusto *et al.*, 1994).

Flow cytometry has been used to separate early progenitor cells with various phenotypes including $CD34^+/CD38^{low}$ (Terstappen *et al.*, 1991), $CD34^+/lin^-/Thy^+$ (Baum *et al.*, 1992), $CD34^+/DR^-$ (Civin and Gore, 1993) and $CD34^+/CD33^-$ (Brandt *et al.*, 1988). Such highly purified populations of cells have been shown to be capable of engraftment of SCID-hu mice when prepared from either bone marrow (Chen *et al.*, 1994) or peripheral blood (Murray *et al.*, 1995). The interest in obtaining very pure populations of early self-replicating stem cell populations is on the increase, as strategies for gene transfer and transplantation of tumour free haemopoietic cells are investigated. Currently, combinations of CD34 enrichment procedures with flow cytometric purification are being used to obtain such cell populations (Korbling *et al.*, 1994; Ziegler *et al.*, 1994). However, the next generation of cell sorters may reduce the need for enrichment procedures, since high speed cell sorting systems are under development that can deal with clinical scale purifications (Reading *et al.*, 1994).

Several markers of murine progenitor population have been identified using monoclonal antibodies (mAbs) including Thy-1, Sca-1 (Spangrude *et al.*, 1988a), A4.1 (Jordan *et al.*, 1990), and, more recently, the murine equivalent of CD34 (Krause *et al.*, 1994). Although highly purified sorted populations of early cells could be isolated using these antibodies, these populations were found to be heterogeneous in their ability to provide long term durable engraftment of irradiated recipients, which is a feature associated with the most primitive of bone marrow cells. Heterogeneity has been revealed in the populations by the use of biological markers such as rhodamine 123 (Visser and de Vries, 1988; Bertoncello *et al.*, 1991), which stains mitochondria, or Hoechst 33342 which stains DNA (Baines and Visser, 1983). Rhodoamine 123 low (dull), non-cycling populations of cells could be isolated from their activated or cycling counterparts (Fleming *et al.*, 1993) and it has been shown that the long term reconstitution activity was found in this population (Wolf *et al.*, 1993). In summary, the techniques to enrich primitive cell populations and long term marrow reconstituting cells are being constantly refined as new antibodies and other tools become available. These populations can then be used to determine the effects of the various growth factors on the survival, proliferation and development of the primitive cells.

STROMAL CELL INTERACTIONS WITH PRIMITIVE HAEMOPOIETIC CELLS

Using the long term bone marrow culture system it has been possible to demonstrate that haemopoietic stem cells can survive *in vitro* when allowed intimate contact with bone marrow derived stromal cells (including adipocytes, macrophages and fibroblasts). Furthermore the stem cells can undergo differentiation and development to form mature haemopoietic cells. Stromal cells have been shown to produce a variety of growth factors *in vitro* including M-CSF, GM-CSF, G-CSF, IL-1, IL-6, IL-7, transforming growth factor β (TGFβ) (Gualtieri *et al.*, 1984; Fibbe *et al.*, 1988; Yang *et al.*, 1988; Henney, 1989) and stem cell factor (SCF) (see Table 1). Many of the cytokines produced from the stromal cells in a soluble form can specifically associate with components of the extracellular matrix and in particular the heparan sulfate proteoglycans (Gordon *et al.*, 1987,

Table 1 Colony stimulating factors which can influence mature progenitor cell development.

Growth factor	Responsive progenitor cell population
Interleukin-3 (IL-3)	CFC-Mix GM-CFC Eos-CFC Meg-CFC Bas-CFC BFU-e
Granulocyte-Macrophage Colony Stimulating Factor (GM-CSF)	GM-CFC Eos-CFC BFU-e*
Macrophage Colony Stimulating Factor (M-CSF or CSF-1)	GM-CFC M-CFC
Granulocyte Colony Stimulating Factor (G-CSF)	GM-CFC G-CFC
Erythropoietin (epo)	CFU-e Meg-CFC
Thrombopoietin (TPO or MDGF)	Meg-CFC

*Denotes limited response.
See text for definition of cell populations.

1991; Siczkowski *et al.*, 1992). The proteolysis of these cytokines may be limited by the interaction between the cytokine and the ECM whilst the appropriate 'niches' for self-renewal or development of haemopoietic stem cells may have as a key determinant the levels and types of cytokines presented to the haemopoietic cell. In this respect a key characteristic of one cytokine found in the bone marrow, SCF, which could be of importance within the stromal cell environment, is its ability to decrease by several hundred fold the amount of a colony stimulating factor required for maximal proliferation and development (Heyworth *et al.*, 1992; Lowry *et al.*, 1992). Thus within the niche (possibly expressing high levels of SCF) concentrations of other cytokines need only be vanishingly small (as has been observed using immunocytochemical approaches to study long term bone marrow cultures) (De-Wynter *et al.*, 1993) for a maximal biological response to be achieved.

Primitive haemopoietic cells can also be maintained on adherent cell lines over a limited period (Breems *et al.*, 1994), during which time they form cobblestone areas consisting of clonal outgrowths from a single cell. The molecular interactions between primitive progenitor cells and these stromal cell lines have still to be characterized, but it is becoming increasingly apparent that there is an essential role for *membrane bound* cytokines. The FLT3/FLK2, c-Kit and c-Fms receptors are closely related receptor tyrosine kinases expressed on primitive haemopoietic cells which can bind membrane bound forms of their respective ligands FLT3/FLK2 ligand, SCF and M-CSF (Rettenmier and Roussel, 1988; Brannan *et al.*, 1991; Hannum *et al.*, 1994). In the case of SCF it has been demonstrated that a failure to produce the membrane bound ligand *in vivo* severely depletes blood cell production (Brannan *et al.*, 1991). Furthermore, Toksoz and colleagues investigated this by generating stromal cell lines from Sl mice (SCF deficient, see below) which were then transfected with cDNAs encoding soluble and membrane bound forms of SCF (Toksoz *et al.*, 1992). They found that optimal progenitor cell proliferation was achieved where membrane bound SCF was present. Membrane bound forms of SCF and M-CSF have both been shown to promote the adhesion of cells bearing the appropriate receptors (Rettenmier and Roussel, 1988; Flanagan *et al.*, 1991).

Another indicator of the importance of cytokines, produced within the stroma, to the primitive haemopoietic cells is the observation that *primitive* progenitor cell cycling in long term cultures exposed to IL-3 or feeder layers producing IL-3 is significantly enhanced (Otsuka *et al.*, 1991b). Set against this is the observation of Zipori and Lee that induced over expression of soluble IL-3 on stromal cell lines did not influence their competence to support progenitor cell proliferation (Zipori and Lee, 1988). Hogge, Eaves and co-workers have considered the impact of increasing the concentration of a range of cytokines on the proliferative status of primitive progenitor cells in long term marrow cultures (Hogge *et al.*, 1994). Their results show a degree of subtlety in the frequency, concentration and method of cytokine addition (either by the use of genetically engineered fibroblasts in the culture or adding soluble cytokine). In summary, G-CSF, IL-3, and IL-6 can induce primitive cell cycling but GM-CSF could not be demonstrated to do so. Importantly (with respect to stem cell expansion *in vivo*, see below) relatively modest effects were seen on the *numbers* of long term marrow culture initiating cells (LTCIC) (Szilvassy *et al.*, 1990; Chapter 13, this book), by adding growth factors: IL-3 + IL-6 and IL-3 + G-CSF both gave increases of only twofold (Hogge *et al.*, 1994).

It has also become apparent that integrins have a role in primitive progenitor cell adhesion to stromal cell layers. The counter-receptors, VLA-4 and VCAM, VLA-5 and fibronectin, β_2-integrin and ICAM-1 potentiate progenitor cell interactions with stroma (Kincade, 1992; Teixido *et al.*, 1992; Dittel *et al.*, 1993). The physiological relevance of this interaction has been demonstrated in primates where anti-VLA-4 antibodies stimulate the mobilization of primitive progenitor cells into the peripheral blood (Papayannopoulou and Nakamoto, 1993). A range of cellular adhesion molecules (e.g. $\alpha5\beta1$, $\beta2$, LFA3) are expressed on CD34 positive human progenitor cells from normal bone marrow but the functional significance of this in stromal cell regulation of haemopoiesis remains unclear. One important point to make, however, is that the cellular adhesion molecules are not simply 'anchors', occupation of these entities by their counter-receptors triggers intracellular signalling. As such the possibility that integrins can interact synergistically with cytokines to determine the developmental and proliferative fate of primitive cells deserves further exploration.

Another example of non-cytokine based interactions between stroma and haemopoietic cells is the role of a 110 kD bone marrow protein restricted to areas of the stroma which are associated with developing lymphoctyes (Jacobsen, K. *et al.*, 1992). CD44 is also involved in haemopoietic cell development (He *et al.*, 1992). Perturbing CD44 binding to haemopoietic cells in long term bone marrow cultures can inhibit myelopoiesis although non-stroma (i.e. free cytokine) stimulated progenitor cell development is unaffected, indicating a role for CD44 in stromal cell mediated progenitor cell growth and development which is independent of the production of cytokines.

Negative regulators of primitive haemopoietic cell proliferation are also endogenously produced by the bone marrow stroma. Both macrophage inflammatory protein 1α (MIP-1α) and TGFβ have been shown to be produced by stromal cells and their levels within the culture can be shown to inhibit primitive progenitor cell cycling (Eaves *et al.*, 1991; Otsuka *et al.*, 1991a) (see also Chapter 13). It is difficult to assess whether the effects observed within a long term culture are due to direct effects upon the primitive cells or mediated by indirect effects via cytokine production in the stroma. However, the relative cycling status of the primitive cells is undoubtedly governed by the balance of growth inhibitory factors and promoters present in the culture (Hogge *et al.*, 1994; Jacobsen, S.E. *et al.*, 1994b).

RECEPTORS TYPES FOUND ON PRIMITIVE CELLS: MOLECULAR CHARACTERIZATION

One of the intriguing questions in primitive cell responses to growth factors is whether the diversity in the structure of haemopoietic growth factor receptors reflects an ability to initiate a range of distinctive intracellular effector mechanisms which govern differentiation and development. Currently there is little evidence of a cytokine-mediated deterministic signal leading to lineage commitment or development in primitive cells. In committed progenitor cells (bipotential granulocyte macrophage colony forming cells) we have recently found that cytokines that promote macrophage development stimulate a chronic increase in phosphatidylcholine breakdown and diacylglycerol (the physiological activator of PKC) production that leads to translocation of PKCα (but no other isoform) to the nucleus within 2 hours (Whetton *et al.*, 1994; Nicholls *et al.*, 1995). No cytokine that promotes neutrophil production has such an effect. This signal for macrophage development may be just one example of the way in which cytokines can influence differentiation. To understand how this might be achieved in primitive cells a great deal of effort has been expended on cloning the cDNA for cytokine receptors and examining the structure which correlates with the function of the various receptors, in order to dissect out the signals associated with survival, proliferation and development, respectively.

The cytokines which are active on haemopoietic primitive and progenitor cells bind to specific cognate receptors. Some of these receptors are members of receptor families which share structural similarities and common subunits. The M-CSF receptor, FLK-2/FLT3 receptor and the SCF receptor are all members of the type 2 receptor/tyrosine kinase family (which includes the PDGF receptor). Similarly IL-2, IL-3, GM-CSF, erythropoietin, thrombopoietin and other cytokines share some common structural motifs in their receptor subunits (and are therefore members of the so-called cytokine receptor family). This family of receptors have no intrinsic protein tyrosine kinase activity but are able, when occupied, to activate a number of distinct tyrosine kinases (Miyajima *et al.*, 1993). In some instances members of this family also share a common subunit. IL-3, GM-CSF and IL-5 all bind to distinct and specific receptors (αIL-3, αGM-CSF, and αIL-5). However, these receptors all share a common β (β_c) subunit that couples with the specific α subunit, the β_c and α subunits both play an essential role in intracellular signalling. IL-5, IL-3 and GM-CSF have very different effects *in vivo* and in soft gel assays with normal bone marrow. Is this a reflection of differential expression of receptor α subunits on different target cell populations, the conformation and signalling capacity of the specific $\alpha\beta$ dimer in question, or of both common and specific signals mediated by the α and the β subunit respectively? The answers to these intriguing questions awaits further molecular dissection of these receptors in appropriate model systems.

Other receptor types are also presumed to be present on primitive cells, based on the ability of their cognate ligands to stimulate or inhibit their proliferation and development. In practice so few receptors are present on primitive cells and the number of primitive cells that can be prepared from normal bone marrow is so limited that the activity of a cytokine in a colony forming cell assay is often the only evidence for receptor expression. Other receptor types found on primitive cells include the IL-1 type which binds a cytokine that acts synergistically with other growth factors to stimulate proliferation and development of primitive cells. The IL-1 receptor is a member of the immunoglobulin superfamily (Sims *et al.*, 1988). Tumour necrosis factor α (TNF-α), can also act on highly purified human progenitor cells to promote growth when added in combination with other cytokines (Caux *et al.*, 1990, 1991). Two distinct TNFα receptors have been cloned and sequenced (p55 and p75) and are members of a growing receptor family, members of which can promote cell death as well as proliferation (Smith *et al.*, 1994).

As well as growth stimulators, a number of molecules having myeloininhibitory actions *in vitro* and in animal models have been characterized (see Chapter 13); these include TGFβ and MIP-1α. The MIP-1α binding site on primitive cells may be related to the MIP-1α receptor whose sequence has recently been determined (Gao *et al.*, 1993; Neote *et al.*, 1993). This is a member of the seven membrane spanning G protein-coupled receptor superfamily. The TGFβ receptor and its relatives are members of an unique family which has intrinsic serine/threonine protein kinase activity (Massague, 1992; Attisano *et al.*, 1994). There also is a complete range of cellular adhesion molecules expressed on haemopoietic progenitor cells which can of course also be counted as receptors (see above). The physical demonstration of receptor expression for specific cytokines on primitive cells is covered in the next section.

EFFECTS OF GROWTH FACTORS ON PRIMITIVE CELLS: *IN VITRO* STUDIES

Initially the colony stimulating factors were characterized by the types of colonies formed from normal bone marrow cells in soft gel assays (Table 1). The more mature progenitor cells (Figure 1) can respond to a single cytokine *in vitro* to produce colonies (>50 cells) containing mature cells. We now know that (in general terms) the cells that were identified using this 'readout' were committed progenitor cells which did not have the features that define the haemopoietic stem cell (see Chapter 13). Further research revealed that combinations of cytokines can act on primitive progenitor cells in a synergistic fashion to promote proliferation and development and assays were established to quantify the numbers and cycling status of such cells (including the high proliferative potential colony

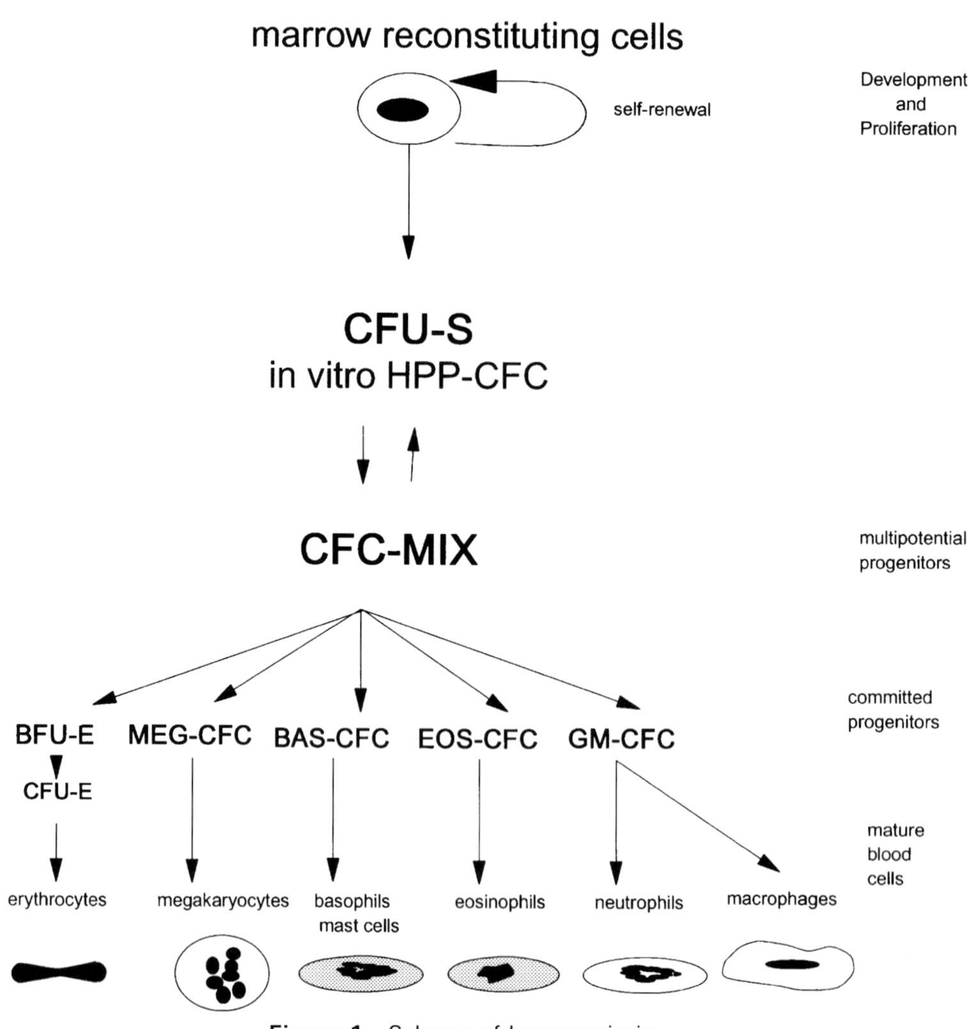

Figure 1 Scheme of haemopoiesis.

forming cell or HPP-CFC assay (Bradley and Hodgson, 1979; McNiece *et al.*, 1986), the blast cell colony forming cell assay (Leary and Ogawa, 1987), the long term culture initiating cell assay (Szilvassy *et al.*, 1990) and the delta assay (Moore *et al.*, 1990)). These assays are described more fully in Chapter 13.

Using these *in vitro* approaches of recombinant growth factors and enriched cell populations (described earlier) it has been possible to assess some of the actions of these growth factors on primitive cells. A list of some of the cytokines which act on haemopoietic cells is shown in Table 2. Details of the synergistic interactions between the haemopoietic growth factors leading to colony formation are also shown. The amplification and development of these cells appears to depend on the number of cytokines present: several groups have shown that, as the number of factors is increased then the amplification of the cell numbers produced is increased (Heimfeld *et al.*, 1991; Haylock *et al.*, 1992). These cells are also sensitive to the inhibitory stimulus of various growth factors such as TGF-β and MIP-1α. The action of both these growth factors is dependent upon the developmental status of the cell, they inhibit the cell cycling of the primitive cells but can act in a growth stimulatory manner on the more mature progenitor cell. Interestingly TGF-β and MIP-1α have been suggested to have slightly different target cells in the primitive progenitor compartment (Keller *et al.*, 1994). However, it is apparent that a balance between these stimulatory and inhibitory growth factors is

Table 2 Growth factors that may interact synergistically on primitive haemopoietic cells.

Growth Factor	References
Interleukin-3 (IL-3)	(Tsuji *et al.*, 1991b)
Granulocyte Macrophage Colony Stimulating Factor (GM-CSF)	(Koike *et al.*, 1987; McNiece *et al.*, 1988; Bot *et al.*, 1990)
Macrophage Colony Stimulating Factor (M-CSF or CSF-1)	(Heyworth *et al.*, 1988; McNiece *et al.*, 1988; Zhou *et al.*, 1988; Bartelmez *et al.*, 1989)
Granulocyte Colony Stimulating Factor (G-CSF)	(Ikebuchi *et al.*, 1988a, 1988b)
Stem Cell Factor (SCF or c-kit ligand)	(Migliaccio *et al.*, 1991; Tsuji *et al.*, 1992)
Ligand for the Flk2/Flt3* receptor	(Small *et al.*, 1994; Ziegler *et al.*, 1994)
Interleukin-1 (IL-1)	(Heyworth *et al.*, 1988; Ikebuchi *et al.*, 1988b; Jacobsen, S.E. *et al.*, 1994a)
Interleukin-4 (IL-4)	(Musashi *et al.*, 1991a)
Interleukin-6 (IL-6)	(Ikebuchi *et al.*, 1987; Jacobsen *et al.*, 1994a)
Interleukin-7 (IL-7)	(Fahlman *et al.*, 1994; Jacobsen, F.W. *et al.*, 1994)
Interleukin-9 (IL-9)	(Lemoli *et al.*, 1994; Sonoda *et al.*, 1994)
Interleukin:10 (IL-10)	(Rennick *et al.*, 1994)
Interleukin-11 (IL-11)	(Musashi *et al.*, 1991b; Tsuji *et al.*, 1992)
Interleukin-12 (IL-12)	(Ploemacher *et al.*, 1993)
Interleukin-13 (IL-13)	(Jacobsen, S.E. *et al.*, 1994c)
Leukemia Inhibitory Factor (LIF)	(Leary *et al.*, 1990; Verfaillie and McGlave, 1991)
Tumour Necrosis Factor Alpha (TNF-α)	(Caux *et al.*, 1991, 1992)
Erythropoietin (epo)	(Migliaccio *et al.*, 1988; Brugger *et al.*, 1993; Rebel *et al.*, 1994)

*Flk2 denotes fetal liver kinase-2. Flt3 denotes fms-like tyrosine kinase-3.

essential for the maintenance of the homeostasis associated with normal haemopoiesis (Hogge *et al.*, 1994; Jacobsen, S.E. *et al.*, 1994b).

Although the synergistic interactions of the stimulatory factors have been extensively studied it appears that the inhibitory factors may also interact in this way (Broxmeyer *et al.*, 1993; Jacobsen, S.E. *et al.*, 1994b). Whether two factors are required to stimulate mutually exclusive signal transduction pathways which act synergistically to promote proliferation, or alternatively one factor promotes survival and the other proliferation, remains to be seen. One current hypothesis is that growth factors such as G-CSF, IL-6, IL-11, IL-12, LIF and SCF trigger the cycling of primitive progenitor cells, promoting their exit from the dormant G_0 or resting phase of the cell cycle. This then allows the primitive cells to respond to a second set of cytokines including IL-3, GM-CSF and M-CSF (Ogawa, 1993). The continuous presence of one or more cytokines from each group is then required to promote proliferation and development of the multipotent cells into mature cells.

How do these cytokines work together at the molecular level to promote proliferation and development of the primitive cells? Possibly by potentiation of cell surface receptor expression. Binding of cytokines such as IL-1α, IL-6 or G-CSF to cell surface receptors can influence the expression of receptors for other growth factors such as GM-CSF or IL-3 (Jacobsen, S.E. *et al.*, 1992; Sato *et al.*, 1993; Testa *et al.*, 1993). Similarly IL-3, GM-CSF and G-CSF can increase type II IL-1 receptor levels on progenitor cells (Dubois *et al.*, 1992; Hestdal *et al.*, 1992).

The description of the role of receptor transmodulation in the synergistic interactions between cytokines can now be approached using fluorescently tagged cytokines. Using the cell purification procedures described above plus a flow cytometric analysis of receptor expression it will be possible to describe receptor transmodulation on subsets of primitive progenitor cells. Wognum and Wagemaker and co-workers have been active in this field since its inception and have now described the progressive increase in GM-CSF receptor expression during monocyte/granulocyte development (Wognum *et al.*, 1994b). Furthermore, $CD34^{bright}$, $HLA\text{-}DR^{dull}$ human primitive progenitor cells have been shown to express the SCF, IL-6 and IL-3 receptors but the GM-CSF receptor is expressed only at low levels (if at all) (Wognum *et al.*, 1994a). It is important to note that the SCF receptor has been detected by others on primitive haemopoietic cells (Orlic *et al.*, 1993; Simmons *et al.*, 1994a) and that this cell surface marker and other receptors can also now be used as markers for purification of primitive cells.

EFFECTS OF GROWTH FACTORS ON PRIMITIVE CELLS: *IN VIVO* STUDIES

The ability of haemopoietic cells to reconstitute blood cell production in syngeneic mice has been a cornerstone of the advances made in stem cell biology and has been covered in chapter 13. Whilst cells which can reconstitute haemopoiesis have been purified (see above) the mechanisms whereby they 'home' to sites of haemopoiesis and 'find' the sites within the bone marrow which promote long term survival and self-renewal remain obscure.

Infusion of exogenous cytokines into mice can give some details of the roles of cytokines *in vivo* (Pojda *et al.*, 1989; Molineux *et al.*, 1991). IL-1 for example can improve haemopoietic recovery from cytotoxic drug damage *in vivo* (Krumwieh *et al.*, 1990; Moore *et al.*, 1990) and also improves the survival of mice after irradiation or bone marrow transplantation (Oppenheim *et al.*, 1989; Neta *et al.*, 1990). However, these effects are probably due in part to the ability of IL-1 to stimulate monocytes, fibroblasts (including bone marrow fibroblasts), endothelial and other cell types (Sieff *et al.*, 1987; Kaushansky *et al.*, 1988; Shannon *et al.*, 1988; Yang *et al.*, 1988) to produce a range of cytokines and growth factors including M-CSF and G-CSF and GM-CSF, making the interpretation of the *direct* effects of IL-1 on primitive cells *in vivo* complex to interpret from such experiments. Some experiments using anti-IL-1 receptor antibodies have suggested that not all the radioprotection effects of IL-1 can be accounted for by the ability to stimulate cytokine production (Neta *et al.*, 1990), it is possible that some of the effects may involve a direct interaction with the primitive cell eliciting, for example, an upregulation of a receptor (Neta *et al.*, 1994). Similarly murine IL-3 *in vitro* is a potent stimulator of primitive progenitor cell proliferation which can put CFU-S (see Chapter 13) into an actively cycling state (Schrader and Clark, 1982; Spivak *et al.*, 1985). Yet when IL-3 is infused into mice initial experiments showed it to have a marginal effect on the bone marrow and a marginal effect on splenic haemopoiesis (Lord *et al.*, 1986; Metcalf *et al.*, 1986). More recently IL-3 has been shown to stimulate haemopoiesis *in vivo* but this is manifested as an increase in peripheral blood cells only when a second cytokine is added in tandem with the IL-3 (Donahue *et al.*, 1988; Krumwieh *et al.*, 1990). Thus the effects of a potent colony stimulating factor *in vitro* are seen to be decidedly subtle *in vivo*.

G-CSF, on the other hand has little colony stimulating activity *in vitro* (when normal bone marrow is plated in soft gel assays with this cytokine small neutrophilic colonies are generally formed, see Table 1) but marked and clinically significant effects *in vivo*. Twice daily injections of G-CSF over 4–5 days into mice led to an increased neutrophil count in the peripheral blood, a large increase in the numbers of splenic CFU-S and granulocyte macrophage colony forming cells (GM-CFC) with a decrease in the quantity of these progenitor cells in the bone marrow. A marked decrease in the duration of neutropenia was also observed in G-CSF treated mice which had been sub-lethally irradiated or treated with cytotoxic agents (Fujisawa *et al.*, 1986; Moore and Warren, 1987; Shimamura *et al.*, 1987). Subsequently clinical trials have thoroughly investigated the effects of G-CSF on neutropenia which occurs as a consequence of cytoreductive treatments in patients with cancer (Morstyn *et al.*, 1994). G-CSF does relieve neutropenia in humans, apparently by reducing cell cycle time and increasing the number of cell divisions in the late granulocyte precursor compartment in the marrow (Lord *et al.*, 1989). However, this will in the future be perhaps less significant than the discovery that infusion of G-CSF (and a number of other cytokines) can mobilize primitive and progenitor cells into the peripheral blood, these mobilized cells are suitable for bone marrow reconstitution and will increasingly figure in the treatment strategies for a number of diseases (Molineux *et al.*, 1990; Pettengell *et al.*, 1993).

In summary, we obtain information on the gross effects of non-physiological doses of cytokines by injecting them into animals and humans and this has had a profound impact on the perception of ways in which cytokines can be used in the treatment of a range of diseases. Because of the complex interactions in the cytokine network where cascades of cytokine production can be initiated by infusion of a single growth factor or cytokine it is not possible to infer a *direct* effect of any agent on stem cells from such experiments. To achieve such an aim the genetic manipulation of animal models is required to generate mice which do not produce specific cytokines.

SCF (or KL or Kit Ligand), binds to the proto-oncogene product c-kit, a transmembrane tyrosine kinase receptor which has been mapped to the mouse W or white spotting locus (Copeland *et al.*, 1990; Matsui *et al.*, 1990; Witte, 1990). SCF is the product of the gene found at the steel (Sl) locus. Mutations at these loci result in defective haemopoietic, melanocyte and gonadal development. Studies using long term marrow cultures demonstrated that the defect in the Sl mutation was due to a stromal cell environment that could not support haemopoiesis whilst the W locus mutations led to defects which are intrinsic to haemopoietic stem cells (Dexter and Moore, 1977). This indicates a major role for SCF–c-kit receptor tyrosine kinase interaction in haemopoiesis *in vivo*. In 1990 several groups reported the functional expression of stem cell factor cDNA (Williams *et al.*, 1990; Zsebo *et al.*, 1990) which was formally demonstrated to be the product of the Sl locus. Infusion of SCF can 'cure' the defective haemopoiesis seen in steel mice (Zsebo *et al.*, 1990). Experiments *in vitro* show that SCF can act on primitive cells to promote their survival, although SCF has to be added in combination with other cytokines to stimulate primitive progenitor cell proliferation and development (Migliaccio *et al.*, 1991; Tsuji *et al.*, 1991a). SCF appears to be synthesized in both a soluble and membrane bound form (Flanagan *et al.*, 1991) and data from mice with the Steel–Dickie mutation demonstrate that the membrane bound form of this cytokine is the essential element for normal haemopoiesis in the bone marrow (Brannan *et al.*, 1991). Furthermore use of the W mutant mouse has shown that the SCF receptor is also important in adhesion of primitive cells to the stroma (Kodama *et al.*, 1994). Thus a genetic approach suggests a pivotal role for this growth factor in primitive cell biology *in vivo*: primitive progenitor cells are presented with several different stimuli which act in a combinatorial or synergistic fashion to promote blood cell development and SCF is a key factor in this process.

Another mouse mutation that has arisen spontaneously has been of great value in elucidating the role of M-CSF (which is structurally related to SCF and binds to a similar receptor protein tyrosine kinase, see above) in stem cell biology. M-CSF was first identified as a growth factor for the monocyte/macrophage lineage by its ability to stimulate the formation of monocytic colonies in soft gel assays, in mice with the osteopetrotic mutation (op) there is a mutation in the M-CSF gene which leads to the production of a truncated, biologically inactive M-CSF molecule (Wiktor-Jedrzejczak *et al.*, 1990; Yoshida *et al.*, 1990). The op mice would appear to be osteopetrotic due to a deficiency of osteoclasts, furthermore the mice are deaf and blind, display poor male and female fertility and lactation is affected (Wiktor-Jedrzejczak *et al.*, 1982; Begg *et al.*, 1993). Yet there is no significant effect on stem cells or primitive progenitor cells, notwithstanding the fact that synergistic interactions between M-CSF and other cytokines are observed in primitive haemopoietic cells (Stanley *et al.*, 1986; Mochizuki *et al.*, 1987).

These experiments demonstrate that the most definitive way in which to describe the role of a cytokine *in vivo* is to genetically modify animals in a directed fashion such that the cytokine in question is no longer produced. Using embryonal stem cells plus an appropriate cytokine targeting vector to delete part of the cytokine gene several gene knockout mice have been established to identify the role of cytokines, other than SCF and M-CSF, which are essential for haemopoiesis. Leukaemia Inhibitory Factor (LIF), for example, has a variety of activities *in vitro* including inhibiting the differentiation of embryonic stem cells and stimulating the survival and proliferation of primitive haemopoietic progenitor cells (see Table 2 and Leary *et al.*, 1990; Fletcher *et al.*, 1991; Verfaillie and McGlave, 1991). LIF deficient mice show decreased numbers of splenic and bone marrow CFU-S and committed myeloid and erythroid progenitor cells, effects which could be reversed by the steady release of LIF into the animals from mini-osmotic pumps. LIF^- mice primitive haemopoietic cells could differentiate normally when injected into irradiated LIF^+ mice demonstrating that the defects observed are not intrinsically associated with the stem cells. Furthermore, infusion of cytokines (IL-6 and oncostatin M) which bind to the same receptor family, employing a common signalling subunit (gp130) could partially compensate for the lack of LIF, demonstrating the redundancy that in part exists in cytokine action which probably accounts for the viability of the LIF^- mice (Escary *et al.*, 1993).

This redundancy is exemplified by the observed pathology of GM-CSF deficient ($GM\text{-}CSF^-$) mice. The peripheral blood cell counts of the $GM\text{-}CSF^-$ mice are normal compared to controls, fertility was unaffected, marrow cellularity was unaltered and slight changes in haemopoietic cell progenitor cell numbers in the bone marrow and spleen (Stanley *et al.*, 1994) were observed: in terms of haemopoiesis there was no significant effect. Yet the $GM\text{-}CSF^-$ mice displayed a bronchovascular invasion of B lymphocytes and a proportion of the animals had inflamed areas in the lung, or pneumonia. This totally unpredicted result of a cytokine gene knockout experiment demonstrates the danger of extrapolating from *in vitro* observations to a likely role for a cytokine *in vivo*.

Perhaps more in line with the predicted effects, G-CSF knockout mice had a granulopoietic defect which depleted peripheral blood neutrophil counts and also the neutrophil reserves in the bone marrow (although the mice were born healthy) (Lieschke *et al.*, 1994a). $G\text{-}CSF^-$ mice also had an impaired ability to fight infection, a result which tallies with the observed effects of cyclic neutropenia in man which can now be treated by infusion with G-CSF to boost neutrophil count and decrease adventitious infection (Hammond *et al.*, 1989). But with respect to the effects of cytokines on stem cells the most significant observation in $G\text{-}CSF^-$ mice is a decrease in the splenic and bone marrow granulocyte precursor cells and colony forming cells.

The next step will be to cross mice strains deficient for one cytokine each respectively to generate double knockouts, to look for combinatorial effects. One such study, for example, has reported that GM-CSF- and M-CSF-deficient mice are osteopetrotic and have lung disease (Lieschke *et al.*, 1994b). Another approach will be to look at the effects of deleting receptor subunits or signal transduction proteins downstream from the receptors. Kishimoto has shown, for example, that gp130 null mice (gp130 is a receptor subunit which functions as a signalling protein for IL-6, LIF and other cytokines) cannot be obtained as they die at the early embryonic stage, the fetal livers contain very few primitive cells (as assessed by the CFU-S assay) which indicates a role for this ligand-activated signal transduction pathway in primitive haemopoietic cells (Kishimoto, 1994) although IL-6 null mice are viable but show an impaired immune response (Kopf *et al.*, 1994; Poli *et al.*, 1994). Yet a naturally occurring mutation in the IL-3 receptor α subunit gene has been identified in several mouse strains, this disallows colony formation stimulated by IL-3 bone marrow from these animals but the animals show apparently normal haemopoiesis (calling into question the need for this specific IL-3 stimulated signal transduction pathway *in vivo* (Ichihara *et al.*, 1995). The identification of a cytoplasmic protein tyrosine phosphatase, SH-PTP1, (Yi *et al.*, 1993) as the product of the 'motheaten' locus, mutations in which give rise to a number of haemopoietic abnormalities (Shultz *et al.*, 1993), suggests a critical role for this phosphatase in haemopoiesis. SH-PTP1 (also called PTP1C, HCP and SHP) binds the phosphorylated EPO receptor (Klingmuller *et al.*, 1995), c-KIT (Yi and Ihle, 1993) and the IL-3/GM-CSF/IL-5 receptor β_c subunit (Yi *et al.*, 1993). Its role is to deactivate occupied receptors, clearly the machinery activated by stem cell growth factors, when perturbed, is equally capable of causing malfunctions which may in the future be found to relate to human diseases such as myelodysplasia or anaemia.

Whilst the perhaps obvious effects of growth factors and cytokines on stem cells are to modulate their survival, proliferation and development, one of the more intriguing and clinically useful *in vivo* effects of certain cytokines is the mobilization of primitive progenitor cells from the haemopoietic organs into the peripheral blood. One of the consequences of increasing the intensity of chemotherapy in treatment of tumours is haematological toxicity, often the limiting factor in treating human tumours. One way of overcoming this is peripheral blood stem cell transplantation, which has been shown to result in successful engraftment after chemotherapy in a number of malignant diseases (Scheding *et al.*, 1994). A full consideration of this approach is beyond the scope of this article but the intriguing question is how do cytokines act on the haemopoietic organs to mobilize into the peripheral blood? G-CSF was one of the first cytokines shown to mobilize stem cells and committed progenitor cells into the peripheral blood (Molineux *et al.*, 1990) and these cells have marrow repopulating ability (Lord *et al.*, 1995). G-CSF can directly promote the entry of primitive cells into the cell cycle but it is unclear whether exogenous G-CSF or other cytokines such as SCF (Molineux *et al.*, 1991) act directly on stem and progenitor cells to induce mobilization. The precise mechanisms of mobilization remain unclear but there may be an involvement of the cytokines in potentiating the cellular adhesion molecules expressed on the surface of the stem and progenitor cells (Simmons *et al.*, 1994b).

MAINTENANCE AND EXPANSION OF STEM CELLS *IN VITRO* FOR CLINICAL USE USING GROWTH FACTORS

Another way to manipulate stem cell numbers is to expand them *ex vivo*. The combination of SCF, IL-3 and IL-6 can at least maintain, and possibly expand, the primitive murine and human cell populations (Bodine *et al.*, 1991a, 1992). This treatment also renders the cells more susceptible to retroviral-mediated gene transfer by stimulating cycling, which will be of key importance in future gene therapy strategies for human disease (Bodine *et al.*, 1989; Luskey *et al.*, 1992). It is of interest that IL-3 plus IL-6 alone do not expand the population of cells with repopulating ability whereas the addition of SCF with these two cytokines can aid in this respect (Bodine *et al.*, 1989, 1991a), perhaps reflecting the fact that the larger the number of cytokines which act on primitive cells included in the culture medium the greater the probability of *ex vivo* expansion of stem cells (Haylock *et al.*, 1992; Brugger *et al.*, 1993). Other approaches to stem cell expansion involve the use of stromal cell layers $\pm$ exogenous cytokines (see above). This area of *ex vivo* expansion studies is under intense study in a number of laboratories: the optimal or near optimal conditions for expanding a population of cells with long term reconstituting capabilities are far from being defined but the essential elements of the 'recipe' will undoubtedly include some of the cytokines described above.

EFFECTS OF CYTOTOXIC TREATMENT ON THE STEM CELL POPULATION

Mammals possess a complement of stem cells vastly in excess of that needed to supply adequate numbers of mature functional haemopoietic cells for a normal lifespan: experiments in mice have established that about 5×10^4–10^5 transplanted normal bone marrow cells are sufficient to rescue the haemopoietic system following potentially lethal irradiation. As the number of haemopoietic cells in the bone marrow of one mouse is of the order of $2–4 \times 10^8$ cells, it can be calculated that the bone marrow of one donor would be able to rescue 2–4,000 mice (Dexter and Testa, 1994). These transplant recipients can in turn be used as bone marrow donors for a second, and maybe even a third sequential generation of ablated recipients. It is, however, difficult to envisage a sound evolutionary reason for this vast reserve capacity. Haemopoietic stress requiring increased cell production (for example, after blood loss or haemolysis, or during infection) elicits a response which involves the recognizable maturing cells in the bone marrow, and according to the intensity and duration of the stimulus, may also involve progenitor cells. There exist several complementary mechanisms, comprising first increased cell mobilization from the marrow to the circulation, and then extra cell divisions in the maturation sequence, shortening of the intermitotic time (Lord *et al.*, 1989) and increases of the proportion of progenitor cells in active cell cycle (Tejero *et al.*, 1988). These mechanisms contribute to increased cell production without necessarily having to involve the very primitive progenitor cells or the stem cell population, which are normally found in a quiescent state. How this quiescent state is controlled is largely unknown, but probably inhibitory cytokines are involved (see above).

Although normal haemopoiesis is polyclonal, experimental evidence indicates that only a few of the available stem cells undergo clonal expansion, involving both proliferation and differentiation, at any given time (Micklem *et al.*, 1975; Abkowitz *et al.*, 1988). Therefore, it can be postulated that in a normal individual, both steady-state and stress haemopoiesis (as described above) will only necessitate a minority of the available stem cells in order to maintain adequate production of mature haemopoietic cells. However, the twentieth century brought the therapeutic use of radiation and of cytotoxic drugs, and the menace of radiation accidents. Possible for the first time, random or even preferential (for example, by busulfan) extensive kill of stem cells will occur. In these circumstances, the haemopoietic system will have to regenerate from a depleted stem cell population, either by endogenous regeneration, or from a usually small number of transplanted stem cells. There is experimental evidence that after acute regeneration has been completed, a new steady state exists, where an adequate output of mature cells can be obtained from a reduced stem cell and progenitor cell compartment (Table 3). Total numbers of progenitor cells are usually found at about half the control values. An even more marked defect may be estimated in the stem cell population, as indicated by their poor capacity (5–10% of normal) to generate progenitor cells in long-term culture (Dexter *et al.*, 1979) or in spleen colony assays (Molineux *et al.*, 1986). However, because

Table 3 New steady state of the haemopoietic system after recovery from cytotoxic injury (% of normal values).

Treatment	Bone marrow				Peripheral blood			Evolution
	Capacity to generate CFC	CFU-S	GM-CFC	Cellularity	WBC	Hb	Platelets	
Mice (a)								
1.5 Gy × 4	—	30	60	100	—	—	—	Stable
4.5 Gy × 4	5	10–30	40–50	80–100	50–100	100	—	Late anaemia†
Busulfan 40 mg/kg × 1	10	50	50	—	—	—	—	Late aplasia
BCNU 30 mg/kg × 4	—	30	48	100	—	—	—	Stable
PATIENTS (b)								
Acute lymphoblastic leukaemia	10–15	—	20–30	100	100	100	100*	Stable†
Hodgkin's disease	10	—	27	100	100	100	100	Stable†
Acute myeloid leukaemia	10	—	45	100	100	100	100	Stable†

(a) Followed until the end of their life span; (b) followed for periods of 1 to 14 years after cessation of treatment; * moderate plaquetopenia in a minority of cases; † instances of secondary malignancy.

Bus, busulfan. BCNU, bis-clonoethyl nitrosourea.

References in text.

of the compensatory mechanisms listed above, mature cell output is, in general, normal. In patients who have been treated for oncological/haematological disease, a similar pattern can be detected for periods of observation that range from 1–14 years (Table 3). Here, very good documentation of the number of cells and platelets in peripheral blood establishes that, except for occasional cases of mild plaquetopenia in children who have been treated for acute lymphoblastic leukaemia (ALL), a normal cell output is maintained. In mice a late anaemia or aplasia is observed in some experimental protocols, but not in others (Testa *et al.*, 1985). However, late haemopoietic failure has not, to date, been reported in patients as a consequence of cytotoxic treatment, and the clinical implications of these observations are unknown.

TO WHAT EXTENT DO STEM CELLS REGENERATE THEIR OWN POPULATION?

From the examples presented in Table 3, it would appear that stem cell self-regeneration after cytotoxic treatment is limited. Indeed, we may question whether, if normal cell output can be maintained for the rest of the lifespan from a greatly diminished stem cell compartment, there is a need for such regeneration.

The advent of recombinant growth factors, however, has generated great enthusiasm for stem cell expansion *in vitro*, mainly for the purposes of transplantation and gene manipulation. While it is clear that the numbers of more mature cells including progenitor cells, can be expanded *in vitro* (Haylock *et al.*, 1992), the numbers of putative stem cells can, at best, be only maintained at present (Henschler *et al.*, 1994).

Reported increases in the numbers of $CD34^+$ cells stimulated *in vitro* by different cocktails of growth factors (see earlier) are not accompanied by increases in LTCIC (which are originally only 0.5–1% of the $CD34^+$ population (Pettengell *et al.*, 1994). Stem cells, however, can undergo cell division in long-term culture, as has been established by studies using murine bone marrow cells with unique retroviral markers, with the newly generated cells assessed for repopulation capacity in transplantation studies (Fraser *et al.*, 1992). However, the total numbers of stem cells decreased in these cultures, implying that stem cell death is higher than the birth of new stem cells.

Long-term cultures of bone marrow cells from patients who had received extensive cytotoxic treatment, showed that addition of growth factors (G-CSF, GM-CSF or IL-3) produced similar responses to those in cultures of normal bone marrow, with stimulation of mature cell production and increase in the numbers of progenitor cells. However, although the relative magnitude of the responses observed were comparable in patients or control cultures, the noted low baseline capacity to generate new CFC in the former (Table 3). Here, very good documentation of the number of cells and platelets in peripheral blood establishes that, except for occasional cases of mild plaquetopenia in
No information is yet available on whether the stem cell compartment is expanded in patients treated with growth factors. On the experimental information available, however, such an event appears unlikely – it may also be argued whether it is desirable: growth factors which act on stem cells, like SCF, may induce stem cell proliferation (Harrison *et al.*, 1994). They may also protect cells from death by apoptosis (Yee *et al.*, 1994) and may thus allow stem cells which may have undergone deleterious mutations during

cytotoxic treatment, to survive. Such cells might subsequently originate second malignancies which are already a measurable risk after some treatment protocols.

DO WE WANT TO PROTECT STEM CELLS BY USING CYTOKINES?

Although no data are available on deleterious long-term effects in the oncological patients treated with G-CSF, GM-CSF and/or IL-3 during acute recovery following cytotoxic treatment, the modification of present protocols has to be considered carefully. Protective effects on haemopoiesis, with marked increase in survival in mice treated with G-CSF before being submitted to irradiation look encouraging (Waddick *et al.*, 1991). However, the interpretation of such data is difficult: whether those are direct effects of G-CSF (and on which cells?), or the results of activation of cytokine cascades that are not well-known. We also know that the pretreatment of mice with SCF before chemotherapy may be deleterious: the lethality of 5-FU is much higher when it follows SCF administration (Molineux *et al.*, 1994). Again, how SCF sensitizes stem cells to 5-FU is unknown. However, these findings may open a new way of marrow ablation without dose escalation, although the effects in other tissues (such as gut, where 5-FU is also more toxic when preceded by SCF) still have to be evaluated. The sensitization of haemopoietic stem cells may also be useful for therapeutic approaches, although the timing of such treatments are likely to be critical.

Protecting stem cells by using proliferation inhibitors that may exert a selective action on their entry into active cell cycle by holding them in G1 is at present being considered. The human equivalent of MIP-1α will shortly undergo clinical trials. If administration of this molecule is followed by cell cycle-active drugs, the stem cells may be protected (Lord *et al.*, 1992). However, if multiple drugs are used, MIP-1α, like SCF, may allow the survival of damaged cells, which might have undergone deleterious mutations. Therefore, the potential advantages of such protocols need to be assessed in more contexts than those which consider only the immediate potential advantages.

ACKNOWLEDGEMENTS

This work was supported by the Cancer Research Campaign and the Leukemia Research Fund. We wish to thank Stella Pearson for her help in preparing this manuscript.

REFERENCES

Abkowitz, J.L., Ott, R.M., Holly, R.D. and Adamson, J.W. (1988). *Blood* **71**:1687–1692.
Attisano, L., Wrana, J.L., Lopez, C.F. and Massague, J. (1994). *Biochim. Biophys. Acta* **1222**:71–80.
Baines, P. and Visser, J.W. (1983). *Exp. Hematol.* **11**:701–708.
Bartelmez, S.H., Bradley, T.R., Bertoncello, I. *et al.* (1989). *Exp. Hematol.* **17**:240–245.
Baum, C.M., Weissman, I.L., Tsukamoto, A.S., Buckle, A.M. and Peault, B. (1992). *Proc. Natl Acad. Sci. USA* **89**:2804–2808.
Begg, S.K., Radley, J.M., Pollard, J.W., Chisholm, O.T., Stanley, E.R. and Bertoncello, I. (1993). *J. Exp. Med.* **177**:237–242.

Berenson, R.J., Bensinger, W.I., Kalamasz, D.F. *et al.* (1992). *Prog. Clin. Biol. Res.* **377**:449–457.
Bertoncello, I., Bradley, T.R., Hodgson, G.S. and Dunlop, J.M. (1991). *Exp. Hematol.* **19**:174–178.
Bodine, D.M., Karlsson, S. and Nienhuis, A.W. (1989). *Proc. Natl Acad. Sci. USA* **86**:8897–8901.
Bodine, D.M., Crosier, P.S. and Clark, S.C. (1991a). *Blood* **78**:914–920.
Bodine, D.M., McDonagh, K.T., Seidel, N.E. and Nienhuis, A.W. (1991b). *Exp. Hematol.* **19**:206–212.
Bodine, D.M., Orlic, D., Birkett, N.C., Seidel, N.E. and Zsebo, K.M. (1992). *Blood* **79**:913–919.
Bot, F.J., Van, E.L., Schipper, P., Backx, B. and Lowenberg, B. (1990). *Leukemia* **4**:325–328.
Bradley, T.R. and Hodgson, G.S. (1979). *Blood* **54**:1446–1450.
Brandt, J., Baird, N., Lu, L., Srour, E. and Hoffman, R. (1988). *J. Clin. Invest.* **82**:1017–1027.
Brandt, J., Briddell, R.A., Srour, E.F., Leemhuis, T.B. and Hoffman, R. (1992). *Blood* **79**:634–641.
Brannan, C.I., Lyman, S.D., Williams, D.E. *et al.* (1991). *Proc. Natl Acad. Sci. USA* **88**:4671–4674.
Breems, D.A., Blokland, E.A., Neben, S. and Ploemacher, R.E. (1994). *Leukemia* **8**:1095–1104.
Broxmeyer, H.E., Sherry, B., Cooper, S. *et al.* (1993). *J. Immunol.* **150**:3448–3458.
Brugger, W., Mocklin, W., Heimfeld, S., Berenson, R.J., Mertelsmann, R. and Kanz, L. (1993). *Blood* **81**:2579–2584.
Brugger, W., Henschler, R., Heimfeld, S., Berenson, R.J., Mertelsmann, R. and Kanz, L. (1994). *Blood* **84**:1421–1426.
Caux, C., Saeland, S., Favre, C., Duvert, V., Mannoni, P. and Banchereau, J. (1990). *Blood* **75**:2292–2298.
Caux, C., Favre, S., Saeland, S. *et al.* (1991). *Blood* **78**:635–644.
Caux, C., Moreau, I., Saeland, S. and Banchereau, J. (1992). *Blood* **79**:2628–2635.
Chen, B.P., Galy, A., Kyoizumi, S. *et al.* (1994). *Blood* **84**:2497–2505.
Civin, C.I. and Gore, S.D. (1993). *J. Hematother.* **2**:137–144.
Civin, C.I., Strauss, L.C., Fackler, M.J., Trischmann, T.M., Wiley, J.M. and Loken, M.R. (1990). *Prog. Clin. Biol. Res.* **333**:387–401.
Copeland, N.G., Gilbert, D.J., Cho, B.C. *et al.* (1990). *Cell* **63**:175–183.
Coutinho, L.H., Will, A., Radford, J., Schiro, R., Testa, N.G. and Dexter, T.M. (1990). *Blood* **75**:2118–2129.
De-Wynter, E., Allen, T., Coutinho, L., Flavell, D., Flavell, S.U. and Dexter, T.M. (1993). *J. Cell Sci.* **106**:761–769.
Dexter, T.M. and Moore, M.A. (1977). *Nature* **269**:412–414.
Dexter, T.M. and Testa, N.G. (1994). *Haemopoietic Growth Factors: Review of Biology and Clinical Potential.* Gardiner-Caldwell Communications Ltd, Macclesfield, U.K.
Dexter, T.M., Schofield, R., Hendry, J. and Testa, N.G. (1979). *Hamatol. Bluttransfus* **24**:73–78.
Dexter, T.M., Coutinho, L.H., Spooncer, E. *et al.* (1990). *Ciba Found. Symp.* 148:76–86.
DiGiusto, D., Chen, S., Combs, J. *et al.* (1994). *Blood* 84:421–432.
Dittel, B.N., McCarthy, J.B., Wayner, E.A. and LeBien, T.W. (1993). *Blood* **81**:2272–2282.
Donahue, R.E., Seehra, J., Metzger, M. *et al.* (1988). *Science* **241**:1820–1823.
Dubois, C.M., Ruscetti, F.W., Jacobsen, S.E., Oppenheim, J.J. and Keller, J.R. (1992). *Blood* **80**:600–608.
Eaves, C.J., Cashman, J.D., Kay, R.J. (1991). *Blood* **78**:110–117.
Escary, J.L., Perreau, J., Dumenil, D., Ezine, S. and Brulet, P. (1993). *Nature* **363**:361–364.
Fahlman, C., Blomhoff, H.K., Veiby, O.P., McNiece, I.K. and Jacobsen, S.E. (1994). *Blood* **84**:1450–1456.
Fibbe, W.E., Van, D.J., Billiau, A. *et al.* (1988). *Blood* **72**:860–866.
Flanagan, J.G., Chan, D.C. and Leder, P. (1991). *Cell* **64**:1025–1035.
Fleming, W.H., Alpern, E.J., Uchida, N., Ikuta, K., Spangrude, G.J. and Weissman, I.L. (1993). *J. Cell Biol.* **122**:897–902.
Fletcher, F.A., Moore, K.A., Ashkenazi, M. *et al.* (1991). *J. Exp. Med.* **174**:837–845.
Fraser, C.C., Szilvassy, S.J., Eaves, C.J. and Humphries, R.K. (1992). *Proc. Natl Acad. Sci. USA* **89**:1968–1972.
Fujisawa, M., Kobayashi, Y., Okabe, T., Takaku, F., Komatsu, Y. and Itoh, S. (1986). *Jpn J. Cancer Res.* **77**:866–869.
Gao, J.L., Kuhns, D.B., Tiffany, H.L. *et al.* (1993). *J. Exp. Med.* **177**:1421–1427.
Gordon, M.Y., Atkinson, J., Clarke, D. *et al.* (1991). *Leukemia* **5**:693–698.

Gordon, M.Y., Riley, G.P., Watt, S.M. and Greaves, M.F. (1987). *Nature* **326**:403–405.
Grimsley, P.G., Amos, T.A., Gordon, M.Y. and Greaves, M.F. (1993). *Leukemia* **7**:898–908.
Gualtieri, R.J., Shadduck, R.K., Baker, D.G. and Quesenberry, P.J. (1984). *Blood* **64**:516–525.
Hammond, W.P., Price, T.H., Souza, L.M. and Dale, D.C. (1989). *N. Engl. J. Med.* 320:1306–1311.
Hannum, C., Culpepper, J., Campbell, D. *et al.* (1994). *Nature* 368:643–648.
Harrison, D.E., Zsebo, K.M. and Astle, C.M. (1994). *Blood* **83**:3146–3151.
Haylock, D.N., To, L.B., Dowse, T.L., Juttner, C.A. and Simmons, P.J. (1992). *Blood* **80**:1405–1412.
He, Q., Lesley, J., Hyman, R., Ishihara, K. and Kincade, P.W. (1992). *J. Cell Biol.* **119**:1711–1719.
Heimfeld, S., Hudak, S., Weissman, I. and Rennick, D. (1991). *Proc. Natl Acad. Sci. USA* **88**:9902–9906.
Henney, C.S. (1989). *Immunol. Today* **10**:170–173.
Henschler, R., Brugger, W., Luft, T., Frey, T., Mertelsmann, R. and Kanz, L. (1994). *Blood* **84**:2898–2903.
Hestdal, K., Jacobsen, S.E., Ruscetti, F.W. *et al.* (1992). *Blood* **80**:2486–2894.
Heyworth, C.M., Ponting, I.L. and Dexter, T.M. (1988). *J. Cell Sci.* **91**:239–247.
Heyworth, C.M., Whetton, A.D., Nicholls, S., Zsebo, K. and Dexter, T.M. (1992). *Blood* **80**:2230–2236.
Hogge, D.E., Sutherland, H.J., Cashman, J.D., Lansdorp, P.M., Humphries, R.K. and Eaves, C.J. (1994). *Baillière's Clin. Haematol.* **7**:49–63.
Ichihara, M., Hara, T., Takagi, M., Cho, L.C., Gorman, D.M. and Miyajima, A. (1995). *EMBO J.* **14**:939–950.
Ikebuchi, K., Wong, G.G., Clark, S.C., Ihle, J.N., Hirai, Y. and Ogawa, M. (1987). *Proc. Natl Acad. Sci. USA* **84**:9035–9039.
Ikebuchi, K., Clark, S.C., Ihle, J.N., Souza, L.M. and Ogawa, M. (1988a). *Proc. Natl Acad. Sci. USA* **85**:3445–3449.
Ikebuchi, K., Ihle, J.N., Hirai, Y., Wong, G.G., Clark, S.C. and Ogawa, M. (1988b). *Blood* **72**:2007–2014.
Jacobsen, K., Miyake, K., Kincade, P.W. and Osmond, D.G. (1992). *J. Exp. Med.* **176**:927–935.
Jacobsen, F.W., Rusten, L.S. and Jacobsen, S.E. (1994). *Blood,* **84**:775–779.
Jacobsen, S.E., Ruscetti, F.W., Dubois, C.M., Wine, J. and Keller, J.R. (1992). *Blood* **80**:678–687.
Jacobsen, S.E., Ruscetti, F.W., Okkenhaug, C. *et al.* (1994a). *Exp. Hematol.* **22**:1064–1069.
Jacobsen, S.E., Ruscetti, F.W., Ortiz, M., Gooya, J.M. and Keller, J.R. (1994b). *Exp. Hematol.* **22**:985–989.
Jacobsen, S.E., Okkenhaug, C., Veiby, O.P., Caput, D., Ferrara, P. and Minty, A. (1994c). *J. Exp. Med.* **180**:75–82.
Jones, R.J., Wagner, J.E., Celano, P., Zicha, M.S. and Sharkis, S.J. (1990). *Nature* **347**:188–189.
Jordan, C.T., McKearn, J.P. and Lemischka, I.R. (1990). *Cell* **61**:953–963.
Kannourakis, G., Johnson, G.R. and Battye, F. (1988). *Exp. Hematol.* **16**:367–370.
Kaushansky, K., Lin, N. and Adamson, J.W. (1988). *J. Clin. Invest.* **81**:92–97.
Keller, J.R., Bartelmez, S.H., Sitnicka, E. *et al.* (1994). *Blood* **84**:2175–2181.
Kincade, P.W. (1992). *Baillière's Clin. Haematol.* **5**:575–598.
Kishimoto, T. (1994). *Stem Cells* **12** (suppl 1):37–45.
Klingmuller, U., Lorenz, U., Cantley, L.C., Neel, B.G. and Lodish, H.F. (1995). *Cell* **80**:729–738.
Kodama, H., Nose, M., Niida, S., Nishikawa, S. and Nishikawa, S. (1994). *Exp. Hematol.* **22**:979–984.
Koike, K., Ogawa, M., Ihle, J.N. *et al.* (1987). *J. Cell Physiol.* **131**:458–464.
Kopf, M., Baumann, H., Freer, G. *et al.* (1994). *Nature* **368**:339–342.
Korbling, M., Drach, J., Champlin, R.E. *et al.* (1994). *Bone Marrow Transplant.* **13**:649–654.
Krause, D.S., Ito, T., Fackler, M.J. *et al.* (1994). *Blood* **84**:691–701.
Krumwieh, D., Weinmann, E., Siebold, B. and Seiler, F.R. (1990). *Int. J. Cell Cloning* **1**:229–247.
Leary, A.G. and Ogawa, M. (1987). *Blood* **69**:953–956.
Leary, A.G., Wong, G.G., Clark, S.C., Smith, A.G. and Ogawa, M. (1990). *Blood* **75**:1960–1964.
Lemoli, R.M., Fortuna, A., Fogli, M. *et al.* (1994). *Exp. Hematol.* **22**:919–923.
Lieschke, G.J., Grail, D., Hodgson, G. *et al.* (1994a). *Blood* **84**:1737–1746.
Lieschke, G.J., Stanley, E., Grail, D. *et al.* (1994b). *Blood* **84**:27–35.
Lord, B.I., Molineux, G., Testa, N.G., Kelly, M., Spooncer, E. and Dexter, T.M. (1986). *Lymphokine Res.* **5**:97–104.

Lord, B.I., Bronchud, M.H., Owens, S. *et al.* (1989). *Proc. Natl Acad. Sci. USA* **86**:9499–9503.
Lord, B.I., Dexter, T.M., Clements, J.M., Hunter, M.A. and Gearing, A.J. (1992). *Blood* **79**:2605–2609.
Lord, B.I., Woolford, L.B., Wood, L.M. *et al.* (1995). *Blood* **85**:3412–3415.
Lowry, P.A., Deacon, D., Whitefield, P., McGrath, H.E. and Quesenberry, P.J. (1992). *Blood* **80**:663–669.
Luskey, B.D., Rosenblatt, M., Zsebo, K. and Williams, D.A. (1992). *Blood* **80**:396–402.
McNiece, I.K., Bradley, T.R., Kriegler, A.B. and Hodgson, G.S. (1986). *Exp. Hematol.* **14**:856–860.
McNiece, I.K., Robinson, B.E. and Quesenberry, P.J. (1988). *Blood* **72**:191–195.
Massague, J. (1992). *Cell* **69**:1067–1070.
Matsui, Y., Zsebo, K.M. and Hogan, B.L. (1990). *Nature* **347**:667–669.
Metcalf, D., Begley, C.G., Johnson, G.R., Nicola, N.A., Lopez, A.F. and Williamson, D.J. (1986). *Blood* **68**:46–57.
Micklem, H.S., Anderson, N. and Ross, E. (1975). *Nature* **256**:41–43.
Migliaccio, G., Migliaccio, A.R. and Visser, J.W. (1988). *Blood* **72**:944–951.
Migliaccio, G., Migliaccio, A.R., Valinsky, J. *et al.* (1991). *Proc. Natl Acad. Sci. USA* **88**:7420–7424.
Miltenyi, S., Muller, W., Weichel, W. and Radbruch, A. (1990). *Cytometry* **11**:231–238.
Miyajima, A., Mui, A.L., Ogorochi, T. and Sakamaki, K. (1993). *Blood* **82**:1960–1974.
Mochizuki, D.Y., Eisenman, J.R., Conlon, P.J., Larsen, A.D. and Tushinski, R.J. (1987). *Proc. Natl Acad Sci USA* **84**:5267–5271.
Molineux, G., Testa, N.G., Massa, G. and Schofield, R. (1986). *Biomed. Pharmacother.* **40**:215–220.
Molineux, G., Pojda, Z., Hampson, I.N., Lord, B.I. and Dexter, T.M. (1990). *Blood* **76**:2153–2158.
Molineux, G., Migdalska, A., Szmitkowski, M., Zsebo, K. and Dexter, T.M. (1991). *Blood* **78**:961–966.
Molineux, G., Migdalska, A., Haley, J., Evans, G.S. and Dexter, T.M. (1994). *Blood* **83**:3491–3499.
Moore, M.A., Muench, M.O., Warren, D.J. and Laver, J. (1990). *Ciba Found. Symp.* **148**:43–58.
Moore, M.W. and Warren, D.J. (1987). *Proc. Natl Acad. Sci. USA* **84**:7134–7138.
Morstyn, G., Foote, M., Perkins, D. and Vincent, M. (1994). *Stem Cells* **12** (suppl 1):213–228.
Murray, L., Chen, B., Galy, A. *et al.* (1995). *Blood* **85**:368–378.
Musashi, M., Clark, S.C., Sudo, T., Urdal, D.L. and Ogawa, M. (1991a). *Blood* **78**:1448–1451.
Musashi, M., Yang, Y.C., Paul, S.R., Clark, S.C., Sudo, T. and Ogawa, M. (1991b). *Proc. Natl Acad Sci USA* **88**:765–769.
Neote, K., DiGregorio, D., Mak, J.Y., Horuk, R. and Schall, T.J. (1993). *Cell* **72**:415–425.
Neta, R., Vogel, S.N., Plocinski, J.M. *et al.* (1990). *Blood* **76**:57–62.
Neta, R., Oppenheim, J.J., Wang, J.M., Snapper, C.M., Moorman, M.A. and Dubois, C.M. (1994). *J. Immunol.* **153**:1536–1543.
Nicholls, S.E., Heyworth, C.M., Dexter, T.M. and Whetton, A.D. (1995). *J. Immunol.* **155**:845–853.
Ogawa, M. (1993). *Blood* **81**:2844–2853.
Oppenheim, J.J., Neta, R., Tiberghien, P., Gress, R., Kenny, J.J. and Longo, D.L. (1989). *Blood* **74**:2257–2263.
Orlic, D., Fischer, R., Nishikawa, S., Nienhuis, A.W. and Bodine, D.M. (1993). *Blood* **82**:762–770.
Otsuka, T., Eaves, C.J., Humphries, R.K., Hogge, D.E. and Eaves, A.C. (1991a). *Leukemia* **5**:861–868.
Otsuka, T., Thacker, J.D., Eaves, C.J. and Hogge, D.E. (1991b). *J. Clin. Invest.* **88**:417–422.
Papayannopoulou, T. and Nakamoto, B. (1993). *Proc. Natl Acad. Sci. USA* **90**:9374–9378.
Pettengell, R., Testa, N.G., Swindell, R., Crowther, D. and Dexter, T.M. (1993). *Blood* **82**:2239–2248.
Pettengell, R., Luft, T., Henschler, R. *et al.* (1994). *Blood* **84**:3653–3659.
Ploemacher, R.E., van-Soest, P.L., Boudewijn, A. and Neben, S. (1993). *Leukemia* **7**:1374–1380.
Pojda, Z., Molineux, G. and Dexter, T.M. (1989). *Exp. Hematol.* **17**:1100–1104.
Poli, V., Balena, R., Fattori, E. *et al.* (1994). *EMBO J.* **13**:1189–1196.
Reading, C., Sasaki, D., Leemhuis, T. *et al.* (1994). *Blood* **84** (suppl 1):399a.
Rebel, V.I., Dragowska, W., Eaves, C.J., Humphries, R.K. and Lansdorp, P.M. (1994). *Blood* **83**:128–136.
Rennick, D., Hunte, B., Dang, W., Thompson-Snipes, L. and Hudak, S. (1994). *Exp. Hematol.* **22**:136–141.
Rettenmier, C.W. and Roussel, M.F. (1988). *Mol. Cell. Biol.* **8**:5026–5034.
Sato, N., Sawada, K., Koizumi, K. *et al.* (1993). *Blood* **82**:3600–3609.
Schattenberg, A., De, W.T., Preijers, F. *et al.* (1990). *Blood* **75**:1356–1363.

Scheding, S., Brugger, W., Mertelsmann, R. and Kanz, L. (1994). *Stem Cells* **12** (suppl 1):203–211.
Schrader, J.W. and Clark, L.I. (1982). *J. Immunol.* **129**:30–35.
Schwartz, G.N., MacVittie, T.J., Monroy, R.L. and Vigneulle, R.M. (1986). *Exp. Hematol.* **14**:963–970.
Shannon, M.F., Gamble, J.R. and Vadas, M.A. (1988). *Proc. Natl Acad. Sci. USA* **85**:674–678.
Shimamura, M., Kobayashi, Y., Yuo, A. *et al.* (1987). *Blood* **69**:353–355.
Shultz, L.D., Schweitzer, P.A., Rajan, T.V. *et al.* (1993). *Cell* **73**:1445–1454.
Siczkowski, M., Clarke, D. and Gordon, M.Y. (1992). *Blood* **80**:912–919.
Sieff, C.A., Niemeyer, C.M. and Faller, D.V. (1987). *Blood Cells* **13**:65–74.
Simmons, P.J., Aylett, G.W., Niutta, S., To, L.B., Juttner, C.A. and Ashman, L.K. (1994a). *Exp. Hematol.* **22**:157–165.
Simmons, P.J., Leavesley, D.I. and Levesque, J. (1994b) *Stem Cells* 12 (suppl 1):187–202.
Sims, J.E., March, C.J., Cosman, D. *et al.* (1988). *Science* **241**:585–589.
Small, D., Levenstein, M., Kim, E. *et al.* (1994). *Proc. Natl Acad. Sci. USA* **91**:459–463.
Smith, C.A., Farrah, T. and Goodwin, R.G. (1994). *Cell* **76**:959–962.
Sonoda, Y., Sakabe, H., Ohmisono, Y. *et al.* (1994). *Blood* **84**:4099–4106.
Spangrude, G.J. (1989). *Immunol. Today* **10**:344–350.
Spangrude, G.J., Heimfeld, S. and Weissman, I.L. (1988a). *Science* **241**:58–62.
Spangrude, G.J., Muller, S.C., Heimfeld, S. and Weissman, I.L. (1988b). *J. Exp. Med.* **167**:1671–1683.
Spivak, J.L., Smith, R.R. and Igle, J.N. (1985). *J. Clin. Invest.* **76**:1613–1621.
Stanley, E.R., Bartocci, A., Patinkin, D., Rosendaal, M. and Bradley, T.R. (1986). *Cell* **45**:667–674.
Stanley, E., Lieschke, G.J., Grail, D. *et al.* (1994). *Proc. Natl Acad. Sci. USA* **91**:5592–5596.
Strauss, L.C., Trischmann, T.M., Rowley, S.D., Wiley, J.M. and Civin, C.I. (1991). *Am. J. Pediatr. Hematol. Oncol.* **13**:217–221.
Sutherland, H.J., Eaves, C.J., Eaves, A.C., Dragowska, W. and Lansdorp, P.M. (1989). *Blood* **74**:1563–1570.
Sutherland, D.R., Marsh, J.C., Davidson, J., Baker, M.A., Keating, A. and Mellors, A. (1992). *Exp. Hematol.* **20**:590–599.
Sutherland, D.R., Keating, A., Nayar, R., Anania, S. and Stewart, A.K. (1994). *Exp. Hematol.* **22**:1003–1110.
Szilvassy, S.J., Humphries, R.K., Lansdorp, P.M., Eaves, A.C. and Eaves, C.J. (1990). *Proc. Natl Acad. Sci. USA* **87**:8736–8740.
Teixido, J., Hemler, M.E., Greenberger, J.S. and Anklesaria, P. (1992). *J. Clin. Invest.* **90**:358–367.
Tejero, C., Hendry, J.H. and Testa, N.G. (1988). *Cell Tissue Kinet.* **21**:201–204.
Terstappen, L.W., Huang, S., Safford, M., Lansdorp, P.M. and Loken, M.R. (1991). *Blood* **77**:1218–1227.
Testa, N.G., Hendry, J.H. and Molineux, G. (1985). *Anticancer Res.* **5**:101–110.
Testa, U., Pelosi, E., Gabbianelli, M. *et al.* (1993). *Blood* **81**:1442–1456.
Toksoz, D., Zsebo, K.M., Smith, K.A. *et al.* (1992). *Proc. Natl Acad. Sci. USA* **89**:7350–7354.
Tsuji, K., Zsebo, K.M. and Ogawa, M. (1991a). *Blood* **78**:1223–1229.
Tsuji, K., Zsebo, K.M. and Ogawa, M. (1991b). *J. Cell Physiol.* **148**:362–369.
Tsuji, K., Lyman, S.D., Sudo, T., Clark, S.C. and Ogawa, M. (1992). *Blood* **79**:2855–2860.
Umemura, T., Papayannopoulou, T. and Stamatoyannopoulos, G. (1989). *Blood* **73**:1993–1998.
Verfaillie, C. and McGlave, P. (1991). *Blood* **77**:263–270.
Visser, J.W. and de Vries, P. (1988). *Blood Cells* **14**:369–384.
Visser, J.W. and Van, B.D. (1990). *Exp. Hematol.* **18**:248–256.
Waddick, K.G., Song, C.W., Souza, L. and Uckun, F.M. (1991). *Blood* **77**:2364–2371.
Wagner, J.E., Donnenberg, A.D., Noga, S.J. *et al.* (1988). *Blood* **72**:1168–1176.
Wagner, J.E., Santos, G.W., Noga, S.J. *et al.* (1990). *Blood* **75**:1370–1377.
Whetton, A.D., Heyworth, C.M., Nicholls, S.E. *et al.* (1994). *J. Cell Biol.* **125**:651–659.
Wiktor-Jedrzejczak, W.W., Ahmed, A., Szczylik, C. and Skelly, R.R. (1982). *J. Exp. Med.* **156**:1516–1527.
Wiktor-Jedrzejczak, W., Bartocci, A., Ferrante, A. Jr *et al.* (1990). *Proc. Natl Acad. Sci. USA* **87**:4828–4832.
Williams, D.E., Eisenman, J., Baird, A. *et al.* (1990). *Cell* **63**:167–174.
Witte, O.N. (1990). *Cell* **63**:5–6.

Wognum, A.W., de Jong, M.O., Egeland, T. and Wagemaker, G. (1994a). *Blood* **84**:16a.
Wognum, A.W., Westerman, Y., Visser, T.P. and Wagemaker, G. (1994b). *Blood* **84**:764–774.
Wolf, N.S., Kone, A., Priestley, G.V. and Bartelmez, S.H. (1993). *Exp. Hematol.* **21**:614–622.
Yang, Y.C., Tsai, S., Wong, G.G. and Clark, S.C. (1988). *J. Cell Physiol.* **134**:292–296.
Yee, N.S., Paek, I. and Besmer, P. (1994). *J. Exp. Med.* **179**:1777–1787.
Yi, T. and Ihle, J.N. (1993). *Mol. Cell Biol.* **13**:3350–3358.
Yi, T., Mui, A.L., Krystal, G. and Ihle, J.N. (1993). *Mol. Cell Biol.* **13**:7577–7586.
Yoshida, H., Hayashi, S., Kunisada, T. *et al.* (1990). *Nature* **345**:442–444.
Zeigler, F.C., Bennett, B.D., Jordan, C.T. *et al.* (1994). *Blood* **84**:2422–2430.
Zhou, Y.Q., Stanley, E.R., Clark, S.C. *et al.* (1988). *Blood* **72**:1870–1874.
Ziegler, B.L., Thomas, S., Lamping, C., Peschle, C. and Fliedner, T.M. (1994). *Blood* **84** (suppl 1):272a.
Zipori, D. and Lee, F. (1988). *Blood* **71**:586–596.
Zsebo, K.M., Williams, D.A., Geissler, E.N. *et al.* (1990). *Cell* **63**:213–224.

15 Haematopoietic stem cells for gene therapy

Jan A. Nolta and Donald B. Kohn

Division of Research Immunology/Bone Marrow Transplantation, Children's Hospital Los Angeles, Departments of Pediatrics and Microbiology, University of Southern California School of Medicine, Los Angeles, CA, USA

INTRODUCTION

Treatment of genetic disease in the haematopoietic system by gene therapy is theoretically possible if the error in blood cell formation or function results from defective activity of a single gene. If a normal counterpart of the defective gene has been cloned, it may be introduced into haematopoietic stem cells from the affected patient, then returned in an autologous bone marrow transplant. Gene therapy via haematopoietic stem cells may ultimately be used to treat sickle cell anaemia, thalassaemia, congenital immune deficiencies (SCID), lysosomal storage disorders, and Fanconi's anaemia. Possible future targets for gene therapy are leukaemia and acquired immune deficiency syndrome (AIDS), although the treatment will be more complex than a simple gene replacement.

Transduction of stem cells is a key requisite for successful, enduring gene therapy of disorders involving haematolymphoid cells. Gene insertion into committed progenitor cells would result in generation of modified blood cells for a limited period. In contrast, transduction of haematopoietic stem cells should result in generation of progeny of all haematopoietic lineages carrying the transgene for the lifetime of the recipient (Williams, 1990; Karlsson, 1991; Kohn, 1995).

A pluripotent haematopoietic stem cell (HSC) is defined as having the ability to give rise to all lineages of mature blood cells and to self-renew. The murine stem cell was identified by its ability to reconstitute a lethally irradiated animal, but the human stem cell has not been fully characterized due to the lack of an experimental transplant system. Although human haematopoietic progenitors may be maintained for several months in culture (Gartner and Kaplan, 1980; Sutherland *et al.*, 1989), the limitation to all *in vitro* systems developed to date is that true pluripotent human HSC are not studied. All cells analysed *in vitro* may be the progeny of committed progenitors rather than pluripotent stem cells. Committed progenitor cells expand in the cultures, obscuring the differentiation of progeny from stem cells which may be present. Another limitation is the inability to support the simultaneous growth and differentiation of both lymphoid

STEM CELLS
ISBN 0–12–563455–2

and myeloid cells in all culture systems developed to date (Schmitt *et al.*, 1993). Research in the field of gene therapy has, therefore, focused on the use of animal models to study transduction of long-term reconstituting stem cells in transplant systems.

MURINE MODELS FOR GENE THERAPY

Early studies in murine bone marrow transplant systems demonstrated that genes can be stably inserted into pluripotent haematopoietic stem cells, which then produce mature blood cells containing the exogenous genetic sequences. The most efficient method for transduction of the stem cells was shown to be the retroviral vector.

Joyner *et al.* (1983) reported the first successful retroviral-mediated transfer of the bacterial neomycin phosphotransferase (*neo*) gene into murine haematopoietic cells. The *neo* gene imparts resistance to the selective agent Geneticin, or G418, to transduced cells. In the study by Joyner, a small percentage of colony-forming units granulocyte/macrophage (CFU-GM, 0.35%) from marrow exposed to the retroviral vector were able to grow *in vitro* in the presence of G418 at a concentration that was completely inhibitory to growth of non-transduced CFU-GM.

Williams *et al.* (1984) next used a recombinant murine sarcoma virus (MSV)-based retroviral vector to transfer the *neo* gene into murine bone marrow cells which were transplanted into irradiated recipients. The *neo* gene was demonstrated in cells derived from colony-forming units-spleen (CFU-S) and in the haematopoietic cells of both primary recipients of the marrow as well as secondary recipients of serially passaged marrow. Thus, the ability of retroviral vectors to genetically modify long-lived murine haematopoietic stem cells was shown.

Murine marrow cells, transduced with the *neo* gene and re-infused, were then used by many groups to study the basic biology of haematopoietic stem cells. These studies identified the features necessary for efficient gene transfer into a high percentage of reconstituting pluripotent HSC: pre-treatment of donor mice with 5-fluorouracil to eliminate cycling progenitors and induce cycling of stem cells, use of retroviral vectors produced at high titre (e.g. $>1 \times 10^6/\text{ml}^{-1}$), co-cultivation of the marrow cells directly upon lawns of irradiated vector-producing fibroblasts and inclusion of recombinant haematopoietic growth factors, notably IL-3 and IL-6 to induce cycling of stem cells (Eglitis *et al.*, 1985; Bodine *et al.*, 1986; Luskey *et al.*, 1992).

Retroviral vectors integrate into different chromosomal sites in each cell they transduce. Each unique vector integrant then becomes part of a distinct restriction fragment detectable by Southern blot. Analyses of the patterns of clonal vector integration were used to demonstrate that retroviral vectors *are* capable of introducing genes into pluripotent stem cells, with ensuing production of genetically-modified progeny of both lymphoid and myeloid lineages bearing the same unique proviral integration pattern.

Dick *et al.* (1985) published one of the first reports using retroviral vectors to act as clonal tags to follow development of haematopoietic cells. The *neo* gene was transferred into murine marrow cells and transplanted into anaemic W/W^v mice. Three months after transplantation, analysis of DNA from marrow, spleen and thymus demonstrated the presence of both unique and common proviral integration sites. This study indicated that the progenitors of both lymphoid and myeloid cells had been transduced by the retroviral vector. Subsequent studies using this elegant system defined the murine stem cell and allowed examination of the properties of self-renewal, lineage commitment and clonal fluctuation (Eglitis *et al.*, 1985; Keller *et al.*, 1985; Snodgrass and Keller, 1987; Lemischka *et al.*, 1986; Jordan and Lemischka, 1990; Capel *et al.*, 1990).

For an introduced gene to have lasting, long-term benefit for the gene therapy recipient, expression must be maintained through the development of mature cells from the transduced stem cell. Initial attempts at retroviral-mediated transfer of a human adenosine deaminase (ADA) cDNA into murine bone marrow cells resulted in expression in cultured cells using an SV40 promoter, but a lack of expression in day 12 spleen colony-forming units (day 12 CFU-S) (Williams *et al.*, 1985). Subsequently, Belmont *et al.* (1986, 1988) obtained high level expression of the human ADA gene, through use of the retroviral long terminal repeat (LTR) promoter to drive transcription. A number of other investigators verified and extended these findings (Lim *et al.*, 1987; van Beusechem *et al.*, 1992; Wilson *et al.*, 1990; Kaleko *et al.*, 1990; Apperley *et al.*, 1991).

Expression of various genes driven by the strong enhancer and promoter in the retroviral LTR has now been widely reported in the murine gene transfer/bone marrow transplant model. In one study, long-term expression of the human ADA gene was maintained in various tissues of primary murine bone marrow transplant recipients. However, when marrow from the primary animals was transplanted into secondary recipients, colonies arising in the spleens at 12 days (2^0 CFU-S) had a frequent lack of expression (Moore *et al.*, 1990). Challita and Kohn (1994) have also observed strong expression of the human glucocerebrosidase cDNA from the LTR in primary murine gene transfer/BMT recipients but a high expression failure rate in secondary CFU-S. They subsequently demonstrated that the LTR of the MoMuLV-based vectors were methylated in the secondary CFU-S to a much higher extent than in the primary recipients. It is not known if the methylation caused the lack of expression, or if it occurred secondary to the LTR being silenced by other mechanisms. At this point, the relevance of LTR silencing in the murine system to potential vector silencing in the human system is unknown. The secondary CFU-S is thought to be a descendant of a cell that was a true pluripotent stem cell at the time of initial gene transfer and thus the observed late expression failure may presage similar problems in clinical situations.

Murine models have been used to study gene therapy for haemoglobinopathies. Dzierzak *et al.* (1988) first used a retroviral vector to carry an intact human β-globin gene into murine stem cells. The vector contained a genomic fragment of the human β-globin gene with the three exons encoding the β-globin protein as well as the 5′ and 3′ flanking sequences and introns which contain the promoter and enhancers which normally regulate β-globin expression. The gene was expressed specifically in the murine erythroid cells, for 4–9 months post-transplantation. Karlsson *et al.* (1988) and Bender *et al.* (1989) subsequently reported similar results. In all of these initial experiments both the levels of gene transfer and gene expression were too low to be clinically useful. The level of expression of the exogenous human β-globin gene was typically at less than 1% of the endogenous murine globin expression.

Studies using transgenic mice have identified the sequence flanking the globin gene which control the lineage and temporal specificity of expression. In addition to the promoter sequences lying immediately upstream from the structural gene, there is a strong enhancer of expression in the second intron. Additionally, sequences lying over 50 kb upstream globally regulate globin expression in erythroid cells. Inclusion of this 'locus control region' (LCR) in transgenic constructs allows high level expression of β-globin, independent of the site of chromosomal integration (Grosveld *et al.*, 1987;

Townes and Behringer, 1990). After the importance of the β-globin LCR was determined for achieving high level expression of globin in transgenic mice, efforts were made to incorporate the LCR into retroviral vectors. The LCR/β-globin vectors did produce physiologically relevant levels of human β-globin in erythroid cell lines and in the murine BMT models, but they suffered from vector instability, making it difficult to achieve efficient gene transfer (Novac *et al.*, 1990; Chang *et al.*, 1992; Plavec *et al.*, 1993). Leboulch *et al.* (1994) have recently reported making extensive modifications in the LCR sequences which maintain their enhancing activity on expression but reduce their deleterious impact on vector titre.

Another strategy currently under investigation for gene therapy for sickle cell anaemia focuses on the expression of fetal hemoglobin. Up-regulation of expression of the fetal γ-globin gene product causes decreased haematolysis and more effective erythropoiesis in patients with sickle cell disease and β-thalassaemia (Ley *et al.*, 1982; Dover *et al.*, 1985; Perrine *et al.*, 1993). The γ-globin gene product does not polymerize with the sickle cell mutant haematoglobin (HbS, $\alpha_2\beta_2^S$) and therefore its expression may ameliorate sickling crises. Theoretically, the γ-globin gene would only need to be expressed at levels of 20–30% of endogenous α-globin to have a beneficial effect. Successful application of gene therapy to the treatment of the relatively common haematoglobinopathies remains a central challenge to the field.

Murine bone marrow transplant systems have also been used to study gene therapy for lysosomal storage disorders, such as the mucopolysaccharidoses (MPS). Retroviral-mediated transfer of the β-glucuronidase gene into bone marrow cells of mice genetically deficient for β-glucuronidase (analogous to human MPS VII, Sly syndrome) resulted in long-term expression of the enzyme in BMT recipients. Although the levels of expression were only 1% of normal, there was partial correction of the disease phenotype (Wolfe *et al.*, 1992). In a similar study, only 5% of the β-glucuronidase-deficient murine bone marrow cells were successfully transduced by the vector, yet enzyme activity was detectable in various organs and lysosomal storage of undegraded glycosaminoglycans was markedly reduced (Marechal *et al.*, 1993). The secretion of active enzyme, with subsequent uptake by non-transduced cells, was responsible for inducing the decrease in lysosomal storage levels.

Gene therapy for Gaucher disease, a deficiency in the lysosomal enzyme glucocerebrosidase, has also been studied in murine bone marrow transplant models by several groups (Correll *et al.*, 1989; Weinthal *et al.*, 1991; Ohashi *et al.*, 1992). Expression of the human enzyme was readily detected in tissues from the primary animals. However, it is not expected that there will be significant cross-correction, in which non-transduced cells assimilate the enzyme, as had been observed with β-glucuronidase gene transfer. Therefore, for gene correction of Gaucher disease, a significant percentage of haematopoietic stem cells must be transduced in order to achieve enduring benefit for human recipients.

LARGE ANIMAL MODELS OF GENE THERAPY

Stem cell transduction in animals larger than the mouse may more closely represent human gene therapy, for both phylogenetic as well as logistic considerations. Therefore, canine and simian systems have been used by several groups, with efforts focusing on attemping to reproduce the efficient gene transduction of stem cells produced in murine models. The first report of successful gene transfer in a large animal model (Kantoff *et al.*, 1987) yielded low but detectable levels of human ADA in peripheral blood cells transiently for a few months after gene transfer/BMT. Further work by the NIH group of Nienhuis and the Dutch group of Valerio led to sustained production of genetically-modified haematopoietic cells lasting for the several years of follow-up (van Beusechem *et al.*, 1992; Bodine *et al.*, 1993). In all of these studies, gene transfer into the CFU-C measured by *in vitro* assays was reasonably successful, with rates between 20% and 40% observed. However, when gene transfer into pluripotent HSC was assessed by analysing the leucocytes produced *in vivo* for many months after BMT, much lower frequencies were seen, on the order of 0.1–1.0%.

Schuening and the group at Seattle (Schuening *et al.* 1989, 1991; Kiem *et al.*, 1994) have explored a variety of methods for transduction and cytoreductive conditioning to increase the level of gene-containing cells in a canine gene transfer/BMT model. Growth of marrow cells in long-term culture with multiple applications of retroviral-containing supernatants, use of G-CSF mobilized peripheral blood progenitor cells and increased strength of cytoreduction all increased the levels of gene-containing cells found in long-term reconstituted recipients.

Carter *et al.* (1992) have also transduced canine bone marrow by growing the cells in long-term culture with addition of retroviral supernatant following weekly media changes. Following 21 days of culture, the cells were reinfused into augologous recipients with and without marrow ablative conditioning. Three months after transplantation, 10% of the marrow cells contained the provirus. When assessed 10–21 months post-BMT, the percentage of marked cells had decreased to 0.1–1%. Interestingly, they reported no difference in the frequency of engrafted gene-containing cells between animals that did and those that did not receive cytoablative conditioning.

In summary, despite extensive efforts to maximize the extent of gene transfer into long-lived pluripotent stem cells in large animals, a frustratingly low level (0.1–1.0%) has been attained.

HUMAN CELLS *IN VITRO*

Initial assessments of gene transfer into human bone marrow progenitor cells used *in vitro* colony-forming assays. Typically, retroviral vectors carrying a drug resistance gene, such as the *neo* gene conferring resistance to the aminoglycoside G418, have been used and the percentage of colonies formed in the presence of drug used as a marker of the extent of gene transduction. Initial studies demonstrated that a low percentage of colony-forming units could be made drug-resistant following retroviral-mediated gene transduction (Gruber *et al.*, 1985; Hock and Miller, 1986; Hogge and Humphries, 1987; Laneuville, 1988; Hughes *et al.*, 1989).

One limitation to the utility of retroviral vectors for gene transduction of haematopoietic stem cells is their requirement for target cell proliferation for viral integration. The majority of human haematopoietic stem and progenitor cells are quiescent and therefore are not susceptible to retroviral-mediated gene transduction. Stimulation of progenitor cells with appropriate cytokines may be able to induce cell proliferation and permit retroviral mediated transduction. The combination of interleukin-3 (IL-3) and interleukin-6

(IL-6) was shown to be uniquely capable of moving quiescent progenitor cells into active cell cycling (Ikebuchi *et al.*, 1987). It was subsequently shown that pre-culture of human marrow in IL-3 and IL-6 increases the extent of gene transfer into clonogenic progenitors, from 10% in the absence of growth factors to 40% (Nolta and Kohn, 1990). The cytokine Stem Cell Factor (SCF) further increases gene transfer rates when used in combination with IL-3 and IL-6 (Nolta *et al.*, 1992). The ability of different combinations of cytokines to increase gene transfer into human marrow progenitor cells has been shown to be proportional to the effects of the factors on progenitor cell proliferation. This observation supports the hypothesis that the quiescence of the cells is the key limit to effective gene transfer.

STROMAL SUPPORT

In initial studies of gene transfer into human marrow progenitor cells, investigators repeatedly found that co-cultivating the marrow cells directly upon the vector-producing fibroblasts produced 4–8-fold higher extents of gene transduction than did transduction using cell-free vector-containing supernatant. Moore *et al.* (1992) reported that efficient gene transduction into human marrow progenitor cells by cell-free retroviral vector supernatant can be achieved if an underlying monolayer of primary marrow stromal cells was present during the gene transfer period. Our laboratory determined that the growth of $CD34^+$ cells on a stromal layer increases the percentage of cells in active cell cycle, similar to the effects of recombinant growth factors.

The necessity for a stromal layer for efficient transduction of marrow which the Belmont group has observed stands in contrast to reports by Breni *et al.* (1992) and Cassel *et al.* (1993), who reported efficient transduction of peripheral blood progenitor cells by cell-free vector in the absence of stroma. This dichotomy may reflect differences in the biology of marrow progenitor cells from that of G-CSF-mobilized peripheral blood progenitors. By direct comparison, we found that the optimal transduction conditions differ for bone marrow and peripheral blood progenitors. We demonstrated that a stromal underlayer is essential for efficient retroviral-mediated transduction of lineage restricted, colony-forming progenitors as well as primitive, long-lived progenitors from human bone marrow. In contrast to bone marrow-derived progenitors, precursors from G-CSF mobilized peripheral blood were effectively transduced without stromal support if cytokines were present (Nolta *et al.*, 1995).

The mechanism by which stromal underlayers enhance gene transfer is not known. Stromal cells may produce and present growth factors, such as membrane-bound c-kit ligand, which stimulate proliferation of progenitor cells. A report by Moritz *et al.* (1994) suggests an alternative role for stroma in the enhancement of gene transfer. They demonstrated that a small fragment from the matrix protein fibronectin may be used to replace stromal cells to augment gene transfer. The fibronectin fragment may either bind and present the retroviral particles to associated haematopoietic cells or may stimulate the progenitor cells by signalling through cell surface receptors. The use of fibronectin-coated culture flasks would be more logistically simple than preparation of autologous stromal layers for clinical gene transfer procedures.

XENOGRAFT SYSTEMS

The limitation to the studies based on colony-forming assays is that they do not measure gene transduction of pluripotent human stem cells. As described above, in the large animal studies, high rates of transduction of progenitor cells, measured by *in vitro* assays, did not predict similarly high rates of stem cell transduction, measured *in vivo* by gene transfer/BMT of cytoablated recipients. Thus, improved model systems are needed to allow study of transduction of the true pluripotent human haematopoietic stem cell.

Recently, several models have been described in which human haematopoiesis may be observed *in vivo* using xenograft systems. One such model has been developed in which human stem cells are transplanted into pre-immune fetal sheep, and multilineage human haematopoietic development ensues (Srour *et al.*, 1993). Human $CD34^+$, Lin^-, Thy^+ progenitors were transplanted into the sheep in one study, and donor cells of lymphoid, myeloid and erythroid lineages were detected in the lambs 10–50 days before birth. The fetal sheep xenograft system was also used to determine that the primitive human $CD34^+$/lineage$^-$/$CD38^-$ population contains stem cells capable of multilineage differentiation (Civin *et al.*, 1993). While this elegant model allows study of the true pluripotent human stem cell, it requires surgical procedures and maintenance of a large animal care facility.

Systems in which human stem cells are studied in smaller animals would be more widely applicable. The transplantation of immune deficient mice with human bone marrow provides a good *in vivo* model of human haematopoiesis. Several strains of mice with different deficiencies have been studied. Mice with the SCID mutation have no functional T or B cell activity, due to a recombinase defect, and have been successfully engrafted with human haematopoietic cells. Systems include surgical implantation of human fetal thymus and liver (McCune *et al.*, 1988), or bone fragments (Kyoizumi *et al.*, 1992). Transplantation of human cord blood into SCID mice leads to multilineage engraftment without a requirement for human cytokine supplementation (Vormoor *et al.*, 1994). T lymphocytes in the cord blood inoculum may secrete enough species-specific cytokines to sustain human haematopoiesis in the mice. However, systems which study haematopoiesis from fetal tissues are of limited use, since they are not clinically relevant to treatment of adult or pediatric haematopoietic stem cells by gene therapy.

A study by Lapidot *et al.* (1992) demonstrated that xenografted adult human marrow survived and proliferated in SCID mice if large quantities of recombinant human cytokines were injected into the mice every 2 days for 2 months. Mice which did not receive human cytokines had very low levels of human cell engraftment and proliferation. Since many cytokines are species-specific, the essential human interleukins may not be present in a xenograft setting and must be supplied exogenously. More recently, this group has reported that mice bearing the non-obese diabetic (NOD) gene with the SCID defect (NOD-SCID) are more receptive to human xenografts, with levels of engrafted human cells as high as 50% achieved without exogenous cytokine administration.

Our group demonstrated that sustained human haematopoiesis in immune deficient mice (*beige/nude/xid-bnx*) can be achieved from transplanted $CD34^+$ progenitors if they are co-injected with human bone marrow stromal cells which have been engineered to secrete human IL-3 (Nolta *et al.*, 1994). The presence of human multilineage colony-forming progenitors, mature myeloid and T lymphoid cells marked with provirus can be demonstrated in the marrow, spleens and blood of the mice for up to 11 months (Nolta *et al.*, 1994, 1995). The major limitation of the system to date is that no human B lymphoid cells have been produced in the mice. Potentially, the continuous exposure to IL-3 may divert the human progenitors away from B lymphoid development, as has been demonstrated by Hirayama *et al.* (1994) with isolated murine stem cells.

Xenograft models may also facilitate investigation of the impact of different transduction protocols on both the efficiency of gene transfer into stem cells and the maintenance of pluripotentiality. We have compared the transduction and engraftment of human marrow and peripheral blood progenitors in *bnx* mice, focusing on the role of stroma on augmenting gene transfer. We confirmed the observation of Moore *et al.* (1992) that stroma increased gene transfer into clonogenic progenitors. Interestingly, when mice were examined 7–11 months after infusion of human progenitor cells, those that received marrow transduced *without* the presence of stroma failed to have significant numbers of human cells present, whereas those that were transplanted with cells transduced *on* stroma showed persistent human haematopoiesis (Nolta *et al.*, 1995). Therefore, the presence of bone marrow stroma has dual benefits in that it increases gene transfer efficiency and is also essential for survival of long-lived human haematopoietic progenitors during the *ex vivo* gene transfer period.

CLINICAL STUDIES USING GENE MARKING

Several clinical trials of haematopoietic cell marking have begun, and have yielded valuable data concerning the development and transduction of human haematopoietic stem cells. Brenner *et al.* (1993a, 1993b) used the *neo* gene to label one third of the bone marrow cells reinfused into patients undergoing autologous bone marrow transplantation for acute myeloid leukaemia and neuroblastoma. These studies had two goals. The first was to determine whether any clonogenic malignant cells present in the transplanted marrow contribute to relapses after transplantation. If malignant cells present in the transplant inoculum were transduced by the *neo* retroviral vector, they would be detectable by PCR in populations of malignant cells found in patients suffering relapse. The second goal was to examine the engraftment kinetics of normal haematopoietic cells which may have been labelled by the vector.

In patients who subsequently relapsed, a fraction of the malignant cells have been shown to contain the *neo* marker. These results confirm that there are clonogenic malignant cells present in these unpurged autologous marrows which play a role in relapse, although there may also be residual cells in the patients which contribute to the relapse. Similar marking studies are now planned to examine the efficacy of various purging protocols to eliminate malignant cells from the transplant inoculum.

Additional information was obtained by studying the normal haematopoietic cells which developed after transplant with the marked marrow. Up to 5% of the normal CFU-GM present for greater than 1 year after transplant contained the *neo* marker, demonstrating that these cells form part of the normal reconstituting elements. The marker gene was found in circulating myeloid and lymphoid cells, suggesting that multipotent progenitor cells had been transduced. The level of labelled cells was somewhat higher than what may have been predicted from animal studies. Only one third of the marrow was exposed to the vector and no attempts were made to stimulate the marrow cells to proliferate by exogenous cytokines or stroma. Potentially, these marrow samples may have contained a higher fraction of proliferating stem cells because they were collected as the patients were recovering from myelosuppressive chemotherapy. These results highlight the fact that clinical trials of gene marking can provide information not available from *in vitro* or xenograft experiments. Currently, similar marking studies are being performed to compare the relative contributions to long-term reconstitution by bone marrow and mobilized peripheral blood after transplant with both cell types.

HUMAN GENE THERAPY TRIALS

The first use of gene transfer for therapy of a human disease was performed in September 1990 by Blaese, Anderson and Culver at the NIH (Blaese *et al.*, 1995). They transduced peripheral blood T lymphocytes from patients with ADA-deficient SCID and re-infused the cells intravenously. After repeated monthly cycles of this procedure, one of the patients has had persistence of circulating T lymphocytes expressing the exogenous ADA gene at levels estimated to be 10–50%. These treated T cells are persisting for at least 2 years after the last cell infusion. Both patients remain on enzyme replacement with PEG-ADA so that the broadness of the immunologic repertoire contained in the transduced cells is not known, nor is their potential duration of survival.

The first successful use of stem cells for gene therapy was performed by Claudio Bordignon and co-workers, in Milan, Italy in March 1992, with two patients with ADA deficiency (Bordignon *et al.*, 1995). They transduced bone marrow $CD34^+$ cells using one ADA vector and, in a protocol similar to that of the Blaese group, used a second ADA vector to transduce peripheral blood T lymphocytes. Cells were administered in nine monthly i.v. injections. Two years later, they detected between 5% and 30% transduced progenitors in the bone marrow of both patients. Initially all vector-positive cells in the blood were derived from the marked T lymphocytes (ave = 8.5% $vector^+$), but later T cell clones arising from marked bone marrow cells began to appear. Continued evaluation of these patients will reveal the relative contributions of mature peripheral blood lymphocytes and bone marrow stem cells to long-term lymphopoiesis.

In May and June of 1993, our group at Children's Hospital Los Angeles treated three newborns with ADA-deficiency by retroviral-mediated gene transfer into their umbilical cord blood $CD34^+$ cells. Serial evaluation has shown persistent presence of mononuclear cells and granulocytes in their peripheral blood for the 18 months of follow-up (Kohn *et al.*, 1995). The frequency of gene-containing leucocytes has been of the order of 0.01%, which is consistent with animal studies showing transduction of 1% of the stem cells in a treated marrow inoculum followed by a further 100-fold dilution of the infused umbilical cord blood cells by the non-ablated normal marrow cells. Examination of bone marrow samples obtained 1 year after the treatment showed the presence of 2–6% G418-resistant CFU-GM. Thus, this study demonstrates that long-lived haematopoietic progenitor cells (stem cells?) from umbilical cord blood can be transduced by retroviral vectors and engraft in non-cytoablated neonates. While limited by the low gene transfer frequency, these results highlight the potential for treating newborns with genetic diseases or HIV-1 infection by transduction of autologous umbilical cord blood.

Other studies which are currently in the review process or imminent include two which will transfer the multi-drug resistance gene (MDR-1) into progenitor cells of cancer patients to confer increased resistance to myelosuppression and three which will transfer the human glucocerebrosidase gene into cells from patients with Gaucher disease. Studies are also under review for insertion of genes which may inhibit HIV-1 replication into stem cells of patients with AIDS. It is likely that many more such studies will follow, although it remains to be proven whether the current retroviral vectors and available gene transfer techniques will lead to transduction of a clinically significant fraction of the stem cells present.

FUTURE PROSPECTS

Cytokines

Cytokines influence the growth and differentiation of haematopoietic cells. Possibly, retroviral-mediated transduction of human stem cells may be enhanced by cytokines which have not yet been defined. The growth factors used to trigger stem cells into cycle to allow integration must not destroy the pluripotentiality of those cells. Cytokine combinations which promote limited self-renewal divisions without forcing lineage commitment and terminal differentiation would be ideal for use in the field of retroviral-mediated gene therapy.

The recently identified cytokine flk-2/flt-3 ligand binds to the most primitive haematopoietic cells and may act to stimulate them, in synergy with other cytokines (Lyman *et al.*, 1993). It remains to be determined whether this cytokine will enhance the gene transfer extent of human stem cells. The growth factors IL-11, G-CSF and basic fibroblast growth factor have been reported to act upon primitive haematopoietic cells, and to increase gene transfer into haematopoietic cells (Lemoli *et al.*, 1993; Dilber *et al.*, 1994). Recently, the cytokines LIF and MIP-1α were reported to cause cycling of primitive cells while blocking differentiation and expansion of commited progenitors (Brandt *et al.*, 1994; Verfaille and Miller, 1994). Retroviral-mediated transduction in the presence of these cytokines might, therefore, allow a higher extent of gene transfer and integration into stem cells. Alternatively, inclusion of antibodies to block cytokines which inhibit cell cycling may enhance the extent of gene transfer, although this approach has not been reported. Inhibitory cytokines which may be present in the transduction media are TGF-β, and TNF-α (Broxmeyer *et al.*, 1986; Cashman *et al.*, 1990).

Pseudotyping

New strategies to increase the efficiency of transduction by retroviral vectors are constantly sought. The rate of gene transfer may be limited by low expression of the receptor for the murine amphotropic virus envelope (env) glycoprotein, which has been used in all of the retroviral vectors used for clinical trials to date (Miller *et al.*, 1994). The cell surface receptor for amphotropic retrovirus (Ram-1) is in low abundance on haematopoietic progenitors (Kavanaugh *et al.*, 1994). Pseudotyping the vector by using an envelope protein in the packaging construct from a virus with a different tropism is one strategy which might allow higher levels of transduction than can be obtained by the amphotropic packaging lines currently in use.

This limitation may be partially overcome by the use of hybrid packaging cell lines which express the Gibbon Ape Leukaemia Virus or murine xenotropic *env* proteins, since haematopoietic stem cells may express a greater number of receptors for these ligands (Miller *et al.*, 1994; Adams *et al.*, 1992). The use of the PG-13 packaging cell line, which uses the Gibbon Ape Leukaemia Virus (Glvr-1) envelope proten, may more efficiently infect human haematopoietic stem cells (Miller *et al.*, 1991).

The vesicular stomatitis virus G glycoprotein (VSV-G) has also been used to pseudotype MoMLV vector particles (Burns *et al.*, 1993). The VSV pseudotyped viral particles are more physically stable than those containing the amphotropic *env* and can be effectively concentrated up to 200-fold. VSV pseudotyped vectors have been shown to mediate gene transfer into clonogenic $CD34^+$ progenitors from human bone marrow (Yang *et al.*, 1994b). It currently is unknown to what extent deficiencies in env/receptor interactions limit gene transfer into human haematopoietic stem cells.

Selection

An alternative strategy for obtaining high frequencies of gene-containing cells would be to select for the cells which have been successfully transduced based upon expression of a transferred gene. Such selection may be performed *ex vivo*, prior to cell re-infusion or *in vivo*, after re-infusion of the cells. The two major approaches which have been studied for selection include the use of a drug resistance gene allowing survival in the presence of a selective agent or the use of a gene which produces a cell surface antigen which can be used to isolate the cells which express it by immunoaffinity procedures.

A variety of drug resistance genes have been studied, including the neomycin phosphotransferase (*neo*) gene encoding resistance to G418, hygromycin resistance, puromycin resistance, dihydrofolate reductase (dhfr) encoding methotrexate resistance and the multi-drug resistance gene (MDR-1) encoding resistance to a variety of chemotherapeutic agents (Eglitis, 1991).

The *neo* gene has been used extensively as a marker of gene transfer into haematopoietic cells by measuring the percentage of colonies which form in the presence and absence of G418. There is evidence that expression of the *neo* gene alters cellular glucose metabolism, and may have adverse effects on haematopoietic cells (Valera *et al.*, 1994; von Melchner and Housman, 1988). In addition, preselection of transduced murine bone marrow cells in the neomycin analogue G418 prior to transplantation reduces their competitive engraftment capacity.

DHFR and MDR-1 represent genes which may be used to select for transduced cells *in vivo*. Methotrexate, taxol, anthracyclines or vinca alkaloids may be administered to patients after transplant of transduced cells, thereby suppressing proliferation of non-transduced cells and allowing those bearing the drug resistance gene to increase in relative frequency. Studies in murine BMT models have demonstrated that this approach does allow the frequency of transduced haematopoietic cells to be increased following transplant of marrow (Williams *et al.*, 1987; Sorrentino *et al.*, 1992). Aran *et al.* (1994) have reported the construction of a vector carrying the MDR-1 gene as a selectable marker and the human glucocerebrosidase cDNA as a 'therapeutic' gene. Transduction of cells followed by selection with taxol led to an increase in the amount of glucocerebrosidase enzyme produced.

An alternative to drug selection for expression of a resistance gene would be selection for expression of a novel surface protein antigen (Strair *et al.*, 1988). The ideal antigen for this purpose would not normally be present on the surface of the cells being analysed, would be biologically inert, and would not be immunogenic if expressed *in vivo*. Recently, uses of a truncated human nerve growth factor receptor or the murine heat stable antigen as selectable antigens have been reported (Traversari *et al.*, 1993; Pawliuk *et al.*, 1994).

Regulated expression

Initial studies of gene transfer have focused on 'housekeeping enzymes' such as ADA and glucocerebrosidase which are ubiquitously expressed in a loosely regulated manner. Expression of these genes under control of constitutive viral promoters, such as the MoMuLV LTR or the CMV or SV40 enhancer/promoters, is likely to result in expression in all transduced cells at variable levels. Such non-regulated expression may be tolerated for these enzymes. However, other genes will require more precisely regulated expression for effective, non-toxic results. The need for precise expression of β-globin for the treatment of haemoglobinopathies has been discussed previously. Other genes which may require regulated expression include those encoding proteins involved in signal transduction pathways, such as Bruton's tyrosine kinase (*btk*) which is defective in X-linked agammaglobulinemia, or the common γ-chain of the receptors for IL-2, IL-4 and IL-12 which is defective in X-linked SCID. For these proteins, both the specific cell types or the cellular developmental state in which they are expressed may need to be controlled or adverse effects on cellular function and proliferation may result. To date, efforts to achieve precisely regulated expression of genes delivered by retroviral vectors has not been achieved.

Targeted delivery

Targeting the retroviral vector to only bind and enter cells of a specific lineage might increase effective transduction levels and overcome the need for tissue-specific expression. By optimizing the number of virions binding to the target cell, the chances of viral entry and integration should be enhanced. Lineage-specific targeting may also allow *in vivo* systemic delivery of vector, avoiding the need for marrow harvest, *ex vivo* transduction and re-infusion. Thus, vectors which specifically transduce human haematopoietic stem cells would be very useful for gene therapy.

Kasahara *et al.* (1994) constructed a chimeric retroviral envelope gene that contained the erythropoietin (Epo) protein in place of a portion of the retroviral ecotropic *env* gene. The resulting Epo-*env* virions efficiently transduced both murine and human cells which expressed the Epo receptor. Hybrid vectors of this type might be employed to specifically target β- or γ-globin genes to the erythroid progenitors of sickle cell patients. One drawback to this type of targeting system is that only maturing, lineage committed cells would be treated with the transgene, so that the gene therapy would need to be repeated on a regular basis.

A similar strategy was used to replace part of the *env* protein with the extracellular domain of the ligand human SCF (Wong *et al.*, 1994). The resulting SCF *env* vectors demonstrate tissue-specific cross-species transduction of human haematopoietic progenitors bearing the c-kit receptor. The use of specific integrin sequences in place of *env*, to target various cell types, is also under investigation (Valsesia-Wittmann *et al.*, 1994).

New vector delivery systems

Despite advances in increasing the affinity of the virion for the target cell, one major limitation to the extent of transduction by retroviral vectors still remains. The cell must undergo DNA replication and nuclear membrane breakdown, to allow the viral genome with its associated proteins access to the chromosomal DNA for integration. However, pluripotent human haematopoietic stem cells are seldom in cell cycle, so the integration rates are low. Alternative gene delivery methods are therefore under development.

Adeno-associated virus

Adeno-Associated Virus-2 (AAV) is a recently developed system with potential advantages over retroviral vectors for gene therapy (Muzyczka, 1992). Like retroviral vectors, AAV efficiently integrate their genome into target cell chromosomes. Wild-type AAV is a non-pathogenic human parvovirus which has strong integration-site specificity. Current AAV-based vectors unfortunately do not display site-specific integration, but the mechanism is under intense investigation. Preliminary studies suggest that AAV can transfer genes into non-replicating cells (Podsakoff *et al.*, 1994) although it has been technically difficult to stringently prove this point (Russell *et al.*, 1994). Due to its small size (20 nM), the AAV virion may enter through nuclear pores without cell-cycle associated breakdown of the nuclear membrane. Transient expression of introduced genes may be achieved with high efficiency using AAV vectors, but the level of stable integration is much lower. No stable packaging systems currently exist for AAV, leading to difficulty producing quantities of virus large enough for clinical studies with consistent high titres and purity. Further work is needed with this newer vector system before its full advantages and disadvantages compared to retroviral vectors are known.

The pathogenic human parvovirus B19 replicates autonomously and infects cells of the erythroid lineage almost exclusively. The B19p6 promoter sequence has been introduced into a recombinant AAV vector (Wang *et al.*, 1994), and causes autonomous replication in human erythroid cells. Modifications of this vector system may be useful for gene therapy of human haemoglobinopathies.

Adenovirus

The 36 Kb double-stranded DNA genome of the adenovirus has been modified for use as a gene delivery vector (Graham *et al.*, 1977; Levrero *et al.*, 1991). Recombinant adenoviruses can efficiently transduce genes into a variety of non-dividing cells *in vivo*. Unfortunately, their use in clinical gene therapy with haematopoietic stem cells is limited, due to the lack of sustained transgene expression, since the vector does not integrate into the host cell genome. An additional disadvantage is the development of an immune response after repeated administration of the vector and ensuing damage to the target tissue (Yang *et al.*, 1994a). Adenoviral vectors may be used to transiently express genes in haematopoietic stem cells to induce cycling (such as homeobox genes, mutant p53 or RB) and thereby allow subsequent stable gene delivery by retroviral-mediated transduction (Mitani *et al.*, 1994).

CONCLUSION

In summary, progress in the use of gene transfer into haematopoietic cells has led to initial clinical trials. Information developed by these early efforts will be used to guide future developments. Ultimately, gene therapy may allow a number of genetic and acquired diseases to be treated, without the current complications from bone marrow transplantation with allogeneic cells.

REFERENCES

Adams, R.M., Soriano, H.E., Wang, M., Darlington, G., Steffen, D. and Ledley, F.D. (1992). *Proc. Natl Acad. Sci. USA* **89**:8981–8985.
Apperley, J.F., Luskey, B.D. and Williams, D.A. (1991). *Blood* **78**:310–317.
Aran, J.M., Gottesman, M.M. and Pastan, I. (1994). *Proc. Natl Acad. Sci. USA* **91**:3176–3180.
Belmont, J.W., Henkel-Tigges, J., Chang, S.M.W. *et al.* (1986). *Nature* **322**:385–387.
Belmont, J.W., MacGregor, G.R., Wager-Smith, K. *et al.* (1988). *Mol. Cell Biol.* **8**:5116–5125.
Bender, M.A., Gelinas, R.E. and Miller, A.D. (1989). *Mol. Cell Biol.* **9**:1426–1434.
Blaese, R.M., Culver, K.W., Miller, A.D. *et al.* (1995). *Science* **270**:475–480.
Bodine, D.M., Karlsson, S. and Nienhuis, A.W. (1986). *Proc. Natl Acad. Sci. USA* **86**:8897–8901.
Bodine, D.M., Moritz, T., Donahue, R.E. *et al.* (1993). *Blood* **82**:1975–1980.
Bordignon, C., Notarangelo, L.D., Nobili, N. *et al.* (1995). *Science* **270**: 470–475.
Brandt, J.E., Sundy, S., Hoffman, R., Tsukamoto, A. and Tushinski, R. (1994). *Exp. Hem.* **22**:Abstract 177.
Breni, M., Magni, M., Siena, S., Di Nicola, M., Bonadonna, G. and Gianni, A.M. (1992). *Blood* **80**:1418–1422.
Brenner, M.K., Rill, D.R., Holladay, M.S. *et al.* (1993a). *Lancet* **342**:1134–1137.
Brenner, M.K., Rill, D.R., Moen, R.C. *et al.* (1993b). *Lancet* **341**:85–86.
Broxmeyer, H.E., Williams, D.E., Lu, L. *et al.* (1986). *J. Immunol.* **136**:4487–4495.
Burns, J.C., Friedmann, T., Driever, W., Burrascano, M. and Yee, J.K. (1993). *Proc. Natl Acad. Sci. USA* **90**:8033–8037.
Capel, B., Hawley, R.G. and Mintz, B. (1990). *Blood* **75**:2267–2270.
Carter, R.F., Abrams-Ogg, A.C.G., Dick, J.E. *et al.* (1992). *Blood* **79**:356–364.
Cashman, J.D., Eaves, A.C., Raines, E.W., Ross, R. and Eaves, C.J. (1990). *Blood* **75**:96–101.
Cassel, A., Cottler-Fox, M., Doren, S. and Dunbar, C.E. (1993). *Exp. Hem.* **21**:585–591.
Challita, P.M. and Kohn, D.B. (1994). *Proc. Natl Acad. Sci. USA* **91**:2567–2571.
Chang, J.C., Liu, D. and Kan, Y.W. (1992). *Proc. Natl Acad. Sci. USA* **89**:3107–3110.
Civin, C.I., Lee, M.J., Hedrick, M. and Zanjani, E.D. (1993). *Blood* **82**:707.
Correll, P.M., Fink, J.K., Brady, R.O., Perry, L.K. and Karlsson, S. (1989). *Proc. Natl Acad. Sci USA* **86**:8912–8916.
Dick, J.E., Magli, M.C., Huszar, D., Phillips, R.A. and Bernstein, A. (1985). *Cell* **42**:71–79.
Dilber, M.S., Bjorkstrand, B., Li, K.J., Smith, C.I., Xanthopoulos, K.G. and Gahrton, G. (1994). *Exp. Hematol.* **22**:1129–1133.

Dover, G.J., Charache, S., Charache, S., Boyer, S.H., Vogelsang, G. and Moyer, M. (1985). *Blood* **66**:527–532.
Dzierzak, E.A., Papayannopoulou, T. and Mulligan, R.C. (1988). *Nature* **331**:35–41.
Eglitis, M.A. (1991). *Hum. Gene Ther.* **2**:195–201.
Eglitis, M.A., Kantoff, P., Gilboa, E. and Anderson, W.F. (1985). *Science* **230**:1395–1398.
Gartner, S. and Kaplan, H.S. (1980). *Proc. Natl Acad. Sci. USA* **77**:4756–4769.
Graham, F.L., Smiley, J., Russell, W.C. and Nairn, R. (1977). *J. Gen. Virol.* **36**:59–72.
Grosveld, F., van Assendeelft, G.B., Greaves, D.R. and Kollias, G. (1987). *Cell* **51**:975–985.
Gruber, H.E., Finley, K.D., Hershberg, R.M. *et al.* (1985). *Science* **230**:1057–1060.
Hirayama, F., Clark, S.C. and Ogawa, M. (1994). *Proc. Natl Acad. Sci. USA* **91**:469–473.
Hock, R.A. and Miller, A.D. (1986). *Nature* **320**:275–277.
Hogge, D.E. and Humphries, R.K. (1987). *Blood* **69**:611–617.
Hughes, P.F., Eaves, C.J., Hogge, D.E. and Humphries, R.K. (1989). *Blood* **74**:1915–1922.
Ikebuchi, K., Wong, G.G., Clark, S.C., Ihle, J.N., Hirai, Y. and Ogawa, M. (1987). *Proc. Natl Acad. Sci. USA* **84**:9035–9039.
Jordan, C.T. and Lemsischka, I.R. (1990). *Genes Dev.* **4**:220–232.
Joyner, A., Keller G., Phillips, R.A. and Bernstein, A. (1983). *Nature* **305**:556–558.
Kaleko, M., Garcia, J.V., Osborne, W.R.A. and Miller, A.D. (1990). *Blood* **75**:1733–1741.
Kantoff, P.K., Gillio, A., McLachlin, J.R. *et al.* (1987). *J. Exp. Med.* **166**:219–234.
Karlsson, S. (1991). *Blood* **78**:2481–2492.
Karlsson, S., Bodine, D.M., Perry, L., Papayannopoulou, T. and Nienhuis, A.W. (1988). *Proc. Natl Acad. Sci. USA* **85**:6062–6066.
Kasahara, N., Dozy, A.M. and Kan, Y.W. (1994). *Science* **266**:1373–1376.
Kavanaugh, M.P., Miller, D.G., Zhang, W. *et al.* (1994). *Proc. Natl Acad. Sci. USA* **91**:7017–7075.
Keller, G., Paige, C., Gilboa, E. and Wagner, E.F. (1985). *Nature* **318**:149–154.
Kiem, H.P., Darovsky, B., von Kalle, C. *et al.* (1994). *Blood* **83**:1467–1473.
Kohn, D.B. (1995). *Current Opinions in Ped.* **7**:56–63.
Kohn, D.B., Weinberg, K.I., Nolta, J.A. *et al.* (1995). *Nature Medicine* **1**:1017–1026.
Kyoizumi, S., Baum, C.M., Kaneshima, H., McCune, J.M., Yee, E.J. and Namikawa, R. (1992). *Blood* **79**:1704.
Laneuville, P., Chang, W., Kamel-Reid, S., Fauser, A.A. and Dick, J.E. (1988). *Blood* **71**:811–814.
Lapidot, T., Pflumio, F., Doedens, M., Murdoch, B., Williams, D.E. and Dick, J.E. (1992). *Science* **255**:1137–1141.
Leboulch, P., Huang, G.M.S., Humphries, R.K. *et al.* (1994). *EMBO J.* **13**:3065–3076.
Lemischka, I.R., Raulet, D.H. and Mulligan, R.C. (1986). *Cell* **45**:917–927.
Lemoli, R.M., Fogli, M., Fortuna, A., Rizzi, S., Benini, C. and Tura, S. (1993). *Exp. Hematol.* **21**:1668–1672.
Levrero, M., Barban, V., Manteca, S. *et al.* (1991). *Gene* **101**:195–202.
Ley, T.J., DeSimone, J., Anagnou, N.P. *et al.* (1982). *N. Engl. J. Med.* **307**:1469–1475.
Lim, B., Williams, D.A. and Orkin, S.H. (1987). *Mol. Cell Biol.* **7**:3459–3465.
Lim, B., Apperley, J.F., Orkin, S.H. and Williams, D.A. (1989). *Proc. Natl Acad. Sci. USA* **86**:8892–8896.
Luskey, B.D., Rosenblatt, M., Zsebo, K. and Williams, D.A. (1992). *Blood* **80**:396–402.
Lyman, S.D., James, L., Vanden, B.T. *et al.* (1993). *Cell* **75**:1157–1167.
McCune, J.M., Namikawa, R., Kaneshima, H., Lieberman, M. and Weissman, I.L. (1988). *Science* **241**:1632–1639.
Marechal, V., Naffakh, N., Danos, O. and Heard, J.M. (1993). *Blood* **82**:1358–1365.
Miller, A.D., Garcia, J.V., con Suhr, N., Lynch, C.M., Wilson, C. and Eiden, M.V. (1991). *J. Virol.* **65**:2220–2224.
Miller, D.G., Edwards, R.G. and Miller, A.D. (1994). *Proc. Natl Acad. Sci. USA* **91**:78–82.
Mitani, K., Graham, F.L. and Caskey, C.T. (1994). *Hum. Gene Ther.* **5**:941–948.
Moore, K.A., Fletcher, F.A., Villalon, D.K., Utter, A.E. and Belmont, J.W. (1990). *Blood* **75**:2085–2092.
Moore, K.A., Deisseroth, A.B., Reading, C.L., Williams, D.E. and Belmont, J.W. (1992). *Blood* **79**:1393–1399.
Moritz, T., Patel, V.P. and Williams, D.A. (1994). *J. Clin. Invest.* **93**:1451–1457.

Muzyczka, N. (1992). *Curr. Top. Microbiol. Immunol.* **158**:97–129.
Nolta, J.A. and Kohn, D.B. (1990). *Hum. Gene Ther.* **1**:257–268.
Nolta, J.A., Crooks, G.M., Overell, R.W., Williams, D.E. and Kohn, D.B. (1992). *Exp. Hematol.* **20**:1065–1071.
Nolta, J.A., Hanely, M.E. and Kohn, D.B. (1994). *Blood* **83**:3041–3051.
Nolta, J.A., Smogorzewska, E.M. and Kohn, D.B. (1995). *Blood* **86**:101–110.
Novak, U., Harris, E.A.S., Forrester, W., Groudine, M. and Gelinas, R. (1990). *Proc. Natl Acad. Sci. USA* **87**:3386–3390.
Ohashi, T., Boggs, S., Robbins, P. *et al.* (1992). *Proc. Natl Acad. Sci. USA* **89**:11332–11336.
Pawliuk, R., Kay, R., Lansdorp, P. and Humphries, R.K. (1994). *Blood* **84**:2868–2877.
Perrine, S.P., Ginder, G.D., Faller, D.V. *et al.* (1993). *N. Engl. J. Med.* **328**:81–86.
Plavec, I., Papayannopoulou, T., Maury, C. and Meyer, F. (1993). *Blood* **81**:1384–1392.
Podsakoff, G., Wong, K.K. Jr. and Chatterjee, S. (1994). *J. Virol.* **68**:5656–5666.
Russell, D.W., Miller, A.D. and Alexander, I.E. (1994). *Proc. Natl Acad. Sci. USA* **91**:8915–8919.
Schmitt, C., Ktorza, S., Sarun, S., DeJong, R. and Debre, P. (1993). *Blood* **82**:3675–3685.
Schuening, F.G., Storb, R., Stead, R.B., Goehle, S., Nash, R. and Miller, A.D. (1989). *Blood* **74**:152–155.
Schuening, F.G., Kawahara, K., Miller, A.D. *et al.* (1991). *Blood* **78**:2568–2576.
Sorrentino, B.P., Brandt, S.J., Bodine, D. *et al.* (1992). *Science* **257**:99–103.
Snodgrass, R. and Keller, G. (1987). *EMBO J.* **13**:3955–3960.
Srour, E.F., Zanjani, E.D., Cornetta, K. *et al.* (1993). *Blood* **82**:3333–3342.
Strair, R.K., Towle, M.J. and Smith, B.R. (1988). *J. Virol.* **62**:4756–4759.
Sutherland, H.J., Eaves, C.J., Eaves, A.C., Dragowski, W. and Lansdorp, P.M. (1989). *Blood* **74**:1563–1570.
Townes, T.M. and Behringer, R.R. (1990). *Trends in Genetics* **6**:219–223.
Traversari, C., Ferrari, G., Bonini, C. *et al.* (1993). *Blood* **82 (Suppl. 1)**:214a.
Valera, A., Merales, J.C., Hatzoglou, M. and Bosch, F. (1994). *Hum. Gene Ther.* **5**:449–456.
Valsesia-Whittmann, S., Drynda, A., Deleange, G. *et al.* (1994). *J. Virol.* **68**:4609–4619.
van Beusechem, V.W., Kakler, A., Meidt, P.J. and Valerio, D. (1992). *Proc. Natl Acad. Sci. USA* **79**:7640–7644.
Verfaillie, C.M. and Miller, J.S. (1994). *Blood* **84**:1442–1449.
von Melchner, H. and Housman, D.E. (1988). *Oncogene* **2**:137–140.
Vormoor, J., Lapidot, T., Pflumio, F. *et al.* (1994). *Blood* **83**:2489–2497.
Wang, X.S., Yoder, M.C., Zhou, S.Z. and Srivatsa, A. (1994). *Blood* **84 (Suppl. 1)**:266a.
Weinthal, J., Nolta, J., Yu, X.J., Lilley, J., Uribe, L. and Kohn, D.B. (1991). *Bone Marrow Transp.* **8**:393–399.
Williams, D.A. (1990). *Hum. Gene Ther.* **1**:229–239.
Williams, D.A., Lemischka, I.R., Nathan, D.G. and Mulligan, R.C. (1984). *Nature* **310**:476–480.
Williams, D.A., Orkin, S.H. and Mulligan, R.C. (1985). *Proc. Natl Acad. Sci. USA* **83**:2566–2570.
Williams, D.A., Hsien, K., DeSilva, A. and Mulligan, R.C. (1987). *J. Exp. Med.* **166**:210–218.
Wilson, J.M., Danos, O., Grossman, M., Raulet, D.H. and Mulligan, R.C. (1990). *Proc. Natl Acad. Sci. USA* **87**:439–443.
Wolfe, J.H., Sands, M.S., Barker, J.E. *et al.* (1992). *Nature* **360**:749–753.
Wong, C., Kasahara, N., Cowan, M.J. and Kan, Y.W. (1994). *Blood* **84 (Suppl. 1)**:254a.
Yang, Y., Nunes, F.A., Berencsi, K., Furth, E.E., Gonczol, E. and Wilson, J.M. (1994a). *Proc. Natl Acad. Sci. USA* **91**:4407–4411.
Yang, Y., Vanin, E.F., Whitt, M.A. and Neinhuis, A.W. (1994b). *Blood* **84 (Suppl. 1)**:358a.

Index